ENCYCLOPEDIA OF AUTOMOTIVE ENGINEERING

ENCYCLOPEDIA OF AUTOMOTIVE ENGINEERING

Volume 4

Part 5: Chassis Systems
Part 6: Electrical and Electronic Systems

Editors-in-Chief

David Crolla†
University of Leeds, Leeds, UK

David E. Foster
University of Wisconsin—Madison, Madison, WI, USA

Toshio Kobayashi
Japan Automobile Research Institute, Tokyo, Japan

Nicholas Vaughan
Cranfield University, Bedford, UK

† Deceased

WILEY

This edition first published 2015

Registered office

John Wiley & Sons Ltd, The Atrium, Southern Gate, Chichester, West Sussex, PO19 8SQ, United Kingdom

For details of our global editorial offices, for customer services and for information about how to apply for permission to reuse the copyright material in this book please see our website at www.wiley.com.

Library of Congress Cataloging-in-Publication Data

Encyclopedia of automotive engineering / editors-in-chief, David Crolla, David E. Foster, Toshio Kobayashi, Nicholas Vaughan.
volumes cm
Includes bibliographical references and index.
ISBN 978-0-470-97402-5 (cloth)
1. Automobiles–Encyclopedias. I. Crolla, David A.
TL9.E5227 2015
629.203–dc23

2014025608

A catalogue record for this book is available from the British Library.
ISBN: 978-0-470-97402-5

Typeset in 10/12pt Times by Laserwords Private Limited, Chennai, India
Printed and bound by Markono Print Media Pte Ltd.

This book is printed on acid-free paper responsibly manufactured from sustainable forestry, in which as least two trees are planted for each one used for paper production.

Editorial Board

Associate Editors

Contents

VOLUME 1

VOLUME 3

PART 5
Chassis Systems

Suspension

Chapter 109

Possibilities of Coil Springs and Fiber-Reinforced Suspension Parts

Joerg Neubrand

Chassis Technologies—Mubea Fahrwerkstechnologien GmbH, Attendorn, Germany

1 INTRODUCTION

As early as over 400 years ago, the first helical compression springs (coil springs) were already used for wheel suspensions of a wagon body, and at the very latest since the invention of the wheel suspension strut (McPherson), they have represented the best spring design for the vertical dynamics of passenger cars. They took over suspension functions and because of their advantages regarding weight and installation space, replaced leaf springs almost completely. The necessary suspension arms to carry the coil springs may be regarded as a disadvantage, but their design enables exceptionally good road holding and safety.

Springs will be deformed elastically and during this process, they take up potential energy, which will be released when relieved. They will have to live up to such duties also during repeated and dynamic loads and considerable deforming processes. For this reason also, steel continues to be an ideal material to make springs.

Over the past few years, resource-saving weight reduction has assumed growing importance. Assuming that the weight of a smaller middle class car was kept constant, the weight of a chassis support spring was reduced by about 55% since 1992 (Figure 1). Such weight reductions were brought about by higher loads, underpinned by optimized manufacturing technologies and new types of steel, without any reductions in robustness.

One example of the developments in manufacturing technology is the high performance process (HPP) developed by the Mubea company, as a result of which the load-bearing strength of helical compression springs was increased by more than 10%. This method was first introduced in 2003, and it has been used as the global benchmark since 2004.

Moreover, in future, there will be increasing demands by car builders as regards CO_2 reductions, lower vehicle weights, and a reduction of unsuspended sections and the robustness of springs, particularly as affected by corrosion. This means that optimum use of materials employed will play a decisive role, underpinned by efficient design. In addition to that, alternative materials, such as composite materials, may be increasingly important when used for suspension purposes.

2 CALCULATION OF COIL SPRINGS

In general, the equations of the German DIN standard 2089 can be used for an initial estimate of the dimensions and stresses of a cylindrical compression coil spring with a constant wire diameter. Figure 2 shows the definition of calculation values for one deformable coil ($n = 1$).

Encyclopedia of Automotive Engineering, print © 2015 John Wiley & Sons, Ltd.
Edited by David Crolla, David E. Foster, Toshio Kobayashi, and Nick Vaughan.
This article is © 2015 John Wiley & Sons, Ltd. ISBN: 978-0-470-97402-5
Also published in the *Encyclopedia of Automotive Engineering* (online edition)
DOI: 10.1002/9781118354179.auto001

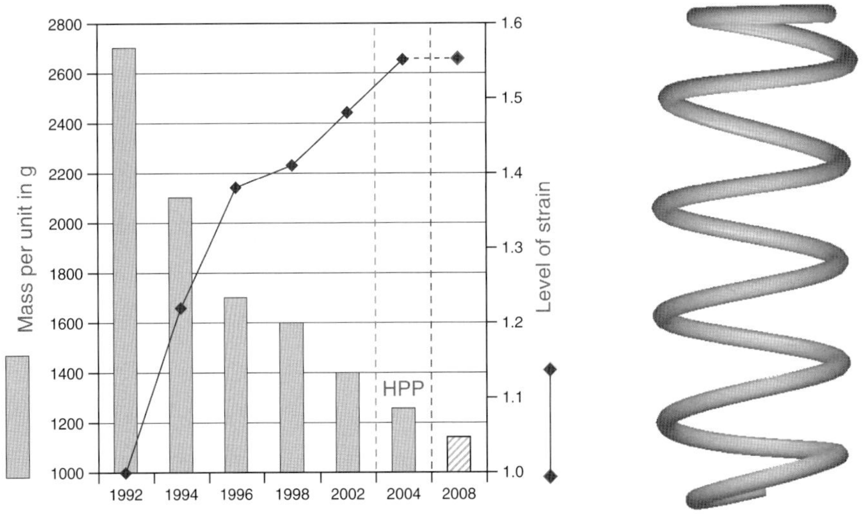

Figure 1. Evolution of weight and load stress levels of suspension springs based on a medium-sized vehicle, given constant vehicle weight. (From Neubrand *et al.*, 2010. Copyright © 2010 SAE International. Reprinted with permission.)

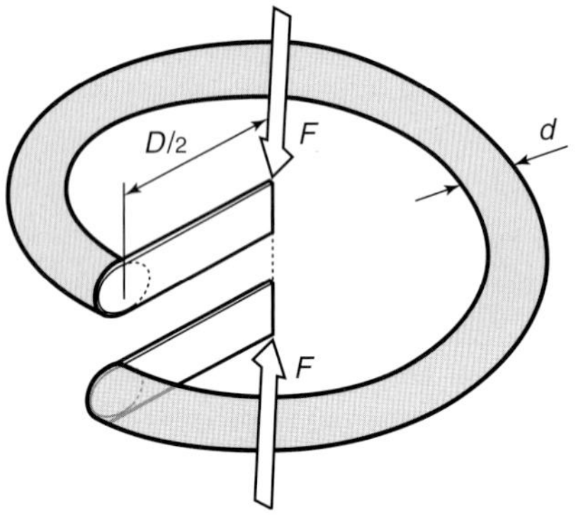

Figure 2. Compression coil spring calculation values. (Reproduced from Carlitz and Neubrand, 2008. With kind permission from Springer Science+Business Media.)

Spring work is defined as:

$$W = \frac{F \cdot s}{2} \tag{1}$$

where s is the spring stroke.

Spring force can be calculated as follows, considering G as the shear modulus:

$$F = \frac{G \cdot d^4 \cdot s}{8D^3 \cdot n} \tag{2}$$

Spring stiffness is given by

$$c = \frac{F}{s} = \frac{G \cdot d^4}{8D^3 \cdot n} \tag{3}$$

Shear stress may be calculated using the following equation:

$$\tau = \frac{G \cdot d \cdot s}{\pi \cdot D^2 \cdot n} = \frac{8F \cdot D}{\pi \cdot d^3} \tag{4}$$

Such stress equation is valid for a straight wire and torsion stress only. In coil springs, the wire is curved and the stresses on the inside are higher than those on the outside of the coil. Torsion and shear forces will have to be added. Moreover, the same force is being applied to a smaller area on the inside of the coil (Figure 3).

Therefore, stress has to be corrected by the factor k, depending on w, the ratio of the coil diameter to the wire diameter.

$$w = \frac{D}{d} \tag{5}$$

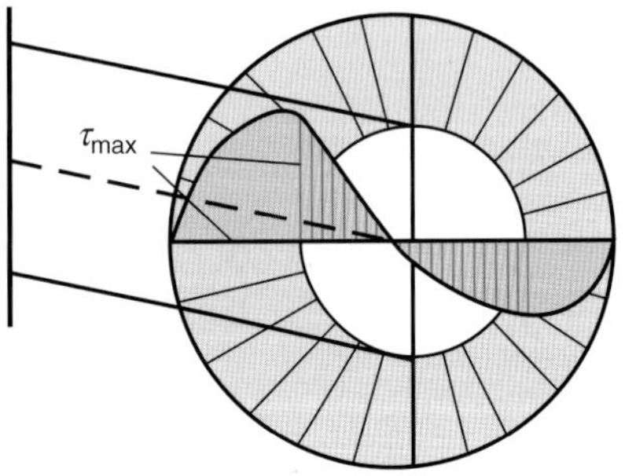

Figure 3. Stress increase on the inside of a coil. (Reproduced from Carlitz and Neubrand, 2008. With kind permission from Springer Science+Business Media.)

$$k = \frac{w + 0.5}{w - 0.75} \tag{6}$$

Corrected stress may then be calculated as follows:

$$\tau_k = k \cdot \tau \tag{7}$$

When designing a spring, the maximum stress τ_{max} is most important for minimizing the weight of a spring. Assuming a uniform deflection without any disturbing side forces, the mass of a cylindrical coil cannot be lower than a certain minimum mass m_{min} (Brandt, Kobelev, and Neubrand, 2007). This mass m_{min} is calculated according to Equation 1 with ρ as the density of the spring wire material and G as the shear modulus. The force F_{max} is the load at full jounce height. The coil spring mass m_{min} is reciprocally related to the maximum shear stress τ_{max}.

$$m_{min} = 2\rho \cdot G \frac{F_{max}^2}{c \cdot \tau_{max}^2} \tag{8}$$

In general, there are two ways to reduce coil spring weight:

1. Shape optimization for decreasing the load stress along the coil spring wire.
2. Increasing maximum allowed stress τ_{max}.

Until quite recently, it was common practice to decrease spring weight by increasing the stress level of suspension springs (Figure 1) and keeping robustness with certain measures regarding material and process optimization.

In real-life springs, load stresses are distributed nonuniformly along the coils. Therefore, it is necessary to calculate springs using a finite element analysis (FEA) as described, for example, in Georges (2000).

Owing to tremendous progress in calculation methods and corresponding production technologies, shape optimization is now possible for increasing the material usage by stress homogenization.

3 SIDE LOAD SPRINGS

The concept and design of a vehicle axle mainly defines the shape of a suspension coil spring. Moreover, the loads occurring and packaging are important factors. Two types of loads can be differentiated. First, the coil over shock, where both ends of the coil spring travel in line with each other, spring and damper form a single device. The second load application is to place one end of the coil spring on a suspension arm. Here, this end travels along a spatial curve, which leads to a nonuniform stress distribution along the spring profile and causes a distortion of the spring body (Carlitz and Neubrand, 2008).

The McPherson strut is a special case of the first kind of load. Owing to the conceptual axle design (Figure 4), unwanted moments occur, resulting in side loads at the

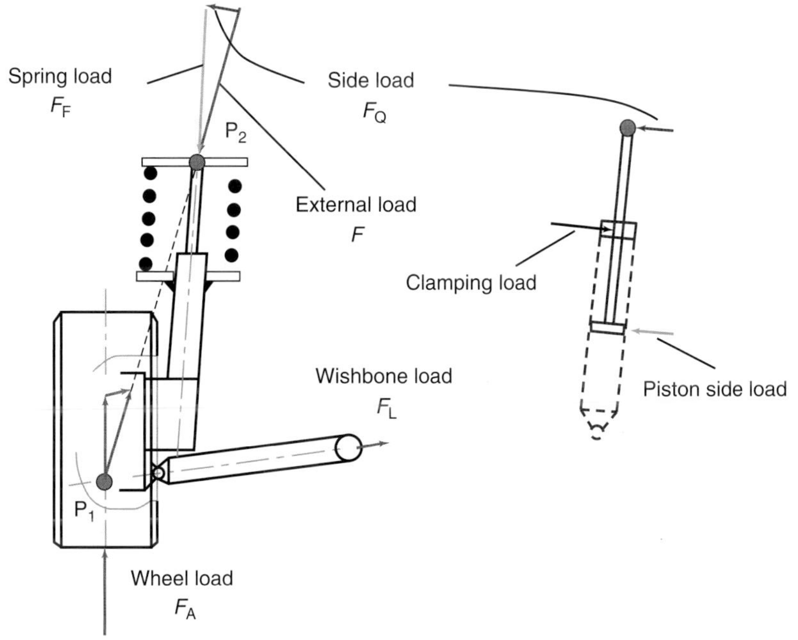

Figure 4. Principal kinematics of a McPherson strut. (From Neubrand *et al.*, 2010. Copyright © 2010 SAE International. Reprinted with permission.)

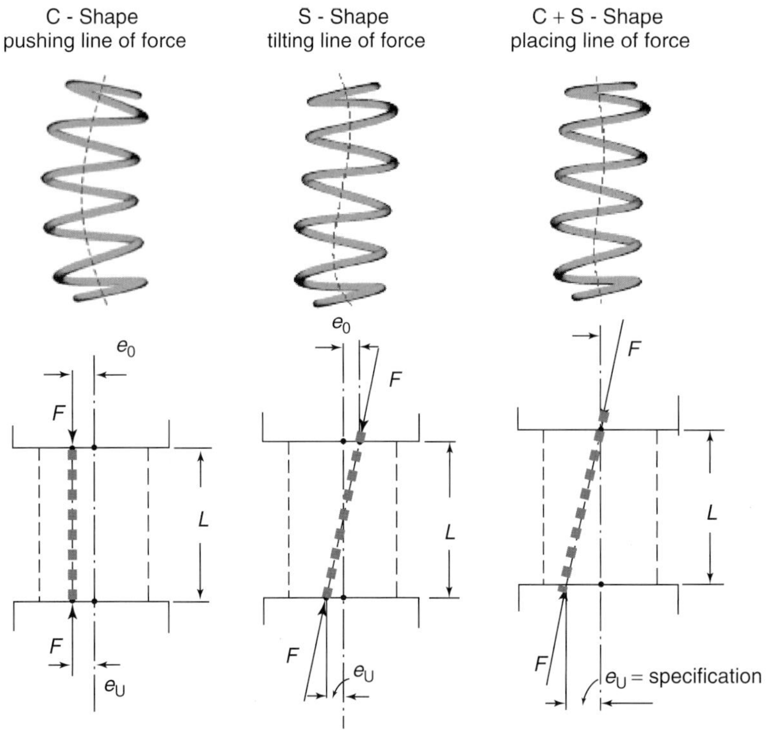

Figure 5. Adjustment of the load vector through spring shape. (From Neubrand *et al.*, 2010. Copyright © 2010 SAE International. Reprinted with permission.)

damper rod and therefore, to an increased friction along the damper rod and damper piston bearings. This friction does not only generate increased wear and tear of the damper piston and seals but driving comfort may also be reduced. A slip–stick effect is noticeable especially at very small damper travels.

Until the late 1980s, this problem had been solved by tilting a huge cylindrical coil spring on a McPherson strut, which offered compensation for such unwanted moments. However, as packaging space for suspension components becomes ever smaller, this solution was soon obsolete because of the voluminous coil springs needed to allow for tilting.

For this reason, the development of side load (SL) springs had been started (Muhr and Schnaubelt, 1991; Kobelev *et al.*, 2002; Brandt and Neubrand, 2001). These counteract the unwanted moments within a small package because of geometrical intelligence. These springs are showing an S-shaped centerline in unloaded condition, which leads in loaded condition to a tilted load vector and fully compensates the unwanted moments around the McPherson strut (Carlitz and Neubrand, 2008; Muhr and Schnaubelt, 1991; Kobelev *et al.*, 2002; Brandt and Neubrand, 2001).

Furthermore, the load vector of these coil springs can be further attuned by adjusting spring geometry, without having to change spring seat designs. Moreover, spring seats can be minimized, because SL coil springs are mostly designed as double-pig-tail springs.

The principle of control of the load vector is shown in Figure 5, and the load vector offset can be "pushed" by a C-shape and "tilted" by a pure S-shape. A combination of these two adjustments leads to a typical SL spring, where the load vector can be "placed."

The technology of the SL spring as explained earlier triggered the development of multiple versions and modifications of spring designs and as of today side load technology is used in more than 70% of all McPherson strut applications worldwide.

Moreover, the ability for controlling the load vector of coil springs could also be used for additional functions, such as height leveling systems as described in Carlitz, Neubrand, and Hengstenberg (2005).

4 NEW GENERATION COIL SPRINGS

Nonparallel coil spring deflection leads to a nonuniform load stress distribution along a coil spring. Even though an SL coil spring is a coil over shock, the deflection is nonparallel. The load vector is inclined relative to the axis

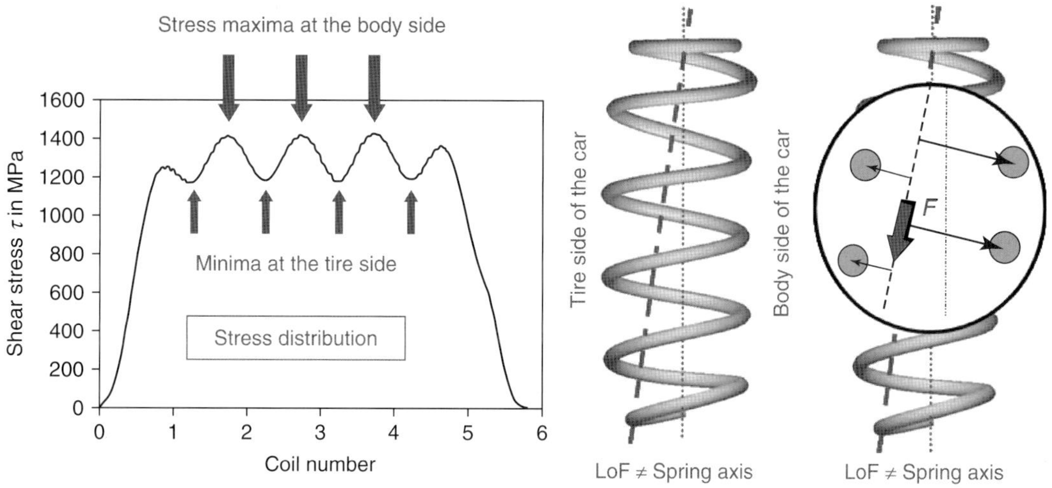

Figure 6. Stress distribution of a conventional SL spring with inclined load vector. (From Neubrand *et al.*, 2010. Copyright © 2010 SAE International. Reprinted with permission.)

of the damper and this is causing a nonuniform distance of coil spring wire to the load vector. Distances from the wheel are smaller than the distances on the opposite side of the load vector (Figure 6).

The distance from the coil spring wire to the load vector also reflects the moment arm of the moment around the coil spring wire centerline. Thus, nonuniform deflection of the spring and an inhomogeneous stress distribution occur under load (Figure 6). This leads to local stress minima at the coil position toward the wheel and local maxima toward the vehicle body. Hence, material utilization is not at its best and spring weight is too high, considering maximum stress τ_{max} as the limiting factor.

Therefore, a significant weight reduction and optimal materials utilization may be achieved by homogenizing stress distribution. Modern FEA methods as well as corresponding manufacturing processes are offering such a chance.

The following equations are used to describe the true shape of a suspension spring (Figure 7; Neubrand *et al.*, 2010).

$$\underline{r_g} = \left[\cos(n)\cdot R(n) \quad \sin(n)\cdot R(n) \quad z(n)\right]^T \tag{9}$$

$$\underline{r_m} = \left[x(z) \quad y(z) \quad z(n)\right]^T \tag{10}$$

$$\underline{r} = \underline{r_m} + \underline{\underline{T}}\cdot \underline{r_g} \tag{11}$$

These equations can be used for a numerical determination of local stresses of the whole spring wire during deflection. Latest FEA software offers an optimized coil spring shape respecting all restrictions such as package, load vector, rate, and maximum stresses and the shape is automatically translated into required computer-aided design (CAD) formats. Automated pre- and postprocessing as well as parametrical definition of the spring shape allow for systematic adjustment of coil spring parameters while solely focusing on the FEA output.

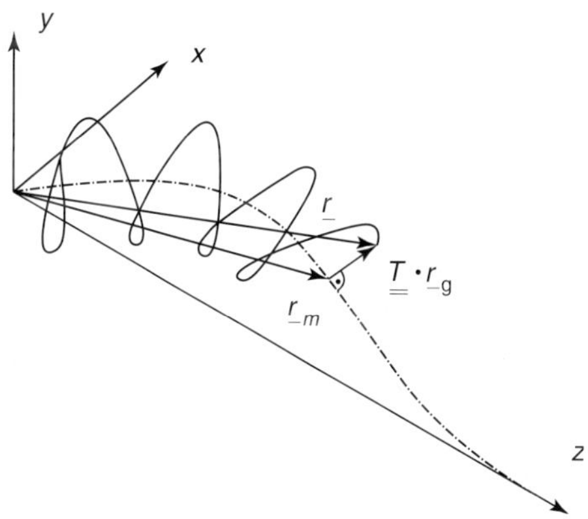

Figure 7. Parametric description of an SL coil spring. (Reproduced from Neubrand *et al.*, 2001. © R. Brandt and J. Neubrand.)

Using this powerful method, it is possible to automatically generate the uniform load stress distribution of a new generation spring (Figure 8). Coil spring geometry is optimized in a way that preserves the defining spring characteristics, while load stress is uniformly distributed along the coil spring profile. Hence, the lowest possible spring mass m_{min} has been numerically calculated.

This new generation spring still offers all advantages of a traditional SL spring regarding side load compensation within small package space and furthermore, a huge weight reduction due to perfect material utilization.

This method of design may also be applied to other kinds of coil springs. The following example shows a rear

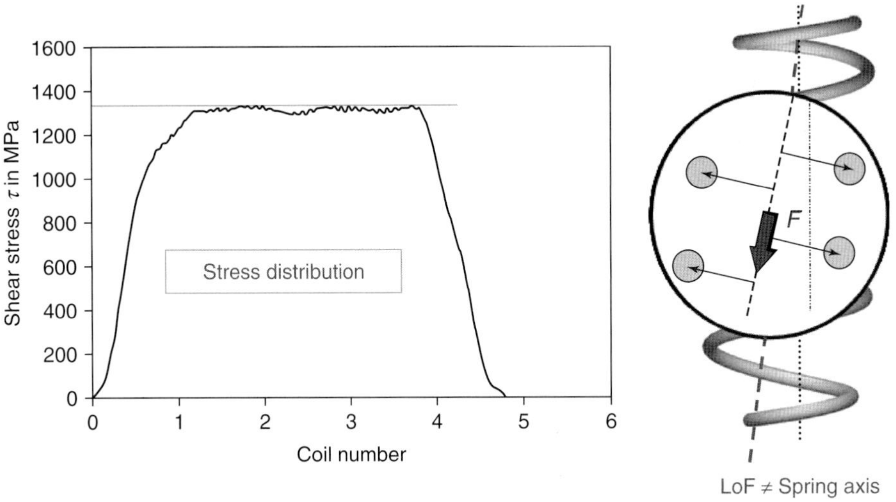

Figure 8. Uniform load stress distribution with Mubea SL new generation spring. (From Neubrand *et al.*, 2010. Copyright © 2010 SAE International. Reprinted with permission.)

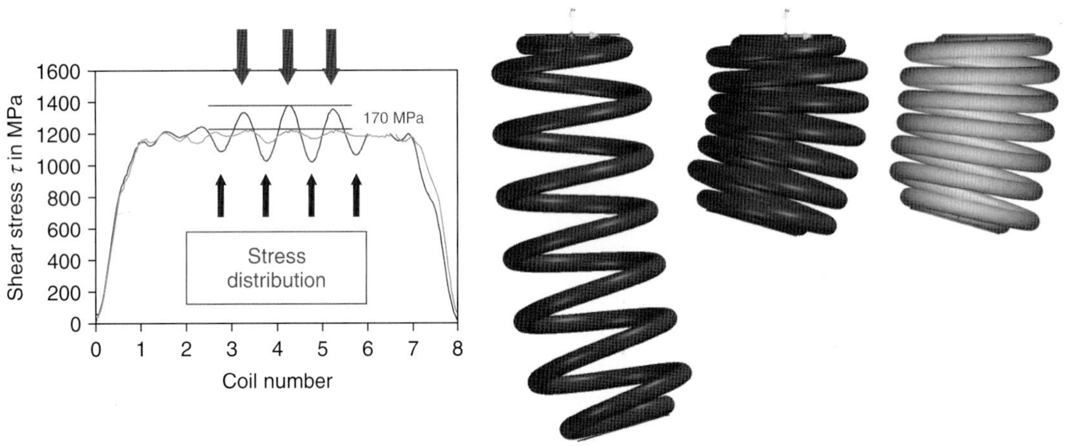

Figure 9. Mubea new generation suspension spring for axles with spatial curve travel. (From Neubrand *et al.*, 2010. Copyright © 2010 SAE International. Reprinted with permission.)

axle coil spring, where the spring seats deflect along a spatial curve (Figure 9). The light gray springs show a conventional cylindrical spring at rebound and full jounce condition, whereas the dark gray spring shows a new generation spring at full jounce condition.

Spring deflection in such a system is highly nonuniform; deflection extremes even change from one direction (at rebound height) to the other side (at jounce height). The result of this nonuniform deformation once more is an inhomogeneous load stress distribution along the coil spring profile (Figure 9). Furthermore, coil clearance issues particularly at jounce height may occur.

Applying the same methods employed for new generation spring design and relying on the same principles as were employed for strut application, uniform deflection, and load stress distribution can be achieved. Just like for strut applications, significantly better steel utilization results in welcome weight reduction while load stress limits have not been increased.

The idea of stress homogenization by new generation design and, therefore, better material utilization offers a weight reduction of more than 10% without increasing maximum stress and/or better robustness.

Coil springs are being produced by hot or cold forming. In recent years, cold forming won significant market shares because of the high flexibility in forming complex shapes. The latest production processes can transfer the FEA design into the forming operation and plastic as well as elastic deformations during manufacturing may be anticipated.

Coiling programming systems have been developed to efficiently produce most complex coil spring designs with the highest possible dimensional capability.

5 MATERIALS AND ROBUSTNESS

Spring materials must have superior properties regarding storing and releasing huge amounts of elastic energy. Therefore, materials with high shear and elasticity modulus are showing the best properties for springs. Furthermore, materials should have high strength and high elasticity to withstand high loads without any plastic deformation (Neubrand, 2004). Properties such as corrosion resistance, sag loss resistance, and dynamic fatigue strength are very important; in addition, superior toughness is also important, which is increasing resistance against notches and cracks.

In general, steel is still fulfilling most of these required properties as best in class. Several alloy elements can be added for improving the steel. Until the late 1980s, CrV steels were commonly used for coil springs. Nowadays, SiCr steels represent a major part of vehicle suspension springs because of their very good properties, such as high toughness at great strength and high settling strength, even at higher temperatures.

Until the late 1980s, material impurities such as inclusions represented a major problem for springs regarding durability, especially in case of high cycle fatigue. Super clean steels were developed as the superior material for valve spring applications, avoiding critical inclusions as far as possible (Neubrand, 2004) by creation of deformable, low melting inclusions.

Continuously decreasing packages, increasing vehicle weights and higher demands regarding component lightweight design caused a significant increase of tensile strength and hardness. The inductive tempering represents a milestone of improvement in this direction, because it provided an excellent fracture toughness for tensile strength higher than $R_m = 2000\,\text{MPa}$. In addition to that, corrosion robustness is assuming ever greater importance. This is the result of combining higher stresses with high tensile strength material under corrosive environmental conditions, when the topic of stress corrosion cracking is assuming ever more relevance. In principle, a distinction is made between two corrosion mechanisms: corrosion in the presence of oxygen (rust), where corrosion scars will be formed by a local dissolving of iron and corrosion under the influence of acids, which is better known as *hydrogen embrittlement*, that is to say, the material will get brittle when cathodic hydrogen will be formed along the grain boundaries (Carlitz and Neubrand, 2008).

These two processes may be counteracted by a number of steps:

- turning the layer of rust into a covering layer by alloying metals such as Ni, Mo, or Cu, thus reducing the speed of corrosion;
- forming hydrogen traps with the help of V (vanadium) or Ti (titanium), where hydrogen will be held up and where there will be less hydrogen embrittlement as a result.
- bringing up fracture toughness by fineness of grain formers such as V, so that tension peaks at corrosion scars can be more easily reduced as a result of plastic distortion.

Using such modifications of materials, such steels as the so-called HPM190 were developed by changing chemical composition on the basis of classical 54SiCr6. Especially in Japan, other alloys were also developed. However, they always limited the strength of steel, so that it was not possible to use the weight reduction potential to the full. Over and above, such modifications of materials using alloy elements are very expensive and they very often result in "special steels." These can only be used for individual spring or car factories and they will not be made in sufficient quantities to gain a sustained foothold in the market.

Over the past few years, a number of very interesting techniques were brought to series production that increase toughness at high strength. Thermomechanical treatment offers one case in point. Here, material will be formed during the austenitic stage in such a way that a very fine grain structure will be produced during the martensitic state that results in very good mechanical properties.

Classical tempering in various types offers additional methods to bring about very small grain sizes. The temperature for austenitization is quite important for the grain size evolution. An adequate method to create a fine microstructure in the rim of a wire could be a heating at high temperatures for hardening the whole wire section, followed by an austenitization at low temperatures just within the rim to create a fine microstructure (Liu, 2011). Other methods such as intensive mechanical working by multiple drawing processes as well as subsequent simple or multiple tempering were investigated. Here, a multitude of nuclei generated by mechanical working will be used to produce small grains. Shaping and tempering parameters need to be very finely balanced for all these methods. Surface layer modification is a procedure of its own (Neubrand and Hartwig, 2009), in that it combines a tough lower strength marginal layer with the hard core of the wire. Giving it some thought, this method can be integrated into inductive wire tempering processes, which makes it very economical (Figure 10).

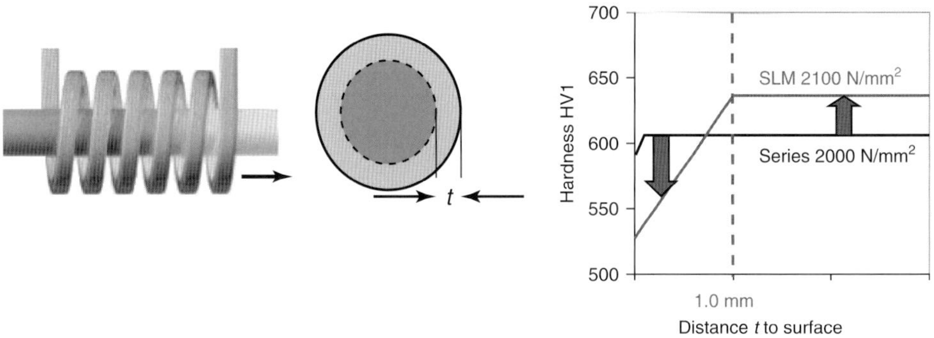

Figure 10. Mubea surface layer modification.

Over the past decade, alternative materials have been investigated to a large extent but could not really move from applications in racing and exotic cars to volume car manufacture. Some particular mass production applications occurred on and off, such as titanium springs for the VW Lupo FSi (Schauerte *et al.*, 2001) and composite leaf springs for the Daimler Sprinter and VW Crafter, but the main reason against greater usage in volume cars can be defined as poor cost competitiveness compared to steel springs. Nevertheless, steel still has some further potential for improvement, but the properties are going to reach a certain limit and therefore, new material will become of greater importance, also in the spring world.

6 COMPOSITE SPRINGS

From the materials point of view, composite materials harbor specific potential. Of practical relevance are combinations of glass and carbon fibers with pressure setting plastics material. Thermoplastics materials are also gaining importance as matrix materials; however, because of technological restrictions, they have been unable so far to meet suspension challenges at high dynamic loads. Glass fibers represent a better material for spring application because their lower elasticity module compared to carbon fibers is favorable regarding high strokes and deforming requirements. Given their high specific strength and the stiffness of composite materials, it becomes in principle possible to achieve weight reductions ranging from 30% to 70%. While reducing unsuspended masses it is also possible to reduce driving dynamics and also as regards noise, vibration, and harshness (NVH) behavior, composite materials may offer advantages when compared to steel, because their material amplitudes are very much higher. In view of high corrosion resistance and resistance against the influence of other environments, surface protection mostly is something that can be discarded, but special protection will be needed when it comes to possible damage by rocks. Studies into the comprehensive energy input for making compound components have shown that their CO_2 footprint is larger than that for steel making (Geuder, 2004). However, this may be compensated for partly or wholly in view of the considerable reduction in quantities required during operational cycles.

However, the potential of this group of materials is also limited by serious disadvantages, which have so far prevented the use of fiber composite materials in large quantities. The load transmission is demanding special designs. Typically, quite frequently high loads come up cross-wise to the main load direction so that the material is not taking up a load in an ideal way and following the direction of the fiber, so that only medium loads may be imposed on the matrix. In addition, allowances will have to be made when it comes to large series production and available manufacturing processes when comparing these with units made of steel. At present, these represent the focal point of research and development efforts all over the world.

Continuous reinforced compound fiber materials as they are used for structural elements in car-making reveal strong anisotropic, that is to say, direction-dependent properties. Fibers employed will be oriented as to the loads occurring so that they take up tension and pressure, if possible. Typical fiber volume percentages for structural elements reach about 60%.

When it comes to suspension, leaf springs made of composite fibers represent an application offering high potential for glass-fiber-reinforced plastics in the chassis section that may bring about industrialization and large quantity manufacturing (Figure 11).

Predominant bending requirements for leaf springs offer a possibility to align fibers in one direction only. When compounded, their elasticity modulus may reach about 45,000 MPa. Compared to conventional multi-leaf springs made of steel, it is possible to score considerable reductions

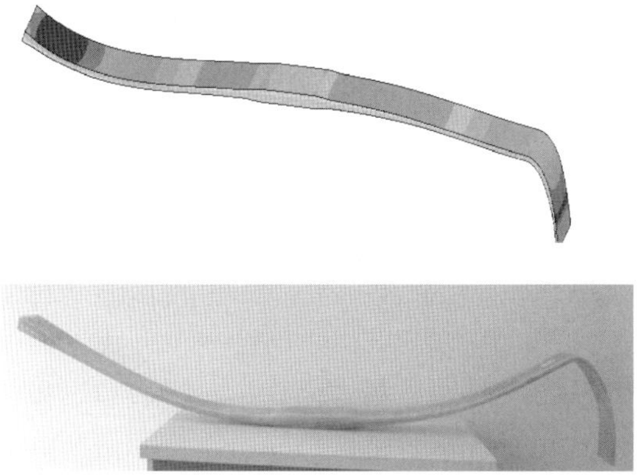

Figure 11. Tension leaf spring made from glass-fiber-reinforced plastics. (Schürmann and Keller, 2010.)

in weight, amounting up to 70%. The number of leaves in a multi-leaf spring may be reduced; sometimes springs may be replaced by a one-leaf spring only. Tension leaf springs (Schürmann and Keller, 2010) represent a specific development, which might replace multi-leaf springs. Over and above reduced weight, it might also be possible to set up progressive spring characteristics.

Using fiber compound materials, it may also be possible to improve damping properties. Contrary to steel leaf springs, spring eyes will not be shaped, they have to be linked to the interfaces with a wedge. A number of different types emerged for steel leaf springs, and they come in differing width and breadth. Such structural shapes may also be offered in glass-reinforced fibers. The simplest shape is a rectangular spring with a constant cross section. Other shapes of cross sections lead to (approximate) constant bending stress, which means a constant use of material employed. Hyperbola-shaped springs may easily be manufactured of glass-fiber-reinforced plastics. In this type of spring, there is a linear reduction of width from the center, whereas the cross-sectional area remains constant. For springs of this type of manufacture, no cut fibers need to be used and it is quite easy to make a blank, or a preform, of various sections. There is a constant width for parabolic springs and their width will be adapted to the load course. Given their varying cross sections, different quantities of fibers will be needed over the course of the spring. This is equivalent to more input in complex manufacture; on the other hand, parabolic springs are weight-optimized structural types, and constant width is in keeping with the installation space situation in a car (Franke, 2004). It is also possible to bring about a compound-fiber compromise between a rectangular and a parabolic spring, by designing a smaller change in width than for the parabolic spring, which may be achieved by a variable fiber volume percentage (Götte, 1989), so that fibers will not have to be cut. Variations in the range 45–65% in the fiber volume would be possible. In practice, however, such low fiber volume percentages may lead to unequal distribution of fibers over the entire cross section.

There is a large number of manufacturing methods for the production of fiber-plastics compounds. Because of the resulting good properties at dynamic loads, the prepreg method stood the test of time for leaf springs. Prepreg processes produce optimum bond strength (Schürmann, 2007).

Component manufacture here first involves impregnating fibers with resin so that subsequent parts of the process may work with strip sections easy to handle (Figure 12).

An impregnated semifinished product represents the first step (preimpregnated fibers). To make it, the resin system and the fiber will be brought together on backing paper while adding some heat. The cross-linking reaction of the resin begins when a prepreg is made. This process, however, may be interrupted by cooling, which means that a weakly cross-linked stage will be reached, and the resulting material may be stored when cooled down. Afterward, prepreg strips will be cut and a blank will be prepared. The blank will then be put into a compression mold and cured at high pressure and temperatures between 110°C and 170°C. Springs will then be subject to mechanical postprocessing.

It is also possible to manufacture leaf springs using a resin injection process. For this resin injection process,

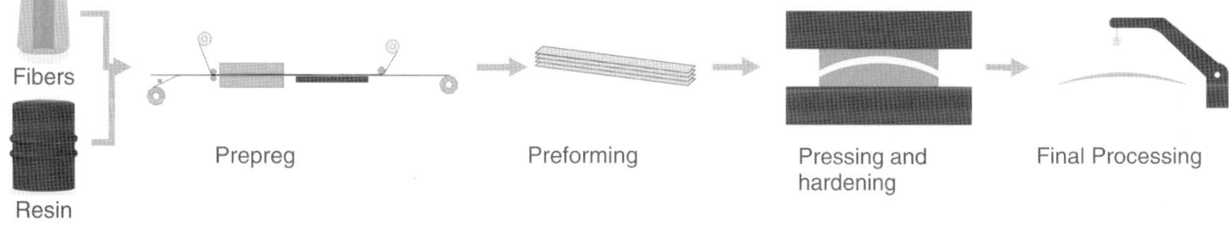

Figure 12. Sequence of operations during spring leaf manufacture, using the prepreg process. (From Müller, 2010. Reproduced by permission of David Müller, Mubea. © Mubea.)

a fiber structure will first be manufactured of the dry reinforcing fibers, which follows the component geometry required. Structural cohesion may be achieved as needed using textile methods, such as sewing processes or involving binder, gluing fibers together. Such fiber structures are known as *preforms*. This preform will then be inserted into a pressing tool, which may be used to inject the resin system. The tool has the jets required and the resin system will be fed into this unit by a mixing device. Resin injection comes in a large number of varieties, distinguished primarily by the type of injection used. Pressure, vacuum, and combined processes are used to achieve the desired results. There are also variations of closing the tool and applying press pressures. The term *resin transfer molding* (*RTM*) is used for the group of pressure-supported resin injection methods (Mitschang and Neitzel, 2004). Compared to prepreg methods, resin injection processes use resins with considerably lower viscosity, as fibers will have to undergo a wetting process as complete as possible within a very short time frame. This is decisive for resultant compound properties and at the same time, this is also the decisive disadvantage for this particular process when manufacturing thick-walled components.

Conventional suspension springs, used as helical suspension springs, offer another field of application for fiber compound materials. Such helical compression springs are mainly made of steel today. Figure 13 shows different types of suspension springs, which are currently investigated regarding feasibility in fiber composites. Specific properties of fiber compound materials are integrated into their design in various types of manufacturing (Müller, 2010). In the case of a mere materials substitution when comparing a steel spring and a helical compression spring (Sardou, 2002), the direction of the fibers must be chosen in such a way that torsion loads may be borne by the spring in an appropriate manner. A classical orientation is to position fiber layers interchangeably in $+45^{\circ}$ and -45° directions. This corresponds to the direction of the main normal tension in cases of pure torsion.

For spring cross sections, both a circular cross section and a circular ring cross section similar to a tube have been proposed. The circular ring cross section not only offers additional weight-saving potential but it also involves more installation space for the finished product. A significantly reduced shear modulus of compound fiber materials using a $\pm 45^{\circ}$ layering when compared to steel results in the fact that the spring geometry of a steel spring cannot be transferred, but that it must be rethought. Composite fiber materials have a low compressive resistance. Therefore, the spring interfaces must differ from those for steel springs, and the design of these must consider the reduction of surface stresses.

A bellows-type spring structure offers one possibility of materials-adapted design (Marquar *et al.*, 2010a).

As opposed to a conventional helical spring, this spring element bears loads by bending and membrane tension. Thus, it is possible to make full use of fiber strength.

A third type of design is the meander-type spring (Kobelev *et al.*, 2008). It is almost exclusively bearing bending loads, which means that it may also be used effectively for compound fiber applications. Subdivided into two elements, this type of spring may also be placed around a shock absorber.

First concepts are also available for designing wheel-bearing units and shock absorbers to be made of compound fiber material. Because of high stiffness demands, carbon-reinforced plastics are given preference here (Marquar *et al.*, 2010b). Compound fiber material is very appropriate for such complex components, because the number of elements may be brought down and additional functions

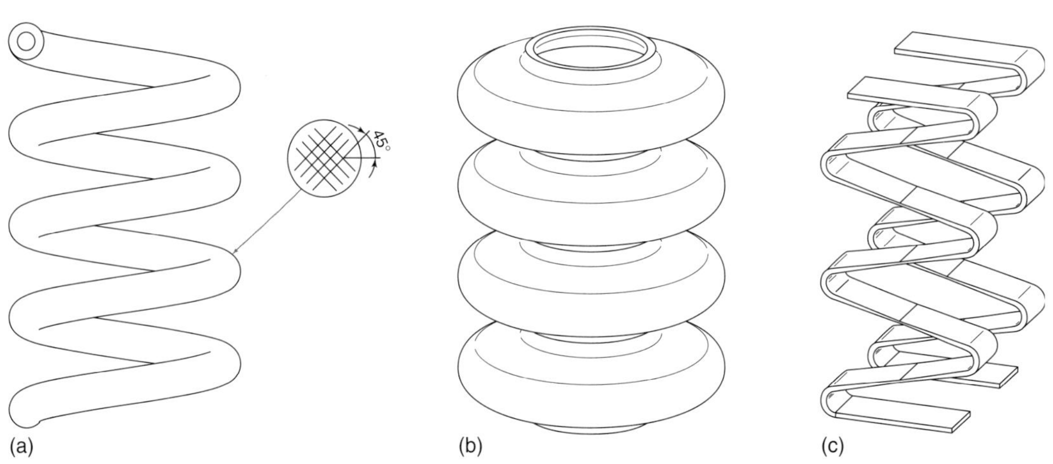

Figure 13. Different types of composite springs: (a) coil spring, (b) bellows-type spring, and (c) meander-type spring.

can be integrated. As already known from air transport, it is for instance possible to integrate sensors directly into such material, so as to protect them against external stress.

A significant disadvantage of compound fiber materials is to be seen in that manufacturing technologies lack sufficient flexibility to represent various forces and constants as taken up by springs. Given a large platform, a great diversity of brands, engine types, and configurations must be implemented for the same installation space with similar connection concepts. This means that quite readily one may need over one hundred different spring force/spring constant combinations for one platform and for one axle at each platform. For the present compound fiber concepts, this would need a new tool for every new component number to produce the exact spring geometry. In case of steel, an isotropic material, this can be achieved fairly easily by varying wire diameters, the number of active windings, or the diameters of coil windings.

REFERENCES

Brandt, R. Kobelev, V., and Neubrand, J. (2007) Simulation und FE-Analyse der Fahrzeugfedern. Presented at the Seminar Fahrzeugfedern, Technische Akademie Esslingen, March 14.

Brandt, R. and Neubrand, J. (2001) Kaltformtechnik für PKW-Tragfedern, Kontrolle der Kraftwirkungslinie und ihr Einfluss auf die Dämpferreibung unterschiedlicher Achssysteme. Presented at Fahrwerke, ihre Komponenten und Systeme, Haus der Technik Essen, February 2.

Carlitz, A. and Neubrand, J. (2008) Federn und Stabilisatoren, in Fahrwerkhandbuch (eds B. Heißing and M. Ersoy), Vieweg Verlag, Wiesbaden. ISBN: 978-3-8348-0444-0

Carlitz, A. Neubrand, J., and Hengstenberg, R. (2005) Radaufhängung mit Federverstellung für Kraftfahrzeuge. Patent EP05025854.0-1264, Muhr und Bender, November 26.

Franke, O. (2004) Federlenker aus Glasfaser-Kunststoff-Verbund—Spannungs- und Festigkeitsanalyse zur Optimierung eines hochbelasteten Bauteils, Shaker Verlag, Aachen.

Georges, Th. (2000) Zur Gewichtsoptimierung von Fahrwerkstragfedern unter besonderer Beachtung des schwingfestigkeitsmindernden Einflusses bruchauslösender Fehlstellen im Halbzeug Federdraht. Dissertation. VDI Verlag, Düsseldorf.

Geuder, M. (2004) Energetische Bewertung von Windkraftanlagen. Diplomarbeit FH Würzburg-Schweinfurt.

Götte, T. (1989) Zur Gestaltung und Dimensionierung von Lkw-Blattfedern aus Glasfaser-Kunststoff, VDI-Verlag GmbH, Düsseldorf.

Kobelev, V. Neubrand, J. Brandt, R., and Lebioda, M. (2002) Radaufhängung. Patent DE10125503C1, December 12.

Kobelev, V., Westerhoff, K., Neubrand, J., Brandt, R., and Brecht, J.D. (2008) Patentschrift EP 2 082 903 B1; May 21.

Liu, Y. (2011) The analysis of the influence in grain refinement through increased dislocation density by cold drawing and heat treatment of Si-Cr spring steel, depending on the degree of deformation. Masterarbeit. Universität Siegen/Muhr und Bender.

Marquar, H., Schuler, M., Renn, J., *et al.* (2010a) Offenlegungsschrift DE 102010040142 A1, Anmeldetag September 2.

Marquar, H., Schuler, M., Renn, J., *et al.* (2010b) Offenlegungsschrift EP 2295826 A2, Anmeldetag September 7.

Mitschang, P. and Neitzel, M. (2004) *Handbuch Verbundwerkstoffe—Werkstoffe*, Verarbeitung, Anwendung, Hanser Verlag, München Wien.

Muhr, K.-H. and Schnaubelt, L. (1991) Radaufhängung mit einem radführenden Federbein. Patentschrift DE3743450C2, March 28.

Müller, D. (2010) Entwicklung eines Prozesses zur Fähigkeitsbewertung von dynamisch beanspruchten Federelementen aus FVW. Masterarbeit. Universität Siegen/Muhr und Bender.

Neubrand, J. (2004) *Entwicklungstendenzen bei Werkstoffen für Fahrwerksfedern, Beitrag zur Tagung Federung und Dämpfung*, Car Training Institute, Düsseldorf.

Neubrand, J. Brandt, R. Junker, C., and Lindner, A. (2010) Light weight suspension coil springs by advanced manufacturing techniques and innovative design definition methods. SAE World Congress.

Neubrand, J. and Hartwig, M. (2009) Gehärteter Federstahl, Federelement und Verfahren zur Herstellung eines Federelements. Patent EP 2 192 201 A1, Muhr und Bender, November 23.

Sardou, M. (2002) Patent FR 000002837250 B1, Anmeldetag March 18.

Schauerte, O., Metzner, D., Krafzig, R., *et al.* (2001) Fahrzeugfedern federleicht. *Automobiltechnische Zeitschrift*, **103**, 654 ff, Wiesbaden.

Schürmann, H. (2007) *Konstruieren mit Faser-Kunststoff-Verbunden*, 2.Auflg., Springer Verlag, Berlin.

Schürmann, H. and Keller, T. (2010) GFK-Blattfeder für Transporter Hinterachse mit progressivem Verlauf der Federkennlinie ohne Federaugen. Patentanmeldung 102010015951.4, TU Darmstadt, March 12.

Chapter 110

Air Suspension Systems—What Advanced Applications May Be Possible?

Friedrich Wolf-Monheim
Ford Motor Company, Aachen, Germany

1 INTRODUCTION

Air suspension systems offer various advantages compared to conventional suspension systems, featuring usual coil or leaf spring elements, leading toward improvements in terms of the ride comfort, the vehicle dynamics performance, and the driving safety. The advantages of air suspension systems in general relate mainly to the practicability of a vehicle body leveling functionality, the technical feasibility of a constant eigenfrequency independent of the loading condition of a vehicle, and the flexibility in terms of the tuning potential of the force versus travel characteristics of the air spring elements by shaping the rolling pistons.

Using air suspension systems, the distance between the vehicle body and the road surface can be varied within the limitations given by the available wheel travels with the aid of the defined supply and exhaustion of pressurized air to and from the air spring elements. On the one hand, the vehicle body can be raised compared to the design height to generate a greater ground clearance between the vehicle underbody and the road surface under severe off-road driving conditions. On the other hand, the vehicle body can be lowered compared to the design height to allow for an unimpeded ingress and egress for all vehicle passengers when the vehicle is standing still. Similarly, the lowering of the vehicle body can enable the reduction of the aerodynamic driving resistance at higher vehicle speeds, for example during highway driving. At the same time, the body lowering improves the road holding abilities of the vehicle, the vehicle handling stability and finally the brake stability of the vehicle. A well-known possibility to reduce fuel consumption and to maintain stability at high speeds is changing the vehicle body pitch angle (lowering the vehicle body height at the front axle only). This functionality is currently used in the Bentley Continental GT.

In the same way, the vehicle body height can be maintained constant independent of the loading condition of the vehicle by varying the air pressures in the individual air spring elements. As a result, also the body eigenfrequency remains unchanged as the air spring stiffnesses can be adapted to the loading conditions. This effect improves the ride comfort for the vehicle passengers.

Furthermore, the geometric design of the rolling pistons can be used to adapt the force versus travel characteristics of the air spring elements to the specific tuning needs over a wide range. Compared to conventional coil or leaf springs, highly nonlinear curve characteristics can be accommodated.

Encyclopedia of Automotive Engineering, print © 2015 John Wiley & Sons, Ltd.
Edited by David Crolla, David E. Foster, Toshio Kobayashi, and Nick Vaughan.

ISBN: 978-0-470-97402-5
Also published in the *Encyclopedia of Automotive Engineering* (online edition)
DOI: 10.1002/9781118354179.auto004

Basically, all air suspension systems consist of air spring modules as primary suspension force elements, a pneumatic compressor in order to ensure the supply of pressurized air, an electronic control unit to handle the control tasks, one or more sensors to detect the ride height and the tilting of the vehicle body as well as pneumatic and electrical connection lines. Some air suspension systems also include a compressed air reservoir in addition to the pneumatic compressor in order to supply the air spring modules with high air mass flow rates if required.

In contrast to the commercial vehicle sector, almost exclusively rolling lobe air spring modules are used for passenger cars. With regard to the ride comfort of air suspended vehicles, the initial dynamic response of the individual air spring modules is particularly important (Wallentowitz, 2005). The contribution of the air spring module stiffnesses with decreasing excitation amplitudes to the total wheel suspension vertical stiffnesses plays a major role in this context. Increasing the wheel suspension vertical stiffnesses results in higher vertical body accelerations with a negative impact on the ride comfort for the vehicle passengers. In the field of automotive engineering, the increase of the wheel suspension vertical rates with decreasing excitation amplitudes is usually referred to as the *harshness effect* (Puff, 2009). Figure 1 illustrates the dynamic air bellow[1] stiffness differences between comfort bellows (one axial fiber layer) and conventional bellows (two crossed fiber layers) with respect to the excitation amplitude.

The frequency dependency starts already slightly above 5 Hz. Thus, the passengers can feel the stiffening of the air bellows. This reduces the driving comfort. Several car companies therefore decided to use comfort bellows and designed air spring systems with externally guided bellows. These differences are explained in more detail in the following: essentially rolling lobe air spring modules can be divided into those that are externally guided and those that are not. From a technical point of view, air spring modules without external guiding can be best executed with cross-ply lobes. Cross-ply lobes normally consist of two or more layers of rubber-coated rayon or nylon cord laid in a cross-ply manner in an angular arrangement with respect to the direction of motion of the air spring unit with an outside layer of abrasion resistant rubber and sometimes an additional internal layer of impermeable rubber to minimize the loss of air (Heisler, 2002). The cross-ply design enables the stability of the geometrical outer shape of the air spring lobe, even under high internal air pressure, without an external guiding tube, as it allows the lobe to carry not only the axial acting force components but also the radial acting force components of the internal air spring pressure. Naturally, cross-ply lobes show greater wall thicknesses compared to axial lobes with external outer guiding tubes. This is the reason for the higher bellow stiffness in Figure 1. In addition, the hysteresis of these bellows during jounce and rebound motions of the suspension is larger compared to the hysteresis of comfort bellows. It is well known that the more friction and hysteresis there is in a suspension system, the less the system deflects for small road irregularities, giving rise to the vehicle riding on its tires rather than on its suspension; this leads to a "busier" secondary ride and to what the Americans call "boulevard jerk". The behavior of former diagonal tires compared to the behavior of radial tires of today is in some way comparable to the behavior of air bellows, which is discussed in this chapter.

The lobes of air spring modules with external outer guiding tubes can be designed with lower wall thicknesses compared to cross-ply lobes without external outer guiding tubes, as the external outer guiding tube carries the circumferential forces. Usually, axial lobes are made from one

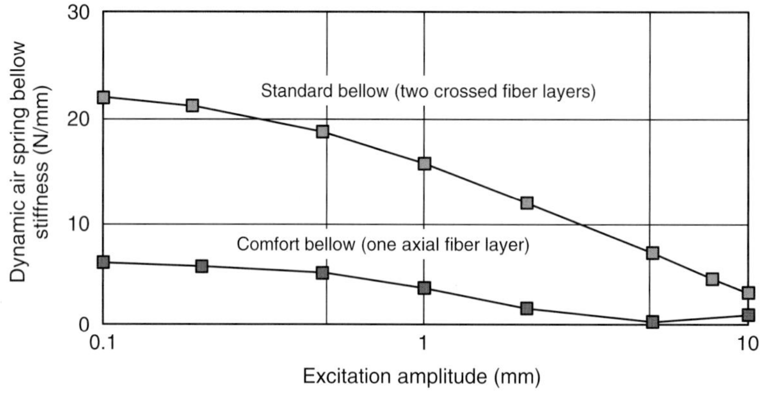

Figure 1. Differences in the dynamic stiffness of one-layer axial air spring bellows and two-layer cross-ply air spring bellows, indicating harshness differences.

layer of rubber-coated rayon or nylon cord in a parallel arrangement with respect to the direction of motion of the air spring unit with an outside layer of abrasion resistant rubber and sometimes an additional internal layer of impermeable rubber to minimize the loss of air (Heisler, 2002). Owing to the lower wall thicknesses of air spring lobes with external guiding tubes, the harshness effect is usually significantly lower compared to air spring lobes without external guiding.

Generally, air spring modules with outer guiding are more limited in terms of the allowable cardanic movements between the upper and lower ends within a given suspension architecture compared to air spring units without outer guiding. This disadvantage can be compensated by alternative guiding concepts of the air spring lobe, for example, the use of pivoted rolling pistons. Next to the design of air spring modules, also the history of the modules (aging) influences the harshness characteristics. In terms of the air spring module design not only the external guiding elements of the air spring lobes but also the upper and lower connecting joints (ball joints, bearings, bushings, etc.) as well as the dimensions of the modules themselves influence the harshness performance. In the area of air spring lobe design, the influencing factors are the fiber orientation of the different layers, the fiber material, the fiber diameter, the elastomeric material and the air spring lobe wall thickness. These are additional contributors to friction and hysteresis. With regard to the aging of air spring lobes, various environmental influencing factors such as ozone in the ambient air, the combination of ultraviolet light and oxygen, reactive metals such as copper, oils and fats play a key role in terms of the harshness effect. In addition, the air spring module loading history (temperature, force and pressure) has an effect on the harshness characteristics (Puff, 2009).

2 STATE OF THE ART—FOUR-CORNER AIR SUSPENSION SYSTEMS

Vehicles equipped with a four-corner air suspension system are characterized by the fact, that all suspension springs are air spring modules. Four-corner air suspension systems can be divided into two distinct classes: Those without switchable additional volumes and those with switchable additional volumes. Already in the 1960s, vehicles as for example the Borgward P100 and the Mercedes 600 were equipped with four-corner air suspension systems. Even today, four-corner air suspension systems are mainly offered in the premium passenger car segment of the upper and middle classes because of the high cost of such systems.

The air spring modules of today's state-of-the-art four-corner air suspension systems are usually equipped with air spring bellows with small wall thicknesses and outer guiding tubes to optimize the ride comfort for the vehicle occupants. Generally, the systems incorporate comprehensive functionality with regard to automatic leveling. This functionality comprises for example the speed-dependent lowering of the vehicle body to reduce the aerodynamic driving resistance or to increase the dynamic driving stability and the hoisting of an SUV body to improve the off-road capabilities. The functions are available to the driver either fully automatic or on demand of the driver.

Figure 2 shows the Jaguar XJ front suspension with the air spring modules of the four-corner air suspension system integrated into a double-wishbone-type front suspension as an example of application. Driven by the front suspension architecture of the Jaguar XJ, air spring modules with outer guiding tubes can be used because no significant bending moments need to be carried by the air spring and damper assemblies. For air sprung strut type front suspension architectures, the bending moments on the damper are usually compensated with the aid of asymmetric shaped air spring bellows without outer guiding elements. Alternatively, it is possible to use rotationally symmetrical air spring bellows that are angularly arranged with respect to the dampers to achieve a side force compensation. Four-corner air suspension systems as in the Jaguar XJ are often combined with adaptive damping systems. The switchable damper units are usually centrally arranged inside the air spring modules to optimize the vehicle package.

Four-corner air suspension systems with switchable additional volumes are semiactive systems. By connecting switchable additional volumes to the individual air spring modules, the air spring rates can be varied depending on the driving condition without changing the pressure levels inside the air spring modules. By opening the pneumatic connections between the individual air spring modules and

Figure 2. Four-corner air suspension system—Jaguar XJ front suspension. (Reproduced by permission of Jaguar Land Rover.)

their corresponding additional volumes, softer air spring rates can be achieved because of the volume increases of the different air spring modules. This results in an improved ride comfort for the vehicle occupants. In the case where the pneumatic interconnection lines between the air springs and their additional volumes are closed, the driving stability and therefore the driving safety during lateral dynamic driving maneuvers are increased because of the resulting higher air spring rates. Therefore, air suspension systems with switchable additional volumes are a potential enabler to defuse the conflict of goals between the driving safety and the driving comfort. Depending on the specific geometric design of the pneumatic interconnection lines, including the electromechanical valves between the air springs and the additional volumes, four-corner air suspension systems with switchable additional volumes can generate damping forces caused by the airflow processes between the air volumes through the interconnection lines, as energy is dissipated by the system. For very high suspension travel velocities and/or unfavorably designed pneumatic interconnection lines between the air springs and the additional volumes, the additional volumes can uncouple dynamically from the air springs. The dynamic uncoupling is driven by the fact that the pressure equalization between the volumes significantly deteriorates because of the dynamics of the suspension excitation and/or the pneumatic interconnection line designs. These decoupling effects can be avoided by a sufficient system design with respect to the fluidic system layout. The general design rule is to use short interconnection lines with large cross-sectional areas. However, interconnection lines with large cross-sectional areas need to be combined with suitable electromagnetic valves with similar cross-sectional areas and adequate air tightness. From a technical point of view, pneumatic valves, which are suitable for use in high volume production, are limited in terms of the maximum switchable cross-sectional area. In addition, the weight and the cost of these valves increase with increasing switchable cross-sectional areas of the interconnection lines. The integration of the individual additional volumes into the air spring modules offers the opportunity to reduce cost and weight of the entire system. At the same time, the lengths of the interconnection lines between the air spring volumes and the additional volumes can be minimized and the airflow between the volumes can be improved accordingly.

In Figure 3, the front suspension air spring module with integrated additional volume of the Porsche Panamera is shown. Air spring bellows with a small wall thickness and an outer guiding tube are used to improve the harshness characteristics of the air spring module. In the upper part of the module, the integrated switchable additional volume (1) is depicted. In the comfort setting of the air suspension system, the additional volume can be added to the base volume (2) by operating a pneumatic valve (3). The base volume is arranged in the lower part of the air spring module. The adaptive damping system with switchable valve unit (4) is coaxially integrated into the air spring module.

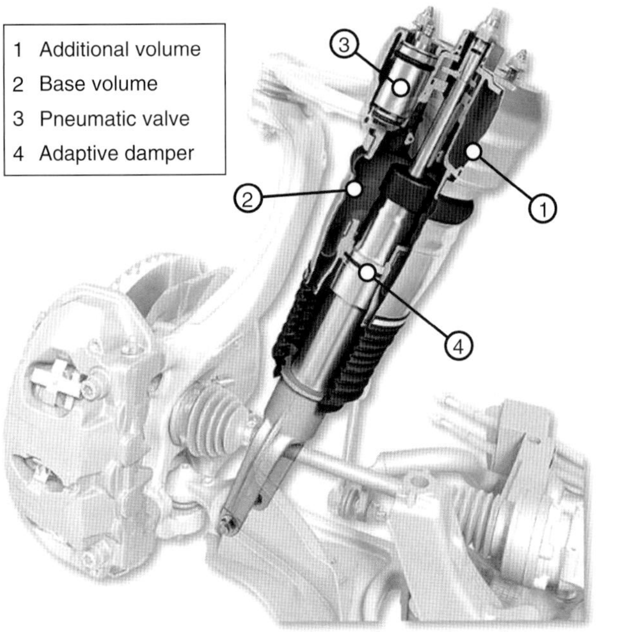

Figure 3. Semiactive four-corner air suspension system—Porsche Panamera front suspension air spring module with integrated switchable additional volume. (Reproduced by permission of Porsche Aktiengesellschaft.)

3 CONTINUOUSLY VARIABLE AIR SUSPENSION SYSTEMS

Usually, four-corner air suspension systems with one switchable additional volume for each air spring module allow only the implementation of two discrete force versus travel characteristics per module. In contrast to these systems, continuously variable air suspension systems enable the individual, seamless transition of the air spring module stiffnesses over a large adjustment range without changing the amount of air in the different air spring modules. In general, two different design approaches for the air spring modules of continuously variable air suspension systems are known. On the one hand, the air spring volumes or the additional volumes can be actively varied to generate continuously changing air spring rates. On the other hand, the effective areas of the air spring modules can be actively increased and decreased over specific roll areas of the air spring bellows over the rolling pistons to generate seamless changing rates of the air spring modules. The interrelationship between the effective area A and the stiffness c of an air spring is shown in Equation 1

$$c = \frac{n \cdot p_{\mathrm{i}} \cdot A^2}{V} \tag{1}$$

On the one hand, the polytropic exponent n and the internal pressure p_{i} cannot be influenced by the air spring design. On the other hand, the volume V and the effective area A can be influenced by the air spring design within the given package constraints.

As an example for a continuously variable air spring module, Figure 4 shows the Active Air Suspension module from Meritor, Inc. The design of the Active Air Suspension module with coaxially integrated adaptive damper unit (3) is similar to the design of a conventional air spring module. In addition to the main air spring bellow (1) of the module, an additional smaller bellow (2) is mounted directly to the rolling piston to enable the continuous variation of the air spring module rate. This second air spring piston bellow has hermetically sealed interfaces with the rolling piston on the upper and lower ends.

The filling of the air spring piston bellow with compressed air is independent from the filling of the main air volume of the air spring module. By varying the amount of air in the air spring piston bellow, the shape of the rolling piston in the roll area of the air spring bellow changes. Owing to the diameter variation of the air spring piston bellow mounted to the rolling piston with the pressure change inside the air spring piston bellow, the characteristic of the effective area as a function of the air spring module travel can be continuously varied. As a result of the variation of the effective area versus travel characteristic, the air spring module rate can be changed continuously. Because of the high sensitivity of the air spring module rate to the effective area, already relatively small variations of the air volume of the air spring piston bellow can generate large air spring module rate changes. As a result of this correlation not only quasi-static but also fast dynamic air spring module rate variations are possible. For example, at a static air pressure of 9 bars inside the main air spring bellow, the air spring module rate can be varied from 16 to 36 N/mm by changing the air pressure of the bellow mounted to the rolling piston from 1 to 14.4 bars.

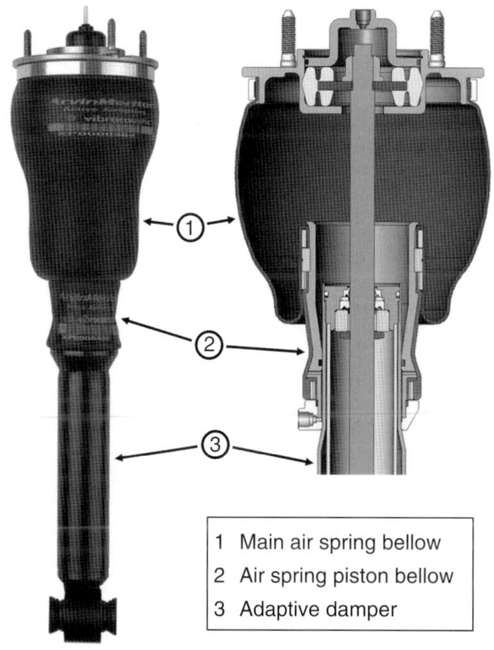

Figure 4. Continuously variable air spring module. (Reproduced by permission of Meritor, Inc.)

4 AIR SPRING DAMPER SYSTEMS

The particularity of air spring damper systems is, that not only the spring forces but also the damping forces are generated with the aid of air as the working fluid. The spring and the damper functionalities are combined in one single functional component. In this manner, the use of conventional hydraulic damper units can be dispensed with. Conventional hydraulic damper modules, used within four-corner air suspension systems with switchable additional volumes, are usually semiactive adaptive units to enable the variation of the damping force versus velocity characteristics of the damper units, in addition to the spring force versus travel characteristics of the air spring modules with switchable additional volumes, depending on the driving situation. However, adaptive hydraulic damper modules are technically more complex in comparison to conventional hydraulic damper units as they require switchable valve elements. In case of air spring damper modules, the damping functionality is achieved with the integration of throttle elements. During the bump and rebound motions of the air spring damper modules, the air is forced to flow through these throttle elements. This leads to the generation of damping forces. Fundamentally, air spring damper units can be distinguished into air spring damper units with two air volumes and air spring damper units with three air volumes. For air spring damper modules with two air volumes, only one maximum of the damping work occurs in the frequency domain that can be aligned with the vertical eigenfrequency of the vehicle body. With the aid of air spring damper modules with three air volumes, two separate maxima of the damping work can be implemented in the frequency range. To achieve the optimum system performance for an air spring damper system, the first maximum of the damping work can be adjusted to the vertical eigenfrequency of the sprung mass respectively to the vehicle body mass and the second maximum of the damping work to the vertical eigenfrequencies of the unsprung masses. Theoretical and practical investigations with regard to air spring damper modules and systems are documented in Kranz (1934) and Gold (1973).

Currently, in the fields of passenger cars, no applications of air spring damper systems are known for series production. In the fields of motorbikes, a series production application of an air spring damper system is known on the single arm rear suspension of the BMW HP2 Enduro. This air spring damper module is supplied by Continental and shown in Figure 5 (Müller *et al.*, 2005).

The module is based on a compact aluminum part with integrated guiding for a hollow aluminum piston rod, which is shown in the upper part of Figure 5. A damper piston with integrated throttle elements is rigidly mounted to the piston rod. In the lower part of Figure 5, a fabric reinforced rolling lobe is shown, which can roll up and down in direct contact to a cone-shaped rolling piston as the air spring damper unit moves in bump or rebound direction.

Figure 5. Cutaway view of the BMW HP2 Enduro air spring damper module. (Reproduced by permission of Continental.)

One bump and one rebound stop made of polyurethane rubber provide the necessary progressivity of the air spring damper module in the extreme ends of the available module travel. Conventional asymmetric shim stacks are used as throttle elements in the rebound and compression motion directions. The shim stacks generate annular gaps as a function of the pressure difference over the damper piston. With the aid of the bypass adjuster, which is integrated into the filling valve, two different settings can be applied to the module to avoid extreme damping force peaks. This damping force limitation enables an improvement of the driving comfort.

During the assembly process of an air spring damper module, the mounting tolerances between the base part and the dividing piston do not need to be as narrow as with conventional hydraulic damping units, because the leakages through the seals are not very sensitive to the generated damping forces. This means, that the leakages on the seals do not immediately lead to a reduction of the generated damping forces. This is based on the fact, that the transported air volumes of air spring damper modules are much higher compared to the transported oil volumes of conventional hydraulic damper units. As a result, the outer diameters of air spring damper modules are usually larger compared to the outer diameters of conventional hydraulic damper units. This leads to an increased package space demand of air spring damper modules, that needs to be considered as a part of the full vehicle layout process.

5 INTERLINKED AIR SUSPENSION SYSTEMS

In Wolf-Monheim (2011), a semiactive interlinked four-corner air suspension system, which is based on a conventional four-corner air suspension system with switchable additional volumes, is analyzed. The system enables an improved ride comfort for the vehicle occupants without neglecting the driving safety aspects. The semiactive interlinked four-corner air suspension system is currently not in series production. To analyze the system performance in terms of the ride comfort improvement for the vehicle occupants, the system was installed into a Volvo XC90 prototype vehicle. The semiactive interlinked four-corner air suspension prototype system consists of four air spring modules with switchable additional volumes as well as two pneumatic pipes routed from the left side to the right side of the vehicle to interconnect the two front suspension air spring modules with one another and the two rear suspension air spring modules with one another. Depending on the current driving condition, the pneumatic interconnection lines between the air spring modules can be switched

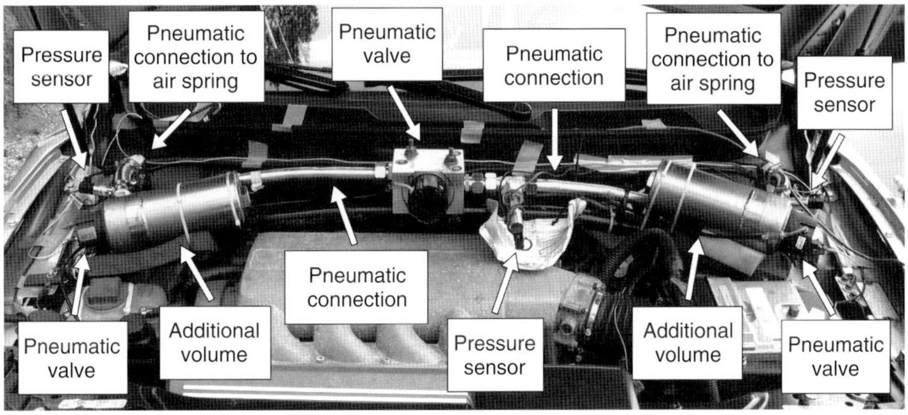

Figure 6. Interlinked air suspension system—Volvo XC90 front suspension. (Reproduced by permission of Wolf-Monheim, 2011. © Wolf-Monheim.)

on and off with the aid of electromagnetic pneumatic valves.

A pneumatic connection between the front and rear suspension air spring modules is not shown in Wolf-Monheim (2011). Within semiactive four-corner air suspension systems, a front to rear connection can be used to reduce the pitch angles of the vehicle body while driving over single impacts and to reduce the pitch angle oscillation accelerations while driving on uneven road surfaces. Within active four-corner air suspension systems, a pneumatic connection between the front and rear suspension air spring modules can be used to allow for an active anti-lift and/or anti-dive pitch angle compensation during acceleration and braking maneuvers.

The achievable improvement in terms of the ride comfort for the vehicle passengers of the semiactive interlinked four-corner air suspension system shown in Wolf-Monheim (2011) results from the fact, that the roll stiffness of the vehicle can be adapted to the individual driving conditions. In this manner, both high roll stiffness during high lateral acceleration driving maneuvers, such as while driving around bends, and low roll stiffness to improve the driving comfort for the vehicle passengers while driving straight ahead on uneven road surfaces can be achieved. Furthermore, semiactive interlinked four-corner air suspension systems can specifically generate roll damping by the appropriate design of the pneumatic interconnection lines and the electromagnetic pneumatic valve geometries as well as by the controlled variation of the cross-sectional areas of the electromagnetic pneumatic valves. Specifically, in combination with switchable additional volumes and continuously variable hydraulic damper modules, the interlinked semiactive four-corner air suspension system can independently influence the roll, pitch and bounce oscillation accelerations of the vehicle body.

In Figure 6, the semiactive interlinked air suspension system with two switchable additional volumes of the prototype vehicle front suspension is shown. Figure 7 shows the equivalent semiactive interlinked air suspension system with two switchable additional volumes installed into the rear suspension of the prototype vehicle. The flows of the pressurized air through the pneumatic interconnection lines between the left and right air spring modules of the front axle and those of the rear axle, respectively, can be interrupted with the aid of electromagnetic pneumatic valves. The additional volumes are connected to the air spring modules with the smallest possible distances and can also be switched on and off separately by means of electromagnetic pneumatic valves. The lengths of the pneumatic interconnection lines of the front and rear suspensions are approximately 1 m. Owing to packaging restrictions, the pneumatic interconnection lines on the front and rear suspensions have some curvatures.

The semiactive interlinked four-corner air suspension system enables a significant improvement in terms of the ride comfort for the vehicle passengers compared to a conventional passive four-corner air suspension system. Compared to a semiactive four-corner air suspension system with switchable additional volumes the ride comfort improvement of the interlinked air suspension system without switchable additional volumes is on a similar level. However, the required package space of the interlinked four-corner air suspension system is significantly lower in comparison to the four-corner air suspension system with switchable additional volumes. Compared to the four-corner air suspension system with switchable additional volumes, the interlinked four-corner air suspension system enables a significantly greater reduction of the body roll oscillation accelerations. The body roll oscillation accelerations are particularly uncomfortable for the vehicle passengers

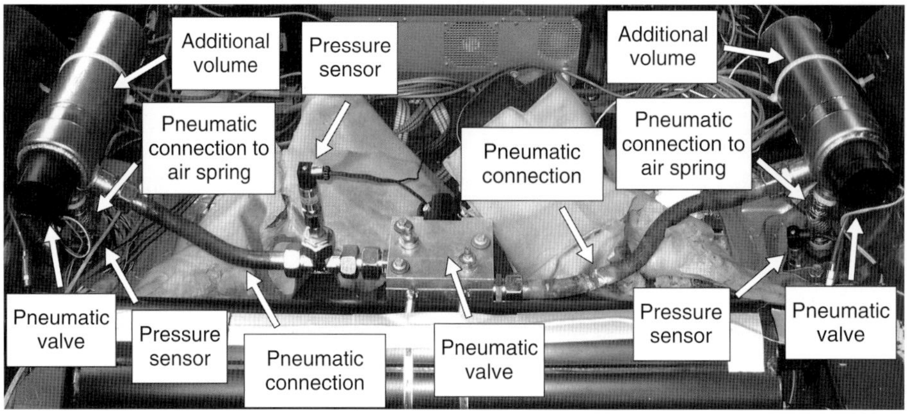

Figure 7. Interlinked air suspension system—Volvo XC90 rear suspension. (Reproduced by permission of Wolf-Monheim, 2011. © Wolf-Monheim.)

compared to the other translational and rotational oscillation accelerations of the vehicle body (Ilgmann, 1979; Kenneth and Griffin, 1978; Parsons, Whitham, and Griffin, 1979).

The vibration intensities introduced to the vehicle body can be further reduced by the combined use of the pneumatic interconnection lines and the additional volumes. Finally, the biggest advantages of the semiactive interlinked air suspension system with switchable additional volumes can be shown in combination with a semiactive adaptive damping system. As mentioned earlier, up to now the semiactive interlinked air suspension system was realized as a prototype system installed into a prototype vehicle only. For future applications of the system under series production conditions, further development work is needed. When the semiactive interlinked four-corner air suspension system with switchable additional volumes is applied to a particular vehicle, it is important that the specific characteristics of the system are aligned with the specific characteristics of the other chassis and suspension systems and components, especially with regard to the driving dynamics. Especially in this context, the tuning of the anti roll bars of the front and rear suspensions should tend toward softer rates whereas the rates of the air spring modules of the front and rear suspensions should tend toward higher rates, as the semiactive interlinked air suspension system allows for a variation of the body roll stiffness. In addition, the air volumes of the pneumatic interconnection lines on the front and rear suspensions should be considered when the air spring modules are designed.

In Wolf-Monheim (2011) various driving test measurement and simulation results on full vehicle bases are presented and analyzed using a prototype vehicle with semiactive interlinked air suspension systems with switchable additional volumes installed on the front and rear suspensions. Initially, simulation and measurement results are analyzed while driving on a ride comfort test track with firstly a sinusoidal profile on one vehicle track only and then secondly with sinusoidal profiles on both driving tracks of the vehicle. During the test drives with only one driving track of the vehicle on the ride comfort test track with the sinusoidal road profile, the other driving track of the vehicle runs on a flat road surface. The results of the test runs with a single track driving on the sinusoidal test track road profile show that the resulting dynamic air spring module stiffnesses can be significantly influenced by the diameters of the pneumatic interconnection lines and by the vehicle driving velocity. Furthermore, the resulting body roll oscillation accelerations and the resulting seat rail vertical oscillation accelerations in relation to the excitation frequencies are studied. The influence of various system designs on the respective effective accelerations is analyzed. With the aid of suitable analysis models to describe the ride comfort as perceived by the vehicle passengers, weighting functions are selected and applied to the simulation as well as the measurement results. In this context, parameters not only for the individual vibration severities but also for the total vibration severities are determined for the various designs of the semiactive interlinked four-corner air suspension system with switchable additional volumes. For the vehicle with opened pneumatic interconnection lines with diameters of 19 mm compared to the prototype vehicle with a reference system design with closed pneumatic interconnection lines and disconnected additional volumes driving with one single track on the ride comfort test track with sinusoidal road profile, the vibration severities of the vehicle body roll oscillation accelerations and the seat rail vertical oscillation accelerations sensed by the vehicle passengers can be reduced by 24%. On various ride comfort test tracks with a wide range of different stochastic unevenness distributions, the potential of the semiactive interlinked air suspension system with switchable additional volumes to reduce the vibration intensities introduced to the vehicle body is also analyzed. In the first place, the potential for variations with regard to the vehicle body control with opened pneumatic interconnection lines of different diameters on the front and rear suspensions, respectively, and with different additional volume sizes is analyzed in the frequency range up to 10 Hz in comparison to the reference vehicle configuration as described earlier.

In this context it becomes evident, that the body roll motions can be influenced by opening the pneumatic interconnection lines to a far greater degree compared to the connection of the additional volumes. The vehicle body roll motions are perceived more clearly by the vehicle passengers compared to other vehicle body motions, such as the vehicle body pitch motions, the vehicle body yaw motions or the vehicle body bounce motions. The reason for this observation can be explained as follows. The road profile excitation content of the left and right driving tracks, primarily leading to vehicle body roll oscillations, results in less dynamic wheel load fluctuations between the two front and two rear wheels, respectively, for the vehicle configuration with opened pneumatic interconnection lines on the front and rear suspensions and disconnected additional volumes in comparison to the vehicle configuration with closed pneumatic interconnection lines on the front and rear suspensions and connected additional volumes.

This is due to the fact that in the case of opened pneumatic interconnection lines on the front and rear suspensions and antiphase travel excitations of the wheels of each axle, the pressure differences between the two front and two rear suspension air spring modules, respectively, can be equalized because of the air mass exchanges through the pneumatic interconnection lines. This results in almost constant forces of the air spring modules of one axle.

In contrast to the aforementioned, the level of the bounce oscillations of the driver's seat as well as the level of the pitch oscillations of the vehicle body can be better varied by connecting or disconnecting the additional volumes compared to switching the pneumatic interconnection lines between the left and right sides of the vehicle. The greater efficiency of the switchable additional volumes, with regard to the bounce and pitch angle degrees of freedom, follows from the fact, that the road profile excitation content of the left and right driving tracks, primarily leading to bounce and pitch angle oscillations of the vehicle body, can only influence the forces introduced into the vehicle body and therefore the vehicle body bounce and pitch angle oscillations in the case, when the additional volumes on the front and rear suspensions are varied respectively switched with the resulting variations of the base rates of the individual air spring modules.

The pneumatic interconnection lines between the air spring modules of the front and rear axles, respectively, have only a minor influence on the resulting air spring module forces, when the road profile excitation content of the left and right driving tracks primarily leads to bounce and pitch angle oscillations of the vehicle body. This can be explained by the fact, that the base rates of the air spring modules are equal for opened and closed pneumatic interconnection lines between the left and right sides of the vehicle for almost identical road profiles of the left and right tracks.

In a similar way, the oscillation acceleration reduction potential in the enhanced frequency range of up to 100 Hz is analyzed for different design variants of the semiactive interlinked air suspension system with the switchable additional volumes, in comparison to the reference vehicle configuration with closed pneumatic interconnection lines and disconnected additional volumes. These tests are based on acceleration measurements on the outer seat rail of the driver's seat and on the steering wheel. In addition to the analyses of the influence of the pneumatic interconnection lines and the switchable additional volumes on the ride comfort for the vehicle passengers, also the influence of a semiactive adaptive damping system is analyzed. The results are evaluated in various individual frequency bands. For all tests, the percentage reductions of the root mean square values of the power spectral densities (PSDs) are determined.

For opened pneumatic interconnection lines, the PSDs of the seat rail vertical oscillation accelerations can be reduced in comparison to the reference vehicle configuration by up to 9.6%, depending on the individual frequency range. The PSDs of the steering wheel oscillation accelerations can also be reduced significantly (by 15.6%) with opened pneumatic interconnection lines in comparison to the reference vehicle configuration. In case the additional volumes are connected, instead of opening the pneumatic interconnection lines, the PSDs of the seat rail vertical oscillation accelerations as well as the PSDs of the steering wheel oscillation accelerations can be reduced by a similar magnitude. On the seat rail of the driver's seat, the PSDs can be decreased by up to 11.3% and on the steering wheel by up to 12.7%, depending on the frequency range.

The combined use of the pneumatic interconnection lines and the additional volumes enables the further increase of the ride comfort perceived by the vehicle passengers. In this case, the PSDs of the seat rail vertical oscillation accelerations can be diminished by up to 15.4%, depending on the frequency band. The PSDs of the translational steering wheel oscillation accelerations can be reduced by up to 19.8%.

A further improvement of the vibration comfort felt by the vehicle passengers is possible if a semiactive adaptive damping system is used in addition to the opened pneumatic interconnection lines between the air spring modules of the front and rear axles, respectively, and the connected additional volumes.

In comparison to the reference vehicle configuration with closed pneumatic interconnection lines between the left and right sides of the vehicle, disconnected additional volumes, and inactive adaptive damping system, this system configuration enables the reduction of the PSDs of the outer seat rail vertical oscillation accelerations measured at the driver's seat of up to 24.6%. In addition, it is possible to decrease the PSDs of the translational steering wheel oscillation accelerations by up to 29.5%.

Subsequent publications referring to semiactive interlinked four-corner air suspension systems are Wolf-Monheim *et al.* (2008a), Wolf-Monheim *et al.* (2008b), Wolf-Monheim *et al.* (2008c), and Wolf-Monheim *et al.* (2009).

6 ACTIVE AIR SUSPENSION SYSTEMS

From a physical point of view, active air suspension systems can be based on either an active air mass change or an active air volume change in the individual air spring modules. Next to the ride height control functionality of conventional semiactive four-corner air suspension systems with and without additional volumes, active four-corner air suspension systems also enable additional functionalities, for example, active roll and/or pitch angle compensations. On the one hand, an active air suspension system based on dynamic air mass change can be realized, for example, by means of high pressure air reservoirs and air compressors. If required, the air masses can be shifted actively from the high pressure air reservoirs to the air spring modules. On the other hand, an active air suspension system based on dynamic air volume change can be made, for example, with the aid of hydraulic, pneumatic, or electromechanical actuators. To minimize the energy consumption of the actuators used within the system layout, the static loads generated by the sprung mass can be carried partly or fully with the help of passive spring elements. This means, that only the requested dynamic force variations are demanded from the actuators.

In Zhang (2006), three different active air suspension systems based on dynamic air volume change are presented and compared with one another with the aid of complex numerical simulation models. All the three systems documented in Zhang (2006) are based on a driven double-acting cylinder, which moves two additional connected mechanical two-chamber cylinders. These mechanical two-chamber cylinders are used to supply pressurized air to one air spring module and remove pressurized air from another. With the help of appropriate valve technology, the active air suspension system can be switched between roll angle compensation mode for lateral acceleration maneuvers and anti-lift and anti-dive pitch angle compensation mode for acceleration and braking maneuvers, respectively, depending on the driving situation. For the fully pneumatic system (Figure 8), a pneumatically operated double-acting drive cylinder is used, whereas a hydraulically operated double-acting drive cylinder is utilized within the hydropneumatic system (Figure 9). The third system is referred to as the *fully hydraulic system* (Figure 10), as in this case, a hydraulically operated double-acting drive cylinder moves two additional connected hydropneumatic two-chamber cylinders. In this system, four additional passive cylinders with one hydraulic and one pneumatic chamber are needed to realize the active air volume change of the individual air spring modules.

For the fully pneumatic active air suspension system, two key advantages in comparison to the other two systems can

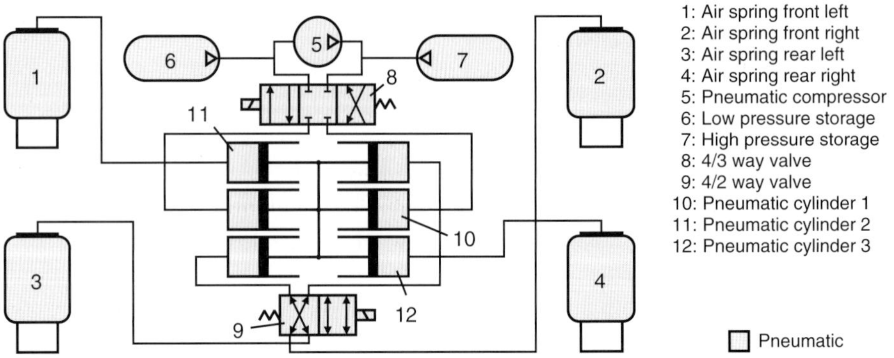

Figure 8. Active air suspension system—fully pneumatic system. (Reproduced by permission of Zhang, 2006. © Zhang.)

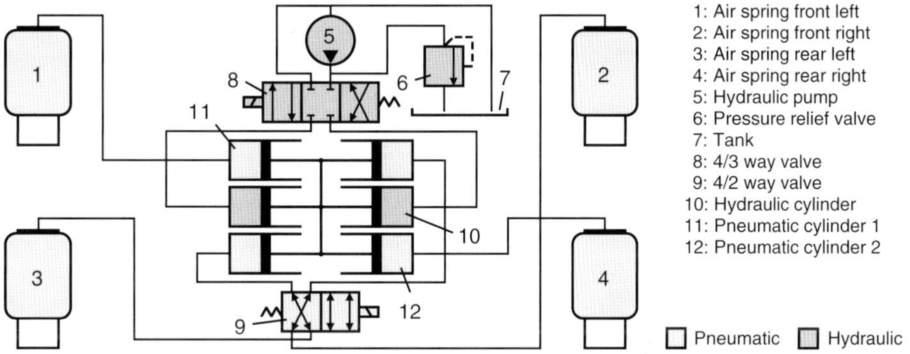

Figure 9. Active air suspension system—hydropneumatic system. (Reproduced by permission of Zhang, 2006. © Zhang.)

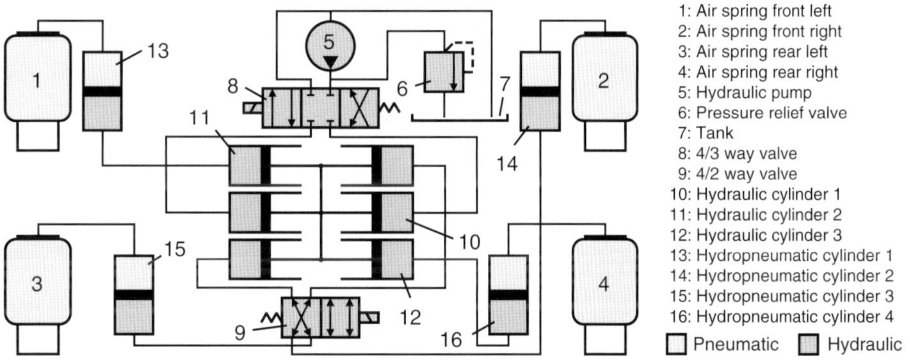

Figure 10. Active air suspension system—fully hydraulic system. (Reproduced by permission of Zhang, 2006. © Zhang.)

be mentioned. On the one hand, only pressurized air is used in this case as the working fluid and hydraulic oil can be avoided to improve the environmental friendliness of the system. On the other hand, the actual design and development of the system is relatively easy. However, the disadvantage of the fully pneumatic active air suspension system is that it tends to oscillate because of the compressibility of the pressurized air as the working fluid. This results in difficulties in terms of the precise position control of the individual actuators.

In contrast to the fully pneumatic active air suspension system, the fully hydraulic active air suspension system can be controlled in a stable way as it features only a small amount of volumes filled with pressurized air. However, it is technically more complex and demands more packaging space compared to the fully pneumatic and the hydropneumatic active air suspension systems. In addition, it is also more expensive compared to the other two systems. The hydropneumatic active air suspension system is implemented and tested in a prototype vehicle.

A further active air suspension system is presented in Kranen (1996). In the system presented here, a volume exchanger is utilized as the actuator to shift the pressurized air between the air spring modules. The power steering device supplies the energy needed to actuate the system.

7 SUMMARY

This chapter gives an overview of advanced air suspension technologies and applications based on current state-of-the-art four-corner air suspension systems. In the introduction, the main advantages and base functionalities of air suspension systems as well as the available design variants of air spring modules are presented.

Section 2 deals with the two key distinct categories of the state-of-the-art four-corner air suspension systems. The first category covers air suspension systems without switchable additional volumes and the second category systems with switchable additional volumes.

Continuously variable air suspension systems presented in Section 3 offer the advantage to vary the force versus air spring module travel characteristics of the individual air spring modules and therefore also the air spring module stiffnesses in a continuous way. For these systems, the stiffness variations can be generated by varying either the effective areas or the air spring module respectively additional volumes.

Air spring damper systems as discussed in Section 4 do not only provide a springing functionality but also a damping functionality based on pressurized air as the working fluid. The air damping is achieved with the aid of pneumatic throttle elements integrated into the air spring damper modules where the pressurized air is forced to flow through while dissipating energy. Compared to conventional hydraulic damping elements, air spring damper units avoid the use of hydraulic oil as a working fluid and therefore improve the environmental friendliness.

In Section 5, interlinked air suspension systems, characterized by pneumatic interconnection lines between the individual air spring modules, are described. Semiactive interlinked four-corner air suspension systems offer enhanced possibilities to tune the vehicle body bounce and roll stiffness properties independently from one another, depending on the current driving condition. In addition, the vehicle body pitch angles can be influenced while driving over single impacts or on uneven road surfaces.

Finally, active four-corner air suspension systems are introduced in Section 6. Active four-corner air suspension systems enable additional functionalities, for example active vehicle body roll angle or anti-lift and/or anti-dive vehicle body pitch angle compensations during lateral acceleration or longitudinal acceleration and braking maneuvers.

ENDNOTE

1. The use of "bellow" should be understood as a flexible bladder or bag containing pressurised air.

REFERENCES

Gold, H. (1973) Über das Dämpfungsverhalten von Kraftfahrzeug-Gasfedern. PhD thesis. RWTH Aachen University, Aachen, Germany.

Heisler, H. (2002) *Advanced Vehicle Technology*, 2nd edition, Elsevier Science, Amsterdam, The Netherlands.

Ilgmann, W. (1979) *Ergonomische Untersuchungen über die Einwirkung rotatorischer Schwingungen, Forschungsbericht aus der Wehrtechnik*, Bundesministerium der Verteidigung, Dokumentationszentrum der Bundeswehr, Bonn, Germany.

Kenneth, P. and Griffin, M. (1978) The effect of rotational vibration in roll and pitch axes on discomfort of seated subjects. *Ergonomics*, **21** (8) pp. 615–625. Taylor & Francis, London, England.

Kranen, H. (1996) *Auslegung und Aufbau einer Steuereinheit für ein aktives pneumatisches Wankausgleichssystem*. Diploma thesis. Institute of Automotive Engineering (ika), RWTH Aachen University, Aachen, Germany.

Kranz, M. (1934) *Luftfederung für Kraftfahrzeuge*. PhD thesis. Technical University of Stuttgart, Stuttgart, Germany.

Müller, P., Reichl, H., Heyl, G., *et al.* (2005) Das neue "Air Damping System" der BMW HP2 Enduro. *ATZ Automobiltechnische Zeitschrift*, **107** (10) pp. 848–857. Vieweg+Teubner Verlag / GWV Fachverlage GmbH, Wiesbaden, Germany.

Parsons, P., Whitham, E., and Griffin, M. (1979) Six axis vehicle vibration and its effects on comfort. *Ergonomics*, **22** (2) pp. 211–225. Taylor & Francis, London, England.

Puff, M. (2009) Entwicklung einer Prüfspezifikation zur Charakterisierung von Luftfedern. Abschlussbericht zu einem vom VDA FAT AK20 geförderten Projekt. Projektbericht Fluidsystemtechnik, Technische Universität Darmstadt, Verband der Automobilindustrie (VDA), Berlin, Germany.

Wallentowitz, H. (2005) *Vertikal-/Querdynamik von Kraftfahrzeugen, Vorlesungsumdruck Fahrzeugtechnik II*, Forschungsgesellschaft Kraftfahrwesen Aachen mbH, Aachen, Germany.

Wolf-Monheim, F., Frantzen, M., Seemann, M., and Wilmes, M. (2008a) Modeling, Testing and Correlation of Interlinked Air Suspension Systems for Premium Vehicle Platforms. *Proceedings of 32nd FISITA Congress, F2008-SC-040, FISITA*, London, England.

Wolf-Monheim, F., Seemann, M., Schommer, M., and Wilmes, M. (2008b) Fahrkomfortoptimierung durch den Einsatz vernetzter Luftfederungssysteme. 17. Aachener Kolloquium Fahrzeug- und Motorentechnik 2008, Forschungsgesellschaft Kraftfahrwesen Aachen mbH, Aachen, Germany.

Wolf-Monheim, F., Frantzen, M., Seemann, M., and Wilmes, M. (2008c) Das Potenzial gekoppelter Luftfederungssysteme zur Verbesserung des Fahrkomforts. 2. Fachtagung Federn und Dämpfungssysteme im Fahrwerk, mic—management information center, Munich, Germany.

Wolf-Monheim, F., Schumacher, M., Frantzen, M., *et al.* (2009) Interlinked air suspension systems—the influence on ride comfort in testing and simulation. *ATZautotechnology*, **9** (3) pp. 58–61. Vieweg+Teubner Verlag / GWV Fachverlage GmbH, Wiesbaden, Germany.

Wolf-Monheim, F. (2011) Fahrkomfortoptimierung durch den Einsatz vernetzter Luftfedersysteme. PhD thesis. RWTH Aachen University, Aachen, Germany.

Zhang, J. (2006) Aktives Luftfedersystem für einen PKW. PhD thesis. RWTH Aachen University, Aachen, Germany.

Chapter 111
The Harshness of Air Springs in Passenger Cars

Andreas Kind and Andreas Rohde
Continental Teves AG & Co. oHG, Hannover, Germany

1 INTRODUCTION

Air suspension systems have found widespread application in the luxury automobile segment in recent years. The advantages of air suspension systems in terms of load leveling, variable spring rates, and comfort have been greeted by a high degree of acceptance among consumers.

Customers in the luxury segment demand and expect constant refinements to driving comfort more than anything else and so the goal is to reduce vibration and noise to a level that is acceptable to passengers. This has resulted in permanent refinement and improvement of chassis components.

The comfort of air suspension can be determined mainly by the curve of the spring rate in relation to the frequency and amplitude of excitation. Harshness in this context refers to the undesirable stiffening of air suspensions in response to small, rapid vibration. This effect is primarily due to the characteristics of the bellows employed. Steps to improve comfort can flow into the development process as a function of the design of the air suspension system, the geometry of the components, and the materials that have been employed.

Encyclopedia of Automotive Engineering, print © 2015 John Wiley & Sons, Ltd.
Edited by David Crolla, David E. Foster, Toshio Kobayashi, and Nick Vaughan.
This article is © 2015 John Wiley & Sons, Ltd. ISBN: 978-0-470-97402-5
Also published in the *Encyclopedia of Automotive Engineering* (online edition)
DOI: 10.1002/9781118354179.auto015

2 HARSHNESS OF AIR SPRINGS

2.1 Ride comfort

The demand for riding comfort in the passenger car segment is becoming ever more pronounced. While in the past, luxury cars were synonymous with excellent ride comfort, nowadays, middle-class automobiles, SUVs, and even sports cars come with expectations of comparable comfort. Most modern electronic chassis systems now offer the choice between comfort and sport modes. Changes in the suspension and dampers occur, depending on the input received by the wheels. These changes then influence the vibratory characteristics and, as a desired result, the impression of comfort. The result is an ideal balance between vehicle dynamics and ride comfort (Figure 1).

Undesired vibrations nevertheless do occur, depending on the type and severity of input to the wheels. The jolts travel from the pavement through the wheel and the chassis into the body. Vibrations and structure-borne noise are thus transmitted into the interior and to the occupants. Ride comfort is a reflection of the occupants' sense of well-being as a function of all vibratory influences. Depending on its frequency, vibration can be experienced as palpable vibration, noise, or a combination of both. The term *NVH* is used to refer to vibrations specific to automobiles. NVH stands for noise vibration and harshness and encompasses acoustical and mechanical vibration as well as people's subjective perception of it (Ersoy and Heißing, 2008).

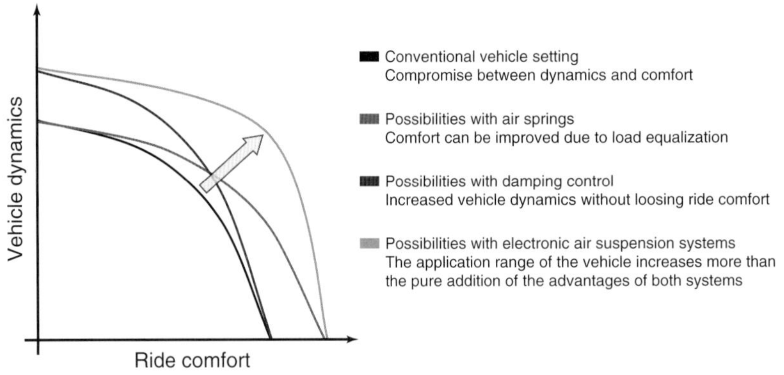

Figure 1. Increase of the chassis application range. (Reproduced by permission of Continental AG.)

2.2 A general definition of harshness

The term *harshness* refers to the intersection of vibrations and noise in a typical range of frequencies between 20 and 100 Hz. Harshness can also be interpreted as a kind of roughness. This roughness is elicited by an undesirable increase in the stiffness of the axle, which may lead to increased acceleration of the bodywork. The harshness effect is magnified at low amplitudes and high frequency excitation. Harshness is an unpleasant experience subjectively because the visual perception of the road often does not coincide with the rough ride the occupants are experiencing. For example, increased harshness can cause a ride over a road with an apparently smooth surface and not even the slightest unevenness visible to the human eye to become an uncomfortable experience because vibration and noise reign in the passenger compartment.

Axle components exert the main influence on harshness. The springs, the dampers, and all other components such as bushings and axle joints are responsible for transmitting vibration. The layout, and the resultant transmission characteristics of the components, thus exerts a direct influence on a vehicle's vibration. In addition, numerous elastomers are used because of the elastic-kinetic properties of the axle. These elastomers play a very important role because of their usually quasi-static and dynamic transmission characteristics and their nonlinear properties. Aside from the material properties, the design of the component is critical. Unevenness and abrupt transitions in the force–travel ratio lead to a marked deterioration of the impression of comfort.

Friction effects are still another leading cause of diminished comfort. On the one hand, there is material friction within the components themselves. On the other hand, friction also arises because of the relative movement of components in contact.

2.3 The basics of air suspension

The actual air suspension element is the air spring bellows. In passenger cars, it appears exclusively as a rolling sleeve-type. Figure 2 depicts a rolling sleeve-type. Metal rings clamp the bellows between the cap and the piston. Under pressure, the bellows roll over the rolling lobe on the outer piston geometry during axial motion.

Load capacity F of the air spring is a function of the pressure p_i inside the air spring and the effective area A_W (Equation 1).

$$F = p_\mathrm{i} A_\mathrm{W} \tag{1}$$

The effective area A_W is proportionate to the effective diameter (Equation 2):

$$F = \frac{\pi D_\mathrm{W}^2}{4} \tag{2}$$

The spring rate c is derived from the load capacity F according to the distance s traveled by the air spring (Equation 3).

$$c = \frac{\mathrm{d}F}{\mathrm{d}s} \tag{3}$$

The rolling sleeve-type consists of thin layers of elastomer between which reinforcement materials in the form of individual threads or mats of fabric have been embedded. Depending on the application, reinforcement material can cross in two layers or can be arranged in one layer along an axis. One is known as a *cross-layered bellows* and the other is known as an *axial bellows*. The reinforcement material can generally only absorb the forces generated in the bellows in the direction of the threads. Cross-layered bellows can therefore also absorb circumferential forces as a function of the angle α between the strands and hence can assume a defined diameter on the application of

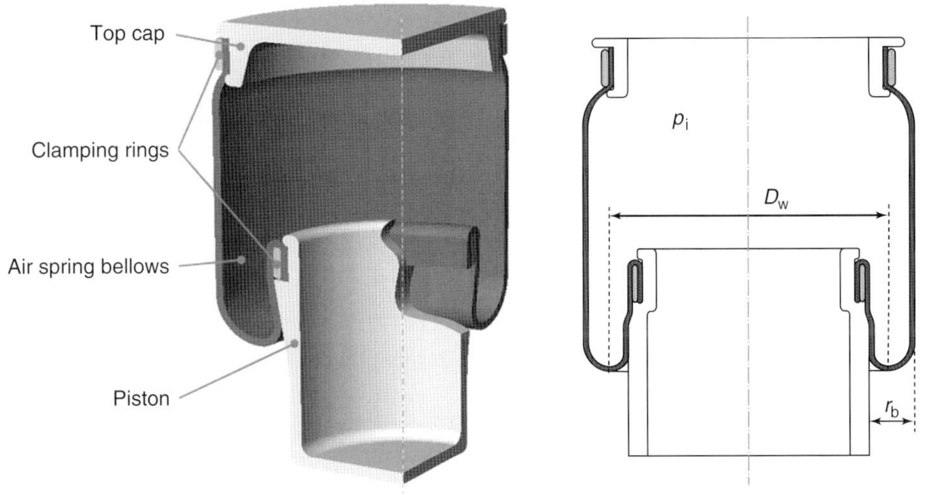

Figure 2. Construction of an air spring bellows suspension unit with pertinent parameters.

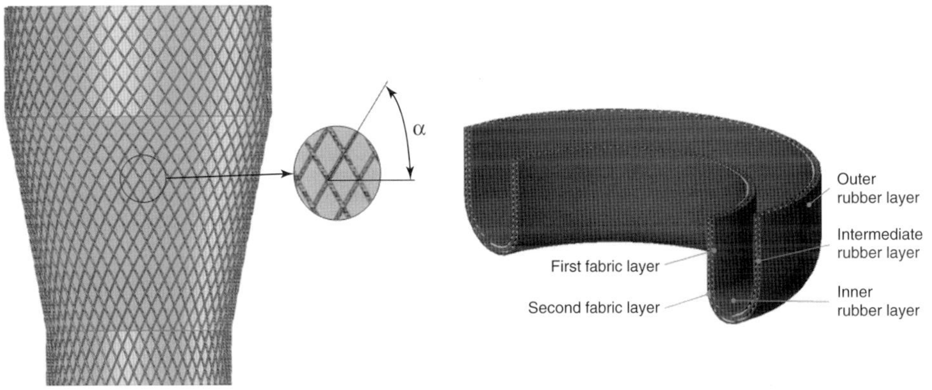

Figure 3. Cross-layered bellows and rolling-lobe shape. (Reproduced by permission of Continental AG.)

pressure. Figure 3 is a schematic diagram of a cross-layered bellows. The five-layer design consists of three elastomer layers—inner, intermediate, and outer rubber layers—and two fabric layers embedded in between. Cross layering results in an angle-dependent rhomboid pattern.

As opposed to axial bellows with a single fabric layer of 90°, the angle of the strands when manufacturing the blank for the cross-layered bellows generally lies somewhere between 45° and 70°, depending on the application. The original angle diminishes as manufacturing proceeds, first because of the bellows widening during vulcanization and shaping and then during pressurization in the static operating state. Here, an equilibrium angle and, as a function thereof, the exterior bellows diameter adjust themselves. The fabric strands stretch under pressure because of their elastic character. During operation, the outer diameter is subject to fluctuations depending on the type of the material used in the fabric. This is due to the magnitude of pressurization. These fluctuations must be considered when designing the component.

During spring motion, the bellows wall does not remain rigid; it fluctuates between the small piston diameter and the large exterior diameter. As the rolling lobe passes through, constant deformations of the bellows wall occur, as do modifications of the strand angle. The pressure inside the air spring causes the bellows wall to roll smoothly down the piston (Voss, 2002).

Alternatively, the air spring can be guided with a fixed external cylinder, which takes up the circumferential forces and diminishes stretching of the bellows wall. This is known as an *exterior-guided air spring*. Both cross-layered bellows and bellows with axial fabric layer find application in guided air springs. The advantages of guided air springs are that there is a lot less wear and tear on the bellows

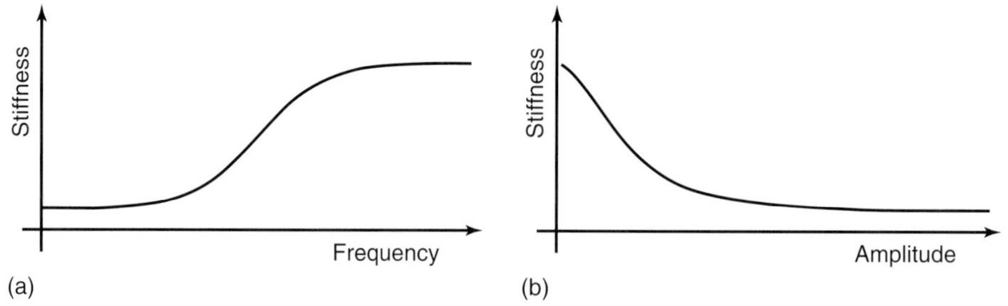

Figure 4. General stiffness curve on an air spring as a function of frequency (a) and amplitude (b).

and they guarantee a fixed diameter for design purposes. The bellows, primarily the reinforcement material and the rubber layers, can be designed much more thinly and delicately.

2.4 The harshness of air suspension

Harshness in connection with air suspension means that spring characteristics stiffen with specific axial movements. Increased stiffness is generally based on two main effects:

- *Frequency-dependent increase in stiffness*. In the direction of high frequencies due to thermodynamic changes in the air.
- *Amplitude-dependent increase in stiffness*. In the direction of low amplitudes due to the rolling resistance of the bellows.

Figure 4 illustrates the general stiffness curves of both effects.

The amplitude-dependent increase in stiffness plays the main part in the actual undesirable harshness of air suspensions. It is predominantly frequency independent and thus significant with every input from the road. The frequency-dependent increase in stiffness, on the other hand, occurs predominantly at frequencies that lie below (i.e., at frequencies below 0.1 Hz) those that are relevant for normal driving conditions (from 0.1 to 2 Hz). Therefore, the stiffness has already increased when the driving relevant frequencies are reached. What follows is a brief explanation of this behavior.

2.4.1 Frequency-dependent increase in stiffness (due to thermodynamic changes)

One can distinguish between static (isotherm) and dynamic (adiabatic) behaviors due to thermodynamic changes in the air. Depending on the rapidity of these changes, the following static equation applies (Equation 4),

$$pV = \text{constant} \tag{4}$$

The conditional equation (5) for the adiabatic case with $\kappa = 1.4$:

$$pV^{\kappa} = \text{constant} \tag{5}$$

The frequency-dependent increase in stiffness (Figure 4a) is marked by a transition from isotherm to adiabatic air behavior. It occurs at what is known as the *cutoff frequency* f' starting at approximately 0.1 Hz. The hysteresis of air suspension assumes a maximum value here. Force–travel curves as a function of different frequencies illustrate the marked degree of hysteresis in Figure 5.

For purposes of identifying the parameters of air suspension stiffness, test specifications thus call for sufficient distance from cutoff frequency f', mostly with test frequencies of 0.01 Hz (static) and 1 Hz (dynamic). These frequencies do not quite attain purely isotherm or adiabatic behavior but have nevertheless proved themselves to be of value for technical reasons.

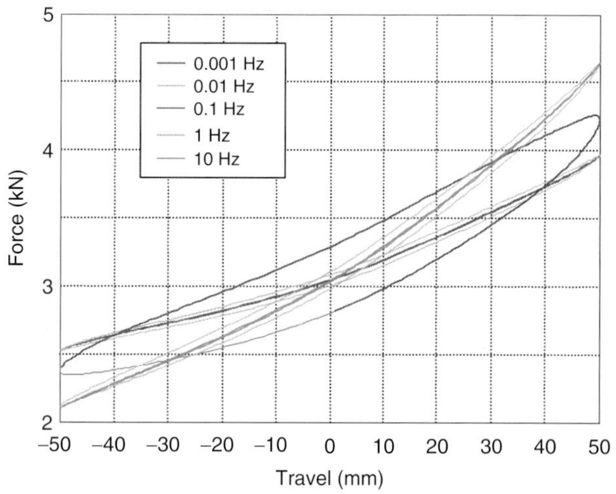

Figure 5. Force–travel curves of air suspension as a function of frequency. (Reproduced from Ilias and Sorge, 2001. Reproduced by permission of Continental AG.)

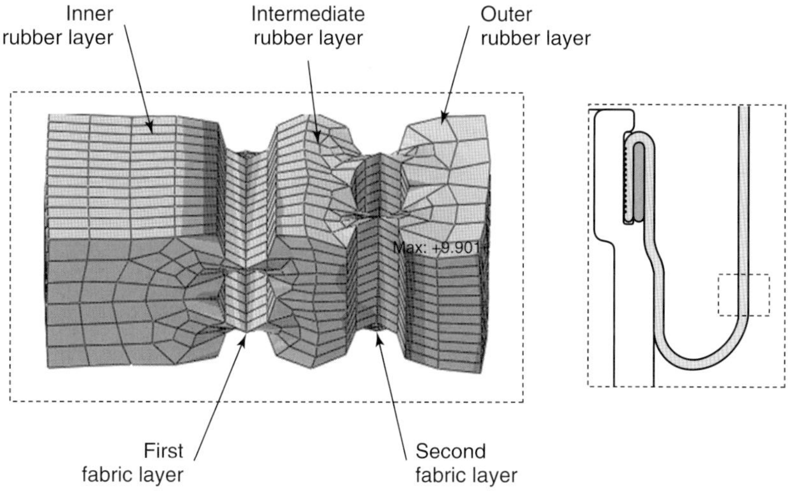

Figure 6. Extended bellows (submodel rolling lobe outside). (Reproduced by permission of Continental AG.)

Incorporating thermodynamic behavior, the value c_{spring} can be calculated as follows in Equation 6;

$$c_{\text{spring}} = p_{\text{i}} \frac{\text{d}A_{\text{W}}}{\text{d}s} + \frac{n(p_{\text{a}} + p_{\text{i}})(A_{\text{W}}^2)}{V} \tag{6}$$

The first term represents the spring value as a function of the change in the effective diameter. The second term in the equation reflects the thermodynamic portion of the spring value. For purely static spring action, $n = 1$ applies, whereas n to the isentropic coefficient $\kappa = 1.4$ results for purely adiabatic behavior. It is thus apparent that the consequence of dynamic spring action is increased spring rate as a function of the spring volume V.

Dynamic spring action is the main factor in a car body's natural frequency and hence in riding comfort. Static spring action, on the other hand, is responsible for pitch and roll. Static spring values are therefore also important for good handling because they have an influence on the car's stability in curves. It is thus worth trying to keep the difference between static and dynamic spring values as small as possible.

2.4.2 *Amplitude-dependent increase in stiffness (due to the rolling resistance of the bellows)*

The actual reduction in comfort that harshness causes results from low amplitude vibration. A classic example of this is rough, uneven asphalt on leaving the highway. However, driving over larger obstacles such as road joints, edges, or asphalt patches may result in an undesirable increase in stiffness. In addition to the harshness being affected by the bellows, often the harshness manifests itself on apparently smooth roads that have perturbations so small that the excitation cannot break through the suspension friction (including the rubber bushings of the suspension arms), resulting in the vehicle bouncing only on the tires in the frequency range of approximately 4–8 Hz and giving what Americans call "boulevard jerk." In the air spring system itself, the reason for the harshness is that the bellows pits a certain rolling resistance against axial motion once it leaves a static resting position. The flexible wall of the bellows must deform during expansion and contraction. As it passes through the rolling lobe, a complex deformation of the elastomer compound and the embedded reinforcement material occurs (Voss, 2002). Figure 6 shows how the rubber matrix stretches at the outer rolling lobe.

Added to the basic stiffness of the air suspension c_{spring} is consequently the portion of stiffness Δc_{harsh} from the rolling resistance of the rolling lobe. The result is the total amplitude-dependent stiffness of the air suspension c^*_{spring} (Equation 7):

$$c^*_{\text{spring}} = c_{\text{spring}} + \Delta c_{\text{harsh}} \tag{7}$$

2.5 Measuring harshness

Objective test methods do a good job of characterizing the harshness of air suspensions. To do this, they record the force–travel curves of an air spring at various amplitudes under harmonic sinusoidal stimulation at a frequency of mostly 1 Hz (Figure 7). The corresponding rates of stiffness are then derived from the observed data via the attendant slopes passing through zero (Figure 8). Amplitudes fall typically in the range 0.1–25 mm.

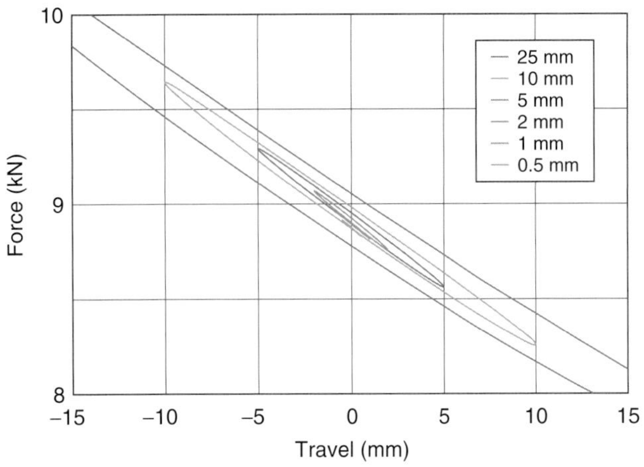

Figure 7. Force–travel curves as a function of amplitude at $f = 1$ Hz.

While the basic design stiffness of air springs regulates at high amplitudes of 25 mm, the spring rate increases in the direction of lower amplitudes. The hysteresis curves rise more steeply as they characterize the harshness effect along the stiffness curve. The harshness effect occurs largely frequency independent (Figure 8).

Measurement using actual air springs, as mentioned earlier, always includes thermodynamic changes. An alternative means of measurement is to attach the metering device to what is known as a *double rolling-lobe arrangement* in order to measure the pure resistance to rolling of the bellows during axial movement.

The device consists of two opposing, geometrically identical air springs with a cylindrical roll-off contour (Figure 9). The volumes of both are linked to each other. The double rolling-lobe arrangement is outwardly force free and allows direct measurement of the rolling resistance of both bellows. A direct, subjective impression "by hand" of harshness or the hysteresis forces is also possible. This arrangement thus permits unadulterated assessment of the components without intrusion by factors influencing the entire module.

Figure 10 illustrates the curve of forces and stiffness on the double rolling-lobe arrangement. While the hysteresis forces diminish as the amplitude declines, the force curve rises more steeply, and stiffness increases.

For the purpose of comparison and obtaining test results that can be reproduced, it is essential to precondition the components extensively and directly before recording any

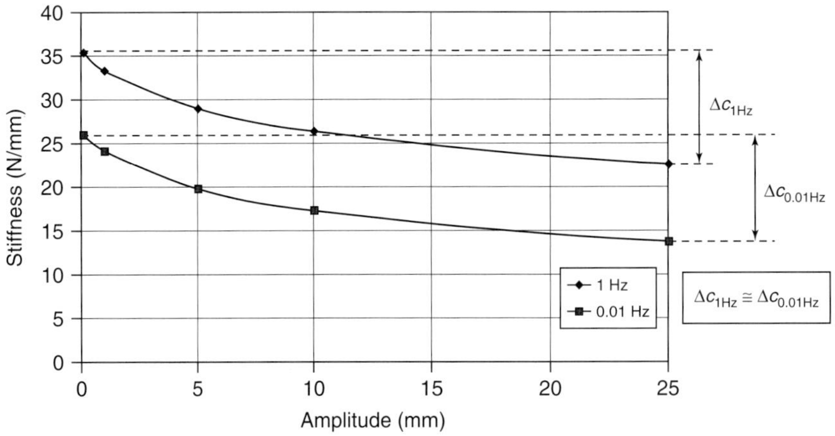

Figure 8. Stiffness as a function of amplitude at $f = 0.01$ and 1 Hz.

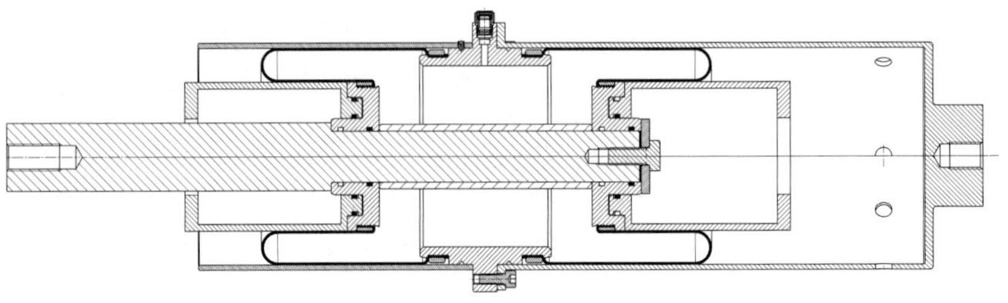

Figure 9. Design for a double rolling-lobe test bench. (Reproduced by permission of Continental AG.)

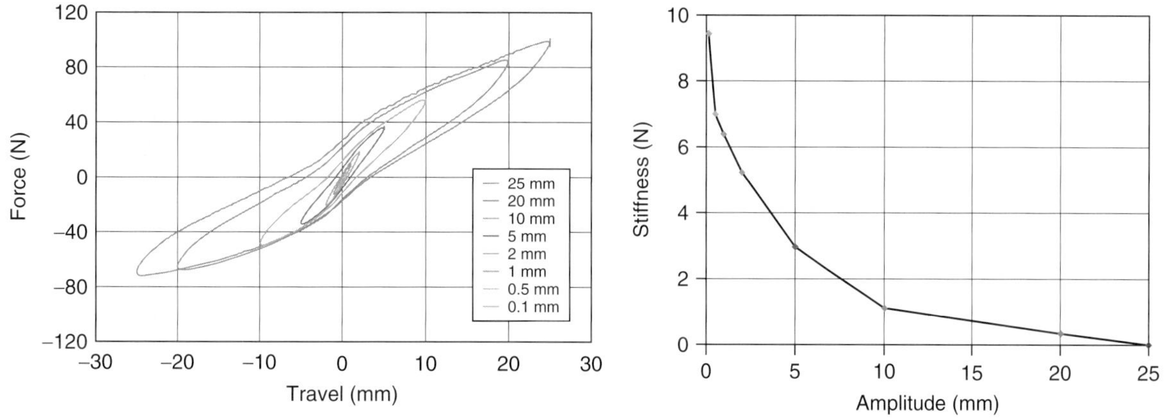

Figure 10. Hysteresis curves with deduced stiffness progress measured on a double rolling-lobe arrangement.

signals. In the case of air springs, springs are usually preflexed to the maximum for a large number of cycles.

2.6 Assessment of harshness

2.6.1 Assessment using harshness coefficients

A coefficient is often used to assess harshness. On the basis of the amplitude response (see Chapter 2.5 "Measuring harshness"), the harshness coefficient $k_{\text{harshness}}$ in Equation 8 sets the relation of increase in the spring rate in relation to the basic spring rate $c_{25\text{mm}}$ of the air spring.

$$k_{\text{harshness}} = \frac{c_{0.1\text{mm}} - c_{25\text{mm}}}{c_{25\text{mm}}} \cdot 100\% \qquad (8)$$

Use of the coefficient, however, insufficiently reflects the actual typical behavior of the component. For example, an air suspension with a high degree of harshness associated with a high basic spring rate would presumably rate better under this equation than a highly comfortable air suspension with an associated very low basic spring rate that might be brought into play by a direct final axle ratio.

The harshness coefficient $k_{\text{harshness}}$ separated from the actual curve of extreme stiffness increase only makes sense in conjunction with the assessment of a specific vehicle or a specific axle.

2.6.2 Assessment in view of the stiffness curve

Simple observation of the curve of amplitude-dependent stiffness leads meanwhile to an optimized, independent assessment of harshness. Two criteria in particular merit our attention:

1. The increase in spring rate.
2. The deflection of the curve.

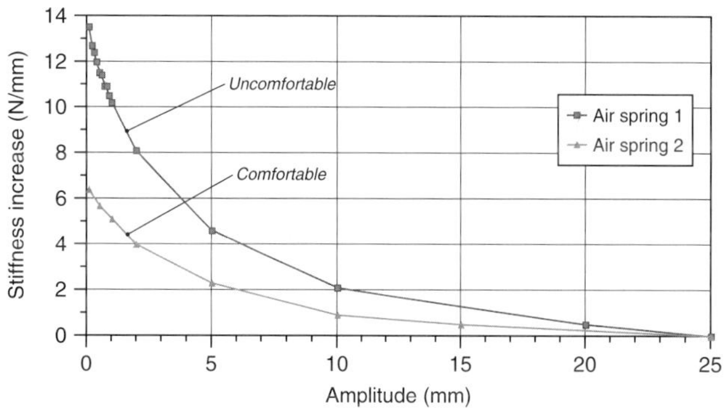

Figure 11. Illustration of amplitude-dependent increase in spring rate Δc_{harsh}.

The increase in spring rate and/or the leptokurtosis of stiffness Δc_{harsh} normalizes the amplitude-dependent spring rate by the value of maximum observed amplitude.

This sets the increase in spring rate at maximum amplitude to zero. This also considers the simple increase in spring rate. This depiction makes it easy to compare air suspensions with varying basic spring rates and facilitates discussion of their curves. Figure 11 illustrates the qualitative progression of a comfortable and an uncomfortable air suspension.

Another important assessment factor is the deflection of the curve in the form of a bulge. A slight increase in spring rate associated with a prominent bulge, that is, a rapid reaching of basic stiffness, is necessary for the comfort of air suspensions.

2.6.3 Correlation to subjective driving impressions

To what extent, there is a correlation to the increase in spring rate observed and its associated curve, in comparison to the subjective driving impression, depends greatly on the total automobile handling. There is thus no single specific answer to this question. Assessment of harshness on the test bench may reliably quantify the behavior of the component on the one hand. On the other hand, though, it can only provide a qualitative indication of how comfortable the vehicle is. The preconditioning performed on the components with large amplitudes (to ensure that the results can be reproduced and compared) does not correspond to the way the component actually behaves in the vehicle. What is missing is permanent high amplitude flexing. Uniform harmonic excitations also occur extremely rarely. It is therefore vital to constantly compare the observed, objective component characteristics with subjective driving impressions in order to accurately assess the comfort of air suspensions. Objective measurement on the vehicle can proceed in parallel with the aid of applied metrology and vibration profiles that can be reproduced. One example would be acceleration measurements on selected test tracks. This testing is expensive, however.

2.7 Causes and influencing factors

2.7.1 Causes of harshness

Characteristic of the increase in spring rate at low amplitudes is the *rolling resistance of the rolling lobe*.

This resistance to rolling that the bellows pits against axial motion results at the macroscopic level from the stiffness of the bellows wall at the rolling lobe. The bellows wall, acting as a stressed membrane under internal pressure, exerts a certain stiffness that influences the characteristic of the force–travel hysteresis. This portion of the bellows wall that stiffens the spring rate is the harshness effect. Among the causes is the mobility of the molecular chain, a characteristic of polymers (Voss, 2002).

To activate the spring, energy is necessary to deform the bellows wall in the rolling lobe. Various force components that occur while passing through the rolling lobe determine the amount of energy needed. Bending/flexing forces (as described by Thurow, 1995) that remain constant throughout the stroke are characteristic on the one hand. On the other hand, elastic and retracting forces occur that determine, as a function of amplitude, how much the material in the rolling lobe stretches. Viewed through a microscope, the rubber mainly deforms between and within the fabric layers during a rolling-lobe cycle. Figure 12 illustrates simulation of a bellows section of an unguided air spring transitioning from the interior to the exterior of the rolling lobe. The rubber stretches much more, particularly near the fabric layer.

Sufficiently large travel amplitudes cause a nearly constant resistance to rolling in the bellows while hysteresis remains constant. As amplitude diminishes, only a partial stroke of the rolling lobe occurs. The associated hysteresis force decreases. By contrast, the hysteresis curve rises more steeply, causing the spring rate to rise. The reason is that diminishing deformations within the elastomer matrix (due to its material properties) elicit

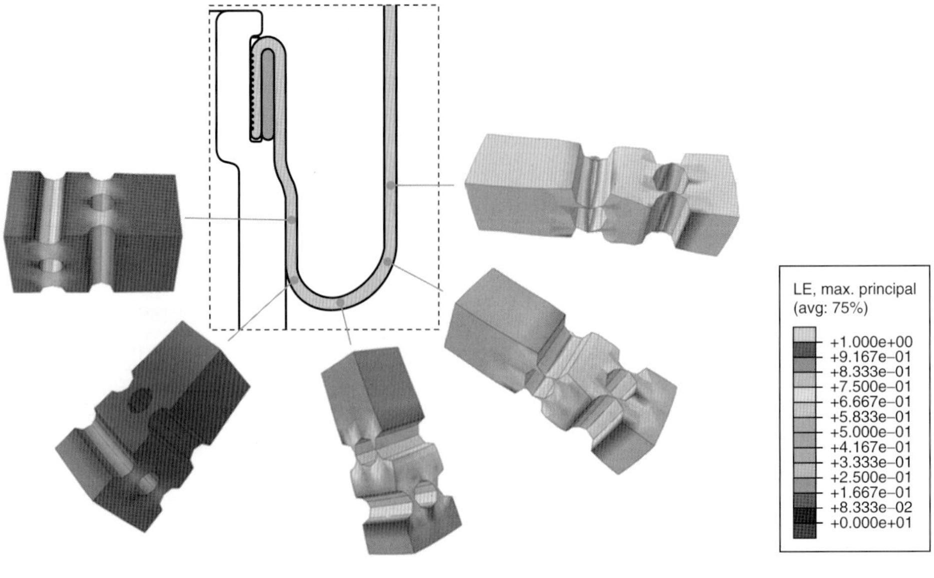

Figure 12. Stretching in the rubber matrix of a bellows submodel during a rolling-lobe cycle. (Reproduced by permission of Continental AG.)

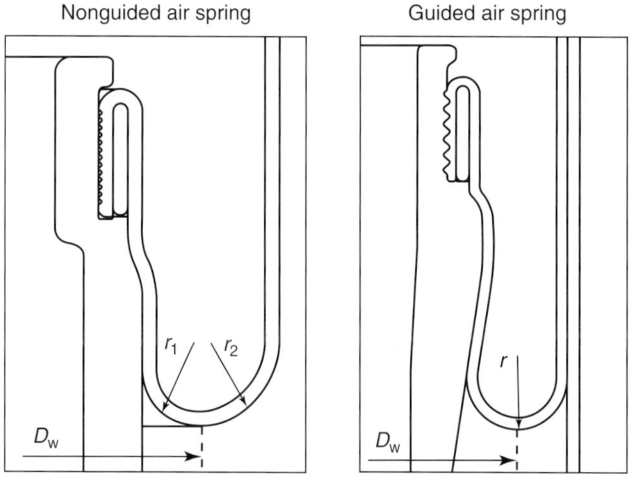

Figure 13. Geometry of the rolling lobe in an unguided and guided bellows.

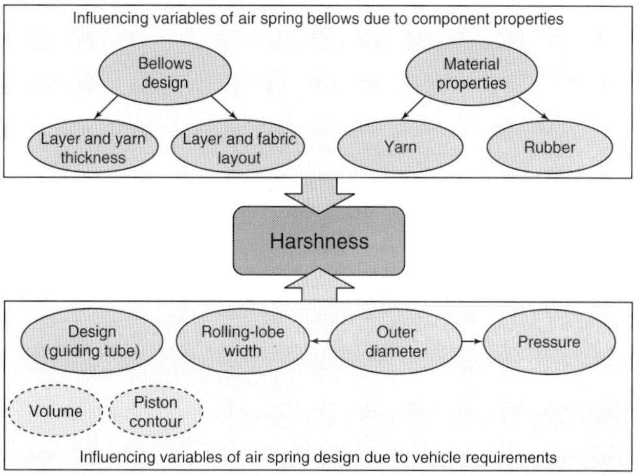

Figure 14. Factors influencing harshness.

the rising, nonlinear changes in force. Moreover, the resistance against the downward rolling motion results in deformation of the rolling lobe's geometry. A shift in the effective diameter may occur, independent of the piston geometry, depending on the stability of the rolling lobe. The result of this effect is an accompanying change in force, which plays a major part, primarily in unguided air springs. Figure 13 provides a detailed view of the rolling lobe's geometry with both a guided and an unguided air spring. The fold factor f_f describes the position of the lowest point of the rolling lobe and thus the effective diameter. With unguided air springs, the fold factor is subject to some variation because of the bellows design selected in conjunction with the material properties of the reinforcement material. The curvatures in the rolling lobe, simply described by the radii r_1 and r_2, do not act constant when rolling motions commence.

With guided air springs, the rigid outer guides largely prevent a change in the effective diameter. The rolling lobe remains more stable in its geometric form, and the characteristic geometry is more capable of retaining a constant curvature with a constant fold factor.

The bellows' resistance to rolling can be accompanied by a certain learning effect after it has stood still for a while. The persistence of the rolling lobe in one position when pressurized, as a function of time, increases resistance to any new motions. The elastomer's material memory "stores" a default position because of deformation in the matrix with the reinforcement material. Age can reinforce this effect. Added to this is the geometric formation of the bellows during vulcanization. A certain number of sufficiently large amplitudes are necessary to break down the resistance of the remembered default position.

Determining the effects of resistance to rolling demonstrated here are mainly the material properties of elastomer and yarn, plus the geometric relation of the whole assembly in conjunction with the kinematics of the rhomboid network in the rolling lobe. As a consequence, it is not impossible to cite a single specific factor for purposes of assessment. A multitude of interactive factors are responsible for the comfort of air suspensions. Figure 14 illustrates the main factors that influence comfort.

As Figure 14 demonstrates, influences from the specific properties of the bellows on the one hand, and influences from the air spring design due to the specific vehicle requirements are chiefly responsible for resultant harshness. The following discussion treats both influence groups in detail.

2.7.2 *Influence of specific component properties of the bellows*

2.7.2.1 Influence of the material properties of elastomer. Influencing the dynamic deformation behavior of an elastomer at low amplitudes are the fillers used in creating it. Fillers such as carbon black or silicic acid greatly improve physical properties of rubber compounds such as strength and durability.

Aside from the basic elastomer properties, the addition of fillers also leads to increased viscoelastic behavior on the part of the material. If sinusoidal deformation occurs, there will be a phase shift between stretching/shearing and tension. The complex shear module G^* can be subdivided into a memory module G' and a dissipation module G'' (Equation 9). The memory module G' is a measure of the stored elastic energy that is regained within a deformation cycle. The dissipation module G'' is the measure of energy

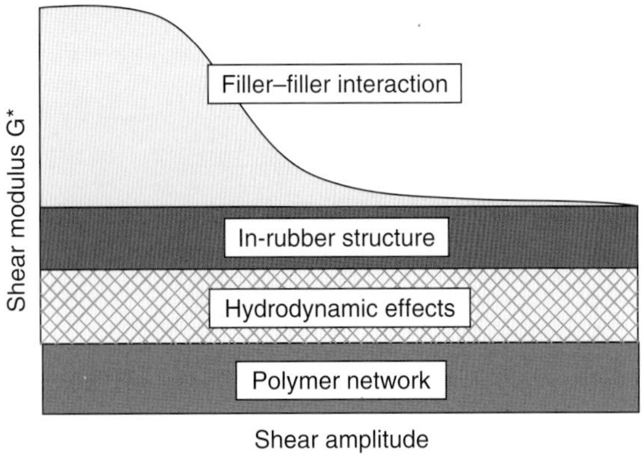

Figure 15. Payne effect. (Reproduced by permission of Joachim Fröhlich, 2007.)

dissipated as heat (Klüppel, 2007).

$$G^* = G' + iG'' \tag{9}$$

The ratio of both factors is known as the *dissipation factor* δ (Equation 10).

$$\tan\delta = \frac{G''}{G'} \tag{10}$$

The Payne effect describes the slope of the dynamic shear module G^* of a filled, vulcanized sample at an increasing amplitude of deformation (Figure 15).

The Payne effect adduces the breakup of the interactive filler compounds as the reason. According to Payne, other interactions between fillers and rubber, hydrodynamic effects of the filler particles or the properties of the pure polymer are largely amplitude independent. When the deformation retracts, these filler compounds rapidly reform. Figure 16 shows the breakdown of the interactive filler compounds using the example of a silica-filled elastomer.

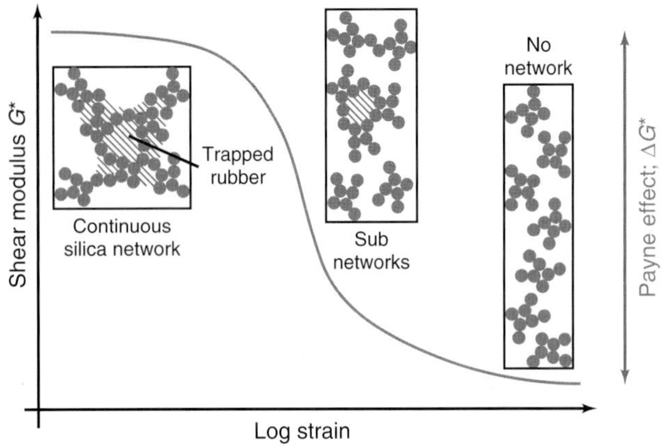

Figure 16. Payne effect regarding silica network. (Reproduced by permission of Joachim Fröhlich, 2007.)

The Payne effect and the characteristic viscoelastic properties become much more pronounced as the amount of filler increases. Figure 17 illustrates the decline of the memory module G' toward large deformations, plus the bulge of the dissipation module G'' as a function of the silica content.

There have been different interpretations of the Payne effect and its causes. The basic explanation of the material behavior of the elastomer ingredients nevertheless follows Payne's original core idea. Interactions occur not only within the filler material but also in the compound of the filler with the polymer matrix (Boehm, 2001). The literature generally speaks also of internal material friction based on a type of breakaway effect analogous to contact–friction phenomena.

Filled elastomers find application mainly in automobile air suspensions. The use of CR (chloroprene rubber), a chlorinated elastomer, is widespread. With the help of suitable ingredients, its properties render it capable of satisfying many of the requirements made on air suspension components in automobiles. Carbon black and silica are the main fillers used in various mixtures. The Payne effect is quite apparent here as well and explains the amplitude-dependent behavior of the increasing spring rate.

Alternative rubber compounds also come under consideration for improving comfort if one is willing to compromise regarding requirements for diverse automotive components. Natural rubber (NR), for example, offers great potential. A combination of various elastomer compounds for the bellows is also possible, depending on the function and the location in the bellows.

2.7.2.2 Influence of the material properties of the reinforcement material. Polyamide (PA) and, to a lesser extent, polyester (PES) and aramid find application as reinforcement materials in the form of individual strands of yarn or rolled fabric mats in automotive air suspensions. PA adheres very well to rubber and is inexpensive but exhibits rather low strength when highly stretched. The strength of PES is comparable to that of PA while its expansion properties are superior, especially with regard to temperature and durability. Aramid, on the other hand, is extremely strong and rather brittle, which somewhat limits its application because of its sensitivity to compression.

The material of the reinforcement plays a rather insignificant role in harshness. This is especially true as the configuration of the bellows leaves little freedom to choose materials because of the strength demands placed on the component. However, a strong fabric made of strong yarn

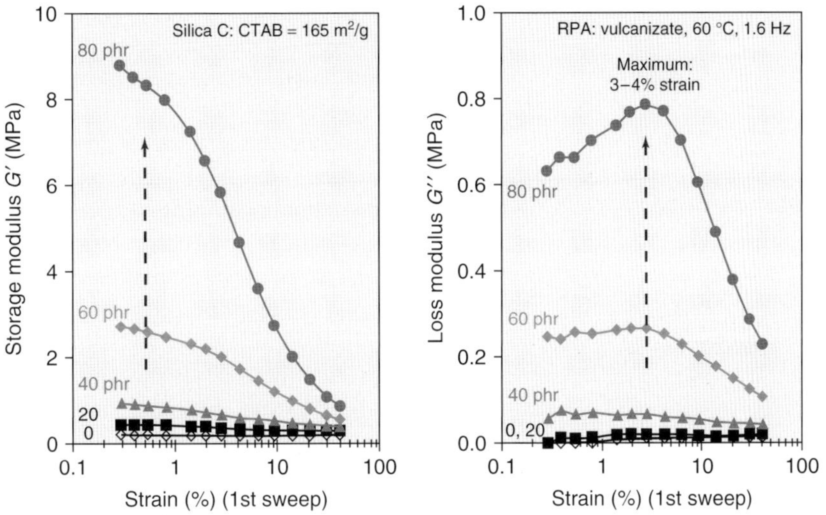

Figure 17. Memory- and dissipation-module curves. (Reproduced by permission of Joachim Fröhlich, 2007.)

and a high yarn density guarantee a stable rolling lobe with regard to less harshness. By contrast, however, the bellows wall may lack flexibility. Material properties may also have an influence on acoustical transmission.

2.7.2.3 Influence of the design of the bellows (wall thickness, yarn angle, yarn density, and thread count).

The resistance to deformation of the bellows matrix not only depends on the materials used in the elastomer and the reinforcement but depends also on the design of the bellows. The main components here are the thickness of the elastomer layers, the diameter of the yarn, the yarn angle, and the thread count and/or the separation between strands. The combination of these parameters gives rise to a linked system that must undergo deformation. In the case of cross-layered bellows, a rhomboid pattern arises that determines the geometry and the deformation of the elastomer within the rhomboid. Figure 18 illustrates the stretch of the material within the rubber matrix as the rolling lobe passes through from the piston to the outer diameter. The example shows a bellows submodel of an unguided spring. The amount of material strain varies as a function of the bellows design.

To reduce harshness, material deformation must be kept low and the geometry of the rolling lobe must be kept stable. An advantageous design in this regard exhibits thin bellows with thin reinforcement material, little material, and slight deformations in the rhomboid pattern, thanks to optimized packing of yarn in conjunction with an optimized yarn angle. Such a high strength networked system, however, on the other hand, may slightly elevate the rigidity of the bellows wall during the rolling-lobe cycle.

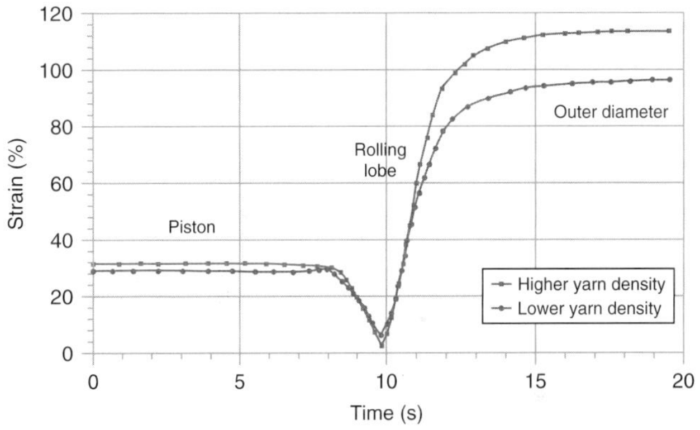

Figure 18. Depiction of the principal tensile strain in the rolling-lobe's cycle as a function of the bellows design.

Reducing the intermediate and outer rubber layer improves harshness markedly with regard to wall thickness. While yarn angle and amount in unguided springs depend essentially on the space available and requirements of engineering strength, the yarn parameters of an externally guided air spring can be put into effect for purposes of comfort under certain conditions. There is thus an ideal yarn angle for every bellows design.

Axial bellows do not exhibit any deformation with the rhomboid pattern because of the single layer of purely vertical yarn. This is the reason why axial bellows enjoyed the reputation until recently as being good for comfort. The entire arrangement, however, also provides less stability against deformation when the rolling lobe is in motion than the cross-layered arrangement. Moreover, thicker yarn must be combined with low rolling-lobe widths in an axial

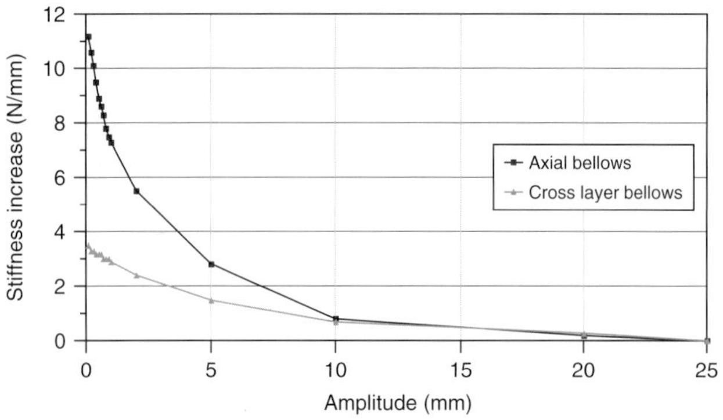

Figure 19. Harshness comparison between cross-layered bellows and axial bellows.

bellows for reasons of engineering strength. In addition, the operating pressure is capable of compressing the rubber material more strongly between the strands of yarn and deforming it even further. The rubber in the rolling lobe consequently experiences very severe deformation during movement. Bellows design is thus subject to severe limitations when it comes to the comfort of axial bellows.

By contrast, thin, guided cross-layered bellows offer the greatest possible leeway in shaping the parameters for improving harshness. Current developmental work has indicated that it is possible to greatly reduce harshness by employing very thin PA fibers and thin elastomer layers of CR in conjunction with the optimized air spring application. Figure 19 illustrates the great potential of a comfortable cross-layered bellows compared to an axial bellows used in the same air suspension of a luxury car. The excessive stiffness of the cross-layered bellows was reduced to approximately 30% of the level of the axial bellows. Its harshness of less than 4 N/mm has been reduced to a minimum.

2.7.3 *Influence of the air spring design due to specific vehicle properties*

The actual configuration of the air suspension exerts a major influence on harshness in conjunction with the properties of the bellows' components. Worthy of mention here are parameters such as air spring diameter with the accompanying pressure level, plus the opportunity to integrate exterior guidance, all the while considering such vehicle requirements as space for installation, load, and the intended spring rates.

2.7.3.1 Influence of air suspension diameter (pressure level and width of rolling lobe). Exploitation of the maximum air spring diameter is reflected in the reduction of the static pressure level. The internal pressure in the air spring stretches the bellows membrane and consequently puts an end to material expansion and the associated deformation (Voss, 2002). Reducing the pressure thus directly mitigates material deformation. The ideal amount for a minimum level of pressure, which would just guarantee the optimum forming of the rolling lobe in every situation, is generally not achievable in passenger car applications because of space limitations. In reality, the basic pressure $p_{\text{stat.rel.}}$ of approximately 8 bar is usual.

If there is sufficient space and clearance for the air springs, there is also a way to optimize the width of the rolling lobe r_b. Rolling lobes that are too small can quickly result in an increase of harshness because of a high rate of material deformation in the bellows wall. Moreover, they have a negative impact on engineering strength. By contrast, rolling lobes that are too wide necessitate large amplitudes for their initial pass-through, associated with a greater change of the yarn angle. The effective diameter (as a function of the rolling-lobe width) changes in parallel with the resultant operating pressure. The effect on harshness is, however, rather negligible.

Therefore, there will always be an ideal rolling lobe for each type of air suspension. Steps for reducing pressure and designing the ideal rolling-lobe width can make a major contribution toward reducing harshness. Figure 20 illustrates the influence of the rolling-lobe width using the example of a guided air spring.

2.7.3.2 The influence of exterior guidance. The expansion of the bellows reduces greatly when exterior guides are used. The rigid outside guide now absorbs the circumferential forces. This permits setting the yarn angle of the bellows generally between 60° and 90°, which then approximates the actual equilibrium angle during operation.

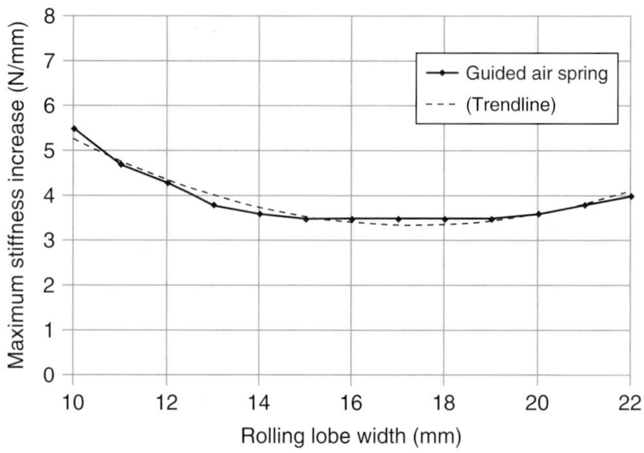

Figure 20. Influence of the rolling-lobe width on harshness.

Reinforcement material and elastomers are thus subject to much less load and deformation. This, in turn, allows for very thin-walled bellows. Above and beyond that, this protects the bellows from environmental influences. This makes it possible to keep the thickness of the outer cap to a minimum. Embedding thinner threads of yarn also contributes to a thinner bellows.

The rolling-lobe geometry remains stable because of the rigid outside guides and reduced deformation.

For purposes of comfort, air suspensions on passenger cars cannot do without exterior guidance. Most externally guided air springs are designed in the form of struts. It is somewhat more problematic to incorporate rigid guides in the case of freestanding air springs.

In contrast to the influences described earlier, the parameters such as air spring volumes and piston-roll geometry have only a negligible effect on harshness. Both generally affect only the basic spring rate.

2.7.3.3 The influence of air spring volume. The change in the air spring's nominal basic stiffness does not result in any increase in stiffness at low amplitudes. Studies of air springs with adjustable additional volume have confirmed this. While the basic spring rate increases as volume reduces, the resulting harshness only changes insignificantly (Figure 21).

On the other hand, the hysteresis forces of the air springs change as a function of the spring volume.

2.7.3.4 The influence of piston contour. The contour of the piston is an important parameter that influences spring characteristics, making it possible to adapt spring characteristics to the requirements of the car. When optimum rolling-lobe widths are considered and the angle of the piston is kept within certain limits, the effect on harshness is negligible. Of course, the piston contour exerts a strong influence on the geometry of the rolling lobe and its deformation, so that it is impossible to exclude an influence, especially for unguided air springs and very large changes in angle such as rises >20°. As long as the rolling-lobe geometry can still "follow" the piston contour (i.e., the effective diameter is proportional to the piston diameter), its influence will be negligible.

Significant interactions among the influencing factors, depicted in Figure 14, will always occur. Each individual factor can change the harshness of air springs but it is necessary to find the optimum balance of all parameters for an individual application. Air spring manufacturers have specific configuration rules depending on the type of spring and bellows design employed. As a practical matter, it would be difficult to quantify values precisely with the aid of simulators and calculations because of the complex behavior of the material in the bellows structure. Values obtained empirically generally serve the development process. Table 1 summarizes the influencing factors and their importance.

It will be necessary to transfer all the influential factors shown here to the design and configuration of any air suspension system during the development process. The vehicle's axle design will determine the type of air spring. Consequently, certain limits already apply when attempting

Table 1. Factors influencing harshness.

Variable			Influence
Bellow design	Rubber	Material	Strong
		Thicknesss	Strong
	Yarn	Material	Neutral/medium
		Thicknesss	Medium
	Matrix layout		Strong
Air spring design	Guiding tube		Strong
	Pressure		Medium
	Rolling lobe width		Strong
	Volume		Neutral
	Piston contour		Neutral

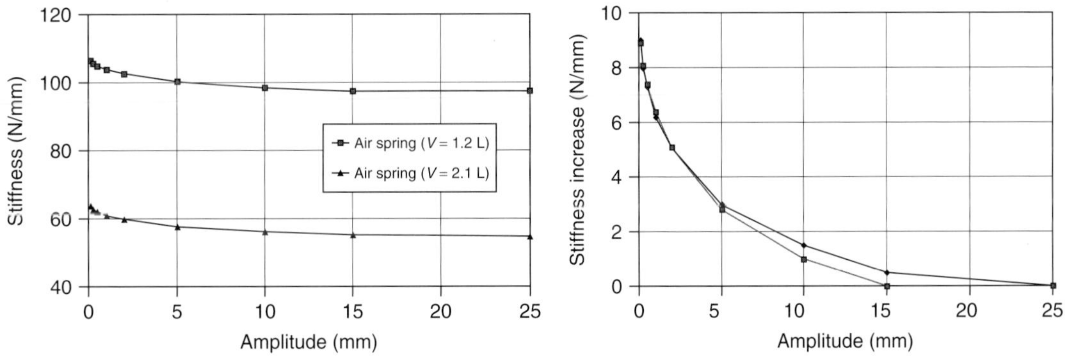

Figure 21. Harshness curves as a function of air spring volume.

to improve harshness due to the type of air suspension chosen.

3 HARSHNESS OF DIFFERENT TYPES OF AIR SUSPENSION

3.1 An overview

Air springs may be configured either as a strut or as a freestanding spring, depending on the type of axle employed. Struts see use in luxury cars predominantly on both front and rear axles while freestanding springs are found mostly on rear axles in middle-class and economy cars. Nevertheless, struts offer the greatest potential for diminishing the effects of harshness (Figure 22). The following spring–rate curves $\Delta c_{\mathrm{harsh}}$ are attainable in air suspensions being mass-produced currently:

- air strut, externally guided: $\Delta c_{\mathrm{harsh}} = 3\text{–}6\,\mathrm{N/mm}$;
- air spring, freestanding, externally guided: $\Delta c_{\mathrm{harsh}} = 6\text{–}10\,\mathrm{N/mm}$;
- air strut or air spring, unguided: $\Delta c_{\mathrm{harsh}} = 10\text{–}15\,\mathrm{N/mm}$.

The differences make it quite apparent that the combination of an externally guided strut on the front axle with a freestanding, unguided air spring on the rear axle elicits a pronounced mismatch of the respective harshness. While the front axle probably provides good comfort, the rear axle may lag noticeably behind. Experience has shown that it is not possible to eliminate this discrepancy. The trend toward modular construction in the automobile industry is leading to uniformity in the axle concepts of middle-class and luxury cars. Consequently, there are efforts afoot to establish the freestanding spring for the rear axle also in the luxury segment.

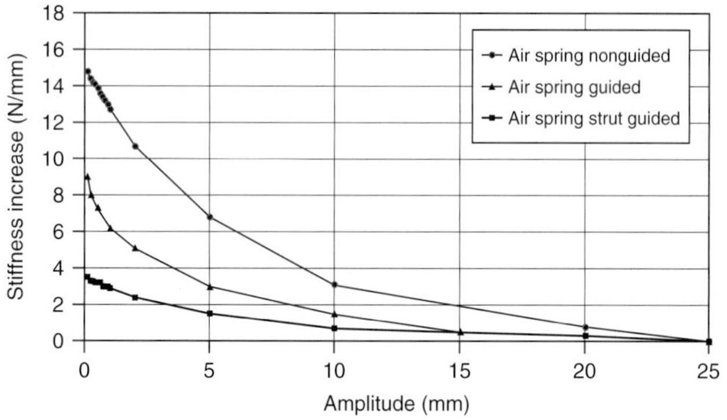

Figure 22. Harshness curves of different type of air suspensions.

3.2 Air struts

In strut architecture, the air spring is arranged concentrically around the damper, just as with a steel spring. The strut contains all the elasticity necessary to compensate for movements from the axle kinematics. Different configurations of air suspensions are possible, depending on the requirements of the vehicle and manufacturer. Figure 23 illustrates three possible configurations. Aside from simple, inexpensive unguided air springs, there are the more commonly used guided designs. The restricted motion of the bellows due to the rigid guide, however, necessitates integration of cardan compensation. Cardan means in this context that the air bellow needs to have certain flexibility into all directions. As will be shown soon, the strongly defined damper movement, defined by the suspension arm, makes the additional movement of the air spring itself necessary, as this is also connected to the top mount which is with the car body.

3.2.1 Externally guided struts

External guidance is indispensable if one is to diminish harshness appreciably. Assuming ideal parameters (see Section 2.7), it is possible to use cross-layered bellows with thin walls between 1.4 and 1.8 mm. Ultrathin PA yarn measuring 235 dtex × 1 serves as reinforcement. The yarn

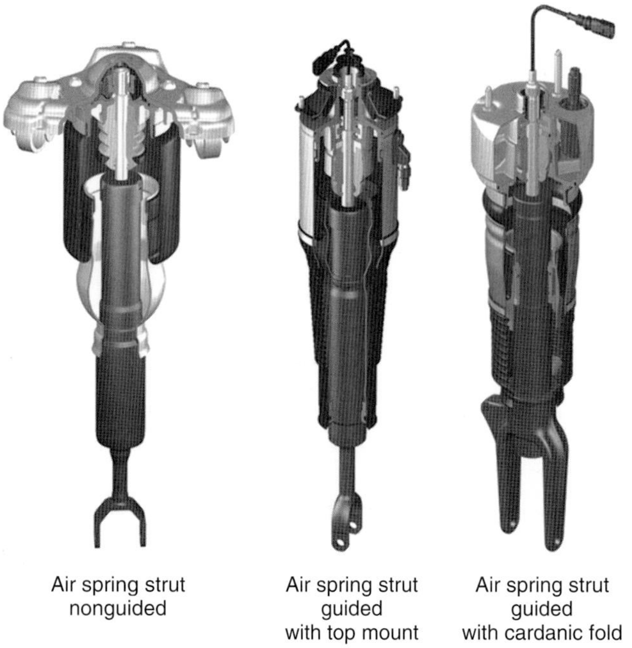

Figure 23. Selected strut configurations. (Reproduced by permission of Continental AG.)

has a diameter of just 0.19 mm. This makes it possible to achieve a minimal harshness $\Delta c_{\mathrm{harsh}}$ of up to 3 N/mm, which is hardly noticeable. The potential here has surely been largely exhausted. Any future improvements will require much effort in manufacturing the bellows when processing the fine yarn and the thin rubber layers. A major focus, moreover, will have to be on improving the material properties of the elastomer.

3.2.1.1 Cardan compensation. It appears much more important to focus on parallel effects that diminish comfort. The architecture of the air suspension can influence these two effects. Aside from the actual harshness of the air springs, the contact friction of the damper-piston rod plays a decisive part. Owing to the cardan motion of the strut in the axle, bending (flexural) moments are transmitted to the damper. They generate lateral forces on the piston rod, which increases friction as a direct consequence. As the strut compresses, it has to overcome a breakaway effect. This breakaway effect diminishes comfort appreciably (particularly at low amplitudes) and masks the harshness of the air springs noticeably (Ilias and Sorge, 2001).

The design of the struts determines the properties for taking up cardan movement within the axle's kinematics. The elasticity in the strut determines the quality of the cardan compensation and the transmission of bending (flexural) moments. Besides the elasticity of the air springs, this also includes the elasticity of the bearings used. A suitable air suspension design is thus capable of supporting optimum cardan compensation. Figure 24 illustrates that an additional elastomer bearing or a cardan fold could find application here. Elastomer bearings in the form of rubber and metal components can also have a decoupling effect because of their transmission behavior, and this also adds to the impression of higher comfort. Simply put, the cardan fold is an exposed, unguided bellows section that is reinforced and possesses a great deal of elasticity. It represents a simple, inexpensive solution.

3.2.2 *Unguided air struts*

Struts with unguided air springs play a somewhat subordinate role when it comes to comfort. The harshness effect is more marked because of their more robust structure and exceeds that of a guided spring many times over. The higher demands placed on the bellows make a thicker yarn necessary (at least PA 940 dtex × 1, yarn diameter approximately 0.33 mm) and a thicker outer cap. The total bellows wall thickness then lies in the range of around 2.0–2.4 mm. An optimum configuration of the remaining influential parameters can confine the spring rate to an acceptable range here. On the other hand, the good cardan compensation, mostly in the form of bellows elasticity, is advantageous.

3.2.2.1 Wheel-locating air struts. There is one promising way to improve comfort for unguided air springs with suspension struts such as the MacPherson strut. By locating the wheel, they transfer higher bending (flexural) movements to the strut, resulting in a greater lateral force with accompanying friction of the piston rod of the damper. An optimum air suspension design and configuration can nearly compensate for the transverse forces that occur. The reduced lateral forces consequently reduce friction of the piston rod and also provide a way

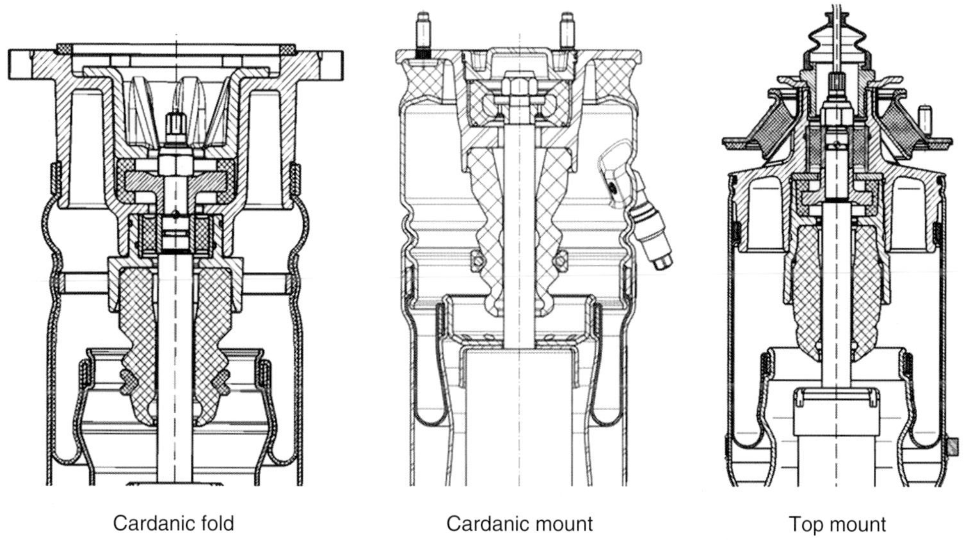

Figure 24. Cardan compensation of guided air struts. (Reproduced by permission of Continental AG.)

to improve the design of the seals and the guides for the piston rod. However, these air suspension concepts are complicated and costly. What is more, the asymmetric arrangement of the bellows places greater demands on engineering strength.

3.3 Freestanding air springs

Rear-axle designs for automobiles call mostly for a spatial separation between dampers and springs. The freestanding springs are located more inboard. Their lower design height allows for a wider loading space in the cargo area. However, the springs' inboard position does result in limited space for installation. It also results in a relatively offset position for the air springs' pistons and cap because of spring motion. The rolling lobe may become more constrained as a consequence. Furthermore, the high spring ratio (up to $i = 0.5$) increases the load-carrying capacity and, consequently, the static pressure level. The high potential for additional load on the rear axle also adds to operating pressure. There is, therefore, a need for air suspension architecture that is capable of accepting cardan movements very well on the one hand and can offer the necessary space for spring travel with very little clearance on the other. The parameters mentioned earlier are counterproductive if one wants to optimize comfort. Using rigid, external guidance is problematic because of cardan motion and the permissible space. Parallel to this, the higher level of pressure requires robust bellows. Various types of freestanding air springs may offer a solution as illustrated in Figure 25. In addition, it is possible to make the lobe rolling upward. A guided version can also be seen in Figure 25.

The differences among the shown versions and the conventional model are the upside-down arrangement, the use of a double rolling lobe, and a flying external guide.

Figure 25. Selected versions of freestanding air springs and a guided one with a double rolling lobe. (Reproduced by permission of Continental AG.)

3.3.1 Externally guided, freestanding air springs

Compared to the front axle, acceptable values for harshness at the rear axle are achievable only with a guided design. To ensure cardan compensation and guarantee freedom of movement, experience has shown that it is necessary to have two rolling lobes and an upside-down arrangement. Inclusion of external guides requires a very compact flying design. This means that the external guide covers only a portion of the bellows and leaves the second rolling lobe largely exposed. All in all, the bellows wall can be made thinner although it must still be robust in the vicinity of the second rolling lobe. An extremely thin wall such as the strut is currently not possible for this reason. As a practical matter, though, it has already been possible to achieve a spring rate that is less than 10 N/mm in a luxury sports car.

There is potential for still more development. The second rolling lobe must be geometrically configured so that the stress sinks, on the one hand, and the harshness effect diminishes, on the other. In view of the above, it would be conceivable to achieve maximum stiffness rates of approximately 5 N/mm and thus to approximate the comfort level of the front axle.

3.3.2 Unguided, freestanding air springs

The use of freestanding unguided air springs in middle-class and light trucks is widespread due in part to cost and space limitations. Comfort is not the top priority in this segment, especially not in the case of rear-axle systems with load leveling. Combined with guided front air springs, the increased harshness occurs noticeably which means that it is necessary to pay attention to this area. As there is only limited freedom in configuring the bellows, any design must be functional and geometric.

An upside-down arrangement in conjunction with a second rolling lobe improves the kinematic movement of the springs and reduces the design height needed. It is then possible to design the geometry of the rolling lobe better and reduce the constriction of the rolling lobe.

What is more, a concept with a second rolling lobe permits placing it at the height of the effective diameter so that the spring rate of the rolling lobe activates in line and consequently sinks. This principle does entail limited freedom of configuration though. Optionally, an additional antiharshness layer in the form of an elastomer or polyurethane (PUR) layer can greatly improve transmission properties, particularly when it comes to acoustics. A simple elastomer layer is generally all that is necessary where the spring joins the bodywork.

The acceptable comfort range for unguided freestanding air springs is approximately 10–15 N/mm of spring rate.

With proper measures, it should be possible to drop the rate to under 10 N/mm in the future.

These harshness numbers do not present a real alternative when it comes to comfort, however. Guided air springs are still indispensable here, ideally as a strut. Added to this is optimum cardan compensation to minimize any additional friction effects.

4 OUTLOOK

Air suspension's market penetration will continue. Air springs bestow a noticeable improvement in comfort, primarily in luxury and middle-class passenger cars. Small cars (which are coming to the market next time) need air springs most, but the cost is (up to now) the largest obstacle. The focus for future work is on improving the characteristics of air suspensions. The harshness effect is a fundamentally undesirable property of air spring bellows and is also an important, quantifiable gage of comfort for automotive components. Objective test results, however, must always be brought into line with subjective assessment.

In view of the various factors influencing harshness, it is possible to demonstrate extremely slight increases in spring rate at low amplitudes using targeted air suspension designs. Externally guided air springs with thin-walled cross-layered bellows and fine yarns of PA are capable of reducing the bellows' resistance to rolling to a minimum. These bellows find use predominantly in strut designs. More development work will be necessary to use thin bellows in freestanding spring designs in order to attain a comparable level of comfort. In addition to empirical experience with comfort-based air suspension, simulation processes will increasingly be used to predict harshness properties.

Beyond that, it will be necessary to analyze and reduce other factors that diminish automotive comfort. Besides the air springs themselves, it is worth mentioning the properties of elastomers where the suspension joins to the body and the response of hydraulic dampers as factors that can generate undesirable rubbing effects. However, the structural stiffness of the bodywork also plays an important part in transmitting undesirable vibrations and noises to the car and to the passengers.

5 SUMMARY

Air spring systems are introduced into the market, but for passenger cars, these systems suffer from harshness. This contribution gives a detailed explanation on the effects of harshness and the reasons for the existence. In addition, there are descriptions for the measurement, the simulation, and the reduction of the harshness effects to modern air spring systems. Detailed information will enable the automotive chassis designer to reduce the harshness effect in the future.

REFERENCES

Boehm, J. (2001) *Der Payneeffekt: Interpretation und Anwendung in einem neuen Materialgesetz für Elastomere*, Dissertation, University of Regensburg, Regensburg.

Ersoy, M. and Heißing, B. (2008) *Fahrwerkhandbuch*, Vieweg Verlag, Wiesbaden.

Fröhlich, J. (2007) *Verstärkung durch Moderne Füllstoffe—Teil II*, Degussa GmbH, DIK, Hannover.

Ilias, H. and Sorge, K. (2001) *Grundsatzuntersuchung zum Thema Harshness, Interner Entwicklungsbericht*, Continental AG, Hannover.

Klüppel, M. (2007) *Hochfrequenzeigenschaften und Reibung von Elastomeren*, DIK, Hannover.

Thurow, G. (1995) *Lebensdauerergebnisse und Federverhalten von Geführten Luftfedern*, ContiTech Luftfedersysteme, Hannover.

Voss, H. (2002) *Die Luftfederung, eine Regelbare Federung für Straßen- und Schienenfahrzeuge*, ContiTech Luftfedersysteme GmbH, Hannover.

FURTHER READING

Betzler, J. and Reimpell, J. (2005) *Fahrwerktechnik Grundlagen*, Vogel Verlag, Würzburg.

Braess, H. and Seiffert, U. (2007) *Handbuch Kraftfahrzeugtechnik*, Vieweg Verlag, Wiesbaden.

Gauterin, F. and Sorge, K. (2001) *Noise, Vibration and Harshness of Air Spring Systems*, Continental AG, Hannover.

Puff, M. (2009) *Prüfspezifikation zur Charakterisierung von Luftfedern*, Technische Universität, Darmstadt.

Chapter 112
Air Supply for Advanced Applications

Hans O. Becher
WABCO, Hannover, Germany

1 INTRODUCTION

The desired characteristic and performance of air suspensions in passenger cars can only be achieved in combination with an air supply unit (ASU). The ASU itself is necessary but undesired: it consumes space, increases weight, consumes energy, creates noise and vibration, and costs money. And exactly they are the key characteristics of an ASU, which are subject to continuous optimization.

However, the ASU is mandatory to generate compressed, dry, oil-free, and clean air to pressurize the air springs. The suspension elements have to be filled with a certain pressure for normal working operations. Natural leakage through bellows, pipes, connectors, and valves has to be considered to be compensated. In many cases, there are additional demands to increase the chassis level by inflating the air springs and decrease the chassis level by deflating the air springs.

Encyclopedia of Automotive Engineering, print © 2015 John Wiley & Sons, Ltd.
Edited by David Crolla, David E. Foster, Toshio Kobayashi, and Nick Vaughan.
This article is © 2015 John Wiley & Sons, Ltd. ISBN: 978-0-470-97402-5
Also published in the *Encyclopedia of Automotive Engineering* (online edition)
DOI: 10.1002/9781118354179.auto016

The compressed air has to be distributed into the air springs, which is mostly done with multiple solenoid valve blocks. To lower the vehicle chassis height, the air springs have to be vented. The ASU provides solenoid valves for this function.

In addition to the above-mentioned criteria, key performance indicators for ASUs are airflow, maximum pressure, duty cycle, electrical current consumption, and peak current.

2 SYSTEM REQUIREMENTS FOR AIR SUPPLY UNITS

2.1 Two-corner applications for rear axles

There are two types of two-corner air suspension systems on the market:

- Leveling systems (combination of steel springs and air springs)
- Full air suspension systems (without steel springs).

Leveling systems are designed to carry the basic vehicle rear axle load (empty condition) with a steel spring and the additional load of passengers and cargo with an air spring. There is just one nominal level requested. Quick lifting/lowering is not needed.

Consequently, the airflow performance requirement is low, typically about 10 norm dm^3/min [where norm dm^3 is the amount of air which fills one dm^3 at 0°C and 1013 mbar at 6 bar. The pressure range however is between 1 (minimum pressure to avoid bellow damage) and 12 bar in the fully loaded condition. The pressure demand is quite high as the diameter of the air spring is limited because of packaging reasons. These systems are popular in the North American market, but the market penetration is decreasing. The system layout is simple (only one common height sensor for right and left side) and the ASU design is very much cost driven.

Full air suspension systems for rear axles, carrying the whole rear axle load (empty vehicle load + passengers + cargo), are more demanding regarding the ASU (Figure 1). While the pressure levels are similar, the airflow requirement is higher (>20 norm dm^3/min at 6 bar). The end-user's requirement is a quick, visible compensation of chassis level drop after loading. Additional functionality is for instance a switch in the trunk area for manual lowering of the chassis for easy loading operation. This lowering procedure has to be quick as well to avoid waiting time. Consequently, the air passage through all system elements has to have a sufficient diameter. Usually, these ASUs have a double solenoid valve block to control left- and right-side air spring individually. This feature allows a right/left-balanced leveling even if the axle load is not centered. In addition, the roll stiffness is increased, as the airflow between right and left air bellows is blocked. Another feature is the load transfer away from one wheel in case of a damaged "run flat" tire. Unloading the damaged tire can increase the allowed driving distance.

Duty cycle demand for ASU's in two-corner systems is low (typically 10% on-time of a 10-min period, environmental temperature 23°C), as there is no normal operation mode, which requires a long continuous compressor run. The longest sequence is given by filling the empty air spring from buffer level to normal ride height and loaded condition. Typically, this operation does not take longer than 60 s.

In some cases [transporters and sports utility vehicle (SUV)], these systems have a reservoir for quick lifting and/or compressor running noise avoidance in certain conditions. Then, the ASU performance in terms of airflow and maximum pressure has to be even higher and on a level of typical four-corner reservoir systems (Section 2.2).

2.2 Four-corner applications

Most of the four-corner systems have a reservoir to enable quick lifting by a high airflow from the reservoir into the bellows. Performance demand for ASU's used in reservoir systems is typically >25 norm dm^3/min at 6 bar, maximum pressure up to 18 bars. Duty cycle requirement is higher (15%) than for ASU's in reservoirless systems, as the compressor run time to fill up a reservoir up to 18 bar is significantly longer than to inflate the bellows directly.

Another advantage of reservoir use is that the compressor must not be operated in the vehicle standstill condition, which is very noise critical (start–stop system, engine off). In closed systems, a reservoir is mandatory, as the air is moved between air spring and reservoir. Usually, a direct bellow filling by the compressor or exhausting of the bellows is not possible.

Some four-corner systems are working without a reservoir. These systems are designed to operate without attracting attention from the end user. Chassis-level changes are expected to be "invisible," thus the demand for the ASU airflow is low. Maximum pressure is similar to two-corner systems (12 bar). The compressor has to be operated also in standstill mode of the vehicle, and so the demand regarding very low compressor running noise is high.

Many four-corner systems are using not only the benefits of air springs in normal operation mode but also the possibility to lift or lower the chassis. Reasons for this include dynamic driving behavior and reduced fuel consumption

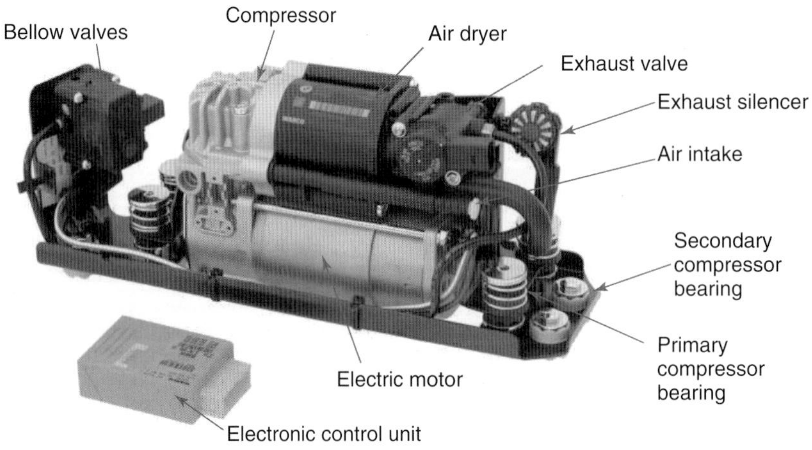

Figure 1. Air supply unit for rear axle air suspension system. (Reproduced with permission from WABCO.)

by low level at high speed, easy entry for passengers, and extended ground clearance for certain driving situations (e.g., off-road). These kinds of systems are demanding higher performance from the ASU: high airflow during lifting and lowering operations, high maximum pressure, and high duty cycle for frequent repeats of level changes. To monitor the duty cycle of a compressor and getting close to the thermal limits, the information about compressor temperature or at least environmental temperature in the area of the compressor is mandatory. Some systems have additional volumes that can be connected/disconnected to the air bellows to vary the spring stiffness. For the ASU, this means more air volume to be transported into and off the bellows. Figure 2 shows a typical example for a four-corner ASU.

2.3 Further demands for compressed air

Having an ASU installed in a passenger car increases the desire to use this energy also for other systems. As an example, a brake booster with compressed air could be packaged much more compactly than a typical vacuum brake booster. However, this concept was never realized. The only option, which was marketed so far, was a tire inflator. The compressor can be used to fill a flat tire or leisure equipment. For some special applications, there are systems on the market, which are able to control the tire pressure during driving. The main reason not using compressed air for other applications than air springs is that many features as the air suspension itself are just optional. One example is a dynamic seat contour, which uses inflatable cushions. An ASU for air suspension systems would be oversized for the seat operation alone. Therefore, these applications have own small compressors, directly installed in the seat.

2.4 General system requirements: future outlook

The load-bearing force of an air spring is as follows:

$$F = p_i \cdot A$$

The demand for smaller air bellow diameters and thus effective cross-sectional area of an air spring A because of packaging reasons in passenger cars leads to increasing bellow pressures p_i. Thus, the main future challenge for ASUs is a higher maximum pressure.

Furthermore, the vehicle manufacturers will specify size, weight, and inrush current limitation more strictly. Noise and vibration levels are also subject to improvement, because hybrid or electric drives in passenger cars allow a very low basic noise level, and the ASU must not stick out during those driving conditions.

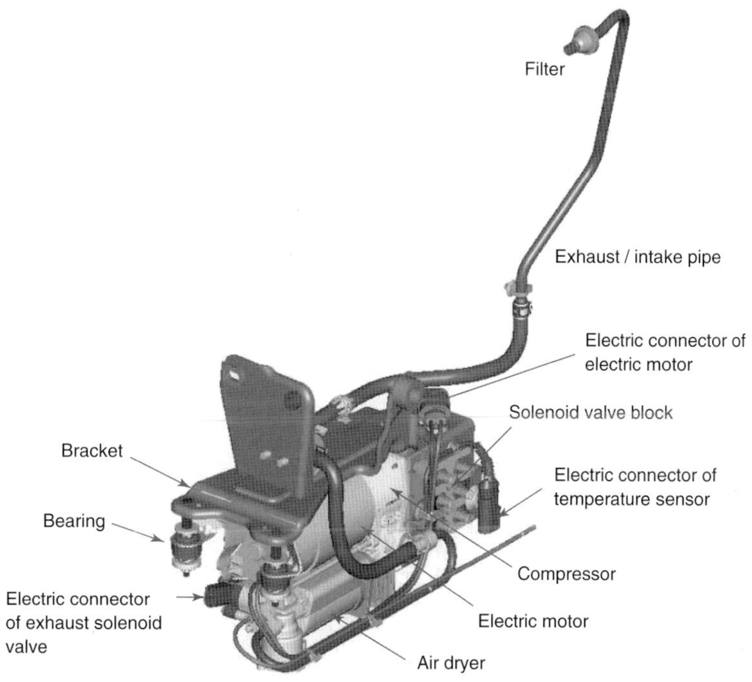

Figure 2. Four-corner air supply unit. (Reproduced with permission from WABCO.)

3 GENERAL FUNCTION OF A COMPRESSOR FOR AIR SUSPENSION

3.1 Compressor

Following the system requirements in Section 2, the preferred solution for an ASU is a dry-running piston compressor driven by an electric motor. Other devices such as vane-type pumps or rotary screw compressors are not able to generate the requested pressure levels (10–18 bar) or are commercially not attractive.

3.1.1 One-stage open type

Open type means the compressor is sucking air directly from the atmosphere, compressing it to the requested level and feeding the air springs and/or reservoir. Exhausting the air from the bellows into the atmosphere lowers the chassis.

A typical 4-corner open system schematics is shown in Figure 3. By far, the most popular and common design is a one-stage piston compressor, open type (Figure 4). The air is taken in from the atmosphere through the air intake into the crankcase. If the piston is moving down,

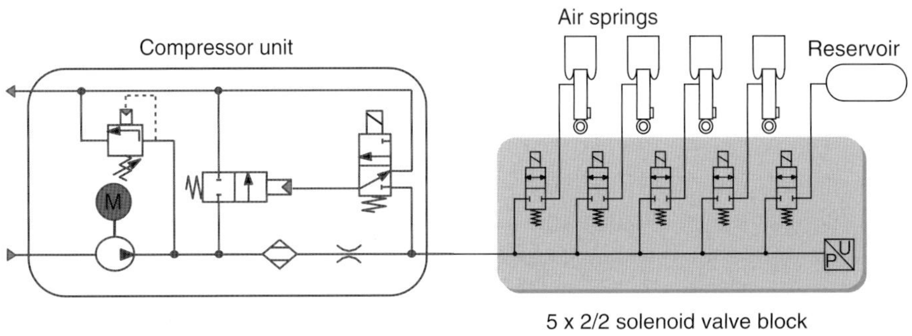

Figure 3. System schematics of a four-corner air suspension system (open system). (Reproduced with permission from WABCO Hitachi.)

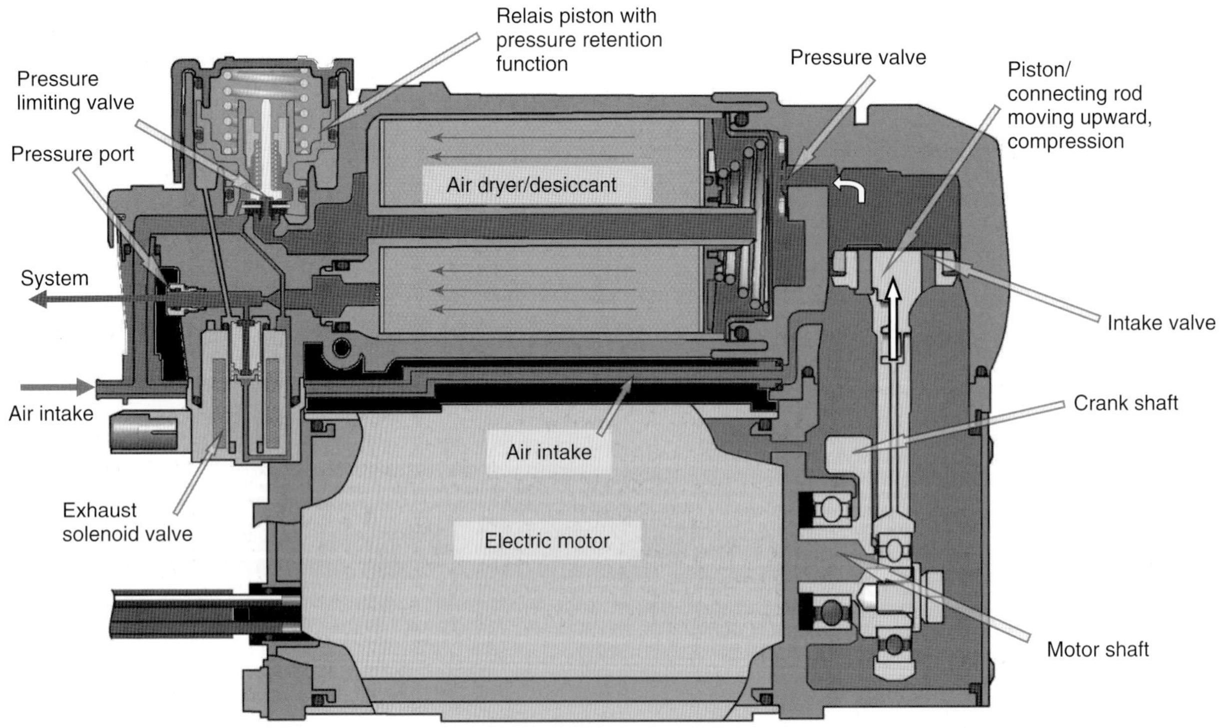

Figure 4. Sectional drawing of a one-stage open type compressor. (Reproduced with permission from WABCO.)

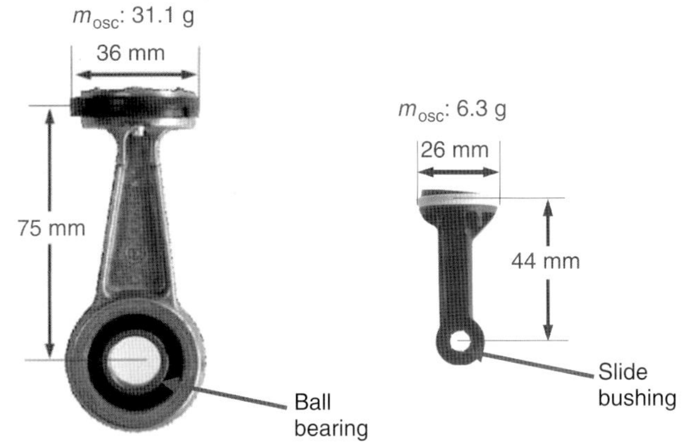

Figure 5. Wobbling pistons with ball bearing and slide bushing. (Reproduced with permission from WABCO.)

the air flows through the intake valve in the piston into the cylinder. If the piston is moving up, the air is compressed and guided through the pressure valve into the air dryer. The compressed air passes the desiccant and is then guided to the pressure port and entering the system.

All known one-stage piston compressors for air suspensions are directly driven by an electric motor. The crankshaft is fixed on the motor shaft and drives the connecting rod. The connecting rod bearing is—depending on the operation conditions and performance range—designed as a ball bearing or slide bushing (Figure 5).

In many cases, the piston and the connecting rod are designed in one piece without articulation between piston and connecting rod. This "wobbling piston" design requires special focus on the piston sealing elements, which have to tolerate an angular motion of the piston (Figure 6).

There are two possible solutions to piston sealing: piston ring or gasket (Figure 7). For both designs, the proved material is PTFE providing a sufficient heat and wear resistance without being lubricated. At the same time, the sealing has also to be able to work under low temperature down to −40°C.

The piston ring (Figure 8) is usually a machined part with a slot for assembly and for wear reasons. This design ensures a permanent defined axial contact pressure between the piston ring and the cylinder wall.

Although the piston sealing should be as tight as possible to achieve maximal volumetric efficiency, the piston ring has a natural leakage under static conditions. Thus, pressure balancing above and below the piston is given within the cylinder, which enables an easy compressor start. In comparison to the piston ring, a gasket is a more simple part, is fixed to the connection rod, and has always close contact to the cylinder. A piston ring has always a certain amount of clearance in axial direction (Figure 7) to allow

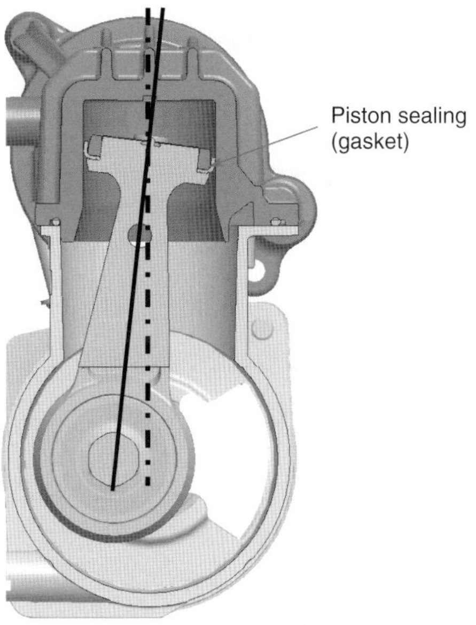

Figure 6. Wobbling piston, maximum angle position. (Reproduced with permission from WABCO.)

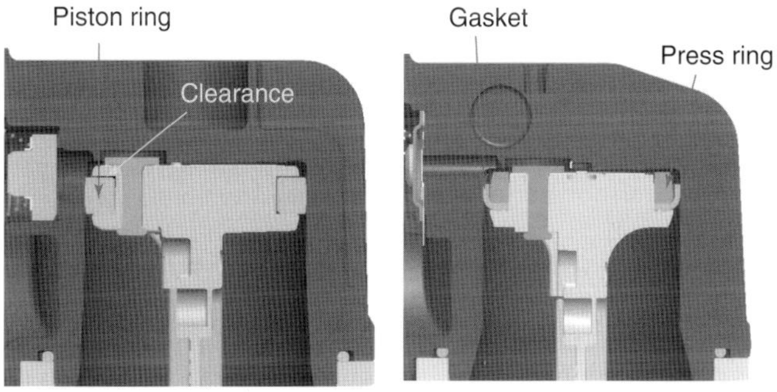

Figure 7. Two different types of piston sealing. (Reproduced with permission from WABCO.)

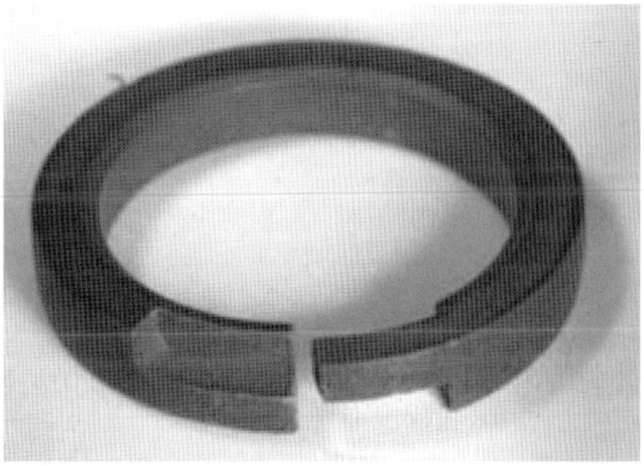

Figure 8. Piston ring. (Reproduced with permission from WABCO.)

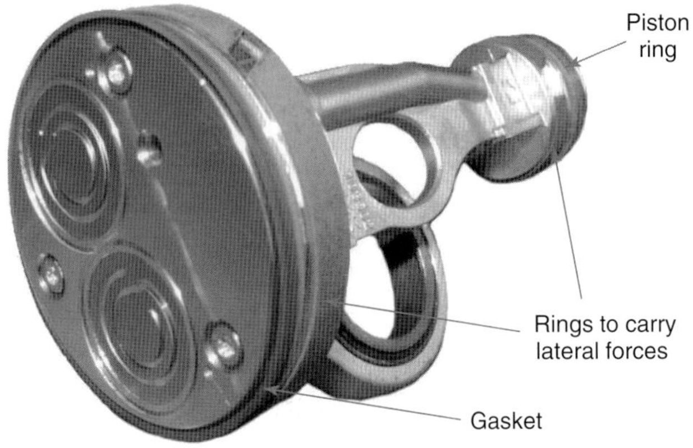

Figure 9. Piston of a two-stage compressor. compare Figure 15. (Reproduced with permission from AMK Automotive GmbH & Co. KG.)

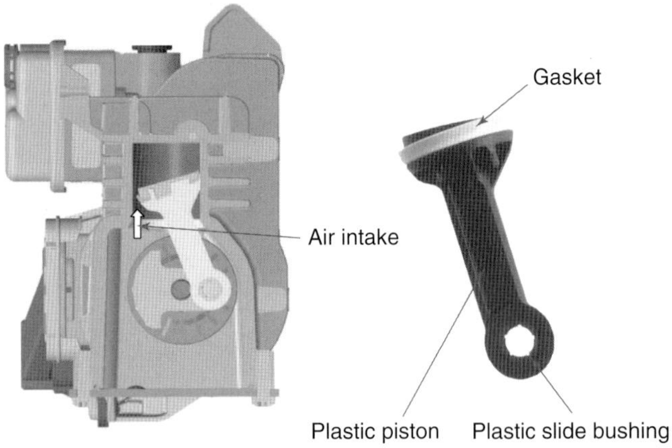

Figure 11. Low power compressor. (Reproduced with permission from WABCO.)

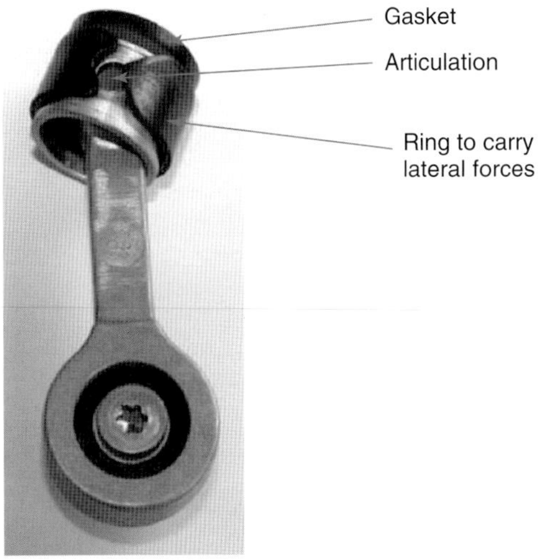

Figure 10. Articulated compressor piston. (Reproduced with permission from Tokico/Hitachi.)

the ring moving. However, this clearance has a negative impact on acoustics.

Another focus lies on lateral piston forces. These have to be minimized to achieve low friction, low wear, and thus a long lifetime. In some cases, an additional ring to carry the lateral forces and unload the sealing (Figures 9 and 10) accompanies the piston sealing.

There are also compressors in the market, which have an articulation between connecting rod and piston (Figure 10). The additional articulation with another bearing and a certain clearance might have a negative impact on acoustics.

Usually, the piston material is aluminum. In the case of lower pressure ranges (and thus lower temperature), the piston can be designed from plastic (comparison in Figure 5).

3.1.1.1 Air intake/suction valve. Air from the atmosphere has to be led into the cylinder. This can be done with an intake valve in the cylinder head or directly in the piston (Figure 4). The lower the stiffness of the intake valve, the higher is the volumetric efficiency. On the other hand, the design has to ensure low tension forces and sufficient strength of the part over lifetime.

In the compressor range of low pressures <10 bar, the intake valve can be omitted by opening a bypass in the lower position of the piston (Figure 11).

3.1.1.2 Pressure valve. The pressure valve is located in the hot area of the compressor, and consequently, it has to withstand high temperatures. In addition, it should have a low delta pressure (low stiffness) for efficiency and a low leakage level. Leakage would allow a flow back into the compression chamber, reducing efficiency and starting capability.

There is a trade-off between those demands: an elastomer pressure valve can achieve low leakage level, but high temperature could be an issue. Another design uses steel leaves as a check valve, which has good robustness against high temperature, but allows a certain amount of leakage. The steel leaf type has a very low mass and thus high dynamic capability. This gives advantages in terms of noise. In the market, both types of pressure valves are known (Figure 12).

3.1.1.3 Dead volume. To optimize the volumetric efficiency, the dead volume has to be minimized. This dead volume is also important for the stabilized pressure, which is achievable. Even at low environmental pressure (high geodetic heights) of about 0.6 bar, the compressor performance must not drop below the specified level. Test results are shown in Figure 13.

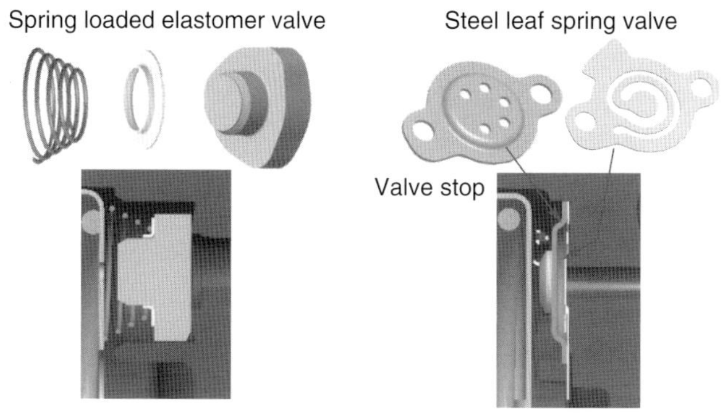

Figure 12. Two different types of pressure valve. (Reproduced with permission from WABCO.)

3.1.1.4 Compensation of oscillating and rotating masses. Figure 14 shows a typical configuration of a one-cylinder compressor. The target is a well-balanced share between rotating and oscillating masses. The force distribution in the *x*- and *y*-directions can be seen in the diagram. The overall compressor position in a car and the insulation between compressor and car chassis has to be considered as well for the layout of the balancing.

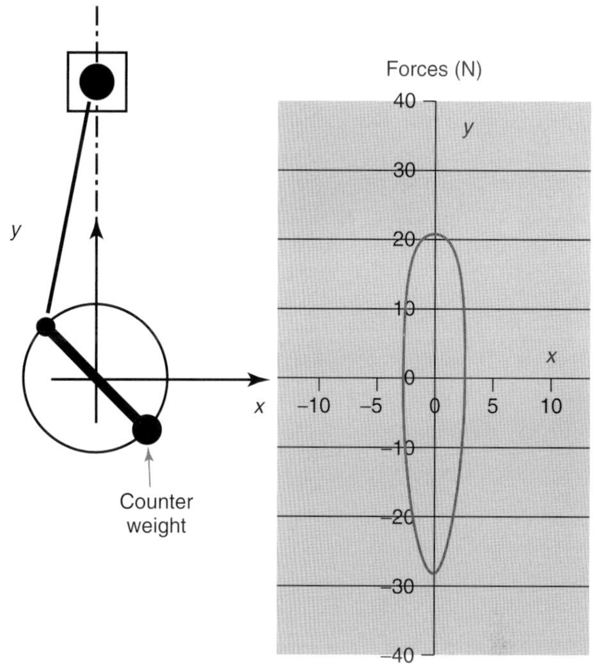

Figure 14. Trade-off between balancing of longitudinal and lateral forces. (Reproduced with permission from WABCO.)

3.1.2 Two-stage open system type

Two-stage compressors have an advantage in generating higher pressures; as a one-stage compressor is limited to maximum 18 bar, the two-stage compressor can provide much higher pressure and the efficiency in higher pressure ranges is better. Furthermore, the required

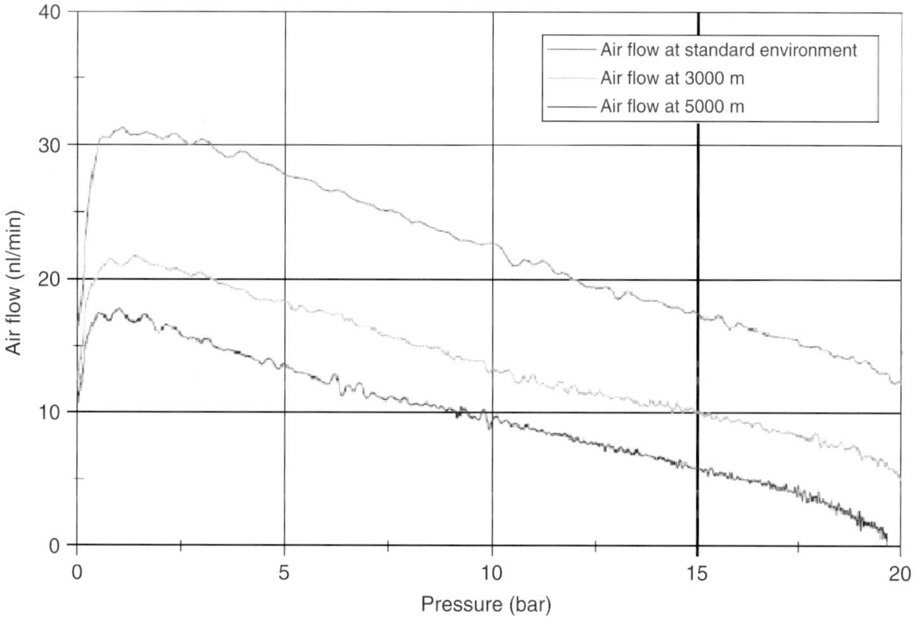

Figure 13. Compressor airflow at different geodetic heights simulated by different intake pressures. (Reproduced with permission from WABCO.)

torque to drive the compressor does not show such high peaks as with a one-stage compressor. The torque is better distributed during one cycle. On the other hand, there are more mechanical parts needed to provide this function, friction is typically higher (two-piston sealing instead of one), and, owing to that fact, the efficiency in lower pressure ranges is worse than for a one-stage compressor.

The principle of a two-stage compressor is shown in Figure 15. Air is taken into the crankcase and then compressed in stage one (at the bottom, piston moves downward). This compressed air is guided through a channel within the double piston into the second stage for compression to the final pressure.

Figure 16 shows the comparison of two medium power compressors. The efficiency advantage of a one-stage compressor in lower pressure ranges is obvious.

However, at higher pressure ranges, the two-stage compressor is showing benefits compared to a one-stage compressor. Figure 17 compares the efficiency and the flow rate of one- and two-stage high power compressors.

3.1.3 Closed system type

Currently, there is one closed system type compressor on the market (Figure 18). The system configuration is more complex than with an open system (Figure 3). The reason is the airflow control valve unit. In a four-corner system,

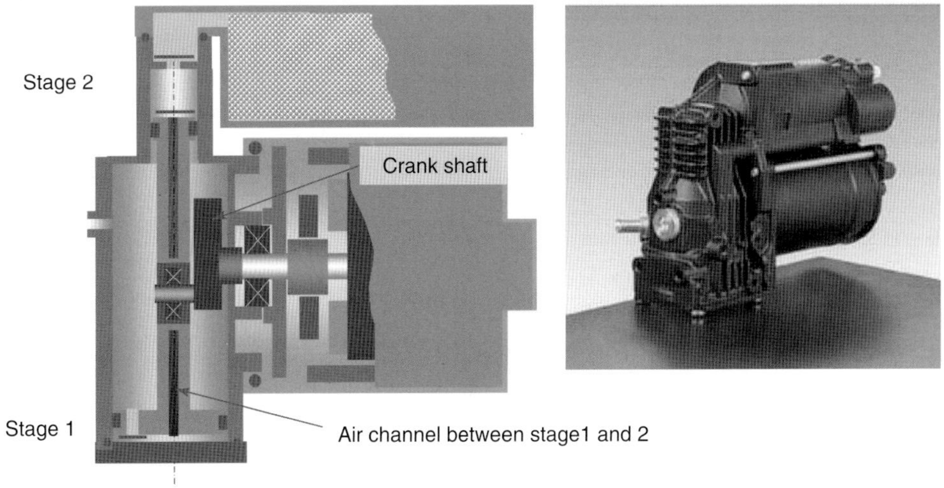

Figure 15. Two-stage compressor. (Reproduced with permission from AMK Automotive GmbH & Co. KG.)

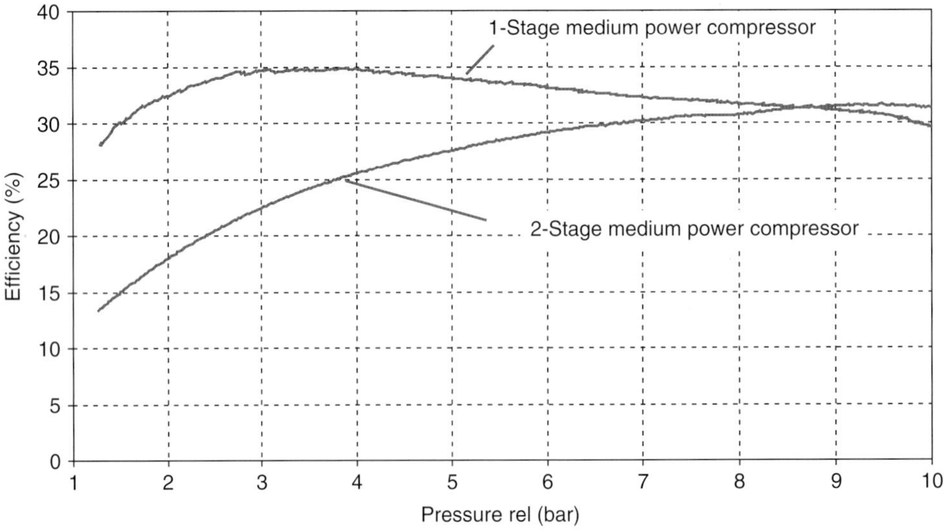

Figure 16. Comparison of a one- and a two-stage medium power compressor. (Reproduced with permission from WABCO.)

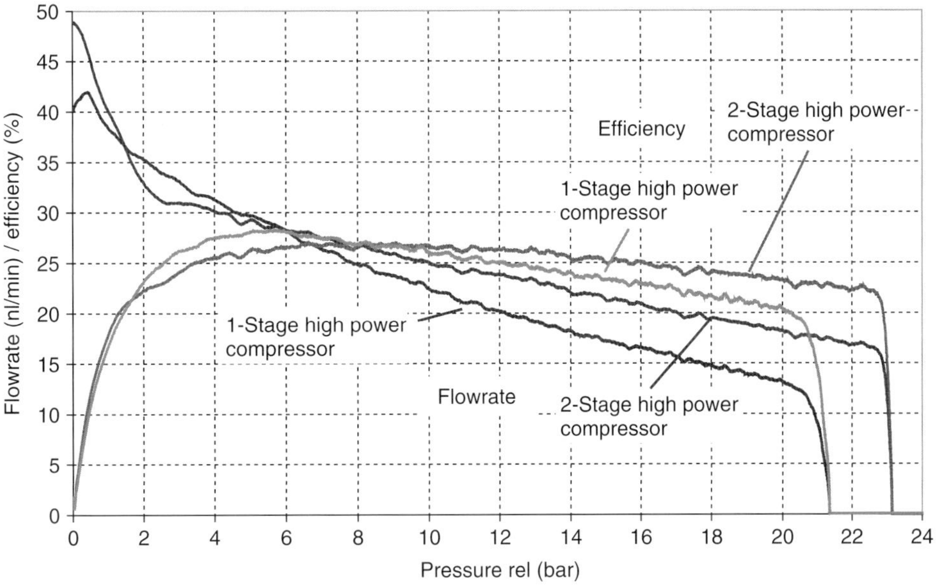

Figure 17. Comparison of a one- and a two-stage high power compressor. (Reproduced with permission from WABCO.)

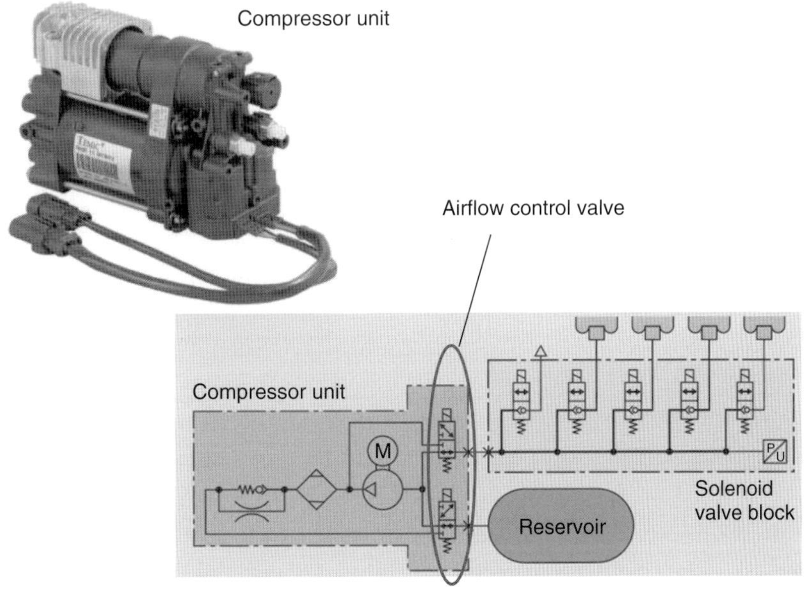

Figure 18. Closed system compressor and pneumatic layout. (Reproduced with permission from Continental AG)

a total of seven solenoid valves are necessary, whereas the open system only needs six solenoid valves.

Compared to an open system compressor, which gets always the atmospheric pressure at the air intake, the closed system type compressor can be precharged. Figure 19 compares the compression cycle of an open system type compressor and a closed system type compressor. The closed system type compressor is able to further compress a certain precharge pressure to the requested end pressure with a smaller stroke than an open type compressor. The area surrounded by the p-V line is equivalent to the work, which is needed to compress the air, or in other words, the requested energy for the compression cycle. In the shown example, the area (=work) is smaller in the closed system type compressor, and this is the reason for the energy-saving advantage of such systems.

The flow rate versus pressure of a closed system type compressor is shown in Figure 20. As a parameter, the

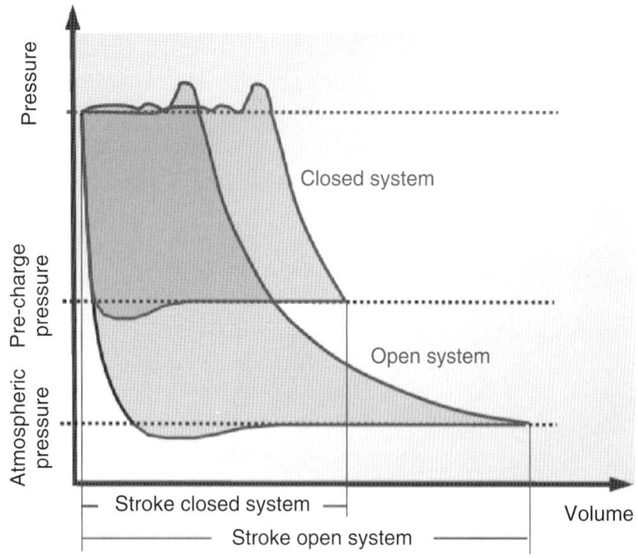

Figure 19. Compression cycle of an open system type compressor and a closed system type compressor. (Reproduced with permission from WABCO.)

precharge pressure is modified between 0 (atmospheric pressure) and 8 bar. This compressor has a smaller displacement than the open system type. Consequently, the airflow is very limited if it is operated in open system mode (e.g., system filling in workshop, see 0 bar line in Figure 20). The performance can be significantly increased if the compressor is charged with compressed air, either from the air springs or from the reservoir. The principle is to pump the air between the reservoir and the air springs or vice versa. The benefit is that the stored energy of the compressed air remains on a higher level and the compressor only has to increase the pressure by a low ratio (Figure 19). Lower energy consumption over vehicle lifetime is the positive consequence. The runtime of a closed system type compressor is roughly 30% of an open system type compressor in a four-corner application with reservoir, if just the pure lifting and lowering processes are considered. This will partly be compensated by necessary purge cycles to regenerate the air dryer: in a special operation mode, the dry air in the reservoir is used to dry the air dryer desiccant. This air loss has to be filled up from the atmosphere. Depending on environmental conditions, the share of compressor runtime for purging can be a significant amount on top of the normal compressor operation for lifting and lowering. Though, the overall compressor runtime in a closed system is about 50–60% compared to an open system.

3.2 Electric motor

Brushed direct current (DC) motors are used to drive compressors for ASU's in air suspension systems. Another option would be a compressor directly driven by the engine. This has been done in earlier times at Daimler Benz (1960s).

Advantages of electric motor for compressor drive:

- The compressor can be operated independently of engine type and vehicle.
- In the case of optional air suspension, the same vehicle engine can be used for both suspension variants.
- Compressor installation position can be anywhere in the vehicle.

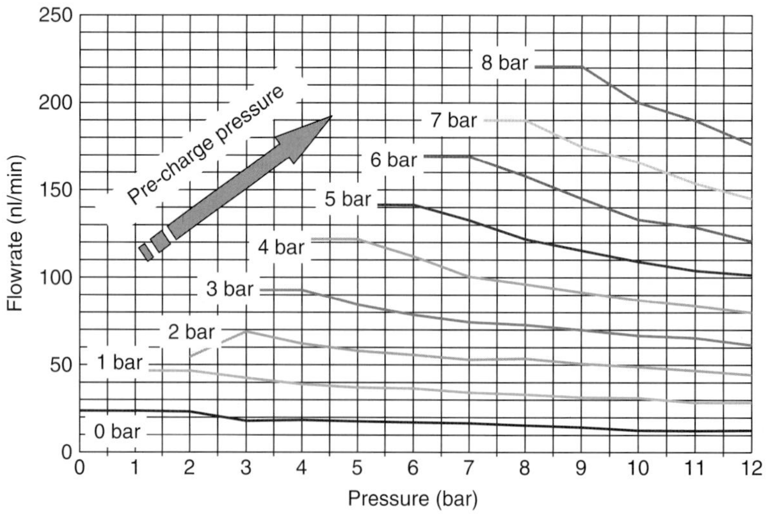

Figure 20. Flow rate versus pressure at various precharge pressures. (Reproduced with permission from WABCO.)

- No issue with varying speed of the engine between idle and maximum speeds (ratio 10 : 1).
- No clutch would be required to couple/uncouple the compressor.
- In case of engine off (hybrid cars), the compressor can still be operated and thus lifting is possible even in "engine-off" condition.

Disadvantage of electric motor for compressor drive:

- Limited performance due to limited battery and generator capacity and electrical wiring dimension.
- Lower total efficiency due to double energy transformation.

In newer applications, for example hybrid cars, the inrush current of a relay-controlled DC motor is not accepted any more. To reduce the peak current and current gradient during the switch-on phase of the compressor, the relay can be replaced by a PWM (pulse-width modulation) control, which provides a soft start. The following picture (Figure 21) shows an example: the relay-controlled peak current of this compressor would be nearly 60 A. Thanks to a PWM control, the peak current can be reduced to 32 A.

In this context, there are also brushless motors considered for compressor drive. Motor life time is not the driver. Typically, the compressor run time is around 750 h in a vehicle life, which is easy to achieve with brushed motors. The main benefit of a brushless motor is the possibility for speed/current control.

The required torque at a certain speed mainly influences size and weight of the electric motor. To reduce the torque but keep the same performance, the speed has to be increased. However, the pneumatic efficiency of a compressor would decrease in case of higher speeds (valve dynamics, etc.) and thus such motors have to be transmitted to lower speeds at the compressor. Another factor, which influences weight and size, is the material of the magnets. Here is the classical trade-off between cost and weight.

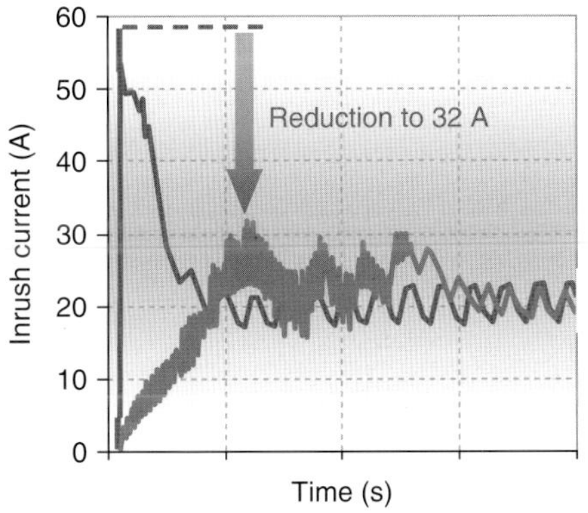

Figure 21. Comparison between relay controlled and PWM controlled compressor activation. (Reproduced with permission from WABCO.)

3.3 Compressor control

Compressors for passenger car air suspension systems are designed for intermittent operation, as a continuous run is not required and would increase weight and cost. Consequently, the compressor has to be protected against overheating. In simple applications, there is a thermal cut-off switch positioned in the electric motor. In systems with electronic control, at least the environmental temperature of the car is used to estimate the working condition. To be more accurate, a dedicated temperature sensor can be used to provide a signal to the controller. This sensor should be placed close to the hot spot of the compressor (e.g., cylinder head) to provide a realistic temperature information to the controller. In some cases, the sensor is placed somewhere in the environment of the compressor to provide at least an indication about the actual environmental condition. For economical reasons, the temperature information can be achieved using the solenoid of the exhaust valve as a temperature-dependent resistor (Bodet and Meier, 2007) (patent no. DE102005062571A1).

Another control feature is the starting and stopping of the compressor against low pressure: to avoid strong torque reactions, the gallery and the air dryer should be evacuated before the compressor is started or stopped.

3.4 Air dryer

By compressing the air taken in from the atmosphere, the natural humidity has to be considered: during compression, the water content remains constant; however, the relative humidity is increasing. This means that at a certain point water droplets are appearing as condensate.

Example: filling of a 5-L reservoir from 0 to 16 bar with air, $t = 30°C$, and 75% relative humidity. Total mass of collected water is 1,767 g, which are about 35 droplets.

Car manufacturers expect a maintenance-free system, so a periodical manual drain of the condensate is not acceptable. Consequently, the air dryer is mandatory to avoid humidity/water ingress into the air suspension system. Without an air dryer, the amount of water transported into the air springs would be more than 5 L (considering average climate conditions in Stuttgart) and 20 L in Singapore over the vehicle lifetime with an open system. This calculation does not consider any leakage. In reality, there is always leakage present. Considering a leakage rate of 5%, an open system would be able to provide dry air over the lifetime, as the air dryer regeneration does not need all air back during exhaust to achieve full regeneration.

In a closed system, the air dryer is only required in case of system filling in a workshop and in case of leakages. 5% leakage rate means 0.25 L water over the lifetime in Stuttgart or 1 L in Singapore. For this reason, even a closed system has to have an air dryer, and special purge cycles have to be fulfilled (Section 3.1.3).

As the air dryer itself has to be maintenance free, it has to provide a self-regenerating function. As desiccant in those applications, special beads of crystalline metal-aluminum silicates are used. These beads are produced synthetically and are highly porous, having a strong affinity to water. Depending on temperature, the desiccant is able to store water mass of 20% of its own weight. Figure 22 shows the principle.

Humid air flows through the desiccant cartridge (1). During this process, water is extracted from the air and stored temporarily in the desiccant beads (adsorption). The dry air leaves the cartridge via a check valve (2) into the pneumatic system.

During each lowering of the chassis level, the dry air that comes from the bellows flows through the air dryer back into the atmosphere. The 3/2 way solenoid valve (3) is activated and opens the relay valve (5). The air passes the solenoid valve and the throttle (4). In the throttle, the relative humidity of the air further decreases due to expansion. While passing the air dryer, the expanded air is regenerating the desiccant, meaning the air takes the water out of the desiccant beads and carries it through the relay valve (5) out into the atmosphere (desorption).

Key for efficient regeneration is the ratio between the nominal width at the entrance of the air dryer (dry air) and at the outlet (humid air). The orifice area at the outlet should be roughly four times bigger than at the inlet. This is to ensure a sufficient expansion of the compressed air. For further reference on this topic, see Section 3.5.

3.5 Exhaust circuits at open systems

The exhaust circuit of an ASU has to fulfill several demands:

- Controllable with electrical energy (solenoid valve).
- Providing sufficient airflow from the system into the atmosphere to achieve high chassis lowering speed.
- Expansion of the compressed air before it enters the air dryer for efficient regeneration.
- Robustness against water/ice in the "wet" area.
- Possible separation of the air dryer volume from the gallery. Advantage: air dryer volume does not need to be pressurized for a pressure measurement in the bellows or in the reservoir, thus the air loss during pressure measurement is minimized.
- Provide low noise level during exhausting.

Depending on the vehicle demands, there are several exhaust circuits on the market (see Figures 23 and 24).

3.5.1 *Standard release circuit*

The standard release exhaust circuit is also shown and explained in detail in Figure 22. To avoid too big and heavy solenoid valves, this circuit uses a small 3/2 solenoid as the pilot valve (3). Thus, the electrical current to operate the solenoid can be minimized. The 3/2 valve operates a relay valve (5) by air, which opens a big orifice to allow high airflow without velocity pressure out of the air dryer

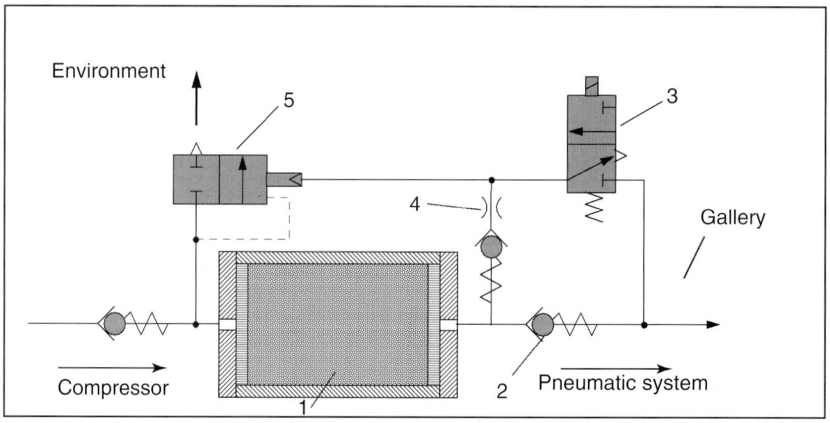

Figure 22. Standard release air dryer circuit. (Reproduced with permission from WABCO.)

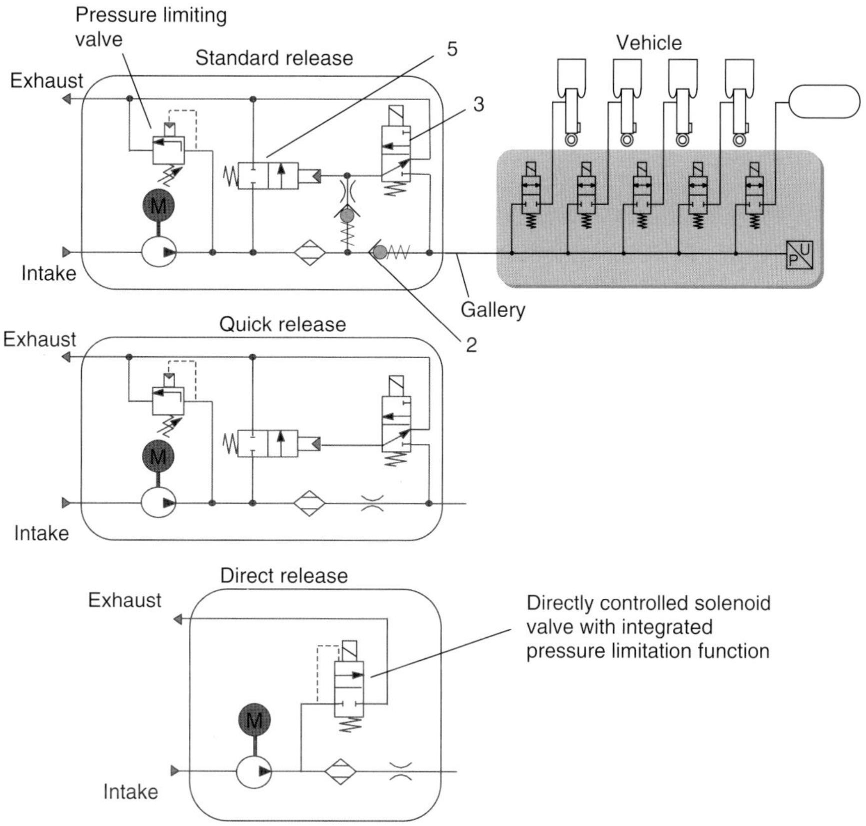

Figure 23. Different air dryer circuits in idle mode. (Reproduced with permission from WABCO.)

during exhausting. The relay valve itself, which operates in the wet area, is robust against water/ice. Separation of air dryer volume and gallery is achieved using a check valve (2). The only disadvantage of this circuit is a limited airflow, as the exhausting air has to pass the 3/2-solenoid valve (3).

3.5.2 *Quick release circuit*

The disadvantage of a limited orifice is not acceptable for vehicles that require quick lowering (SUV's). For this reason, the standard release circuit was modified and called a *quick release* circuit. Here, the exhausting air is not passing through the solenoid valve, and a bigger orifice is possible. To achieve this, the check valve (2) does not exist, which creates one disadvantage that the gallery and air dryer volume are always connected. In case of pressure measurements in the bellows or the reservoir, the air dryer volume has always to be inflated that consumes air and increases the chilling time for a pressure measurement.

3.5.3 *Direct release circuit*

For economical reasons, the relay valve/pilot solenoid valve is being replaced by one bigger solenoid valve. This circuit is called *direct release* (Figure 25). The solenoid valve is positioned behind the air dryer, and the exhausting air has to pass the solenoid valve. Special focus is on the robustness against water/ice, as the solenoid valve is located in the wet area. Pressure limiting function is integrated in the solenoid valve. Further advantages are achieved by integrating the filter/silencer into the air dryer body.

3.6 Pressure retention

In some cases, air suspension systems require a minimum pressure within the bellows, for example, if the car is lifted in a workshop. Without pressure retention, the bellows might be exhausted completely and would be damaged in case of lowering the car. The pressure retention function is provided either by dedicated valves in each bellow or by one central valve in the air dryer of the ASU. In case of standard and quick release circuit, the relay piston,

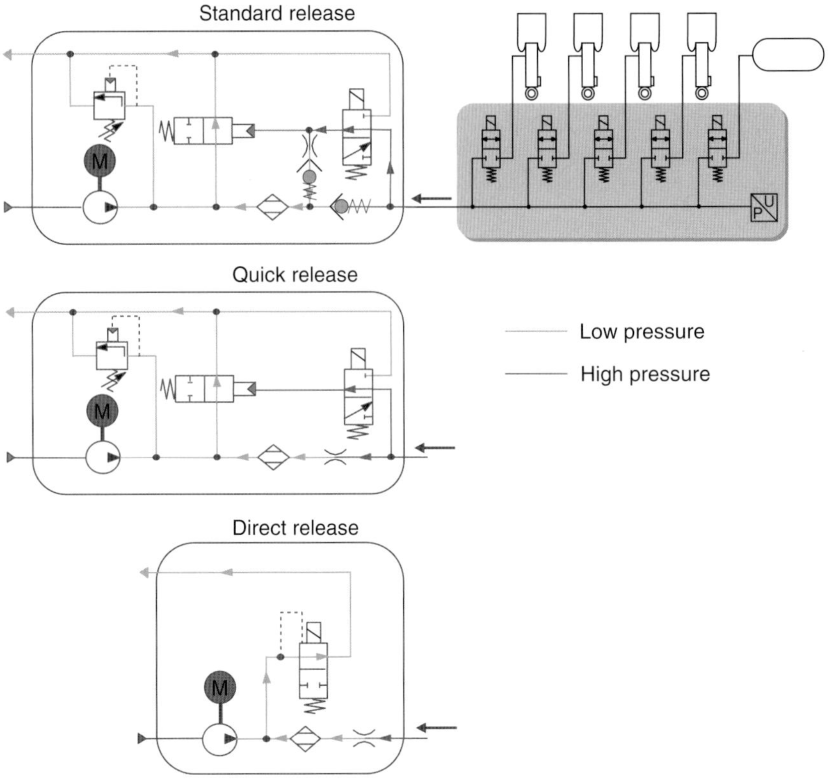

Figure 24. Different air dryer circuits in exhaust mode. (Reproduced with permission from WABCO.)

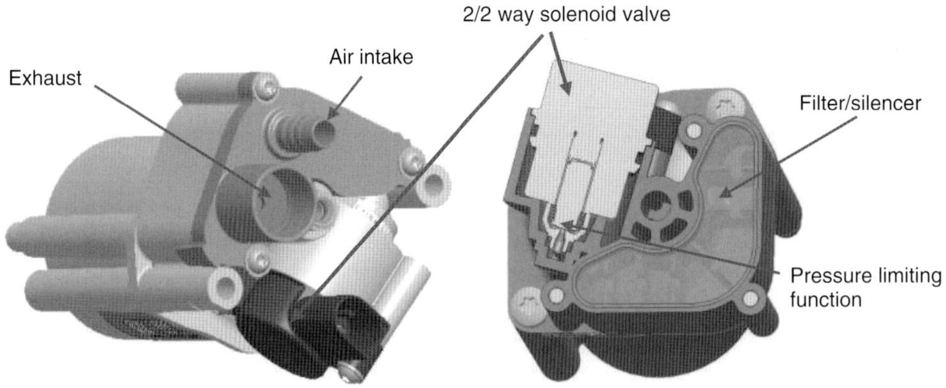

Figure 25. Direct release air dryer with integrated filter/silencer. (Reproduced with permission from WABCO.)

which closes by help of a spring, provides the pressure retention function if the system pressure drops below a specified level (Figure 4). Typically, this is between 3 and 1.5 bar.

If the air suspension system requires lower bellow pressures, this cannot be achieved with such a circuit. However, the direct release air dryer is able to exhaust the system down to atmospheric pressure.

3.7 Pressure limiting

To mitigate a risk that comes mainly from system FMEA (failure mode and effects analysis) considerations, a pressure-limiting function has to be integrated in the air suspension system. Typically, the ASU provides this function. This feature is only required if one other component has a failure. As an example, a sticking relay that controls the electric motor/compressor might occur. As the compressor cannot be switched off anymore, the pressure in the compressor and the gallery would increase up to the natural stabilization pressure, depending on compressor performance and height above sea level. The desired cut-off pressure should be above the normal working pressure with sufficient margin. In reservoir systems, the maximum operating pressure is about 16–18 bar. Considering technical tolerances, the blow off valve should not open before 17.5–19.5 bar.

Pressure-limiting function is provided by a valve in the relay piston (Figure 3 and 4) or by the 2/2-way solenoid valve in the direct release air dryer (Figure 23). The bellow valves must be safe against this pressure level (ratio between gallery and bellow pressures); otherwise, the compressed air could pass through the solenoid valve and might lift the chassis unintentionally.

Although the pressure-limiting valve avoids damage in the system caused by too high pressure, the compressor might run until it overheats or empties the vehicle battery. Owing to this risk, some applications have electronic "relays" in both supply and ground line. A safe shut off of the compressor is always possible.

3.8 Air distribution

3.8.1 Air intake

The compressor has to get clean air from the atmosphere. For this reason, an air filter is installed in the air intake (Figures 2, 25, and 26). Particle sizes smaller than 10 μm are acceptable. The position where the air is ingested has to be in a dry environment (e.g., inside the car or at a place above the wade line) and where splash water has no access.

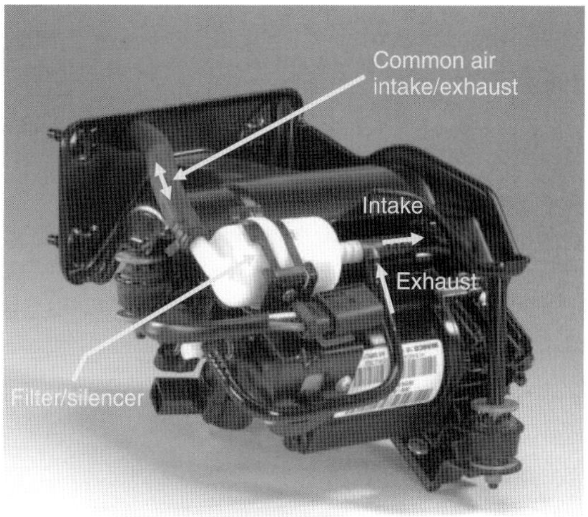

Figure 26. Air supply unit with common air intake/exhaust. (Reproduced with permission from WABCO.)

3.8.2 Exhaust

The exhaust port is noise critical. Therefore, a position outside the cab is favorable. A silencer is mandatory in many cases. Often, the air intake and exhaust of the compressor is combined into one port. Advantage is that the air filter can be used as silencer and ease of assembly. An example is shown in Figure 26.

A disadvantage of a combined air intake/exhaust is the trade-off between dry position and exhaust noise. In addition, the air-drying capability can be diminished. In the case of the exhaust process, humid air is guided through the comment pipe/filter. There might be condensation effects if the pipe temperature is lower than the exhausted air. During the next compressor run, the humid air—or even

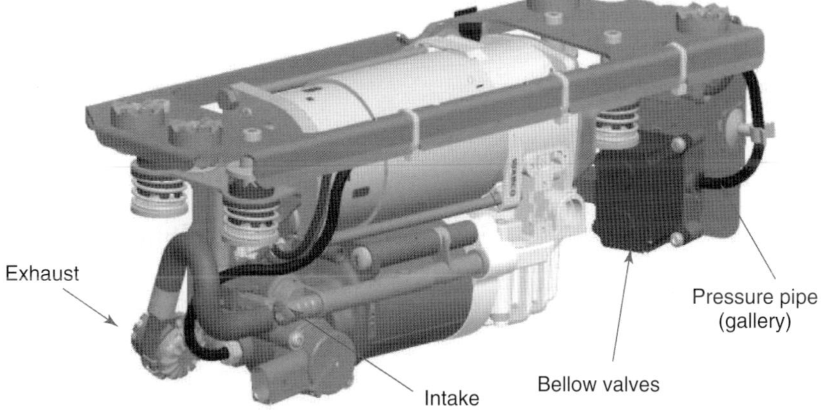

Figure 27. Air supply unit for rear axle air suspension, air intake, and exhaust separated. (Reproduced with permission from WABCO.)

water drops—in the pipe are ingested into the ASU. In case of longer air intake/exhaust pipes, a separation is recommended (Figure 27).

In many cases, the solenoid valves to control the air springs (bellow valves) and reservoir are assembled close to the compressor on one bracket (Figure 27). The connection between the multiple solenoid valve block and the compressor (gallery) should be as short as possible to reduce the volume inside this pipe. The diameter has to be selected depending on the nominal width of the bellow valves. Typically, it is a pipe 6 × 1.5 mm (outside diameter 6 mm, inside diameter 3 mm, and wall thickness 1.5 mm), in some cases, 4 × 1 mm is also used (inside diameter 2 mm).

A typical 5 by 2/2-solenoid valve block is shown in Figure 28.

A single pressure sensor is integrated. By connecting it to the bellows and the reservoir in a sequence one by one, the individual pressure can be measured. Here, the disadvantage of the quick release and direct release circuit becomes obvious, because in each case the air dryer volume has to be pressurized during a measurement. This takes a certain time and the pressure value can only be taken after a chilling time. After each measurement, the air dryer volume has to be exhausted.

Air distribution from the multiple solenoid valve block to the bellows is usually done with PA tubes 4 × 1 mm. This dimension avoids expensive preforming of the tubes, which would be required in case of bigger tubes (e.g., 6 × 1.5 mm).

Depending on the exhaust circuit, special attention is required on the pneumatic layout and dimensioning. In case of low bellow pressures and large pressure drops due to long pipes with small orifices, the relay piston in the air dryer might not receive sufficient pressure levels to stay fully open. In this case, the relay piston would swing within an intermediate position and would not be able to open the complete nominal width. This results in a bad regeneration of the air dryer, as the expansion of the compressed air cannot be provided as specified.

At rear axle systems, an integration of the bellow valves into the compressor can be advantageous. An example is shown in Figure 29.

4 COMPRESSOR PACKAGING/ACOUSTIC INSULATION

4.1 Inside mounting

Inside mounting is advantageous in terms of low demand for protection against water, corrosion, and so on. On the other hand, acoustic (air borne noise) and vibration (structure borne noise) of the ASU is very demanding. The

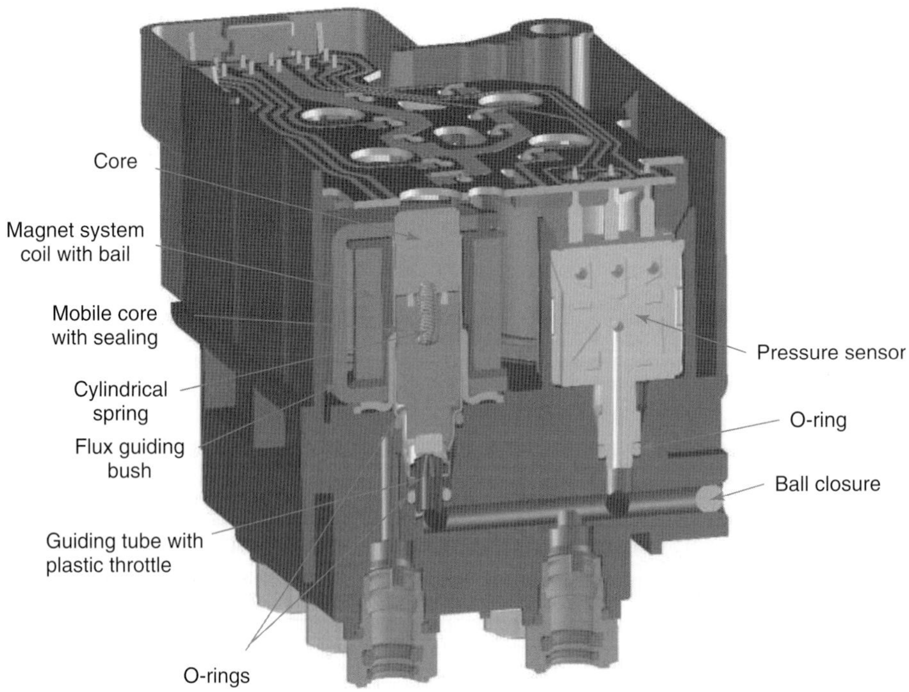

Figure 28. 5 by 2/2-solenoid valve block with integrated pressure sensor. (Reproduced with permission from RAPA GmbH.)

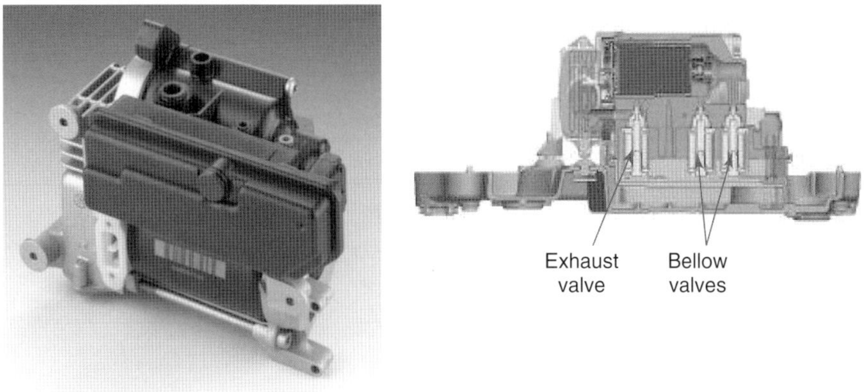

Figure 29. Air supply unit for rear axle air suspension with integrated bellow valves. (Reproduced with permission from WABCO.)

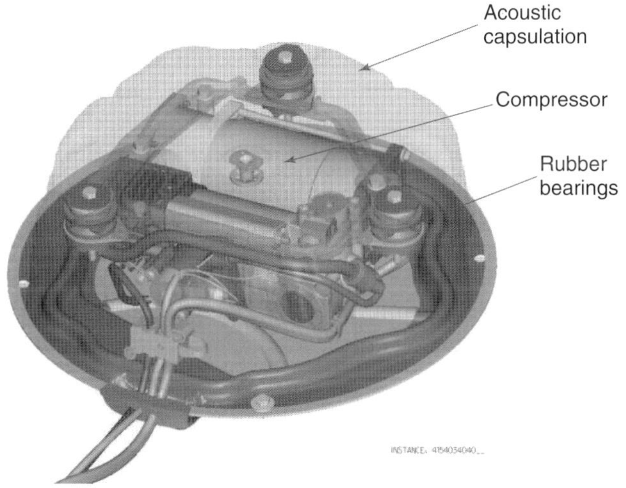

Figure 30. Capsulated air supply unit. (Reproduced with permission from WABCO.)

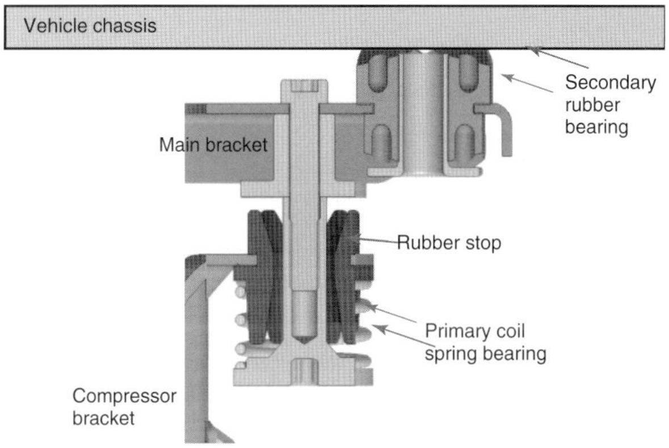

Figure 31. Typical insulation design between compressor and vehicle chassis. (Reproduced with permission from WABCO.)

compressor has to be suspended on flexible elements such as rubber bearings as shown in Figure 30.

Best insulation is achieved with coil springs (Figure 31). The tuning of the mass/spring system is typically subcritical, meaning that the eigenfrequency of the compressor/spring combination is below the first order of eigenfrequency of the compressor vibration if it is running. A well-balanced load distribution between the bearings (3 or 4) is key for NVH (noise, vibration, and harshness) reduction. As the coil springs have to be soft and damping is just done by a certain amount of friction, stroke limiters are required. To avoid the compressor structure hitting the bracket, rubber buffers are used. One reason for compressor movement is vibration caused by the vehicle on uneven roads. Another reason is the torque reaction if the compressor is switched on or off.

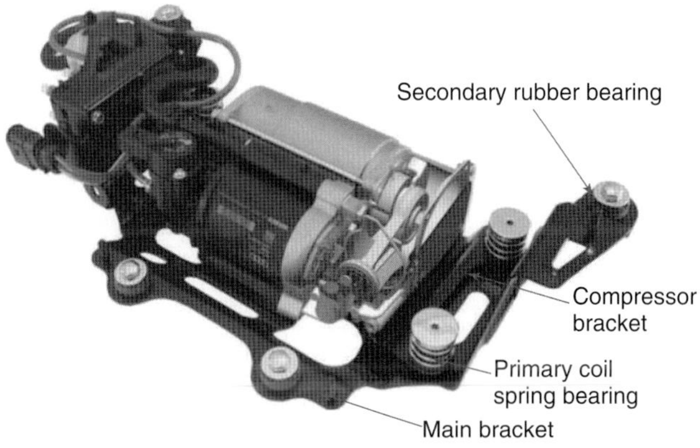

Figure 32. Air supply unit for four-corner air suspension. (Reproduced with permission from WABCO.)

4.2 Outside mounting

Corrosion resistance has to be provided following ISO 9277 NSS, typically 480 h. Protection class requirement is

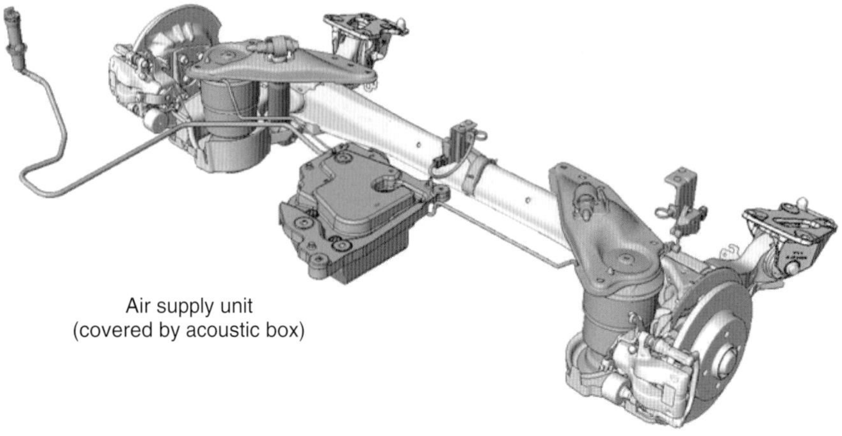

Figure 33. Rear axle installation air supply unit with acoustic box. (Reproduced with permission from WABCO.)

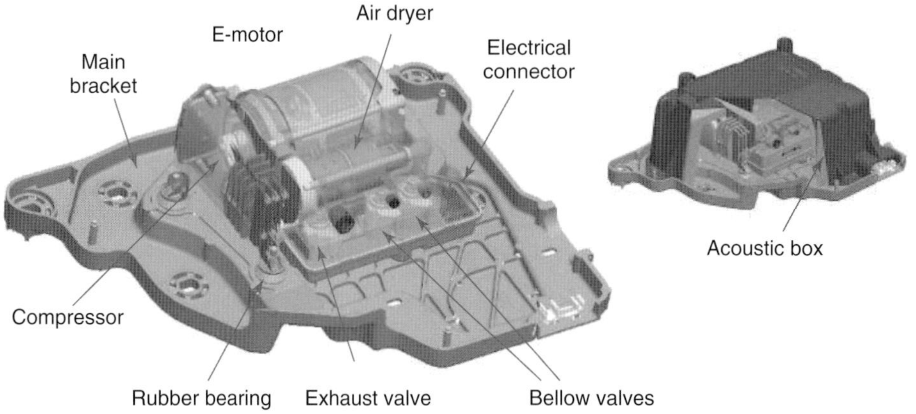

Figure 34. Details of a rear axle installation air supply unit with acoustic box. (Reproduced with permission from WABCO.)

typically IP6K6K, IP6K7, and IP6K9K according to ISO 20653.

Compared to the inside mounting, the air-borne noise is not that critical. In many cases, the ASU is not capsulated (Figure 32). A preferred installation space is the engine compartment. However, packaging and the neighborhood to hot elements might be critical. In many cases, the ASU is installed below the chassis in the rear area of the car. There might be hot elements (exhaust muffler) in the neighborhood as well.

In the following application, the ASU is covered by a case, which reduces airborne noise and protects the ASU against stone chipping and ground contact (Figure 33).

Details of the ASU for this rear axle solution are shown in Figure 34.

These examples show that—depending on the OEM (original equipment manufacturer) demand—there are various solutions how and where to place an ASU in a car. There is no standard solution, and the engineering effort to package a compressor into a car is significant both on OEM side and on supplier side.

5 SUMMARY

The purpose of ASUs in passenger cars is to feed the air springs with compressed dry air. Piston compressor technology is standard on the market, all driven by electric DC motors. System pressures vary between 7 and 20 bar, depending of the vehicle application. There are one-stage compressors (for open and closed systems) and two-stage compressors (for open systems) used. Closed systems are pumping the air between the air springs and a reservoir back and forth, avoiding an exhaust of the pressurized air into the atmosphere. By keeping the pressurized air widely in the system, the overall energy balance has an advantage compared to open systems. In addition, the repeatability of lifting operations is better than with open systems. However, closed systems are more complex and a reservoir is mandatory. Furthermore, the maintenance-free air drying capability of an ASU requires purge cycles at closed systems, which worsens the energy balance. Open systems are less complex and are able to be operated without reservoirs, which also have a positive impact on overall system costs. One-stage compressors are simpler and have a better efficiency in low pressure ranges than two-stage compressor, whereas two-stage compressors have physical advantages in higher pressure ranges.

Besides those differentiators at the compressor, the major variation is packaging of compressors into the vehicle. As there is no "standard" installation space and environment, each new vehicle requires an individual design of bracket, insulation, piping, and wiring harness.

Future challenges for ASUs are weight and size limitations, acoustic performance, current consumption, peak current, and increasing pressure demand for smaller air springs.

REFERENCE

Bodet, M.-M. and Meier, J. (2007) Verfahren zur Ermittlung einer Kompressorumgebungstemperatur und Kompressoranordnung zur Durchführung des Verfahrens. Offenlegungsschrift vom 28.06.2007, DE102005062571 A1.

FURTHER READING

Becher, H.O. (2004) *Steuergeräte und Kompressoren für Luftfedersysteme*. CTI Fachkonferenz Federung und Dämpfung im Fahrwerk, Stuttgart July 06, 2004.

Becher, H.O. (2008) *Rear axle air suspension for compact class and transporters*. 4th International CTI Conference Suspension & Damping Stuttgart, April 23, 2008.

Fitch, B. (2008) A new high power compressor helps to realize high-performance for air suspension systems. Vehicle Dynamics Expo North America, October 23–25, 2008.

Meier, J. (2006) Steuerung von Luftfedern in Straßenfahrzeugen. Technische Akademie Esslingen, Oktober 25, 26, 2006.

Westerkamp, H. (2007) Air suspension enters the European compact car segment. Vehicle Dynamics Expo Stuttgart, May 8, 2007.

Chapter 113

Technology of Active Shock Absorbers and Benefits in Regards of Ride and Handling

Heinrich Schürr, Thomas Kutsche, and Christian Maurischat

ZF Friedrichshafen AG, Schweinfurt, Germany

1 INTRODUCTION

1.1 General tasks of dampers in chassis technology for passenger cars

- To damp post-vibration of vehicle body caused by uneven roads or driving conditions
- To quickly settle road-induced wheel and axle vibrations

Dampers provide a constant tire to road contact and subsequently ensure good traction, braking, and acceleration performance and minimize the vibration transmissibility for the passengers. This causes the positive effect to ensure maximum comfort to the passenger. As known, the human body is sensitive especially to vibrations between 4 and 8 Hz, that is, in the range between the natural frequencies of the vehicle spring mass (0.9–1.6 Hz) and the wheels (10–18 Hz) (Heissing, Ersoy, and Giess).

Encyclopedia of Automotive Engineering, print © 2015 John Wiley & Sons, Ltd.
Edited by David Crolla, David E. Foster, Toshio Kobayashi, and Nick Vaughan.
This article is © 2015 John Wiley & Sons, Ltd. ISBN: 978-0-470-97402-5
Also published in the *Encyclopedia of Automotive Engineering* (online edition)
DOI: 10.1002/9781118354179.auto018

1.2 Damping principles

In the past, many different damping principles for vehicle applications were studied. In general, mechanical, pneumatic, and hydraulic systems can be used. In today's automobile applications, the hydraulic-mechanical damping system prevails. The most widely used principle is the telescopic shock absorber (twin-tube shock or strut, monotube shock, or even named damper). Using hydraulic valve elements, a velocity-dependent, tuneable ratio between damping force and damper velocity can be realized (Causemann, 1999). Figure 1 gives an impression about the widespread damping forces with respect to the damper velocity, which are in use. The borderlines are comfort and stability. The rebound forces are nearly two times larger than the compression forces. The changes within these damping forces are done by applying different current to the damper valves.

A speciality of hydraulic-mechanical damping systems are variable, electronically controlled dampers. Main target of this technology is to adapt the necessary damping force as good as possible and as quick as possible to the actual driving situation, for example, to eliminate as much as possible the conflict between driving comfort and driving safety. The typical damping characteristics, shown in Figure 1, can be realized within a few milliseconds for every velocity. This is true for all possible damping forces between limits in rebound and compression.

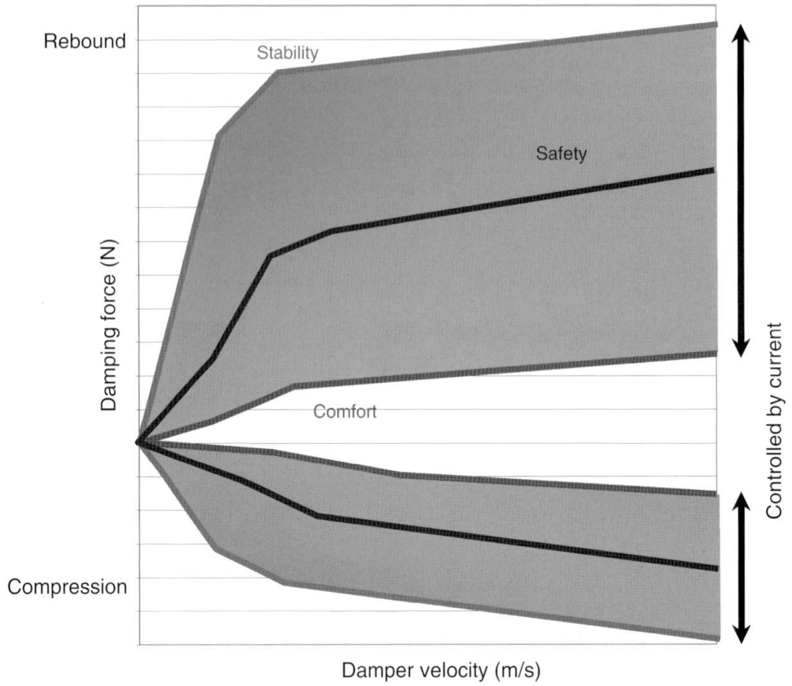

Figure 1. Variable damping characteristics. (Reproduced by permission of ZF Friedrichshafen AG, Germany.)

1.3 History of variable damping systems

Electronically controlled damping systems have been designed and manufactured since the early 1980s. They range from simple hand-operated electro-motorized adjustments all the way to fast, electromagnetic systems with two or more discrete damping levels (Causemann, 1999). State of the art today is continuously adjustable systems of a third generation, which are based on proportional acting damping valves (Kutsche and Raulf, 2001) (Figure 2). Alternative systems based on electro- or magneto-rheological fluids also appear in the market since some years ago.

In general, a variable damper should allow optimum adjustments for each of the three following criteria:

1. Best body damping
2. Low wheel load variations
3. Best possible isolation for small road inputs

1.4 Existing technologies

Valve principles with external or internal designs are used. Most of the technologies allow adapting the valve characteristics to the needs of the car manufacturers as an individual fingerprint (tuning). More details are given in Section 2. In regards of systems, different system architectures are known in the market, most based on Skyhook control and different sensor configurations, often combined with air spring applications. Details will be explained more in detail in the following chapters.

2 TECHNOLOGY

2.1 Damper technology

On the basis of the standard damper design of telescopic shock absorbers, variable dampers were developed. Thereby, both kinds such as mono-tubes and twin-tubes are used. Differences between the different solutions are the used principles of variation of damping forces.

In general two main principles of variation exist:

1. Variation of hydraulic resistance
2. Variation of the hydraulic values (viscosity) of the fluid

2.1.1 Variation of hydraulic resistance

By using these principles additional valves at or in the shock absorber are placed to modify the hydraulic resistance of the oil flow from the upper to the lower working chamber or counter wise.

The easiest way is to use a valve with two positions. In this case the oil flows through the standard configuration

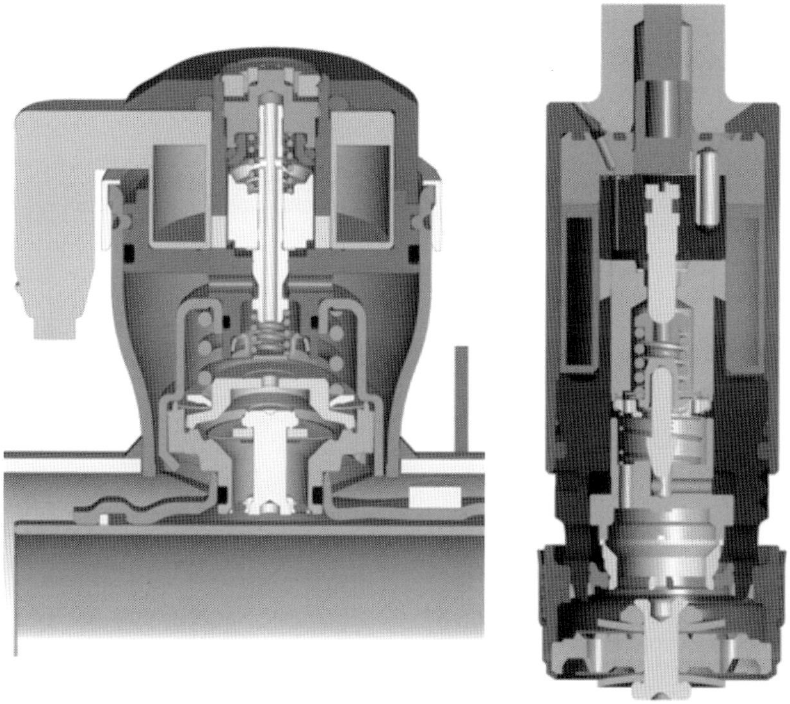

Figure 2. Cross-section of a proportional control valve; (a) external valve (b) internal valve. (Reproduced by permission of ZF Friedrichshafen AG, Germany.)

of base valve and piston valve in the firm position. By acting the additional valve a bypass to these conventional damper valves is opened and the resistance is reduced generating softer damping forces. Differences could be found in the structure of these valves. The bandwidth reaches from just open an additional bypass by using a bigger whole (orifice variation see Figure 3), modify just the pressure relief valve (Figure 4) to valves with two-stage structure. Here, the orifice and the pressure relief characteristic of the shock absorber are modified by the valve (Figure 5).

These types of variable dampers were used for example at the Ford Mondeo in 1993 or the Ford Lincoln Continental in 1994. To generate more than two characteristics, a second valve has to be used. Then, up to 4 characteristics (valve 1 and 2 closed=> firm; valve 1 open and valve 2 closed=> medium; valve 1 closed and valve 2 open=> soft; valve 1 and 2 open=> maximum soft) could be achieved by differences of the single valve orifices. Mostly the valves influence the hydraulic resistance in rebound and jounce. The dimensions of the damper cylinder and the rod define the ratio between rebound and jounce. Only the solution used in the S-class by Mercedes or Grand Cherokee SRT has one valve for rebound and one for jounce.

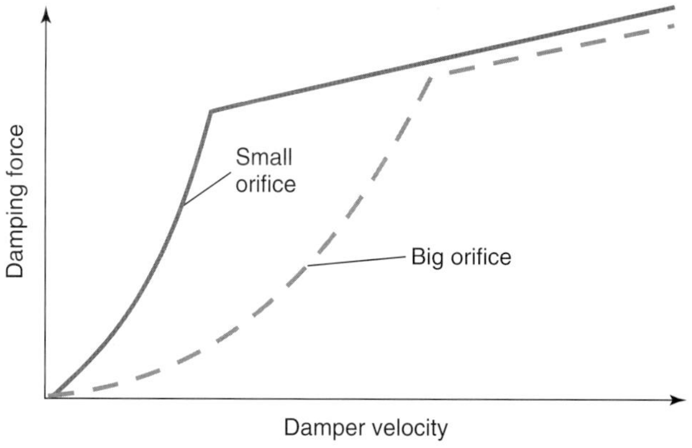

Figure 3. Opening an additional bypass by using a bigger whole.

To improve the function of these types of variable dampers and to reduce the production cost the "on/off"-valves were substituted by just one proportional valve. By changing the control current of this valve type the damping force can be controlled continuously between a fix soft and a fix firm characteristic.

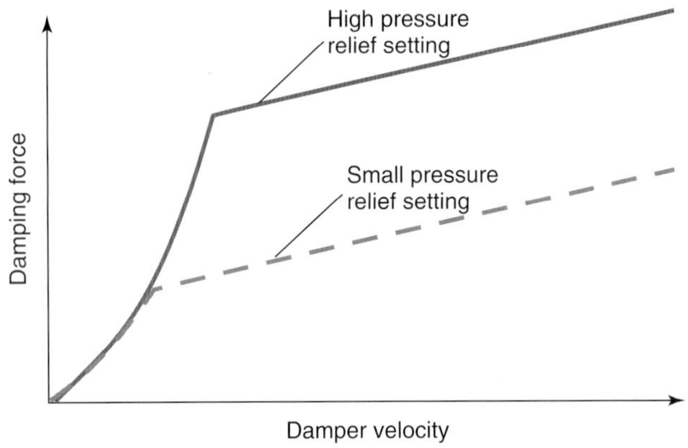

Figure 4. Modifying just the pressure relief valve.

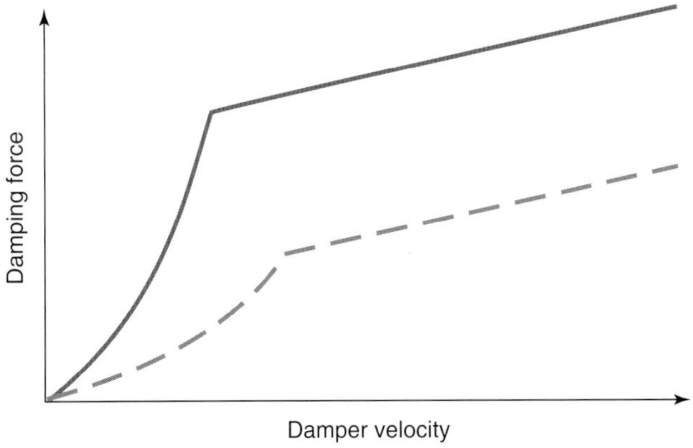

Figure 5. The valve modifies orifice and the pressure relief characteristic of the shock absorber.

Figure 6. Continuous damping control CDC® shock absorber of (a) GM Insignia and (b) Porsche Panamera. (Reproduced by permission of ZF Friedrichshafen AG, Germany.)

With the step to the proportional valves the spread between the soft and the firm characteristic was increased at the same time.

These types of dampers are for example used in the Mercedes C-class, Volvo V70, VW Golf, BMW X3 produced by Tenneco, or in the BMW X5, GM Insignia and Astra, AUDI A8, A4, Porsche Cayenne, Panamera produced by ZF Sachs (Figure 6), Porsche 911 produced by Bilstein, or SYMC Chairman (Figure 7) produced by Mando.

The top solution of this kind of variable damping is the solution used in the BMW 7 and 5 series produced by ZF (former ZF Sachs). Here, two proportional valves are used to control rebound and jounce independent continuously (Figure 8). One continuously controlled valve is responsible for the damping forces in rebound and the other valve controls the compression forces. In this configuration the necessary control of the valves is easier to handle because of the reduced control current frequency (body frequency vs wheel frequency). In addition to this benefit the possible compression spread of this type of shock absorber is higher compared to a shock absorber with only one valve because of the different hydraulic flows. Disadvantages are a higher weight of the component and additional cost.

2.1.2 Variation of the viscosity of the hydraulic fluid

In contrast to the principle described in Section 2.1, these shock absorbers use special fluids instead of oil. Until now, two different fluids are used:

1. Magneto-rheological fluid
2. Electro-rheological fluid

In the case of the magneto-rheological fluid a mixture of oil and iron particles substitutes the damper oil. The valve itself, located at the piston of the shock absorber, is just an electromagnetic solenoid, which has holes to connect the lower with the upper working chamber (Figure 9).

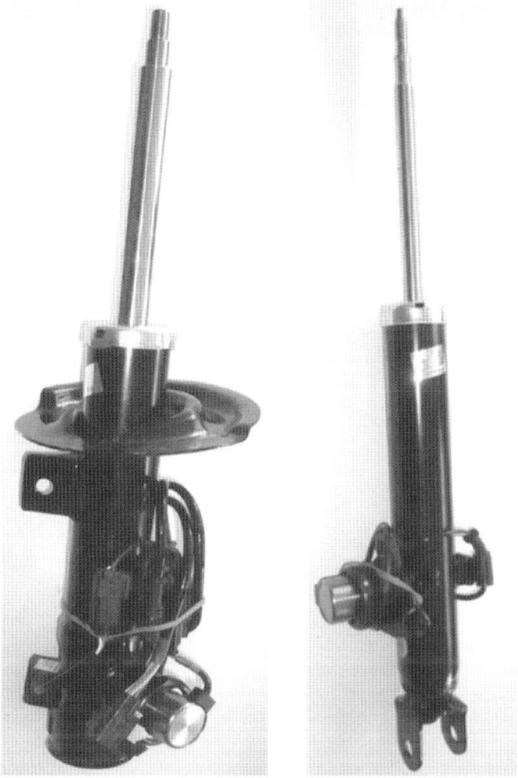
(a) (b)

Figure 7. Shock absorber of (a, b) Kia K7. (Reproduced by permission of MANDO Corporation, Korea.)

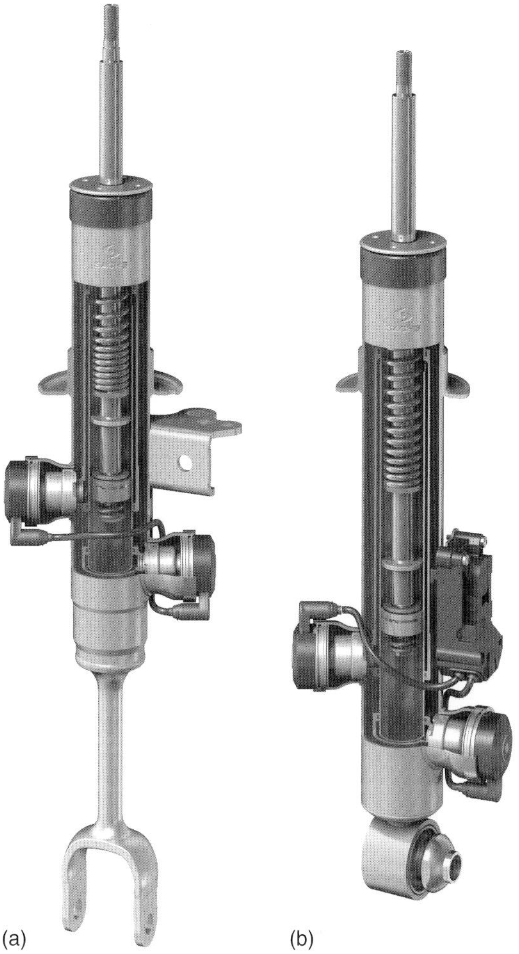
(a) (b)

Figure 8. CDC® shock absorber of (a) front, (b) rear of BMW 7 series. (Reproduced by permission of ZF Friedrichshafen AG, Germany.)

By increasing the magnetic flow through the valve (piston) the fluid inside the holes changes its viscosity. This effects increase of flow resistance partly and generates a pressure change inside the shock absorber in the active (compressed) working chamber. The viscosity inside the working chamber themselves is not changed and keeps the same physical values as, that is, compressibility.

Shock absorber acting with this principle are used in Ferrari, Audi TT, R8; GM Corvette and STS, Landrover Evoque, all produced by BWI.

Very similar to the principle mentioned earlier, the electro-rheological fluids work. The technology is based on a silicon oil fluid with particles of a polymer. This, together with an electric field, generates an effect similar to the magneto-rheological one described before. The only difference is that the viscosity changes by controlling an electric field inside the piston instead of the magnetic field. Because of the needed electric field of 1–2 kV and the challenges of isolation, this principle is not used until now in series car applications. Both principles are limited to mono-tube applications because of the same restrictions of the fluids themselves.

2.2 System

The easiest way of a system for adjust variable dampers is the manual control. This control gives the driver the possibility to select the soft or the firm damping forces (Causemann, 1999). Here, the system contains an electronic control unit (ECU), which just activates in a constant way the related valves. Systems like that are limited to using dampers working with "on/off"-valves (Section 2.1.1).

The next step of improvement was to use in addition one vertical body accelerometer, one lateral accelerometer, and one longitudinal accelerometer to detect the motion of the car. With this information a simple "threshold"-control (Section 2.3) was calculating the needed damping for the car and activating the valves automatically. The driver had a selection between two different control programs (Sport, Comfort) (Causemann, 1999). This type of control was used

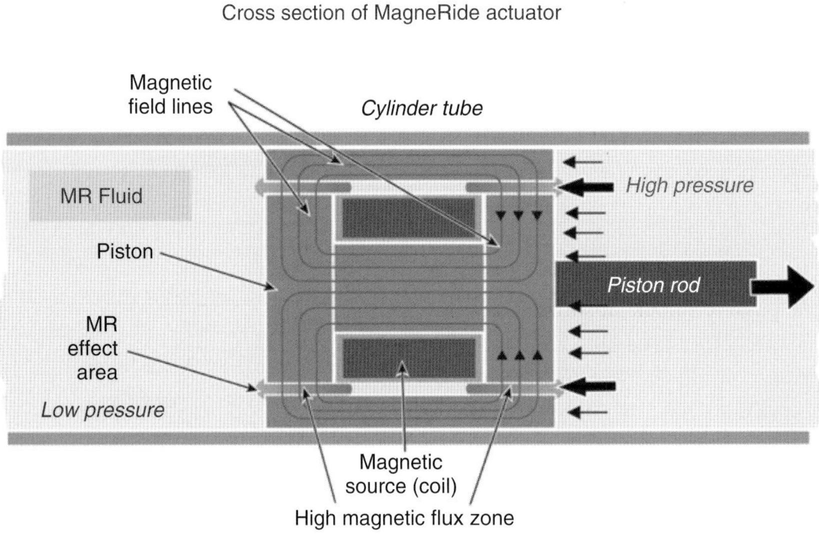

Figure 9. Principle of MagneRide shock absorber. (Reproduced by permission of BWI Group, China.)

for "on/off"-valves and the first generation of proportional valves too. Nowadays, a lot of information is available in the cars by CAN or Flexray-Bus and the additional needed information for control systems of shock absorbers are limited to up to three vertical body accelerometers to detect the body motion roll, pitch, and heave, and some sensors at the wheels. Here, displacement sensors are used in combination with xenon light control and/or height leveling (air spring) or accelerometers at the shock absorbers. The ECUs are now able to control each shock absorber at each wheel independent and fast. Mostly a separate ECU is used or a combination with the ECU for air leveling. Special at BMW X5 and 7 series each shock absorber has a single ECU onboard at the damper, which controls the damping forces locally depending of the demands from a supervisor ECU.

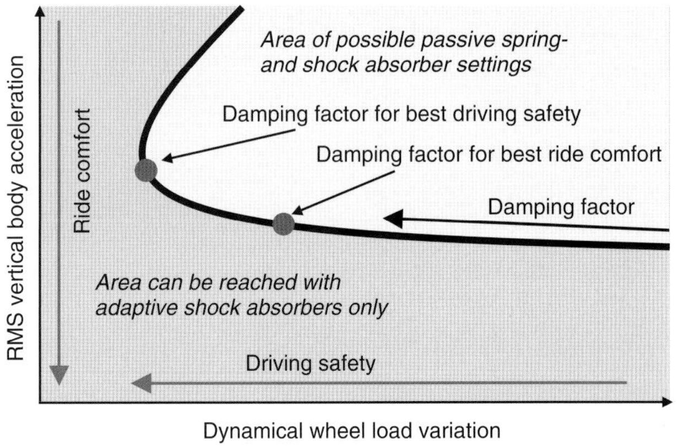

Figure 10. Target conflict between driving safety and ride comfort.

2.3 Control strategy

The simplest way to control adjustable shock absorbers is to use a switch to change between the possible damper settings.

But this kind of manual control by the driver resigns most of the possible benefits of adaptive damping systems.

If we take the body acceleration—as a measure for the ride comfort—in dependence on the dynamic wheel load variation as a measure for the driving safety, a limiting curve of the physically possible adjustments can be derived (Figure 10).

Owing to the fixed damping factor a vehicle with passive suspension can only cover one point on the envelopment curve, which will be either in favor of ride comfort or of driving safety, depending on the vehicle philosophy.

A control strategy automatically reacting to the actual driving situation is able to go some way toward resolving this target conflict.

The first active shock absorber systems entering the market in the 1980s used threshold control strategies, partially including elements of fuzzy logic.

At that time only slow switching valves with merely a few discrete damping curves were available. The development of fast reacting continuous damping valves and the availability of control units with much more computing power helped to establish more complex control strategies.

In today's serially produced vehicles usually control strategies for adaptive damping systems based on the Skyhook algorithm (Causemann, 1999) are used. Further known control strategies are Groundhook (Valasek and Novak, 1996) (aim: maximal safety due to minimized wheel load variation) or strategies based on H-infinity methods (mathematical procedures of system analysis and controller synthesis from the field of robust control algorithms).

The basic idea of Skyhook control is to reduce body movements by placing a passive shock absorber between sky and vehicle's body. To meet the same objective between vehicle's body and wheel, a variable shock absorber is used. As an example of an active shock absorber control strategy the ZF Sachs CDC® (Continuous Damping Control) algorithm is described later. The ZF Sachs CDC® control strategy is also based on the Skyhook algorithm. However, during more than 20 years of development, the simple Skyhook idea evolved into a complex control strategy able to cover both, optimal ride comfort and driving safety.

Standard input signals of ZF Sachs CDC® are three vertical body acceleration sensors and two vertical wheel acceleration sensors. If available, either body- or wheel-acceleration sensors can be replaced by wheel-travel sensors. Additional input signals are read in via the vehicle communication bus. Kind and number of these signals depend very much on the real car configuration, and at least driving speed and steering angle are needed. Figure 11 shows the used signals and the controlled driving situations.

In a first step, measured body movements of the vehicle are analyzed and fragmented into three so-called modal velocities: heave, roll, and pitch. Each modal velocity can be tuned separately by means of an advanced Skyhook algorithm.

Resulting damping force requests for each shock absorber are converted into valve current requests. Variable current limiters restrict these current requests.

Current limitation is controlled by a multiplicity of signals specifying the actual driving situation. This enables the control strategy to change the focus permanent and continuous between ride comfort and driving dynamics (Figure 11).

3 RIDE AND HANDLING IMPROVEMENT

Owing to the ability to adapt to any possible driving situation, an active shock absorber system is able to improve, in addition to ride comfort, driving dynamics, and driving safety.

3.1 Ride comfort

To characterize the ride comfort of a passenger car the power density spectrum (PDS) became widely accepted.

Figure 12 presents a comparison of three PDS of body acceleration front right, measured with one and the same car, serially equipped with conventional damping.

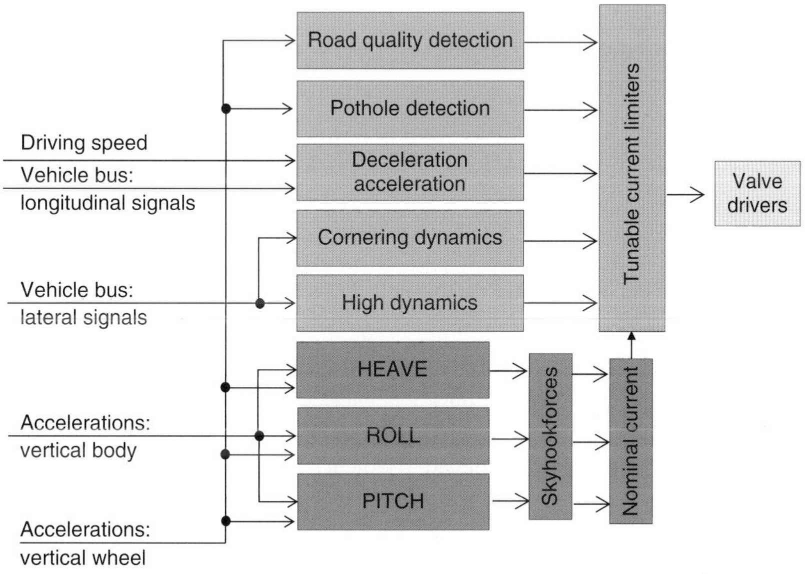

Figure 11. Overview design of ZF Sachs CDC® control strategy. (Reproduced by permission of ZF Friedrichshafen AG, Germany.)

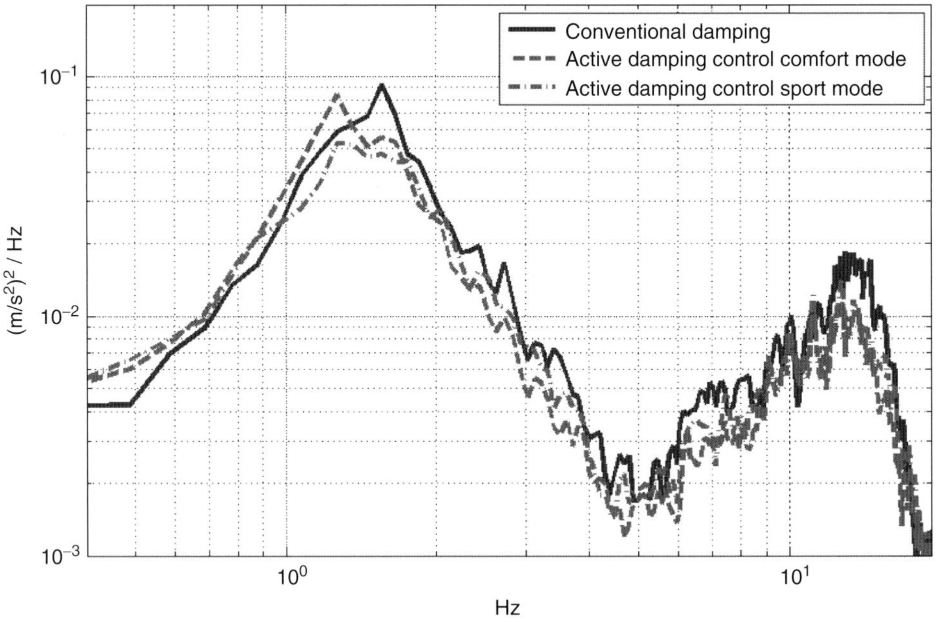

Figure 12. PSD body acceleration front: conventional damping, CDC® comfort mode, CDC® sport mode. (Reproduced by permission of ZF Friedrichshafen AG, Germany.)

The solid curve displays the initial state. The test vehicle then was equipped with active shock absorbers. Dashed and dot-dashed curves display PSDs of the two switchable modes for CDC®-control.

The advantages of active shock absorbers are clearly visible.

Accelerations near eigenfrequency of the body are reduced by 8% in comfort mode and even 40% in sport mode.

In addition, at higher frequencies up to 10 Hz, the accelerations are significantly lower most notably in comfort mode as a result of substantially better isolation against high frequency road inputs.

3.2 Driving dynamics

At the same time, driving dynamics are clearly enhanced because of better body control and more stable yaw-behavior. Figure 13 shows the roll angle during a double lane change maneuver in comparison between a conventional damped car and a car equipped with active shock absorbers.

Not only the absolute values of roll angle are reduced with CDC®, but also the temporal variation is visibly steadier.

As a result of minor body movements the possible velocity to perform the standard double lane change increased by 5% with active shock absorbers.

An additional feature of active shock absorber control is the possibility to influence the self-steering behavior of the vehicle. By increasing damping forces of the rear axle in cornering situations an over-steering tendency can be created. The agility of the car is noticeably enhanced.

Figure 14 shows the influence of CDC® to yaw behavior. The driving maneuver was a moderate steering step input at 114 km/h (71 mph). The stationary lateral acceleration was 6 m/s^2. In this situation active damper control could increase the initial yaw-velocity by 15%.

3.3 Driving safety

The influence of CDC® to the self-steering behavior described earlier can also be used to stabilize the car in critical driving situations and thereby support the driver.

To demonstrate the mode of action, lane change maneuvers were performed using a steering robot. Electronic stability program (ESP) function was switched off. The error between target yaw-velocity calculated by ESP software and measured yaw-velocity was evaluated. Figure 15 shows the yaw-error as a mean value of three measurements each.

During the high dynamical change back to the first lane at 4.5 s the yaw velocity error was reduced by CDC® by 28%.

In order to achieve this goal, the damping control system permanently monitors steering angle, driving speed, measured yaw-velocity, and lateral acceleration to detect

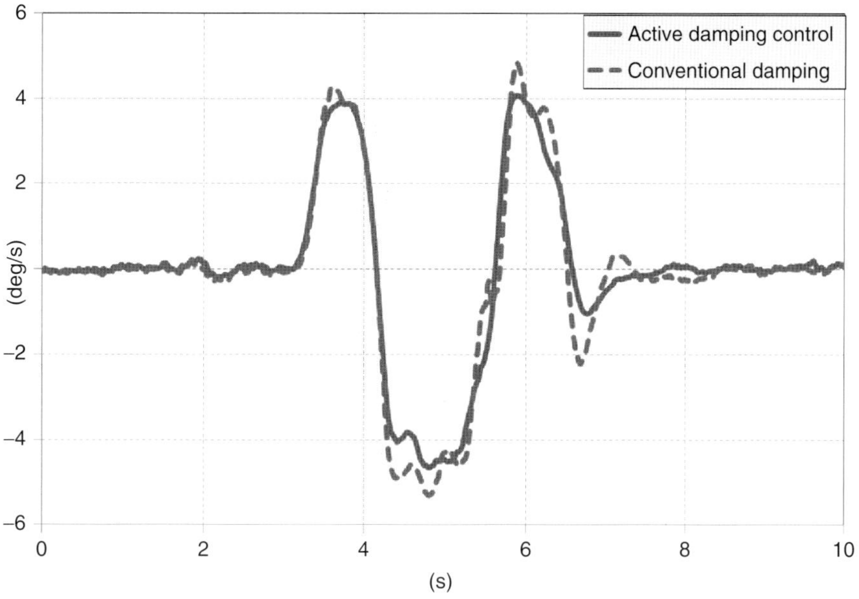

Figure 13. Roll angle during double lane change maneuver—conventional damping versus CDC®. (Reproduced by permission of ZF Friedrichshafen AG, Germany.)

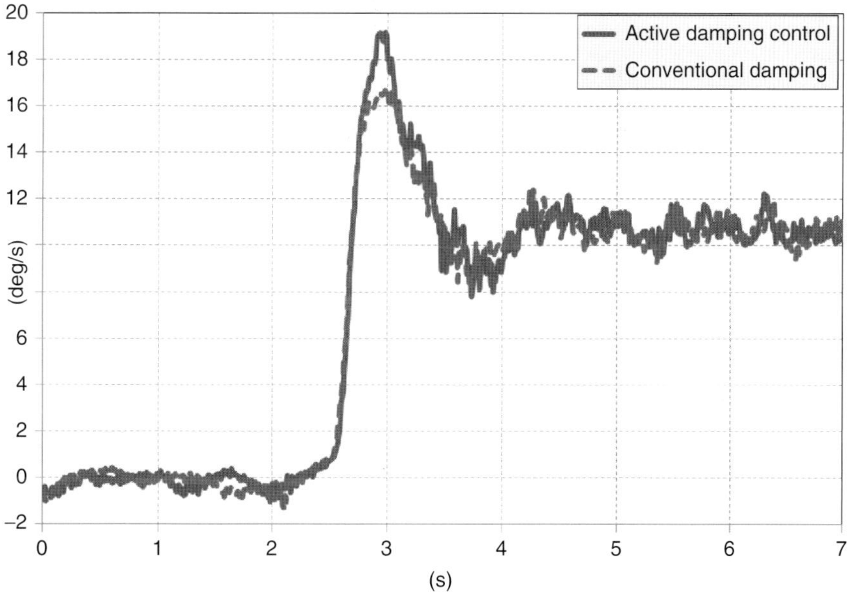

Figure 14. Yaw velocity during steering step maneuver—conventional damping, CDC®. (Reproduced by permission of ZF Friedrichshafen AG, Germany.)

over-steer and under-steer tendencies. In case of an over-steer tendency detected, the damping force of the rear axle should be reduced.

Figure 16 shows the working of over-steer/under-steer control during a double lane change. The lower the valve current is, the higher are the damping forces generated.

In the beginning, normal shock absorber control is working. At 2.2 s, a high dynamical situation is detected and the control strategy changes to over-steer/under-steer control. Valve current of the front axle is reduced to 0 mA to provide maximum damping force with the aim to reduce roll angle and to improve steering response. The valve current

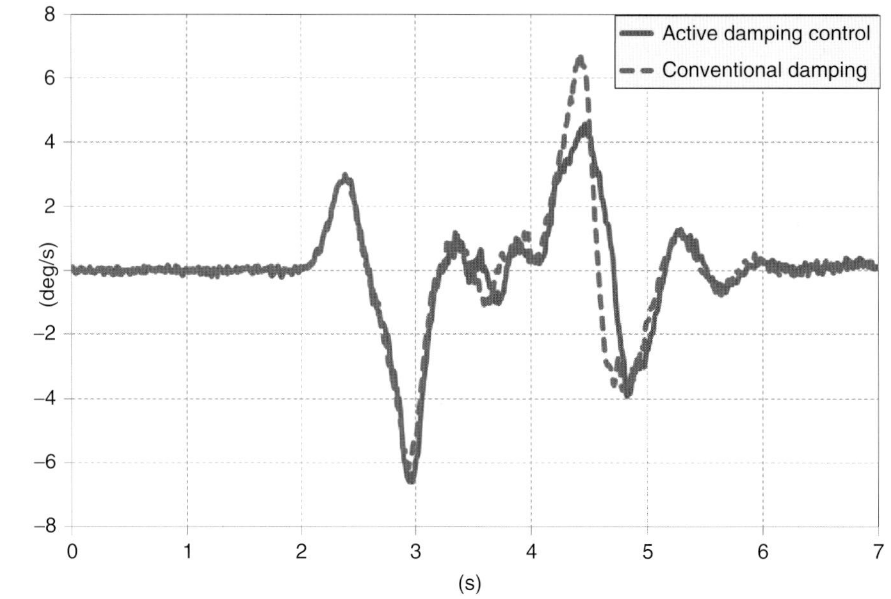

Figure 15. Yaw velocity error during double lane change maneuver—conventional damping, CDC®. (Reproduced by permission of ZF Friedrichshafen AG, Germany.)

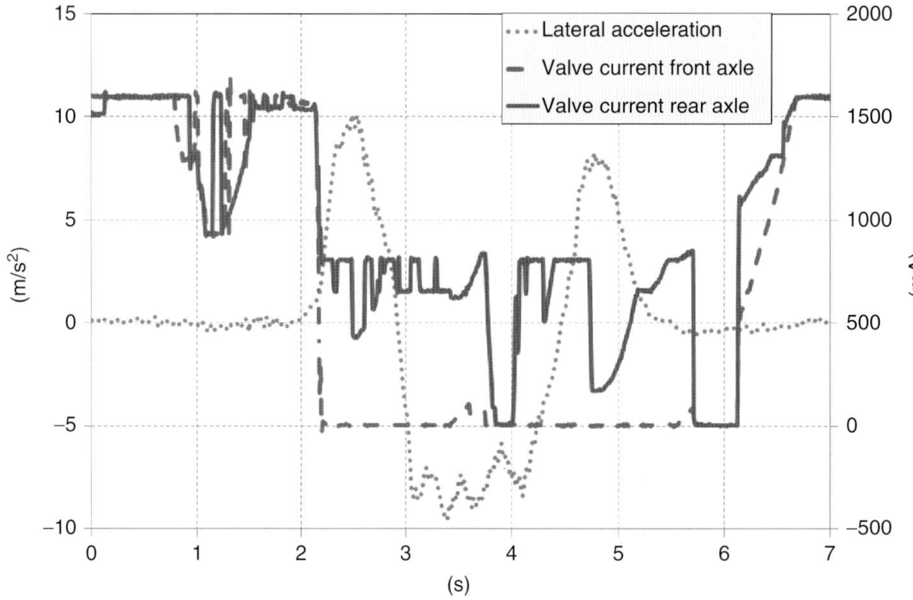

Figure 16. Over-steer/under-steer control: lateral acceleration, current front axle, current rear axle. (Reproduced by permission of ZF Friedrichshafen AG, Germany.)

and so the damping force of the rear axle changes dependent of over/understeer calculation/reaction of the car. Thus the rear shock absorbers are tuning/influencing the degree of understeering/oversteering.

To achieve a braking deceleration as large as possible wheel oscillations should be minimized (Niemz, 2007; Niemz *et al.*, 2007). Usually the control strategy of active shock absorber systems aims to achieve maximal ride comfort. Changing the objective of control strategy in case of heavy braking from comfort to low dynamic variation of wheel forces has a great potential for influencing braking performance.

Comprehensive investigations carried out in cooperation with Technische Universität Darmstadt have shown that, in fact, controlling wheel load variations with active damping leads to significantly shorter braking distances up to 2% (Figure 17).

4 OUTLOOK

In the meantime variable damping systems (active shock absorbers), coming from upper and luxury car segments, have meanwhile reached the lower middle class and smaller car segments. Cost pressure in these segments demand new solutions and optimizations. Actual visible solutions offer simple bi-stage switchable shock absorbers or propose the use of fully variable dampers on the rear axle only, the so-called rear axle control, to bring down total system cost (Nikiel and Schürr, 2011). In the meantime first applications appear in the market. Besides passenger cars, this last proposal could also be interesting for applications for vans

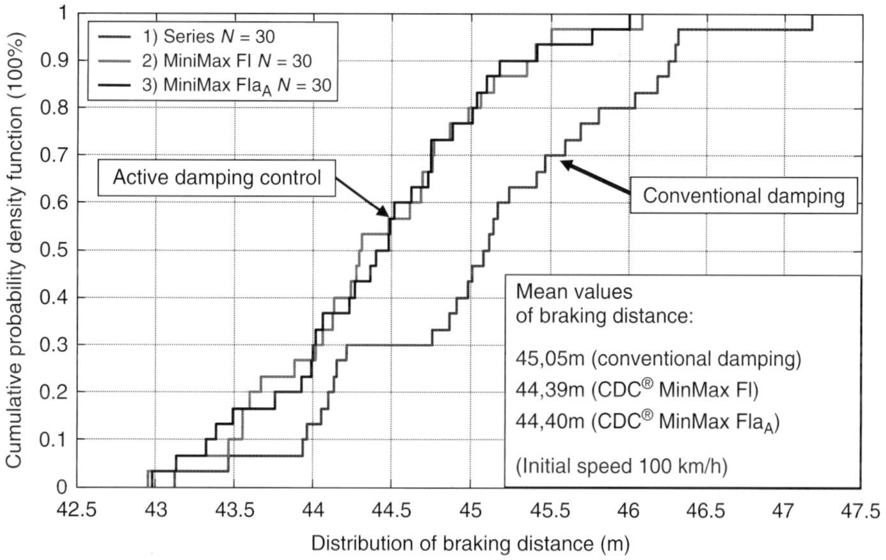

Figure 17. Active damping control can shorten braking distance significantly. (From Reul, 2011. Reproduced by permission of VDI Verlag GmbH.)

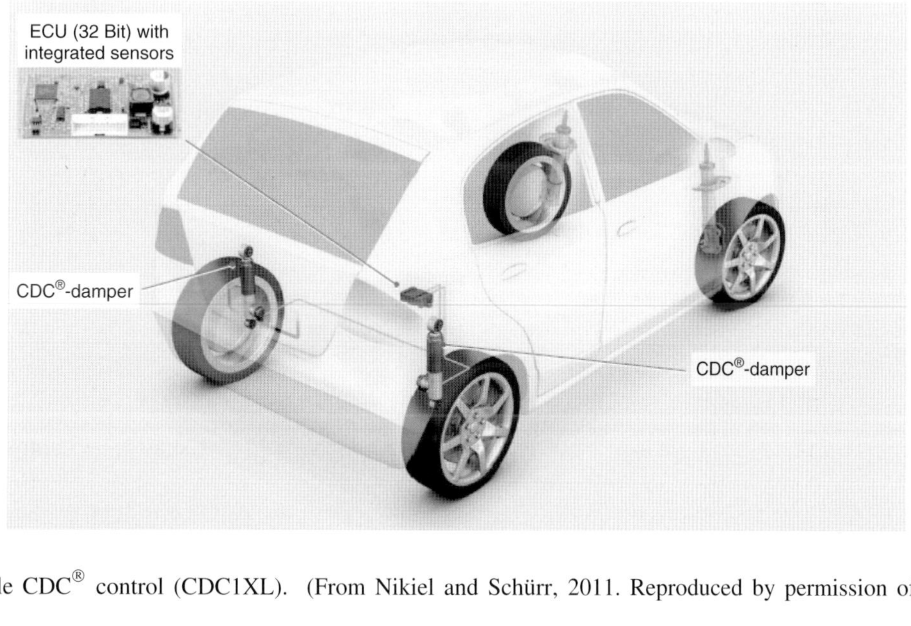

Figure 18. Rear axle CDC® control (CDC1XL). (From Nikiel and Schürr, 2011. Reproduced by permission of ZF Friedrichshafen AG, Germany.)

and light commercial vehicles (LCVs). Figure 18 gives an impression of such a design.

5 SUMMARY

This chapter shows a technological overview of existing damping principles. The main focus is directed on fluid-based, electronically controllable dampers. An historical back view on damper technologies in general is given. Existing solutions and functional principles are explained and application examples are shown to modify the hydraulic resistance. This can be realized by adjusting holes in the hydraulic circuit of the damper or vary the viscosity of the fluid (based on magneto-rheological/electro-rheological principles)

The total electronic system level-based explanations focus mainly on the Skyhook related control strategy.

A basic description of a general system topology and control strategy is given and explained.

Possible improvements in regards of ride and handling (driving comfort, driving dynamics, and driving safety) are described and some detailed examples/explanations from real applications are shown.

RELATED ARTICLES

Volume 4, Chapter 110
Volume 4, Chapter 127
Volume 4, Chapter 128
Volume 4, Chapter 130
Volume 4, Chapter 131

REFERENCES

Causemann, P. (1999) *Kraftfahrzeugstossdämpfer: Funktionen*, Bauarten, Anwendungen, Verlag Moderne Industrie, Germany.

Heissing, B., Ersoy, M., and Giess, S. (2007) *Fahrwerkhandbuch: Grundlagen, Fahrdynamik, Komponenten, Systeme, Mechatronik, Perspektiven*, Friedrich Vieweg & Sohn Verlag (ATZ/MTZ-Fachbuch), Wiesbaden.

Kutsche, T. and Raulf, M. (2001) Optimised Ride Control for Passenger Cars and Commercial Vehicles. ATZ, Germany.

Niemz, T. (2007) *Reducing Braking Distance by Control of Semi-Active Suspension*, Fortschritt-Berichte VDI Reihe 12 Nr. 640, Düsseldorf. ISBN: 3-18-364012-6

Niemz, T., Kutsche, T., Schürr, H., and Winner, H. (2007) Verbesserung des Bremsverhaltens von Pkw durch Einsatz variabler Dämpfer. In: TÜV SÜD Automotive GmbH (Hrsg.): chassis.tech.

Nikiel, R. and Schürr, H. (2011) Rear axle CDC-System, an approach to improve driving comfort and driving dynamics in smaller car segments. Chassis. Tech Munich.

Reul, M. (2011) Bremswegverkürzungspotential bei Informationsaustausch und Koordination zwischen semiaktiver Dämpfung und ABS. Dissertation Technische Universität Darmstadt.

Valasek, M. and Novak, M. (1996) *Ground Hook for Semi-Active Damping of Truck's Suspension*. Proceedings of CTU Workshop 96, Engineering mechanics, CTU Prague, Brno.

Chapter 114

Suspension Arms, Steel versus Aluminum, Where are the Benefits?

Dirk Adamczyk, Markus Fischer, and Klaus Schüller

ZF Friedrichshafen AG, Friedrichshafen, Germany

1 INTRODUCTION

Suspension arms or links connect the wheel or the wheel carrier with the vehicle body, usually via a subframe, and allow with respect to the vehicle chassis the controlled motion of the wheel. They have to transmit the loads from braking, cornering, vehicle acceleration, and impacts from the road surface. There are even strong crash requirements for the suspension arms. Crash or controlled buckling and durability are key targets for suspension arms. In special axle types, the suspension arm has to transmit the loads from vertical acceleration and "carry" the vehicle load. In this case, the spring is directly or indirectly assembled to the arm (Heißing, 2011).

Owing to the fact that the control arm is assembled to the wheel carrier, the mass of the arm is at least partially unsprung mass. On the basis of this, lightweight is one of the most important design targets for a suspension arm as it supports two design targets on vehicle level:

1. General low vehicle weight to reduce CO_2 emissions and
2. To improve the vehicle performance, for example, improved vehicle acceleration and improved vehicle safety and comfort by reducing the unsprung mass (Adler, 1991).

Suspension arms are highly integrated in the overall vehicle package. Main drivers are the required wheel or tire clearance and the ground clearance. As on the one hand, wheels are getting bigger and bigger with every new vehicle model, and on the other hand, engines and especially hybrid engines require more and more package space, the allowable space to package suspension arms gets more and more limited.

Beside all functional and packaging requirements, there is a very strong focus on costs and the manufacturing/assembly process. As suspension arms have a critical function for the vehicle, it is essential to have a strong focus on quality and robust production processes that have to be globally available for worldwide vehicle platforms.

1.1 Where are the benefits?

We have to take aluminum to get the best weight and steel to get the best cost.

Of course, it is not as simple as this. Although for many applications, the sentence is true.

Considering the material data (Table 1), it is easy to imagine that there is no simple answer to the question: Steel versus aluminum – where are the benefits?

Encyclopedia of Automotive Engineering, print © 2015 John Wiley & Sons, Ltd.
Edited by David Crolla, David E. Foster, Toshio Kobayashi, and Nick Vaughan.

ISBN: 978-0-470-97402-5
Also published in the *Encyclopedia of Automotive Engineering* (online edition)
DOI: 10.1002/9781118354179.auto002

Table 1. Mechanical properties of steel and aluminum.

	Steel	Aluminum
Young's modulus	210 Gpa	70 GPa
Density (ρ)	7850 kg/m^3	2700 kg/m^3
Tensile strength (typical high end)	800 MPa	Up to 380 MPa

Some thoughts about ratios between the material properties of aluminum and steel:

$$\frac{\text{density (steel)}}{\text{density (alu)}} = \frac{7.85\frac{\text{g}}{\text{cm}^3}}{2.70\frac{\text{g}}{\text{cm}^3}} = 2.91 \quad (1)$$

$$\frac{\text{young's modulus (steel)}}{\text{young's modulus (alu)}} = \frac{210\,\text{Gpa}}{70\,\text{Gpa}} = 3.00 \quad (2)$$

$$\frac{\text{Tensile strength (steel)}}{\text{Tensile strength (alu)}} = \frac{800\,\text{MPa}}{380\,\text{MPa}} = 2.11 \quad (3)$$

One of the main drivers for using aluminum in automotive business is its weight reduction potential. The density of aluminum is 2.91 times lower than for steel (1). At the same time, the tensile strength is lower as well, but only 2.11 times lower (3). This means, if tensile strength is essential for the design, it is still possible to achieve a major weight reduction.

An example that considers previous material facts is calculated (Figure 1), based on a simple tension rod. A tension load of 40 kN was chosen, as it is an average force in today's mid-sized passenger cars.

This simple assumption of a rod tension loaded by a one-axis static load shows the aluminum weight-saving potential of 27.7%. In both cases, the tensile strength of the respective material is on the high level of the forging alloys that are typically used in the automotive industry for chassis components (380 vs 800 MPA).

If the stiffness is essential for the design, with the Young's modulus being the relevant key parameter, then there are some slight advantages for steel regarding weight, as the Young's modulus of aluminum is even 3.00 times lower than the Young's modulus of steel (3).

For most of the control arms, the required buckling load is important for design and weight. A basic factor for the buckling load is as well the Young's modulus. Taking a simple straight control arm, the following ideal example shows that the steel version is even lighter than the aluminum version – at least the middle section (Figure 2).

For the buckling calculation, there are different buckling conditions (Figure 3), but for typical suspension arms, the mode (2) has to be used.

For the calculation in the example, the following extended equation will be used. The equation includes bending/offset (Figure 4) to cover, for example, tolerances.

$F_{\text{Buckling}} = \frac{\delta_{\text{steel/alu}}}{\frac{e+f}{W}+\frac{1}{A}}$; based on offset [$e$]; additional displacement/bending [f] under load [F]; maximum stress for steel or alu [δ]; resistant torque [W]; cross section [A].

F_{max} = 40 kN (Load applied to the rod)

l = 300 mm (Total length of the rod)

$R_{m\,steel}$ = 800 MPa (Ultimate tensile strength)

$R_{m\,alu}$ = 380 MPa (Ultimate tensile strength)

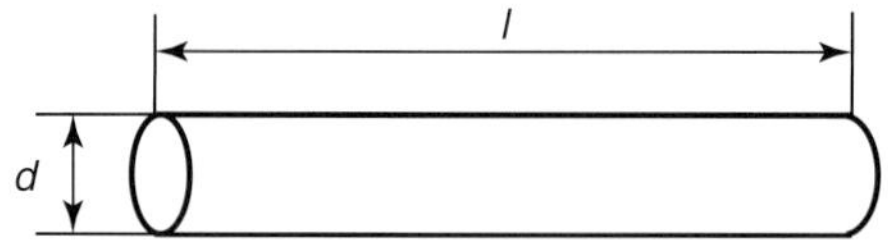

$$\sigma = \frac{F}{A} = \frac{F}{\frac{\pi d^2}{4}} \quad \text{(Stress)}$$

$$d = 2\sqrt{\frac{F}{\sigma\pi}} \quad \text{(Required rod diameter)}$$

$$m = \frac{d^2\pi}{4}\, l\, \varrho \quad \text{(Mass of the rod)}$$

$$d_{steel} = 2 \times \sqrt{\frac{40\text{ kN}}{800\text{ MPa} \times \pi}} = 8.0\text{ mm};\ m_{steel} = \frac{(0.8\text{ cm})^2 \times \pi}{4} 30\text{ cm} \times 7.85\frac{\text{g}}{\text{cm}^3} = 118.4\text{ g}$$

$$d_{alu} = 2\sqrt{\frac{40\text{ kN}}{380\text{ MPa} \times \pi}} = 11.6\text{ mm};\ m_{alu} = \frac{(1.16\text{ cm})^2\,\pi}{4} 30\text{ cm} \times 2.70\frac{\text{g}}{\text{cm}^3} = 85.6\text{ g}$$

Figure 1. Example: a mass comparison of a steel rod and an aluminum rod under the same load requirement.

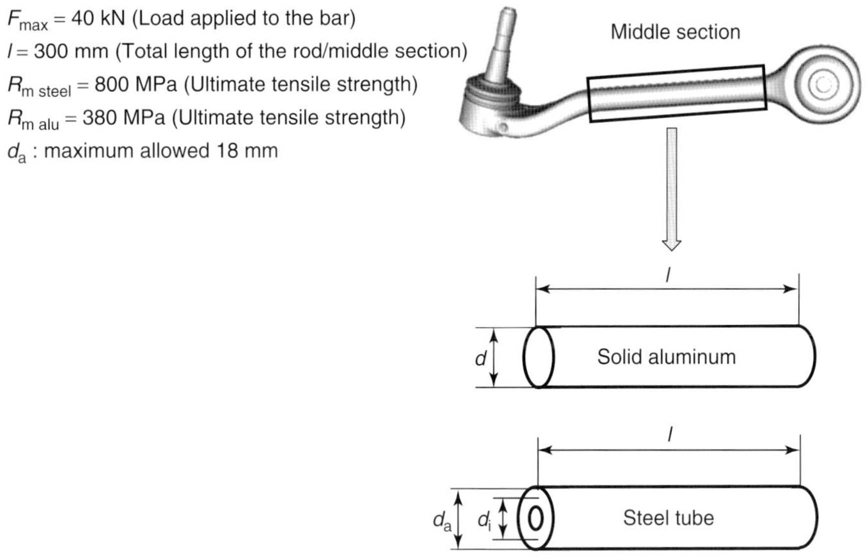

Figure 2. Example: middle section of a straight control arm.

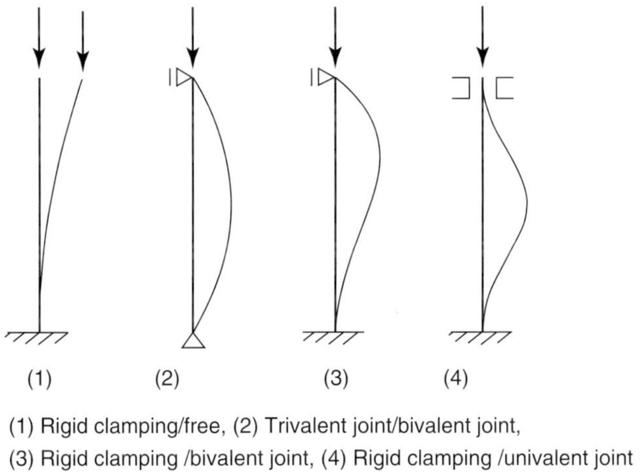

(1)
(2)
(3)
(4)
(1) Rigid clamping/free, (2) Trivalent joint/bivalent joint,
(3) Rigid clamping /bivalent joint, (4) Rigid clamping /univalent joint

Figure 3. The four Euler buckling modes. (Reproduced from Beitz and Grote, 2001. © Springer Science+Business Media.)

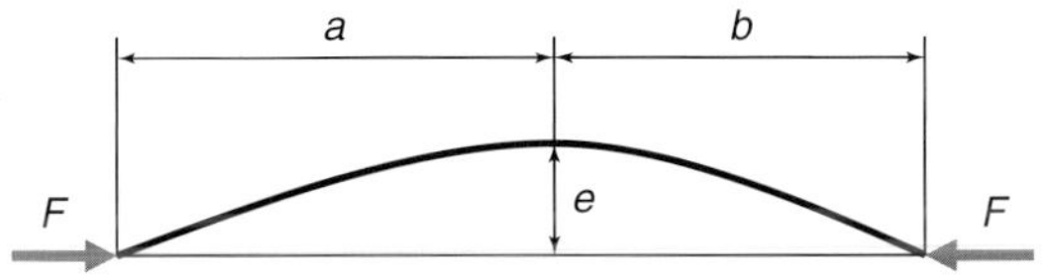

Figure 4. Buckling including offset.

$$f_{\text{buckling}} = -\left[\frac{e+W\left(\frac{1}{A}-x\right)}{2}\right] \pm \sqrt{\left(\frac{e+W\left(\frac{1}{A}-x\right)}{2}\right)^2 + W\cdot e\cdot x};$$

displacement [f] under buckling load [F]; $x_{\text{buckling}} = \frac{\delta_{\text{steel/alu}} a\cdot b}{3\cdot E\cdot I}$; distance [$a$]/[$b$] (Table 2).

The simple comparison (Figure 5) shows that aluminum does not necessarily lead to the best weight solution. In this case, the steel design is approximately 11% lighter than the aluminum design. This example is also a very simplified case based on only one requirement – in reality, further boundary conditions such as interfaces and packaging requirements between suspension arm, subframe, and knuckle have to be considered. Considering all the packaging requirements, the picture will likely change and a weight advantage for the aluminum design can be expected.

There are additional aspects that further influence weight (and costs). In typical sheet metal applications, there are weld seams that reduce the material properties significantly. For forging and casting parts, the tolerances of the raw part (e.g., thickness) have to be considered.

The existing applications show a typical weight-saving potential of 10–50% by the use of aluminum. The more complex a part is and the more it is effected by fatigue, the higher the weight-saving potential is. For the cost side, it is more or less the opposite.

The following sections show the typical influencing factors for the design and the choice of the best material and the manufacturing process for suspension arms.

2 KEY FACTORS FOR THE DESIGN OF SUSPENSION ARMS

2.1 Weight and environmental aspects in suspension design

Looking at today's vehicle fleet, vehicle simulations indicate that fuel consumption is reduced by 0.25 l/100 km or

Table 2. Annulus.

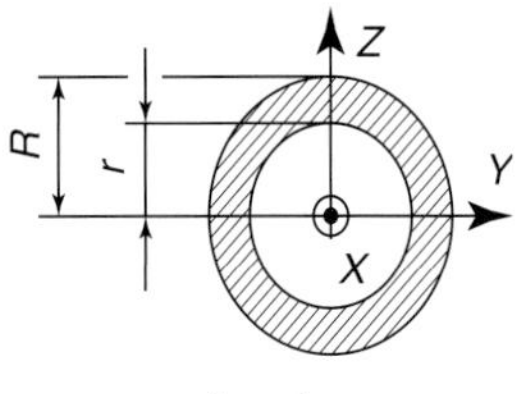

Annulus

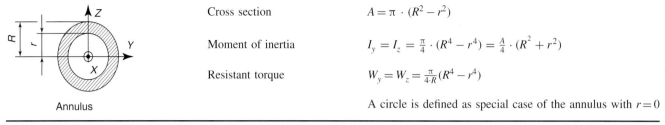

Cross section	$A = \pi \cdot (R^2 - r^2)$
Moment of inertia	$I_y = I_z = \frac{\pi}{4} \cdot (R^4 - r^4) = \frac{A}{4} \cdot (R^2 + r^2)$
Resistant torque	$W_y = W_z = \frac{\pi}{4 \cdot R}(R^4 - r^4)$
	A circle is defined as special case of the annulus with $r = 0$

Given geometry data:
s length = 300 mm $a = b = 150$ mm $e = 0.5$ mm $d_{max} = 18$ mm ($R_{max} = 9$ mm)

Given data for a steel tube:	Given data for an aluminum rod:
$R_{steel} = 9$ mm	$R_{alu} = 9$ mm
$r_{steel} = 7.5$ mm	$r_{alu} = 0$ mm
$A_{steel} = 77.7$ mm^2	$A_{alu} = 254.5$ mm^2
$I_{steel} = 2668$ mm^4	$I_{alu} = 5153$ mm^4
$W_{steel} = 593$ mm^3	$W_{alu} = 1145$ mm^3
$R_{e\ steel} = 680$ MPa (δ_{steel})	$R_{e\ alu} = 330$ MPa (δ_{alu})

Results steel tube:	Results aluminum rod:
$x_{steel} = 0.0091 \frac{1}{mm^2}$	$x_{alu} = 0{,}0069 \frac{1}{mm^2}$
$f_{steel} = 0.77$ mm	$f_{alu} = 3{,}87$ mm
$F_{steel} = 45$ kN	$F_{alu} = 43$ kN

Conclusion:	
The mass comparison for aluminum and steel version: ($F_{buckling}$: F_{alu} » F_{steel})	$m_{steel} = 183$ g $m_{alu} = 206$ g Δm » 11%

Figure 5. Calculation including offset.

even up to 0.4 l/100 km and 0.5 l/100 km (an improvement of ca 1 mile per gallon) for light trucks per every 100-kg weight reduction (Cheah, Heywood, and Kirchain, 2010; Goede, 2007). This figure shows the increasing need for the reduction of weight and in consequence fuel for all vehicle types – vehicles with combustion engine, vehicles with hybrid engine, or electric vehicles. This requirement of weight reduction puts a lot of pressure on new designs and challenges current production processes regarding improvements and optimizations.

The goal of weight reduction nowadays can be realized by the utilization of high strength steel, standard and high strength aluminum, higher strength ductile iron (DI), and the advancement of plastics. Nevertheless, the most common material for weight-optimized links is still aluminum. There are solutions for nearly all imaginable control arm applications, using forging, casting, and performance casting processes as well as cast–forge with various alloys. Only strongly limited package spaces in combination with the need to maintain a certain maximum load level

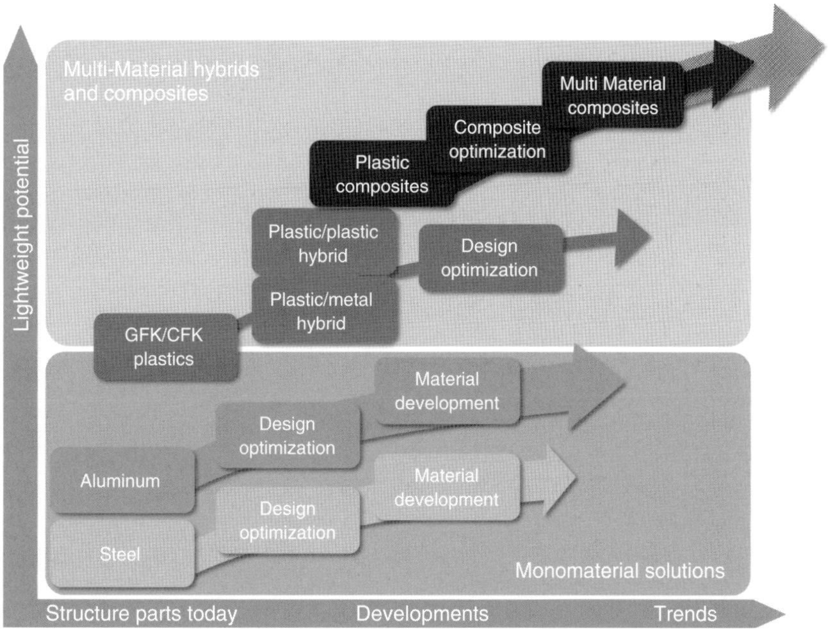

Figure 6. Material mixture as trend for lightweight concepts. (Reproduced with permission from ZF.)

would not allow using the light metal, from a functional point of view.

On the other hand, the energy balance producing aluminum generally speaks against this alloy: the amount of energy used – throughout the process from the raw material to the final product – is higher compared to that used for the processing of steel products.

Nowadays, not only the amount of energy used to produce an alloy is considered but also, seeing the global warming, a more holistic approach is chosen: more and more attention has to be paid to the overall CO_2 emission – not only during the usage of the vehicle but also under cradle-to-grave aspects. A study on a front end of a 2007 Cadillac CTS, for instance, came to the result that the aluminum design achieves the break-even distance from energy use and GHG (greenhouse gas) emission perspectives earlier within the vehicle lifetime (Dubreuil *et al.*, 2010). This achievement in overall environmental friendliness has further potential for improvement. It can be reached by considering the capability of aluminum to be recycled indefinitely, which requires only 5% of the original energy to be put back into use (Heidi and Thomas 2011).

Apart from the well-known disadvantages of steel arms compared to aluminum – such as higher weight and the need for corrosion prevention – the steel producers and manufacturers push toward high strength material and smart material mixes. This is to improve weight and fight the losses caused by aluminum and upcoming composite applications. Steel control arms using sheet metal, forging and cast iron alloys, and processes are still being used widely; especially in mid- and low-sized vehicles with standard engines, they show an increasing market share. Reason for this is the focus on cost in this market segment as main driver for design selection, and other considerations have lower priorities.

Figure 6 shows an overview about trends in lightweight solutions similar to the trend we see for the vehicle body (Goede *et al.*, 2008). Beside momo-material solutions, multi-material solutions will offer further lightweight potential for chassis.

2.2 General design aspects

Control arms are typically designed for two load conditions – fatigue load cycles and misuse load cases: load duty cycles (fatigue load or "dynamic" load): this is a combination of a number of single load events or block cycles. These loads are generated during "normal" driving conditions.

Single load events (the so-called quasi-static):

(a) Loads that appear frequently: these can be taken out of a load spectrum, as the load level is below the maximum, which means there are no special events or maximum loads included. The criteria are no crack or fracture, no or only minimal plastic deformation.
(b) Loads that appear just a few times over the vehicle life, such as special events and misuse. These load

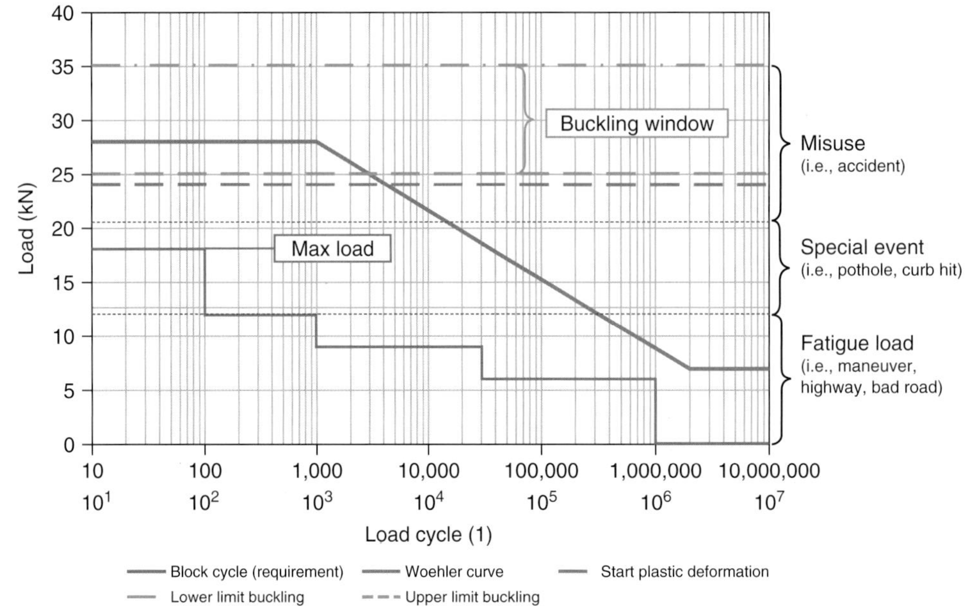

Figure 7. Relation between the load cases fatigue load, special event, and misuse.

cases are usually given through "Nastran-load-decks" (an input data format for finite element calculations) or "Adams-loads" (a software to simulate the kinematic and kinetic vehicle/component function) if not available out of real road load data measurements with test vehicles. They cover various driving maneuvers according to certain load philosophy rules such as "brake over railway tracks," "obstacle overdrive," and "washboard braking" just to name a few possible events. These loads are generated or simulated/calculated through pothole braking, curb impacts, crash (like through an accident) events, and some more. The load level reaches up to the maximum load expected from all relevant events. Depending on customer and design targets, it might be expected that after a special event full fatigue life is still ensured.

The relationship between the different load cases is shown in Figure 7.

Simulation tools check the dimensional layout of control arms before physical prototypes are built. As a standard, linear and nonlinear finite element analysis (FEA) calculations are used for the "quasi-static" cases and fatigue life FEA if needed. In most cases, it is not required to perform a fatigue life FEA because design rules for maximum strength will cover it. Structural or surface-related iteration tools such as "Optistruct" as one example support optimizations of complex structural parts considering the load requirement optimizations.

At the beginning of the design process, the following pieces of information are typically available.

The material selection or –request (material or weight target), load data and calculation requirements (i.e., for instance "Abaqus Skin"). FEA parameters such as bearing constraints (i.e., "RBE" = rigid body elements). Assessment requirements such as a maximum plastic elongation under a permissible load, maximum damage rate for fatigue life or stiffness and buckling requirements.

For suspension arm design, the following aspects are to be considered if applicable:

- Structural related: concept model constraints, special tolerances, target weight, specific section modulus or cross-sectional constraints, and stiffness. A control arm package is sometimes given as a CAD model with all applicable component contacts in worst condition, by minimum distance faces or a kinematics model (Figure 8). The kinematics model contains the kinematics points of all components and distances, as well as parameters for jounce, rebound, and steering, including stiffkinematic (structural components without elasticity content) and elastokinematic (elasticity added, such as from rubber bushings). The kinematics analysis should also consider interim movement positions to catch all possible positions that could touch minimum package requirements. Some defined steering overtravel might be required as well. In addition, it is closely to be judged and decided whether full jounce and rebound of the strut should be reduced to a reasonable amount.

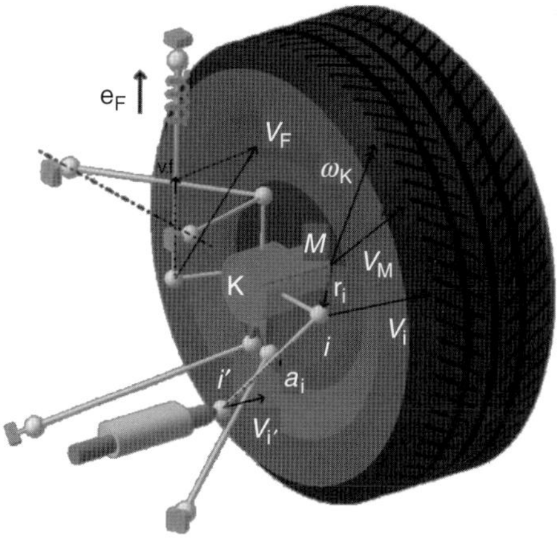

Figure 8. Kinematics model. (Reproduced with permission from ZF.)

Special care should be given to not create any stress raiser notches. Not only radii of main structures can be designed too sharp but also tight transitions to bushing bosses, mold part sections, mold part markings, and surface imperfection due to manufacturing process constraints. This is especially important not only to fulfill the fatigue lifetime requirements but also at fastener connections and for the application of coatings. In addition, pockets that can collect dirt or water have to be avoided or drainage solutions have to be found. Material raw part properties impact the function of a control arm through their mechanical, chemical, and grain conditions; material tolerance ranges; work hardening stiffening; and fracture behavior, which is different from push to pull direction. Furthermore, the geometrical tolerance chain from the raw part over the machining contour and the assembly components has to be considered in the common locator scheme at all production stages to avoid unnecessary machining or assembly effort or conflicts.

Smooth and "flowing" blended transitions are intended to achieve a consistent stress distribution under load. Special attention has to be given to avoid flat, straight, and even control arm structures, which would provide an unintended so-called close-to-Euler buckling case that reacts sensitively – small changes in typical tolerance windows of manufacturing processes would lead to large buckling load variations. Especially, this design rule contradicts the target of minimum weight because "leaving" the direct kinematics line will add weight. As a compromise, structural pinching dedicated to certain load directions can be a solution for cases that cannot allow for "buckling friendly" bending.

In some rare cases, it might be required to design a minimum natural frequency (nuisance avoidance) or minimum energy consumption (crash characteristic).

Table 3. Examples for aluminum.

	High Performance Aluminum Casting	Aluminum Forging Design
Material	SAE A 356-T6	SAE J 454-6082-T6
Yield strength	220 MPa min	240 MPa min
Ultimate strength	290 MPa min	290 MPa min
Elongation	8% min	10% min

Table 4. Examples for steel or iron.

	Ductile Iron Casting	Steel Forging Design
	Example Middle Range	Typical
Material	EN-GJS-500-7	30MnVs6 + P
Yield strength	320 MPa min	560 MPa min
Ultimate strength	500 MPa min	820 MPa min
Elongation	7% min	14% min

- Load related: maximum allowable stress for specific load cases, misuse requirements such as lateral and longitudinal buckling windows (minimum and maximum permissible buckling load) considering push and pull direction. This can be in conjunction with a required defined minimum plastic deformation limit and the direction to ensure the detection of a misuse event by a steering misalignment for instance or a defined travel of control arms relative to other components. In the case of a front impact, the requirement can be to provide a certain limited wheel travel, which must not block the doors for instance. Consequences of plastic deformations in case of hits or accidents may apply to the ball joints as well. It is to be considered that even a defined control arm bend would not separate the ball joint – in order to provide permanent wheel and/or steering control.

Example: Typical mechanical material properties of different manufacturing processes (Tables 3 and 4).

The expression "high performance casting" contains various processes such as counterpressure casting (cpc), vacuum riser-less casting (VRC), pressure riser-less casting (PRC), and cast forge (Cobapress) with high mechanical properties.

2.2.1 *Optimization of suspension arms*

Figure 9 shows the typical design process for forged and cast control arms starting with package model and the specification (step 1), initial optimization, for example, with

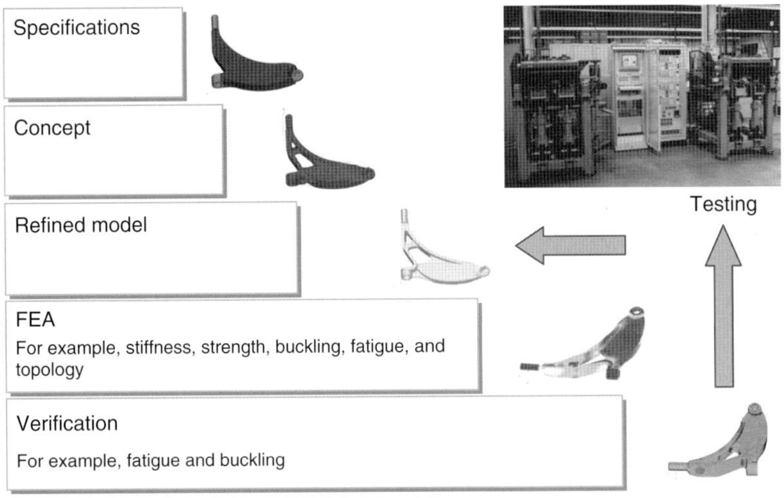

Figure 9. Typical development optimization process via simulation.

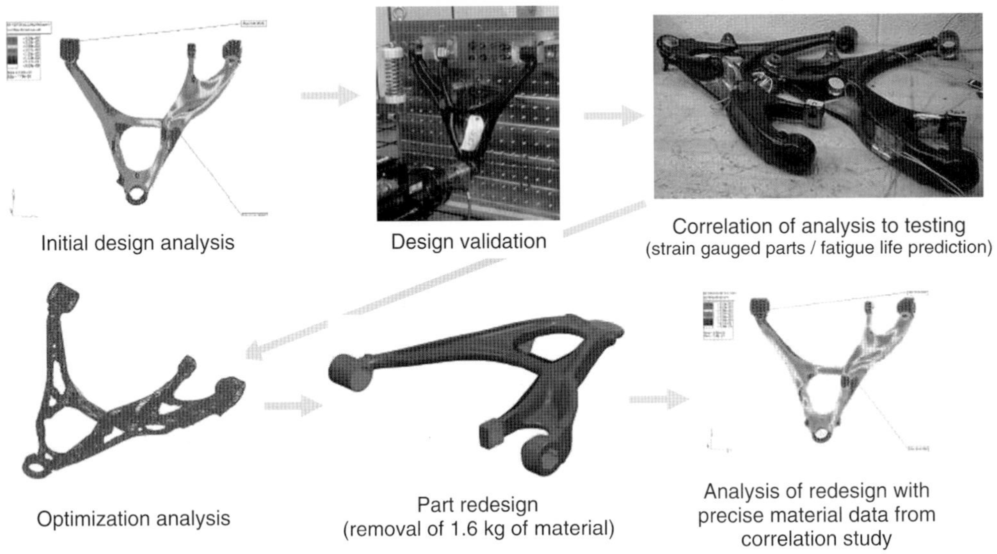

Figure 10. Development optimization including correlation study.

Optistruct® (step 2), optimized design (step 3), FEA (step 4), verification by fatigue simulation (step 5), and finally verification by testing (step 6). In some cases, a shape optimization will be done on top to further optimize some design details. In Figure 10, there is a special process shown to even consider design- and process-specific material behavior based on initial verification tests.

The above-mentioned process takes the optimization process even one step further. Real part material properties depend not only on the alloy but also on the raw part manufacturer and their production processes and variation of mechanical properties. On the basis of this knowledge, parts have been produced, strain gauged and real stress in critical areas of the part have been measured. These data have been used in FEA to further optimize the design based on a selected foundry process and tool design. This approach could reduce the weight by another 1.6 kg over the original FEA with typical material properties for the shown example of a front lower control arm (ZF).

2.3 Buckling and failure chain

Compact and light axle arrangements lead more and more to extraordinary challenges for the design to cover multiple

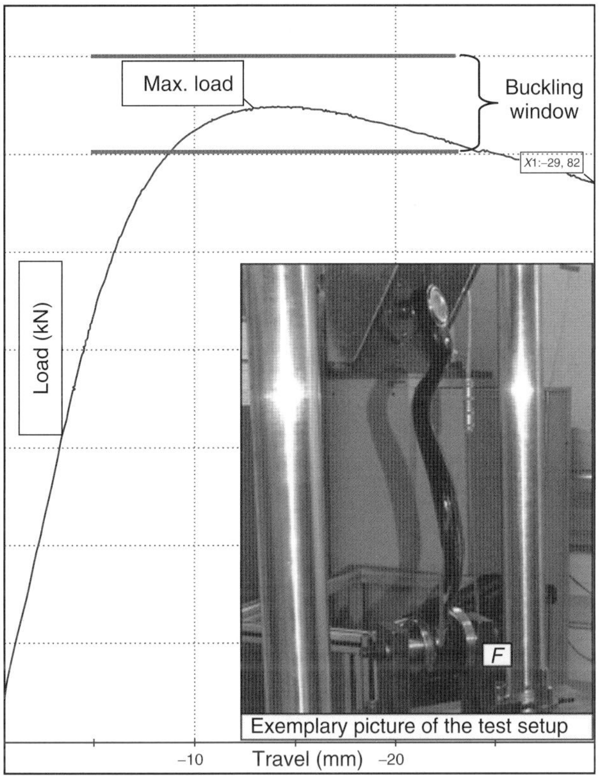

Figure 11. Typical buckling curve measured in compression direction.

load requirements for various load directions. On top of that, the so-called damage chain is defined for the vehicle. The damage chain describes the cascade of misuse load through the chain of components linked to each other, leading to a recognizable failure of single defined components. This follows the principle "leak before break" and means that a driver for instance recognizes a wheel misalignment after hitting a curb with too high speed.

The damage chain would be cascaded down from the major vehicle weight and ability requirements down to each component. For a side impact at the front end, the tie rod needs to deform first to protect the steering gear, and typically a control arm needs to buckle before the knuckle or subframe is damaged. The suspension arms are usually relatively easy to be replaced and therefore lead to low repair costs in case of minor misuse cases. However, this depends also on the manufacturing damage chain philosophy. These components are therefore the so-called sacrificial or victim parts, which act like a mechanical fuse. The damage chain defines the maximum permissible load for structural parts and for link attachments and screw connections. Figure 11 shows an example for a control arm. The buckling window describes the range for the maximum permissible load.

3 SUSPENSION ARM TYPES

3.1 Front axles

Requirements toward a chassis are high and complex and have to be a compromise among but not limited to driving dynamics, ride comfort, safety, weight, cost, reliability, durability, allowable space, crash behavior, and environmental friendliness. Owing to these requirements, the following suspension principles and modifications within those are typical for modern vehicles and SUV as front suspension (Figure 12; Elbers *et al.*, 2004)

3.1.1 Suspension arms for McPherson axles

The McPherson suspension type (Figure 13) is the preferred solution on passenger cars up to the luxury segment – thanks to the simplicity of design, associated cost effectiveness, and cross-car installation space. Nevertheless, this type of suspension has still limitations due to the fact that the shock has to transfer additional loads besides the vertical ones, which can add friction to the system and therefore influences the performance of the suspension. In today's cars and crossover vehicles, the typical design is a rear or front facing L-arm, manufactured out of stamped steel (single and double shell), DI,

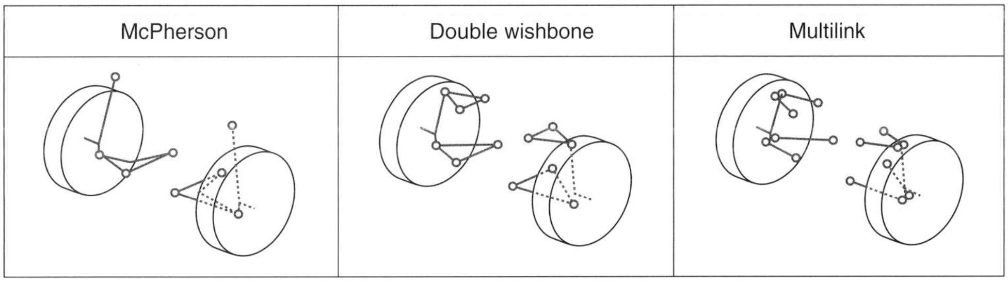

Figure 12. Typical front axle concepts.

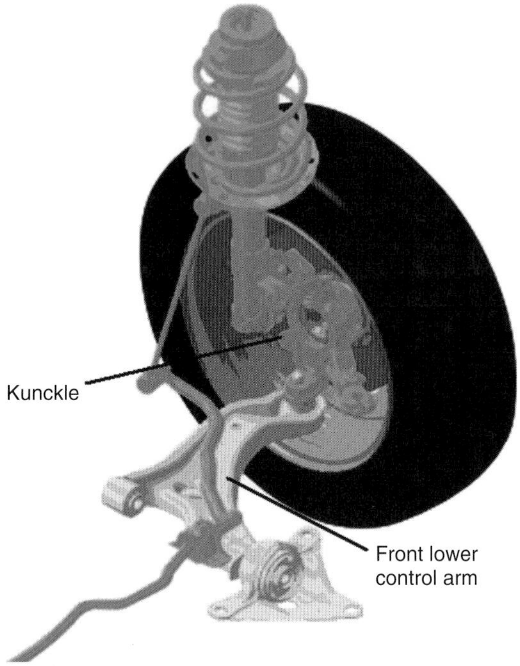

Figure 13. McPherson axle.

aluminum forging, and aluminum casting (Figure 14). From the design perspective, vertical loads are not too high, but the fore-aft and cross-car loads plus the crash performance have to be considered during the design as well as the space requirement for tire envelope, ground clearance, and stabilizer bar.

Figure 15. Strut front suspension with two lower suspension arms.

The figures above show examples for the different technologies for L-shape suspension arms. Application in the vehicle depends on requirements such as loads, packaging space, and cost. Therefore, all of the above-mentioned samples are feasible and production intend.

3.1.2 Two separated arms

To improve the freedom to position the lower hard point (virtual hard point) and therefore also to move the upper shock mount further out, the lower control arm can be replaced by two links (Figure 15 – links are marked red), typically called *tension* and *lateral* or *compression link*. These arms are typically designed out of forged steel or forged aluminum, but, if clearance and deformation requirements allow casting or tubular, weld designs are also feasible (Figure 16).

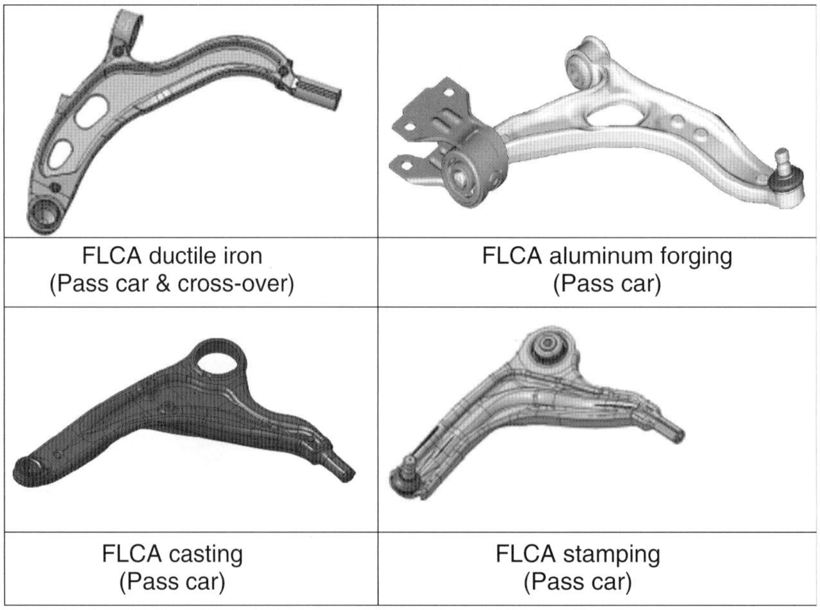

Figure 14. Examples: front lower controls arms.

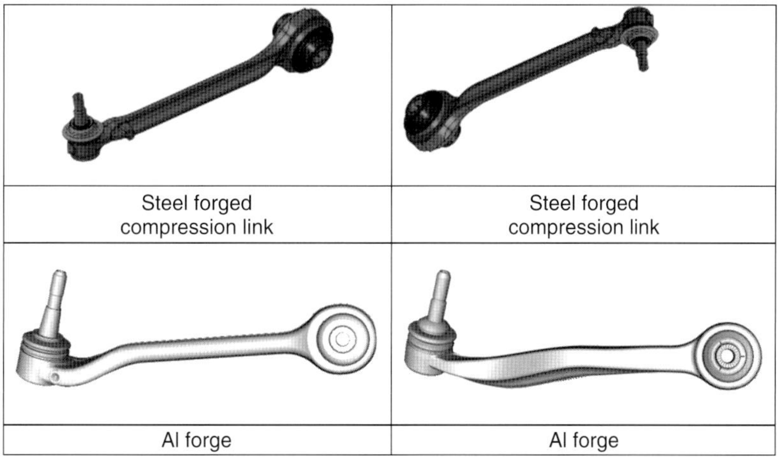

Figure 16. Examples: front lower compression and tension links.

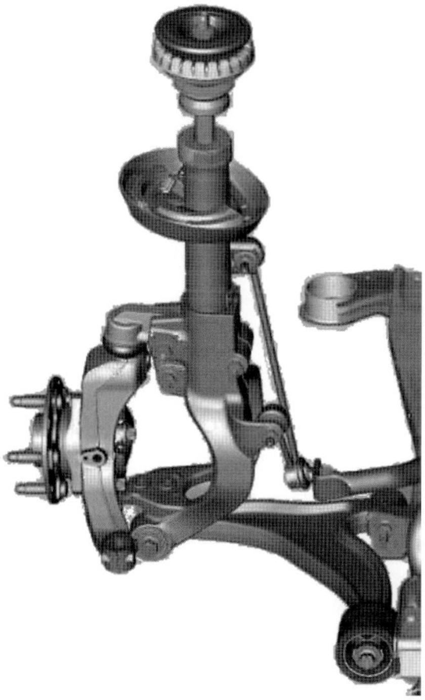

Figure 17. McPherson suspension with revolute joint.

A special modification of the McPherson suspension is the revolute joint (Figure 17), which improves the scrub radius and the king pin offset; it therefore can be used for high powered front-wheel-drive vehicles.

3.1.3 Suspension arms for double wishbone

The double wishbone suspension or SLA (short long arm) is typically used on SUVs, pickup trucks, and luxury passenger cars. The big advantage of this suspension is that the shock absorber is decoupled from the wheel guiding function. The lower control arm carries the vertical forces in addition to driving loads and the upper control arm on an SLA suspension has to provide clearance for the strut mount.

3.1.4 Lower triangle arm, A-arm, and V-arm

Lower arms can be made out of DI, steel forging, aluminum casting, or complex multipiece stamping. All of those have to provide attachments for shocks and springs, which can be torsion bars or coil springs (Figure 18).

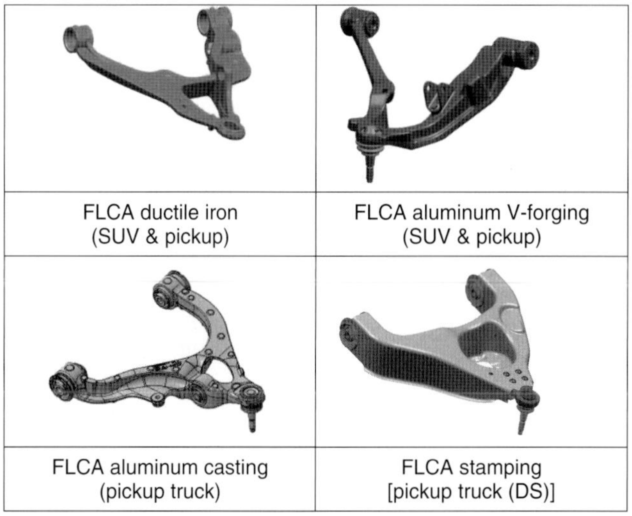

Figure 18. Example: SUV front lower control arms of double wishbone suspension.

3.1.5 *U-shape arm*

The upper control arm of a double wishbone suspension (Figure 19) has a typical "U-shape" to provide clearance for the strut assembly, which makes it sometimes difficult to achieve the required stiffness values. Owing to the position of the arm, the vehicle loads are low, but the arm has to be built short to allow space for the engine compartment. Therefore, the ball joint and the bushings have high requirements regarding working angles especially on SUV and pickup trucks.

3.1.6 *Separated arms (preloaded arm and wheel-guiding arm)*

Similar to the strut suspension with two arms, a double-wishbone axle can be designed with two arms on the lower and/or upper level for the same reason (Figure 20).

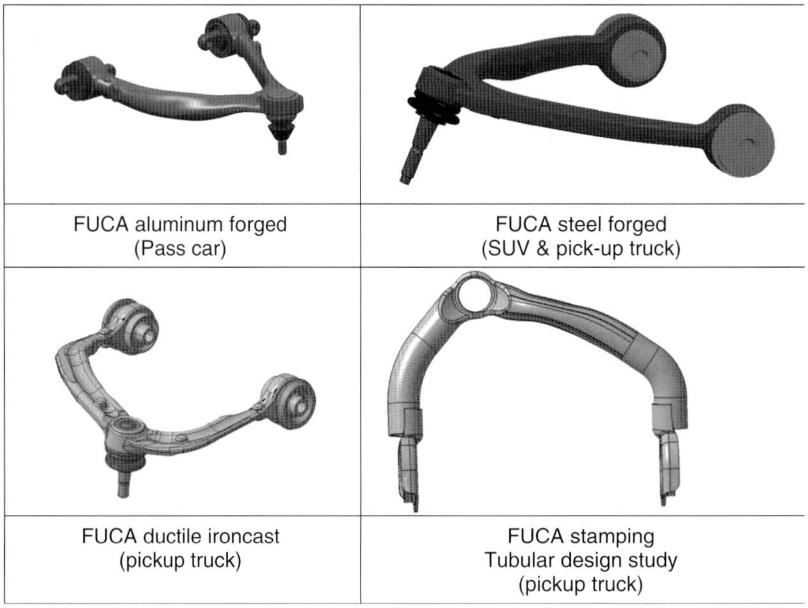

Figure 19. Example: front upper control arms.

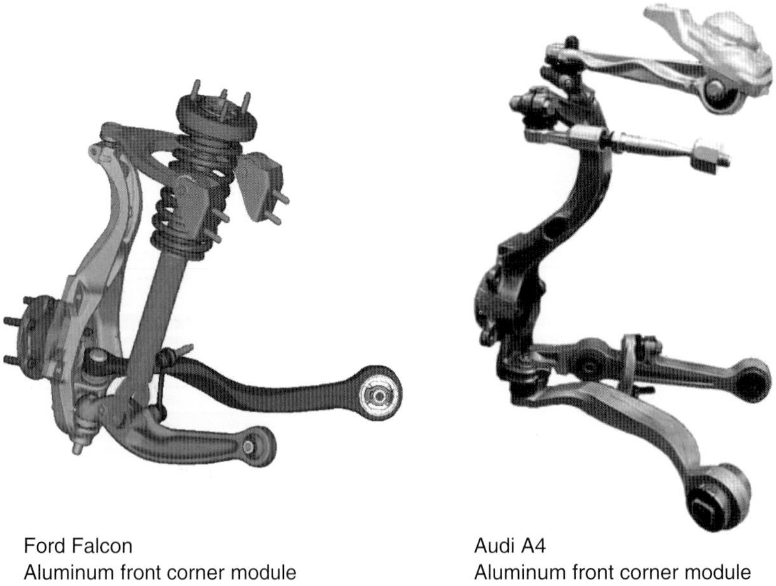

Figure 20. Examples: multilink front axles.

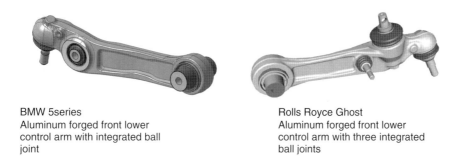

BMW 5series
Aluminum forged front lower control arm with integrated ball joint

Rolls Royce Ghost
Aluminum forged front lower control arm with three integrated ball joints

Figure 21. Examples: preloaded suspension arms.

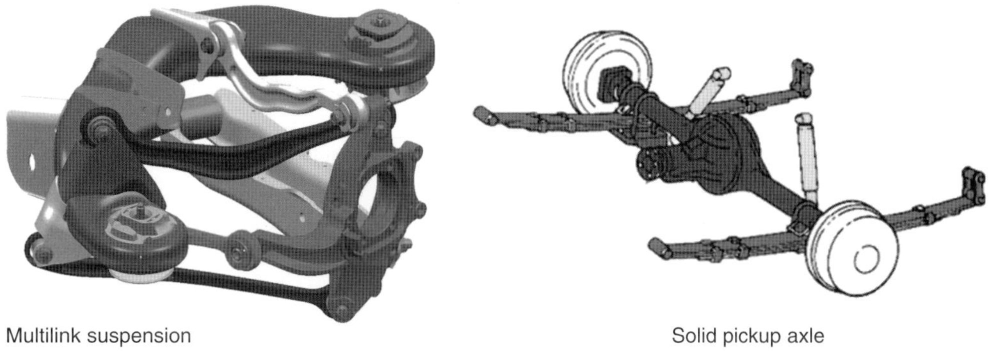

Multilink suspension

Solid pickup axle

Figure 22. Rear axle types.

3.1.7 Preloaded suspension arms

For multilink axles, one of the control arms has to carry the load of the spring/strut as a permanent static preload (Figure 21).

3.2 Rear axles

The typical modern suspension can be a multilink rear suspension, for smaller vehicles a twist beam axle and for pickup trucks – due to high payload – a rigid rear axle (Figure 22).

3.2.1 4-point suspension arm

Especially, for premium rear axles with higher loads, we can often find suspensions with a 4-point suspension arm. Owing to the size of the arm, there are a lot of aluminum solutions available for this kind of control arms. Owing to high stiffness requirements and complex load conditions, the hollow cast solution offers the best weight-saving potential (Figure 23).

4 SUSPENSION ARM TECHNOLOGIES

Today's independent passenger car and truck suspensions use either ferrous material or aluminum. Within those materials, forging, casting, and stamping are the prime choices in ferrous design; casting and forging, potentially extrusion, are the choices in aluminum design. Castings, especially new riser-less technologies, provide freedom to design sections on the suspension with a minimum of unused material (sprue and flash). Forging and sheet metal can provide higher strength materials, but typically the rate of nonutilized material is higher than with casting technologies (flash, trim-off, and raw part net weight vs weight of the forged part).

The different technologies stay in constant competition; the technologically and commercially best solution has to be found for each application, based on vehicle goal and target requirements (Figure 24).

4.1 Integrated ball joint

What is special about suspension "die-based" technologies such as forging and casting is the possibility to integrate the ball joint directly into the structural component element. Although this is a common technology, ZF (Figure 25)

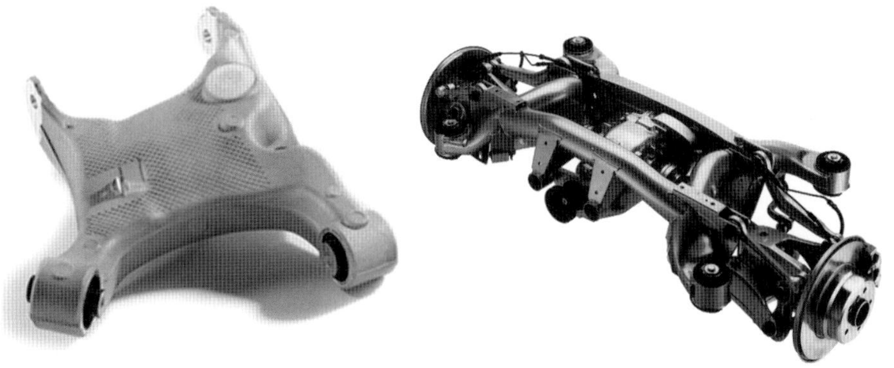

Figure 23. Four-point suspension arm. (Reproduced with permission from ZF.)

	Weight	Costs	Package
Forged aluminum	+++	O	+
Cast aluminum, LPPM	++	+	O
Special cast aluminum processes	++	++	+
Forged steel	O	++	++
Ductile iron (*)	O/–	+++	++
Sheet metal integrated control arm (SMICA)	+	++	+
Sheet metal CA + press-in ball joint	+	++	O
Sheet metal CA + flange ball joint	+	+	+

(*) Crossover, SUV, pickup truck / passcar

Figure 24. Technology matrix.

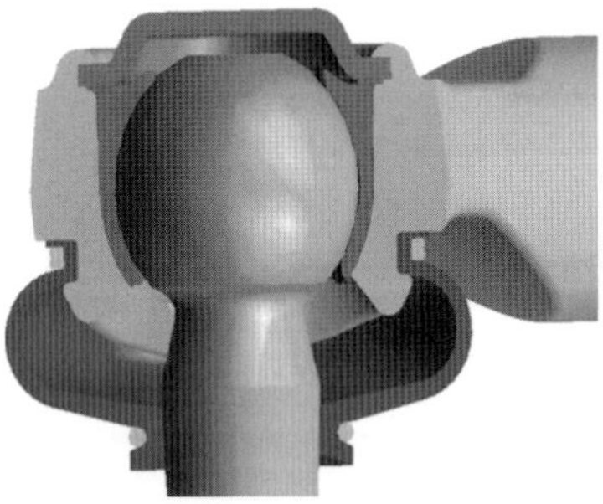

Integrated ball joint advantages:
Packaging, weight, cost, low tolerances, and low torque

Figure 25. Integrated ball joints (ZF Friedrichshafen AG). (Reproduced with permission from ZF.)

has improved the functional performance of such joints to implement low or, if needed, high motion torques with small tolerance range on the one hand and lifetime wear resistance in each driving condition on the other hand. Furthermore, the position of the hard points to each other is kept extremely tight with minimum tolerances. Special ball joint types cover different needs such as suspension- or compression-loaded variations.

In general, this is different to sheet metal, which requires a separate bolted, riveted, or pressed-in ball joint solution. It is obvious that an integrated ball joint has advantages compared to others. Pressed-in ball joints can be just right from a functional performance point of view but, when assembled, the press fit can have negative impacts on the motion torque of the ball joint. This is caused by the design of the press fit with the respective tolerances and the selected stiffness of the hub. A press fit has to be chosen to maintain a minimum of press-out to withstand vehicle loads, which can as a result double or triple the motion torque of the ball joint.

To provide the advantages of integrated joints to sheet metal control arms as well, ZF has developed and patented

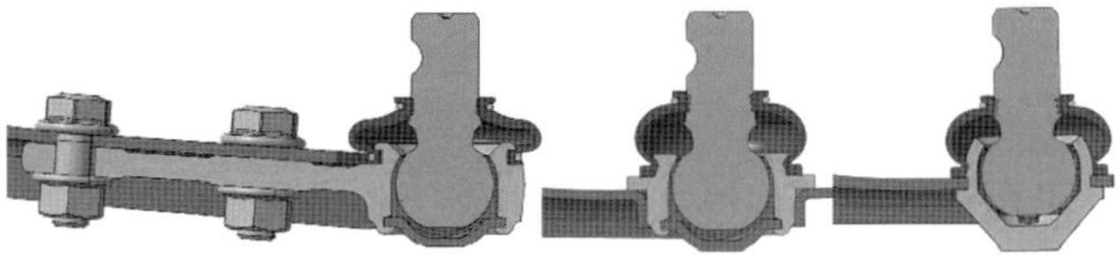

Figure 26. Ball joint types for sheet metal control arms.

the so-called sheet-metal-integrated control arm (SMICA) solution. The unique solution with the robustness of a conventional integrated joint contains a complete ball joint with a steel housing, welded by laser as a cartridge into the arm. The advantages of this technique are a smaller package space to the knuckle, less parts, and less weight compared to all other solutions; this brings the steel solutions a little closer to the weight saving that light metal solutions can offer (Figure 26).

4.2 Ferrous material

4.2.1 Castings

Like all casting technologies, DI casting gives a lot of freedom in design (shape, change of cross sections, and potentially coring). The freedom to distribute material where it is needed in a shape that is most efficient to support loads or the stiffness requirement is the big advantage of the casting process. All this is possible without the need of putting too much material in flash that has to be trimmed, which results in a good material usage. Furthermore, small draft angles can be realized and portions can be cored to reduce the raw material usage.

The typical DI alloy used in suspension components is DBC 450 (Rm = 450 MPA), optionally DBC 550 (Rm = 550 MPA), which both offer a compromise between strength and ductility.

SiboDur® is a material for use in chassis components subject to high levels of stress as typically seen in control arms and steering knuckles. SiboDur has been developed by Georg Fischer Automotive on the basis of spherical graphite cast iron. The name SiboDur is derived from the two materials silicon and boron and from the English word durability. SiboDur makes it possible to produce lightweight cast iron parts.

Tempered ductile iron (ADI) is a heat-treated form of cast DI. It improves the strength of the DI and maintains the ductility to a certain degree. Sizes of the parts and packaging densities in the heat treatment furnace have to be considered to develop a cost-effective part.

Overall, with its flexibility in design, its potential to core certain areas, and considering its mechanical properties, DI offers a high potential to design chassis components such as control arms and knuckles efficiently and cost effectively.

4.2.2 Forgings

4.2.2.1 General. Steel forging had been the most common production process for control arms for decades until the weight factor became more and more demanding. Even today, it still can be efficiently used in axles with specifically tight package areas and high load requirements leading to mostly locally concentrated yield strengths way above 360 MPa or ultimate 400 MPa and up.

The damper and spring as well as the spring tower for the upper arm and the lower arm, especially with separated arms on the outbound ends of the control arms – because of the additional ball joint and the special kinematics with a virtual center for the steering axis, limit the package on double wishbone corners. At the inbound end of a control arm, it is sometimes tight for the bushing bosses in combination with large bushing diameters driven by desired stiffness and allowable space, limited by available space to the subframe. Further examples are on lower L-arms with a spike pin bushing attachment; the transition from the pin to the structure of the arm is a weak point for bending stress. This can be handled more easily with forgings. In general, it is a big benefit if the packaging and the loading are specified correctly with little potential of change during the program: so a decision for a particular material and process selection can be made with high confidence.

Even though – from a functional standpoint – quenched and tempered material is generally more desirable for a control arm design because of its high ductility and more preferable impact behavior, it has been nearly completely eliminated from the control arm market. As the 1990s, microalloy steel has been the only option in use. Inventions on these specific material types were driven in the 1970s already, especially because of their cost efficiency for crankshafts. The cost efficiency lies in the forge–temper–rework process chain: the final raw part reaches its mechanical properties just by controlled air cooling on a conveyor after forging and trimming with least thermal. Additional advantages are a delayed tendency to crack under fatigue life, a far better machinability because of its specific grain structure and a smaller border-to-core hardness decrease compared to quenched and tempered material. However, the disadvantages are a faster crack progress, once a crack is induced, and a worse notched-bar impact works (Wegener, 1998). So, it is possible to meet the demanded combination of best function and least possible cost.

The most popular alloy in Europe is 30MnVS6 +P, in the typical range 560-MPa yield and 800-MPa ultimate

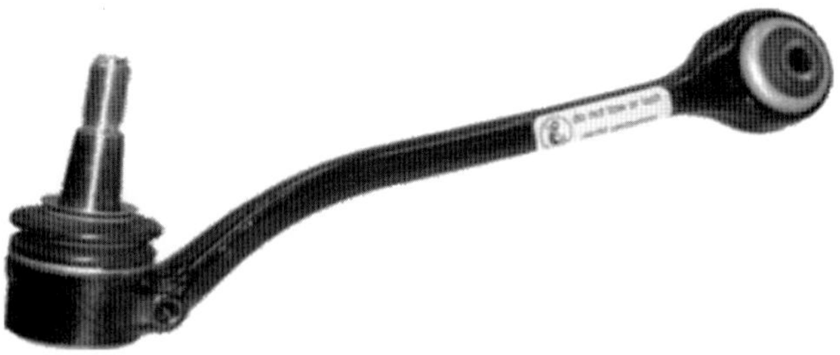

Figure 27. Two-point link in steel forging technology for BMW (ZF Friedrichshafen AG). (Reproduced with permission from ZF.)

strength. The alloy is also available on the Asian and American market or can be purchased with very close similar chemical properties such as 15V24.

The process and the cost with some major aspects drive the forging design. Two-point arms with straight and easy routing as well as small cross sections with easy geometry are the most desirable ones. The more complex the geometry gets and the wider the projected area in forging direction is, the less likely it is to find a cost-effective design (Figure 27). If steel forging is needed in particular sections of the arm, then large arms will be hybrids out of sheet metal or aluminum-plus-steel combinations.

The material consumption on a steel-forging arm surely depends on the arm's size and complexity, but there is a general disadvantage with this process. The flash extension required to fill the cavity of the mold properly is large compared to that required in casting processes. It is extraordinary on flat arms with a wide extent but significantly lower on straight thin arms. A rate factor larger than 2 for the operational weight in proportion to the final part weight is borderline in terms of cost efficiency. In addition, the larger and more complex the arms are, the bigger is the height and mold shift tolerance impact with its effect on the function of the part, like buckling. Aluminum forgings of a similar size and complexity are easier to control, compared to steel (Figure 28).

4.2.3 Sheet metal (1-shell, 2-shell)

Sheet metal is widely used in vehicle suspensions and on all types of vehicles from passenger cars over SUVs up to pickup trucks (Figure 31). Depending on load case, the stampings can be single shell, double shell, or fabricated and welded together out of multiple different sheet

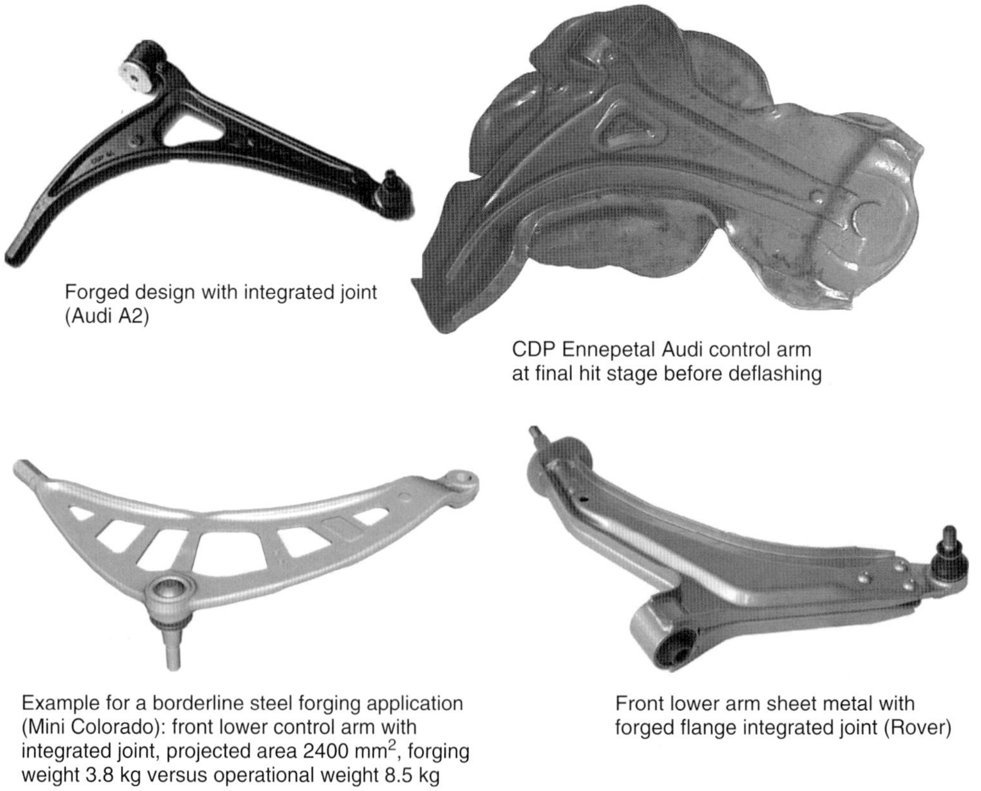
Forged design with integrated joint (Audi A2)
CDP Ennepetal Audi control arm at final hit stage before deflashing
Example for a borderline steel forging application (Mini Colorado): front lower control arm with integrated joint, projected area 2400 mm^2, forging weight 3.8 kg versus operational weight 8.5 kg
Front lower arm sheet metal with forged flange integrated joint (Rover)

Figure 28. Technology range for suspension arms.

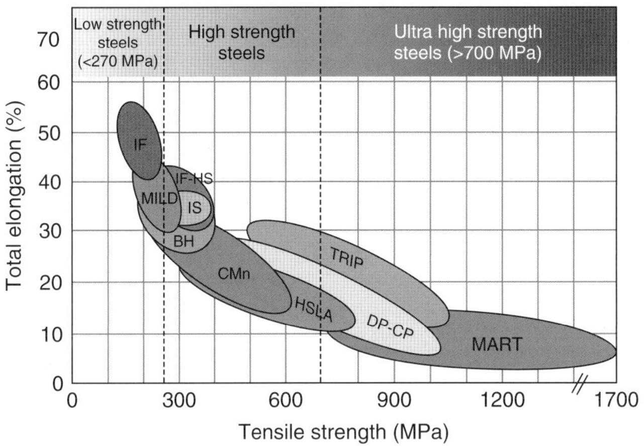

Courtesy - American iron and steel institute

IF	Interstitial free
MILD	low carbon steel, for example, 1008 and 1010
IF-HS	High strength interstitial free
BH	Bake hardenable
CMn	Carbon manganese
HSLA	High strength low alloy
TRIP	Transformation-induced plasticity
DP	Dual phase
CP	Complex phase
MART	Martensitic

Figure 29. Sheet metal technology range. (Adapted from Shaw, 2003. Reproduced by permission of the American Iron and Steel Institute.)

metal parts and reinforcements. The more complex the stamping gets, the more efficient it might be to substitute it by a casting. During the past years, new higher strength sheet metal alloys with good formability have been developed to enable lightweight stampings (Figure 29). This development was feasible because of dual and multiphase steel alloys that provide a soft ferrite grain matrix with a bainitic or martensitic second-phase deposit at the grain borders, which work hardens through the plasticization of the grain (ThyssenKrupp Steel AG, 2008). Although such applications are in production, there are still some huge disadvantages: high steel cost, lack of steel market volume, and missing availability on the global market. In view of increasing global platforms, it appears to be more advisable to accommodate widely used, common alloys of the range up to 600-MPa UTS.

ZF investigated the possibility to produce a single-shell L-arm in a high strength steel grade for vehicles in the C/D class. The aim of this study was to avoid any welds with the exception of welds needed for the ball joint integration. The study showed that single-shell control arms can be utilized up to typical loads for C-segment vehicles; formability of

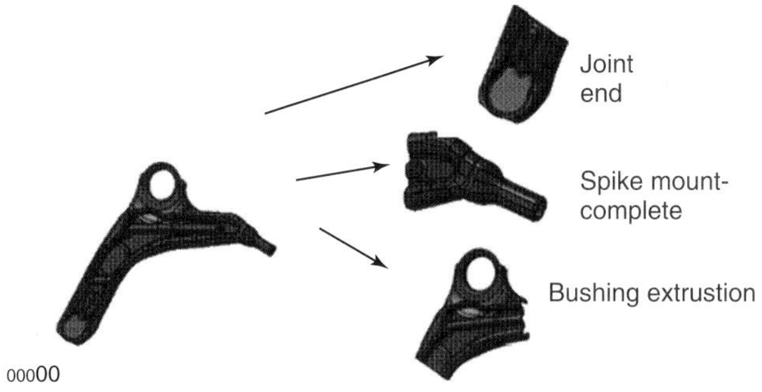

Figure 30. Study of a single-shell L-arm.

the boss hole for the vertical bushing plus the forming of the pin for the rear facing bushing could be realized. Furthermore, the production process could be optimized so that the part can be coated after stamping and then all additional processing can be done in the production line (ball joint integration and bushing assembly). The welding method to integrate the ball joint was developed, so that the ball joint can be integrated at precise position without any negative effects on the ball joint torque and performance and with minimum space requirements to the brake disk and knuckle.

Figure 30 demonstrates the steps of the study: to assure that the single-shell L-arm was feasible, the most critical areas regarding the degree of deformation such as ball joint end, spike mount, and bushing extrusion have been developed as sections. After those successful trials, the whole control arm was formed to demonstrate design feasibility (Figure 31).

4.3 Aluminum

4.3.1 Aluminum casting

Despite intense research and in contrast to sheet metal technology, the alloy used for aluminum suspension parts has not changed over the past 10 years and is out of the 3xx.x series with the elements silicon, copper, and/or magnesium; commonly A 356 is used with the T6 heat treatment (solution heat treated and artificially aged). Even though the casting alloy for control arm applications has not changed, a development process took place on the casting production process side: a variety of performance casting processes has been developed such as squeeze casting, VRC/PRC, Cobapress™, Vacural®, CPC, and others. These performance castings could be used in the chassis application as they provide low porosity parts with good mechanical properties and sufficient elongation values (T6 heat treatment).

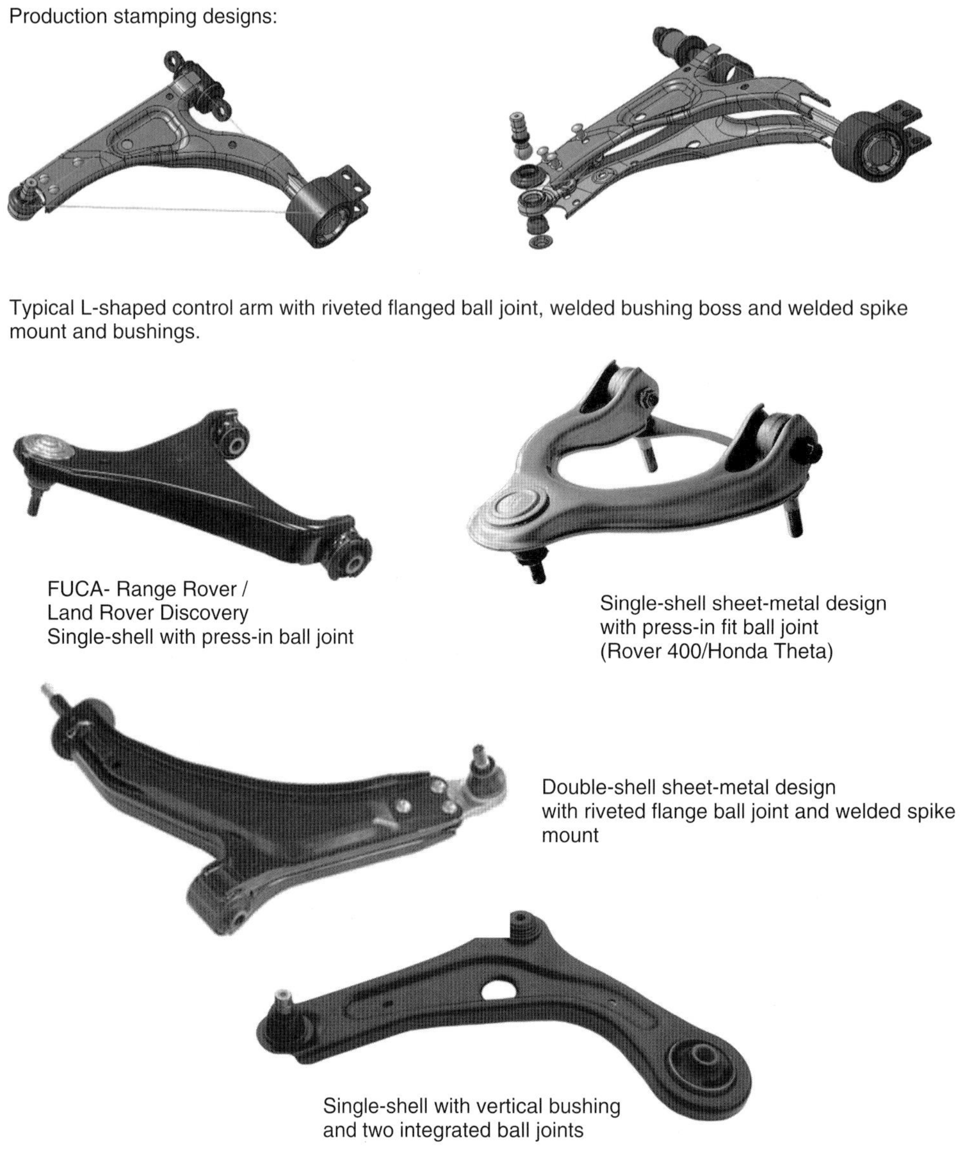

Production stamping designs:
Typical L-shaped control arm with riveted flanged ball joint, welded bushing boss and welded spike mount and bushings.
FUCA- Range Rover / Land Rover Discovery Single-shell with press-in ball joint
Single-shell sheet-metal design with press-in fit ball joint (Rover 400/Honda Theta)
Double-shell sheet-metal design with riveted flange ball joint and welded spike mount
Single-shell with vertical bushing and two integrated ball joints

Figure 31. Examples for sheet metal control arms.

Owing to the need to further reduce weight, an increasing focus on the potential for hollow cast is seen; currently, different chassis parts such as control arms and knuckles (Figures 32 and 33) are under development. The usage of cores allows a box section in the part that gives an ideal shape for stiffness (Figure 34); in addition, local reinforcements can be casted into the part. Furthermore, reduced heat transfer to the core may allow thinner walls and may reduce mold complexity as no mold pulls are required; designs may be possible that cannot be realized with pulls. In addition, a core can be designed without a draft angle or even undercut, which would create mold lock in a typical casting process.

4.3.2 Aluminum forging

4.3.2.1 General. Aluminum forgings do not share the fate with steel forging, which is more and more decreasing. As weight has become more important and therefore strength and ductility, aluminum is in favor. The market share of aluminum forgings is still increasing; an aluminum forging is the best functional solution with the best process capability. Despite the fact that aluminum forgings come with relatively high costs, they are still very popular at OEMs especially for front axle applications, as casting processes require expensive quality controls such as X-ray checks.

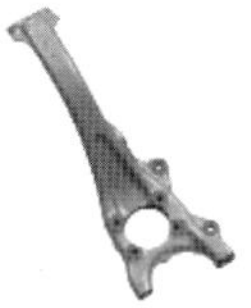		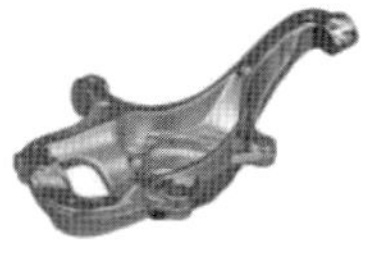	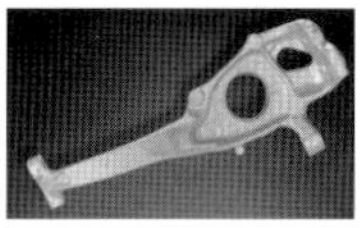
Coba press	Squeeze casting	Vacuum riser-less casting / Pressure riser-less casting	Counter pressure casting

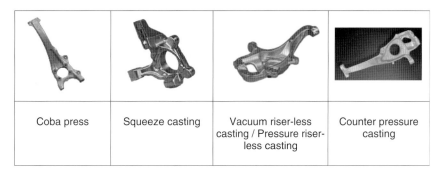

Figure 32. Example Knuckles in Al casting.

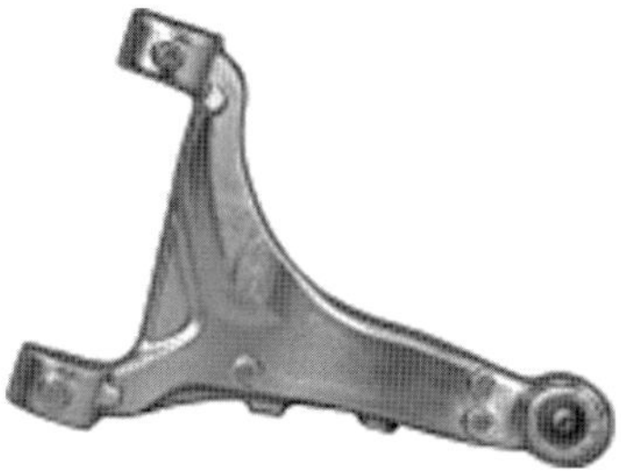

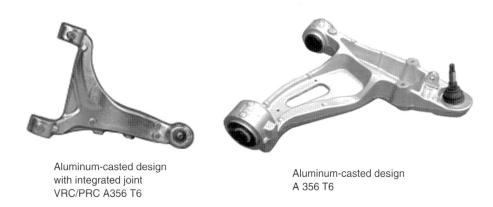

Aluminum-casted design
with integrated joint
VRC/PRC A356 T6

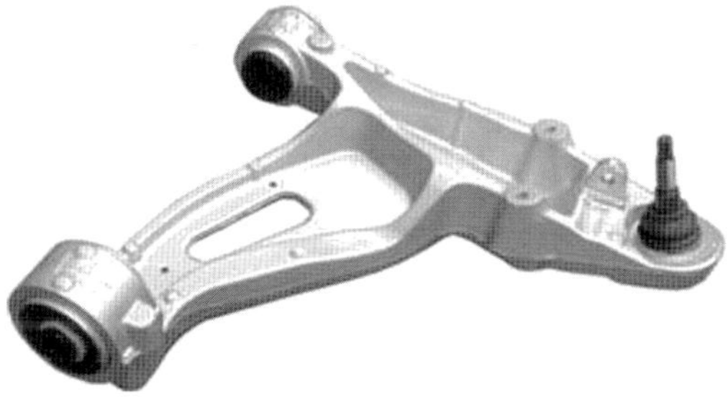

Aluminum-casted design
A 356 T6

Figure 33. Examples: suspension arms in cast technology.

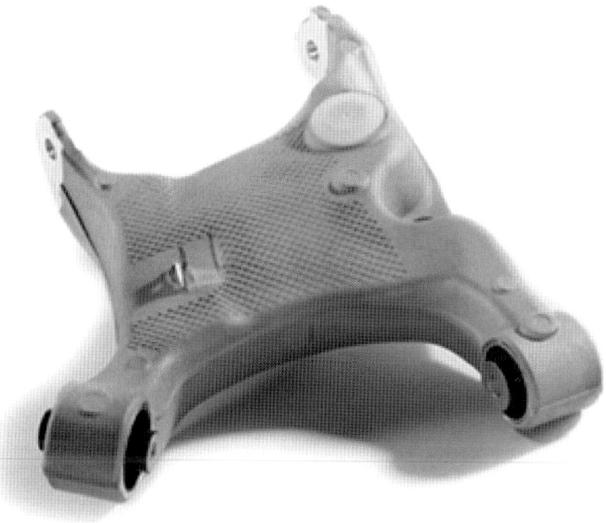

Figure 34. Example: BMW X5 RLCA: VRC/PRC with sand core (hollow cast).

If the design constraints described in the steel chapter allow for softer material, then aluminum forge will be the best choice of all, if the comfort ability of best fit and function can be afforded. What has been said about steel forging is also valid for aluminum: the production process is very stable and reliable; high end forging is for instance controllable by fully automated high speed lines.

Compared to steel, aluminum material provides some additional advantages beside weight reduction, which support the decision to use it. Very obvious is the optical appearance, a dull or even shiny silver like surface with smooth roughness. The functional aspect of the smooth surface is the lack of roughness notches acting as stress raiser on fatigue life. However, the disliked effect with aluminum is that it does not show a constant long life fatigue strength level. Therefore, assumptions need to be made like setting a minimum cycle number of 2×10^6 for long life criteria.

Similar to casting materials, forging material ranges of the alloy SAE J 454–6061 (yield 260 MPa, UTS 300MPA, and elongation 10%) and SAE J 454–6082 (yield 310 MPa, UTS 340MPA, and elongation 10%). Depending on the specific company alloy composition and heat treatment, a level of yield 390 MPa with UTS 410 MPA is possible especially for two-point links.

It is very important for aluminum arms to maintain the rule of providing as little coarse grain as possible.

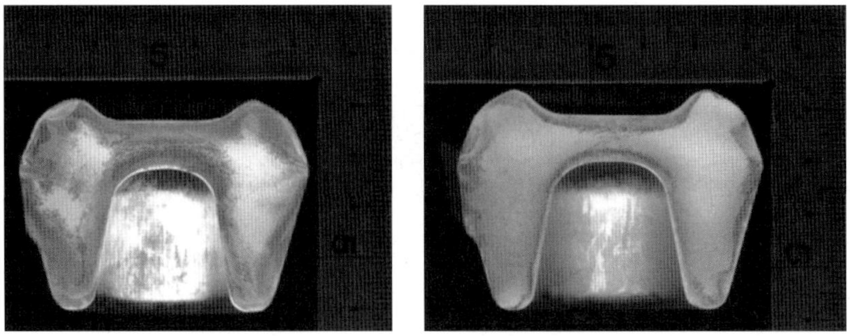

Example for coarse grain deep through the cross section.

Example for coarse grain mostly at the skin only.

Figure 35. Cross section of forged aluminum suspension arms (grain structure).

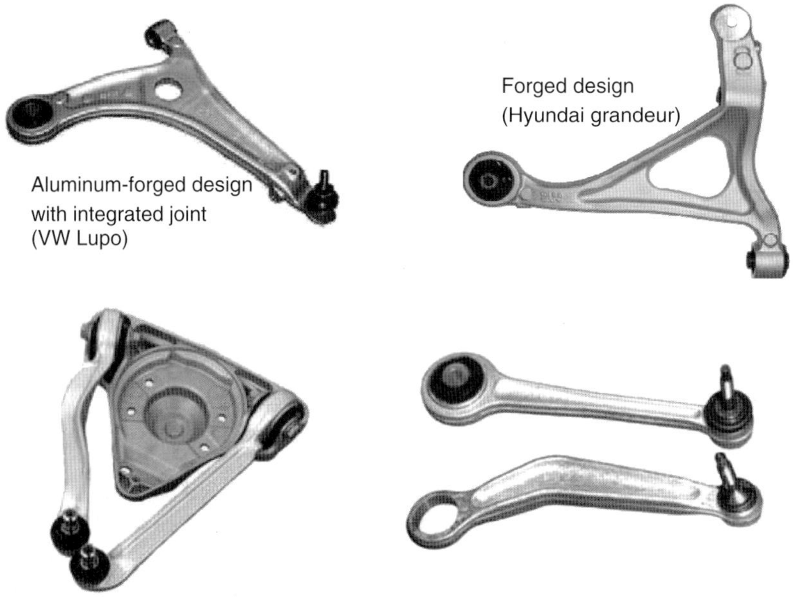

Front upper control arm module in aluminum-forged design (and casting hub)

Rear axle control arms in aluminum forging

Figure 36. Examples: aluminum forging.

On typical arms, only a thin layer of coarse grain is allowed at the skin to enable the desired deforming and fatigue life behavior as well as the formability after machining – allowing the integration of ball joints (Figure 35).

In addition to the description in the design chapter, the aluminum forging process generally provides the same restrictions as the steel forging process. The advantages versus steel are hidden in details such as smaller draft angles or sharper radii. The loss of material in the flash extension is less (thinner), and, as a rule, the scrap material is completely used for company-internal recycling, as mentioned in chapter 2.1.

5 DESIGN MATRIX (JUDGMENT OF TECHNOLOGIES FOR DIFFERENT SUSPENSION ARM TYPES)

5.1 L-shape front lower control arm for McPherson suspension

L-shape control arms are currently the preferred FLCA (front lower control arm) for McPherson suspensions; they are typically designed to lateral and traversal stiffness, ground clearance, crashworthiness, clearance for tire envelope and drive shaft, stabilizers, and so on – Owing to those requirements with regard to space and performance,

Stress: $\sigma_{max} = \frac{M}{W_x}$

Section modulus for A, B, C: $W_x = \frac{1}{6H}(BH^3 - bh^3)$

Profiles considered with reference to x-axis

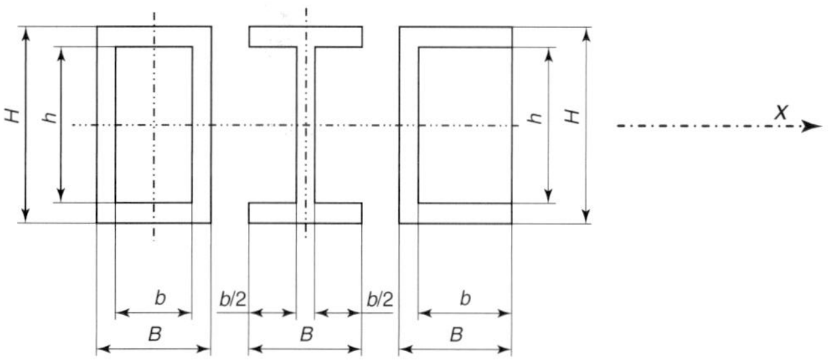

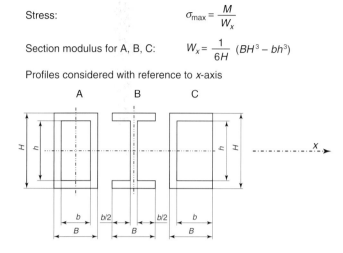

Figure 37. Shape (cross section) of control arms.

not only one material or process is applied. L-arms are rather produced in aluminum and steel forged, stamped, or casted parts.

The following arm was designed as double-shell control arm during early development stages with regard to fore-and-aft stiffness and lateral stiffness. Considering the fore-and-aft stiffness in a simplified way, the design engineer, while keeping the bending stress in mind, has to find the best section modulus and therefore the best cross-sectional shape suitable for the intended manufacturing process.

The different cross sections in Figure 37 have the same sectional modulus (based on reference to the x-axis) and therefore the same maximum stress and the same weight. This means that the H profile, typical for casting processes, and the box profile, typical for double-shell stampings, can be designed competitively for the relevant processes.

The following front lower control arm (Figure 38) shows a production design for a high volume application. The numbers below show the ranking – "1" for best in this category down to "4" worst in this category. The design was in competition with a double-shell stamping. Before deciding for this aluminum performance casting, the following other options have been considered by the vehicle manufacturer.

Fulfilling customer requirements for stiffness, packaging, and durability, the above table shows the performance of the different materials. In this case, a performance aluminum casting process (vacuum riser-less casting/pressure riser-less casting) was chosen for series production, as it offered the customer a weight advantage at a reasonable cost.

5.2 Front upper control arms

Front upper control arms for a double wishbone suspension with a long knuckle (goose neck) are currently the typical design, because of space and kinematic advantages over the McPherson strut suspension. The upper arm is typically designed to allow space for the strut module assembly; it provides short length to limit cross car space and for stiffness. Furthermore, the arm has to allow for jounce and rebound travel and tire clearance. The following example is taken from a design competition against a steel forging.

For the new vehicle, requirements have been updated regarding lateral and fore-and-aft stiffness; different designs in stamping, DI, and aluminum have been analyzed including a ranking (see above) (Figure 39).

In this case, none of the investigated parts could fulfill the upgraded stiffness requirements – even not the incumbent

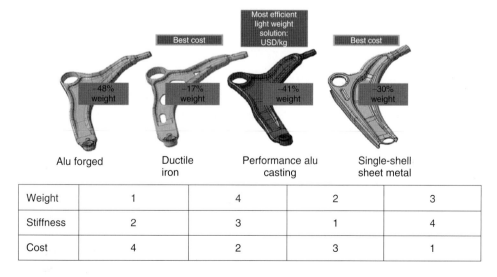

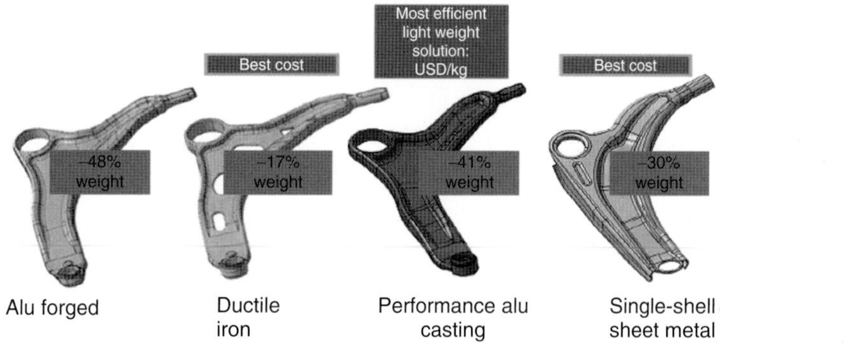

Weight	1	4	2	3
Stiffness	2	3	1	4
Cost	4	2	3	1

Figure 38. Technology range for FLCAs.

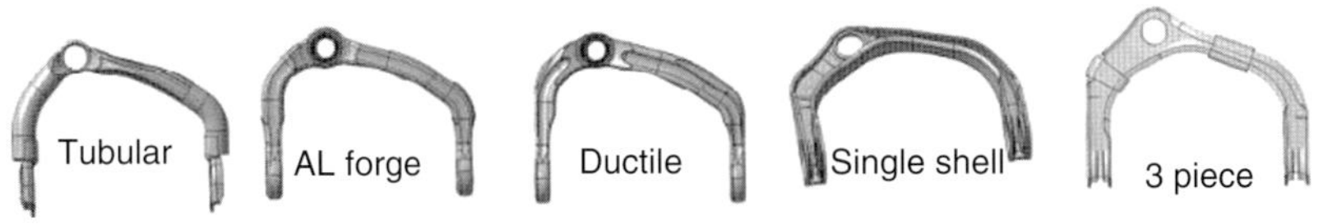

Weight	2	1	3	2	4
Stiffness	2	3	1	4	2
Cost	4	2	1	3	5

Figure 39. Technology range for FUCAs.

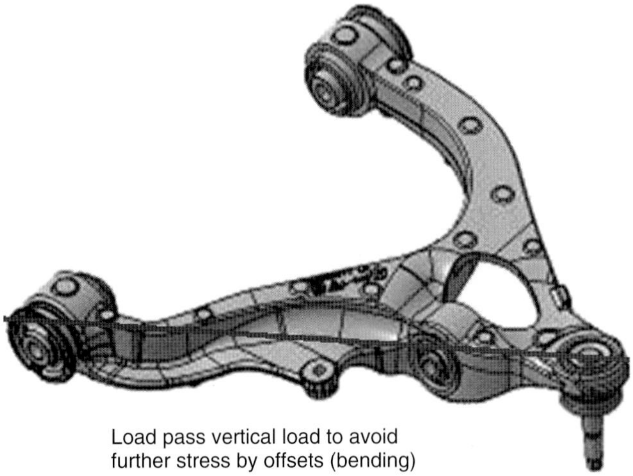

Load pass vertical load to avoid
further stress by offsets (bending)

Figure 40. Example: A-shape FLCA.

forged steel design. With reduced stiffness, it appears that a stamping or an aluminum cast could provide the best compromise between cost and desired performance.

5.3 A-shape front lower control arms (FLCA) for double wishbone suspension

These are in the market in all different materials and processes. Besides all the requirements previously discussed for the L-arm, FLCAs of a double wishbone suspension have to additionally carry the vertical force. The following suspension arm (Figure 40) was developed for a pickup truck. The first generation had a multiple piece double-shell stamping design, which was – for cost reasons – replaced by a DI control arm with torsion bar attachment. The next model changed from a torsion spring to a normal strut-mounted coil spring (due to crash performance), and the design was carried out in DI as the previous arm. During the last refresher, weight and fuel economy came into focus – a lighter design had to be developed for nearly identical requirements as to loads, clearance to tires, and drive shaft. As the DI cast part had already been optimized for maximum allowable packaging space and clearance, the aluminum casting could only be designed in the same space. Moreover, material could

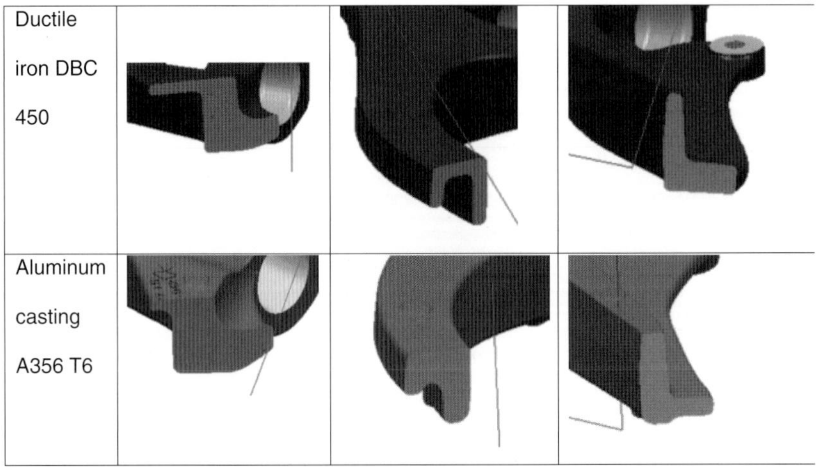

Figure 41. Cross-sectional variation possibility in casting process – ductile iron versus aluminum casting.

only be added in areas that are not as efficient to create force-carrying capability (Figure 41).

The development history shows that for similar applications different processes and materials have been in production, based on overall vehicle goals, which resulted in aluminum casting with the main focus on weight at reasonable costs. With the replacement of DI – which had been the cost effective solution – by aluminum casting, a weight saving of 5.3-kg per control arm was achieved, resulting in an overall weight saving of ca. 40% for the complete assembly.

6 SUMMARY AND OUTLOOK

Suspension arms, steel versus aluminum, where are the benefits? Coming back to the initial question, this chapter shows the overall tendency that aluminum applications offer the best weight whereas steel offers best cost. Owing to the fact that designs for suspension arms are often driven by stiffness and/or buckling requirements, the individual solution might lead to different results – so in the end, it depends on the application. For other chassis parts, for example, knuckles, the direction is more constant.

Sheet metal solutions already have a big market share and, based on ultrahigh strength steel, there is further weight reduction potential. Key for a good sheet metal design is the design of the weld seem. Or, the other way round, the best material properties do not support a lightweight design if we have to weld too much in critical areas. Owing to cost advantages and the possibility of weld reduction, one-shell solutions (if feasible) seem to be most interesting for sheet metal solutions.

Steel forging has more and more become a niche process for volume products. For heavy and small packages and high loads (heavy vehicles) it will remain an important technology for suspension arms.

DI solutions are often seen as old-fashioned heavy solutions. This chapter has shown some examples where DI is the best option or compromise among stiffness, package, and cost. There are alloys on the market that even offer very good elongation properties to fulfill the buckling/bending requirements of automotive industries. Depending on the market situation, DI can lead to the best-cost solution at a competitive weight.

Aluminum forging seems to offer low weight but high cost. Competitive costs can be reached for designs with a good material utilization of the aluminum preform to avoid as much machining as possible and as much aluminum scrap due to flash as possible. The machining of aluminum is cheaper than the machining of steel, and in some chases, this more or less compensates for the higher material costs for aluminum. There are materials on the market with much higher strength than standard steel.

High performance aluminum cast solutions and combined cast forge solutions are becoming more and more interesting. The market offers an increasing number of special production technologies that support a good stable quality of the material and material properties with high strength. This leads to best-cost lightweight solutions. Depending on the axle concept, we have to pay special attention to the overall crash concept because of limited elongation properties of cast solutions.

Owing to the upcoming endless range of fiber-reinforced plastic applications especially for vehicle bodies, the interest in plastic applications for chassis components has increased, too. There is still some skeptics regarding plastics for chassis application, but it becomes a more open discussion. For some chassis parts, there are already solutions in series production, for example, the leaf spring for GM Corvette or Volvo 940 and the anti-roll bar links for several BMW and GM vehicles. The complex structure of suspension arms with their high load and crash requirements make it much more difficult to substitute traditional suspension arms. So to overcome hurdles and find different axle layouts that better support plastic applications will take some time.

In other words, the question of "Suspension arms, steel versus aluminum, where are the benefits?" will survive some more years.

REFERENCES

Adler, U. (1991) *Robert Bosch, Kraftfahrtechnisches Taschenbuch*, VDI-Verlag, Düsseldorf.

Beitz, W. and Grote, K.-H. (2001) *Dubbel, Taschenbuch für den Maschinenbau, C28, C43*, Springer, Berlin, Germany.

Cheah, L., Heywood, J. and Kirchain, R. (2010) *The Energy Impact of U.S. Passenger Vehicle Fuel Economy Standards*. http://web.mit.edu/sloan-auto-lab/research/beforeh2/files/IEEE-ISSST-cheah.pdf (accessed 08 January 2014).

Dubreuil, A., Bushi, L., Das, T., *et al.* (2010) A comparative life cycle assessment of magnesium front end autoparts. SAE Paper 2010-01-0275. Society of Automotive Engineers: PA, USA.

Elbers, C., Ersoy, M., Hausfeld, G., *et al.* (2004) *Automotive Chassis Technology*, Vmi, Landsberg am Lech, Germany.

Göde, M., Stehlin, M., Rafflenbeul, L., *et al.* (2008) published online by European Conference of Transport Research Institutes (ECTRI).

Goede, M. (2007) *Contribution of light weight car body design to CO_2 reduction (Karosserieleichtbau als Baustein einer CO2-Reduzierungsstrategie)*. Presentation at 16th Aachener Kolloquium Fahrzeug- und Motorentechnik 2007, 9th October, Aachen, Germany.

Heidi B. and Thomas B. (2011) *Annual Report – The Aluminum Association*, http://www.aluminum.org (accessed 08 January 2014).

Heißing, E. (2011) *Chassis Handbook*, Springer Science+Business Media, Berling, Germany.

Shaw, J. (2003) *ULSAB-Advanced Vehicle Concepts ULSAB-AVC*, presentation to the American Iron and Steel Institute. http://www.autosteel.org/~/media/Files/Autosteel/Great%20 Designs%20in%20Steel/GDIS%202003/02%20-%20ULSAB% 20Advanced%20Vehicle%20Concepts.pdf (accessed 19 February 2003).

ThyssenKrupp Steel AG (2008) Dualphasen-Stähle DP-W® und DP-K®: Für die Herstellung komplexer hochfester Strukurelemente. http://www.thyssenkrupp-steel-europe.com/ upload/binarydata_tksteel05d4cms/39/88/78/02/00/00/2788839/ Dualphasen_Staehle_de.pdf (accessed 08 January 2014).

Wegener, K.W. (1998) Werkstoffentwicklung für Schmiedeteile im Automobilbau, *ATZ Automobiltechnische Zeitschrift*, **100** (12), 918–927.

Chapter 115

Performance Target Conflicts in Normal Tires and Ultrahigh Performance Tires

Reinhard Mundl[1] and Burkhard Wies[2]

[1]*Technische Universität Wien, Vienna, Austria*
[2]*Continental, Hannover, Germany*

1 INTRODUCTION

Few people are aware of the fact that we are surrounded in our daily life almost everywhere by tires. Although they are often perceived as commodity, we all rely strongly on the high quality of their performance. Beside the very basic safety-related tire properties of durability and robustness, there is a bundle of the so-called additional tire performances such as grip on all surfaces, mechanical and acoustic comfort, safe handling properties, and, last but not least, a long lifetime. In recent times—also driven by legislation such as the new EU tire label—additionally, low rolling resistance has become important as a means to reduce fuel costs and climate change.

Unfortunately, these entire tire properties cannot be improved simultaneously. Each change in the tire's design that leads to an improvement in one specific tire property often leads to a trade-off in another one. It is now the demanding job of the tire engineer to come up with a well-balanced optimum for all the tire performances to satisfy the needs of the customers. Intensive research in tire technology and a deep know-how is needed to identify innovations to shift conflicting tire properties to a higher level. This chapter describes in the following these conflicts and provides the underlying rubber physics and tire mechanics.

2 PRODUCT DEVELOPMENT STRATEGY BY ENHANCING TARGET CONFLICTS

At the very beginning of the development of a new tire line stands the definition of a performance requirement book, which is based on competition analysis, marketing needs, and customer—more specifically, original equipment—requirements. This "book" covers tire performances, which are in most cases listed in relation to a reference tire. The differences are quantified by the relative deviation in percentage. More than a dozen of these tire performances define the strengths and weaknesses, which represent the individual character of a tire line on the market.

As an example, see the multidimensional presentation of tire competitors in a simplified, the so-called, spider graph in Figure 1. It shows on each ray the respective relative

Encyclopedia of Automotive Engineering, print
Edited by David Crolla, David E. Foster, Toshio Kobayashi, and Nick Vaughan.
 ISBN: 978-0-470-97402-5
Also published in the *Encyclopedia of Automotive Engineering* (online edition)
DOI: 10.1002/9781118354179.auto012

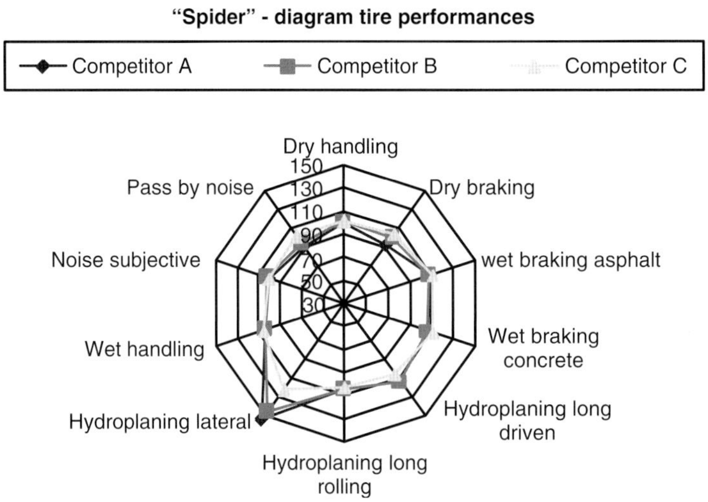

Figure 1. Multidimensional presentation of selected tire performances.

tire performance, thus delivering a specific picture of each competitor. The improvement of a single tire performance conflicts often with another one, making it very difficult for the tire developer to find a new optimum for all of them on a higher level. This sort of multitasking challenge can be easier overseen if one tries to achieve an improvement within a so-called performance conflict, neglecting the interactions to other performances. Owing to tire physics, conventional design changes of specific tire components cause only a shift along a so-called performance conflict line, thus improving only one performance but diminishing the conflicting one. This sort of adaption to changing targets is state of the art for an experienced tire designer, whereas innovative measures improve both performances at once and lead to a higher level perpendicular to the conflict line (Figure 2).

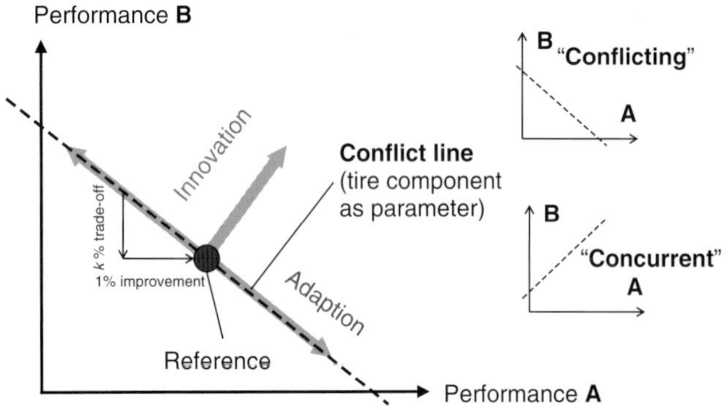

Figure 2. Schematic tire performance conflict diagram, distinguishing oriented measures of adaption and innovation and classification of dependencies.

3 SYSTEMATIC IDENTIFICATION OF TARGET CONFLICTS BY MULTIPLE PERFORMANCE ANALYSIS

From a mathematical viewpoint, if one has a number of n tire performances under investigation, a number of $n(n-1)/2$ target conflicts are possible. Regardless of the underlying physics, one can analyze the results of basic tire programs, where the influence of the change of a specific design parameter on a set of tire performances has been tested. This approach allows for rules of thumb to quantify the existing target conflicts. In particular, this leads to a ratio describing the decrease of performance B of $k\%$, when performance A is increased by 1%. In terms of mathematics, this ratio stands for the inclination k of the so-called conflict line to be seen in Figure 2. This conflict line can be extracted from the test results by applying linear regression analysis and is equivalent to the regression line.

Making these regressions, we can distinguish pairs of tire performances, which are *conflicting*, meaning that the inclination k is negative, and pairs of tire performances, which are *concurrent*, defined by a positive inclination, see again the visualization in Figure 2. It is noteworthy to mention that the described conflicting behavior has always an underlying change in a specific design parameter that can be positioned on the conflict line.

This sort of data mining has been done by Continental utilizing test results collected by the winter tire predevelopment department over almost one decade to get an overview on existing target conflicts for tread pattern variations (Figure 3). As mentioned at the beginning of this chapter, from the 13 tire performances investigated, 78 dependencies

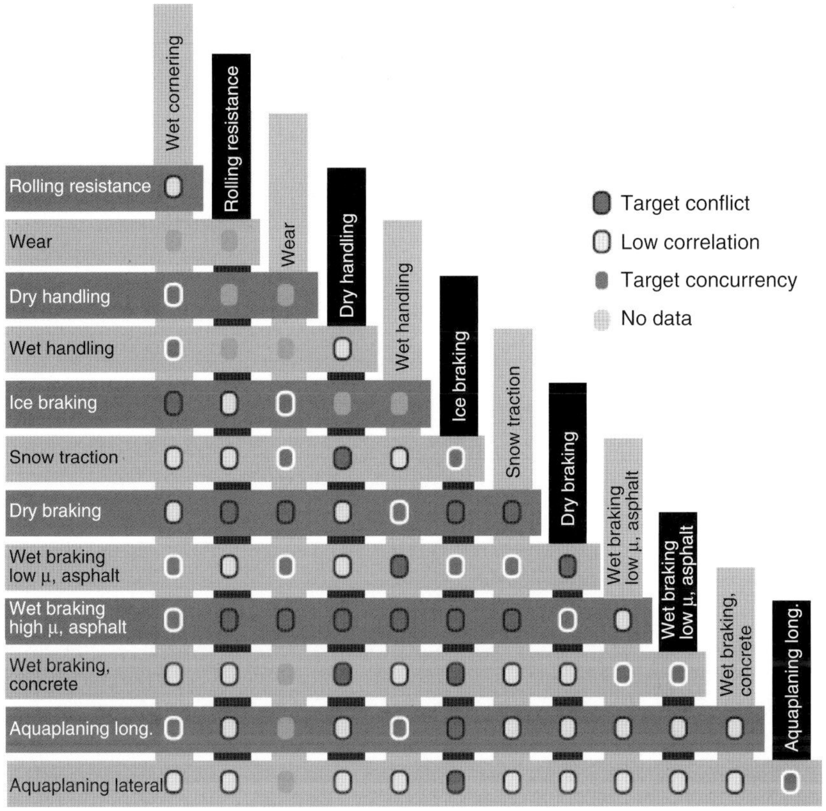

Figure 3. Overview of tire performance conflicts/concurrencies for winter tires.

were analyzed, out of which 18 are conflicting. An example for such a (pattern-inherent) performance target conflict, which can be explained by tread pattern mechanics, will be given in the following chapter.

4 DESIGN POTENTIAL OF MAIN TIRE DOMAINS AND RELATED TARGET CONFLICTS

Within a tire design, one can distinguish four main domains with significant influence on tire performances, the tread pattern, its compound, the outer shape of the cross section, named contour, and the tire carcass including the belts and cap plies (the latter domain herein summarized as construction). A qualitative estimation of the performance potential of design changes within these domains can be found in Heißing and Ersoy (2007) and is to be seen more detailed in Figure 4.

4.1 Tread pattern

From Figure 4, it can be seen that the tread pattern most influences grip on different surfaces and noise. Especially with media on the road, such as water, snow, or ice, the edges within a pattern, from blocks and tire sipes, contribute significantly to tire grip. In a large basic program, published in Doporto *et al.* (2003), a series of patterns ranging from ultrahigh performance (UHP) summer tires to highly siped Scandinavian tires have been tested in their interactions with a similar wide range of tread compounds.

From these results, we take as an example the dependency of wet and dry braking on the density of edges in the patterns structure (Figure 5). The rule of thumb for this *pattern-based conflict between wet and dry braking* consists in the same amount of trade-off for dry braking when one tries to improve wet braking by adding sipes to a pattern structure. This conflict can be explained by different frictional effects of the rubber edges: on a wet surface, the leading edges improve braking by wiping away the lubricating water film; on a dry surface, the trailing edges lift off because of higher frictional forces thus reducing the local contact area and in consequence dry braking.

For the tread *compound*, one sees a *concurrency between wet and dry braking*, where the improvements in dry braking are about only a third of the ones in wet braking. Both performances are concurrent because they are both

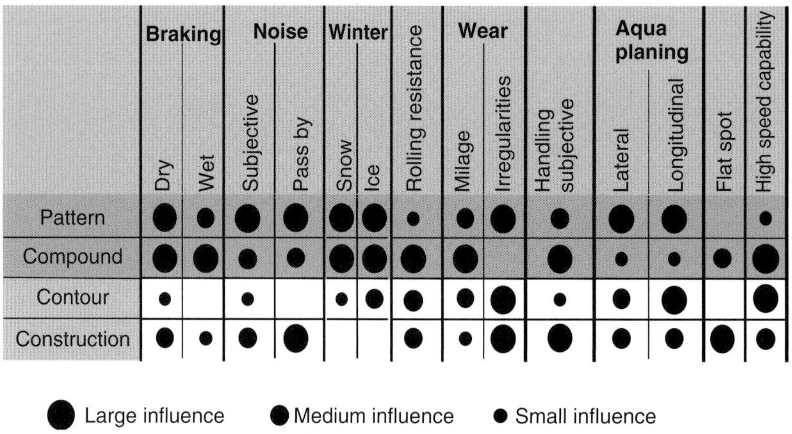

Figure 4. Design influence potential of tire domains on tire performances. (Reproduced from Heißing, 2007. With kind permission of Springer Science+Business Media.)

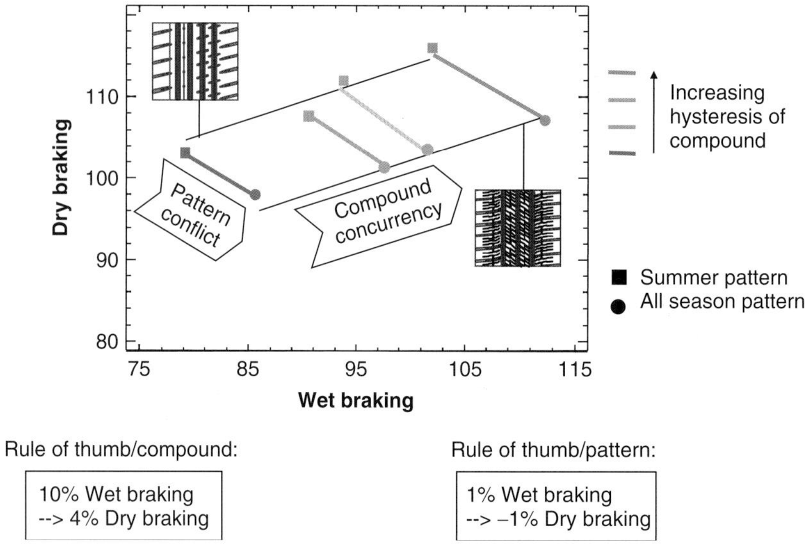

Figure 5. Example for tread induced performance conflict/concurrency and related rules of thumb.

caused by the traction physics of the compound's hysteresis. Its larger contribution to wet braking might be explained by a larger spread in hysteretic compound behavior at lower local contact temperatures because of the cooling effect of the intermediate water film.

Another extensive basic study, investigating the effect of pattern void on all relevant tire performances, was published in Mundl, Roeger, and Wies (2009). In Figure 6, one can see all major and minor conflicts/concurrencies visualized in a kind of balance: conflicting performances can be found on opposite sides of the balance and concurring performances on the same side. The lengths of the levers indicate the change in a specific property while increasing the void of the tire pattern by 1.5%. By void, one understands the empty space in a pattern realized dominantly by grooves in relation to the tread volume of the respective smooth tire. Opposing levers of a pair of properties signal a performance conflict and quantify the related rule of thumb by the levers' lengths.

The main conflict to be influenced by *void* is the one between *hydroplaning and tire noise*. Increasing of void delivers more space for water for drainage, thus increasing the critical speed at lift off of the tire's contact patch by a water wedge. By contrast, when larger grooves contact the road surface, more air volume is stimulated to vibrate and to emit noise. To a lesser extent, we encounter also the above-described conflict between braking on wet surfaces and dry ones. Less known is a conflict for winter

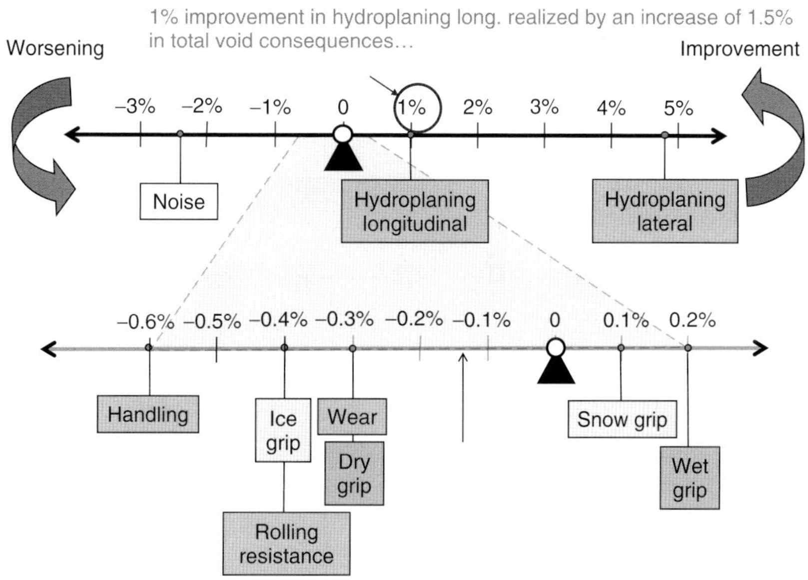

Figure 6. Pattern void based quantified conflicts/concurrencies. (Reproduced from Mundl, Roeger, and Wies, 2009. © The Tire Society.)

tires, namely between traction on snow and ice. Owing to increased sinking into the snow, additional void is better for interlocking between the soft snow and the tread pattern, whereas on stiff ice, the contact area is relevant for braking. Finally, more void weakens the tread's structure and, in consequence, cornering stiffness, which is relevant for handling. Less void means also having more rubber available to wear thus increasing the rolling distance over lifetime. On the other side, the higher vertical deformations of the smaller rubber blocks under contact pressure induce higher hysteresis loss and increase rolling resistance.

It has to be mentioned that some of these results are inconsistent with those in Figure 3. The results, documented in Figure 3, are elder ones and mainly based on variations in sipe technology at constant void. Obviously, one has to be careful with general statements on performance conflicts and has to consider always in detail which influencing parameter is the dominant one in steering the conflict!

4.2 Tread compound

Figure 4 shows the dominant influence of the tread compound on braking, winter performance, rolling resistance, wear, and handling. Four of these capabilities can be found in the so-called "magic triangle," which visualizes the challenges for a rubber chemist (Figure 7). Centrally located is the wet braking behavior, being also most important from a driving safety viewpoint. It conflicts not only rolling resistance but also winter performance and wear.

Especially the conflict between *wet braking and rolling resistance*, which according to Societe de Technologie

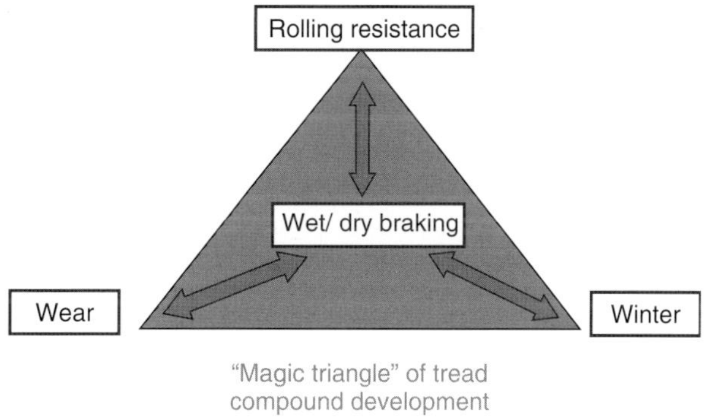

Figure 7. Schematic presentation of main tire performance conflicts influenced by tread compound.

Michelin (2005) causes about 20–30% of cars fuel consumption, must be dealt with in depth. For the reason of this conflict, both performances will be made transparent to the consumer by help of a future tire label classifying these performances by characters from A to G [see the related legislation from the European Union in des Europäischen Parlaments und des Europäischen Rates (2009)]. The material property causing this conflict is the hysteresis of rubber, measurable by the so-called loss angle δ. This loss angle significantly depends on the temperature, as shown in Figure 8. As rolling resistance is 50% because of the tread compound, we have to consider its operating temperature around 50°C and its operating frequency of deformation of about 10 Hz, which is also the excitation frequency of the measurement in Figure 8. Therefore, it can be concluded that the indicated dark gray

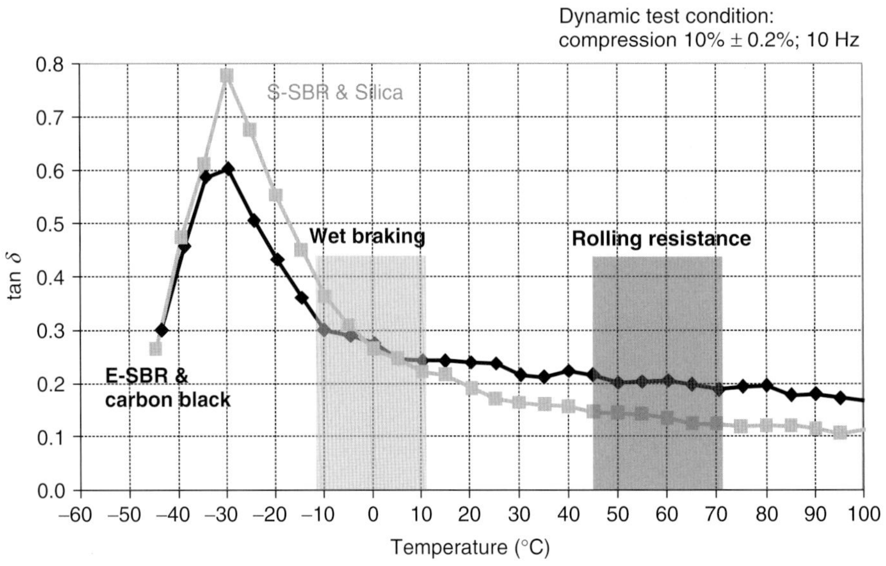

Figure 8. Loss angle as function of test temperature with correlating temperature ranges to rolling resistance and wet braking.

temperature range is the range of correlation in ranking between measured tan δ and measured rolling resistance. The light gray correlation range for wet braking is located around 0°C because of the underlying hysteresis friction mechanism, described in Kummer (1966), which operates around excitation frequencies of 10^4–10^7 Hz, depending on the roughness of the road and the sliding speed of the tread blocks. To make the conclusion complete, one has to consider the equivalence between temperature and dynamic excitation for rubber (Clark, 1982): qualitatively spoken, this stands for the phenomenon that rubber behaves equally when a colder operating temperature is compensated by a higher loading frequency and vice versa. This fact explains the shift of the correlation range for wet braking in regard to rolling resistance to lower temperatures.

One can now overcome the conflict between wet braking and rolling resistance only by changing rubber chemistry in a way that tan δ is increased at lower temperatures and decreased at higher temperatures. This happened with the introduction of silica as substitute for carbon black in the mid-1990s, shown by the light gray curves compared to the black curve in Figure 8. The tremendous progress in reducing this conflict since then, achieved by the rubber industry, is shown in Figure 9. It gives hope that, by use of new rubber materials, this most important conflict will be reduced significantly in future.

The conflict between *wet braking and winter capabilities* can be explained by a shift of the graph for the loss angle δ to negative temperatures. This shift is necessary to achieve winter capabilities by maintaining, under winter temperatures, the flexibility of rubber that is responsible for enhanced traction on snow and ice (Mundl, Wiese, and Wies, 2011). Doing so, one can deduce the lower ranking of those compounds in the temperature range for wet braking at 0°C. The conflict between *wet braking and wear* can be illustrated by the molecular model of the internal damping of rubber: it can be imagined as the energy consuming sliding of the molecular chains of polymers attached to the graphite crystal surfaces of the filler material, carbon black, by van der Waals forces. Hindering this internal sliding increases the internal cohesion and in consequence wear resistance but it lowers hysteresis necessary for wet braking as well.

4.3 Tire contour

A tire contour can be characterized by two main parameters: the radius of the tread and its width. These parameters span a two-dimensional plane qualitatively limited by the terms small/large and narrow/wide, wherein the optimal location of several tire properties are marked qualitatively (Figure 10). With round and narrow treads, the drainage of water is eased because of laterally shorter flowing distances for expressing water and a *hydrodynamic* better suited oval shape of the contact patch for plowing through the water layer. *Noise* is reduced because a narrower leading and trailing edge of the contact patch sees less impacting and snapping out of tread structures. The rounder shape of the contact patch has probably a more uniform and therefore less lateral noise radiation.

These tire performances *conflict* now with less *rolling resistance* and *wear* with a flatter curvature. This fact can

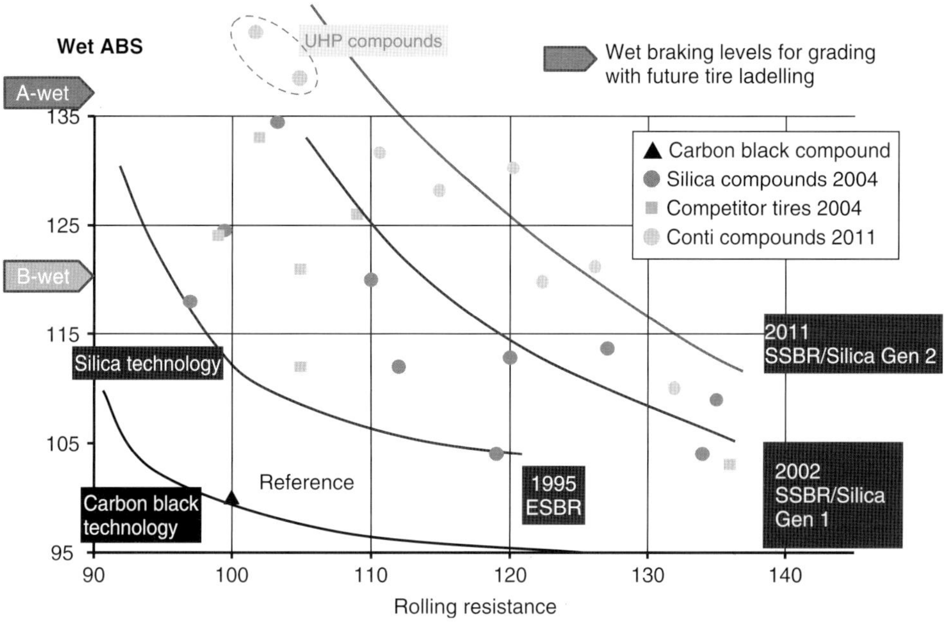

Figure 9. Evolution of the performance conflict between wet braking and rolling resistance from 1990 to 2008.

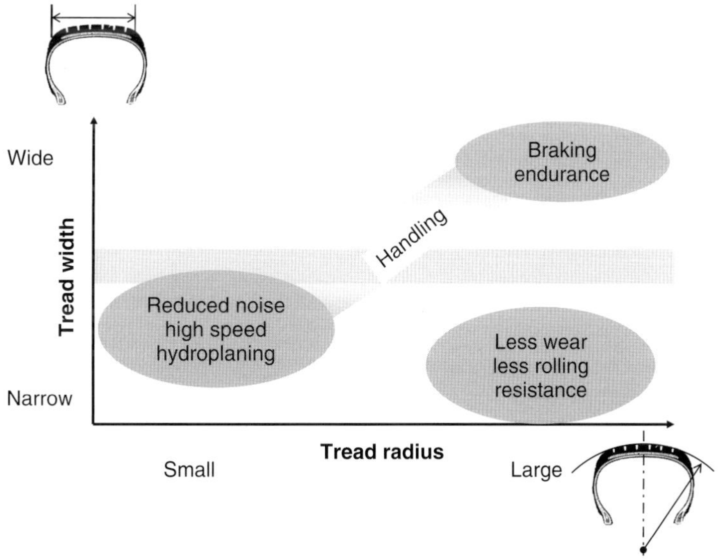

Figure 10. Target conflicts, resulting from tire contour variations.

again be explained with less energy input for flattening the tread structure resulting in a lower amount of internal damping and less frictional energy, which is responsible for abrasion.

They *conflict* also with *braking* capability, which improves with wider treads. This fact can be explained with a shorter contact length, which reduces opposed shear forces in the longitudinal direction, diminishing the overall friction potential.

4.4 Tire construction

Tire construction is first responsible for the endurance and durability of a tire. Secondly, it serves to optimize handling properties. It holds also for about 50% of rolling resistance. The dominant performance conflicts are therefore the ones between *durability/handling versus rolling resistance*. To make, for example, the tire sidewall more resistive against punctures under bad road conditions, a thicker rubber

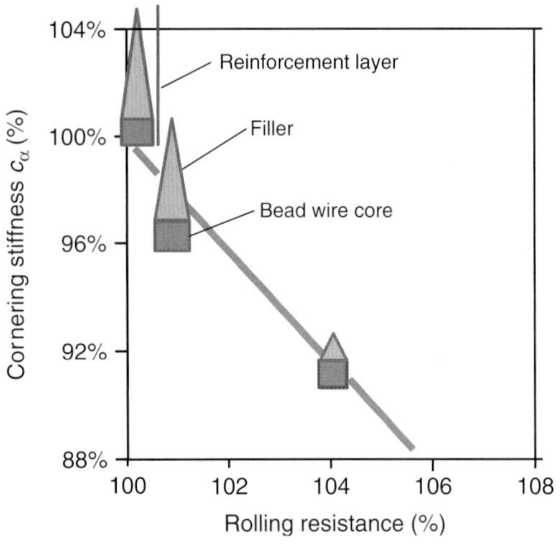

Figure 11. Conflict between handling and rolling resistance controlled by bead stiffness (various schematic bead designs).

coating and an additional ply are necessary to protect the reinforcing cords. This of course increases rolling resistance because more energy is needed to deform the thicker sidewall. For better handling, stiffening of the bead area by a higher apex often helps. Again more material is needed and increases rolling resistance, as can be seen in Figure 11. Therein the different bead constructions are indicated by symbols. The dark gray area stands for the bead wire layers, the light gray area for the apex rubber, and the gray line for the textile reinforcement layer. The textile reinforcement is only influencing the cornering stiffness, not the rolling resistance. Figure 11 shows that 2% increase in cornering stiffness leads to a change of about 1% in rolling resistance.

5 TARGET CONFLICTS OF UHP TIRES

5.1 Description of virtual development process

The development of UHP tires is dominated by the improvement of the three following performances: handling, high speed capability, and wear on race tracks. To shorten the development process and to save costly testing efforts, a major part of the development process has become virtual, which means to achieve handling requirements by simulation of virtual prototypes. Following Fischer *et al.* (2010) and visualized in Figure 12, the interface between tire evaluation and tire design is defined by tire target characteristics. In particular, a corridor in cornering stiffness for front axle (FA) and rear axle (RA) and increased lateral grip limits has to be achieved by the proposed tire variants under investigation. Both properties can be identified in the side force/sideslip angle dependency shown in Figure 13. The cornering stiffness is the inclination of above-mentioned dependency for slip angles up to 3°, whereas the grip limit is achieved at high slip angles. On the one hand, the cornering stiffness results only from stiffness properties of the tire, in particular the tread pattern stiffness and tire body stiffness connected in series. On the other hand, the grip limit is dominated

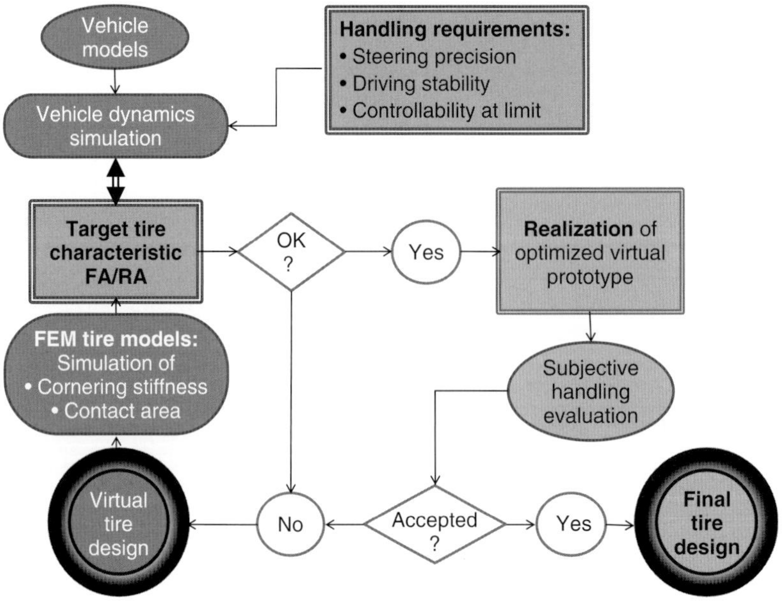

Figure 12. Virtual development process of UHP tires.

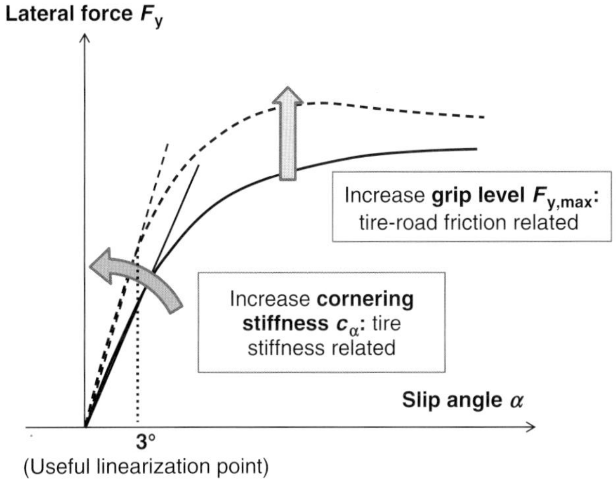

Figure 13. Main parameters of lateral force/slip angle behavior.

by the frictional properties of the sliding tread influenced by tread pattern structure and tread compound in reaction to the road surface. In a first loop of the virtual development process, the proposed tread pattern variants are simulated in their lateral stiffness; this shows the potential for improvement, the cornering stiffness being relevant for steering precision and driving stability. The related simulation tool is discussed in Mundl *et al.* (2008). The optimal tire pattern variant is realized in tire prototypes and checked against a base tire in its subjective handling performance on test tracks. These subjective evaluated results are associated with objective measurements of vehicle dynamics such as vehicle slip angle versus lateral acceleration (Figure 14).

In a second loop, proposed changes in tire body design are modeled and evaluated by finite element analysis in cornering stiffness and contact area. Optimization criteria are again maximal cornering stiffness and additionally maximum contact area under severe lateral forces with the consequence of an as low as possible contact pressure distribution for a maximum coefficient of friction. Again, a final vehicle handling evaluation is performed for the realized best variant to ensure the achieved targets in handling capability in serial production.

5.2 Main target conflict: grip limit versus cornering stiffness

As described earlier, it is essential for UHP tires to design them in a way that excellent handling on sports cars is achieved. This means that both cornering stiffness and lateral grip limit have to be increased and that in a way that the balance between FA and RA is maintained for the purpose of stable cornering. In general, driving at the limit may become also more challenging because of a smaller range of maximum grip (Figure 13). Inexperienced drivers may feel this as unpleasant. As grip is mostly affected by measures in the tread, a balance has to be found predominantly for the tread compound stiffness. The softer the tread compound is, the more the road asperities penetrate into the surface of the rubber blocks and interlock with them, thus producing enhanced grip. On the other hand, these softer blocks produce less lateral contact shear forces at a specific slip angle and therefore a lower cornering stiffness, as visualized qualitatively in

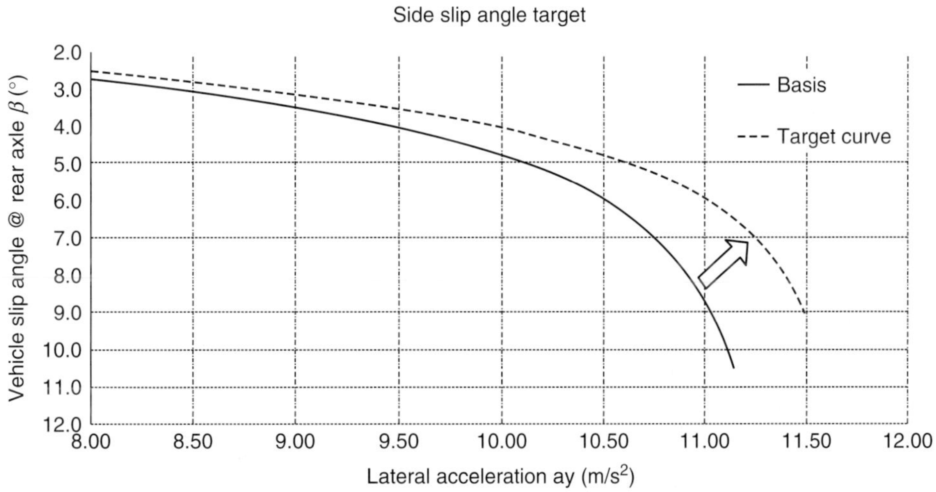

Figure 14. Slip angle at rear axle as indicator for improved grip at the limit. (Reproduced from Fischer *et al.*, 2010. With permission from Aachen Colloquium Automobile and Engine Technology, 2010, M. Fischer, J. Ehlich, C. Schroeder, K. Peda and B. Wies, Continental.)

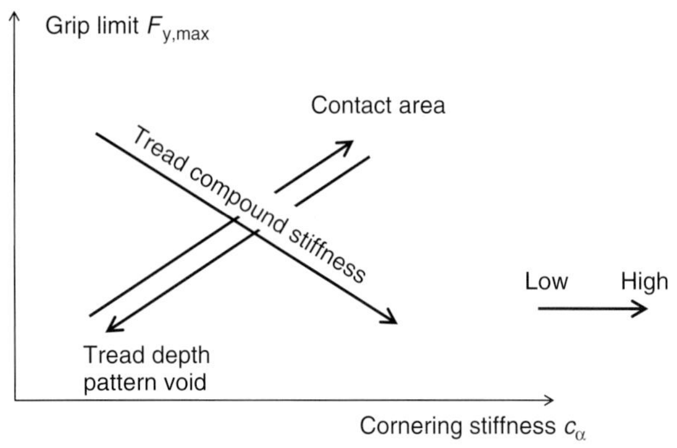

Figure 15. Qualitative influence of tire design parameters on grip/cornering stiffness conflict.

Figure 15. By chance, other design measures in the tread pattern, such as increased positive or a lower height of the pattern blocks, are positive for grip and cornering stiffness simultaneously (Figure 15). The same is true for a larger overall contact area of the contact patch, which can additionally be controlled by tire contour and tire body design.

5.3 Secondary target conflicts: grip limit versus high speed capability and wear

With their higher maximum speed, sports cars do need tires with a raised high speed capability. This can be in conflict to lateral grip as well (Figure 16). The hysteresis of the tread compound dominates this conflict. On the basis of the model of hysteresis rubber friction, see again the theory in Kummer (1966), a high internal material friction leads to increased grip, whereas hysteresis causes heat buildup in the rubber blocks, thus limiting maximum rolling velocity by depolymerization of the tread compound. Going back to Figure 10, we recognize therein that the contour is improving grip when it is flat but worsening high speed at the same time. The reason for this behavior is that a flat contour has a tendency to larger contact areas and better grip as mentioned earlier but causes higher pressure peaks in the shoulder area because of the dynamic growths induced by centrifugal forces and, in consequence, larger deformations and additional heat buildup. A high void in the pattern is a countermeasure for this, because it helps cooling but again negatively influences grip (Figure 16).

It has to be mentioned that resistance against wear of the tread compound conflicts also its grip capability. The explanation is that, again, higher hysteresis of the tread compound causes increased heat buildup, thus making the compound softer and less resistant to wear. The explanation is that under traction forces onto the driven axle, softer

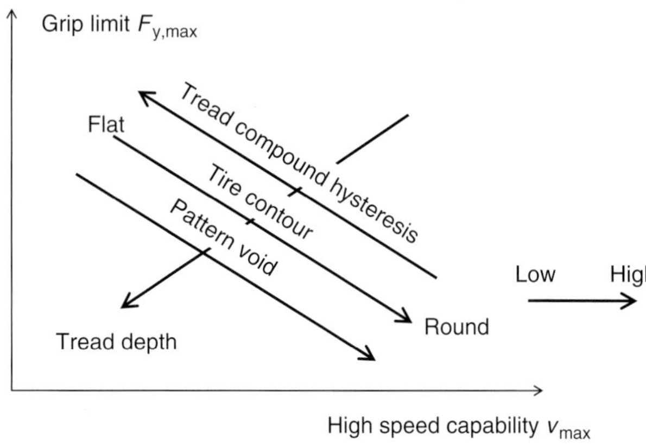

Figure 16. Qualitative influence of tire design parameters on grip/high speed capability conflict.

tread structures suffer from larger deformation energy in the contact area that is transformed into frictional energy at the trailing edge, resulting in increased wear.

6 OUTLOOK FOR FUTURE IMPROVEMENTS OF UHP TIRES

A conflict analysis of tire performances is helpful in giving hints where future innovations may take place. In the field of UHP tires, potential can be seen in two ways to improve the conflict between enhanced steering properties and increased grip level dominated by future compound aspects: an anisotropic compound behavior, such as higher shear stiffness and lower radial stiffness, may be achieved by oriented microfibers within the compound matrix, see also the physical reasoning in Chapter 4.2. Another vision is a deeper interaction with tire properties and advanced vehicle control systems using active steering, thus improving steering precision and handling stability even at lower cornering stiffness of the tires and leaving development space for ultrahigh grip levels.

REFERENCES

Clark, S. (1982) *Mechanics of Pneumatic Tires*, US Government Printing Office, Washington, DC, p. 24.

Doporto, M., Mundl, R., Wies, B., *et al.* (2003) Zusammenwirken zwischen Profil und Laufflächenmischung. *Automobiltechnische Zeitschrift*, **105**(3), 238–249.

Verordnung Nr. 1222/2009 des Europäischen Parlaments und des Europäischen Rates (2009), Über die Kennzeichnung von Reifen in Bezug auf die Kraftstoffeffizienz und andere wesentlichen Parameter, Brüssel, November 25.

Fischer, M., Ehlich, J., Schroeder, C., *et al.* (2010) Virtual based development process for UHP tires with target setting for an excellent driving experience. *Aachener Kolloquium Fahrzeug- und Motorentechnik*.

Heißing, B. and Ersoy, M. (2007) *Fahrwerkhandbuch*, 1st edn, Vieweg Verlag, p. 346.

Kummer, H.W. (1966) *Unified Theory of Rubber and Tire friction*, Pennsylvania State University, 94.

Mundl, R., Fischer, M., Strache, W., *et al.* (2008) Virtual pattern optimization based on performance prediction tools. *Tire Science and Technology*, **36**(3), 192–210.

Mundl, R., Roeger, B., and Wies, B. (2009) Influence of pattern void on hydroplaning and related target conflicts. *Tire Science and Technology*, **37**, 187.

Mundl, R., Wiese, K., and Wies, B. (2011) An analytical thermodynamical approach to friction of rubber on ice. Presented at the 2011, Tire Society meeting, Akron, Ohio; also in), *Tire Science and Technology*, **40**, 124–150.

Societe de Technologie Michelin (2005) *The Tyre, Rolling Resistance and Fuel Savings*, 2nd edn, Karlsruhe, p. 35, CD-Rom, Abt. Öffentlichkeitsarbeit.

Chapter 116
Tire Pressure Monitoring Systems

Victor Underberg[1], Thomas Roscher[1], Frank Jenne[1], Predrag Pucar[2], and Jörg Sturmhoebel[2]

[1]*AUDI AG, Munich, Germany*
[2]*NIRA Dynamics AB, Linköping, Germany*

1 INTRODUCTION

The tire represents the link between vehicle and road surface. As such, the tire characteristics determine the transfer of forces between the vehicle and the road. One major factor influencing the actual tire properties with respect to vehicle handling, braking, fuel consumption, and tread wear is the tire pressure. Only with the vehicle-manufacturer-recommended tire pressures, the optimum of all the above-mentioned characteristics can be guaranteed. The recommended tire pressures do depend not only on the specific vehicle/tire combination only but also on anticipated driving conditions as vehicle load (partly or fully loaded) and travel speed (low or high).

Already in 1923, Bosch patented a mechanical device—the "Bosch-Glocke" (Bosch-Bell)—which was designed to warn the driver of under-inflated tires (Figure 1). The concept behind the "Bosch-Glocke" was that the deformation of an underinflated tire running through the contact patch activates a wheel-mounted lever, which, in turn, will ring a high frequency bell to alert the driver.

Today's tire pressure monitoring systems (TPMSs) are designed to monitor the current tire pressure in all four wheels in service—in some cases also the spare wheel—and to provide a warning to the driver if the tire pressure in at least one the wheels falls below a predefined threshold level. In general, there are three different pressure loss scenarios (Figure 2). First, there can be a tire blow out, where the tire loses all its air in just seconds because of sudden and severe tire damage, for example, by hitting a pot hole (TPMSs are not designed to warn drivers in the blow out scenario, as the driver will be alerted immediately by a sudden change in vehicle response). The second scenario is an air leakage or punctures with pressure loss rates of approximately 10 kPa (equiv. to 1.45 lb/in^2) per minute up to a week, usually caused by a puncture (nail, screw, etc.), a leaking valve, or improperly mounted tires. In most cases, a tire blow out or a leakage only affects one tire at a time. However, tires in service also naturally lose air through a process called *diffusion*, also known as *permeation*. In this case, air molecules pass through the tire, as the tire inner liner does not guarantee total impermeability. This scenario affects all four tires in the same way with an approximate pressure loss rate of about 10 kPa per month and will cause significant under inflation if tire maintenance is neglected by the driver, that is, tire pressure is not checked and adjusted over a long period of time.

At first, TPMSs were introduced as a comfort feature in the late 1980s. The model year 1987, Porsche 959

Encyclopedia of Automotive Engineering, print © 2015 John Wiley & Sons, Ltd.
Edited by David Crolla, David E. Foster, Toshio Kobayashi, and Nick Vaughan.
This article is © 2015 John Wiley & Sons, Ltd. ISBN: 978-0-470-97402-5
Also published in the *Encyclopedia of Automotive Engineering* (online edition)
DOI: 10.1002/9781118354179.auto014

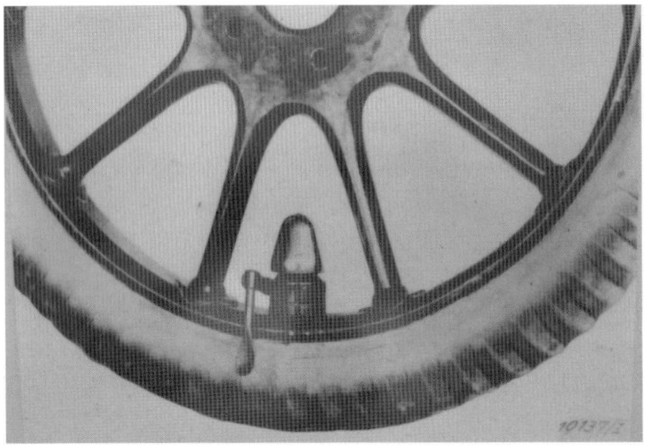

Figure 1. 1923 "Bosch-Glocke" (Bosch-Bell).

was the first series production car to feature a TPMS (Figure 3). The main components of the Porsche 959 TPMS, pressure switch and high frequency unit, are represented in Figure 4. After the turn of the century, having the correct tire pressure with the effect of maintaining low rolling resistance and by that increasing fuel economy shifted into focus. The first TPMSs were "direct" measuring systems, meaning tire pressures, and often temperatures, were measured within the wheel and the information was transmitted via radio frequency to a receiver inside the vehicle. The so-called indirect systems use the information embedded in the wheel speed signals of the antilock brake or electronic stability control system (ABS/ESC) to

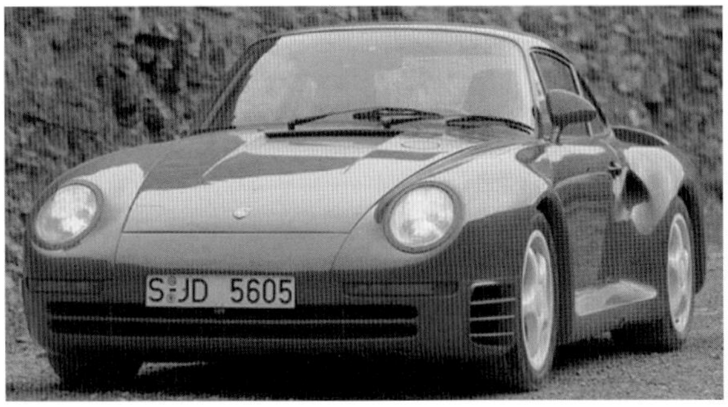

Figure 3. 1987 Porsche 959.

determine under inflation. Indirect measuring systems of the first generation were able to detect puncture scenarios only at one tire, whereas current state-of-the-art second-generation indirect systems are able to detect puncture and diffusion scenarios in up to all four tires. Indirect systems of the first generation became widely used in the early 2000s as a cost efficient alternative to complex and, back then, error prone direct TPMSs. The increasing market share of run-flat tires (others say "extended mobility" tires) around that time also caused the increasing market share of first generation iTPMS (indirect tire pressure monitoring system) as puncture detection systems. Especially when using run-flat tires, drivers have to be provided with underinflation information, as a flat tire will not be detected by visual inspection.

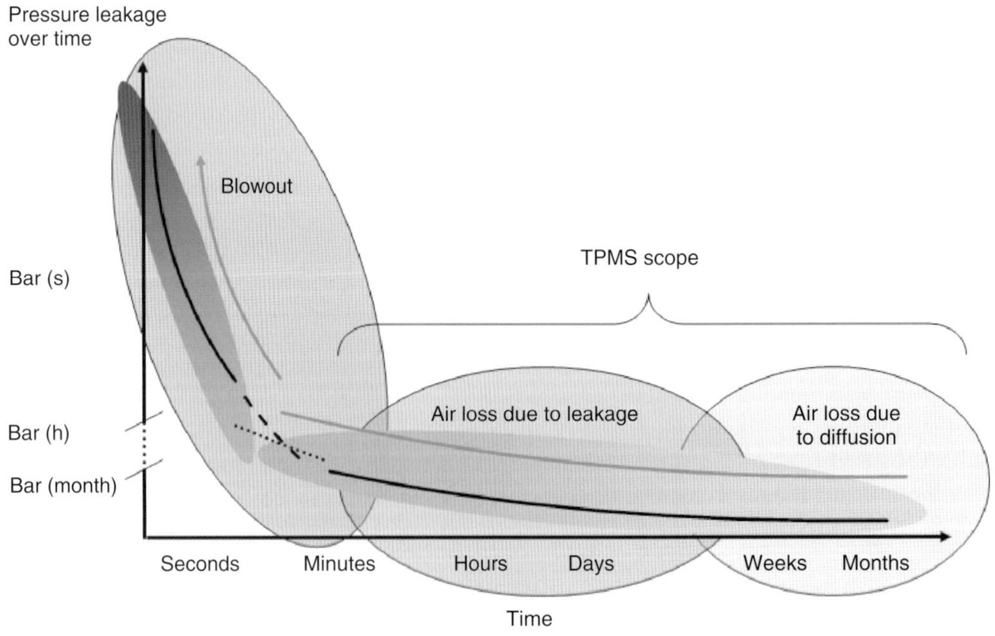

Figure 2. Pressure loss scenarios.

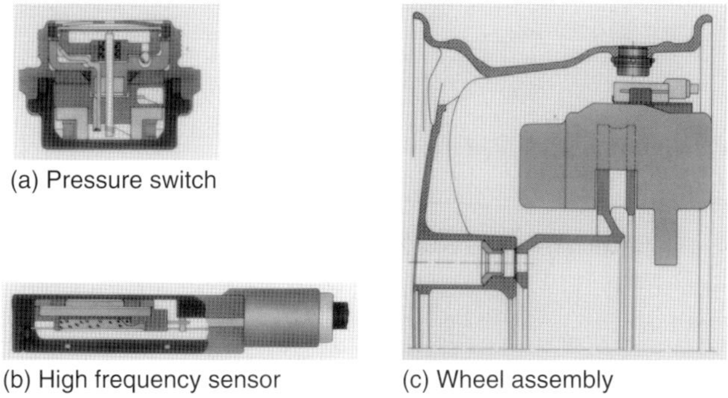

Figure 4. (a–c) 1987 Porsche 959 TPMS components.

In the year 2000, a series of fatal accidents related to tire tread separation on Firestone tires mounted on Ford Explorer sport utility vehicles triggered the US congress to pass the so-called TREAD Act (Transportation Recall Enhancement, Accountability and Documentation Act). Subsequently, in 2005, the new Federal Motor Vehicle Safety Standard No. 138 (FMVSS 138) was enacted, requiring all new passenger cars, multipurpose passenger vehicles, trucks, and buses with a gross vehicle weight rating (GVWR) of 4536 kg (10,000 pounds) to be equipped with a TPMS. Starting with a phase-in period in 2005, by September 2007, all new vehicles had to be fitted with a TPMS conforming to FMVSS 138 requirements. With that the United States was the first country to have a regulation specifying the installation and performance criteria for TPMSs. Targeting the fuel efficiency in addition to the safety aspect of correct tire pressure, in 2009, the EU parliament mandated ECE-R 64, a regulation requiring all new vehicle types to be equipped with a TPMS starting November 2012 and all new vehicles to be registered from November 2014.

2 DIRECT TIRE PRESSURE MONITORING SYSTEMS

Direct pressure monitoring systems are based on the principle of measuring tire internal pressure (and temperature) by wheel units fixed to the rim. The sensors send their data by radio frequency to vehicle body fixed antenna(s). The antenna(s) provides the data to an electronic control unit (ECU) for evaluation. The ECU communicates with the car's network and issues messages that warn the driver in case of a pressure loss via telltales and messages in displays in the instrument cluster and/or human–machine interface (HMI).

The functional principle introduces system complexity to the vehicle design. The parts required for the direct pressure monitoring functionality are the following:

1. wheel units
2. antenna(s)
3. ECU and wiring harnesses
4. optional parts and displays
5. development.

The following sections discuss the function of each system component.

2.1 Wheel unit

This component is mounted at least in each driving wheel. It measures at least tire pressure and temperature. In addition, data such as acceleration and rotational direction may be collected. The wheel unit sends the data wirelessly on standardized carrier frequencies of 433 or 315 MHz to a vehicle body fixed antenna.

Figure 5 shows a block diagram of an HUF Electronic sensor (formerly known as *BERU Electronics*). The low frequency (LF) module enables wake-up and data requests from outside the tire for diagnosis or functional purposes. The "over temperature protection circuit" allows survival for temperatures up to 150°C for several minutes. At 120°C, the wheel unit will fall into a protection mode to safeguard electronic parts at these temperatures. A lithium-based consumer battery, providing 3.0 V output, provides energy. In general, state-of-the-art sensors transmit data:

1. Continuously, when activated once (e.g., one transmission/minutes in regular intervals).
2. Roll switch activated at a certain threshold acceleration.
3. Outside requested, for example, by a diagnostic tool or TPMS trigger (initiated by an LF signal).

As the inner tire environment is hostile because of temperature variations (−40 to +140°C), humidity, and high acceleration (up to 2000 g), the plastic-housed electronics are protected by a potting compound. The opening for pressure measurements is either at the top of the housing or at membrane shielded and integrated into the potting.

For mounting the wheel unit to the rim, several different techniques are used. The most common solution is the attachment of the wheel unit to the valve as snap in or fixed by a screw to an alloy valve (Figure 6). In Figure 7, a valve-mounted wheel unit is shown in the wheel unit assembly and in a cross section. Besides that, metal bands around the rim are also in use.

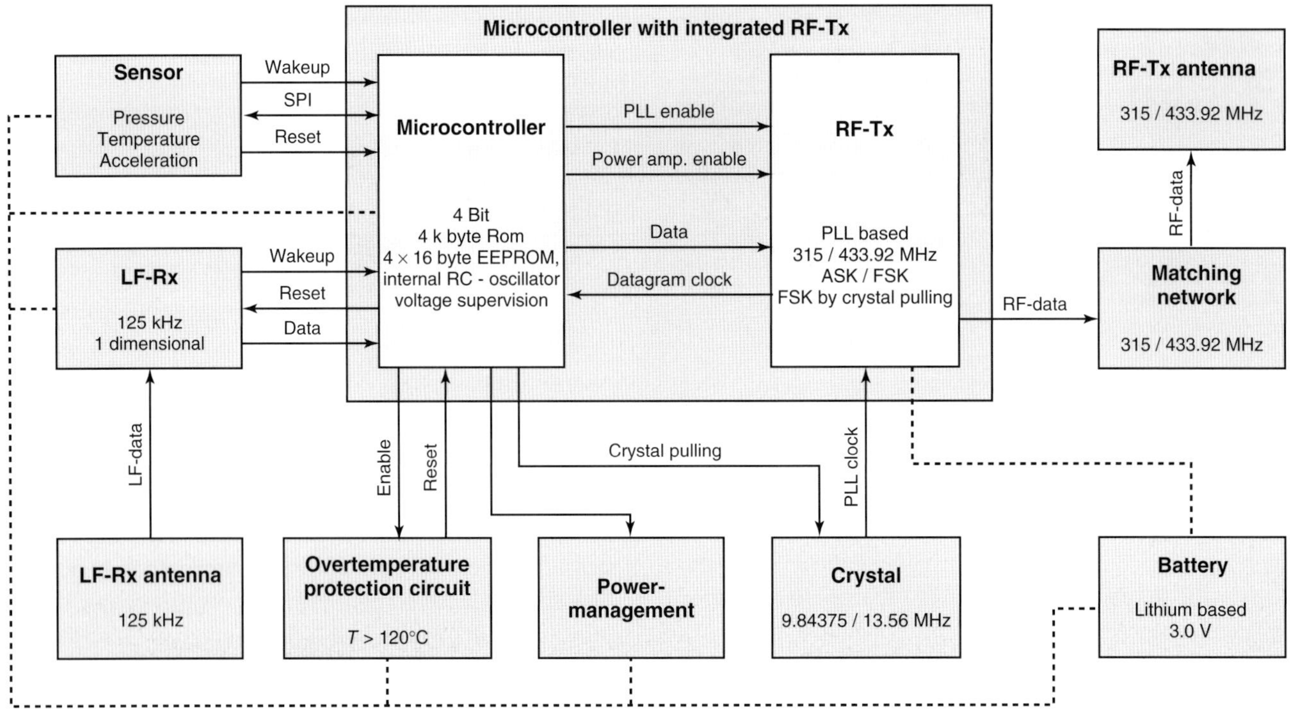

Figure 5. HUF Electronics sensor, block diagram.

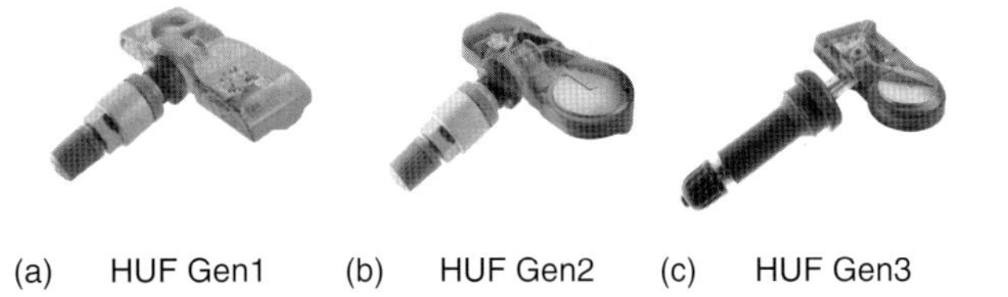

Figure 6. HUF Electronic wheel units, generation 1 (a), 2 (b), and 3 (c).

2.2 Antenna

As to the antenna, receiving data from the wheel unit, several solutions are available:

1. passive external antenna
2. active external antenna
3. integrated antenna.

Passive antennas only receive the radio signals without processing the data. In this case, wire connections transmit the analog signals to a receiver, implemented, for example, into a TPMS ECU or a body computer. During the advent of TPMSs, receiver technology was unable to filter out disturbances effectively from the use-signal. For this reason, these systems were prone to electromagnetic compatibility (EMC) issues. An example of a passive antenna implementation is shown in Figure 8. As receiver technology progressed, and being less susceptible to EMC issues, this cost-effective solution was base for the so-called integrated solutions. Integrated in this sense means sharing receiver and antenna with other vehicle systems (e.g., keyless entry) and implementing the TPMS software into the body computer.

Active antennas represent the next step for improved receiver performance. In this setup, the receiver is integrated into this antenna module. An example of an active antenna component is shown in Figure 9. It decodes the received signals and transmits them in digital format via wire to the TPMS ECU where the analysis takes place. Figure 10 shows an example of an active antenna system implementation. This antenna generation is more sensitive to radio signals but more resistant against EMC issues. It assures easier application and modular system configuration. Another technology leap is represented by the so-called intelligent antenna, which means that TPMS ECU, receiver, and antenna are combined in one housing to reduce parts (Figures 11 and 12).

2.3 Electronic control unit (ECU) and wiring harness

The TPMS ECU processes and evaluates the data received by the antenna or additional data from other ECU; an

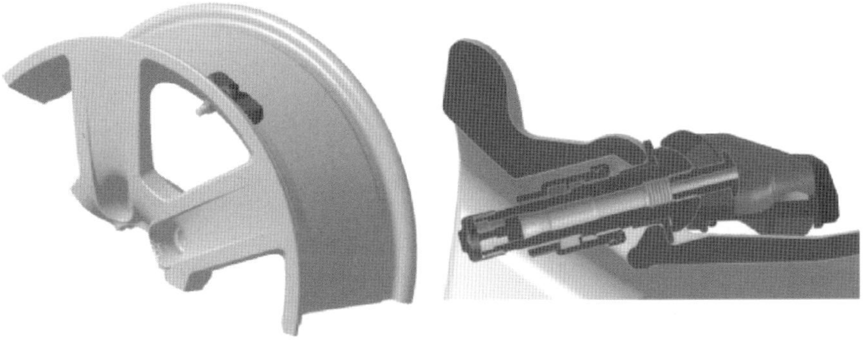

Figure 7. Wheel-mounted pressure sensor with cross-sectional view.

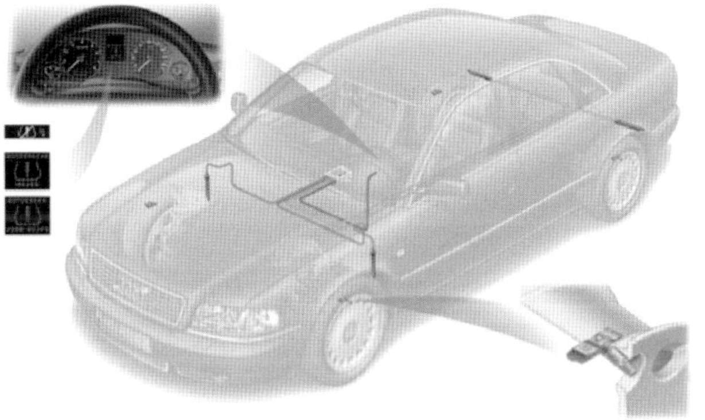

Figure 8. 2001 Audi A8 (D2) with passive antennas.

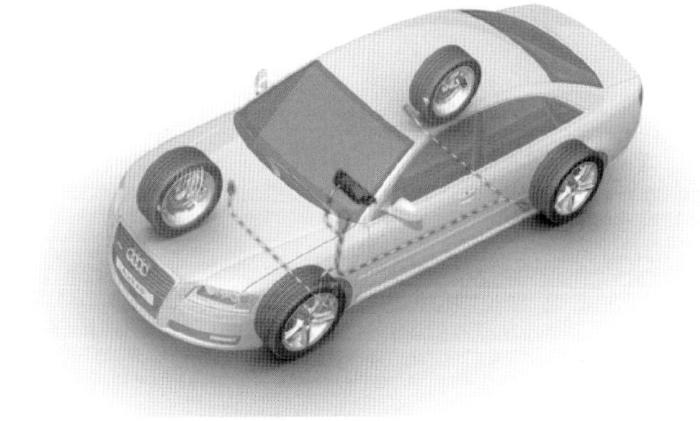

Figure 10. 2004 Audi A8 (D3) with active antennas.

Figure 9. Example of a digital antenna component.

Figure 11. 2007 Audi Q7 modular high TPMS.

example of a TPMS standalone control unit is shown in Figure 13. It manages warnings, allocation of wheel units, and internal and external system diagnoses and provides data and messages for displays and menus. The ECU is connected to the car's wiring harness and communicates as part of this network with other ECU such as gateway, dashboard, and MMI. Wiring depends on TPMS structure and function. As minimum, there is connection to controller area network (CAN) bus, power supply, and ground. Additional components may be linked and organized by local internal network (LIN) or direct linked.

Figure 12. 2007 Audi Q7 modular low TPMS.

Figure 13. Stand-alone TPMS electronic control unit.

Learning and management of wheel units may be done via diagnostic tools at a car or tire dealer or by the TPMS software itself via auto learning and auto location. Auto learning and auto location is more or less static evaluation of information such as number of received tele messages, their signal strength, acceleration in combination with vehicle speed, and turning direction.

2.4 Optional parts and displays

Optional parts may be, for example, a 125 kHz transmitter, the so-called trigger, an example is shown in Figure 14. With this, the TPMS is able to build up a bidirectional communication. The transmitters, one placed in each wheelhouse behind the mudguard, are connected to the ECU. If the wheel unit receives the LF signal from this transmitter, it sends its data on 433/315 MHz to the TPMS antenna no

Figure 14. 125 kHz Transmitter unit, trigger.

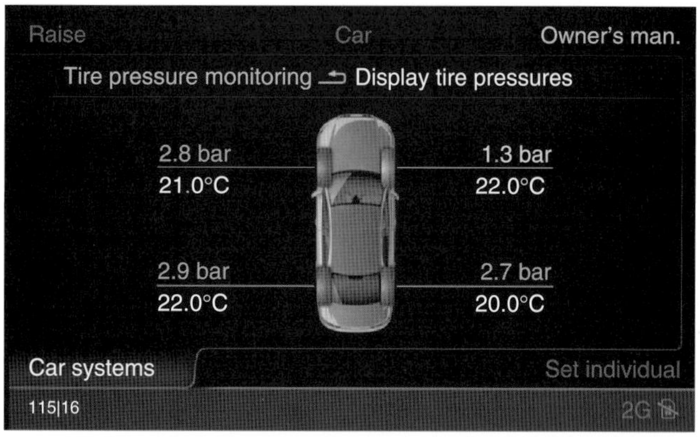

Figure 15. TPMS information in Audi MMI.

matter if the car moves or is stationary. The ECU is always able to request data, for example, when data are lost or for comfort purpose at door open as a status request. In addition, using this additional communication does allocation of sensors quickly and accurately.

Warning telltales and text messages in case of pressure loss or malfunction are displayed in a display in the dashboard. Additional information, such as pressure/temperature status, may be shown either on displays in the dashboard or suitable HMIs (Figure 15).

For operations such as resetting the system, start learning new wheel units, or storing tire pressures after adapting them, a push button connected directly to the ECU, or a menu either in the driver information system in the dashboard (DIS) or another HMI, provides this function to be executed by the driver. As the placard pressure may vary with mounted tire size, velocity to be driven, load and other characteristics TPMSs can provide various parameters to be chosen.

2.5 Development

Development of TPMS depends on OEM (original equipment manufacturer) functional, HMI, and legal requirements. In general, TPMS development can be grouped in the following categories:

1. hardware development
2. location/placement of hardware in the vehicle body (application)
3. software development.

2.5.1 Hardware development

All components of TPMS, except the wheel units, are subject to automotive industry standards depending on the location/placement of the different parts. For example, high temperature conditions can occur in the dashboard area. In addition, outside the passenger compartment, placed components need to be shielded against high humidity influences. In some circumstances, special protection against stone chipping is also needed. Electrical tests, for example, cover over/under voltage, voltage ramping conditions, EMC, or quiescent current topics.

As mentioned earlier, wheel units have to withstand the hostile environment inside a tire. In a joint effort, German car manufacturers (AUDI, BMW, Daimler, Porsche, and VW) and TPMS supplier DODUCO (later BERU, HUF Electronics) have developed detailed requirement specifications that describe the functional testing and validation of TPMS wheel units and attachments to the wheel by Alligator-developed alloy valves. Especially important were considerations about accuracy over component lifetime, minimal temperature dependency, resistance to high acceleration (up to 2000 g), temperatures higher than 120°C, and temperatures as low as −40°C. In addition, realizations of reliable functionality withstanding the high mechanical stresses were challenges mastered.

2.5.2 Placement of hardware at car body (application)

Positioning the receiver is crucial for TPMS functionality, as it strongly depends on the signal strength and reception rate of the wheel unit sensors at the location of the receiver. When placing LF transmitters in the wheelhouse, positioning of these components also requires that the wheel unit can be reached in all possible wheel positions (360° wheel rotation and all steering positions of wheel).

Investigations on digital mockup (DMU) adduce the first positions of components chosen based on available space and experiences from former vehicle projects. Subsequent tests on prototype vehicles give first impressions of the component functionality at the predefined position. To evaluate the receiver placement, the wheel unit sensor is set into a steady-state sending mode. The signal strength at the receiver is measured and evaluated for a full 360° rotation of each wheel. In addition, at the front axle wheels, this procedure needs to be carried out at full steering wheel angle, left and right turns. The objective is to have best possible reception and, if they cannot be avoided, only short "black spots," that is, positions where the signal strength falls below predefined threshold values during one revolution of the wheel. The same principle test needs to be performed for 125 kHz transmitter positions at each wheel position. The objective here is evaluating the flux density at the 125 kHz antenna of the wheel unit sensor during one full turn of a wheel. In addition, here, no or only minor black spots are to be accepted.

The reception rate of wheel unit sensor tele messages is measured while proving ground driving at different steady-state speeds, for example, 100, 150, and 200 km/h. Usually, two test conditions are analyzed and verified: confirmation of HF reception, when triggers are deactivated, and confirmation of the LF communication, when trigger mode is activated. Subsequent driving tests in all customer-relevant scenarios (high and low temperature, snow, rain, etc.) need to confirm proper functionality. As specific tire constructions may influence signal reception (signal damping effect), close cooperation with tire development is also necessary. EMC of the TPMS components/system within the vehicle environment needs to be verified as well, meaning that the TPMS is not influencing other electronic systems in the vehicle and vice versa (Figures 16 and 17).

2.5.3 Software development

TPMS software components can be grouped into base software, diagnostic and functional software. The base software ensures basic communication between the TPMS ECU and other car components via the vehicle network. Functional software contains warning algorithms and algorithms defining the appearance to the driver. To function within a vehicle network, data exchange via signals and protocols are standardized. Diagnostic behavior for defect and malfunction analysis in the technical service environment (e.g., workshop) also needs to be covered and defined to ensure fast and effective troubleshooting.

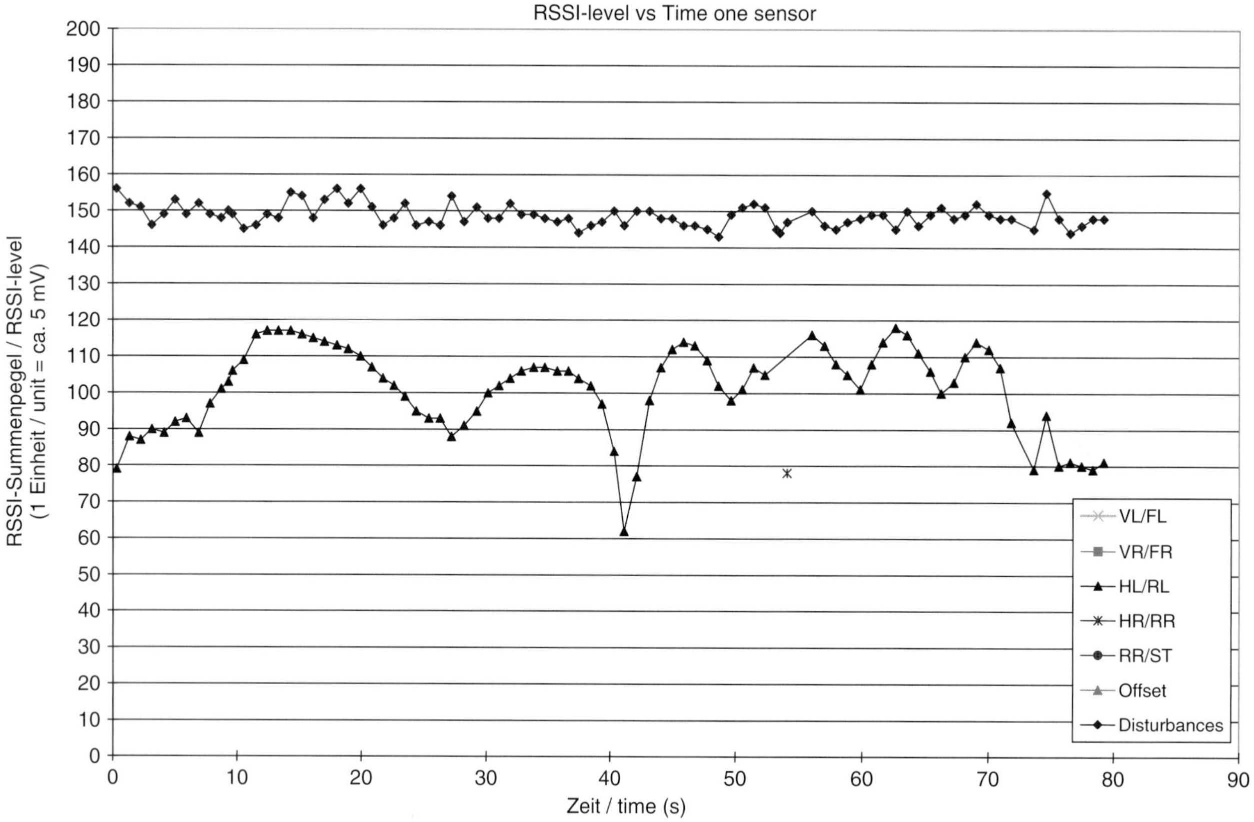

Figure 16. Signal strength diagram for a 360° rotation.

The functional software part defines warning strategies for different pressure loss scenarios, learning procedure of new sensors, learning of sensor positions, and the timing and sequence of telltale warnings and displays. Functional software setup depends on OEM requirements and philosophy as well as legal requirements for different markets. The warning strategy itself defines the pressure thresholds for signaling underinflation to the driver in case of a pressure loss because of puncture or diffusion pressure loss scenarios. As puncture usually means fast pressure loss (10–20 kPa/min), the warning needs to be displayed immediately when detected. Because of its long-term nature, detection of diffusion pressure loss need not necessarily be signaled immediately to the driver. This scenario is usually low pass filtered and often not displayed to the driver in the current ignition cycle but signaled at the next ignition on to assure cold tire pressure when the driver checks the pressures. In general—for customer acceptance purposes—it has been seen that warning thresholds too close to the recommended cold tire pressure have to be avoided. When signaling a warning at a pressures slightly below the cold recommended pressure, warning may be seen as nuisance by the customer and TPMS functionality is questioned.

Software verification is performed in hardware-in-the-loop (HIL), software-in-the-loop, and simulated network tests and in whole vehicle environment. Compliance to legal requirements and endurance is the main focus when doing approval vehicle tests.

3 INDIRECT TIRE PRESSURE MONITORING SYSTEMS

3.1 Tire as sensor

Directly measuring pressure sensors usually have components whose properties are dependent on the tire pressure and can therefore be used for measuring the pressure. Applying this basic principle in a generalized way leads to the approach of using the tire itself as a sensor. This is possible as various properties of a pneumatic tire are strongly pressure dependent and some of these properties are measurable using sensors being common and already installed for other purposes in most vehicles nowadays.

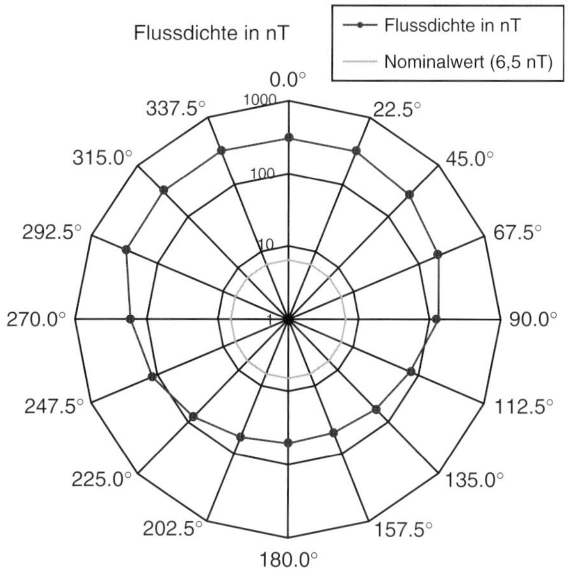

Figure 17. Flux density diagram.

The overall approach has various benefits as it avoids the installation of separate sensors and parts for the sake of monitoring tire inflation pressure. In this way, cost, weight, complexity, service, and logistics are efficiently avoided. In its most consequent application, it enables implementing a purely software-based tire pressure monitoring function into the ABS/ESC control unit as long an ABS or ESC system is fitted to the vehicle and can be utilized as a host for the TPMS software.

On the other hand, tires are complex products, which differ significantly in sizes, dimensions, materials, and technologies, as do their properties related to inflation pressure. The main challenge for iTPMS is to deliver consistent and reliable system behavior independent of the mounted tires. This variance in tire properties consequently is the main reason for iTPMS being relative by nature, meaning that they cannot monitor or display quantitative inflation pressure values but always need a reference state from which a deviation can be detected. The reference state can be viewed as an externally set reference point. The process where the user informs that a reference state is present is usually referred to as *reset* that is essential for the iTPMS. A reset is always required when inflation pressure has been adjusted, tires changed or rotated between different positions on the vehicle. In current applications, the reset is carried out actively by the driver (Figure 18). However, other options such as tire pressure gauges interacting with the vehicle could also enable other solutions in the future.

Subsequent to the reset, the iTPMS will analyze the pressure-dependent tire parameters and store the current parameters as reference data. This process is referred to as the *learning phase*. Once the learning phase is completed, the pressure-related tire parameters can be continuously monitored and compared to the stored reference data. If the current parameters deviate from the reference data in characteristics, according to predetermined patterns and/or exceeding certain thresholds, low tire inflation warnings are issued.

3.2 First-generation indirect TPMS

The first iTPMSs on the market, the so-called first generation iTPMSs, were only roll-radius or rolling-circumference-based systems. The basic principle is that the effective rolling radius of a tire is inflation pressure dependent and will decrease when the tire loses pressure. At a given vehicle speed, any inflation pressure loss can be detected by comparing and monitoring the individual rotational wheel speeds with each other (Figure 19).

The most important input signals are the individual wheel speeds. The wheel speed signals are obtained from measurements of number of teeth of the ABS rotor passing a certain point per time interval. The toothed rotors are fixed to each wheel and this way the rotational speeds used for ABS and ESC applications are calculated. This approach enables the effective and reliable monitoring of up to three out of four wheel positions but cannot detect same-rate

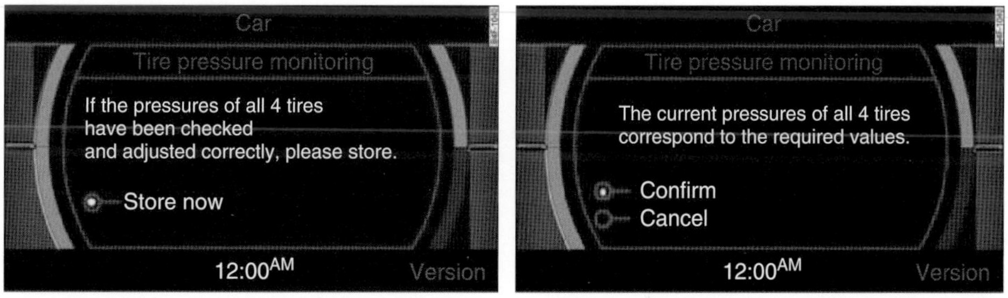

Figure 18. Guided reset procedure.

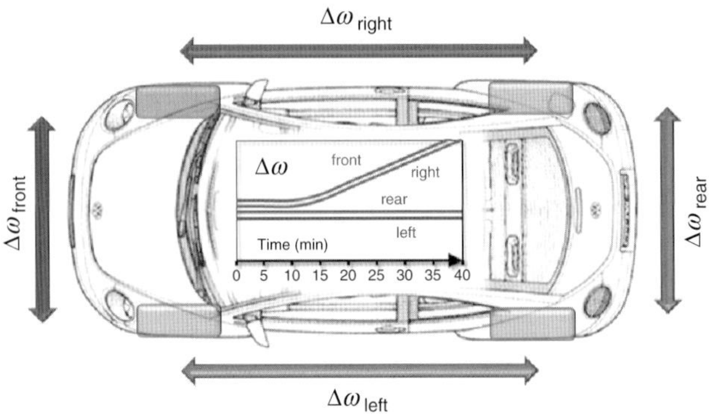

Figure 19. Four-wheel speed differences for rolling-radii-based indirect TPMS indicating a puncture front right.

inflation pressure losses on all four-wheel positions. The reason for not being able to detect the four-wheel deflation is that during the learning phase the relation between the wheel speeds is set. If all four wheels deflate in the same way, the relative difference will remain the same. Puncture detection is therefore possible with first-generation iTPMS but the diffusion/permeation detection due to slow natural air losses in all four wheels is not. Several approaches to overcome this limitation using wheel radius information only have been investigated, or are under investigation, but have not led to series applications yet.

Irrespective of the limitations, first-generation iTPMSs were mainly used and developed as the so-called run-flat-warning systems in combination with run-flat tires (see also ECE-R 64 requirements) as they are cost effective, reliable, and easy to integrate into ABS/ESC systems.

As the rolling radius is not only influenced by inflation pressure, other influencing parameters have to be considered of which the most influential one is the driving speed, which through radial forces influences the effective rolling radius and by that the wheel speed. It is common practice with iTPMS that they work in individual speed intervals for which reference values are stored. The current values are then continuously compared to the reference values stored for specific speed intervals. Other parameters that influence the rolling radius are cornering, braking and accelerating, load, and lateral or longitudinal road inclination. To eliminate the influence of these parameters, additional information such as engine torque, longitudinal and lateral acceleration, or the steering wheel angle are analyzed and used to compensate for the unwanted effects. On the basis of these auxiliary signals, the vehicle load can be estimated and compensated for. Advanced load compensation is possible using signals directly related to vehicle load, for example, air suspension pressures (when so equipped) or axle height signals commonly used with xenon front lighting systems.

3.3 Second-generation indirect TPMS

The wheel assembly is constantly excited by the road surface and oscillating in numerous vibration modes. In the course of trying to overcome the limitations of first-generation iTPMS, the oscillation properties of the wheel assembly have come into the focus of development.

Some of these vibration modes excited are highly inflation pressure dependent and some are possible to monitor with common sensor signals as the wheel speed sensors (Figure 20). The advantage of this approach is that the oscillation behavior of a wheel assembly can be monitored wheel individual and in absolute terms and not only relative to the other wheel positions as is the case for rolling radius. Realizing of monitoring the oscillation behavior therefore enabled the enhancement of iTPMS to a TPMS, comparable to direct systems, which are also capable of monitoring all four tire pressures individually. Indirect systems utilizing both the rolling radius and the oscillation behavior are called *second-generation iTPMS*. With this enhancement, the second-generation systems are able to comply with current and legal requirements that require both puncture and diffusion detections.

There are different options to utilize the vibration behavior but most common is to process the wheel speed signals and analyze them in a narrow frequency band of approximately 30–60 Hz as depicted in Figure 20 to reduce computational effort. Other approaches using vertical and/or translational oscillations have been investigated but not implemented, as additional sensors necessary are not as commonly used.

Figure 20. Spectrum behavior as function of tire pressure.

As with the rolling radius, the vibration behavior is influenced by other parameters than the inflation pressure alone. Vehicle driving speed is the most important and therefore the oscillation analysis is usually divided into different, consecutive speed intervals. The tire itself influences the oscillation behavior through their spring rates, effective mass, and inertia and so does the excitation by the road or effects of vehicle loading. One subset of these influences is compensated for through the reset procedure and learning phase, and the other subset can be actively compensated for by using additional signals such as ambient temperature or the signals also used for performing the rolling-radius analysis.

State-of-the-art second-generation iTPMSs use both rolling-radius and vibration informations and combine them by advanced signal processing to enhance the detection performance and increase robustness against disturbing factors.

3.4 Indirect tire pressure monitoring fundamentals

The so-called tire pressure indicator or TPI, an iTPMS and trademark of NIRA Dynamics, was the first iTPMS to be introduced to the US market in the 2009 model year Audi A6 to comply with regulation FMVSS 138 (Figure 21). In the following, some of its fundamentals will be discussed. Second-generation iTPMSs are also developed and marketed by Dunlop Tech (as deflation warning system or DWS) or Continental (as deflation detection system plus or DDS+).

TPI consists of two main modules; one monitoring the rolling radius called *wheel radius analysis* (*WRA*) and the other wheel spectrum analysis (WSA) is monitoring the oscillation or spectrum behavior.

The WRA and WSA modules calculate common properties or indicators, which describe the rolling-radius ratios of the wheels relative to each other as well as the wheel spectrum behavior. Having compensated for various disturbances, these properties can be used for a number of the so-called detectors, which have the purpose of monitoring certain deflation scenarios. For example, there are one-wheel-puncture detectors, two-wheel-puncture detectors, and several multi-wheel-diffusion-pressure-loss detectors. These indicators work independent from one another. Each of the indicators can, when triggered, issue a low pressure warning. Puncture detectors are typically characterized by their ability to react quickly due to the immediate safety relevance of the underlying pressure loss scenario. Diffusion detectors are typically low pass filtered to exclude influences of short-term disturbances. Detectors in TPI use both rolling-radius and spectrum informations simultaneously, such that the spectrum information is used as an attenuator or amplifier for base information obtained from the rolling-radius information. One exception is the four-wheel diffusion detector, which is purely based on spectrum information. Figure 22 shows the analyzed wheel speed differences and spectrum properties for a puncture scenario. As can be seen, the wheel radius difference and the spectrum are affected by the one-position-only pressure loss case. On the other hand, when all tires are affected at approximately the same pressure loss rate, no wheel speed

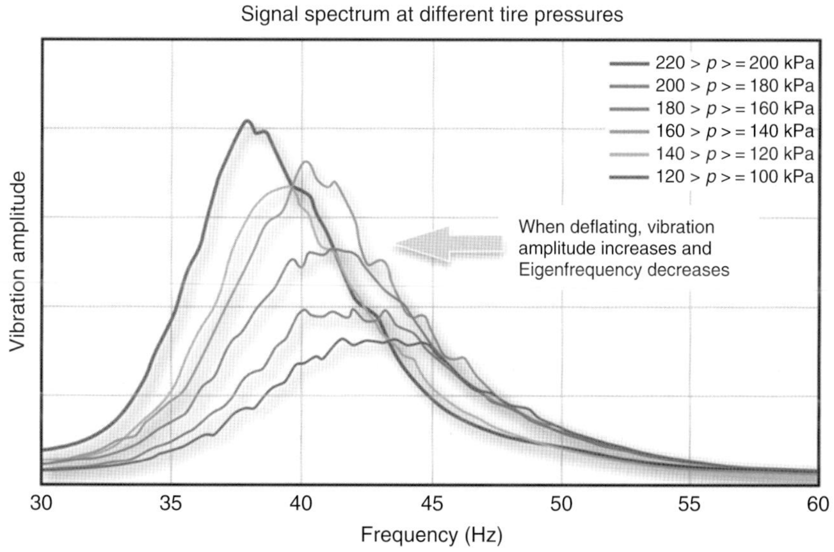

Figure 21. 2009 Audi A6.

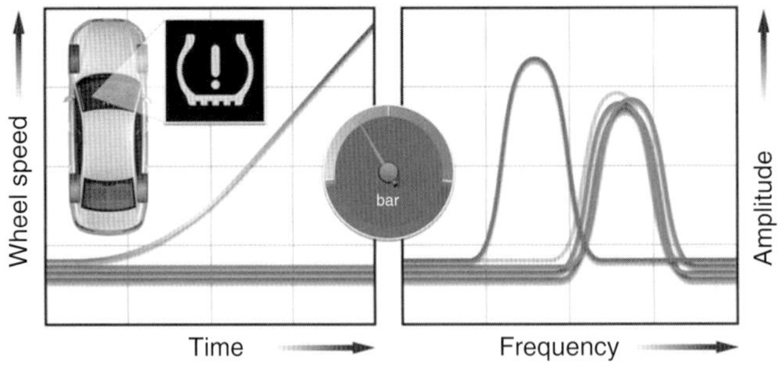

Figure 22. One-wheel puncture scenario.

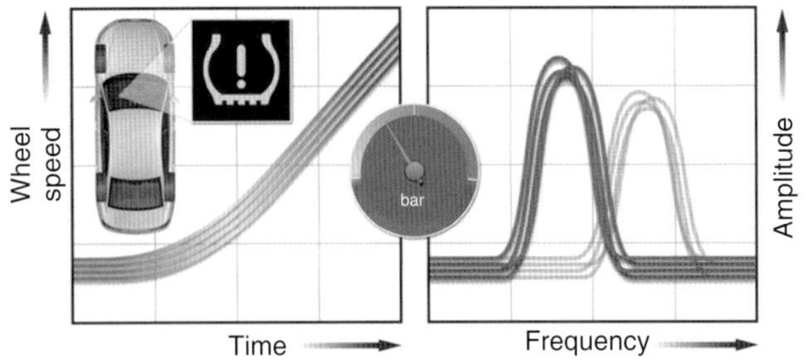

Figure 23. Four-wheel diffusion scenario.

Figure 24. Under inflation warning in cluster.

difference can be seen and only wheel spectrum behavior changes detect under inflation (Figure 23).

This system design is aimed to increase robustness toward short-term disturbances and make it independent to individual tire sensitivities. The use of wheel-individual detectors also enables the so-called isolation functionality that describes the ability of a TPMS to deliver position information about affected wheels, resulting in the possibility to display this information to the driver (Figure 24). Another advantage is the possibility to parameterize and adjust—or even deactivate—detectors individually and by that adapt the system specifically to the target vehicle and its operational/display concept as well as different functional requirement specifications. Additional subfunctions cover self-diagnosis capabilities, for example, in case of invalid or erroneous input signals.

Figure 25. System configuration for second-generation indirect TPMS in an 2012 Audi A6.

3.5 Software integration versus standalone solution

Indirect tire pressure monitoring does not need additional hardware as it only consists of a software module, which is to be executed in the vehicle. Early applications of iTPMS required one additional stand-alone control unit. Examples for that setup are the current 2007 Audi TT or 2008 Audi A6. As this setup does not support one of the most important advantages of iTPMS—no additional physical parts—the integrated solution is preferred. The setup of an indirect system as software module integrated in the ESC software is shown in Figure 25.

As the wheel speed signals are essential for any iTPMS, the selection of suitable host systems is limited to systems, which supply the wheel speed signals in sufficient quality and reliability. Wheel slip control systems therefore represent the natural choice to host iTPMS as they are also highly reliable and provide sufficient hardware resources, software structures, and interfaces. Adoption of standards such as AUTOSAR (automotive open system architecture) is becoming common for both iTPMS and slip control system and simplifies the software integration process. The iTPMS software module is normally embedded or wrapped into a so-called middleware with the main task to control iTPMS execution and transfer input and output signals to and from the iTPMS module. It also controls the use of shared resources such as RAM (random access memory), nonvolatile memory (EEPROM, and electrically erasable programmable read only memory), and diagnostic functions such as system state handling, malfunctions, and diagnostic trouble codes (DTCs).

The integration of an iTPMS is closely linked to the HMI, operational and display concepts, and requirements of the target vehicle. For example, the system integration needs to consider whether position information of underinflated tire is signaled to the driver via additional text messages or if the reset procedure is conducted via button or interactive interface.

3.6 Reset and learning phase

iTPMS as relative measuring systems require the external setting of a reference point, usually by the driver. As soon as relevant parameters such as tire pressure, mounting positions, balancing, or tire types have changed the indirect system needs to be reset to function properly. To prevent the reset function from being executed unintentionally, physical switches (buttons) are usually placed in locations where they cannot be activated without driver awareness (e.g., glove compartment). Additional measures such as a confirmation procedure via interactive menu guidance or plausibility checks, with a minimum vehicle standstill to allow resetting, are common practice.

Once the reset request is accepted by the iTPMS, the existing reference data are erased and the iTPMS starts collecting reference data in the speed intervals visited while the vehicle is driving. Collected reference data are stored in nonvolatile memory and continued learning takes place over several driving cycles if necessary until the learning phase is completed. Different strategies on when to issue full iTPMS functionality, depending on the learning phase completion, are common. Most iTPMS gradually become active depending on the amount of collected reference data and the current speed interval. Full warning sensitivity will be reached after approximately 20 min of driving. Complete learning, covering all speed ranges usually takes about 1 h driving, assuming all speed intervals have been covered. Learning progress also depends on driving style and external road conditions. Additional functionalities, for example, speed extra- and interpolation, allow detection capabilities in speed intervals not yet visited. The reset of the iTPMS is necessary for proper performance and also eliminates numerous disturbing influences such as tire wear, tire aging, seasonal temperature changes, and/or the use of different tire types.

3.7 Application of indirect TPMS

As iTPMS use the tire as a sensor and the fact that tires are not standardized regarding their sensitivity to underinflation, covering a wide range of different tire characteristics, one iTPMS setting is the major application task. Current research activities are under way to take tire sensitivity to under inflation into consideration already when designing tires. This will generate uniform responses over a certain tire program and thus considerably reduce application effort of iTPMS. In addition, chassis or driveline influences can have to be considered when doing iTPMS application. It is therefore important to adjust and adapt the iTPMS in the target vehicle toward the target tire program. As iTPMSs are following the open-loop concept, they can effectively be tested using vehicle test recorded data files covering all input signals. These data files can be replayed off-line using different iTPMS software versions and parameter settings without repeating the actual vehicle test (Figure 26). The application of an iTPMS normally consists of one or more structured data collection campaigns, where relevant target vehicle model variants and tires are tested under various conditions and the input data being collected (detection and robustness tests, winter- and high temperature testing, road surface screenings, etc.). The data files are then replayed off-line and complemented with further vehicle tests on demand. As soon the database status makes it possible, algorithms and parameter settings are iteratively optimized until the test results meet the OEM requirement specifications. Using automatic test environments drastically reduces optimization loop time.

4 LEGAL ENVIRONMENT

After a series of fatal accidents, which were related to tire tread separation on certain Firestone tires mounted to the Ford Explorer in 2000, the US Congress enacted the so-called TREAD act and subsequently a new FMVSS 138, requiring new light vehicles [passenger cars, multipurpose passenger vehicles, trucks, and buses with a maximum

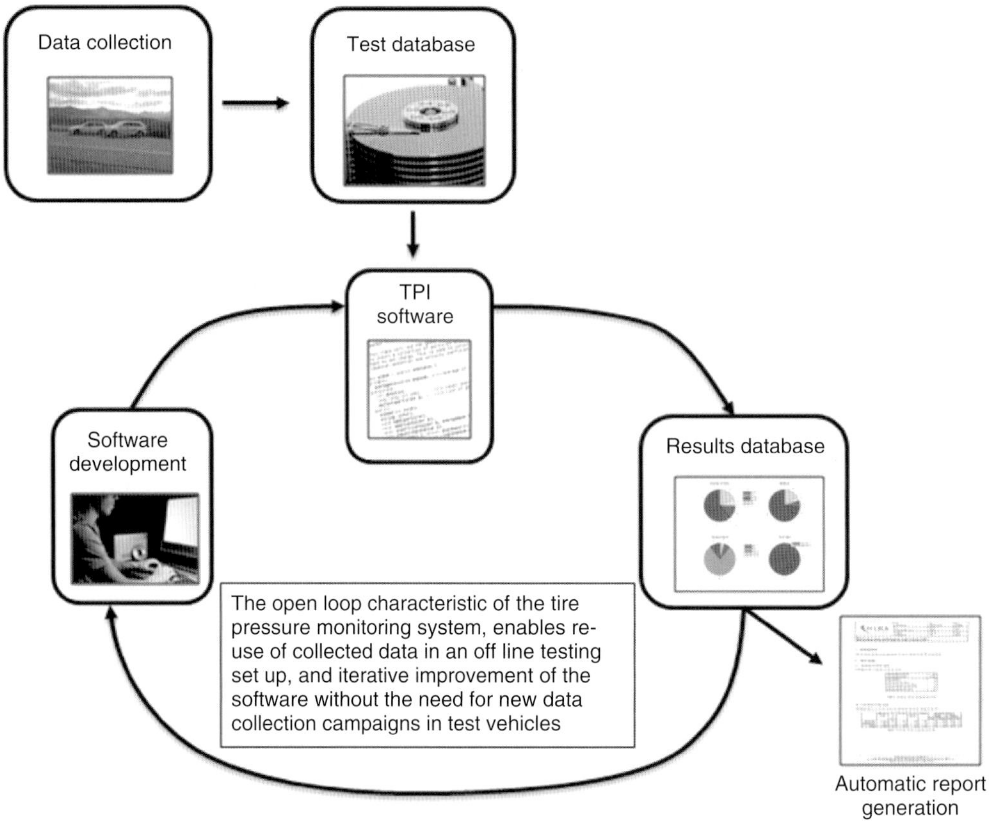

Figure 26. Representation of open-loop concept for iTPMS application.

gross vehicle mass rating of 4536 kg (10,000 pounds) to be equipped with a TPMS]. After a phase-in period starting in 2005, from September 2007, all new vehicles sold in the United States need to be equipped with TPMS conforming to FMVSS 138.

The safety goal of this regulation was to reduce accidents caused by tire failures that in turn were caused by significant underinflation. FMVSS 138 requires a warning to the driver, when the tire pressure falls below 25% of the cold recommended pressure level in one up to all four tires. In general, FMVSS 138 is a technologically neutral drafted performance standard, that is, it does not prescribe specific technology for fulfillment. Besides underinflation detection, FMVSS 138 also requires TPMS malfunctions to be signaled to the driver. A low tire pressure condition will be displayed by illumination of the standardized low tire pressure telltale (Figure 27). A number of OEMs chose to amend the minimum telltale requirements of FMVSS 138 with additional text messages to inform the driver about the situation and point him or her toward corrective actions.

In addition to the safety aspect of tire pressure monitoring covered in FMVSS 138, in 2009, the EU parliament decided mainly with focus on reducing CO_2 emissions to require TPMS for all new vehicles with the new regulation ECE-R 64. The regulation applies to M1 and N1 vehicles (passenger vehicles and vehicles designed to carry goods

Figure 27. TPMS telltale in cluster.

with a maximum mass not >3.500 kg). From November 2012, all new types and, from November 2014, all newly registered vehicles need to comply with ECE-R 64. ECE-R 64 requires a low tire pressure warning at 20% pressure loss with respect to the "warm tire pressure." Depending on the pressure loss scenario, ECE-R 64 requires different warning times, 10 min for a puncture on one wheel and 60 min for simulating the diffusion on all four wheels. The telltale and display requirements are mainly carried over from FMVSS 138, and the malfunction detection time has been reduced to 10 min.

Compared to US regulation FMVSS 138, the European regulation ECE-R 64 represents a substantial reducing of warning thresholds, as not only the warning level has been decreased to 20% but also the pressure reference has changed from the recommended cold tire pressure to the in-service, operational or the so-called warm tire pressure. Actually, ECE-R 64 almost cuts the warning threshold in half compared to FMVSS 138. Figure 28 shows the pressure trace for FMVSS 138 and ECE-R 64 test procedures. Once the recommended tire pressure is adjusted, the vehicles need to be driven to allow the TPMS to learn current tire/pressure characteristics (calibration or learning phase). During driving the calibration phase, the tire will heat up and the pressure will be above the initially adjusted cold inflation pressure. Pressure build up is dependent on different influences such as driving style and axle load but usually reaches values of about 10–15% of the cold tire pressure. After the calibration phase, the pressure will be adjusted to the warning level (25% for FMVSS 138 and 20% for ECE-R 64), whereas the reference pressure for adjusting is the cold tire pressure (FMVSS 138) and the in-service pressure (ECE-R 64). Subsequently, the vehicle is driven in the detection phase, where the detection time requirements vary depending on the exact scenario tested. In FMVSS 138, the warning must be issued at the latest 20 min after deflation independent of the pressure loss scenario (deflation in one up to all four wheels), ECE-R 64 requires detection in 10 min for the one-wheel puncture and 60 min for the four-wheel diffusion scenario. Another important difference between US and ECE requirements is that the United States realizes a self-certifying regulation, whereas ECE-R 64 calls for a type approval/homologation process done by the OEM to allow for registration of vehicle.

All tire pressure-monitoring systems—except first generation indirect systems—have been shown to be able to fulfill current US and ECE regulations. A performance comparison between direct TPMS and iTPMS is shown in Figure 29.

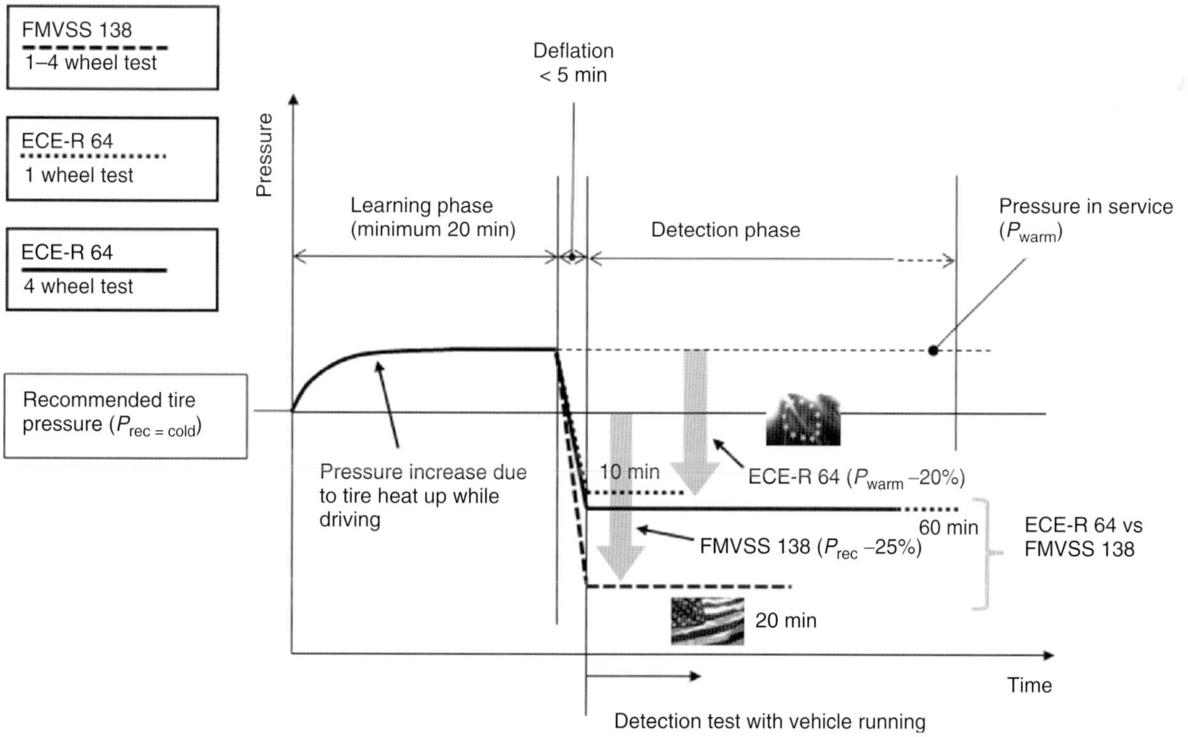

Figure 28. FMVSS 138/ECE-R 64 test procedure.

	Direct measuring TPMS	Indirect measuring TPMS	
		1st generation	2nd generation
Puncture detection	✓	✓	✓
Diffusion detection	✓	✗	✓
Display of tire pressure	✓	✗	✗
Identifying puncture position	✓	✗	✓
FMVSS 138 (U.S.)	✓	✗	✓
ECE-R 64 (E.U.)	✓	✗	✓

Figure 29. Performance matrix of direct/indirect TPMS.

5 FUTURE TRENDS AND DEVELOPMENT

Enacting legal requirements on tire pressure monitoring has and will further increase the market penetration of TPMS. After 100% mandating TPMS in the US market via FMVSS 138 since 2007, in 2014, the EU market will require 100% compliance with ECE-R 64.

On the side of direct systems, there is a clear trend toward reducing the number of components, complexity, and costs, most important for the sake of customer acceptance. US market experience has shown that, especially, the after-market has severe challenges to serve to the vast variety of direct systems currently available. Therefore, there is a clear trend toward standardization of hardware components and data transfer protocols, so systems from different OEMs/TPMS suppliers can interact in the future. It has also been understood that the issue of sensor replacement for direct systems over the lifetime of the vehicle poses a serious customer acceptance concern: market research has shown that only one-third of customers will replace sensors with empty batteries. Extensive research in the field of energy harvesting is on the way to eliminate the need for batteries in the pressure sensors of direct TPMS. Currently, the most promising approach for energy harvesting in the TPMS environment seems to be the deployment of the piezo-electric effect, but as of today, no energy harvesting system is in series production.

The topic of energy harvesting goes strongly connected to the implementation of the tire-integrated sensor (Figure 30). By attaching the sensor to the tire via gluing or curing, more functionality besides tire pressure monitoring could be realized. If permanently fixed to the tire, tire-specific data such as tire type (summer, winter, and run-flat) or speed and load index could be stored to the tire sensor, allowing for additional benefits: speed warnings could be issued when exceeding the speed index specified speed, or driver information could be provided if the tire type does not fit current weather conditions (summer tire running in winter conditions). In addition, the tire-integrated sensor could determine the size of the contact patch between road and tire, information that can be utilized when optimizing rolling resistance or determining wheel loads. Nevertheless, with regard to the tire-integrated sensors, not all issues are resolved yet. For example, the permanent sensor attachment necessary for realizing benefits beyond tire pressure monitoring has to address the issue of tire replacement when a tire is damaged or the tire life is reached because the tread has worn out. In this case, not only the tire but also the sensor will be disposed, which economically can only be done when sensor prices decrease substantially. On the other hand, making the tire-integrated sensor replaceable will not allow for realizing all possible benefits, because the stored tire information may not be correct because of tire changes. It is

Figure 30. Integrated tire sensor.

expected that the tire-integrated sensor will start its market introduction in the high price sports car segment, where the tire program is usually limited and special customer needs justify the special effort and price. At the moment, no tire-integrated sensor solution is currently available in series production.

As it has been proved that indirect systems fulfill legal requirements FMVSS 138 and upcoming ECE-R 64 legislation, the market share of iTPMS is expected to grow substantially over the coming years. The benefits of providing a robust tire pressure monitoring functionality with reduced hardware and complexity in combination with no follow-up costs over the vehicle lifetime will appeal to mid-segment, cost-conscious, and premium-segment OEMs alike. For example, Audi has implemented the indirect system into almost all the vehicles in its model line up.

ACKNOWLEDGMENTS

We would like to thank Ralf Kessler and his team of HUF Electronics Bretten for significant contributions to the segment of direct systems. Another special thanks goes to Volker Rühr of Dr. Ing. h.c.Porsche AG and Dr. Andreas Köbe of Continental Engineering Services GmbH who supported on the topic of tire pressure monitoring system introduction and history.

RELATED ARTICLES

Volume 4, Chapter 115
Volume 4, Chapter 117
Volume 5, Chapter 146
Volume 5, Chapter 149
Volume 5, Chapter 152
Volume 5, Chapter 153
Volume 5, Chapter 161

FURTHER READING

Basbantu, G., Pellicciari, M., and Andrisano, A. (2004) On the tire monitoring systems temperature compensation. SAE-paper, 2004-01-110.

Bosch Kraftfahrttechnisches Taschenbuch (2007) *Reifendruckkontrollsystem*, S. 810/811, 26. Auflage, Friedr. Vieweg & Sohn Verlag, Wiesbaden, Germany.

Fischer, M. (2003) *Tire Pressure Monitoring, Die Bibliothek der Technik*, vol. **243**, Verlag moderne Industrie, Landsberg.

Folger, J., Riedl, H., and Wallentowitz, H. (1989) Electronic Tire Pressure Control. *International EAEC Conference*, Strasbourg.

Greenly, C. and Beverly, J. (2005) Concerns related to FMVSS No. 138: tyre pressure monitoring systems and potential implementation of a similar standard on commercial vehicles. SAE 2005-01-3517.

Kowalewski, M. (2004) Monitoring and managing tire pressure, *IEEE Potentials*, **23** (3), 8–10.

Marshek, K. and Cudermann, J. (2002) Performance of anti-lock braking systems equipped passenger vehicles – past 111: braking as a function of tyre inflation pressure. SAE 2002-01-0306.

Minf, K. (2001) A smart tire pressure monitoring system, *Sensors*, **18** (11), 40–46.

NHTSA (2005) Federal Motor Vehicle Safety Standards—Tyre Pressure Monitoring Systems, FMVSS No. 138, Final Regulatory Impact Analysis.

Paine, M., Griffiths, M., and Magedara, N. (2007) The Role of Tyre Pressure in Vehicle Safety, Injury and Environment, Road Safety Solutions, Caringhah, NSW, Australia.

Persson, N. and Gustafsson, F. (2002) Indirect tyre pressure monitoring using sensor fusion. SAE 2002-01-1250.

Pohen, F.-H. (2009) Entwicklung einer radgebundenen Reifendruckregelanlage für landwirtschaftliche Fahrzeuge. Dissertation RWTH Aachen, 2009 Forschungsbericht Agrartechnik Nr. 482, Shaker Verlag, Aachen.

Umeno, T., Asano, K., Okashi, H., *et al.* (2001) Observer based estimation of parameter variations and its application to tyre pressure diagnosis, *Control Engineering Practice*, **9**, 639–645.

Wallentowitz, H. and Reif, K. (eds) (Hrsg.) (2006) *Handbuch Kraftfahrzeugelektronik*, S. 513 f, Vieweg Verlag, Wiesbaden, Germany.

Williams, R. (1992) DWS – ein neues Druckverlust-Warnsystem für Automobilreifen, *Automobiltechnische Zeitschrift (ATZ)*, **94**, 336–340.

Braking Systems

Chapter 117
Brake Systems, an Overview

Christoph Drexler and Ralf Leiter
TRW Automotive, Koblenz, Germany

1 INTRODUCTION

Today, some highly powered vehicles are on the road, often without it being clear what the required capacity of the brake system of these vehicles is. Thus, for example, when braking a modern vehicle of the upper middle class with an engine output of 130 kW (178 HP) to a complete stop from a speed of 220 km/h, the brake force must be more than three times the engine output. In addition to reliability and safety of the brake system, it is also required that this process takes place with absolutely no noise, is easy for the driver to apply and especially easy to control, as well as being comfortable for all passengers.

Even though the vehicle brakes have reached a very high level of development, more than 50% of all technical defects in motor vehicles are caused by the brake system.

Encyclopedia of Automotive Engineering, print © 2015 John Wiley & Sons, Ltd.
Edited by David Crolla, David E. Foster, Toshio Kobayashi, and Nick Vaughan.
This article is © 2015 John Wiley & Sons, Ltd. ISBN: 978-0-470-97402-5
Also published in the *Encyclopedia of Automotive Engineering* (online edition)
DOI: 10.1002/9781118354179.auto024

The first vehicles (including the means of transportation of antiquity) were already equipped with devices to stop the vehicle from rolling while stationary. This was the start of parking brakes. Thus, the vehicle brake is surely as old as the wheel itself. As these vehicles increased speed (Roman chariot races), an operating brake was developed that was required to stop the vehicle from rolling when stationary, as well as for decelerations, when driving downhill or when stopping.

2 TYPES OF BRAKES

Generally, the brake is a device to slow down the vehicle:

- until it stops
- to secure the parking vehicle
- to control (reduce) the speed of the vehicle.

During the brake application, kinetic energy is converted to heat. The parking brake prevents rolling of the vehicle. When operating during standstill, this brake works without wear or heating. It can be executed as a friction brake or also as a locking device like in an automatic transmission gear box.

According to the physical work principle, vehicle brakes can be subdivided into three main groups:

- mechanical friction brake
  by solid state friction (wear) of two rigid bodies.
- hydrodynamic brake
  - by inner friction of a medium: water, oil, and fluid friction.
- electrodynamic brake
  - by magnetic effect of the electrical field: generator and eddy brake.

2.1 Friction Brake

In this contribution, only friction brakes are considered, which are working on the principle of mechanical solid friction. They can be used as a parking brake as well as a dynamic brake (stop the car, control the speed). In the case of dynamic braking, it works strictly as a wearing brake. Disk, drum, belt, or cable brakes can be used. Here, only disk and drum brakes will be dealt with.

2.2 Disk brakes

The disk brakes usually consist of a disk (gray iron, cast aluminum, carbon fiber pellets, and ceramic reinforced with carbon fibers), which is mounted to the shaft. The disk brake is predominantly made as a partial disk brake. The friction lining only covers a sector of the disk ring, and not the entire perimeter. The brake calliper encloses the disk such as a gripper. Assisted by one or several hydraulically operated pistons, steel or aluminum segments with applied friction material that are axially movable are pressed against the disk. Very high surface press forces are possible due to the opposing tensioning force (Figure 1).

With increasing wear of the friction linings, the pistons extend further and readjust the wear path. The brake clearance, which must be guaranteed after each operation, remains constant. Additional fluid volume required for this flows from a reservoir. Protective caps on pistons and guide bolts prevent the brake from seizing due to dirt and corrosion. Despite high local temperatures, few fluctuations in adhesion factor occur, compared to drum brakes. This will be shown later.

Disk brakes have good cooling from air flow, as long as the installation space in the vehicle permits air access. With high thermal load, inner ventilated and/or perforated disks are used. The heat expansion is not critical, as it works toward the tension forces. Cracks in the brake disk are considered as production or design faults.

Figure 1. Disk brake. (Reproduced with permission from TRW.)

The disk brake provides a simple and easy accessible device for maintenance and checking for wear. The design of the disk brake is further explained in the example of the brake Colette, a frequently used floating calliper brake. The brake (Figure 2) consists of a housing part, which can be adjusted using two guidance bolts, and a stationary support. The brake linings are hold on the support (4) in this type of brake, and they are held in position with springs (1). These springs mainly prevent noisy rattles and clicks (2 and 3). The lining material that is glued to the back plate is between 10 and 15 mm thick and should be replaced when there is residual thickness of 2 mm. Between the lining back plate and the piston, most disk brakes have a noise damping panel.

The floating calliper brake has proved to be well-suited for small installation spaces in vehicles with negative king pin offset. The housing only contains (one up to three) pistons on the disk side. When operating, the hydraulic pressure works on the pistons and the housing evenly. The piston moves out and presses the neighboring lining against the disk. At the same time, the floating housing, perpendicularly arranged to the disk, is moved in the other direction, so that the second lining is pulled against the other disk side. The holding of the brake must be rigid, which limits the realizable disk diameter, as the rim determines the maximum available space. The movement of the floating calliper is arranged by the guide bolts. The bolt on the disk inlet side positions the housing to the support, whereas the second provides a fixed seat with its rubber coating and prevents rattling in the bore while driving. The relationship of lining size to design size is very high compared to other disk brake designs.

To change the lining (example component, Figure 3), merely release the hex screw. Afterward, the housing can be folded away and the linings can be pulled out away from the disk (upward). In addition, the support also carries the linings in circumferential direction, so that no circumferential forces must be transferred via the guide bolts.

New types of designs have a lining without retainer springs on the upper lining edge. One Nirosta spring sheet on each side position the lining in nonbraking condition and increases its lateral mobility to the disk. The response time of the brake improves, and the residual friction torque of the linings are reduced when the brakes are not in operation.

The movements of the lining parallel to the disk are highly limited. A groove together with the spring sheet on both ends prevents unscrewing while braking. This also minimizes the lining travel to the disk for braking and clicks (as well as squeaking) no longer occur when changing

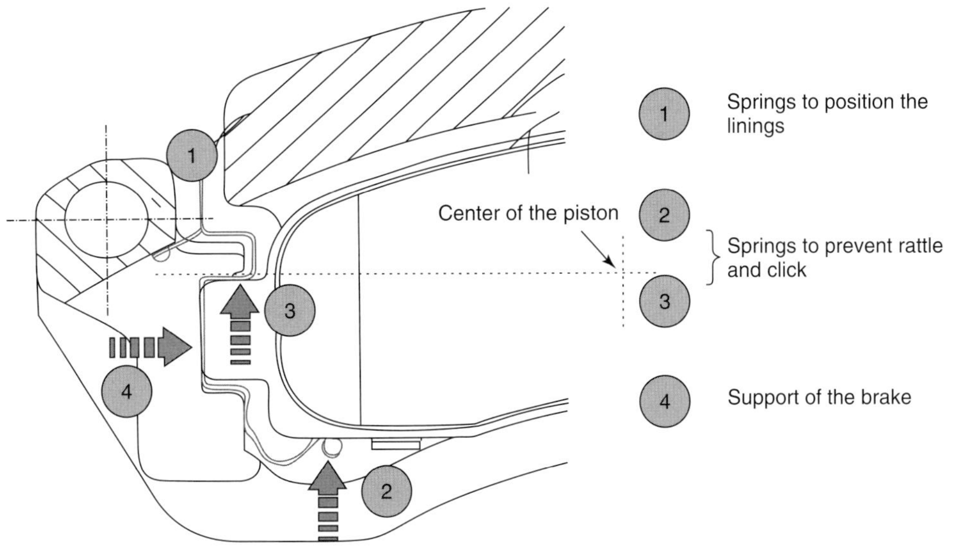

Figure 2. Floating caliper brakes Colette II. (Reproduced with permission from TRW.)

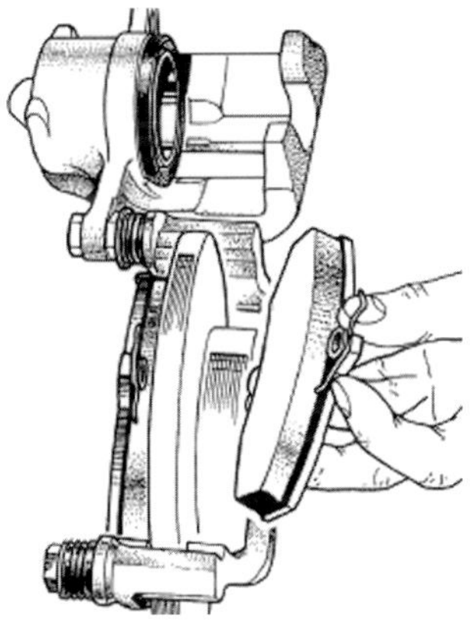

Figure 3. Changing linings Colette I. (Reproduced with permission from TRW.)

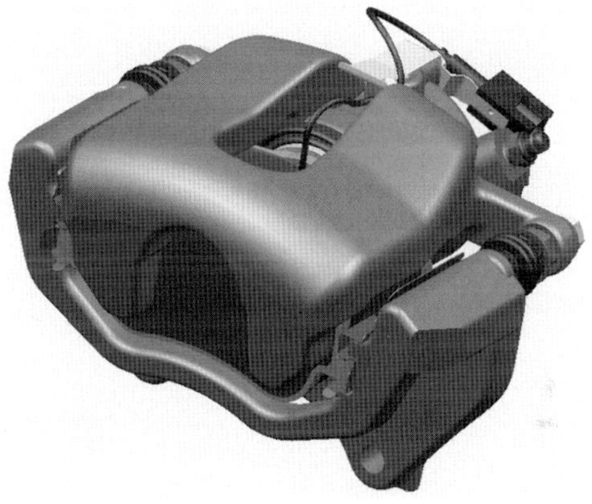

Figure 4. Asymmetrical disk brake. (Reproduced with permission from TRW.)

direction of rotation. On uneven roads there is no rattle as well.

To improve lining wear, the latest brakes (Figure 4) no longer have symmetrical designs. The piston is offset opposite the lining back plate, which also identifies the milling groove necessary on the wheel side for processing the housing. The brake plate is only reinforced on the outlet side, which is also where most of the forces are to be transferred. This leads to additional weight reduction.

The brake lining is eccentrically placed on the back plate corresponding to the piston axle. Opposite the back plate, the piston is also radially offset, using a partial sheet that is simultaneously a damping plate, or the piston is displaced opposite the lining's center of gravity. This sheet, in bare or rubberized execution, can be placed or glued to the back plate. This dampens lining vibrations that result during the braking process in the friction path. Minimizing sounds is becoming more and more significant in modern brakes. For this reason, special design measures are even being introduced for damping noise (dynamic vibration absorber). The friction process is the trigger for all brake noises. Each wear optimization that affects the friction process also leads to noise reduction, as the origins of specific noises are prevented.

The frame brake (Figure 5) is a significantly more rigid execution of a floating calliper brake. Design and function are similar to a floating calliper brake. In the RC 5424/13 design with a high performance spectrum, the housing was shifted to the outside and the carrier with linings to the inside. This makes the housing bridge especially

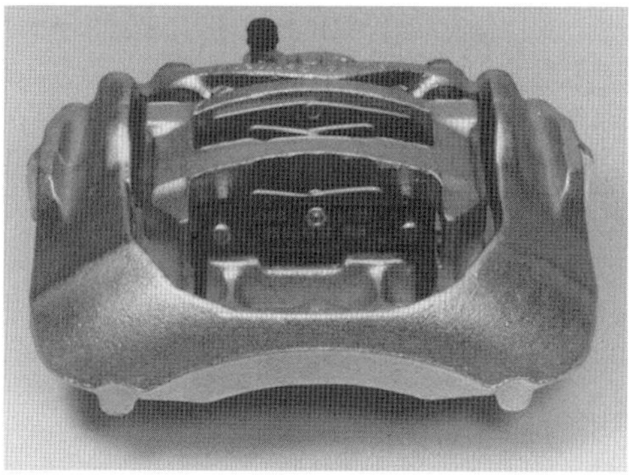

Figure 5. Frame brake. (Reproduced with permission from TRW.)

Figure 6. Semi-integrated disk brake. (Reproduced with permission from TRW.)

rigid. The tension forces on the disk are transferred in the plane of force (piston axle), so that there is no radial bending up. However, the benefit of a thinner bridge is offset by width, and the disk diameter can be designed to be larger. The cross member must be very rigid (solid), which increases the weight. The linings are carried by the support in circumferential direction and they are positioned with central springs in the center of the lining. The stable lining pressure is radially and tangentially well supported. Despite the large size, the lining size is somewhat small. This brake is installed on the front axle in larger vehicles like, for example, the VW Transporter with permitted total weight up to 2.8 t.

The FIS16 brake (Figure 6) is a semi-integrated floating frame calliper brake of the middle performance spectrum (semiabutment). Semi-integrated means that only the outer lining is supported in the housing and the housing guide bolts must transfer these circumferential forces. In contrast, the inner lining is directly supported by an integrated steering knuckle carrier. Fully integrated brakes with support of both linings in the housing are also in series.

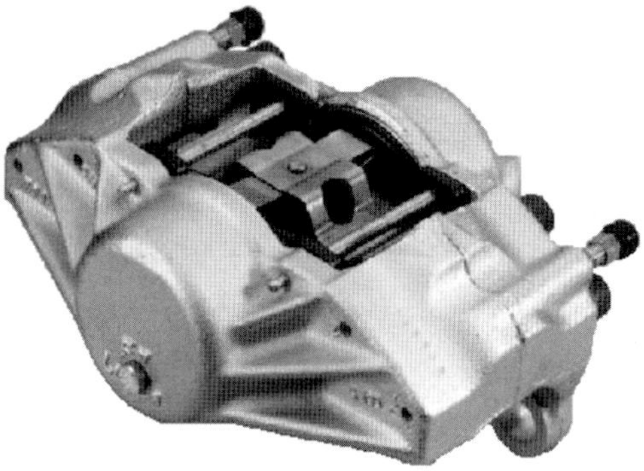

Figure 7. Two-piston-fixed caliper brake. (Reproduced with permission from TRW.)

Today, all brakes are provided with surface protection against corrosion. This can consist of galvanizing followed by yellow chromate conversion coating, of special paint coats or similar.

Another design is the fixed calliper brake (Figure 7). Instead of a sliding housing, a stationary housing is used. There is no support. The tension forces are applied by the opposite pistons. Fixed calliper brakes are built with up to eight pistons per side. All pistons are hydraulically connected with each other. For this, it is necessary to bring brake fluid through the bridge over the disk to the other side of the calliper. If there is longer braking (e.g., in the mountains), or with many successive hard stops, the hot brake disk can heat up the brake fluid in the bridge up to steam formation. As steam is compressible, the brake system loses effect heavily (fluid fading).

The design, which is extremely solid because of the fixed connection between housing halves, allows very large tension forces with minimal housing deformation (high performance spectrum). The response behavior is especially good due to this. The two or three hydraulically connected pistons per lining are individually optimally pressed. The lining wear is thus very even, (no diagonal, tangential, or difference wear). The noise production is also decreased. The clearance when releasing the brake is the same for both linings (inner and outer) and is also quickly deployed. The difference wear is also less when the brakes are not applied. Application noises do not occur.

It is not possible to install fixed calliper brakes for vehicles with negative kin pin offset because of lack of space. The ball joints of the wheel suspension and the brake disk slide too closely together or work on the same spot in the wheel disk.

A completely different version of the disk brake is the wrap-around brake. It has the great advantage of moving the friction process to the largest possible radius, and thus to achieve a higher brake torque than outer disk brakes. It is a high performance brake. However, as there are disadvantages from difficulties in manufacturing and also in assembly and maintenance (due to poor accessibility), this brake may not be considered in this context. As a summary, this type of disk brake is very expensive and technically too complex for use in series production.

2.3 Drum brakes

The drum brake (Figures 8–10) is predominantly designed as an inner shoe brake. By spreading apart radially, the brake shoes are pressed against the inner side of the rotating brake drum. The shoes can be housed on the carrier in different ways: we differentiate between brake shoes with single or double pivot, members or friction pads. The drum brake shown here contains friction pads with glued-on linings. Earlier drum brakes exclusively had riveted linings.

Figure 9. Simplex drum brake. (Reproduced with permission from TRW.)

Figure 8. Drum brake. (Reproduced with permission from TRW.)

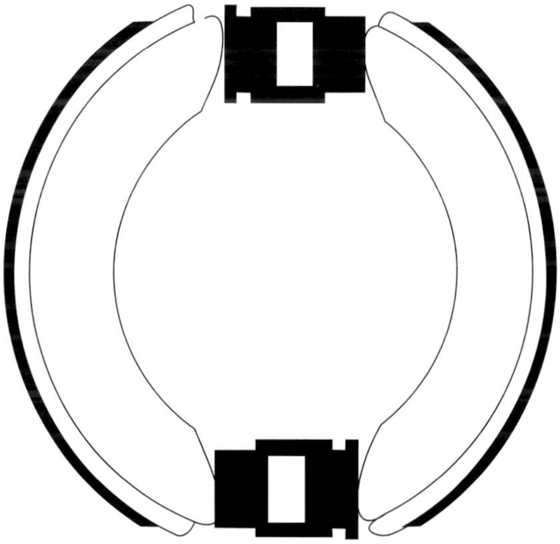

Figure 10. Duo duplex brake. (Reproduced with permission from TRW.)

The linings generally have a very long service life; despite this, a resetting device is required for wear, because of the longer application path. Drum wear is also minimal. Different designs are dependent on the mounting and operation of the brake shoes. The parking brake function is easily realized with an additional mechanical operation. The brake factor is high due to self-reinforcing effect; however, it can vary significantly because of the adhesion factor fluctuations during operation. Altogether, the drum brake is a cost-effective design.

Today, the simplex brake is the most commonly used drum brake. Figure 9 shows the light-weight version, with the anchor plate, the support and the brake wheel cylinder made of aluminum. When the drum turns counterclockwise, the left brake shoe becomes the leading shoe with self-reinforcing effect (accumulating) and the right becomes the trailing shoe with reduction (self-attenuation). By a hand brake lever and a pressure rod, that simultaneously contain the automatic resetting device, the drum brake can be operated mechanically using a cable pull (parking brake). The upper and lower return springs pull the linings away from the drum after releasing the brake and provide clearance again. The retaining spring provides the correct position of the brake shoes along with the lower support.

During operation, high adhesion factor fluctuations can occur, which effects the brake characteristic C*.

The longer application path of the shoes due to lining wear is automatically reset. With the aid of an adjusting lever, a pinion is operated over the application path, which steadily spreads the brake shoes apart. The necessary clearance is always maintained.

If there is a longer application process, the drum expands as a result of the heat. If there is an automatic clearance reset now, the brake would jam after cooling off. To prevent this, a thermal-bimetal (thermoclip) compensates for the longer application path because of temperatures, and the resetting is suspended.

The cooling application of the drum brake, especially those of the brake shoes, are not as good as with a disk brake. The thermal expanding of the drum negatively effects the braking behavior. This results in an extension of the application path of the brake shoes, which is compensated with increased piston travel in the wheel cylinder. This causes a greater volume intake of brake fluid and with it a longer pedal travel. (If the drum is already heated, the driver first steps into emptiness at the start of operation.) The incorrect application of the shoes in the drum decreases the actual lining application length and decreases the brake lining parameter. This results in uneven brake behavior (temperature fading). For this reason, a maximum temperature of of 400°C is permitted at the drum brake during application. Depending on the manufacturer, the disk brakes are between 750 and 1000°C.

Different designs of the drum brake are dependent on the mounting and operation of the brake shoes. Here, the five most important types of drum brakes are more closely described.

The simplex brake (Figure 9) was already described. Both rotation directions have the same total brake force.

Duplex brakes have two single wheel cylinders and two leading shoes; however, they only provide a double reinforced brake force when driving forward. When driving in reverse, the brake force is significantly less (decreasing two times).

The duo duplex (Figure 10) brake is similarly built as the duplex brake. However, it is equipped with two double wheel cylinders and two leading shoes for both rotating directions. This ensures the same strong brake force each time. When changing rotation direction, usually application noises occur.

The power brake consists of a double wheel cylinder and the floating housing of two accumulating sliding shoes when driving forward due to the support force of the primary against the secondary shoe. This leads to a very high self-reinforcement, however only when driving forward. When driving in reverse, it corresponds to a simplex brake. Here, application noises also occur when changing rotation direction.

The duo servo brake is designed similar to the power brake. Dependent on the rotation direction, the two shoes are supported by a support bearing. With this, there are two leading brake shoes for both rotation directions with very high self-reinforcement. Application noises occur when changing rotation direction. This type of brake is frequently used with a mechanical expander lock as a parking brake.

2.4 C* value of brakes

The brake torque is the product of the average friction radius r_{Bm}, the piston surface A_K, of the hydraulic pressure p_{hydr} and the brake lining parameter C*. This parameter describes the relationship between the circumferential force and the tension force, which is dependent on the design and the adhesion factor (Figure 11). As the braking torque should, if possible, remain the same during the brake process, a constant lining-friction parameter (linig adhesion factor) is required.

This requirement is best fulfilled by disk brakes. Here, the C* value is equal to twice the adhesion factor. The brake lining parameter depends on the self-reinforcing factor of the drum brake (geometry). It reacts very sensitively against adhesion factor fluctuations. Figure 11 shows the dependence of the brake lining parameter C* on the lining adhesion factor with different brake designs. As adhesion factor fluctuations are unavoidable, and always occur, but the torque created by the brake should be as even as possible, only the simplex brake is still considered for practical application in modern cars, except for the disk brake. With correct design and moderate self-reinforcement, the simplex brake only differs from the disk brake a little bit. The remaining drum brakes with high self-reinforcements are unsuitable because of the difficult dosing, but were used in the past. The advantage has been that no brake booster

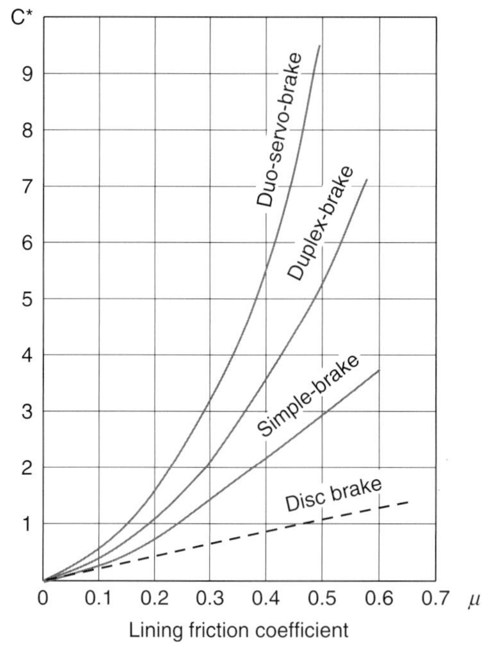

Figure 11. C* values of different brakes. (Reproduced with permission from TRW.)

have been necessary. Today, the brakes with the minimal C* values require additional reinforcement of the driver's foot force through a separate brake booster.

2.5 Light-weight design

In addition to the main task of light-weight design, to reduce the vehicle weight in total and to decrease the fuel consumption, there is especially the requirement to reduce the unsuspended masses on the axles of a vehicle. This includes the brakes and the brake disks as significant parts. Assuming a proved design, the parts containing iron with high density (e.g., steel and cast parts) are replaced with other materials. As a 1 : 1 substitution is usually not possible because of strength, wear, or thermal reasons, the components must be modified in design, corresponding to the requirements of the new material. This often leads to compromises, and the maximum possible weight reductions (with Al 62%) are not reached, as the modified components have more volume.

The following materials are used or under development as light-weight materials:

- Aluminum in various combinations, predominantly as ALMMC (aluminum metal matrix composite), a composite that is reinforced with 20–50 vol% silicon-carbide. This improves the mechanical characteristics, and the melting point increases.
- Ceramics as porcelain from silicate, exclusively used for pistons. Heat-resistant technical ceramics made of silicon-nitride is too expensive by 10-fold.
- Plastic, made of venyl and epoxy resin with good mechanical and thermal characteristics.
- In addition, pistons made of Nirosta sheet metal or anodized aluminum and magnesium are under development.

The modifications due to the new materials are, for example, as follows:

- The brake disk, previously manufactured from gray iron, has a larger thickness in the friction circle so it does not fail thermally. However, in many applications, this material cannot be used, as the thermal stresses overall are too high. Application area is the rear axle.
- On the floating calliper housing, previously made of spherical graphite iron, the bridge strength was increased by ribs when made of aluminum, to keep the expansion during operation to a minimum.
- The wall of the piston made of plastic, ceramics, or aluminum is also thicker than in the previous steel piston, to keep the pressures at 160 bar.
- The weight of the lining rear plate is additionally reduced by bulges. In the former steel plate, there were still reserves here. In addition, the connection with the lining material is improved and the form closure is improved.

The high thermal stress of the brake with the aluminum brake disk is opposite the low melting point and less strength of the material. Therefore, only the use of fiber or particle-reinforced alloying such as ALMMC is possible. The maximum operating temperature of 450°C must be absolutely maintained; above that, the material loses its strength and already reaches its melting point at 550°C. Owing to the lower density, with the same volume, ALMMC only has a very low absolute heat storage ability compared to gray iron, despite the higher specific heat capacity.

The time history of the temperature during a stop, calculated using the finite element method, indicates the distribution of energy within the disk (Figure 12). The temperature in the friction path center increases as the first and quickest; in the disk head, it does not increase until the end or after the stop. The heat released through convection during the stop (radiation or heat conduction into the disk head) can be ignored, independent of material, as during the highest case it only amounts to 5% of the energy conducted in only a few seconds during braking. Therefore, the brake

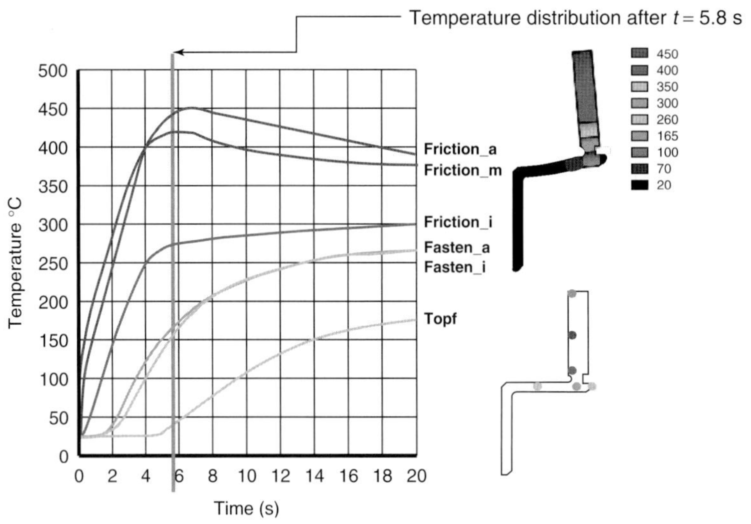

Figure 12. Heating of the disk during a brake process. (Reproduced with permission from TRW.)

disk during a stop is to be considered purely as a heat accumulator, which gets filled.

After braking, in the cooling phase, a material with good heat conduction brings advantages. The temperature profile within the disk circle is better balanced, which causes a quick convective heat release and more convenient heat conduction into the head. The heat conduction into the head must be limited to prevent excessive heating of the wheel bearing; the grease would leak out in fluid form.

Owing to the temperature distribution, different expansions occur, which leads to deformations and tensions of the disk. Figure 12 shows the deformations in a high performance test with successive stops, a vehicle is braked several times in a row with a high deceleration of 70% of gravity and is then accelerated as fast as possible. The first five stops are performed from 90% of the maximum speed to 80 km/h and the last stop to 0 km/h.

Comparing gray iron to an iso-volumetric ALMMC disk, it becomes obvious that the aluminum disk is quickly heat saturated. The temperature in the gray iron disk continues to increase. Owing to the lower storage capacity of the ALMMC, the temperature difference at the same stops is larger. Owing to the higher heat conductivity of the ALMMC, the cooling between stops occurs better than with gray iron.

The temperature of the disk head near the wheel always increases after a stop, an indication that the heat conductivity during a stop can be ignored. The maximum temperature of 180°C clearly shows that the neighboring parts, such as wheel bearing or rim also experience heating, which also occurs more quickly especially with longer braking, and therefore prevails longer than with the gray iron disk.

Test bench tests with ALMMC disks have shown that the thermal loads together with the mechanical loads can be managed with aluminum. The strength of such disks, especially against abrasion, is ensured by silicon carbide particles. A black lining layer forms on the disk surface, so that the lining material rubs against itself. The brake wear is also heavily reduced by this. There is also less wear on the brake linings that are specially developed for aluminum disks.

The tests performed for rear wheel brakes (high performance and fading successive stop brakes as well as the simulation of a slow alpine descent: long-term braking for 45 min) were successful. When exceeding the performance (in this case with double vehicle mass), the disk does not completely fail. On the surface, the melting temperature is reached and deep ridges result. The protective coating is destroyed. An iso-volumetric exchange ALMMC for gray iron will not be possible in most cases. This material can only be used when still additional installation space will be available. Highly loaded brake disks, as are necessary for high powered heavy vehicles on a front axle, cannot be made of ALMMC with the same dimensions.

Brake disks in sandwich construction are under development for use in highly loaded front axle brakes. To decrease the high thermal and wear-inflicted loads of the disk surface, an aluminum brake is protected at the corresponding spots with resistant layers. This principle is already used in brake drums, in which the friction surface is embedded in the inside drum with a gray iron ring. Thus, known friction pairs can be used. Aluminum is then usable without particle reinforcement and is also easier to recycle. The achievable weight reduction, however, is less.

By half-empirical iterative layer calculation, a thermotechnical pre-design was performed, and the transient temperature profiles for all applications were determined. Such a sandwich disk is only safe for operation, if the maximum temperature in the contact surfaces of both materials is under the maximum allowable temperature of the aluminum, and if between both materials there is no thermal contact resistance. The gray iron layer must not overheat. During cooling, a part of the heat can flow in the groove; however, for convection, the protective layers are insulating. Owing to thermal tensions and aging effects, chipping of the protective layers can occur. For most of the planned application areas, the thermal load is at a critical range. However, the sandwich disk is not yet finished in development. The long-time effects through time, temperature, and corrosion aging have not yet been researched. For this, to evaluate the multiparametric operating load collectives, the correlations, and the specific test procedures, tests must still be developed for.

Increasing engine power and increasing cost pressure have led to a decline in light-weight construction in recent decade. The tire diameters grew as fast as the brake disk diameters. More lining volume for higher operating life required broader partial disk segments and thus broader and heavier brakes. The CO_2 laws and increased efforts to reduce fuel consumption will turn this trend back. Light-weight construction is gaining significance again, and people are willing to pay for it.

2.6 Brake circuit design

To prevent the loss of brake power by leaks in the brake system, the system is divided into two separate circuits for safety reasons. From all the possibilities, only the best-known five brake circuit divisions are presented here:

TT. Simple two-wheel brake system (front, rear axle division or also named: black, white division). Each circuit supplies the brakes on one axle. This is a simple system design, there are no special requirements for the axle design—the typical division for upper middle class and upper class cars. In case of a failure in the rear axle circuit: only small brake performance derease, in case of a failure in the front axle: large decrease in deceleration, as the rear brakes are not very powerful. There will not be a moment around the vertical car axis.

X. Diagonal two-wheel brake system (diagonal distribution) per circuit. One front wheel and one opposing rear wheel are braked by one circuit. This is also a simple design. If there is a failure in one circuit, this results in a moment around the vertical axis of the vehicle. In case of a cicuit failure, the brake power is always cut in half. The moment around the vertical car axis can be independently compensated by designing the axle with a negative king pin offset, otherwise the driver must countersteer. This is the usual division of small vehicles up to middle class.

HT. Two-wheel brake system. The front axle requires a four-piston fixed calliper or a two-piston floating calliper or a piston of the floating calliper is connected to a different brake circuit. If there is a circuit failure, only the half brake force can be applied to the front axle with constant brake pressure and the same piston surface.

LL. Three-wheel brake system. Each circuit effects half of the front axle and one rear wheel. In case of failure, the front axle brakes half and on the rear axle there is always one nonbraked wheel that can take the side forces.

HH. Four-wheel brake system. Here, as with the front axle, four-piston fixed callipers or two-piston floating callipers must be installed on the rear axle. Each circuit works on the half front and the half rear axle brakes. The effect on both individual circuits is identical.

The design of the brake circuits depends on, which cause of failure is significant and whether an antilock braking system (ABS) is to be installed. As the standard-production application of ABS and ESP (electronic stability program), the HH division is considered too complicated, the LL division is inconvenient. The HT division has poor emergency brake characteristics (over braking of the rear axle) if there is a failure at the half front axle. All systems with two circuits on the front axle are only convenient with good thermal design, the brake fluid must be connected over the disk to the other pistons if there are fixed callipers, which makes these easy to overheat by the hot disk (evaporation). This leads to failure of both circuits.

Heavy and power vehicles of the upper class with rear wheel drive and motor in front have at least a TT division. For front wheel drive with front motor, a vehicle is equipped with floating callipers because of the (necessary) negative king pin offset and the diagonal division (X) is used.

2.7 Parking brake

The parking brake must work independent of the hydraulic service brake. By adding a mechanical actuation, the parking brake is integrated into the service brake. The primary task of the parking brake is to hold the parked car on a slope. By law, it is required that this is still possible with a slope of 18% in front or rear position of the fully loaded vehicle (ECE in Australia 30% slope).

Parking brakes are predominantly installed on the rear axle of the vehicle, as these brakes are less heavily stressed due to the brake force distribution (which is why they are also smaller) and an installation is simplified by the wheels that are not steered, in contrast to the front axle. The integration of a parking brake into a drum brake was already explained earlier. Operated by a bowden cable, the brake shoes are pressed against the drume with the aid of a lever. Integration is more difficult in disk brakes because of the smaller free space ratio. The installation of a separate smaller duo servo drum brake in the head of the brake disk (drum in disk) provides one possible solution that is exclusively used as a parking brake. Figure 13a shows an example for such a small servo drum brake. Figure 13b shows the arrangement of the drum brake in the disk brake.

Another solution is the integration of a mechanical device into the housing and the piston of a floating calliper brake. From the outside, this parking brake can only be identified because of the hand brake lever and the loop of the cable support (Figure 14a). In Figure 14b, there is the floating calliper housing with the hand brake lever and the cable support. The piston, which brings up the tension forces to the friction linings and the disk, is operated either hydraulically by the pressure of the brake fluid or mechanically by the hand brake lever using the integrated mechanism.

For hydraulic operation by the service brake, the piston is moved and pressed against the lining. The reaction force moves the housing away from the disk and presses the outer lining against the disk. The hydraulic pressure generates the tensioning forces. For mechanical operation, the main shaft is rotated, which is activated by the hand brake lever. Instead of the hydraulic pressure, a ball rolls up a ramp and spreads the piston shaft away from the housing with the piston. Both linings are pressed against the disk. In contrast to hydraulic operation, now the ball generates the tension forces within the ramp-shaped groove. The mechanical operation is independent of the hydraulic operation. However, both effect the same operation of the brake, exactly as with the drum brake.

In the inner part of the piston, there are resettings necessary for the mechanical operation. It is activated during one of each brake procedures by both operating types

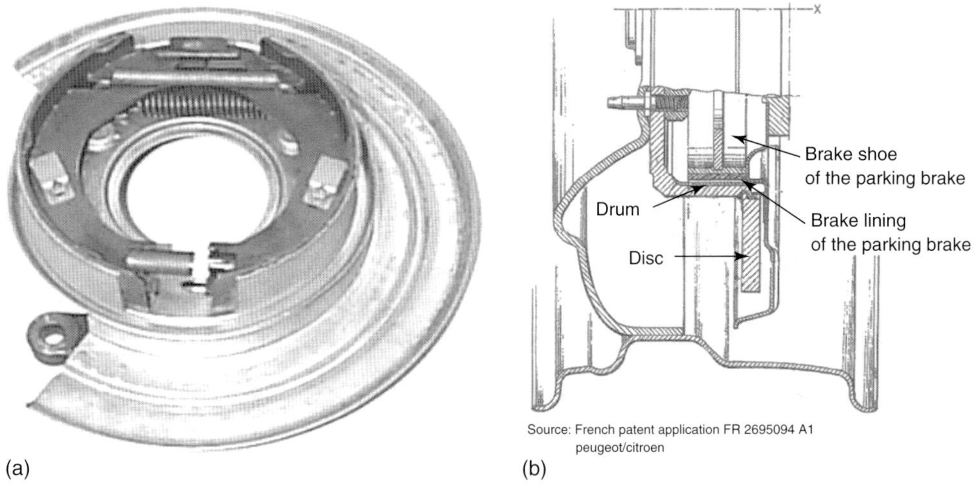

Figure 13. “Drum in disk” parking brake. (a) Drum brake as part of drum in disk. (b) Combination of drum brake and disk brake. (Reproduced with permission from TRW.)

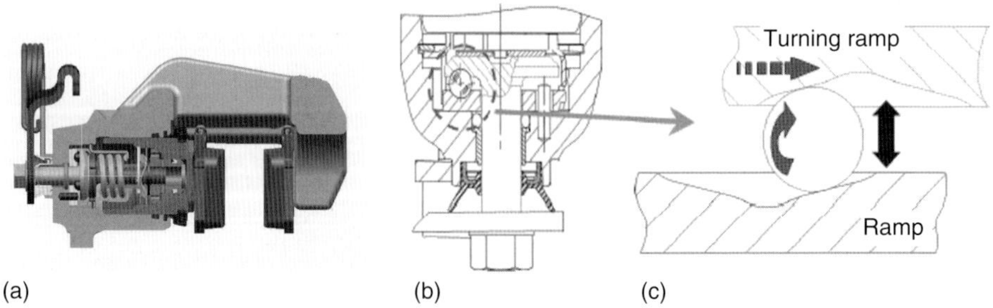

Figure 14. (a–c) “Ball in ramp” parking brake with three balls and the two ramps. (Reproduced with permission from TRW.)

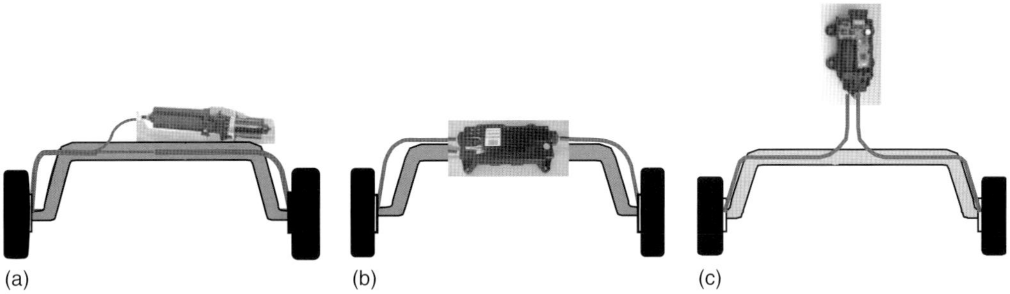

Figure 15. (a–c) Various cable systems for operating parking brakes. (Reproduced with permission from TRW.)

(hydraulic and mechanical) and extends the piston out of the housing mechanically with decreasing lining thickness (due to wear). Owing to the adjustment, there is a constant base clearance at any time between the piston and the mechanical operation when the brake is released. As the mechanically generated base clearance of the piston is larger than the hydraulically created clearance from roll back, the clearance between disk and linings is the same as with a disk brake without parking brake.

The parking brake devices work when operating the cable using hand lever, foot lever, and an electromechanical actuator. The cable puller operation is available for reaction force cable systems (a: also called *conduit systems*), transverse force cable systems (b: also called *cross-pull*), and parallel pull cable systems with force balance bars (c: also called *forward pull*) (Figure 15).

The integrated electric parking brake (EPB) replaces the cable operation using a small electric motor with transmission at the brake and the adjustment device via software. This brake is an additional step to decrease operating forces and to obtain design space in the vehicle interior. The electric operation allows the use of minimal forces, which not only physically weak and disabled people appreciate.

Instead of ball ramp and the automatic adjustment device, a spindle is inserted in the EPB, on which there is a compression nut, which—secured from rotating—is pressed against the piston head or is pulled back from there again. The motion thread is self-locking. A maximum tension force of 23 ± 2 kN can be generated with the device. With this tensioning force, a "slip stick" effect begins in the thread of the spindle. These high tension forces must be considered when developing brake disks (Figures 16 and 17).

The spindle is supported with the flange against the axial bearing on the bottom of the brake housing. The flange has an end stop that prevents the pressure nut from tensioning against the flange. This is important for manual and electric lining change and for the end of the line—calibration (calibration on assembly line).

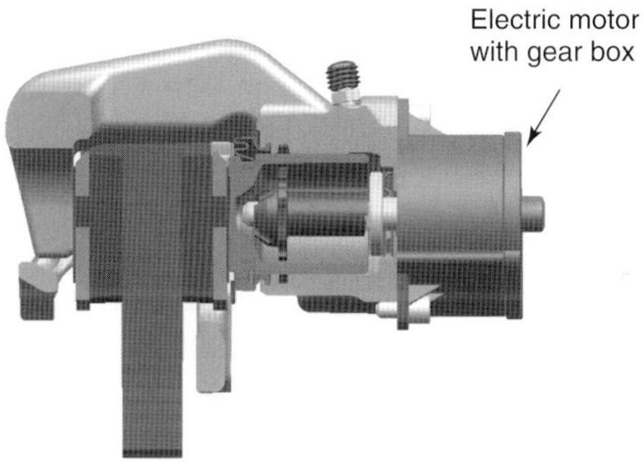

Figure 16. EPB (electric parking brake) in cross section. (Reproduced with permission from TRW.)

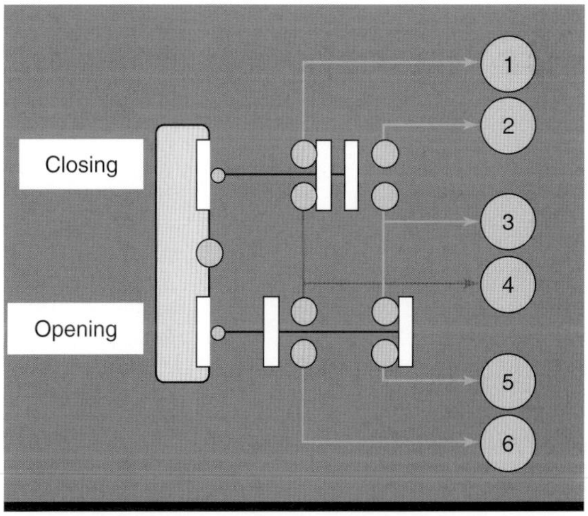

	1/4	1/6	2/3	2/5	3/5	4/6
Neutral	1	0	0	0	1	0
Opening	1	1	0	0	0	1
Closing	0	0	1	1	1	0

Figure 17. Wiring of the electric park brake. (Reproduced with permission from TRW.)

For emergency release, the actuator can be unscrewed and the brake is opened using a Torx wrench—that is a factory solution. Generally, before an emergency release, it must be considered, how and where the vehicle will next be securely parked even if no power is available.

The variant produced by TRW is operated using dual buttons with neutral setting, mounted in vehicle longitudinal direction. The front position (press) closes, and the rear position (pull) opens. If it is not operated, the switch stays in neutral position. The contacts are redundant, each with a normal open (NO) and a normal closed (NC) contact per switch position. Using the NO contact, a capacitor is placed so that the NO connections can also be constantly monitored with a pulsed test voltage. Thus, a single fault can be securely and immediately detected.

The processor is a Star12 from Motorola with on-chip flash memory that is widely used. The motors are each switched to one FET (field effect transistor). Changing rotation direction is realized using two relays. The real flowing currents are measured with each one shunt, likewise the voltage level and polarity. The control unit (Figure 18) can have up to four end levels for direct control of pilot lights.

For ignition "on," the control unit is in active mode and continuously performs self-tests. After ignition "off" and a shut-off delay, the control falls into a power-saving sleep mode (maximum 200 μA), from which it awakens operating either by a switch or by ignition. A second processor that simultaneously executes the monitoring function controls this and independent of the main processor can lock the end levels. The "end of line" configuration and the diagnostics occur using the CAN (controller area network) bus.

For vehicles with integrated starting assistance, an additional inclination sensor can optionally be provided in the control unit. If this is installed, it can differentiate between the static and dynamic mode by evaluating the vehicle movements during the drive even with failure of the CAN velocity signals.

The software runs in a 20 ms main loop, from which all function modules are continuously addressed.

This is strictly a status machine, whereby the different modes can only be exited using predetermined events (mode control). Subroutines, such as the composition of switch commands with other system signals (demand calculator/clamp force controller) or the motor control (motor controller), are set up as status machines.

The main function is the static closing of the brake with ignition on. The switch must be operated for that. The software checks whether the ignition is switched on and whether the wheel speeds are zero. After 100 ms (5 cycles), the command is accepted and transferred to the motor control. This enables all functional motors in tension direction and monitors the tension flow. This measures the voltages. While driving over the clearance, the motor current is determined and considering the voltage, the motor temperature is estimated. Both voltage and temperature are used for the formation of a correction value for the cut-off current that is then added to the no-load current. A fully loaded vehicle on a 30% incline

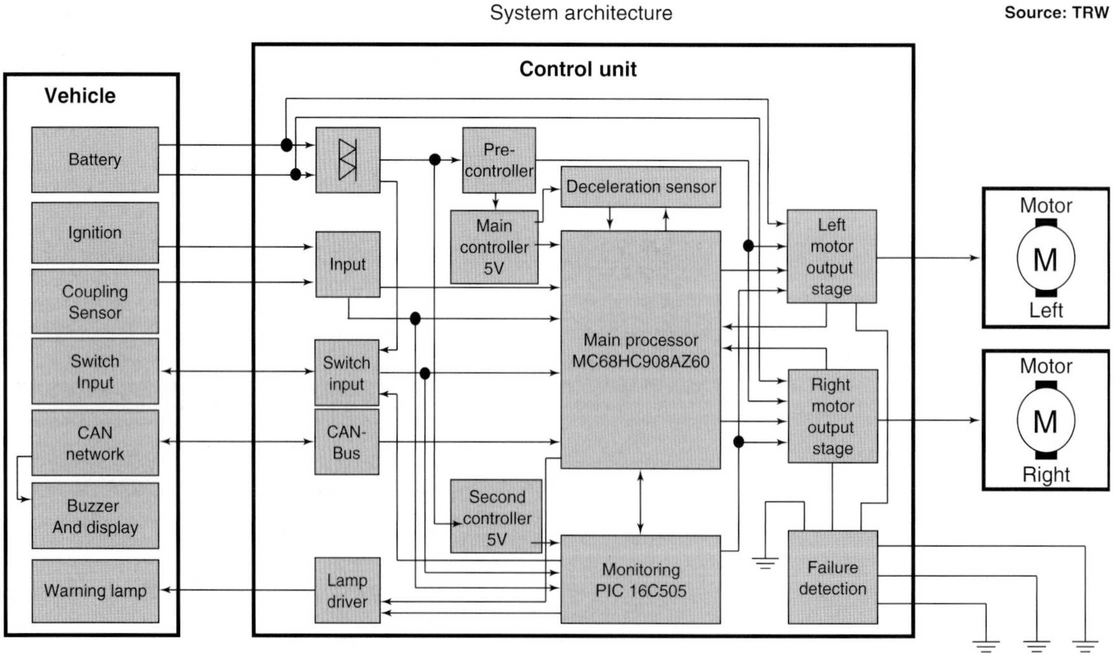

Figure 18. Block diagram of control unit. (Reproduced with permission from TRW.)

is assumed to define the cut-off current. Figure 19 shows the time history of a tensioning and a release process.

During the drive, a temperature model runs for the rear brake disks to detect excessively hot disks. By repeated tensioning, the system provides the opportunity after approximately 3 min to realize a one-time (cycle-related) force super elevation beyond the nominal tension force. Rollaway detection can also trigger multiple tensions, which lead to high tension forces, especially, if there is not enough energy available because of undervoltage.

To open the parking brake, the relays are switched in release direction and the FETs are energized. As soon as the current curve has fallen to the no-load current while opening, this point is retained as the new relative zero point, and after clearance adjustment time, the motor is switched off. This time is determined by declamp-voltage and declamp-current (to estimate motor speed).

If the ignition is off, and the switch is operated, only tensioning the parking brake is possible, but not opening (child safety lock) It becomes difficult for drivers that switch off the ignition while driving downhill; wheel speed signals are no longer available from the CAN bus. A CAN follow-up function ended by the EPB ensures that the electric brake does not close until the vehicle is standing. Even if wheel speeds fail, the brake is not closed until the ignition is off and vehicle standstill has been detected.

The total software overview is shown in Figure 20.

If the service mode is activated with ignition on via the diagnostic function of the service mode, both pressure nuts move back to their end stop. Then the linings can be replaced.

The dynamic deceleration is passed to the operating brake system by the EPB using CAN message, which then executes the braking using the ESP system and takes over again with a speed <7 km/h.

Probably, the most convenient function of the EPB is the automatic release when starting. Using the integrated incline sensor, the required engine torque is determined and when reaching this threshold, the brake is opened. For manual transmission, a clutch position sensor is engaged. From the gradients of the clutch position (so the release speed), the grip of the clutch is calculated and the brake is released when there is sufficiently high torque.

The starting assistance is also important for automatic vehicles. Today, less and less creep torques already allow automatic vehicles to roll back on smaller inclines (12%) If the parking brake is not used, semiautomatic vehicles generally roll back when moving the foot from the brake pedal to the accelerator. Using the starting assistance, the rolling-back can reliably be prevented. Undesired starting is prevented by additional signals, for example, starting with the starting assistance is only permitted for a driver whose seatbelt is buckled. Manually releasing EPB is only possible

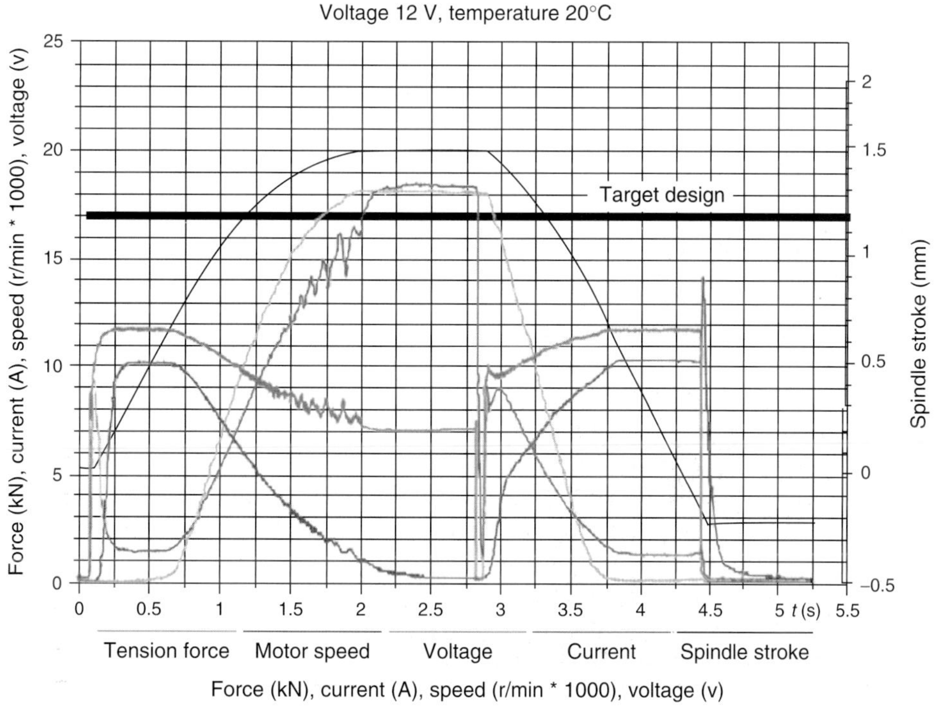

Figure 19. Tensioning and release processes of the EPB. (Reproduced with permission from TRW.)

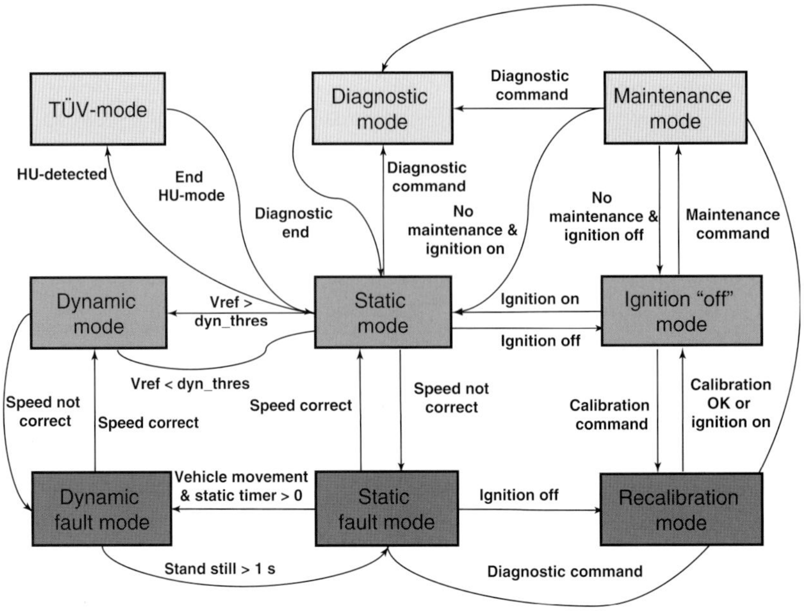

Figure 20. Software condition diagram. (Reproduced with permission from TRW.)

by pressing a button if either the brake or the accelerator is pressed down.

With the typical techniques such as continuous ROM test, RAM test, external watchdog, and second processor, the security of the control unit is ensured up to ASIL D corresponding to ISO 26262. In addition to voltage supply, the redundant operating element, the fault lamp, and the actuators are also monitored for proper function. Using a constant current source (approximately 2 A), the performance output stages are tested for switching on and off. The relays for pole change can only be tested one time when the ignition is switched on, owing to their service life.

All signals are subject to a plausibility test as much as possible before processing. If the ignition is switched on, the CAN must also be there, if the EPB motors are switched on, shortly thereafter the voltages must fit with correct polarity. If the brake opens, the current must decrease; if the brake closes, the current must increase; if the driving speed decreases heavily, there is either an incline or the brake light switch is on. Changes to the drive speed must synchronously go with the signals.

If there are missing signals, if possible, replacement parameters are used. If the vehicle speed is missing via the CAN, the corresponding filtered incline sensor signal is used to detect standstill. If terminal 15 (voltage in the system) is missing, CAN is present and notifies the switched-on ignition, this is used.

The faults are classified as follows:

1. casual faults (e.g., CAN loss): the EPB system continues to work and indicates the faults;
2. interfering faults (e.g., communication with the motor control unit interrupted): the function (here, drive away) is switched off, and the EPB system continues to work and indicates errors;
3. channel fault (e.g., channel to motor defective): the EBP system tries to open the brake with the next release command and switches the affected channel off, the other channel continues to work, and the fault is indicated;
4. system fault (e.g., processor fault): the EPB system switches off, and the fault is indicated.

With this system, the parking brake function is also electrified now. In the coming years, an integration of the electronic steering in the ESP control unit will happen—many functions and signals are "related," the safety requirements are similar.

3 THE BRAKE PROCESS

The stopping distance required during the brake process is composed of

- rolling distance: the vehicle continues to roll without braking; only air and rolling resistance are working—no motor brake torque. The amount of the rolling distance depends on

  - time to turn toward view
  - reaction base time
  - time to implementation
  - response time.
- threshold path: braking begins, the brake torque increases
  - threshold time.
- braking distance: the vehicle is decelerated
  - time of braking (complete or partial braking).

All distances together cover the stopping distance.

The introduction and the execution of a braking process in traffic occurs due to visual perceptions. This process is composed of three phases:

- Objective reaction summons: start of visibility of the object
- perception of the object: in general peripheral, on the edge of the field of vision of the driver, only rarely in central field of vision of the driver (time to turn toward view)
- Object fixation: Recognizing the risk that originates from the object, start of muscular reaction

The fact that conscious decisions and appropriate reactions are never made based on peripheral perceptions has fundamental significance. Conspicuous objects in the peripheral field of vision always first trigger a glance. The foveal area generally only has an opening angle of 1°. If there is more than a 0.5° deviation from the driver's field of vision, a view adjustment is made. Object fixation and a conscious reaction are only possible after the *time to turn toward view* (Figure 21).

The *reaction time* begins after turning toward the view. The reaction base time lies between the object fixation and the start of muscular reaction. It is a process with the same behavioral patterns for all individuals. The time span is independent of outer boundary conditions (e.g., brightness and weather conditions) and independent of the situation.

Afterward, with the start of muscular reaction, the *implementation time* begins. The accelerator is released; thus, the first appropriate movement of the foot happens.

The first contact with the brake pedal is the start of the application or *response time*. The brake shoes/lining are applied. The implementation and application times are short compared to the reaction time.

As the brake pressure increases, the *threshold time* begins, where the increase of the hydraulic pressure is significant for the gradients of the deceleration. Normally, the wheels of the vehicle lock before reaching the maximum pressure. In modern passenger cars, the ABS activates beforehand and prevents locking, keeps the wheels in optimal slip range and secure the maneuverability of the vehicle during the brake application.

The following is the *brake time*, which lasts as long until the entire emergency brake process ends with the vehicle at standstill. It is dependent on the output speed, the degree of deceleration, and the achievable friction values at the tires (at 250 km/h and highest possible vehicle deceleration, the process takes approximately 12 s). This time span is solely determined by the capacity of the brake system and the vehicle, assuming that the driver does not reduce the brake pressure.

In contrast, the *reaction time* depends on human movement sequences, physical–technical laws, and the abilities and conditioning of the driver. Additional factors affecting the driver are of special significance: road and weather conditions, traffic volume, distraction, high degree of information, and stress.

The brake process starts as soon as the driver identifies the reason that requires braking. From this time up to the rise of brake pressure, the vehicle practically continues to roll without brakes. At this point, the modern driver assistance systems start—a shortening in reaction time by 0.5 s reduces the probability of an accident up to 80%. The brake effect from air, rolling, or motor resistance can be ignored here, as their effect is very minimal compared to the power of full braking.

In emergency or sudden braking, the driver applies a pedal force of 250 N and moves the brake pedal for approximately 80 mm. The brake master cylinder, connected to the brake pedal, generates a hydraulic-volume shift in the direction of the brake system because of this pedal movement. Owing to the compressibility of the brake linings and the deformation of the brake callipers, the entire brake system absorbs a volume of approximately 8 ml. At the same time, a brake pressure of 80 bar occurs. This leads to a deceleration of approximately 0.9 g (90% of gravity). In panic situations, more than 500 N pedal force and more than 160 bar hydraulic pressure were measured—however, the deceleration did not increase further from this.

The deceleration, with which an average driver brakes, lies in the range 0.01–0.4 g. Only in absolute emergency or exceptional situations are up to 0.8 g achieved. During their entire driving practice, most drivers never brake above 0.5 g. However, the brake system is designed for up to 1.2 g.

4 BRAKE DYNAMICS

The control of the braking process is decisive for the availability of a vehicle in practice, to be able to ensure safety in traffic. A vehicle must be designed in a way that it does not produce unforeseen reactions even during panic

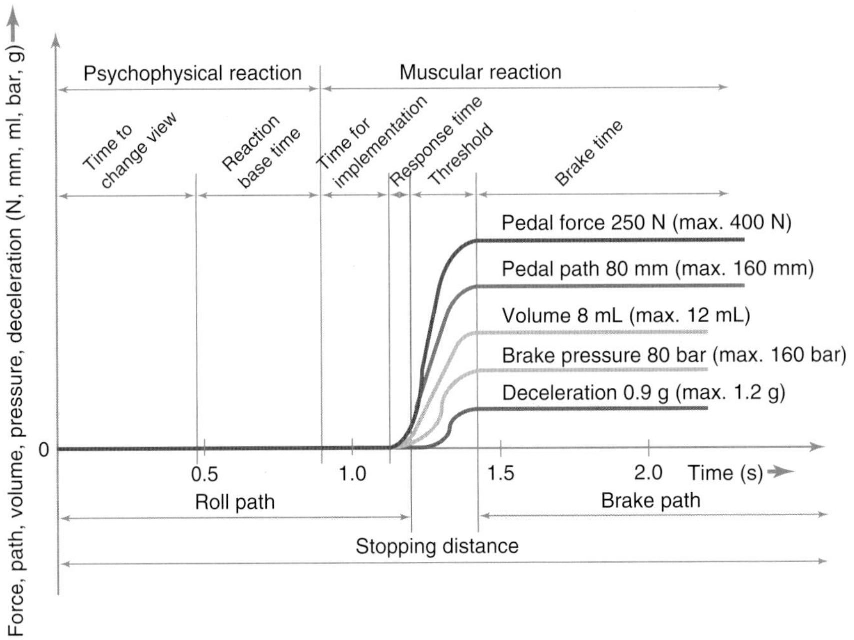

Figure 21. The brake process. (Reproduced with permission from TRW.)

braking and always behaves like the driver expects from his or her experience. Only then the vehicle is safe.

The knowledge of the outer brake dynamics of the vehicle is of special significance for designing a brake system. This is dependent on the brake force distribution that follows from the design parameters of the vehicle. For installation or for the demand-oriented change of the brake force distribution, corresponding control devices are installed in the brake system.

The brake system must be able to brake the vehicle with maximum deceleration. The driver must be able to do this with an application of average forces. The existing limits are the contact area between the wheels and the road, which are only the size of a palm of the hand, and the adhesion coefficient of the tires of $\mu_B = 1.1$ at most. With an assumed friction coefficient of $\mu_B = 1.0$, a maximum vehicle deceleration of $1\,g$ ($=9.81\,\text{m/s}^2$) can be reached. In theory, with specific rubber, the μ-factor may even be higher (race cars).

The brake application must be stable, the vehicle must not become unstable or even worse, that the rear axle swerves. A short braking distance naturally occurs with the highest possible deceleration. The processes during braking will not have any reactions on the driver or the behavior of the vehicle (no noises, no rubbing, no steering wheel torsional vibrations, and no pulsing of the pedal). The response of the system improves with only minor volume absorption in the callipers, which also decreases the brake pedal travel (operation, roll back, and widening the brake calipers). Pedal travel going against zero is not appropriate because of ergonomic reasons.

The brake force will be well dosable, so that with light pedal operation, no strong deceleration occurs. A slight decrease in the liner friction coefficient while braking, which forces the driver to step on the pedal again, is preferred. The support by brake boosters should allow every driver to perform locked braking with average use of force in emergencies. On the other hand, the increase must not lead to poor dosing with minimal deceleration. The booster devices are realized with progressive, degressive, or linear characteristics, depending on the manufacturer. In addition, all brake actions will never generate noise (inside and outside of the vehicle) from the liner and the disk wear. Traffic light systems with signals for blind people had to be changed from beeping to clicking because there was confusion with brake noises.

This especially includes the following:

- rubbing: vibrations due to speed fluctuations from differences in thickness of the brake disk;
- squeaking, chirping, and muh: average to high frequency noise or friction fluctuations;
- rattling and clicking: from the moveability of the linings, with pot holes, when applying the linings.

For the quality of a vehicle in practice, the stability is decisive in case of emergency braking. Only with sufficient brake stability, the vehicle is prevented from skidding, and the steerability is ensured with good cornering forces. Regardless of size, type of motor, or design of any vehicle, the rear wheels may not lock before the front wheels lock (legal requirements are setting a limit with minimum 8 m/s^2 deceleration). When the front wheels lock, the vehicle can no longer be steered; however, it stays in its track. Because modern, electronically controled brake balance no longer has these mechanically secure fallbacks, only partial brakes are allowed for (displayed) failure.

The brakes (brake linings) may not lead to strong friction coefficient changes (high temperature fading), as in addition to the loss in brake power, especially a change in the brake balance by fading of front to rear axle occurs, which can endanger the stability of the vehicle. The same can happen because of evaporation of brake fluid (fluid fading).

The tire characteristics must also be adapted to the vehicle. A vehicle that stably brakes with approved tires (first the front axle locks) can become unstable during emergency braking by installing tires with higher maximum adhesion coefficient on the front axle. Then the rear locks first. If the rear axle skids due to premature locking of the wheels, it is impossible to react to this using steering movements. The vehicle immediately goes out of control (Figure 22).

To calculate the vehicle behavior, a rigid two-wheel model is used, and the method with minor disturbance is applied. In doing so, the vehicle is to be considered a rigid body, deformation of the axles or the wheel suspensions are ignored. Owing to a disturbance, the vehicle has the sideslip angle ß between the longitudinal axis of the vehicle and the driving direction. This driving condition generates sideslip angles at the front and the rear wheels (same amount like ß) and then side forces are acting between the road and the tires. The masses and the lateral force affect the vehicle's center of gravity. These counteract the deceleration forces and the side forces on front and rear axles.

The yaw angle or the yaw acceleration can be determined from the balance of forces in *x*- and *y*-directions and the moment balance. A positive value means an increase in the sideslip angle. This can become unstable, finally a skidding of the vehicle may happen. A reduction of the sideslip angle points toward stability. The vehicle body will returned to the driving direction. The smaller the moment of inertia J_z, the faster the vehicle turns (e.g., when the engine is in the center of the vehicle).

By relating the brake forces to the vehicle mass, the yaw acceleration in dependence of the deceleration can be determined. The simulation results show immediately under what conditions the vehicle becomes unstable (for instance in Figure 23: 70% *g*). The additional load of the vehicle is also of great influence, as not only the overall weight, but also the brake balance changes. The behavior of the rigid two-wheel model with ideal tires is a straight line (neutral steering behavior).

For the evaluation of a vehicle, the deviation from the straight line is decisive.

At a deceleration of more than 80% of *g*, the empty vehicle begins to skid (in the example of Figure 23). The yaw acceleration is positive and increases extremely

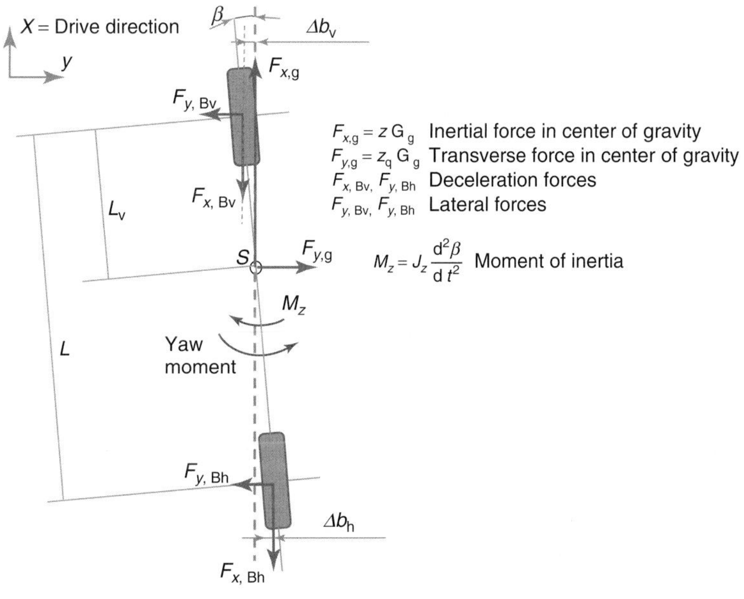

Figure 22. Simple calculation model. (Reproduced with permission from TRW.)

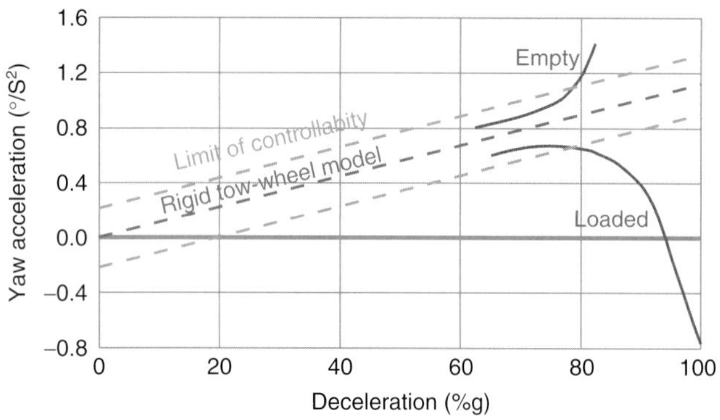

Figure 23. Vehicle stability model.

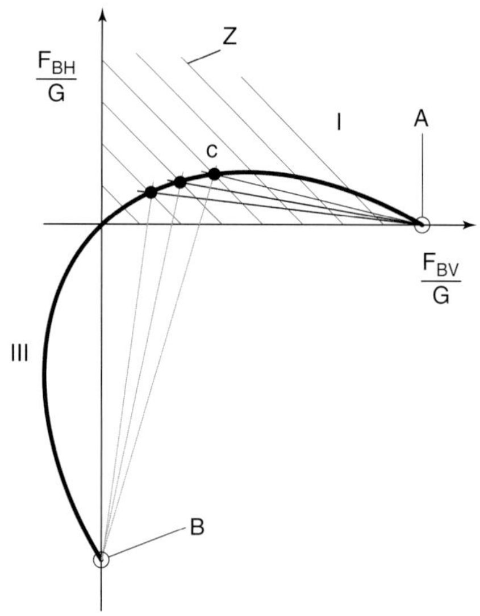

Figure 24. The brake force distribution diagram.

quickly. A yaw acceleration of $1°/s^2$ corresponds to a yaw angle of approximately 57°/s after 1 s and an increase of the angle by 30°. A normal driver cannot handle this. An acceptable value for yaw acceleration is $0.25°/s^2$. Then, the braked vehicle can still be stabilzed by countersteering.

In real vehicles, the elasto-kinematic design of suspensions and stabilizers delivers inherently stable behavior. For the control of the vehicle by a normal driver, the brake behavior in case of circuit failure or with differing low adhesion surfaces (right/left, μ—split) is decisive. At standstill, the weight force in the center of gravity and the wheel loads at the front and the rear axles are acting on the car. The center of gravity has the distance l_1 to the front axle. The wheel base is indicated as l, and the hight of the center of gravity above the road is named h.

During the brake process, the mass forces zG_g (equivalent to vehicle mass multiplied with the deceleration) and the brake forces are added to the front and rear axles. This assumes that neither the position of the vehicle to the road nor its center of gravity change.

The dynamic balance in driving direction leads to the deceleration factor z ($z = a_x/g$ with a_x as longitudinal deceleration) . As a result of the height h of the center of gravity and the inertial force in the center, a dynamic axle load distribution $\pm\Delta G$ results from the rear axle to the front axle. The equations for dynamic axle load arise from the equilibrium of moments around the wheel contact points of the front or rear axles. These equations can be found in Kane (2008).

Assuming the same utilization of traction (that means the same μ) on the front and the rear axles, the equations can be used to generate a diagram that shows the ideal distribution of the brake forces (means: the same friction coefficient at the front and the rear axles during deceleration). This is a diagram (Figure 24) with the brake force related to the total weight, at the front axle as abscissa and at the rear axle as ordinate.

On this parabolic curve, the ideal case of the same adhesion utilization on front and back axles is always realized. It is solely described by the data: center of gravity height, center of gravity position between the axles, total vehicle mass, and wheel base, and it is existing in the first (braking) and third (acceleration) quadrants of the coordinate system.

The deceleration factor z can be drawn in as lines of constant braking at an angle of 45°. On the parabola of the ideal axle force distribution, $\mu = z$ always applies. The lines of same coefficients of adhesion of the tires are also straight lines that go through the "point of ideal operation" (point C).

There are two useful physical explanations for no brake forces at the rear axle:

1. $z = 0$, there is no braking that means no deceleration (coordinate origin).
2. $F_{BH}/G = 0$, the rear axle is completely released (point A).

No brake forces at the front axle can also be explained: The same applies to acceleration, here z is negative

1. $z = 0$, there is no braking that means no deceleration (coordinate origin).
2. $F_{BV}/G = 0$, the front axle is completely released (point B). This means that the car is accelerated (not braked) and therefore the related deceleration z is negative.

The deceleration factor z must be drawn in Figure 24 to be able to plot the green lines (starting in B) and the red lines (starting in A) of constant friction adhesion between tires and road surface. All red straight lines of constant friction adhesion between rear axle and road must cross point A (lifting the rear axle). There is no more friction at the rear axle in this point. Point B is crossed accordingly by straight lines of constant friction at the front axle. During acceleration, the front axle will be lifted under the conditions in Point B.

In point C, the straight lines of friction utilization meet each other. The parabola defines the ideal brake distribution that means for instance the full utilization of friction at the rear axle (may be 0.4) and at the front axle (may also be 0.4). The parabola is valid for $z=\mu$.

Practical statement: The installed brake force distribution in the first quadrant counts as stable, in case the installed distribution line is below the ideal distribution until $z=1$ (by legal requirements $z=0.8$). If a friction coefficient of 0.4 is assumed (the same value at front and rear axles), then the brake force P can increase along the line of the installed brake force distribution until the line crosses the green line of the friction utilization of the front axle (P_1 in Figure 25). The front axle locks at this point and the deceleration can be seen (Figure 25).

In case the brake pressure is still increased, the brake force at the front axle is no longer increasing. The braking point is going upward along the green line of the constant friction utilization of the front axle until the red line for the friction utilization of the rear axle (same value like at the front) is crossed. In Figure 25, this is shown as point P_4. Now the rear wheels are also locked. The gradient from P_1 to P_4 is not vertical, as the lines for the constant friction coefficient at the front axle are crossing in point B (comp. Figure 24).

For the evaluation of the brake behavior of a vehicle, the ideal distribution in the first quadrant is used. The installed brake force distribution is considered with respect to the ideal distribution.

As an example for a vehicle, the ideal brake force distribution is shown in Figure 26. This is like the characteristic in the first quadrand of Figure 24. This ideal distribution is also named as outer dynamics of a vehicle. The brake force distribution installed in the car can be regarded as information about the inner dynamics of the vehicle.

The brake force distribution installed provides information about the inner dynamics of the vehicle. The distribution is evaluated from the parameters of the brakes on front and rear axles and it normally represents a straight line. The intersection point of the straight line with the line of the ideal distribution indicates the deceleration (z-value) at which both axles look at the same time. As long as the line for the installed brake distribution is below the ideal one, the front axle will lock first if the road friction will not be sufficient. If the brake operates above the ideal distribution line, the rear axle will lock first and the vehicle will be immediately unstable. The intersection point of the installed distribution with the ideal brake force distribution indicates the so-called critical braking z_{kr} by which the two axles lock simultaneously.

For safety reasons, most vehicles are designed so that the critical braking (locking of all four wheels) is at $z=0.9-1.0$. As the drivers normally avoid these high deceleration values, the vehicle normally will not become unstable by braking.

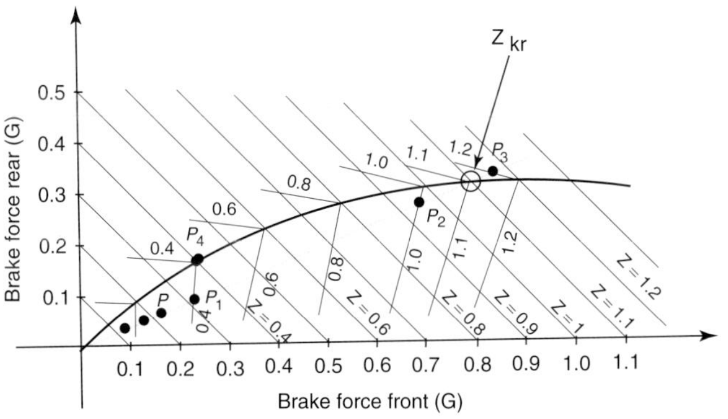

Figure 25. Brake pressure increase and locked wheels at $\mu=0.4$.

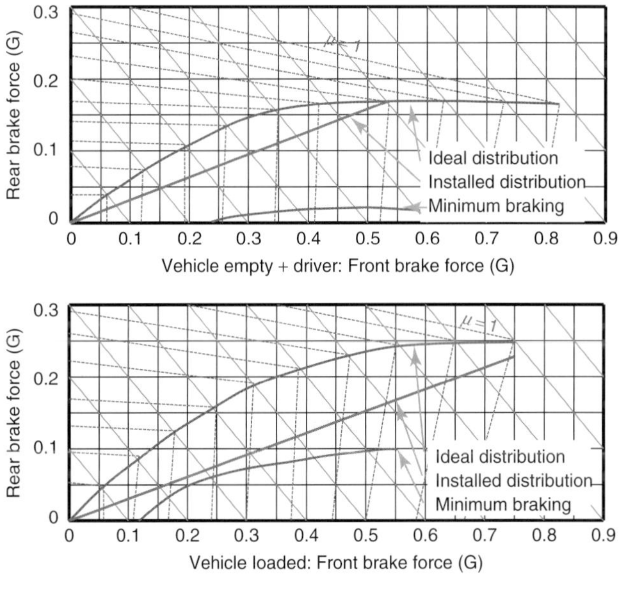

Figure 26. Brake force distribution diagram with vehicle empty + driver and vehicle loaded.

In addition, the minimum brake force distribution given by law is drawn in this diagram. With these minimal values, the requested deceleration will be reached. However, they are far too much low for practical driving.

To prevent the rear axle wheels from locking, which means to keep the installed brake force distribution below the ideal distribution, valves to reduce the brake pressure at the rear axle have already been installed for many years. These components respond to specific brake or vehicle data.

If an increasing pressure is given to the brake system via the master cylinder, the pressure-dependent valve only allows no or only a small increase in pressure to the brakes of the rear axle if a specific brake pressure has been exceeded.

The load-dependent valve works like the pressure-dependent valve. It is mainly used in vehicles with front wheel drive with front motor and low rear axle load, as the brake force distribution can significantly change there because of the vehicle payload. The spring deflection of the body is used to adjust the valve, thus the pressure switching point changes according to axle load (dynamic). In reality, the load dependence is a travel dependence (spring path dependence).

For valves with an acceleration-dependent switching point, the brake force reduction is dependent on the vehicle deceleration. High operating speeds of the brake pedal distort the mode of the valve's action. For diagonal brake circuit distribution, it is necessary to control both circuits in parallel, which is very inconvenient. These valves can also be installed between the wheels of an axle to react to lateral accelerations. For driving in extreme curves, this can prevent locking of the inner wheel.

For a vehicle without brake force control devices or ABS, the limit of brake stability is therefore achieved shortly before the locking limit of the rear wheels at the highest possible deceleration. Figure 26 shows a fixed distribution that enables such a design; it already cuts the ideal distribution with braking of 0.7.

The development is clearly moving from brake pressure control devices and going to real control systems. The antilock system that today predominantly works on all four wheels separately or on both wheels of an rear axle (select low) provides reliable brake stability with almost optimal vehicle deceleration and gives still steerability of a car, even during severe braking.

Together with an electronic brake force distribution, the braking power can ideal be used in every emergency situation, independent of load condition and adhesion coefficient of the tires, as long as the driver only demands the brake system sufficiently, generates corresponding high pressure, and also keeps the pressure while braking.

If the engine power increases with a constant vehicle weight, then generally more powerful brakes are required. If the vehicle weight increases with the same engine power, then the brakes can usually be retained from the point of view of brake force distribution and subsequent braking behavior.

5 THE FRICTION PROCESS

The friction process results from the resistance of the relative movements of two bodies. With relative movements without lubricant, dry friction or solid friction results. The amount of the resistance depends on the size, the texture, and characteristics of the contact surfaces. This includes

- the material combination (strength and deformation behavior, heat conduction, and chemical characteristics);
- the design combination (shape, surface profile, and roughness);
- the operating conditions (kind of stressing: duration, time history, pressure, temperature, and relative speed);
- the environmental medium (chemical affinity between friction partners and environmental medium in connection with the mechanical–thermal energy in friction contact);
- the type of manufacturing process;
- the degree of deformation;
- the formation of reaction layers; and
- the degree of movement (the relative speed and static and dynamic adhesion factor).

Owing to the roughness of technical surfaces, the contact of fixed bodies is always only "discrete." Compared to the nominal, in reality, there is a much smaller contact surface, as the contact only occurs on the roughness peaks. The true contact surface is formed from the sums of these partial surfaces (percentage: contact area). The result is a much higher surface load than the surface pressure formed by the nominal contact surface. This results in elastic and plastic deformations of the roughness and the surface profile, which increase with normal force. In contrast to the nominal specific surface pressure, however, the real surface pressure does not increase proportionally with the normal force. Deformations and wear lead to hardening of the surface layer, a increase of the contact ratio and cause a change in the friction characteristics (embedding).

The friction force is composed of a deformation and a shearing part: the elastic part of deformation causes vibrations of the roughness peaks affected by the friction process. These vibrations are irreversibly converted into inner energy (heat). The softer body, whose roughness peaks have a lower spring stiffness, receives the largest portion of the energy. Whether it also retains and depends on the heat conductivity and the heat capacity of both bodies (heating- up).

The shear force portion of the friction force is due to adhesive force from cold welding and adhesion forces between the oxide layers. The adhesive forces are dependent on the mixture of the oxide layers that in turn are dependent on the mechanical and chemical characteristics of both surfaces, as well friction energy density. The friction energy density can locally lead to very high flash temperatures of several thousand degrees centigrade in the friction path. This stimulates the formation of oxide layers with lower hardness and lower shear resistance that results in a decrease of the adhesion factor (lining fading). This process is a great disadvantage for the vehicle brake, as the brake performance is directly dependent on the adhesion factor. However, a certain continuous renewed oxide layer by wear, also called *friction carbon layer*, is necessary for a constant, stable adhesion factor (Figure 27).

Brake linings consist of a formula of varying materials that are bound with phenolic resin. During manufacturing, the mixture substance is pressed onto the lining back plate either cold or warm. After the installation into the vehicle, the linings do not yet deliver the target adhesion factor. Not enough friction contact surfaces lead to high local friction temperatures with low adhesion factors. After a little while, the linings are ground in and optimal adapted to the disk and to the movements of the brake housing (embedded). In addition, the disk surface did also run in. In general, the adhesion factor also improves after the lining has experienced higher temperatures once during the braking processes and has hardened due to outgassing. However, during the entire lifetime of a lining (approximately 50,000 km), the adhesion factor does not in anyway remain constant.

The adhesion factor depends on the contact pressure, the temperature, and the speed of the vehicle and the disk (with constant vehicle weight). After violent braking, which affects most of the decrease in adhesion factor, the brake linings have the ability to regenerate in the course of the stops following. They are stable against extreme adhesion factor fluctuations (fading) and are not sensitive to wet conditions and salt. The wear of a brake lining should be as low as possible to realize a high lifetime and a narrow design. The friction process, however, leads to oxidation processes that accelerate the mechanical abrasion as well as wear. Often, a specific wear is necessary to ensure comfort characteristics.

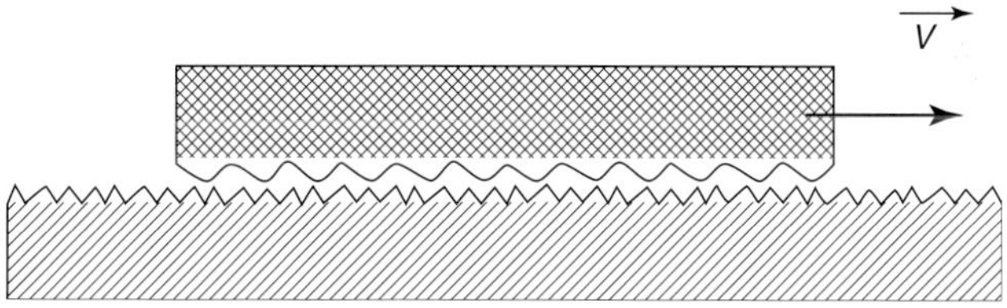

Figure 27. Schematic presentation of the friction process. (Reproduced with permission from TRW.)

The temperature sensivity of the wear from the lining material should be small. Tears or edge breakage due to thermal stress can be ignored in today's series linings. In this context, the coordination of friction partners is important, so the brake disk or the brake drum on the lining. Structural conditions and hardness can shift the wearing process from "lining *eats* the disk" to "disk *eats* the lining." In general, the disk or drum lifetime is twice that of the lining lifetime. The characteristics of the lining materials are as follows:

- high breaking strength against chipping
- high shear strength against loosening from the back plate
- low natural frequency with good damping characteristics
- an average compressibility for good comfort, which, however, keeps the brake from becoming too soft
- low heat conduction to isolate against the brake fluid
- small heat expansions to prevent rolling away while using the parking brake
- no formation of residue on lining or disk/drum
- corrosion resistance.

Today, comfort behavior is more and more important. Thus, a buildup of friction vibrations is critical, if it leads to rattles and steering wheel vibrations. The response and threshold behavior of the brake depends significantly on the friction and compressibility. Compressibility leads to increased pedal travel, and adhesion factor changes effect pedal force changes. A modern lining must not make any noises out of the friction combination and the entire brake system during the whole lifetime; the lining must rathehave damping effects on the vibration system.

6 SUMMARY

Modern vehicles have disk brakes for performance reasons and drum brakes for cost reasons.

Large energy storage devices are made out of gray iron for performance reasons and energy storage devices are made out of alternative materials for weight reasons (aluminum, fiber reinforced ceramics, and carbon fibers). The latter can only store a clearly lower amount of energy with the same design space. Friction brakes ensure safe braking up to standstill, and these brakes can securely hold the vehicle mechanically and without energy conduction.

The secure design of the brake systems is optimized by electronic control systems (ABS, ESP, EVB—electronic brake force distribution, BA—brake assistant). Comfort and safety functions (EPB, ACC—adaptive cruise control—speed and distance contol unit) reduce operation force and support the driver while driving and braking.

The man–machine interfaces are more and more understood (response behavior, pedal feel, and indicator lights).

Everything that the driver experiences in brake technology in the vehicle is a compromise decided by the manufacturer between costs, lifetime, and a performance, which usually goes far beyond the legally required levels.

RELATED ARTICLES

Volume 4, Chapter 118
Volume 4, Chapter 119
Volume 4, Chapter 120
Volume 4, Chapter 130
Volume 4, Chapter 131

REFERENCE

Haken, K.-L. (2008) *Grundlagen der Kraftfahrzeugtechnik*, Hanser Verlag, Germany.

FURTHER READING

Bill, K.-H. and Breuer, B.J. (2012) *Bremsenhandbuch*, Springer Verlag, Germany.

Bill, K.-H. and Breuer, B.J. (2008) *Brake Technology Handbook*, USA, SAE International.

Limpert, R. (2011) *Brake Design and Safety*, 3rd edn, USA, SAE International.

Chapter 118

Carbon-Fiber-Reinforced Silicon Carbide: A New Brake Disk Material

Andreas Kienzle and Hubert Jäger

SGL Carbon GmbH, Meitingen, Germany

1 INTRODUCTION

Brake disks in cars are safety relevant parts, which require high reliability and a high lifetime. The task of a brake disk is to transform the kinetic energy during braking by friction into heat, which is absorbed by the brake disk and the brake pads. The absorbed heat must be dissipated in the surrounding area by thermal conduction, heat radiation, and convective flow of heat. A brake system of this kind is accordingly limited by the friction characteristics of the brake materials and its ability to store and remove the heat. In general, brake materials must have very good thermomechanical properties, high and constant friction characteristics, and good resistance against abrasion. With the development of the first cars in the late *nineteenth* century, the kinetic energies are still very low but never the less, the cars must be stopped safely. Therefore, there was an increasing demand in equipment, which allows their safe stopping. The first cars used the running surface of the wheels directly for braking. Brake linings out of wood or leather were pressed mechanically by a lever on this running wheel surface. This brake was substituted by the brake drum made out of metal. Brake disks for braking need around another 50 years before they were used in cars. Nevertheless, the first patent regarding to brake disks was granted in 1902 to Lanchester (1902): This patent describes the first time the idea to use a brake disk, which is mounted on the wheel hub. The brake linings were pressed mechanically on the disk surface during braking. The first brake disks out of gray cast iron were used for race applications (Jaguar winning the 24-h race of Le Mans in 1953). Alfa Romeo introduced the first brake disks in road vehicles. The BMW 502 3.2l was the first German car equipped with disk brakes on front wheels, which can be ordered as an option followed by Mercedes 1961 in the 300 SL.

In addition, the development in the material optimization of gray cast iron and in the production process technologies for the production of ventilated disks allowed reducing cost and improved the quality constancy of the material. Nevertheless, the main disadvantages of the gray cast iron disks, that is, like high weight due to the high material density of 7.2 gcm^{-3}, corrosion sensitivity by oxygen and water, limited high temperature stability, and high wear rates, could not be solved. In addition, the speeds, which are attained nowadays by such vehicles, are constantly increasing. Since 1950, the top speed of upper class

Encyclopedia of Automotive Engineering, print © 2015 John Wiley & Sons, Ltd.
Edited by David Crolla, David E. Foster, Toshio Kobayashi, and Nick Vaughan.

ISBN: 978-0-470-97402-5
Also published in the *Encyclopedia of Automotive Engineering* (online edition)
DOI: 10.1002/9781118354179.auto026

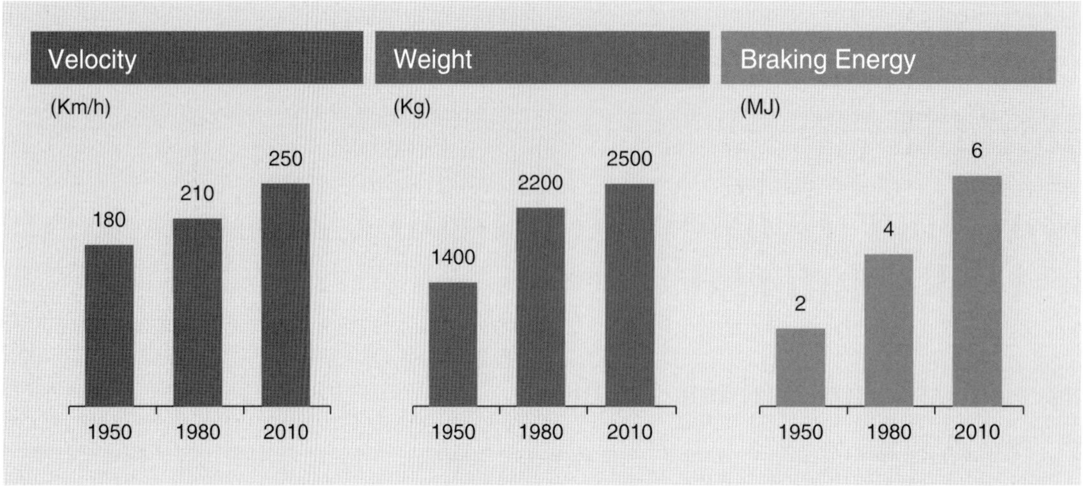

Figure 1. Increase of kinetic energy of upper class limousine shown here since 1950 up to the year 2010 leads to a demand of the automotive industry to look for new brake disk materials. (Reproduced by permission of SGL Carbon GmbH.)

limousine increased from around 180 to 250 km/h in the year 2010. The 250 km/h was a limit decided by the car manufacturer (OEM) for tire reasons. In the same time, also the car weight increases from around 1400 to 2500 kg. This leads to higher kinetic energy from 2.38 MJ in the year 1950 to 6.0 MJ in the year 2010 (Figure 1). This high energy results in a lower lifetime of the metal disks because of the increasing thermal loads during braking. This heat can also negatively result in a reduction of the friction coefficient during braking, called *fading*, with the result of longer stopping distances.

This leads to an increasing demand of the automotive industry to look for new brake disk materials and systems, which are stable under these rising braking conditions.

2 CARBON FIBER AND C/C BRAKES

With the development of the carbon fibers in the early 1960s, a new type of composite material, the carbon-fiber-reinforced carbon (CFC) material, which is made out of these fibers, leads to a new high temperature stable material class. These materials show very low weight because of their low density and found fast interest for high energy braking systems such as race cars or aircraft brakes and railway applications. The used carbon fibers are produced out of polyacrylnitrile polymer (PAN) or pitch-based fibers by carbonization at temperatures up to 1800°C under inert atmosphere. They show very high mechanical strength and high stiffness and additionally they show a very high temperature stability >2000°C in inert atmosphere without losing their mechanical performance. The disadvantages of these fibers are their behavior against oxygen. Owing to the oxidation reaction of carbon with oxygen under the formation of CO and CO_2, they are only stable up to 500–600°C under air. Nevertheless, the C/C brake disks and CFC pad materials developed for race car applications show excellent high friction coefficients that are stable even under extreme loads. The disadvantages of these disks are the low friction coefficient on cold disks surface and also in the wet state of the disk. In addition, the high wear rates especially in the cold disks combined with the high production costs limit this disk material mainly for race or aircraft applications. These disadvantages avoid a broad entrance of this CFC material class in the area of the normal road vehicles.

3 C/SiC BRAKES

The problems of the C/C brake disks could be solved by the development of a new type of brake material: the fiber-reinforced SiC-based ceramic brake disks (CF/SiC). Here, the carbon fibers are embedded in a ceramic matrix made out of SiC. Normally, ceramic materials suffer from their brittle behavior under mechanically and thermally induced stresses, which limit their applications. Different ways to improve fracture toughness of ceramics were developed in the past decades. The toughening of SiC ceramic by C-fibers or ceramic fibers such as SiC-fibers is the most effective way to reduce such brittle behavior (Bader, 1993; Evans, Zok and Davis, 1991) and opens the wide field for the use of this material. First trials to use these types of CF/SiC materials started in the past decade of the last century, based on public-funded projects in Germany for the development of ceramic brake disks for high speed trains.

First patents for CF/SiC brakes were granted for Krenkel and Kochendörfer (1994). They used woven fabrics for the C/C performs production. After the infiltration of this preforms with liquid silicon, they get the C/SiC material, which could be used as brake disk. The commercialization of CF/SiC brake disks started with the development of the short fiber-reinforced CF/SiC materials. This type of material was first developed at SGL and closely parallel at Daimler Chrysler and described by Gruber and Heine (1997) (SGL) and Haug *et al.* (1997) (Daimler Chrysler) in the patents. This material shows good friction coefficients in cold and wet disk states. Owing to the high hardness and low porosity of the material, it shows additional low wear rates in cold and hot disks. Their low weight (more than 50% reduction of unsprung mass compared to cast iron), high hardness and high stability of resulting friction coefficients, high corrosion resistance, and long lifetime are the main advantages for using CF/SiC ceramics as brake disk material or clutch material for automotive applications and their increasing demands. At the Frankfurt Motor Show in 1999, the carbon-ceramic brake disk was shown the first time to the public. In 2001, Porsche AG was the first car producer who installed the carbon-ceramic brake disk as series equipment into the 911 GT2. Since that time, also other premium cars use the advantages of CF/SiC brake disks. At present, for nearly all high end sport cars and luxury limousines for instance from Ferrari, Porsche, Audi, Bentley, Bugatti, and Lamborghini, CF/SiC brake disks are available (Figure 2). Ferrari equips today all new cars by series with the CF/SiC brake systems. In the year 2011, in total around 80,000 CF/SiC brake disks were produced.

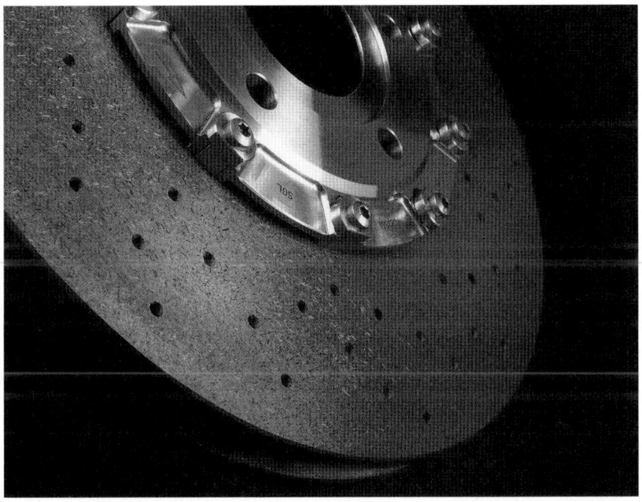

Figure 2. CF/SiC Brake disk. (Reproduced by permission of SGL Carbon GmbH.)

4 PRODUCTION WAYS TO CF/SiC MATERIALS

There are different ways available for the production of carbon-fiber-reinforced silicon carbide, which are summarized in Figure 3. These technologies show different ways to bring in the ceramic matrix in a porous carbon fiber preform under the formation of C/SiC. The technologies used are chemical vapor infiltration (CVI), liquid silicon infiltration (LSI), and the polymer infiltration and pyrolysis (PIP), which bring the ceramic SiC matrix in a carbon fiber preform. These technologies can be used alone or in combination and are explained in the following sections.

4.1 Chemical vapor infiltration

Starting from a porous carbon fiber preform, the ceramic matrix can be impregnated by a gas-phase deposition reactions chemical vapor infiltration (CVI) at temperatures >1000°C. To control the interface of the fiber to the ceramic matrix, a special interface layer must be deposited on the fiber surface. In most cases, this is a thin carbon layer (several micrometer thickness) deposited by CVI. Subsequently, the types of gases are changed and the ceramic matrix is deposited. Very slow infiltration rates are necessary for this reaction to keep the infiltrating channels open. After several days of infiltration, the surface must be brushed to open closed porosity on the surface. These together with the slow infiltration rates lead to long production cycle times of several weeks and to the high

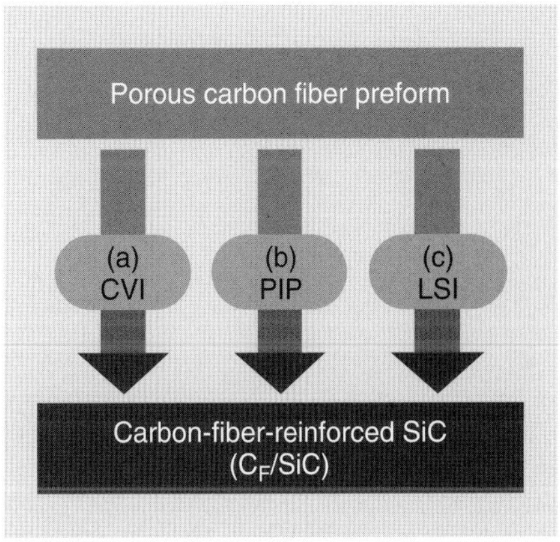

Figure 3. Production ways for carbon-fiber-reinforced silicon carbide materials starting from a porous carbon fiber preform infiltrating the ceramic matrix by (a) chemical vapor infiltration (CVI), (b) polymer infiltration and pyrolysis (PIP), and (c) liquid silicon infiltration. (Reproduced by permission of SGL Carbon GmbH.)

costs of this technology. The resulting parts produced by this technology show excellent mechanical behavior and a matrix porosity up to 15 vol%.

4.2 Polymer infiltration and pyrolysis

The impregnation of the porous carbon fiber preform with inorganic silicon-based polymers such as polysilans or polycarbosilanes used in the polymer infiltration and pyrolysis (PIP) technology is another way to build up the SiC ceramic matrix around the coated carbon fiber preform. The ceramic composition after the pyrolysis depends on the used inorganic polymer. These polymers are infiltrated under vacuum followed by a pressure cycle in the preform. After the pyrolysis of such impregnated preforms at temperatures up to 1000°C under inert atmosphere, the polymer decomposes and is transformed into a porous amorphous ceramic SiC material. To get a mechanical stable material, these impregnation and pyrolysis cycles must be done several times. Each cycle increases the density and closes the residual open porosity and raising the mechanical stability of the resulting CF/SiC material. Normally up to 5–6 cycles are necessary to get enough density, stability, and porosity lower than 10%. Production times for the PIP route are therefore 1–2 weeks. High process and polymer costs and the polymer availability limit currently this production way. To reduce the infiltration cycles, additional fillers in the preform can be introduced. Use of inorganic polymers such as polysiloxanes or polysilazanes lead to other ceramic matrix materials such as ceramics in the ternary systems Si-C-O and Si-C-N, which are currently not used for brake disk applications. This can be an option for the future.

4.3 Liquid silicon infiltration

The fastest way to produce CF/SiC brake disks is the reaction way via the LSI process. Here, the liquid silicon is infiltrated at temperatures higher than the melting point of pure silicon (>1420°C), in the porous C/C preform. To increase the infiltration speed, the infiltration is done under vacuum. The liquid silicon is infiltrated by capillary forces in several minutes into the porosity of the preform and reacts with the carbon matrix in a strong exothermic reaction under the formation of β-silicon carbide (Figure 4).

The residual porosity of the material is filled with the unreacted excess silicon. This LSI production way allows a fast production of C/SiC parts. The resulting material shows a low porosity of <1%. A small dimensional change only during the transfer from the CFC part to the ceramic part is a great advantage of this advanced technology.

For the CF/SiC brake disk, the LSI process is currently the mainly used process for production.

5 ELEMENTS OF A CERAMIC BRAKE DISK

The ceramic C/SiC brake disk of today is normally built up out of the following elements (Figure 5):

- The CF/SiC carrier body is the main element of the brake disk. It has the task to transfer the forces and to store and transport the heat, produced during braking. Therefore, the material of the carrier body must have a high mechanical stability and ductility and a good

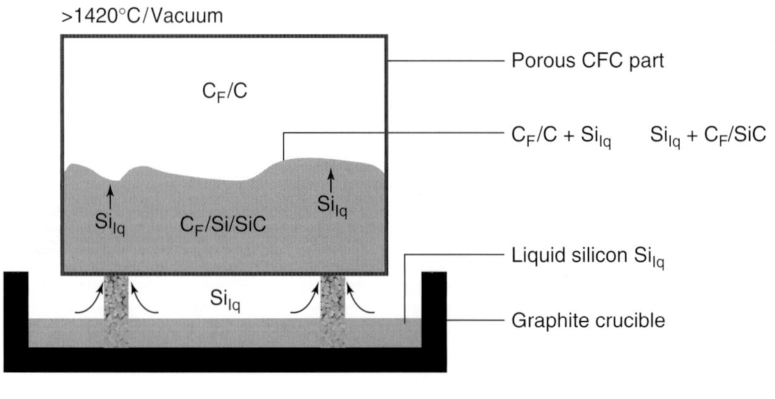

Figure 4. Schematic drawing of the liquid silicon infiltration process of CFC (w = porous carbon wick). (Reproduced by permission of SGL Carbon GmbH.)

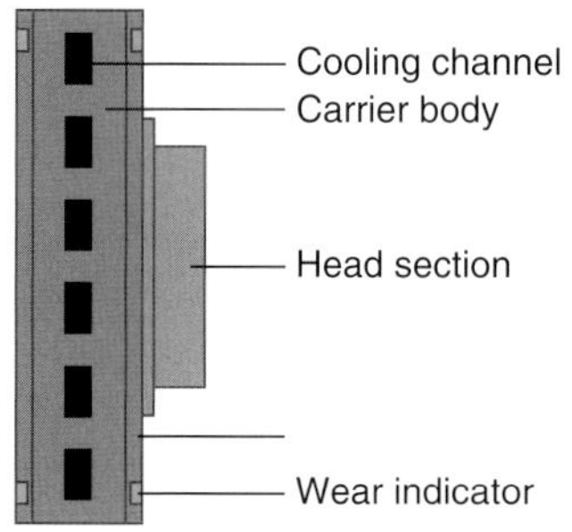

(a)

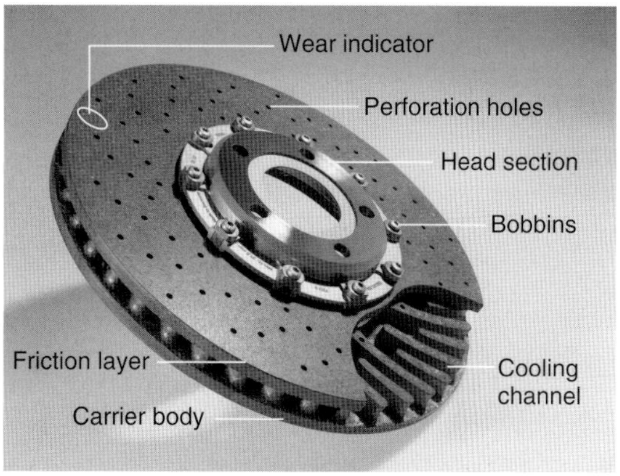

(b)

Figure 5. (a,b) Assembly up of a typical C/SiC brake disk. (Reproduced by permission of SGL Carbon GmbH.)

heat conductivity to transport the heat from the friction layer to the cooling channels. C/SiC material of the carrier body has typically higher carbon fiber content than the separate produced friction layer. Moreover, longer carbon fiber bundles are used than in the other disk parts because of the higher strength needs.

- The cooling channels in the carrier body have the task to allow a fast cooling of the brake disk. Therefore, the heat has to be transported as fast as possible out of the disk by the air passing through the cooling channels. The design of the cooling channel can be optimized to the demands coming from the car such as top speed and weight of the car.
- The friction layer is the contact area of the brake disk to the brake linings. Friction coefficient, wear, and lifetime of the disk are depending on the interaction of the friction partners. The friction layer material has a higher hardness than the carrier body material and typically a low carbon fiber content. If carbon fiber bundles are used, they are smaller and shorter than in the carrier body. As the C/SiC material has already low wear rates, brake disks can also be applied without a special friction layer.
- The head section today is typical out of metal. It is mounted on the brake disk by special bobbins, which allow compensating the different thermal expansion of the metal head section and the ceramic disk during braking. The head section is the contact area of the disk to the hub. The materials used for the head section are stainless steel or aluminum.

The dimensions of the produced C/SiC brake disks today start with outer dimensions of 350 mm up to above 420 mm and a thickness of above 32 mm, depending on the car weight and maximum speed.

6 PRODUCTION OF C/SiC BRAKE DISKS

6.1 Protection of carbon fiber bundles against reaction with liquid silicon

To reduce the direct reaction of the carbon fiber with the liquid silicon during the infiltration step, the carbon fibers must be protected. Therefore, the production of short fiber-reinforced CF/SiC brake disks starts with the preparation of the short fiber bundle with a special protection against liquid silicon. During the past decades, different technologies were developed to protect the fiber bundles against the reaction of liquid silicon (Gruber and Heine, 1997; Haug *et al.*, 1997; Krätschmer *et al.*, 2004). One promising way to prevent the carbon fiber bundles siliconization is the filling of the porosity in the bundles with various types of carbon. One example is the infiltration of the fiber bundles with phenolic resin or pitch followed by a pyrolysis step (Figure 6).

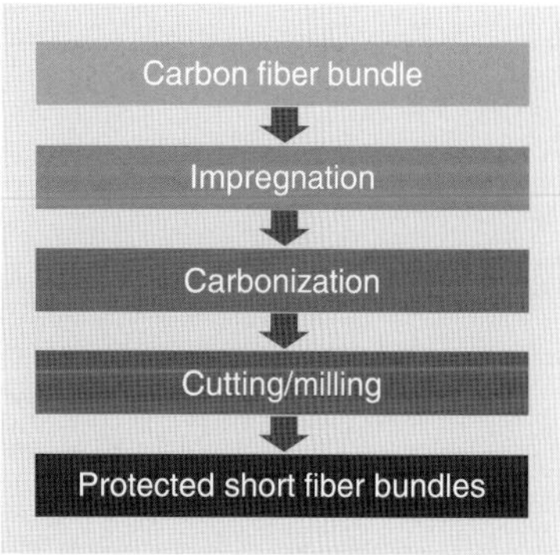

Figure 6. Production of carbon fiber bundles protected against liquid silicon. (Reproduced by permission of SGL Carbon GmbH.)

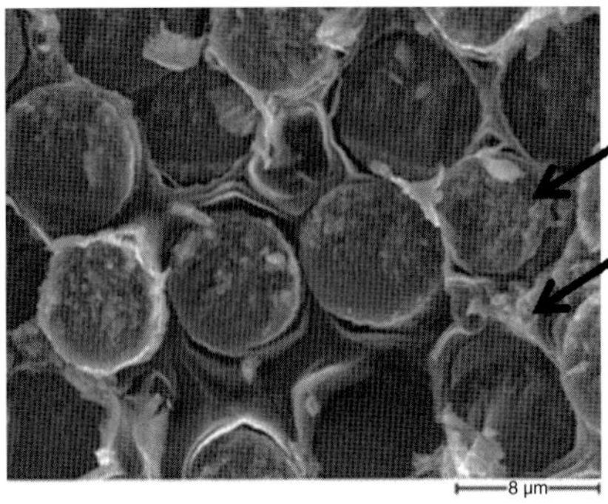

Figure 7. Cutting surface of a densified carbon fiber bundle by impregnation with pitch followed by a carbonization step. (Reproduced by permission of SGL Carbon GmbH.)

The carbon fiber bundles are used in the form of fiber rovings, woven fabrics, or pressed short fiber plates. The bundles are filled with phenolic resin or pitch and then carbonized up to temperature of 1200°C. Additional impregnation and carbonization cycles lead to a decrease in porosity in the bundles and an increase in fiber bundle stability and integrity. Figure 7 shows the SEM image of the cutting surface of an impregnated and compacted C-fiber bundle. The space between the single C-fiber filaments of the fiber bundle is filled with a dense carbon layer of pitch-based carbon. These dense carbon structures in the bundle give a good protection of the carbon fiber filaments against the reaction with liquid silicon during the siliconization process. The schematic production way to a ventilated C/SiC disk with friction layer is shown in Figure 8.

6.2 Mixing and molding step

For the carrier body of the disks, these stabilized fibers are mixed with phenolic resins as binder and further carbon-containing fillers such as graphite powder or coke powders.

After the mixing step is finished, the press masses are ready for the filling step. In the case of a production of solid disks without cooling channels, the press mass is filled directly in the mold and densified under pressure to the necessary green density. The plates therefore are heated up to 200°C where the phenolic resin starts to cross-link. After the cross-linking of the resin, the final CFRP disk is removed out of the mold and ready for further processing.

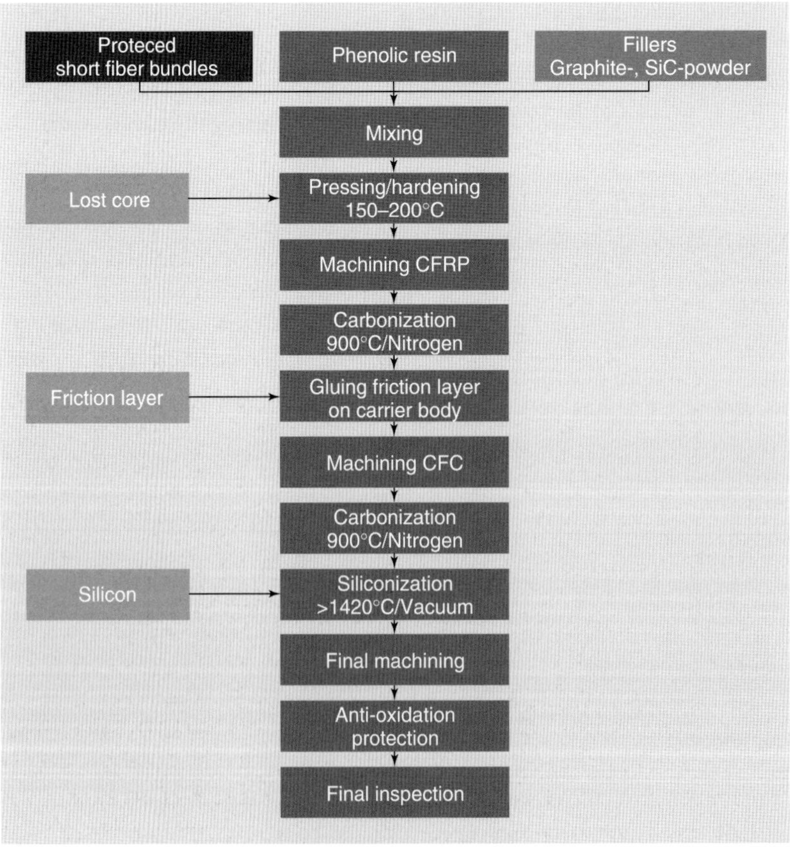

Figure 8. Production process of a ventilated C/SiC brake disk with friction layer. (Reproduced by permission of SGL Carbon GmbH.)

For the preparation of ventilated disks, more effort is necessary. Here, different technologies were developed. For cooling channels with an easy and radial geometry, the solid disks could be machined after the molding step. The disadvantages of these machining of the cooling channels are the high material loss and the high machining effort. To avoid these machining steps and also to allow more complex cooling channel geometries, three different production technologies are used today:

(a) The disk could be produced out of two parts with ribs on each part (Martin, 1999). The cooling channels of the ventilated disk are formed after gluing the two disks together in the contact areas of the ribs.
(b) For basic geometrical radial cooling channel, geometries drawable cores out of metal are used (Pacchiana and Goller, 2001). After the pressing step, the cores are removed mechanically out of the CFRP disk.
(c) For complex formed cooling channels, the lost core technology is used (Bauer *et al.*, 2002). Therefore, a core is produced with the geometry of the designed cooling channels. This core material is decomposed during the heat treatment of the disks completely. A disk with near-net-shaped cooling channels remains for further processing.

6.3 Carbonization

The pressed CFRP disks are removed out of the molds and are machined with conventional machining tools in the outer and inner diameter and thickness. After this premachining step, the disks are carbonized at temperature up to 1200°C under inert gas atmosphere. During the heating up to 1200°C, the organic binder material is thermally decomposed and transformed in porous glassy carbon such as carbon material under evolution of organic species with lower molecular weight. The weight loss of the used phenolic binder systems is typically in the range 40–60 wt%. The main weight loss of the binder is thereby in the temperature area between 400 and 600°C. These temperatures must be carefully crossed to avoid damages of the ceramic fiber-reinforced ceramic (CFRC) disks.

6.4 Friction layer

For disks with friction layer, the friction layer is glued on the porous CFRC carrier body. The friction layer itself is produced in a separate way and a special recipe is used. Therefore, a mixture of carbon fiber bundles, fillers, and phenolic resin as binder is mixed to form a press mass analogous to the carrier body press mass. In difference to the carrier body typically lower, carbon fiber bundle content is used. In addition, the used fiber bundles are smaller in their dimensions than in the carrier body material (Gruber, Heine and Kienzle, 2001). The friction layer is molded equal to the carrier body, on a press at temperatures up to 200°C. The produced CFRP parts were carbonized in the next step under inert atmosphere and then glued on the porous carrier body by a separate pressing and heating step. After the fixation of the friction layer, the final green machining step starts (perforation holes, etc.) and the disks get their finished design. After an additional carbonization step to transform the glue to carbon, the disks are then ready for the siliconization step.

6.5 Siliconization

The siliconization process of the CFC disks is performed in graphite crucible on wicks out of porous carbon material. The crucibles are filled with the calculated amount of solid silicon and are heated up to temperatures of 1700°C. The heating is done under a vacuum atmosphere in special siliconization furnaces. At a temperature of 1420°C, the silicon starts to melt and the molten silicon is absorbed by the wicks and transported by capillary forces into the porous CFRC disk. The infiltration into the body starts immediately and, in a highly exothermic reaction, the CFC disk is infiltrated completely and the matrix carbon and also some fiber filaments on the surface of the fiber bundles react with liquid silicon under the formation of SiC. The formed layer around the fiber bundles protects the fibers in the inner part of the bundle against further reaction with the liquid silicon. After the infiltration is completed, the furnace is cooled down to room temperature. Owing to the lower density of liquid silicon compared to solid density, some of the excess silicon is pressed out of the disks and forms some bigger Si drops, which must be removed after cooling. The density of the resulting C/SiC disk after siliconization is in the range 2.2–2.4 gcm^{-3} and depends on the used recipe and the CFC density of the disks.

In the next production step, the C/SiC disks are machined in inner and outer diameters and on the friction layer using diamond tools. The holes for the fastener elements are drilled in the C/SiC state to guarantee the necessary high precision. To protect the carbon fiber bundles in the disk against oxidation at temperatures >500°C during braking, the disks are further impregnated with an anti-oxidation solution. The C/SiC disks are then assembled with the metal head section using connecting elements made out of stainless steel and the disk is finally machined to the defined tolerances of parallelism between the head section/hub area and the friction area. The assembly is balanced by a groove, machined in the coverage of the brake disk. Every finished disk is quality controlled via sound test to detect internal defects. The dimensions of the disks are measured and additionally the thickness of the friction layer is controlled by measurement. The disks are then ready for shipping.

7 MATERIAL BEHAVIOR

7.1 Microstructure

An overview of the microstructure of a typical short fiber CF/SiC material from the carrier body of a brake disk is shown in the micrograph of Figure 9. The fibers are oriented perpendicular to the press axis z. In the xy-plane, the fibers are distributed with a random orientation or depending on the filling step. The fiber bundles are embedded in a SiC matrix (gray color). The porosity of the porous CFRC structure is filled with silicon (white areas).

Owing to the fiber protection of the bundles, the reaction with the silicon takes place only on the surface of the fiber bundle under the formation of a dense silicon carbide layer. This dense silicon carbide layer protects the fiber against further reaction during the infiltration. The different thermal expansion coefficients of the carbon fiber bundles, the silicon, and silicon carbide matrix lead to the formation of microcracks formed during the cooling of the material from 1420°C to room temperature. The microcracks are stopped in the neighboring fiber bundles. In addition, there are some areas of unreacted amorphous carbon matrix in the microstructure, which are also embedded in a formed SiC layer.

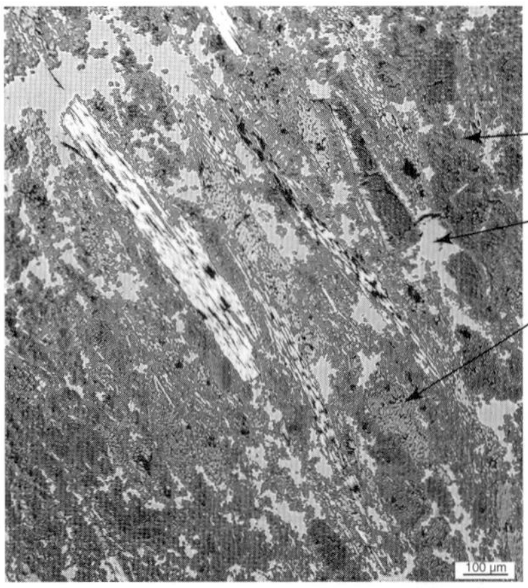

Figure 9. Micrograph of carbon-fiber-reinforced SiC. Sample of carrier body from a brake disk material. (Reproduced by permission of SGL Carbon GmbH.)

7.2 Material properties

In Table 1, the material data of typical short fiber CF/SiC materials in comparison to gray cast iron are summarized. In addition, the typical values of 2D reinforced CF/SiC material are mentioned. This material is produced by the siliconization of 2D CFC material out of woven C-fiber phenolic prepregs by lamination and carbonization. Short fiber-reinforced material shows a lower mechanical stability (strength, modulus, and elongation at maximum stress) than the long fiber-reinforced CF/SiC material. The higher density of the short fiber-reinforced material and the higher content of silicon and silicon carbide lead to increases in heat conductivity through the material. The density of the CF/SiC materials with values of 2.3 gcm^{-3} is significantly lower than the metal with densities of >7.2 gcm^{-3}. This lower density gives a significant weight advantage. If, on the front and rear axles, the CF/SiC brake disk is used instead of metal disks, weight reductions of up to 20 kg by car can be achieved for disks with 400 mm diameter. The differences in physical properties of the two types of CF/SiC materials are summarized in Table 1 together with gray cast iron GG-20.

The specific heat capacity of CF/SiC material is around 40% higher than cast iron. The combination of the high heat capacity, low Young's modulus, and high heat conductivity results in a high thermal shock resistance of the CF/SiC material, compared to the GG 20.

Table 1. Summary of the material data of short fiber- and long fiber-reinforced CF/SiC.

	I C/SiC *Short Fiber*	II C/SiC *Long Fiber*	GG-20
Density (g/cm^3)	2.3	1.9	7.2
Four point bending (MPa)	70–80	200	220
Strength modulus (GPa)	30	70	110
Elongation at break (%)	0.3	0.45	0.3–0.8
Heat conductivity z-axis (W/(mK))	35	10	54
Thermoshock resistance (W/m)	$> 27,000$	$> 27,000$	< 5400
Maximum operating temperature (°C)	1400	1400	700
Specific heat capacity/ weight (kJ/kgK)	0.8	0.8	0.5
Thermal expansion (*10^{-6} 1/K)	2.5	1.5	9.0–12
Phase content			
Si (wt%)	13	10	
C (wt%)	32	48	
SiC (wt%)	55	42	

(Reproduced by permission of SGL Carbon GmbH.)

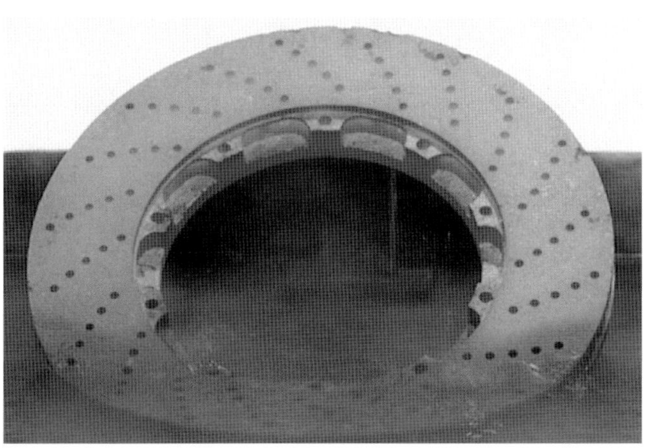

Figure 10. Thermal shock test of C/SiC brake disk. (Reproduced by permission of SGL Carbon GmbH.)

Figure 10 shows a disk, which is heated up to 1000°C and then dipped into water with one-half. While the side in the water is cooled down below 100°C, the non-dipped part is still red hot. Silicon carbide ceramic without fiber reinforcement would not survive such extreme thermal shock conditions. Another big advantage of the CF/SiC materials is the increase in mechanical stability with increasing temperatures (Figure 11). During the heating up to temperatures of 1200°C, the mechanical strength of CF/SiC increases up to 80% higher values, compared to respective room temperature values. In addition, the stiffness of the material is also increasing. In comparison to CF/SiC, the gray cast iron loses at temperature of 700°C around 40% of the mechanical room temperature strength and also stiffness.

Under dynamic load tests, CF/SiC materials show a reduction in stiffness in the range 20% but without reduction in the mechanical strength of the material (Thielicke, 2005). Owing to the heterogeneous material structure of the CF/SiC and the formed microcracks resulting from the different thermal expansions of the materials during the cooling step, critical stresses are reduced. In addition, newly formed cracks from an applied load are bridged or bypassed by carbon fiber bundles and the cracks are stopped. These effects also result in a low notch sensitivity of the CF/SiC materials. The high hardness compared with the good mechanical behavior and the increasing mechanical behavior with increasing temperature makes the short fiber material an ideal candidate for the use as brake disk material. The CF/SiC ceramic material combines the beneficial behavior of a hard ceramic material, which is responsible for the low wear, good tribological behavior, and long lifetime of the brake disk with the advantages of fiber-reinforced material that shows high fracture toughness and damage tolerance. The resulting quasi-ductile properties of the ceramic composite material ensure the resistance to high thermal and mechanical stability. Limitations are coming from the oxidation of the carbon fibers at temperature over 500°C under air. This oxidation can result in a decrease of the mechanical strength of the CF/SiC and may have a negative impact in the lifetime of the

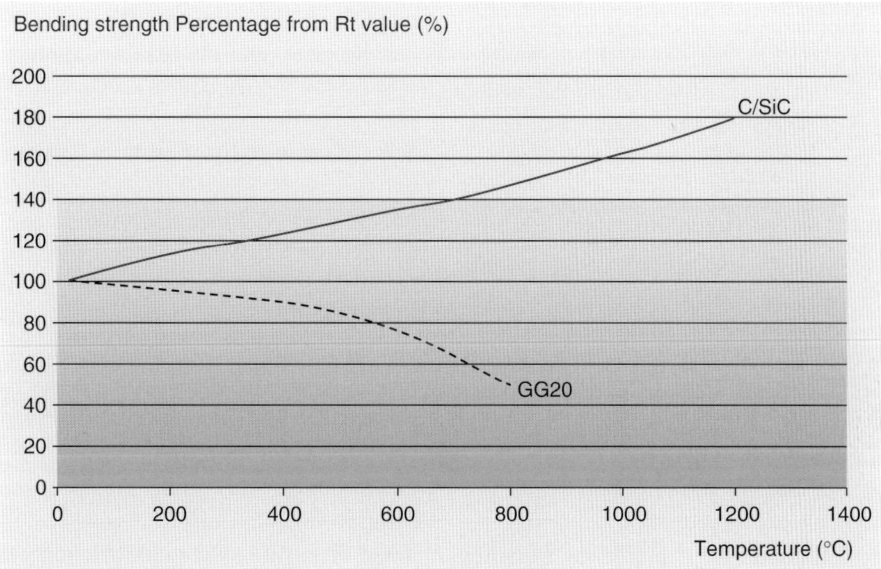

Figure 11. Bending strength behavior of GG 20 and C/SiC with increasing temperature up to 1200°C as percentage of the room temperature value. (Reproduced by permission of SGL Carbon GmbH.)

disks. Anti-oxidation agents such as aluminum hydrogen phosphates are preventing long time any negative oxidation effect during application.

8 C/SiC AS BRAKE DISK MATERIAL

8.1 Dimensioning and design of the disk

For the design of a CF/SiC disk for a new car, identical tests must be performed as for gray cast iron disk. Main control parameters are the maximum speed and the maximum load and their distribution on front and rear axles and the time to stop the car from maximum speed. The brake disk layout is largely determined by the response during full braking at v_{max} when the tensile stresses hit their peak because of the superimposition of the maximum braking and centrifugal forces.

The resistance of the brake disk mainly depends on their design of the chamber interface and interior disk rim as well as on the shape of the webs and cooling ducts. In addition, the cooling conditions of the disks are important factors. In usage, the brake disks are mainly strained by centrifugal loads and braking forces. Therefore, they must withstand lateral forces as well and further consideration must be given to thermal loads and stresses occurring at the interface between the brake disk and the brake disk chamber. Calculations for assembled carbon-ceramic brake disks include the design of the head section connection. The necessary dimensions of the disks are further calculated to ensure all possible mechanical load situations during the brake use over the lifetime of the car. This prevents that neither the disk itself nor any other component in its direct neighborhood is exposed to excessive thermal loads. To remove the heat as fast as possible out of the disks, the cooling channel geometry must be optimized using numerical methods for each car model. First prototypes are produced from the optimized disks and cooling channel dimensions and tested on a test bench also under extreme load conditions (Figure 12). The mechanical stability, braking performance, and heat dissipating behavior of the tested brake disks are the main results from these tests.

These results are used to optimize the simulated and calculated behavior of the disks and consequently modifications in the construction are transferred. The loads over the lifetime of the car are simulated on the disks and their mechanical stability is tested. These tests are done with disks directly after the production and also after surviving the simulated lifetime, for example, by bursting tests. Finally, right after perform dynamometer tests and analyzing the results, the tests on the car are done. They cover not only high speed runs on a test circuit but also mountain passes, descents, and road tests in normal traffic under winter and summer conditions. During these test runs, the test driver evaluates the brake behavior, the braking performance, and the braking comfort.

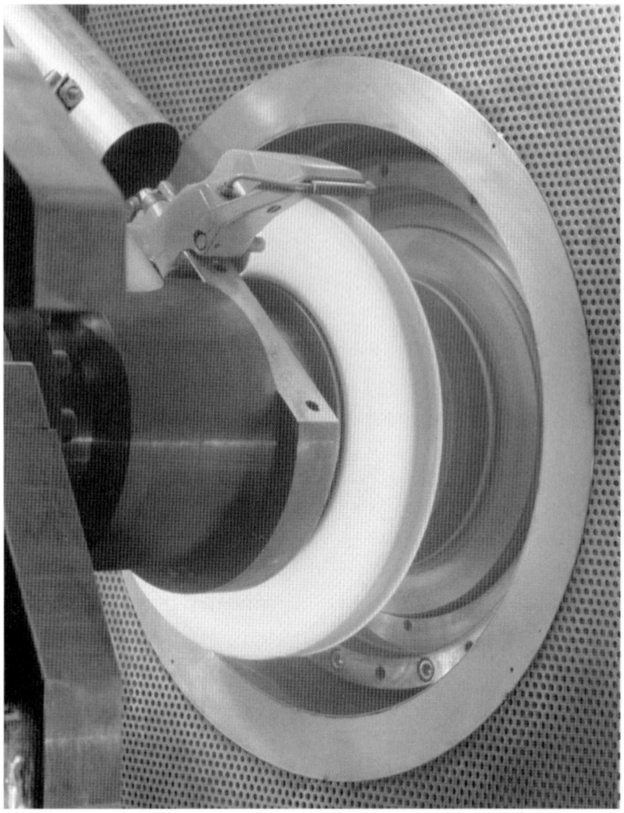

Figure 12. Overload friction test of a C/SiC disk on a test bench. (Reproduced by permission of SGL Carbon GmbH.)

8.2 Brake caliper and brake pad

For the CF/SiC brake disks today mainly fixed caliper with up to four opposing piston pairs are used. The pistons clamp the CF/SiC disk during braking between the pad materials. To avoid overheating of the hydraulic fluid of the caliper, the pistons are equipped with a layer with low heat conductivity between the brake pad and the piston. At present, mainly organic bounded pads are used as brake pad. The material is matched to the friction behavior of the ceramic surface of the disks. Metal bounded pads and ceramic bounded material pads were tested for the C/SiC brake disks and show good friction performance. Owing to comfort problems, such as noise evaluation during braking, various improved pad materials are still under development and testing.

8.3 Advantages C/SiC brakes

In comparison to the standard metal brake disk, this ceramic brake disk has several advantages:

Low Density. The low density of around 2.3 gcm^{-3} leads to a significant reduction in weight of the brake disks compared to gray cast iron (−30% to 70%).

High Abrasion and Corrosion Resistance. Gray cast iron disks show a reduction of the disk thickness during lifetime by wear and corrosion, which can be measured easily. In comparison to this, the CF/SiC brake disk shows only low wear rates because of the high hardness of the SiC ceramic and consequently only a low reduction of disk thickness during lifetime of the car is measurable. The significantly lowered wear of disk and pad material offers as additional advantage the reduction of break dust. This is a very positive contribution for the environment.

Noise during Braking. High thermal loads during cooling after high energy breakings at temperatures over 700°C form cracks and deformations of the metal disks. These deformations result in noise during braking and lead to disk changes. Owing to the low thermal expansion and the high temperature stability of the C/SiC disk, there is no significant change in dimensions or geometries of the disks measurable during braking, which could reduce the performance of the disks.

Friction Behavior and Fading Stability. With increasing temperatures, the CF/SiC disks show additionally a slight increase of the friction coefficient during braking in comparison to the gray cast iron disks. This friction behavior results in an improved fading stability of the CF/SiC disks compared to a metal disk. The fast build up of the friction coefficient and the higher possible values of the friction coefficients of >0.45 compared to gray cast iron (from $\mu = 0.3$ up to 0.45) lead to an excellent positive pedal feeling for the driver and results in significantly shorter stopping distances especially from high road speeds.

8.4 Other automotive friction applications of CF/SiC materials

Owing to the high strength and ductility of the long fiber CF/SiC material and the high hardness of this material, it can be applied as material for clutch disk systems. This type of disk was for the first time used in the high end sports car Porsche Carrera GT. The main advantages for CF/SiC in clutch applications are the low wear rate under extreme load, the low weight, and the high and stable friction coefficient. The disks can be designed in smaller dimensions and allow lowering down the drive train and the center of gravity of the vehicle with the better performance of the driving characteristics. Currently, this ceramic clutch material is mainly used in high end race cars.

9 OUTLOOK

The current CF/SiC disks are mainly used in high end sports and luxury road vehicles because of high costs of the disks. To enter the market of normal cars, the price of the disks must come down. This will happen with the development of new brake systems allowing easier disk geometries and resulting in a less complex production. Of ongoing importance is the increase of the automation of the production processes. Therefore, new production technologies must be developed, implemented, and qualified. With their high lightweight potential, the C/SiC brake disk in combination with their discussed braking performance, the CF/SiC material is the right brake disk material for cars in future.

10 SUMMARY

Since 1950, the top speed of upper class limousine increases from around 180 to 250 km/h in the year 2000. In the same time, the car weight also increases from around 1400 to 2500 kg. This leads to higher kinetic energy, which results in a lower lifetime of the metal disks and fading problems, with the result of longer stopping distances.

In the past decade, the carbon-fiber-reinforced silicon carbide brake disk material is developed. The material is produced by liquid siliconization of porous CFC perform with liquid silicon under vacuum. The resulting material shows a low weight because of the low densities of around 2.3 gcm^{-3} and high hardness. At the Frankfurt Motor Show in 1999, the carbon-ceramic brake disk was shown the first time to the public. At present, for nearly all high end sports and luxury cars, CF/SiC brake disks are available. In this contribution, the production processes of the CF/SiC disks are described and an overview of the CF/SiC material and brake disk behavior is given.

RELATED ARTICLES

Volume 4, Chapter 117
Volume 4, Chapter 119

REFERENCES

Bader, M.G. (1993) Reinforcing fibers: the strength behind the composites. *Materials World*, **1**, 22–26.

Bauer, M., Gruber, U. Heine, M. Huener, R. Kienzle, A. Rahn, A., and Zimmermann-Chopin, R. (2002) Process of Manufacturing Fiber Reinforced Ceramic Hollow Bodies. EP 1300376 B1.

Evans, A.G., Zok, F.W., and Davis, J. (1991) The role of interfaces in fiber reinforced Brittle Matrix composites. *Composites Science and Technology*, **2**, 3–24.

Gruber, U., Heine, M. and Kienzle, A. (2001) Friction or Slip Body Comprising Composite Materials, Reinforced with Fibre Bundles and Containing a Ceramic Matrix. EP 1216213B1.

Gruber, U. and Heine, M. (1997) Mit Graphitkurzfasern verstärkter Siliciumcarbidkörper. DE1971010105.

Haug, T., Kienzle, A., Schwarz, C., Stöver, H., Weißkopf, K., Dietrich, G., and Gadow, R. (1997) Verfahren zur Herstellung einer faserverstärkten Verbundkeramik. DE 19711829 C1.

Krätschmer, I., Kienzle, A., Wuestner, D., Domagalski, P., and Haeusler, A. (2004) Polymer Bound Fiber Tow. EP 1645671B1.

Krenkel, W. and Kochendörfer, R. (1994) Verfahren zur Herstellung einer Reibeinheit mittels Infiltration eines porösen Kohlenstoffkörpers mit flüssigem Silizium. DE 4438455 C1.

Lanchester, F.W. (1902) Improvements in the Brake Mechanism of Power propelled Road Vehicles. GP 26407.

Martin, R. (1999) Bremsscheibe aus Faserverbund Werkstoff. DE 19925003 A1.

Pacchiana, G.P. and Goller R.S. (2001) Mold and Procedure for Manufacturing of a Braking Band with Ventilation Ducts in Composite Material. EP 1412654B1.

Thielicke, B. (2005) Mechanische Charakterisierung von Faserkeramiken für Bremsscheiben. *Konstruktion*, **9**, 20–21.

Chapter 119
The Development of Alternative Brake Systems

Bernd D.M. Gombert, Martin Schautt, and Richard P. Roberts
RG Mechatronics GmbH, Seefeld, Germany

1 INTRODUCTION

Brake system development has always been characterized by the search for and development of new technologies. Although this initially concentrated on the optimization of systems driven by pure muscle power, over the past 150 years, there has been a steady increase in the importance of powered brakes, as a response to both heavier vehicle weights and higher speeds. There has also been repeated interest in self-reinforcement as a means of lowering brake forces, either for the driver or for the actuation system. Critical to the success of such developments has been the provision of good controllability, repeatability, and comfort for the driver. Economic viability and a well-considered safety strategy are of course also essential to any such system. Such hurdles have resulted in many promising systems not being brought to the mass market, but either being restricted to niche applications or disappearing altogether. Some ideas have then reappeared as the technological boundaries have expanded, for example, due to the introduction of powerful but inexpensive microprocessors.

The future of the automobile will be significantly shaped by networked sensors and electronics and increasingly powerful driver assistance systems. These will exert even more influence on the major vehicle controls than is currently the case, intervening in the drive train, brakes, and steering. Starting with ABS (anti-lock braking system) and ESC (electronic stability control), it has been possible to control the vehicle in ways, which are simply not possible for the driver, who has neither the information available (e.g., individual wheel speed sensors), nor the ability to intervene appropriately (e.g., braking of individual wheels for ESC), nor the necessary speed of reaction. This trend will continue in the foreseeable future, with the continued improvement of existing systems, more requirements from more electric vehicles (e.g., blending of regenerative and friction braking), and the introduction of new lower bandwidth systems that exploit the possibilities offered by affordable new sensors.

For existing vehicle systems, such as ABS, TCS (traction control system), and ESC, the braking system must be able to control the force at each wheel individually and autonomously from the driver input, which can be zero. This will continue to be the minimum requirement for future braking systems. A simple electronic interface will increasingly be required between the many vehicle level systems and the high performance brake system. At the same time, the overall use of energy should be reduced to minimize the fuel consumption and production of CO_2. A significant step in this direction can be made using components that only use energy when they are actually required, which typically means a move toward more electric systems. At the same time, the desire to increase vehicle safety will ensure that the

Encyclopedia of Automotive Engineering, print © 2015 John Wiley & Sons, Ltd.
Edited by David Crolla, David E. Foster, Toshio Kobayashi, and Nick Vaughan.
This article is © 2015 John Wiley & Sons, Ltd. ISBN: 978-0-470-97402-5
Also published in the *Encyclopedia of Automotive Engineering* (online edition)
DOI: 10.1002/9781118354179.auto025

performance requirements will tend to increase rather than reduce.

As a response to these requirements, the automobile industry has already introduced a variety of systems that can be signaled electrically and include some form of power supply, which is independent of the driver's mechanical energy. The simplest form of such a system is the pump in an ESC system, which permits braking pressure to be produced without intervention from the driver. In order to proportion the braking forces between friction and regenerative braking, a smart booster may be used, allowing a relatively straightforward modification to existing braking systems. At the high performance end of the spectrum is the electrohydraulic brake (EHB), where, in normal operation, the pedal only functions as a simulator, which generates an electronic demand for the system. The hydraulic pressure is supplied from an accumulator and the pressure in the individual brake cylinders is controlled by means of proportional servovalves. These examples illustrate one of the key features of "by-wire" systems that the control and energy paths are separate, in contrast to a standard mechanical brake. The signaling is electronic, but the power supply can be hydraulic (EHB), pneumatic (smart booster), or electric. Electric-powered solutions can be either "wet" or "dry." An example of the former is using an electric-driven pump to provide hydraulic pressure, either

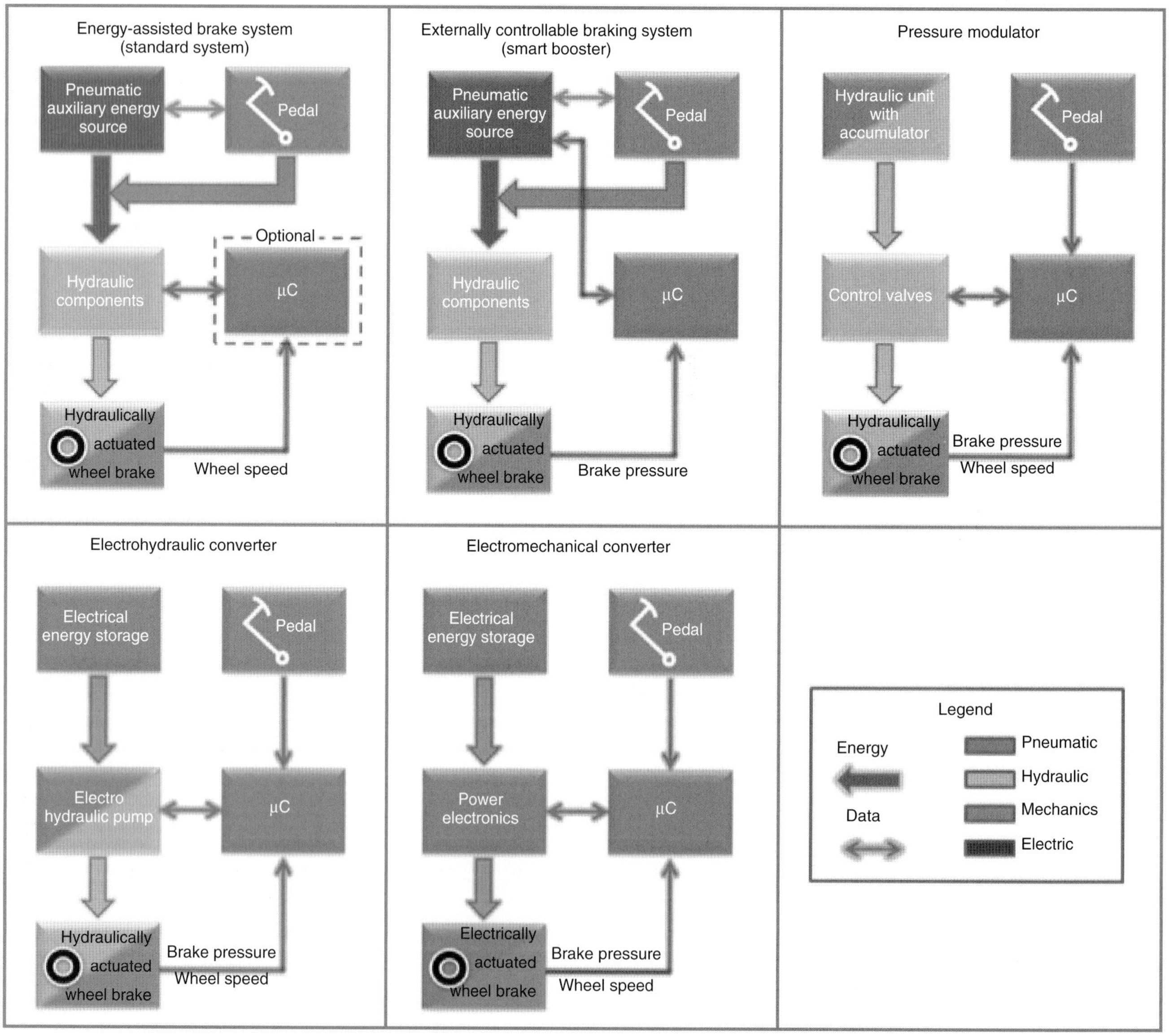

Figure 1. Signaling and power options for brake systems.

on demand or for storing in an accumulator. In "dry" systems, the power is transmitted by purely mechanical means, obviating the need for hydraulics.

An overview of typical possibilities for powering an automobile brake system is provided in Figure 1. Of these five architectures, three can be considered by-wire systems: the pressure modulator, the electrohydraulic converter, and the electromechanical converter. The first two of these use hydraulic wheel brakes, whereas the last one is the only "dry" system illustrated.

Hydraulic systems have the great advantage of being able to store energy at a high energy density over long periods of time without significant losses. This is both an advantage for the system energy supply (a hydraulic accumulator can store much more energy in a smaller volume than a battery) and for the actuation itself. Thus, a constant wheel force can be held simply by closing a valve, without requiring a constant input of new energy, except for that, needed to overcome leakage. This is clearly true of the EHB solution (pressure modulator). Other solutions (electrohydraulic converter), where the hydraulic system can be thought of as a form of gearbox, do not share this advantage, as they have to actively hold the brake pressure. For booster systems, this requires maintenance of a certain amount of suction from the engine, in order to support the driver.

Practical electromechanical systems do not share this advantage and also have to actively hold the brakes closed. This requires current and hence electric power. For a classic electromechanical brake (EMB), this static requirement can be minimized using higher gear ratios, but this conflicts with the actuator dynamics required for ABS systems. Both the achievable rate of change of brake force and the actuator acceleration can be an issue here. This sets an upper bound to the possible gear ratio. The lower bound is effectively set by volume and weight constraints for the brake actuators. Even motors with modern rare earth magnets have a limited torque density and a direct drive system is simply not feasible. Hence, some form of gearing will always be required.

The designer of a classic EMB is therefore faced with a dilemma: if the gear ratio is too low, the long-term power requirements will be higher, which will have an impact on the long-term fuel consumption of the vehicle (not to mention higher requirements on the power electronics, etc.); should it be too high, then achieving the necessary dynamics for effective ABS control will demand high power peaks. These will have very little impact on the fuel consumption directly but will significantly affect the dimensioning of the vehicle power supply. Confronted with this problem, a general consensus has developed within the automobile industry that conventional EMBs are a possible solution for the rear axle but that a 42 V power supply is required, if they are to be used for the front brakes. Consequently, conventional EMBs are still under consideration within the industry. However, alternative approaches have been sought in order to meet all requirements with 14 V power supplies.

Within this chapter, we consider various approaches to the problem of dry brake-by-wire (BBW) actuators. In nearly all cases, an attempt has been made to reduce the required static actuator force. This means that a lower gear ratio is required between motor and brake pad and that the dynamic requirements can consequently be met more easily. The additional benefit is that many components can be reduced in size, providing actuator cost and weight savings.

2 POTENTIAL BENEFITS

The first question that arises is why such developments could be of interest. After all, the performance of existing brake systems is already quite impressive. Some points have been touched on in the introduction but we will expand on them here. It needs to be stated in advance that there is no single overriding reason why such systems have to be introduced. As a result, it will be a judgment on the part of OEMs and suppliers as to when/whether a switch is made. However, when this point is reached, none of the major players can afford to be left behind.

Firstly, it is clear that the coming years will see increasing electrification of vehicles. A wide variety of systems is possible, with mild hybrid systems at one end of the spectrum and electric or fuel cell vehicles at the other. The former will become common in the mass market, whereas the latter will probably be restricted to small production runs in the near future. For all vehicles, there will be a trend to lower fuel usage and reducing CO_2 output, as a response to rising fuel prices and as part of a political solution to achieving climate goals. Within the European Union, this has resulted in the regulatory requirement to reduce fleet average CO_2 production to 95 g/km by 2020. The significant savings offered by hybrids and electric vehicles in city traffic, where a significant proportion of drivers live and commute, will mean that more electric vehicles have an important role to play in meeting such goals. Such developments are also being pushed politically.

As a result of this increasing electrification, more electric energy (and power) will be available in vehicles. Minimizing its use is important both for reducing CO_2 output for more conventional vehicles and for ensuring maximum range for minimum battery size in electric vehicles. Efficient use of the available energy and power requires the following:

- *Minimizing* quiescent power losses (power on demand)
- *Maximizing* the efficiency of the individual components
- *Controlling* power consumption via
  - Intelligent power/battery management
  - Intelligent consumers.

The consumers themselves have to be prioritized so that essential systems (e.g., steering and braking) are always provided with sufficient power, whereas comfort systems, especially those with a large power draw (e.g., seat heating), can be either controlled or disabled, should the need arise. This requires significant information flow around the power management system and demands that the critical systems can signal their needs before the power situation becomes critical.

Thus, we can see that

- the primary signaling interfaces are *already* electronic;
- the primary power interface is *becoming* electric;
- energy-efficient solutions are required, both in operation and when "off";
- more central intelligence will be required for power management;
- more local intelligence will be required to enable the power management to function properly;
- more communication will be required between "better integrated" vehicle systems, to permit optimum use of the available power and energy.

All of these factors point to more automotive by-wire solutions in the future. The main advantages of electrically powered systems are that, if correctly implemented, they only need to draw power when it is actually needed (e.g., electric power-assisted steering) and packaging, installation, and maintenance can be simplified. They are also clearly compatible with situations such as a hybrid vehicle driving in electric mode.

The issues discussed earlier are valid for any vehicle system. However, what would be the advantages of a dry BBW system? The first obvious issue is removing the hydraulic fluid itself. This saves time and hence costs during the production process and in servicing, particularly as there is no longer a need to bleed the brake system. It is no longer necessary to replace and dispose of the fluid several times during the vehicle lifetime. The brake booster and vacuum pump can be removed from the engine compartment, saving considerable volume. The lack of hydraulic architecture permits more flexibility in the system design and contributes to a more modular system, which is easier to assemble.

Perhaps the most important gains can be obtained in the pedal region. The pedal is reduced to a feel simulator system with sensors. This removes any sort of direct connection with the brakes themselves. This has significant benefits for the passive safety and for reduced noise transmission into the passenger compartment. Pedal force and travel is no longer constrained by the requirements of the hydraulic brakes themselves, so the travel can be reduced without any significant effect on the system. The feel can then be made exactly the same for a range of vehicles and can be potentially be tuned via software to individual requirements (e.g., 5% female vs 95% male). Mounting the pedal unit should be simpler and the location of the supporting structure is less constrained by other components. Finally, there will be no pulsing of the pedal during ABS interventions—some see this as an advantage and some as a disadvantage.

As regards performance, electric systems are by nature more linear than hydraulic systems and less sensitive to temperature (viscosity of hydraulic fluid). Individual wheel braking is of course implemented as standard. Experience suggests that it is possible to make an all-electric braking system perform as well as a very "highly tuned" hydraulic system and significantly better than most conventional systems. Traction control interventions in particular can be made smoother than existing systems. Tuning can be performed by means of software, which makes the process quicker and more flexible. Thus, the same hardware can be "branded" differently by adapting the software. The additional electronics provide new possibilities for system diagnosis and fit well with the concepts of intelligent power management.

3 BRIEF REVIEW OF ALTERNATIVE SYSTEMS

Two separate issues are addressed in this section:

- The various proposed brake systems
- Proposed vehicle configurations.

Different developers have taken different routes to realizing the goal of a dry BBW system, all probably with slightly different motivations. As yet, none of these systems has been applied to a mass production vehicle, although some have come close. Our review here is based on published information, which means that some developments are probably excluded, either for lack of such information or because it has not come to the attention of the authors. Many suppliers and OEMs choose to be very discrete about their developments.

3.1 Brake actuators

We will split the brake actuators into two groups: those, which do not use self-reinforcement and those, which do. While self-reinforcement offers significant benefits, the brake response can easily be influenced by varying friction—coefficient between pad and disk, requiring higher gain controllers to ensure repeatable performance. There are also other risks, which we will address later. As a result, much development has focused on more conventional actuators, which were seen as lower risk solutions.

3.1.1 Actuators without self-reinforcement

3.1.1.1 Conventional electromechanical brake. Conventional EMBs consist of a motor, gearbox, and a ball- or roller screw, which replaces the hydraulic piston. They have been investigated by a large number of organizations, and one sample of collaboration between industry and academia is given in the article by Schwarz *et al.* (1998). As discussed earlier, a consensus has developed that a 42 V supply is required to power such actuators for the front axle of a passenger vehicle. Development of such systems has stalled; therefore, the prospect of a vehicle using pure EMB has receded.

Nevertheless, there is still interest in these actuators. The main reason is that it is relatively straightforward to include a latch mechanism in the gear train and so to produce an integrated parking brake. The actuator can be sized for the maximum parking brake force and used on the rear axle of the vehicle with a conventional 14 V supply. As braking forces when driving are generally relatively low on the rear axle because of stability considerations, the actuator dynamics are adequate for existing ABS/TCS applications.

3.1.1.2 Maximum torque brake (Delphi). The maximum torque brake (MTB) was an attempt to solve the problem of using an EMB on the front axle of the vehicle. The idea is relatively simple—instead of one brake disk, the MTB uses two. This doubles the number of friction interfaces and hence approximately halves the actuator force required compared to the EMB. Assuming the same motor, sized on long-term torque requirements, the gear ratio can therefore be halved. As nearly all the inertia is in the motor itself, this change results in approximately double the rate of change of brake force and four times the acceleration of the conventional solution. This provided sufficient dynamics to allow the MTB to be installed on the front axle with a conventional 14 V power supply (Smith and Hudson, 2003).

The most difficult engineering challenge here is in ensuring that all the sliding components move correctly relative to one another under all conditions and throughout the vehicle lifetime.

While development effort on the MTB seems to have ceased, this approach is clearly one way of overcoming the limitations of the conventional EMB. The concept can also still be powered by hydraulics if required, perhaps offering some economies of scale. In addition, it is certainly possible to make such concepts work: aircraft brakes have long relied on stacks of rotors and stators to provide the necessary energy absorption. These systems are also now moving in the direction of electric actuation.

3.1.2 Actuators with self-reinforcement

A second path to reducing the actuator forces and improving the dynamics is to use self-reinforcement. This can ideally result in the condition that no actuator force is required to produce any given braking force. While this was seen as a hard boundary for hydraulic actuators (which were only designed to push), for electromechanical actuators, this is actually the point of neutral stability. When the self-reinforcement goes beyond this point, the actuator is unstable and requires a controller to stabilize it. This means that, in the event of a failure, some independent means is required to ensure that the brake fails open.

A second issue arises with self-reinforcement, namely what happens when it disappears, typically due to the vehicle stopping. This is particularly critical on a slope. For example, if a car is braked to a halt going up a hill, then, once it has stopped, it will try to roll backward. Thus, the force, which should normally be assisting the brake, acts against it and tries to open it. Some strategy is required to deal with this problem smoothly and safely.

3.1.2.1 Electronic wedge brake. Considerable development work was conducted on the EWB by Siemens VDO in the period between 2004 and 2007. It was designed to use self-reinforcement in both the stable and unstable regimes, so as to obtain the maximum benefit from the technology (Hartmann *et al.*, 2002). Adjustment for wear, a parking brake function and a fail open function were implemented in an additional mechanism to the main brake actuator. It was possible to demonstrate vehicles containing all the main modern brake control functions (ABS, TCS, and ESC), which could operate using a 14 V supply.

At this point, it is worth considering why this can be the case. It has been noted by other authors that the energy required to expand the caliper is not that high, with the implication that self-reinforcement is redundant. We will examine the benefits by means of calculations based on simple assumptions.

Firstly, for a conventional EMB, we will assume a linear caliper stiffness of 20 kN/mm and a maximum clamping force of 35 kN. The energy required to achieve the maximum clamping force is therefore

$$\begin{aligned} E &= 0.5 \times F_{\text{Max}} \times \Delta x_{\text{Cal}} \\ &= 0.5 \times 35{,}000 \times \frac{35{,}000}{20 \times 10^6} = 30.6\,\text{J} \end{aligned} \tag{1}$$

Assuming the maximum force is for a fading case and that this needs to be achieved in 0.2 s, then the power required per brake to reach this force is an average of approximately 150 W over this 0.2 s, ignoring the mechanical losses.

For a wedge brake with actuation in the plane of the disk, the actuator requires lower forces to achieve the same clamping force but has to move further. Hence, if α is the wedge angle and μ is the coefficient of friction between pad and disk, then

$$\begin{aligned} E &= 0.5 \times F_{\text{Max}} (\tan\alpha - \mu) \times \frac{\Delta x_{\text{Cal}}}{\tan\alpha} \\ &= \left(1 - \frac{\mu}{\tan\alpha}\right) \times 30.6\,\text{J} \end{aligned} \tag{2}$$

Thus, excluding the mechanical losses, the wedge brake will always require less energy than the conventional solution as long as $\mu < 2\tan\alpha$. For high friction coefficients, the energy works the opposite way round to a conventional brake: the EWB requires energy to open the brake rather than to close it.

To consider the dynamic case, we have to make some more assumptions. We will assume that the wedge brake is constructed with $\tan\alpha = 0.35$ and that it is designed to operate normally for $0.15 \le \mu \le 0.55$. This means that the actuator force is reduced to 20% of that required by an EMB. Assuming we use the same motor, dimensioned so that it can hold the maximum steady torque indefinitely for both applications, then the EWB requires a gearing factor 5 less than the EMB for the same application. If the EWB must rotate at 250 r/s to meet the requirements for rate of change of force, then the EMB must rotate at

$$\omega_{\text{EMB}} = 5\tan\alpha \times 250 = 437.5\ \text{r/s} \tag{3}$$

We will assume that the motor has an inertia of 25 kg mm^2 and that this dominates the inertia of both systems. Hence, the kinetic energy required to achieve the same dynamics for the two brakes is

$$\text{KE}_{\text{EMB}} = 0.5 \times 25 \times 10^{-6} \times 437.5^2 = 2.39\,\text{J} \tag{4}$$

$$\text{KE}_{\text{EWB}} = 0.5 \times 25 \times 10^{-6} \times 250^2 = 0.78\,\text{J} \tag{5}$$

If we now consider that the maximum speed must be reached in the order of 10 ms to produce good ABS control, then the average mechanical power required for the EMB is 239 W whereas that for the EWB is 78 W. The peak power will of course be considerably higher than this.

Thus, the main advantage of the wedge brake is in providing superior dynamics for the same power. This has been demonstrated in dynamometer and vehicle tests. The penalties are as follows:

- More complicated mechanics (forward/backward braking, adjustment, and emergency release functions)
- More complicated control.

3.1.2.2 Cross-wedge brake (Mando). Mando has also been developing a wedge brake to allow braking with a 14 V supply (Kim, Kim, and Kim, 2009). The so-called cross-wedge mechanism has been used, with the aim of producing even pad wear and avoiding the use of rollers, which were felt to be vulnerable to contamination. High force amplification was not a goal, and a subjective view of the wedge design based on the publications seen suggests that the brake should normally operate in the stable regime ($\tan\alpha \approx 0.75$). This would obviate the need for an emergency release mechanism (assuming the friction coefficient can be guaranteed). However, the use of a worm gear to change the direction of the drive through 90° raises the question of whether the system can be back-driven open. An electric parking brake (EPB) function has been integrated into the brake (Kim *et al.*, 2010), which is thought to be implemented by disabling the motor and allowing the worm gear to hold the force. If so, this would cause problems with the system safety requirements, which should demand that a faulty brake is opened if the vehicle is moving.

Results have been presented, which show that the brake can be used for ABS and ESC control, although the authors admit that the performance is not yet equivalent to a conventional brake system (Kim *et al.*, 2010).

3.1.2.3 VE brake. The Vienna Engineering group has proposed a solution, which relies on a nonlinear lever mechanism to provide a limited degree of self-reinforcement (Putz, 2010). Because of the mechanism, both the overall mechanical advantage and the effective wedge angle vary with position. It is designed to produce a higher force ratio at higher forces but should not become unstable. Thus, no additional release mechanism is required, as long as this can be guaranteed. A parking brake function can be integrated in the brake, although its precise implementation has not been specified in the published literature. At the moment, only dynamometer results have been published, but it is clear that vehicle tests will follow (Vienna Engineering, 2011).

In the view of the authors, the main challenge here is that the bearings in the lever mechanism must be able to support the forces acting on the active pad and rotate freely for the lifetime of the brake.

3.2 Vehicle concepts

A second aspect of the development of advanced brake systems is the configuration to be applied in the vehicle. As mentioned earlier, none of the systems has yet reached series production, so all of the solutions mentioned here must be regarded as provisional.

3.2.1 Rear EMBs only

In this configuration, the front axle of the vehicle retains an existing hydraulic brake system, whereas the rear axle uses EMBs, ideally with integrated parking brakes. As a means of introducing the BBW systems to the market, this approach has several advantages:

- The design requirements for the hydraulic system are simplified.
- The rear axle suits the strong points of the EMB and permits a 14 V power supply.
  - No additional parking brake required.
  - Simple integration of hill-holder type functions within the brake system.
  - Fail-silent and adjustment functions without any additional mechanisms.
- The safety concept can build on that for conventional systems.
- Service experience with such a hybrid concept will help generate reliability data for later systems.

Continental has, for some years, been developing an electrohydraulic combi (EHC) brake based on this approach (Neunzig and Linhoff, 2009).

3.2.2 Front and rear EMBs

At the start of interest in dry BBW systems, this was naturally the favored configuration. As discussed earlier though, it was determined that such an approach is only practical when a 42 V supply is used. Consequently, at the time of writing, such configurations do not appear to be being seriously investigated.

3.2.3 Rear EWBs only

Mando is investigating a vehicle with their cross-wedge EWBs on the rear axle (Kim *et al.*, 2010). It is not known whether this is intended to be representative of a production configuration or it is only intended for prototype testing. In the opinion of the authors, it must be the latter, because this configuration gains the minimum advantage from self-reinforcement while being most susceptible to its disadvantages (see, e.g., the following section).

3.2.4 Front and rear EWBs

This configuration was successfully investigated by Siemens VDO. The performance of the system was extremely good and more than lived up to expectations.

The main challenges, which arose, were mainly the result of changes in direction of, or loss of, self-reinforcement. As discussed earlier, on coming to a halt forward on an up slope, the direction of the self-reinforcement changes. If nothing is done, the EWBs will draw a high current and may not be able to hold the vehicle stationary. The solution adopted was to split the braking directions of the EWBs just before the vehicle stopped: the front axle remained in the forward direction whereas the rear axle was set to brake backward. Should the direction of the self-reinforcement change, then there is always one set of brakes, which will hold the vehicle. The overall braking force was held approximately constant while this occurred. On stopping, a brake current limit was introduced to prevent the system using too much power. Because of the high degree of self-reinforcement, the brakes, which were acting in the "correct" direction, could always hold the vehicle using this reduced current.

This works well in most circumstances, but it is more tricky in the case of rear-wheel drive cars, particularly automatics. Here, the driven axle can continue pushing against the closed brakes, increasing the current required to hold them closed. Should traction control be required at low speed, then there is also the danger that the rear brakes will continually be changing direction to allow for the (potentially) different directions of self-reinforcement. Switching round the directions of the braking on the front and rear axles is not attractive because the system is then always trying to brake with the "wrong" axle at low speed.

An additional motor was required for the EPB function. To avoid producing a conventional EMB for this case, the

EPB force was reduced and distributed between all four wheels. This is not ideal for the critical case of holding the vehicle on a slope with a loaded unbraked trailer attached.

Such issues will be relevant to all the proposed systems that use self-reinforcement. Clearly, the lower the level of the self-reinforcement is, the less critical the problem becomes.

3.2.5 Front EWB and rear EMB

MOBIS has prepared a prototype vehicle in which the axles each have a different dry by wire brake (Cheon *et al.*, 2010). It appears to be a well-thought-out solution, because it uses the actuators where they show most potential:

- The EWBs on the front axle require only a 14 V supply, as with the Siemens vehicles.
- The EMBs on the rear axle implement the parking brake function and are not sensitive to changes in direction.

Using this type of concept, it would be possible to increase the braking force on the rear axle once the vehicle has stopped, so that there is no danger of the vehicle moving.

While for production systems, the two different types of actuators might be regarded as a cost disadvantage, in terms of the system safety, the possibility of common mode failures is significantly reduced.

4 CHALLENGES

It is clear from what has already been written that there are benefits to be had from introducing new BBW systems and that there are solutions, which are technically feasible. However, while there has been done considerable research on this topic during the past decade and a half, there is no such system on the mass market yet. This alone demonstrates that there must be significant hurdles to overcome. So, what are they? We will list some of the major issues.

4.1 Somebody has to be first

This category contains a range of issues, for which we could think of no better title. Because no such system exists, there is no single set of requirements, which can be applied to it, so there is no well-defined set of rules to work by. Within ECE-R13H, it can be seen that there are passages, which have been added to accommodate the EHB. For any developers of future EHBs, it is therefore possible to follow these guidelines and it should be possible to obtain certification for a large number of markets. While the first one to market a dry BBW system may have the advantage of being able to help write these regulations, this requires considerable extra effort and liaison with a number of safety authorities. There is also the possibility that competitors may not only offer constructive criticism in such cases!

Any such system will now be subject to ISO/DIS 26262. This also represents a challenge, because the standard is still new and not all the associated documentation is yet available (mid-2011). In some cases, it may still be necessary to fall back on the IEC 61508, because not all the processes are completely clear, especially for such a complex system as this. This will again require additional effort and specialized personnel.

The fact that there is no internationally agreed path to certification is a major obstacle. Until such a process is agreed, there will remain a residual risk of having to repeat tests and certification work, even if the development is almost complete.

Assuming these issues are overcome, there is then the issue that the OEMs may require two independent suppliers for such systems. This is obviously not likely to be the case, and this means that at least one supplier will have to have an OEM firmly "on-board" before deciding to go ahead with production.

4.2 Costs

Costs will always be a challenge, especially now that even relatively high technology brake systems are more-or-less sold "by the kilogram." A supplier has to be willing to invest a lot of money up front to produce a system and must be very sure that there will be sufficient return on it. Quantifying exactly the scale of investment is also not easy, because of the issues mentioned earlier.

In uncertain economic times, this requires some courage, not least because existing brake systems will not stand still, neither in capability nor in price. Moreover, it will be more difficult to leverage low labor costs for a new, high technology system than for well-understood, conventional systems. Finally, competitors will be certain to make life more difficult when the new product is ready to be launched.

Thus, as always in the automobile industry, cost control will be a key factor in a successful development process.

4.3 Reliability/availability

A system containing many electronic components will have to be very well designed and tested to achieve reliability rates comparable with a "dumb" hydraulic system. There is simply more that can go wrong. In addition, some of the electronic components are likely to be located in harsh environments (wheel well, with shock and vibration, and large temperature changes), which will not make this any easier.

4.4 Conclusion

In general, it is clear why there is a marked reluctance to be first in the field, despite the potential of dry BBW systems. It is perhaps not surprising that an "outsider," such as Siemens VDO, has to date been the most willing to push their development: they have the most to gain and the least to loose. Existing suppliers are often satisfied with the status quo, which is more predictable and which (still) produces reasonable margins, especially for "higher level" brake functions. In Germany, in particular, there is also the memory of the introduction of the EHB, which has made suppliers and OEMs alike more risk averse. At the same time, nobody wants to be left too far behind should the market take off.

Both these factors perhaps explain Continental's development of the EHC brake, which represents a stepping stone in this process. There is an understandable desire to minimize risk and also to ensure that the hardware is really ready for production before introducing it.

5 EXAMPLE: ELECTRONIC WEDGE BRAKE

Here, we discuss some of the experiences gathered by the authors during the development of the EWB. We split the section into subsections devoted to a particular topic.

5.1 Mechanical design

A good mechanical design is clearly the basis for any successful product. In the case of the wedge brake, there were several difficult problems to be solved in a relatively confined space.

The first obvious issue is that the brake must be actuated in two directions. This demands motion perpendicular to the direction of travel of the roller screw. The solution chosen was to attach the roller screw to the wedge with reinforced leaf springs (to prevent buckling). The forward actuation was designed to be in the plane of the wedge angle, so that normally no perpendicular motion was required. In the reverse direction, the motion was proportionally quite large, but the travel was reduced because smaller braking forces were needed. A view of this design is provided in Baier-Welt and Schmitt (2007).

The next major issue was the force sensor. Here, a new development was undertaken to produce a sensor integrated in the caliper using cheap Hall sensors to measure the relative deflection of a loaded and unloaded part. The noise level from this sensor was noticeably lower than that of the load cells used in the early prototypes, but it was more sensitive to drift. Some detail about this development is provided in Baier-Welt (2007).

The most complicated single development was that of a mechanism to implement the following three functions:

- Emergency release (fail silent)
- Pad wear adjustment
- Parking brake.

This was built around a second motor, a solenoid, and a cam. During normal operation, a roller sat on the cam slope, supporting the brake clamping force. The cam was actively locked by means of the solenoid. In the event of a loss of power, a spring opened the solenoid and the cam released the brake force. The pad wear adjustment relied on interaction between the solenoid and the secondary motor and could not be conducted under load. Finally, the parking brake used a stable depression in the cam surface to hold load without power. Illustrations of these functions are provided in Baier-Welt (2007).

5.2 Actuator control

Control was obviously a major issue in the development of the EWB, simply because the wedge can be unstable. Any controller has to be able to address this fact. The taken approach was to design a controller, which could stabilize the brake, even if the friction coefficient between pads and disks was 1. This also required an estimate of this friction coefficient, so that the motor could not be overpowered and pulled in.

Two different approaches were taken, which were to some extent dictated by the mechanical design. The early Beta prototypes were designed with two motors so that drive-train backlash could be actively removed (Roberts *et al.*, 2003, 2004). They were also stiff in the axial direction. The latter property allowed the controller to be based on a conventional cascaded design, with an inner current loop, followed by a rate loop, and an outer force control. The motors were controlled either to remove backlash if the friction was close to the ideal value or to cooperate if more axial force was required. Brief details are provided in Ho *et al.* (2006) and Lang *et al.* (2006).

The preproduction prototypes were all based around a single motor. On the one hand, this simplified some of the control, but on the other hand, the mechanism required for forward and backward brakings was much more flexible in the axial direction than in the earlier hardware. For this reason, it was necessary to use a state "feedback" controller. The system was stabilized for a μ of 1.0 and a conventional PID controller was then used to achieve the desired performance. This process is discussed in Fox *et al.* (2007).

5.3 Vehicle control

Vehicle level systems were designed and optimized with considerable use of simulation tools. This, together with the well thought out structure, enabled a comprehensive system to be produced in a relatively short time. It also simplified the process of adapting the control to different vehicles.

A hierarchical structure was built up, with slip controllers at the lowest level to manage ABS and traction control interventions, and a vehicle dynamics controller at the level above this. This implemented the brake force distribution, ESC, and some wedge-specific functions (e.g., direction control). This structure helps produce relatively smooth interventions, which were not only of benefit to passenger noise and comfort but also suit electromechanical actuators in general, where the motor inertia is an important influence on performance. More details are provided in Semsey and Roberts (2006).

5.4 Functional safety

A description of the approach to functional safety is provided in Schaffner *et al.* (2006). The brake system requirements were taken as far as possible from ECE-R13H, whereas IEC 61508 was used to cover the electrical and electronic components. At this stage, ISO 26262 was still a working draft and it was mainly used for categorizing risks into ASILs (automotive safety integrity levels). Three major categories of failures were found to lead to ASIL-D (most severe):

- Unintended braking
- Insufficient braking
- Braking (or lack of it) leading to vehicle stability problems.

After consideration of a number of potential architectures, a solution was proposed based on.

- three pedal sensors (fail-operational)
- two EPB button sensors (fail-silent)
- a central electronic control unit with three microcontrollers (fail-operational)
- two energy management systems, each with an associated backup battery, attached to the main vehicle power supply. Each power channel feeds one brake diagonal, the central electronic control unit, and the pedal unit.

In the case of a single fault, the objective was to achieve either fail-operational or fail-degraded behavior (reduced performance). For a second fault, it had to be possible to bring the vehicle to a controlled stop. For a failure where the next failure could lead to a critical loss of performance, an Auto-Stop function was proposed. This would have led to additional hazards of its own.

5.5 Electronics

The design of the electronics followed the functional safety requirements. A fail-operational central control unit was designed and produced. The wheel unit electronics featured an independent "safing control" processor and were developed in two phases: firstly an external version to check the circuitry and then an integrated version. Considerable thermal modeling and testing was done to support the latter development. It was found that, in order to meet the EMC targets, integration within the actuator was a "must." More information can be found in Zelger (2007).

6 CONCLUSIONS

In this contribution, we have tried to lay out some of the background to the development of dry BBW systems. The technical motivation for such systems belongs to an increasing development over the past few years: more and more manufacturers look seriously at hybrid vehicles and electromobility. Although the level of research being conducted is not currently as high it perhaps has been, there are still a number of organizations working on the problem. There are a variety of different solutions under investigation, both in terms of actuator technology and vehicle architecture. There has even been some new actuator concepts been developed in recent years, relying on limited self-reinforcement.

It is nevertheless easy to see why such systems have not yet been brought to market. There is nothing simple about introducing a revolutionary new product into an established market for a safety-critical application. However, the authors are still in the firm opinion that it is only a question of when this breakthrough occurs, not if. The "fit" between this technology and the trends in automobile development is simply too good. Once this occurs, the very obstacles, which have discouraged many from taking this step, will act to secure a competitive advantage for the successful organizations.

GLOSSARY

ABS	Anti-lock braking system
BBW	Brake-by-wire
EHB	Electrohydraulic brake
EHC	Electrohydraulic combi brake
EMB	Electromechanical brake
EPB	Electric parking brake
ESC	Electronic stability control
EWB	Electronic wedge brake
OEM	Original equipment manufacturer
MTB	Maximum torque brake
TCS	Traction control system

REFERENCES

Baier-Welt, C. and Schmitt, B. Electronic Wedge Brake. *IQPC 3rd Annual Innovative Braking Conference*, Frankfurt am Main, Germany, 26th April 2007.

Baier-Welt, C. *High Performance Electrical Wedge Brake Actuator*. Vehicle Dynamics Expo, 5th May 2007.

Cheon, J.S., Jeon, J.H., Kim, J.S., *et al.* (2010) Brake by wire system configuration and testing using front EWB (electric wedge brake) and rear EMB (electro-mechanical brake) actuators. SAE Paper 2010-01-1708.

Fox, J., Roberts, R., Baier-Welt, C., *et al.* (2007) Modeling and control of a single motor electronic wedge brake. SAE Paper 2007-01-0886.

Hartmann, H., Schautt, M., Pascucci, A., and Gombert, B. (2002) eBrake—the mechatronic wedge brake. SAE Paper 2002-01-2582.

Ho, L.M., Roberts, R., Hartmann, H., and Gombert, B. (2006) The electronic wedge brake—EWB. SAE Paper 2006-01-3196.

Kim, J.G., Kim, M.J., and Kim, J.K. (2009) Developing of electronic wedge brake with cross wedge. SAE Paper 2009-01-0856.

Kim, J.G., Kim, M.J., Chun, J.H., and Huh, K. (2010) ABS/ESC/EPB control of electronic wedge brake. SAE Paper 2010-01-0074.

Lang, H., Roberts, R., Jung, A., *et al.* (2006) The road to 12V brake-by-wire technology. *VDI-Berichte Nr.*, **1931**, 55–71.

Neunzig, D. and Linhoff, P. Electro Hydraulic Combi Braking System (EHC). *IQPC 5th Annual Innovative Braking Conference*, 2009.

Putz, M. (2010) VE mechatronic brake—investigations of a simple electro-mechanical brake. SAE Paper 2010-01-1682.

Roberts, R., Gombert, B., Hartmann, H., *et al.* (2004) Testing the mechatronic wedge brake. SAE Paper 2004-01-2766.

Roberts, R., Schautt, M., Hartmann, H., and Gombert, B. (2003) Modelling and alidation of the mechatronic wedge brake. SAE Paper 2003-01-3331.

Schaffner, J., Doerricht, M., Hartmann, H., and Gombert, B. *Approach to a Functional Safety Concept for the Electronic Wedge Brake*. BremsTech, 2006.

Schwarz, R., Isermann, R., Böhm, J., *et al.* (1998) Modeling and control of an electromechanical disk rake. SAE Paper 980600.

Semsey, A. and Roberts, R. (2006) Simulation in the development of the electronic wedge brake. SAE Paper 2006-01-0298.

Smith, A.C. and Hudson, S.M. (2003) A new, high torque brake design using sliding discs. SAE Paper 2003-01-3309.

Vienna Engineering. *TechnischeUniversität Braunschweig bestellt brake-by-wire System*. www.vienna-engineering.com, 2011.

Zelger, C. (2007) Electronic Wedge Brake—Actuator Integrated Electronics. *IQPC 3rd Annual Innovative Braking Conference*, Frankfurt am Main, Germany, 26th April 2007.

Chapter 120

The Cooperation of Regenerative Braking and Friction Braking in Fuel Cell, Hybrid, and Electric Vehicles

Zhuoping Yu and Lu Xiong
Tongji University, Shanghai, PR China

1 INTRODUCTION

It is well known that when the brakes are applied in a car to slow it down, some energy is wasted. The kinetic energy converts into heat and becomes useless. However, the energy crisis is becoming more and more serious at the moment and carbon emission is reaching a peak. Therefore, energy must be used as efficient as possible.

The beginning of the twenty-first century could very well mark the final period in which internal combustion engines are commonly used in cars. Now, automakers are trying to apply new energy technologies, such as pure electric, hybrid, plug-in hybrid, hydrogen fuel cell, and other alternative fuel energies such as biofuel. At the same time, automotive engineers have beaten their brain out to wring the maximum efficiency out of aerodynamic streamlining of the bodies and use of lightweight materials. Among them, regenerative braking is one of the most important technologies.

When driving a car, you have to hit the brakes occasionally to stop the car or adjust the speed. For a conventional car, about 80% of its energy converts into heat through friction. However, with a system that can recapture much of the car's kinetic energy and convert it into some other restorable and reusable energy, regenerative braking is able to capture as much as half of that wasted energy and put it back to work. Therefore, the fuel consumption can be reduced by 10–25%.

Figure 1 shows the energy conversion by braking system. The conventional braking systems convert the kinetic energy into friction heat and acoustic energy, for example, braking squeal, which cannot be reused. To convert the kinetic energy into some storable and useable energy, a lot of various schemes of regenerative braking have appeared. For instance, the kinetic energy of the vehicle can be converted into electrical energy by the electric motor and stored in the battery or supercapacitor or mechanically converted to the kinetic energy of a flywheel (e.g., KERS in some F1 racing car), etc.

The regenerative brake with electric motors is the well-developed one than other variants of regenerative braking systems. It is used in electric vehicles, primarily pure

Encyclopedia of Automotive Engineering, print © 2015 John Wiley & Sons, Ltd.
Edited by David Crolla, David E. Foster, Toshio Kobayashi, and Nick Vaughan.
This article is © 2015 John Wiley & Sons, Ltd. ISBN: 978-0-470-97402-5
Also published in the *Encyclopedia of Automotive Engineering* (online edition)
DOI: 10.1002/9781118354179.auto028

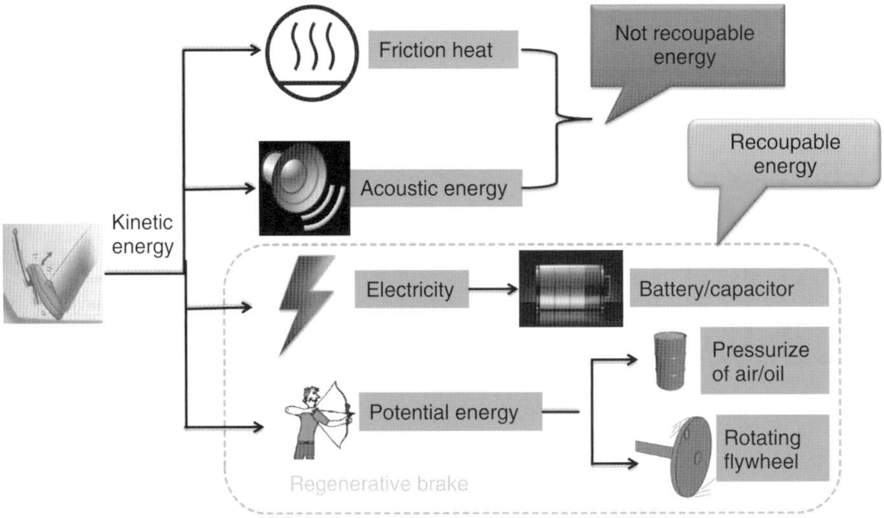

Figure 1. Energy conversion by braking systems.

electric vehicles and hybrid electric vehicles, whose battery can be used for longer periods of time without the need to be recharged by external charger or get gas service. The traction motor or the electric machine would work as a generator when using the regenerative braking, it could produce a brake torque, meanwhile producing electrical energy and stored it into energy storage components, namely, battery or supercapacitor.

2 HISTORY OF THE REGENERATIVE BRAKING SYSTEM

The origin idea of regenerative braking dates from the nineteenth century. In the most common form, there is an electric motor working as an electric generator. In electric railways, the generated electricity is put back to the supply system, whereas in battery electric and hybrid electric vehicles, the energy is stored in a battery or bank of capacitors for later use.

2.1 The Krieger electric landaulet

Louis Antoine Krieger (1868–1951)'s horse-drawn cabs were incipient examples of regenerative braking system. There is a drive motor with a second set of parallel windings in each front wheel. The motors are free to revolve a short distance either way around the large spur wheel and drive the wheels through spur gearing. Under the driver's seat is the main battery and under the passenger seat is an additional one. While descending a hill, there is certain provision for charging the battery (Figure 2).

Figure 2. Krieger electric landaulet. (Reproduced from Harris & Ewing, 1906. Library of Congress, Prints & Photographs Division, photograph by Harris & Ewing.)

2.2 "Regenerative control" in tramway

In order to reduce electricity consumption, tramway operators brought the Raworth system of "regenerative control" into use in the early 1900s in British cities such as Devonport (1903), Rawtenstall, Birmingham, Crystal Palace-Croydon (1906), and many others. While slowing down the car or keeping its speed in the downhill, the motors worked as generators, braking the car and capturing the kinetic energy at the same time. There were also wheel brakes and track slipper brakes in the tramcars in case of failure

Figure 3. Train carrying iron ore transported between Kiruna and Narvik. (Reproduced from Gubler, 2009. © David Gubler.)

of electric braking systems. The tramcar motors were shunt wound in several situations, and the systems on the Crystal Palace line utilized series-parallel controllers. However, an embargo was promulgated on this form of traction in 1911, because of a serious accident at Rawtenstall. Fortunately, the regenerative braking system was reintroduced 20 years later.

2.3 Application of regenerative braking in railways

Regenerative braking has been widely used on railways for many decades. The advantages of the regenerative system are the reduction of energy consumption, the reduction of wearing of brake shoes and wheel tires, and the consequent lowering of maintenance costs. The disadvantage is that the motors are larger and more costly. There are many examples of remunerative regenerative braking abroad in mountain railways and on lines where the gradients are heavy. From Riksgränsen on the national border to the Port of Narvik, the trains use only 20% of the regenerated energy. This regenerated energy is sufficient to power the empty trains back up to the national border. Any excess energy from the railway is pumped into the power grid to supply families and businesses in the region, and the railway is a net generator of electricity (Figure 3).

3 COOPERATIVE WORKING OF REGENERATIVE AND FRICTION BRAKING SYSTEMS

In a traditional braking system, brake pads produce friction with the brake rotors to slow or stop the vehicle. Additional friction is produced between the slowed wheels and the surface of the road. This friction is what turns the car's kinetic energy into heat. With regenerative brakes, on the other hand, the system that drives the vehicle does the majority of the braking. When the driver steps on the brake pedal of an electric or hybrid vehicle, these types of brakes put the vehicle's electric motor into reverse mode, causing it to run with reverse torque, thus slowing the car's wheels. While running with reverse torque, the motor also acts as an electric generator, producing electricity that is then fed into the vehicle's batteries. These types of brakes work better at certain speeds. In fact, they are most effective in stop-and-go driving situations. However, in a cooperative regenerative braking system, there still exist two sub-braking systems, that is, traditional friction braking system (usually hydraulic braking system) and regenerative braking system based on electric motors. Generally, the reasons are as follows:

1. The regenerative braking torque is not large enough to cover the required braking torque; furthermore, when the motor speed rises, regenerative braking torque will decrease because of the flux-weakening control.
2. The regenerative braking cannot be used for many reasons such as high state of charge (SOC) or high temperature of the battery to increase the battery life.
3. Regenerative brake cannot brake to stop (because of its work principal).

The relationship between regenerative and friction brake system is shown in Figure 4.

At the beginning of braking, the motor (light gray line) could meet the driver's rapidly rising demand for the total braking torque (black line), when the motor reaches its full potential, the hydraulic braking (dark gray line) begins to intervene. Continue with the braking, vehicle speed decreases, and the motor torque raises a little (light gray line). The motor torque declines rapidly (light gray line) when the vehicle slows down, and the braking torque is fully provided by the hydraulic brake (dark gray line) when close to stopping.

3.1 Three ways to cooperate the two sub-braking systems

There are three ways to realize the cooperative working of friction and regenerative braking systems (Zheng, 2010; Zhang and Ning, 2009).

The first way is to add the regenerative braking system directly onto the original hydraulic braking system without changing the arrangement of the original hydraulic braking system (Figure 5). The advantage of this way is simplicity.

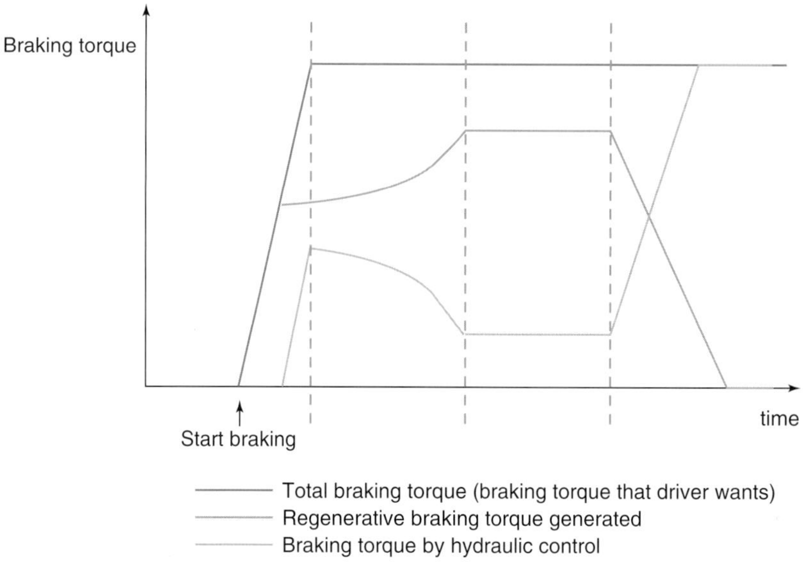

Figure 4. The relationship between regenerative and friction brake system.

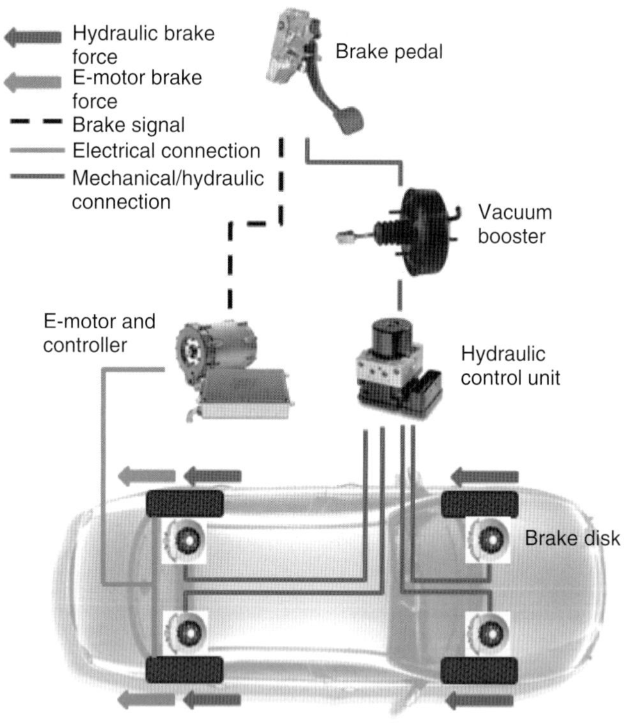

Figure 5. Add the regenerative braking system directly.

It is not necessary to rebuild a braking system. In this way, the structure is simple and what need to be changed can be minimized. However, the disadvantage is the low energy recovery efficiency because of the current braking force distribution design (Zhang and Ning, 2009).

The second way is brake-by-wire technology, where many of the functions of brakes that have traditionally been performed mechanically will be performed electronically (Zhang and Ning, 2009) (Figure 6). Here, electronic braking pedal and integrated sensors are used. When drivers step down the brake pedal, the sensor records the force, displacement, velocity, and other necessary signals to identify drivers' brake intention. The controller would give command to the pedal displacement simulator in order to supply the designed friction brake force. In this way, the friction brake force can be controlled actively to maximize the energy recovery efficiency. At present, different automakers have come up with different circuit designs to handle the complexities of regenerative braking. However, in all cases, the most important part of the braking circuitry is the braking controller.

The third way is redesigning the current hydraulic braking system by a large margin (Zhang and Ning, 2009). By adding and controlling the valves on the hydraulic braking pipelines, the hydraulic braking pressure can be controlled in time to generate the designed friction brake force. In terms of theory, this design of cooperative regenerative braking system can achieve the highest energy recovery efficiency.

3.2 The structure of the cooperative regenerative braking system

Regenerative braking system is used in the vehicles driven fully or partly by electric motors. Electric motor supplies the regenerative brake force. One of the most interesting

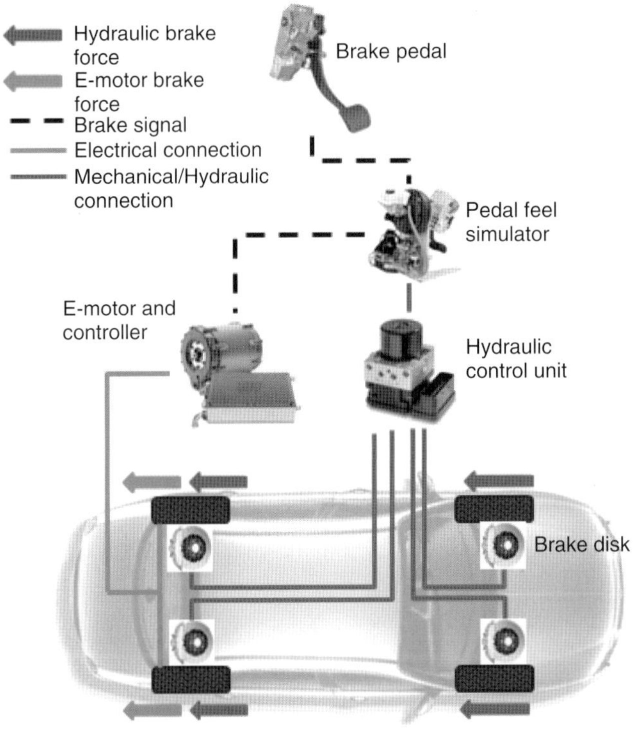

Figure 6. Brake-by-wire technology.

properties of an electric motor is, when it is running in one direction, it converts electrical energy into mechanical energy that can be used to perform work (such as turning the wheels of a car), but when the motor is running in reverse mode, a properly designed motor becomes an electric generator, converting mechanical energy into electrical energy. This electrical energy can then be fed into a charging system for the car's batteries, shown in Figure 7.

A cooperative regenerative braking system includes a frictional braking device for applying a friction brake torque to the wheels, and a regenerative braking device for applying a regenerative brake torque to drive wheels of the vehicle. Here, frictional braking device means a system which operates on the principle of friction mechanism to generate brake force regardless of using disk brake structure or drum brake structure, hydraulic or pneumatic actuating device (usually it is hydraulic braking system). The regenerative braking device means the system which operates to transform the vehicle kinetic energy to electricity energy, usually it is electric motor. A cooperative regenerative braking system is usually used in the following three kinds of new energy vehicles, such as pure electric, hybrid electric, and fuel cell vehicles.

3.2.1 Regenerative braking in battery electric vehicle

Battery electric vehicles use the motor as a generator when using regenerative braking: it is operated as a generator during braking and its output is supplied to electrical loads and it provides the braking effect. This energy can be saved in a storage battery or a supercapacitor and can be used to propel the motor (Figure 8).

3.2.2 Regenerative braking in hybrid electric vehicle

A hybrid vehicle is a vehicle that uses two or more distinct power sources to propel the vehicle. The term most commonly refers to hybrid electric vehicles (HEVs), which combine an internal combustion engine and one or more electric motors. Hybrid system is divided into series hybrid, parallel hybrid, and series–parallel hybrid.

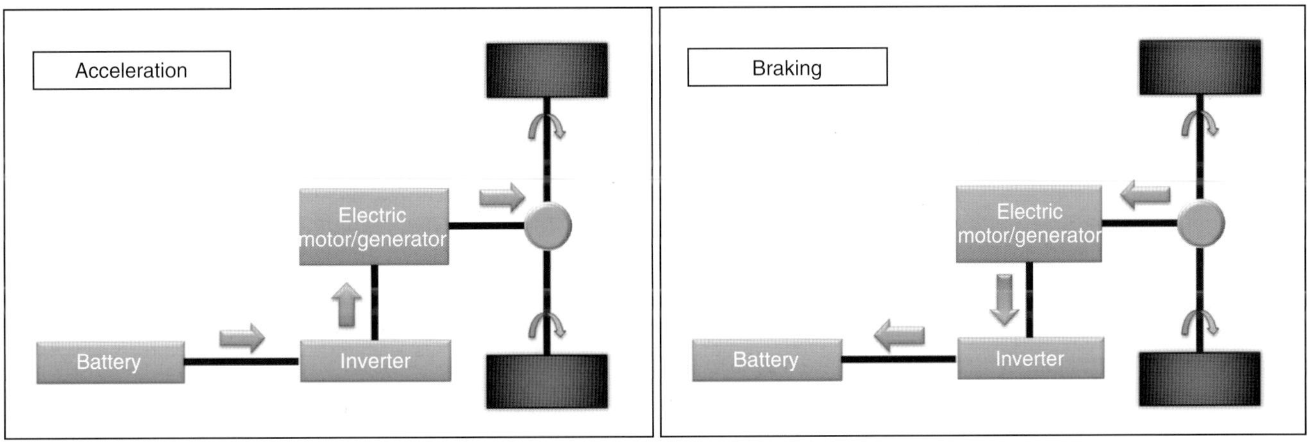

Figure 7. The working property of electric motor.

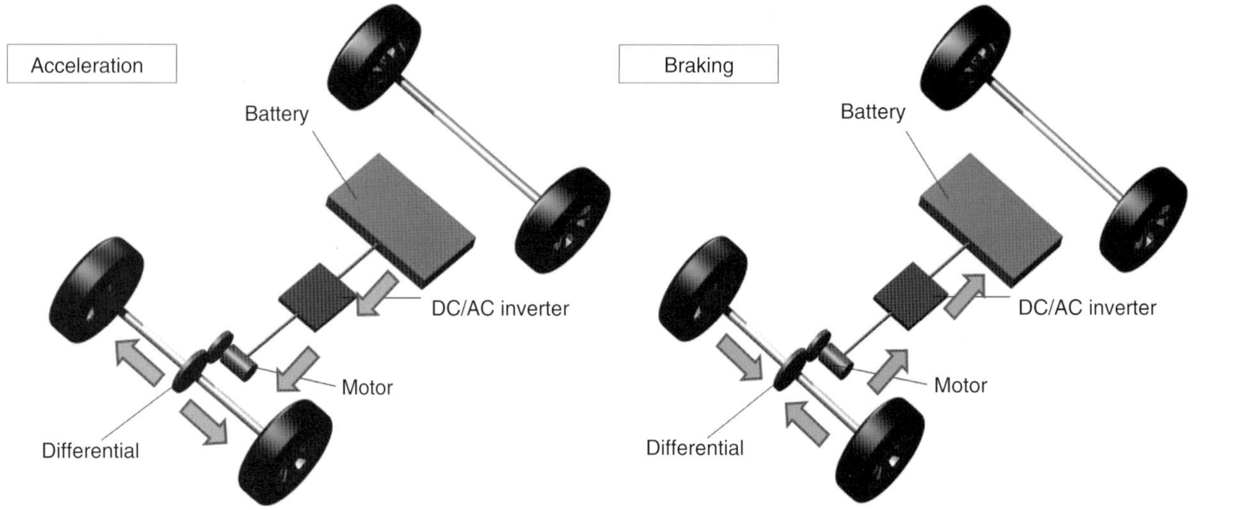

Figure 8. Cooperative regenerative braking system of the pure electric vehicle.

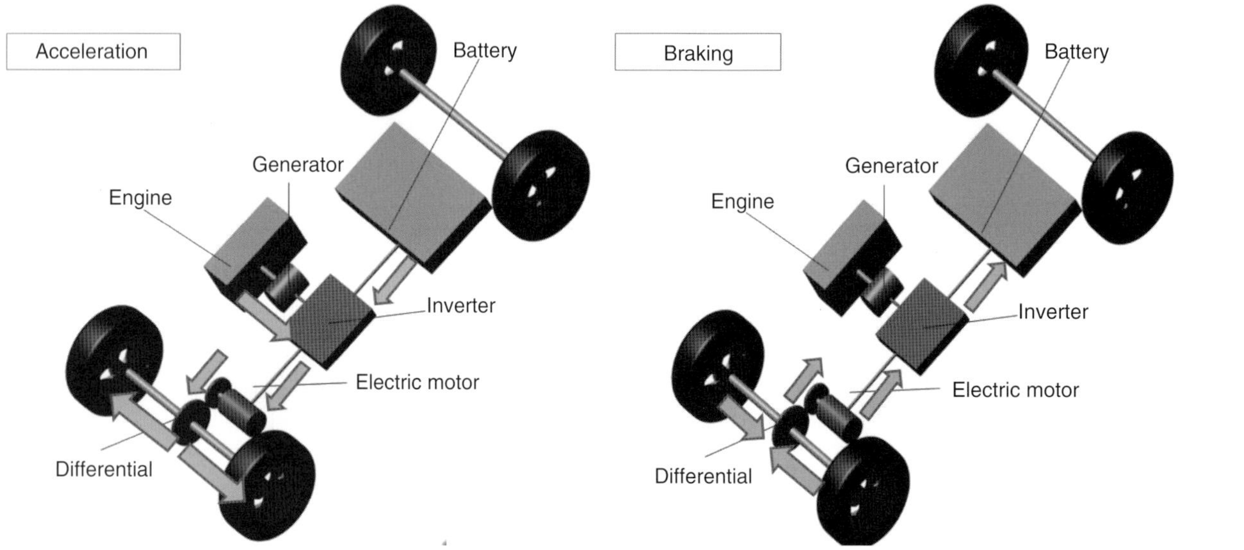

Figure 9. Cooperative regenerative braking system of the series-hybrid vehicles.

Series-hybrid vehicles are driven by the electric motor with no mechanical connection to the engine. Instead, there is an engine tuned for running a generator when the battery pack energy supply is not sufficient for demands (Figure 9).

In a parallel hybrid, the single electric motor and the internal combustion engine are installed so that they can power the vehicle either individually or simultaneously. In contrast to the power split configuration, typically, only one electric motor is installed. Most commonly, the internal combustion engine, the electric motor, and the gearbox are coupled by automatically controlled clutches. For electric driving, the clutch to the internal combustion engine is open, whereas the clutch to the gearbox is engaged. While in combustion mode, the engine and motor run at the same speed (Figure 10).

In a series-parallel hybrid electric drive train, there are two motors: an electric motor and an internal combustion engine. The power from these two motors can be shared to drive the wheels via a power splitter, which is a simple planetary gear. The ratio can be from 0% to 100% for the combustion engine, 0% to 100% for the electric motor, or an anything in between, such as 40% for the electric motor

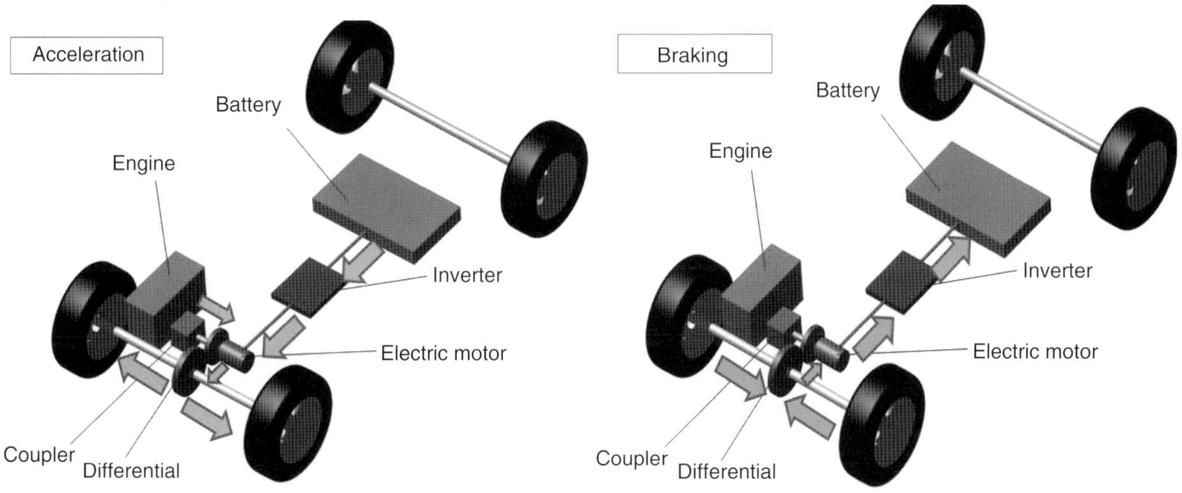

Figure 10. Cooperative regenerative braking system of the parallel-hybrid vehicles.

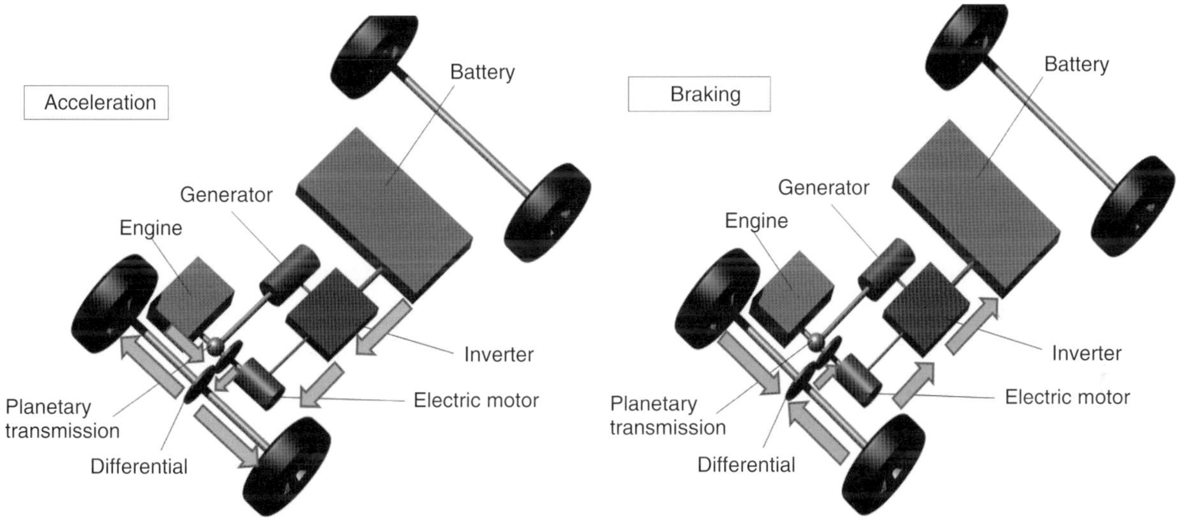

Figure 11. Cooperative regenerative braking system of the series–parallel hybrid vehicles.

and 60% for the combustion engine. The electric motor can act as a generator charging the batteries (Figure 11).

3.2.3 Regenerative braking in fuel cell vehicle

A fuel cell vehicle is a type of hydrogen vehicle that uses a fuel cell to produce electricity to power its electric motor. Fuel cells in vehicles create electricity to power an electric motor using hydrogen and oxygen from the air. The electricity produced by the fuel cell is delivered to the electric drive system in the vehicle, which converts electric power into the mechanical energy and drives the wheels of the car (Figure 12).

4 THREE PARTS OF THE COOPERATIVE REGENERATIVE BRAKING SYSTEM

From the point of view of control system theory, the cooperative regenerative braking system is made of three parts: signal part, control part, and actuator part (Zhang and Ning, 2009), shown in Figure 13.

1. *Signal Part*. The signals include the brake pedal position, vehicle speed, energy storage (e.g., battery) status, hydraulic pressure in the pipeline, pneumatic pressure in braking valves, etc., shown in Figure 14.

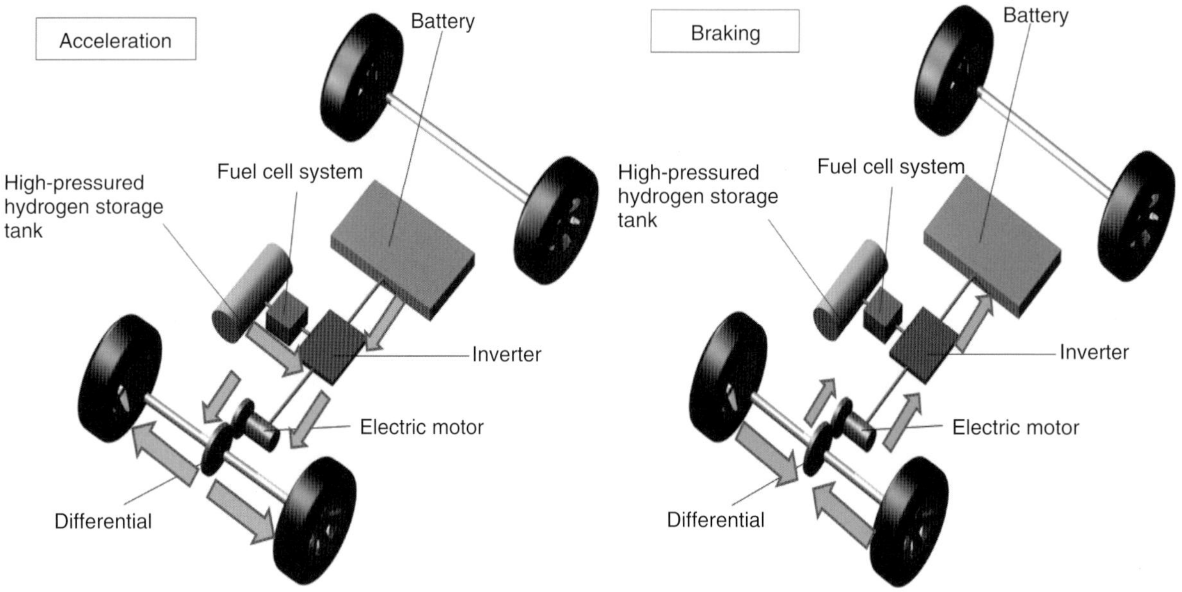

Figure 12. Cooperative regenerative braking system of the fuel cell vehicles.

Figure 13. Three parts based on control system.

The most important task in this part is to identify the driver's right intention in time.

2. *Control Part.* Electronic devices that control the regenerative braking function remotely by transmission parts and electronic control unit (ECU).
3. *Actuator Part.* Considering that there are two sub-braking systems, hydraulic or pneumatic braking system parts belongs to the friction braking system. Electric motor belongs to the regenerative braking system.

4.1 Signal part

The vehicle can be regarded as a closed control circuit consisting of driver, vehicle, and traffic condition. The driver gives the command based on the current traffic condition and personal intension. By steering, shifting gears or stepping down gas pedal or brake pedal, the driver can speed up or slow down the vehicle to adapt to the current outside environment and meet personal demands. Braking intention can be shown as how the driver controls the brake pedal.

4.1.1 The categories of braking intention

The braking intention can be divided into three categories: mild braking, moderate braking, and emergency braking (Zhang *et al.*, 2009), shown in Figure 15.

Mild braking means the driver brakes, slows down, or stops the vehicle actively. It is common to see this kind of braking intention on the cross road when the driver is turning around the corner. In this case, the pedal angular velocity is small and the hydraulic pressure in the pipeline increases slowly (Yu *et al.*, 2005).

Moderate braking means driver brakes passively. In this situation, the pedal angular velocity is big and the hydraulic pressure changes sharply. It is common to see this kind of braking intention on the highway when the front vehicle suddenly brakes or in case the vehicle needs to be stopped within a short brake distance. Sometimes, this kind of braking would be withdrawn very soon.

Emergency braking means the driver has stopped the car rapidly till the car completely stopped.

4.1.2 The ways to identify braking intention

In the cooperative regenerative braking system, by collecting signals such as brake pedal step force, pedal displacement and velocity, or hydraulic pressure change

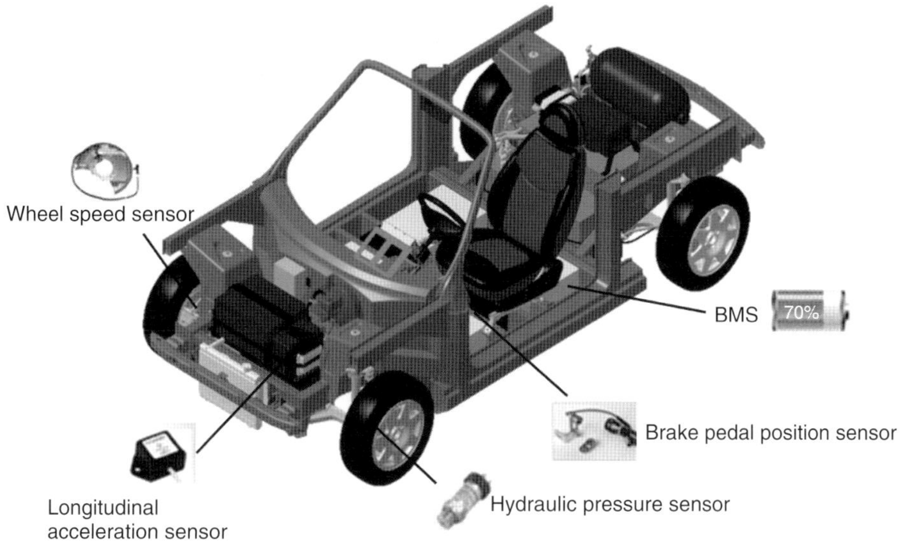

Figure 14. Signal parts.

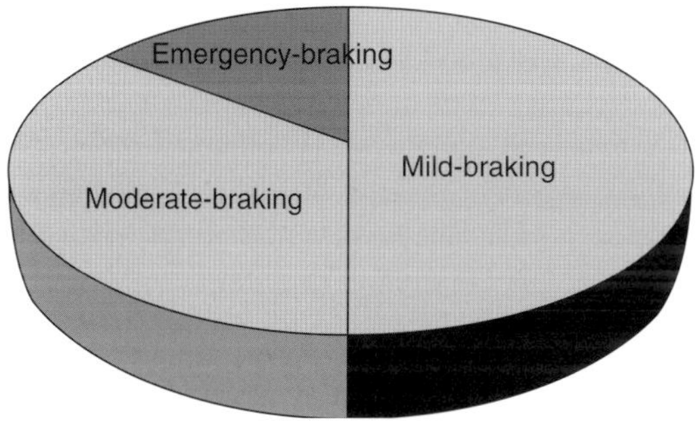

Figure 15. The categories of braking intention.

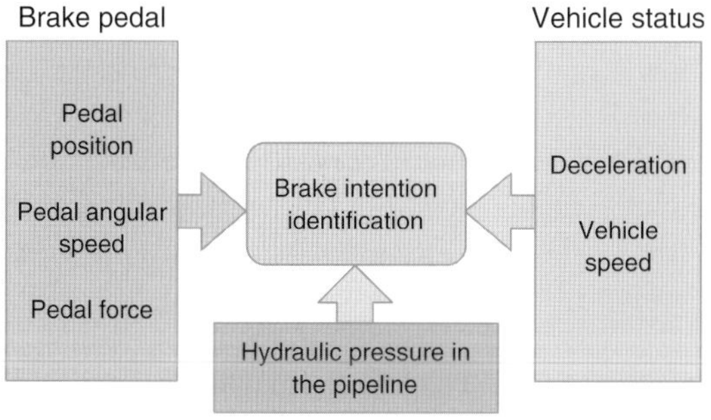

Figure 16. The signals that are used to identify the driver's braking intention.

rate in pipeline, shown in Figure 16, the driver's braking intention can be identified (Zhang *et al.*, 2009). At the same time, the rotation velocity of each wheel should also be monitored, to assure the vehicle is under safe condition.

4.2 Control part

4.2.1 Requirements for the braking system

Because there are two sub-braking systems in the cooperative regenerative braking system, its complexity is higher than the traditional braking system. The requirements for the cooperative regenerative braking system are as follows (Xiao, 2009):

1. sufficient brake force according to the driver's intention supplied;
2. the recovery efficiency of braking energy maximized;
3. the road adhesion coefficient best exploited to achieve the highest braking efficiency;
4. the driver's requirement about braking comfort met.

4.2.2 The role of the controller

The regenerative braking controller monitors the rotational speed of the wheels and the difference in that speed from one wheel to another. In vehicles that use these kinds of brakes, the brake controller not only monitors the speed of the wheels but also can calculate how much torque (rotational force) is available to generate electricity to be fed back into the batteries. During the braking operation, the brake controller directs the electricity produced by the motor into the batteries or capacitors. It not only makes sure that an optimal amount of power is received by the batteries but also ensures that the inflow of electricity is not more than the batteries can handle.

The most important function of the brake controller, however, may be deciding whether the motor is currently capable of handling the force necessary for stopping the car (Lampton, n.d). If it is not, the brake controller turns the job over to the friction brakes, averting possible catastrophe. In vehicles that use these types of brakes, as much as any other piece of electronics on board a hybrid or electric car, the brake controller makes the entire regenerative braking process possible.

4.2.3 *Influence of the position of the electric motor*

As mentioned in Section 3, there are three types of cooperative regenerative braking system according to where the regenerative brake force acts. So obviously the position of electric motor has effect on the braking efficiency and vehicle stability. When the regenerative brake force acts on the front wheels, it can achieve better vehicle stability and braking efficiency because it can make better use of the road adhesion coefficient (Yu *et al.*, 2008). When the regenerative brake force acts on the rear wheels, vehicle's stability is worse and it is not good for energy recovery. Therefore, Section 4.2.4 control strategy is based on the principle that electric motors supply regenerative brake force on the front wheels at first. When brake force is not enough, regenerative brake force would act on the rear wheels. In the case of emergency, electric motor and hydraulic braking system would work together to supply brake force.

4.2.4 *Control strategy*

For the vehicles, on which electric motors are just added to the original friction braking system, the control strategy is to control the electric motor to generate the designed brake torque that equals to the insufficient value between the realistic braking force distribution curve and ideal curve.

For the vehicles, on which the cooperative regenerative braking system is realized by "brake-by-wire" technology, the control strategy is more complicated. With the clutch connecting, the backward drag force from combustion motor should be considered. The following control strategy is under the circumstance without considering the backward drag force from combustion motor. In the case of mild or moderate braking, recovering kinetic energy is regarded as the priority goal. Therefore, in the case of mild braking, all the brake force would be supplied by electric motor. In the case of moderate braking, electric motor would supply brake force at first. The hydraulic braking system serves as a "backup" in case that the brake force is not enough. In the case of emergency braking, braking efficiency would be regarded as priority goal. In such circumstance, electric motor and hydraulic braking system would work together to supply the maximum brake torque within the shortest time (Zhang and Ning, 2009). The interconnectedness and mutual influence of electric motor braking and hydraulic braking should be considered. In this case, it needs help from ABS to keep the wheels unlocked. Considering system control simplification and motor protection, the motor's regenerative braking function of composite brake systems is usually turned off when ABS works (Figures 17 and 18).

4.3 Actuator part

Usually, for the friction braking system, hydraulic braking system is the actuator part whereas for the regenerative braking system, it is electric motor.

4.3.1 *Electric motor*

According to the PWM control theory, by switching the arm of the inverter conduction power tube, the armature current can be changed in opposite direction under the same magnet poles to drag the electric motor reversely. In this way, the power battery can be charged while part of the electricity is consumed by motor internal resistance in forms of heat. The energy stored in the battery can be used in idling, start, acceleration condition to achieve the purpose of energy saving and pollution reducing.

4.3.1.1 The limitations of electric motor regenerative braking.

1. Regenerative braking relies on the electric motors. Therefore, this kind of braking system cannot apply on the wheels that are not driven by electric motors (Zhang and Ning, 2009);
2. Because brake torque supplied by electric motors cannot exceed the maximum electric motor torque at current rotational speed, if pure regenerative brake torque cannot meet the current braking requirement, friction braking system should supply the rest brake torque (Zhang and Ning, 2009);
3. The energy regenerative efficiency relies on the battery SOC. The upper and lower limits of SOC depend on the internal resistance of battery. When SOC is too high, the power battery should not be charged in order to keep its normal operational life cycle. The recovery power of braking energy should not exceed the charge power of the battery (Zhang and Ning, 2009).
4. In the case of regenerative braking, when the rotational speed of electric motor is lower than the rated speed, the electric motor works with the rated brake

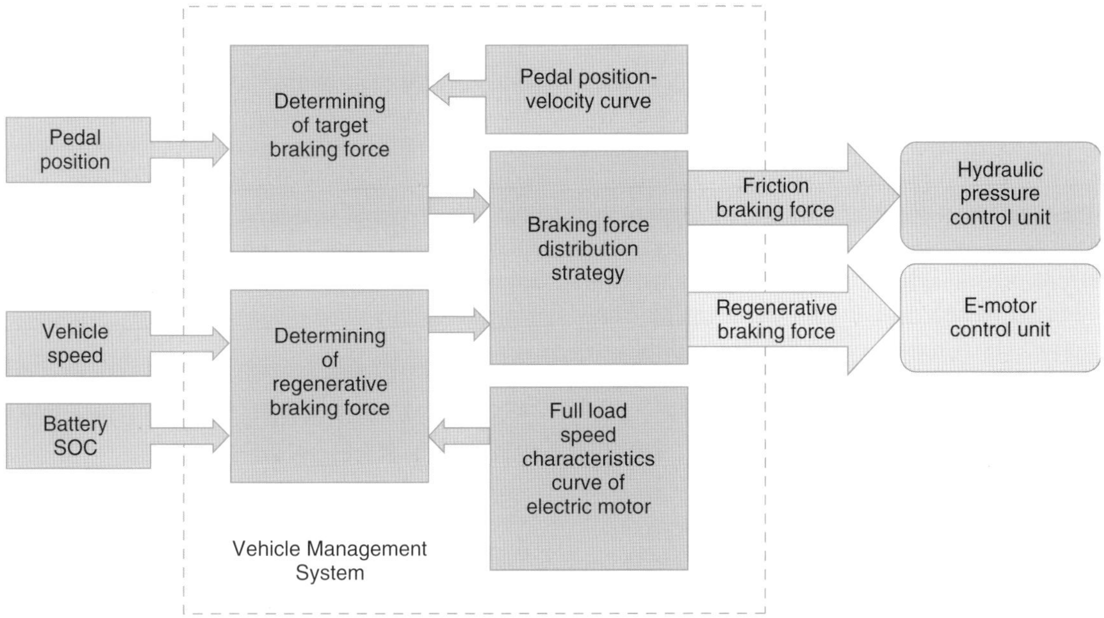

Figure 17. Control strategy.

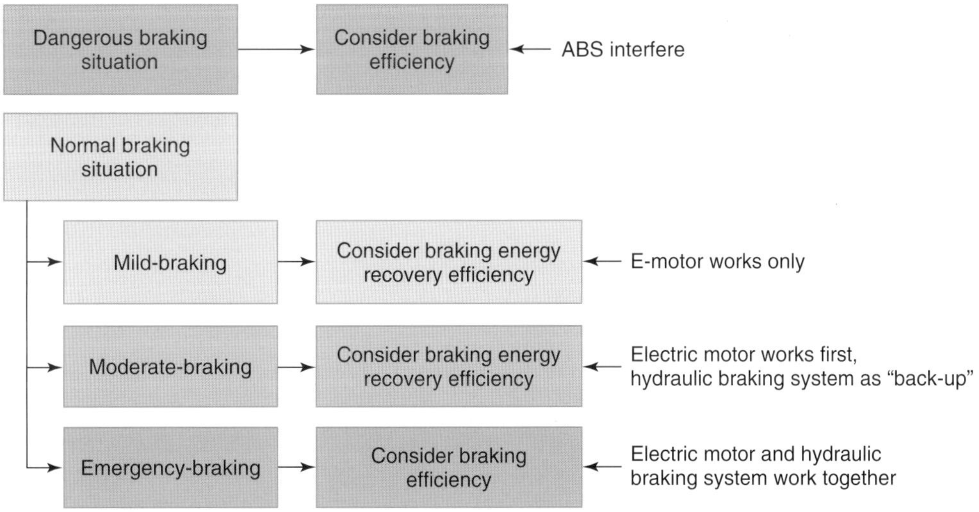

Figure 18. The control strategy under different braking intentions. (Modified from Zhang and Ning, 2009.)

torque. When the rotational speed of the electric motor is higher than the rated speed, the electric motor works under the rated power. When the rotational speed of the electric motor is too low, the regenerative braking fails to work because of the too low electromotive force. In this case, the regenerative braking force decreases to zero immediately (Zhang and Ning, 2009).

5. By emergency braking, the required braking torque may be many times higher than which the traction motor can supply. In this case, the conventional braking system is indispensable for a compensation of the braking torque to achieve the desired deceleration.

4.3.1.2 Different kinds of the electric motors. Most commonly used electric machines in electric vehicles are permanent magnet synchronous motors (PMSM), AC induction motors (IM), and switched reluctance motor (SRM). IM is simple-structured, rugged, low cost, and relatively mature in speed vector control technology. However, compared to the permanent magnet motor, it has a lower efficiency and power density.

PMSM uses permanent magnet to replace its excitation system, and it has a higher power density, a higher efficiency, and a wider speed range. Vector control is often used when the drive system works in a low speed range, whereas the flux-weakening control is used in a high speed range.

SRM system features a compact and solid motor, suitable for high speed operation, with a simple and low cost drive circuit, performance reliable in a wide speed range with relatively high efficiency, and it can easily realize four-quadrant control. The disadvantage is the torque ripple and noise. In addition, by contrast to PMSM, the power density and efficiency is lower.

4.3.2 Hydraulic braking system

The traditional hydraulic braking system can be divided into two categories, with variable β and with fixed β (Zhang and Ning, 2009). Here, β is the braking force distribution coefficient. With variable β, the realistic brake force distribution curve is much closer to the ideal distribution curve. The technique of variable β is called *electric braking distribution* (*EBD*, shown in Figure 19). At present, with the application of ABS, wheel lock cases can be avoided. However, the biggest disadvantage is that ABS cannot realize regenerative braking. All braking energy is wasted.

In the cooperative regenerative braking system, if the hydraulic braking force can be controlled, the realistic brake force distribution curve can be much nearer to the ideal curve by adjusting the hydraulic brake force. Yet, currently in many cases, the cooperative regenerative braking system is realized by adding electric motors directly onto the original hydraulic braking system, the frictional brake force cannot be adjusted. In this case, the electric motor regenerative brake force can be adjusted in order to get near to the ideal distribution curve.

5 RELATED RESEARCH FIELDS IN THIS REGARDS

The cooperative regenerative braking system brings us more energy efficient vehicles, but at the same time, it brings more challenges for automotive engineers to ensure the vehicle safety, control stability, and riding comfort.

5.1 Brake feel

Brake feel (Zhang and Ning, 2009) can help the driver make the judgment, which includes the brake pedal displacement, the resistance force of brake pedal, and the vehicle braking performance which the driver can feel (Figure 20).

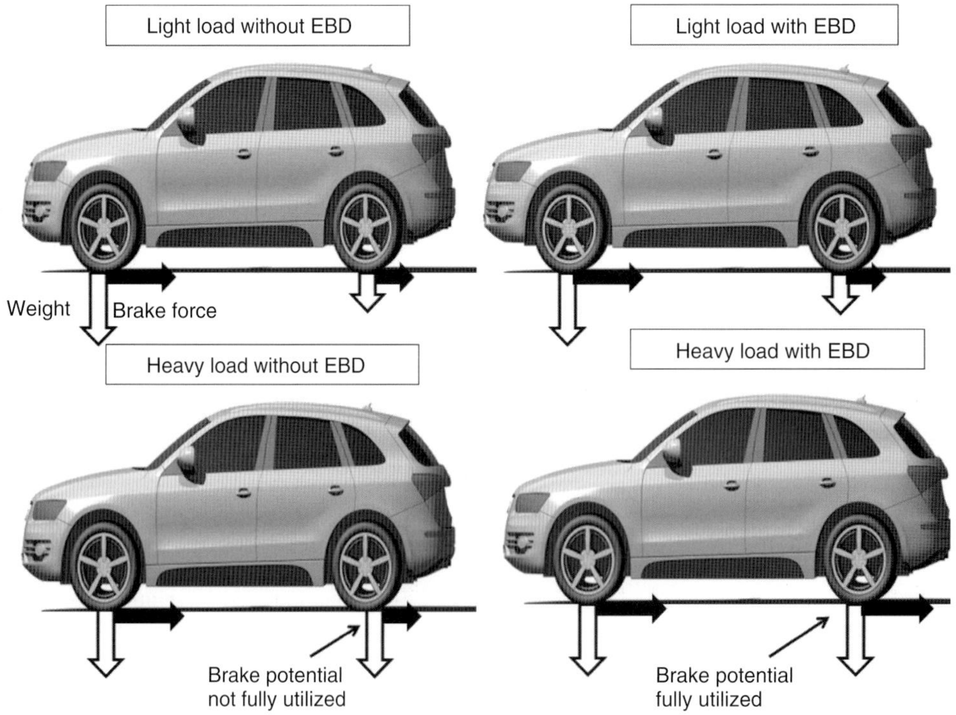

Figure 19. The influence of using EBD.

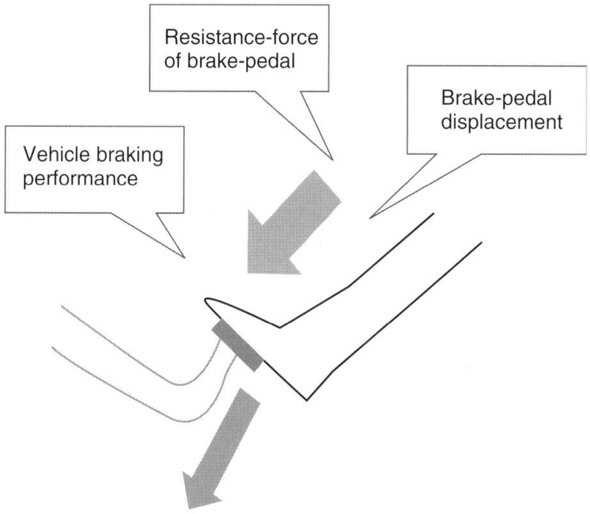

Figure 20. Brake feel.

Currently, some cooperative regenerative braking systems are realized by adding the electric motor systems directly onto the traditional friction braking system. Compared with hydraulic braking system, the electric motors respond faster. However, it brings a new problem, different brake feel. Therefore, if it needs to make sure of the same brake feel, automotive engineers have to postpone the active time point of the electric motor. Another problem is how much brake force should electric motor supply. The uncertainty increased with the change of the brake deceleration. To solve those problems, it is common to install an electric motor under the brake pedal to simulate the feedback of brake pedal force and displacement for the driver.

In addition, in the case of regenerative braking, brake fluid cannot flow to the wheel brake cylinder because of valve resistance. It results in a smaller brake pedal displacement. As the regenerative braking changed to friction braking because electric motor cannot supply torque when the rotational speed is too low, the valve switched from closed to open, it brings a sudden change in brake pedal displacement. At this point of time, definitely, it brings a sudden and significant bake feel change.

5.2 The influence of the cooperative regenerative braking system on ABS

To maximize the braking efficiency, hopefully the wheels would be in a lock/unlock state. ABS is a system that adjusts the wheel slip rate to make sure that the wheels in the right state. Because electric motor has advantages in fast response and easy control, usually the electric motors would act at first to supply the regenerative brake force (Zhang and Ning, 2009). If it is not enough, the traditional friction brake system would work as a "backup."

The motor's regenerative braking function is usually turned off when ABS works in order to simplify motor control and protect motor. For four-wheel-driven vehicles, in-wheel electric motors can work together with hydraulic system to realize the function of ABS to adjust each wheel's slip rate by controlling motor current and the wheel's brake torque in order to keep the wheel's ideal status.

For central electric-motor-driven vehicles, electric motor can only take the job of EBD. The function of ABS can only be realized by the hydraulic brake system. In the vehicles on which the electric motors are just directly added onto the traditional friction braking system, controlling the pressure of the brake fluid can adjust the wheel's slip rate.

5.3 The energy recovery sufficiency

Electric motor plays the role of energy transform in the cooperative regenerative braking system. There should be some parts to store the retrieved energy such as battery. Basically, there are three energy storage devices: battery, supercapacitor, and flywheel. At present, battery and supercapacitor would be the prior choices. Either one or some different kinds of devices can be used together in vehicles (Figure 21).

Energy recovery efficiency depends on the battery property, electric motor working property, the charging speed, etc. In the cooperative regenerative braking system, over charging and fast charging happen quite often. It makes the working environment of electric motor and battery very bad. In this case, the key point to get higher energy recovery efficiency lies in the electric motor control strategy as well as the battery energy storing technology.

6 CASES ABOUT THE COOPERATIVE REGENERATIVE BRAKING SYSTEM

Although the regenerative braking technology was first used in trolley cars, it has subsequently found its way into such unlikely places as electric bicycles and even Formula One racing cars. At present, these kinds of brakes are primarily found in hybrid vehicles, such as the Toyota Prius, and in fully electric cars, such as the Tesla Roadster. Followings are some cases showing how the cooperative regenerative braking system works in the hand of different automakers.

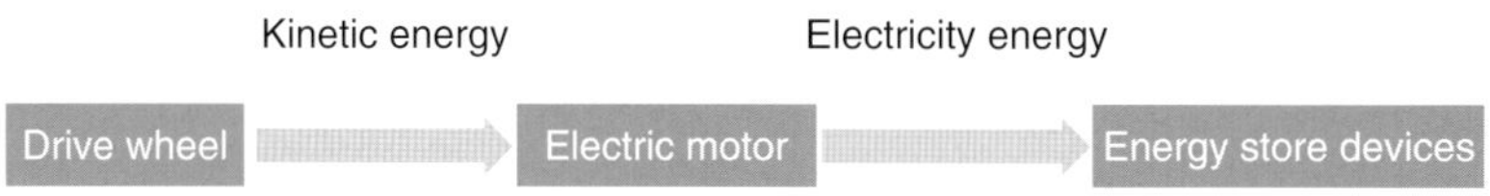

Figure 21. The theory of regenerative braking system.

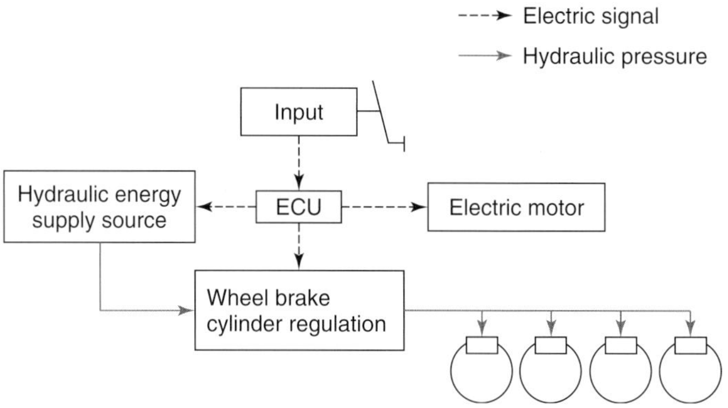

Figure 22. The regenerative braking system of Prius.

6.1 Toyota Prius

It is well known that Toyota Prius is the most popular hybrid electric vehicle in the world with the highest degree of industrialization and largest market share. The cooperative regenerative braking system in Prius is working based on the linear electromagnetic valve developed by Toyota, shown in Figure 22 (Lampton, n.d).

The design is based on the concept "brake by wire" (Zhang and Ning, 2009). Because the input from brake pedal controlled directly by the driver is separated from the brake fluid pressure, the pressure in the wheel brake cylinder can be controlled completely without the influence from the movement of brake pedal. When the driver steps down the brake pedal, the pedal displacement simulator transmits the information to the ECU instead that the pedal pushed the piston in brake main cylinder directly. The ECU decides how much brake force should the hydraulic braking system supply. With this feedback, the piston in the brake main cylinder would be pushed forward to generate the designed pressure.

6.2 Continental

Engineers in Continental developed a regenerative braking system for electric and hybrid vehicles (http://www.conti-online.com/generator/www/de/de/continental/engineering_services/themes/brakes_chassis/download/one_pager_regenerative_braking_pdf_uv.pdf). The design is also based on the concept "brake by wire." The components in this brake system include ESC with eGap (generator), electric vacuum pump, simulated brake actuation, and electromechanical brakes (Figure 23).

Cooperation between brake and power train system include

1. monitoring of driver brake request;
2. decouple driver demand from friction braking;
3. recuperation interface between brake and power train.

Different solutions for regenerative braking is as follows:

1. brake actuation with pedal force simulator (full by-wire braking);
2. Use of electrical brake caliper (by-wire braking at one axle);
3. eGap (the generator) realized by electronic brake system (conventional hydraulic brake system).

7 SUMMARY

Although there is still a long way to go for the cooperative regenerative braking system to achieve the same degree of the success and popularity as ABS, one thing seems certain: the application of regenerative braking is an irreversible trend. As hybrids with electric motors and regenerative brakes can travel considerably farther on a gallon of gas, some has achieved more than 50 miles per gallon at

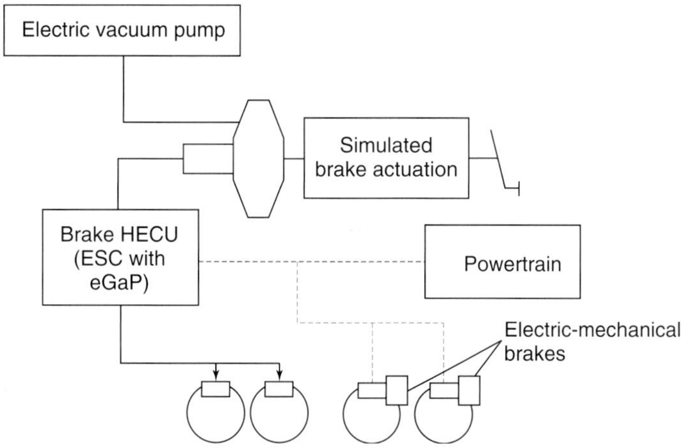

Figure 23. The components in the Continental regenerative braking system.

this point. That is something that most drivers can really appreciate.

RELATED ARTICLES

Volume 2, Chapter 51
Volume 2, Chapter 63
Volume 3, Chapter 74
Volume 3, Chapter 75
Volume 3, Chapter 76
Volume 4, Chapter 119

REFERENCES

Gubler, D. (2009) IORE beim Tornetiäs, http://en.wikipedia.org/wiki/File:IORE_beim_Tornetr%C3%A4sk.jpg (accessed 9 October 2013).

Harris & Ewing, Inc. (1906) Senator Wetmore in Automobile, http://it.wikipedia.org/wiki/File:SenatorWetmoreInAutomobile_retouched.jpg (accessed 9 October 2013).

Lampton, C. (n.d) How Regenerative Braking Works in *HowStuffWorks*, http://auto.howstuffworks.com/auto-parts/brakes/brake-types/regenerative-braking.html (accessed 9 October 2013).

Xiao, K. (2009) Algorithms research on braking force distribution of cooperative braking system *Beijing Vehicle*, **32** (2), 42–46.

Yu, Z., Xiong, L., and Zhang, L. (2005) Matching research on electrohydraulic cooperative brake *Automotive Engineering*, **27** (4), 455–462.

Yu, Z., Zhang, Y., Xu, L., *et al.* (2008) Brake force coordination distribution methods simulation in cooperative braking system *Train Technology*, **5**, 1–5.

Zhang, Z. and Ning, G. (2009) Vehicle mechanical-hydraulic cooperative braking system *Shanghai Vehicle*, **11**, 42–46.

Zhang, Y., Yu, Z., Xu, L., *et al.* (2009) Cooperative braking system brake force distribution strategy research based on the brake intentions *Automotive Engineering*, **31** (3), 244–249.

Zheng, H. (2010) An research on anti-lock braking control method of 4WD electrical vehicle with in-wheel motors based on variable structure control. Tongji University Graduation Thesis for Master Degree, 3.

Axles

Chapter 121
Designing Twist-Beam Axles

Veit Held and Rüdiger Hiemenz
Adam Opel AG, Rüsselsheim, Germany

1 INTRODUCTION

When the twist-beam axle was introduced in 1974 in the Audi 50, in the VW Golf, and later in the Scirocco, it became a rapid success. Historically, this concept can be seen as a further development of the trailing arm axles that were widely used in front-wheel-driven vehicles. The immediate forerunner was the non-driven rear suspension that was launched in the DKW Junior, 1959–1962 (Figure 1). While this axle connected the trailing arms with a slotted axle tube, which did already work as a stabilizer, the longitudinal control arms were still flexible and a Panhard rod was needed to take the lateral forces.

In the Audi 50, we find all the typical elements of a twist-beam axle that work in the way, as we know it today (Figure 2).

Encyclopedia of Automotive Engineering, print © 2015 John Wiley & Sons, Ltd.
Edited by David Crolla, David E. Foster, Toshio Kobayashi, and Nick Vaughan.
This article is © 2015 John Wiley & Sons, Ltd. ISBN: 978-0-470-97402-5
Also published in the *Encyclopedia of Automotive Engineering* (online edition)
DOI: 10.1002/9781118354179.auto003

Today, twist-beam axles or, as they are also referred to, compound crank rear suspensions, have by far the largest volume on the global passenger car market of all rear suspension types (Table 1). It can be expected that, with the growth of the small and compact car segment, the demand for the twist-beam axle type will increase.

Twist-beam rear suspensions dominate the mini to lower medium passenger car segments (Table 2). Typical models in the mini category are the Toyota IQ and Peugeot 107. In the small car segment, we find the Opel Corsa, the Ford Fiesta, and the VW Polo, and in the lower medium category, we find vehicles such as the Honda Civic (EU) and the Renault Megane.

The reasons for this widespread usage have been summarized in numerous textbooks, for example, in Heißing, Metin Ersoy, and Gies (2007). The main advantages and disadvantages of this simple, but very effective axle are listed in Table 3.

2 THE FUNCTIONAL PRINCIPLE OF A TWIST-BEAM AXLE

Apart from the usual design elements that apply to any axle (such as the positions of dampers and springs), the twist-beam axle has the following characteristic components that define the specific properties of this concept (Figure 3):

1. the torsion beam
2. the front bushing or, as it will be called in the remainder of this chapter, the "A-bushing"
3. the left and right trailing arms

This highly integrated design takes all forces and moment which are applied to the tire and wheel during driving. Its mechanical parameters allow tuning both ride and handling

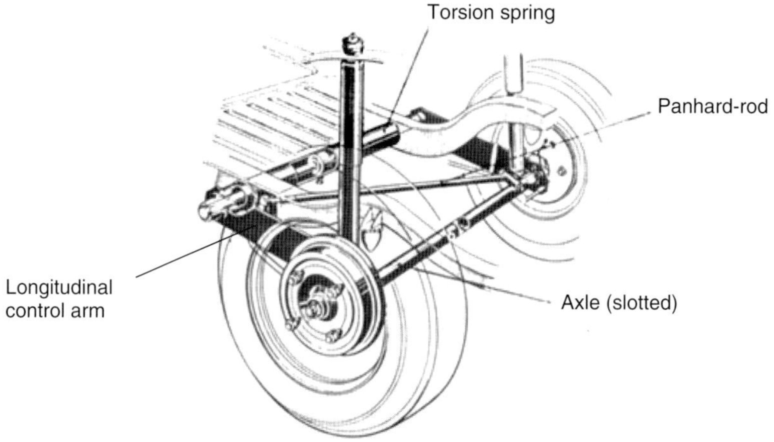

Figure 1. The forerunner of the twist-beam axle was a non-driven rear suspension with elastic longitudinal control arms. (DKW Junior, 1959–1962. Reproduced from Audi company archives, with permission from Audi AG.)

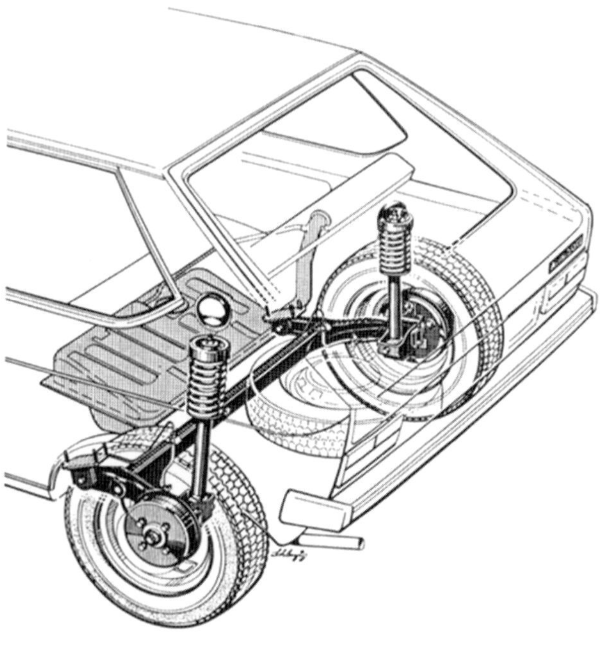

Figure 2. Non-driven rear axle with stiff longitudinal control arms and torsional elastic beam, Audi 50, 1974–1978. (Reproduced from Audi company archives, with permission from Audi AG.)

Table 1. Rear suspensions, worldwide light passenger vehicles around 2007 (%).

	RWD	FWD	4WD	Total
Twist-beam	0.0	31.9	—	31.9
Semi-trailing	0.5	12.1	—	12.6
Solid beam	10.9	10.8	2.1	23.8
Double wishbone	2.2	3.7	0.2	6.1
Multilink	4.0	18.1	2.2	24.3
Strut and others	0.3	0.9	—	1.2
Total	17.9	77.5	4.5	99.9

The automotive suspension systems report (2013), Data taken from SupplierBusiness Ltd., 2007.

Table 2. Twist-beam penetration by segment, worldwide light passenger vehicles (%).

	Twist-Beam
Mini	2.8
Small	11.9
Lower medium	9.3
Medium	4.5
Upper medium	0.9
Luxury and sport	0.1
Off-road	0.1
MPV	2.2
Transport	0.2
Pickup	<0.1

The automotive suspension systems report (2013), Data taken from SupplierBusiness Ltd., 2007.

behavior as well as the noise isolation of this suspension. The small number of components requires each of them to serve several functions, unlike the links in a multilink suspension that have dedicated functions and dominating load directions.

The torsion beam (1) is the key element that differentiates this axle from all other concepts. All side forces that act on the wheels try to turn the whole body of the axle around a vertical axis. This creates internal stresses in the structure, particularly in the crossbeam, which has to provide a stiff coupling of the wheels. The higher the bending stiffness of the profile, the better is the control of the pivoting moment by tuning the elasticity of the A-bushings. With respect to the roll response during cornering, the torsion profile has exactly the same role as the stabilizer bar in independent suspensions.

Table 3. Advantages and disadvantages of twist-beam axles.

Advantages of the Twist-Beam	Disadvantages
• Very simple design—essentially, it takes one body plus two rubber bushings • Excellent packaging, that is, needs little space and has a very flat shape (exceptions: housing the dampers and accommodating the vertical movement of the crossbeam during suspension movements) • Easy to assemble • Stabilizer function included • Small unsprung mass • Enables damper ratios around one (damper can be placed vertically at the wheel center) • Roll steering independent of the load • Good compensation of pitching during braking • Small changes of track width under load	• Load peaks at the joints between the rigid trailing arms and the crosslink • Lateral force steer needs toe-correcting bushings • Limited side force stiffness due to deflecting moments in torsion beam and control arms. Compensating measures (e.g., toe-correcting bushings) affect road harshness • Difficult to enable all wheel drives • High axle loads difficult to achieve because of undue stress on welds • Conflicting targets of ride comfort, noise isolation, and handling are difficult to meet because all forces have to be taken by the front bushings

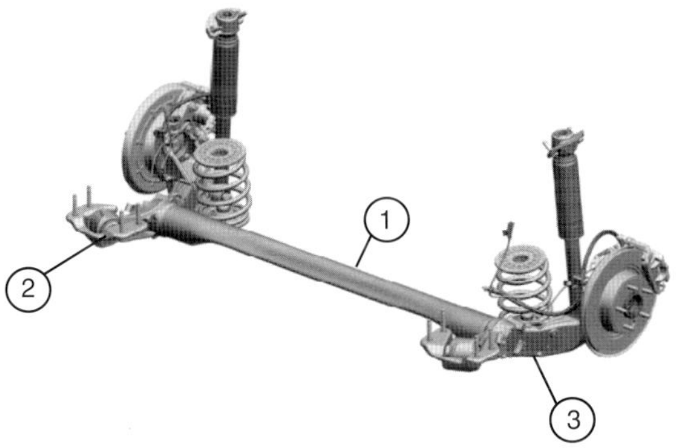

Figure 3. Key elements of the twist-beam axle. (Reproduced by permission of Adam Opel AG, Germany.)

The A-bushing (2) takes all lateral and longitudinal forces. The orientation and the spring rates of the bushing have a huge impact on the steering behavior during lateral loading.

The trailing arms (3) carry the brake force, the side loads during cornering, and the resulting moments, which are mainly the bending moments around the lateral and the vertical axis plus the twisting moment around the roll axis. Therefore, the stiffness of the trailing arms in all directions and around all axes controls the deflection and orientation of the wheel and tire under loads. Stiff welded sheet metal structures or cast profiles are established design solutions.

These three components will be discussed in more detail in the following sections.

2.1 The torsion beam

Understanding the twist-beam axle means understanding the torsion beam. In the case of most other concepts, the movement of the wheel is defined by the geometry of the bushings and the links. In the case of the twist-beam axle, the roll center, the camber angle, and the toe-in are largely defined by the twisting behavior of the beam.

When driving over a wide obstacle such that the left and right wheels move in parallel, the twist-beam axle works exactly like a rigid axle. The situation is more complex under lateral loads or when the car is rolling. If one wheel is raised while the other one is lowered, then the beam gets twisted around its lateral axis, which induces shear forces in the profile. As a consequence of these internal forces, the beam will twist around the shear center of the profile (Figure 4).

If the left and right wheels are moved vertically in opposite directions, then the shear centerline will pivot around its center point in the middle of the car (Figure 5). This means that the intersection of the shear centerline with the middle of the car defines one pivoting point of the left and right half-axles; the second one is the respective A-bushing. In Figure 5, the dashed line marks the axis around which the left wheel turns upward. Apparently, a twist-beam axle behaves like a semi-trailing arm axle with the same A-bushing as the twist-beam and a central bushing at the intersection of the shear centerline with the middle of the vehicle. Figure 5 depicts this relationship in the top view.

The designer can position this "virtual center bushing" by moving the beam backwards or forwards. This "virtual bushing" can also be moved up or down by choosing the profile and the orientation of the beam such that the shear center is located at the desired height. Moving the shear center above the A-bushing will result in a roll understeer behavior, moving it below will result in roll oversteer behavior (Figure 6).

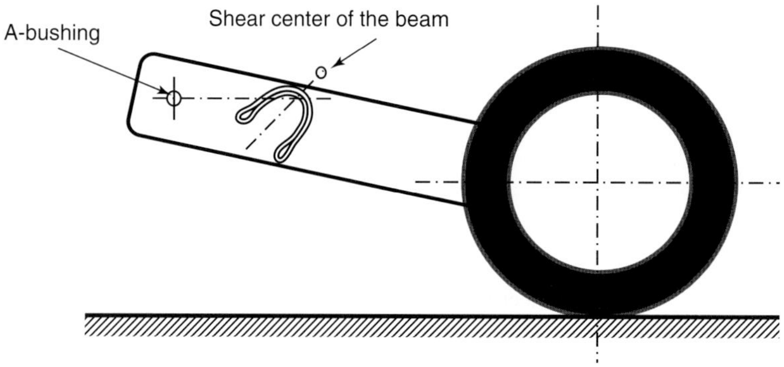

Figure 4. The torsion of the beam is defined by the shear center of the profile.

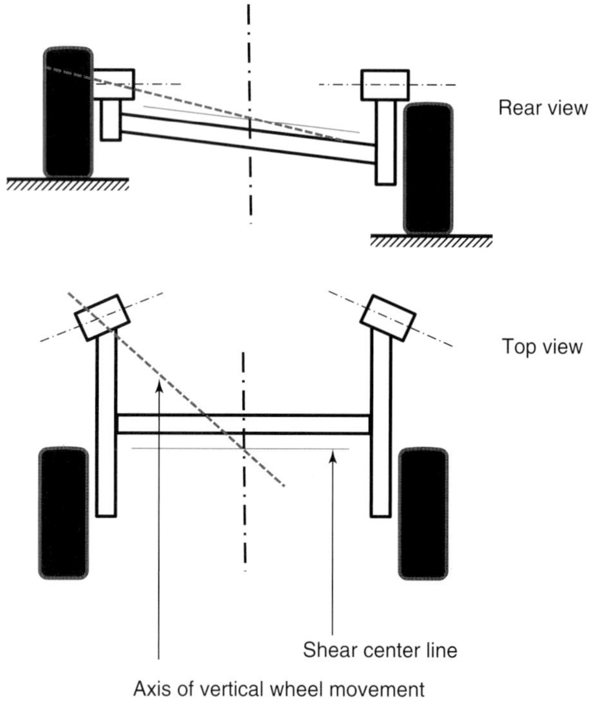

Figure 5. Wheel movements when the vehicle is rolling.

To define both the torsional rate and the position of the shear center, the designer can select the material and the shape of the profile (Table 4).

2.2 The A-Bushing

The A-bushing of the twist-beam has to resist longitudinal and lateral forces and the resulting moments. There are multiple conflicting design goals that make the design of the A-bushing critical. On one hand, the bushing should be soft to absorb the noise and vibrations from the road and to reduce the impacts from obstacles. On the other hand, the bushing needs to be stiff to create a precise handling behavior and determine the desired over- or understeer behavior due to side forces or rolling movements. To meet these requirements, the designer needs to define the vertical and horizontal spring rates of the bushing and the orientation.

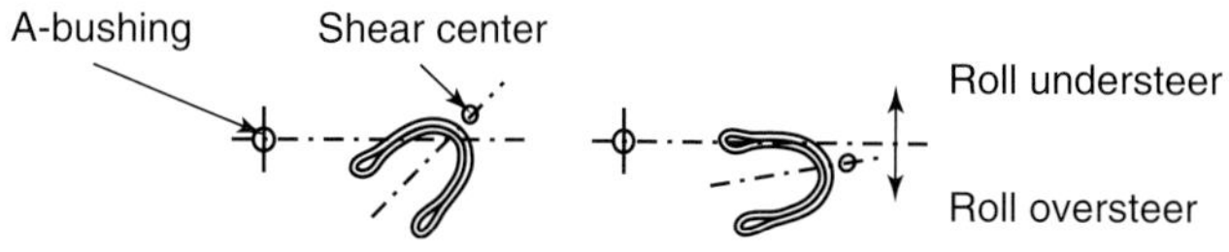

Figure 6. Positioning the shear center will define roll under- and oversteer.

2.3 The trailing arms

Because all side forces have to be transferred to the A-bushing, these arms have to be very rigid. Most commonly, sheet metal arms are chosen to save cost and mass. Another option is to use cast aluminum, which allows for topology optimization. One of the biggest challenges is to design the link between the trailing arm and the torsion beam, in particular, if different materials have to be combined.

3 THE PROCESS OF DESIGNING A TWIST-BEAM AXLE

In the previous chapter, the main design elements of the twist-beam axle were introduced. In the following chapter, the main steps will be summarized to develop a specific twist-beam axle for a particular carline.

The first step in the design of any rear axle is to do the "concept selection", which means to find the right axle type for a given vehicle. In this phase, all alternative concepts

Table 4. Commonly used profiles to define torsional stiffness and shear centerline.

Shape	Design	Property
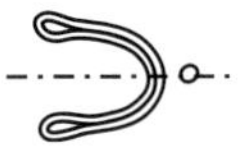	Closed section in U- or V-design, made from tube	High torsional/roll stiffness
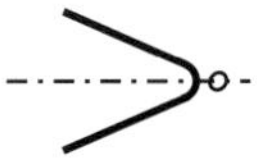	Open section in U- or V-design, made of sheet metal	Low torsional/roll stiffness
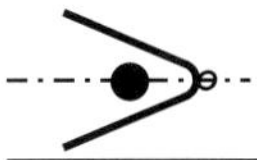	Open section in U- or V-design, made of sheet metal	Additional stabilizer bar for increased torsional/roll stiffness

will be evaluated. Apart from the traditional chassis requirements, there are many other boundary conditions to be met. Some of the primary drivers are the packaging constraints, which are imposed by the required trunk space and fuel tank volume, and the ground clearance. All these requirements have to be evaluated against the allowable system cost.

3.1 Essential design parameters and system characteristics

The twist-beam axle has a specific set of design parameters that can be selected or tuned to achieve the desired handling behavior, ride comfort, and noise isolation. These parameters are marked in Figure 7 and the impact on the targeted properties is listed in Table 5.

Table 5. Design parameters of twist-beam axles.

No.	Parameter	Impact
(1)	Position of the crossbeam Torsional stiffness	Roll center height, Stabilizer rate
(2)	Position of the shear center	Roll center height, Roll steer
(3)	Angle of the A-bushing	Lateral force steer
(4)	Compliances in x- and z-direction, Compliance in y-direction	Impact harshness, Lateral force steer, Road noise, Lateral force compliance
(5)	Side view swing arm angle	Wheelbase change and hence impact in harshness Anti-dive angle

3.2 Understeer contribution

The performance of a twist-beam axle depends on how well the conflicting requirements for the overall steering behavior of the axle and the ride and noise refinement can be balanced within the given package constraints. A typical design iteration process starts with the understeer contribution that the rear axle has to deliver in the context of the desired overall vehicle handling trim.

The rear axle understeer contribution is the sum of roll steer, lateral force compliance steer, and vehicle roll rate fore/aft distribution. Each element relates to a design factor with its own package constraints that limits the amount of total steer that can be derived from the single element. Consequently, the design process of a twist-beam

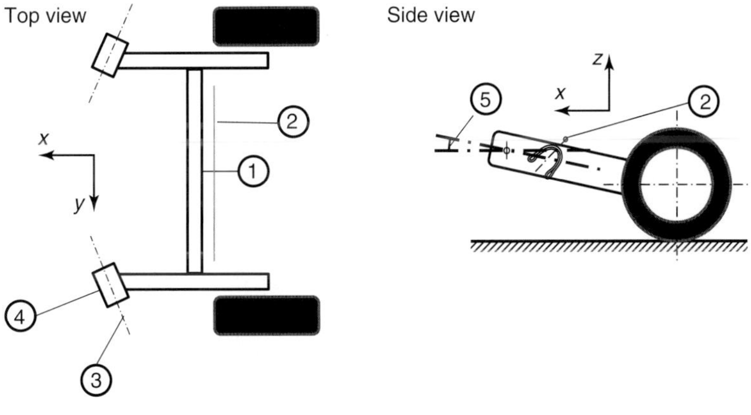

Figure 7. The design parameters of the twist-beam axle.

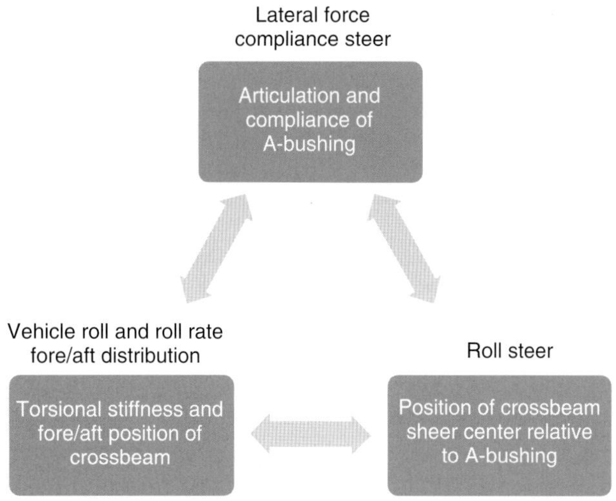

Figure 8. Main elements that contribute to the total axle steer.

axle makes best use from all three factors in combination (Figure 8).

The design evolution starts with the definition of the total vehicle understeer and its distribution to the front axle and rear axle.

3.2.1 Lateral force compliance steer

By its nature, the twist-beam axle goes into toe-out (oversteer) during cornering. Oversteer is caused by the moment that is generated by the side forces acting on the tire contact point longitudinally behind the A-bushings. The moment twists the axle into the direction of the acting side force, that is, oversteer (Figure 9a).

Mainly, three factors control the amount of inherent oversteer. To reduce the amount by which the axle is turned into toe-out by cornering loads, one can shorten the distance between the tire contact point and the A-bushing, choose a wider track width, or increase the fore/aft stiffness of the A-bushing. Of the three main factors, one can assume track width to be a program-given constraint that cannot be used as a design variable to control steer.

The control arm length can be chosen more freely. However, if the arm becomes too short, then the twist-beam is too close to the tire contact point in the fore/aft direction. As a result, the twist-beam is predominantly bent by the loads and not twisted anymore. In this case, the axle behaves more like a rigid axle—with all the known disadvantages. The fore/aft position of the twist-beam is furthermore restrained by the underfloor package where tank, muffler, rails, and foot space leave a small window for the suspension.

The fore/aft stiffness remains the main free parameter that can be tuned to control oversteer. Unfortunately, the fore/aft stiffness not only controls side force oversteer but also affects the noise isolation. Hence, the fore/aft stiffness is a compromise to meet both requirements. The acceptable fore/aft rates alone do not allow to provide sufficient oversteer control.

For this reason, a kinematic trick has become standard in modern twist-beam axles. The pivot line of the A-bushing is not perpendicular to the longitudinal axis of the car, but is angled towards the front.

If the bushing has sufficient axial compliance, then the cornering force pushes the axles sideward up the ramp on the side with the higher side force and down the ramp

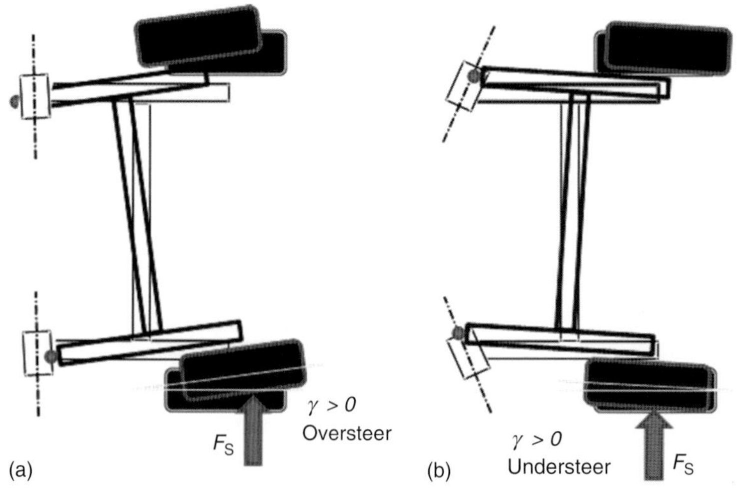

Figure 9. (a) Typical side force oversteer effect caused by the rotation of the axle within the fore/aft constraint of the bushing. (b) Partial compensation by twisting the A-bushing such that the axle slides along the ramping angle, thus inducing a turn.

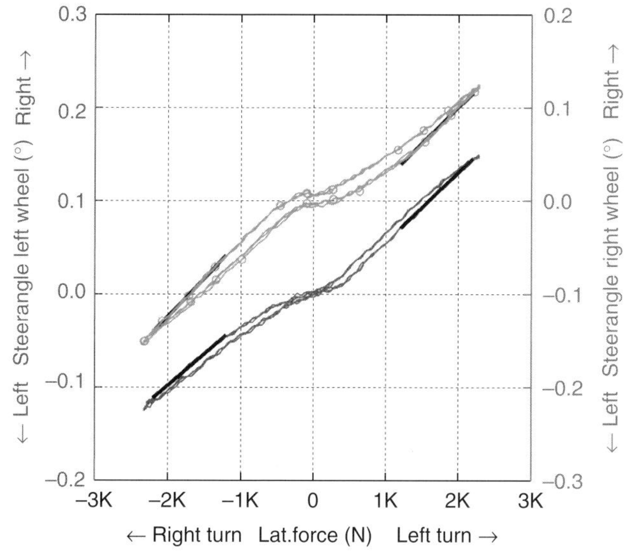

Figure 10. Typical side force steer curves.

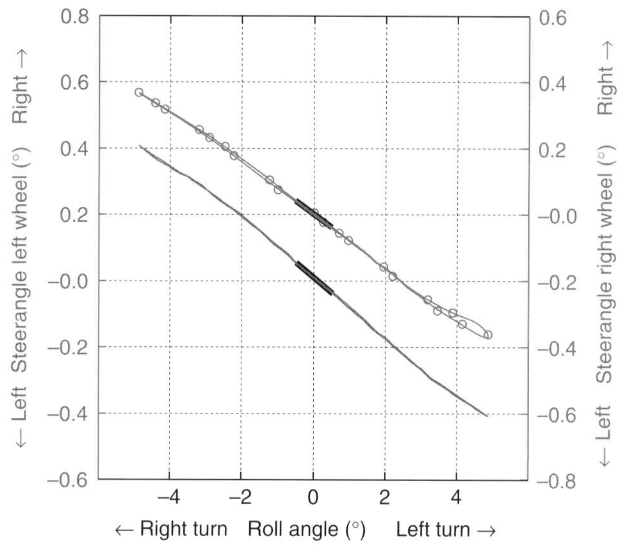

Figure 11. Typical roll steer curve.

on the other side. Effectively, the whole axle is turned into understeer (Figure 9b). The exploration of this effect is limited by the need for directional control of the rear axle that is compromised by the lateral compliance in the bushings. Figure 10 shows a typical curve for the steering angle of the rear axle as a function of the lateral force.

3.2.2 Roll steer

As shown in the last chapter, a particular characteristic of the twist-beam axle is the dependency of roll steer from the position of the shear center of the torsion profile relative to the pivot line of the axle. To further compensate for the remaining compliance oversteer tendency, most passenger cars require a roll understeer contribution from the rear axle (Figure 11).

Maximum roll understeer is gained if the open sections of U- or V-shaped profiles point downwards such that the shear center is above the pivot line of the axle. The optimal vertical position of the A-bushing would be low for understeer; however, for impact smoothness a high position is desirable. The amount of understeer to be gained is limited mainly by packaging restrictions and conflicting performance requirements.

The more the A-bushing is raised, the higher the torsion profile must be situated to provide understeer. Naturally, the underfloor height and the exhaust routing limit the vertical packaging space for the torsion beam. Hence, roll steer will always be a compromise with the influencing factor "harshness".

Twist-beam axles with a tubular interface to the trailing arm offer a marvelous opportunity. With identical crossbeams and trailing arms, the location of the shear centerline can be positioned simply by turning the beam and welding the joint at different angles to the crossbeam (Figure 12). This feature allows tuning the roll steer of an axle to the mass and the mass distribution of the vehicle.

3.2.3 Vehicle roll and roll rate fore/aft distribution

The roll rate of a twist-beam rear axle is defined by the crossbeam torsion stiffness, possibly by the twist rate of an additional stabilizer bar and by the suspension spring rate. The roll rate distribution front to rear together with the axle load distribution front to rear affects the overall steering behavior of the vehicle. The effective roll rate of the crossbeam is a product of its diameter and wall thickness and the fore/aft position of the beam.

By selecting an appropriate effective torsion stiffness as a function of the axle load distribution, the overall steer behavior of the vehicle can be tuned to meet the specification. Furthermore, the steering behavior can be maintained constant for all model variants despite different axle loadings.

The concept of the crossbeam connecting both rear wheels has earned the twist-beam axle the label "semi-independent". In respect to its "copying behavior" of excitations from one side to the other, this notion is misleading. The copying effect is no different to that of an independent suspension that utilizes a stabilizer bar to create roll stiffness. This stabilizer bar does exactly the same thing

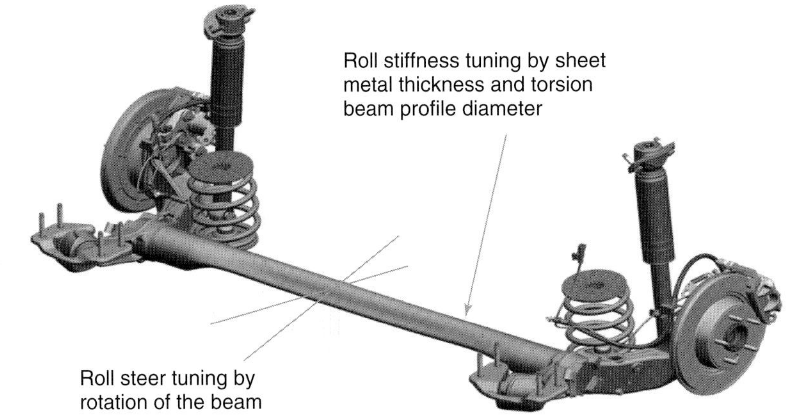

Figure 12. Tuning flexibility of a twist-beam axle with tubular crossbeam.

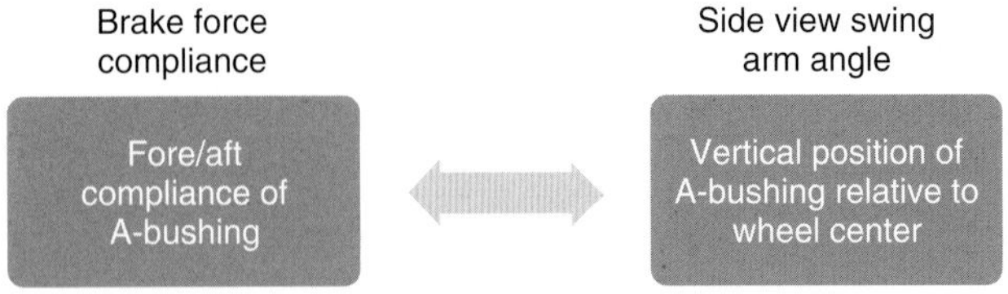

Figure 13. Main elements that contribute to the impact compliance.

as does the crossbeam of a twist-beam axle: it transfers loads and excitations from one side to the other regardless whether intended (to control body roll) or unintended (if caused by road irregularities). A certain fore/aft transfer of road excitations, however, is inherent to the twist-beam, yet far less apparent than the mutual vertical interference that is common to suspension with a high roll stiffness.

3.3 Impact compliance

The main factors that influence the ability of a wheel to elude road irregularities have already been mentioned in connection with understeer control (Figure 13).

3.3.1 Brake force compliance

Forces that are induced by braking have the same effect as any other forces that are generated from an uneven road. In both cases, the tire/wheel moves backwards against the effective compliance of the A-bushing. High fore/aft compliance of the A-bushing allows the wheel to swing back rather freely while contacting a road input such as a crack or a bump.

Although high fore/aft compliance of the A-bushing is desirable for impact isolation, it is undesirable for side force oversteer as shown in the previous section.

3.3.2 Side view swing arm angle (SVSAA)

Side view swing arm angle (SVSAA) is the angle between the horizontal and the line through the wheel center and the A-bushing. The influence of the SVSAA can be easily envisaged by picturing the overcoming of a step with a wheelbarrow. Pushing the wheelbarrow forward into the step results in a high impact, whereas pulling the wheelbarrow instead, while holding the handles high, results in a smaller impact and a more smooth negotiation of the step. This is because the wheel is kinematically guided away from the step, instead of being pushed into it.

Likewise, the A-bushing should be above the wheel center (positive SVSAA) for minimal impact sensitivity. The optimal solution regarding impact compliance, however, is very often not feasible due to interference with the body rails or the foot space in front of the rear seats. In this case, minimal negative SVSAA should be considered.

Since fore/aft compliance is limited by side force oversteer, and SVSAA is limited by packaging constraints, the achievable impact comfort is a compromise. This drawback is inherent to conventional twist-beam axle concepts.

3.4 Roll camber/camber stiffness

Common multilink suspensions have the property that the upper point of the wheel moves inboard when the wheel gets pushed upwards. This implies that the camber angle deteriorates when the vehicle rolls in a corner and the tire can take less side forces. This effect is almost negligible in the case of twist-beam axles.

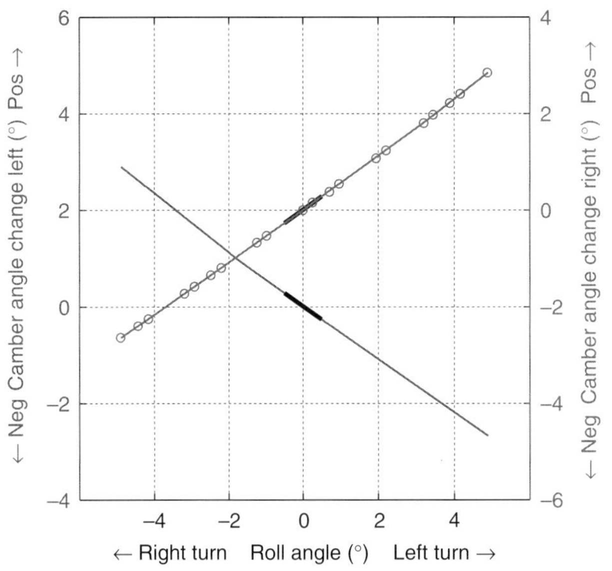

Figure 14. Typical roll camber curve.

Kinematic camber loss is minimal due to the nature of the connection of the axle to the body by the A-bushing that constrains the axle in roll and heave like a hinge with one rotational degree of freedom. Figure 14 shows a typical curve of the camber angle as a function of the roll angle.

Compared to multilink suspensions that suffer from higher kinematic camber loss, the twist-beam axle has significantly better camber compensation.

The second relevant aspect of camber is side-force-induced camber loss. The side force during cornering creates a moment that tilts the wheel in the direction of positive camber. The amount of this effect is predominantly controlled by the torsional stiffness of the trailing arm. Modern cast iron or box-type sheet metal designs provide excellent torsional resistance; hence, twist-beam axles have minimal side-force-induced camber loss.

Sufficient camber compensation and high camber stiffness lead to high residual camber during cornering. This helps grip and rear axle stability.

3.5 Summary: typical design parameters of a twist-beam axle

The said mutual interdependencies and restrictions of the design parameter of twist-beam axle as well as the packaging constraints define a solution space. Most twist-beam axles in the compact and small car segment fall into the bandwidth shown in Table 6.

4 FURTHER ADVANCEMENTS OF THE AXLE CONCEPT

Lately, the usage of the twist-beam axle has been extended to applications that go beyond the conventional performance expectations or use cases of this suspension. New materials and further enhancements to its mechanics show that the limits of the twist-beam axle have not yet been fully explored.

4.1 Twist-Beam with Watts linkage

It has been shown that the twist-beam axle has big advantages to other axle types with respect to camber stiffness, camber compensation, and jounce hysteresis. Lateral force steer and fore/aft compliance, however, are known weaknesses. The A-bushing cannot control the latter three

Table 6. Typical performance attributes of existing twist-beam axles.

Performance Attribute	Phys. Unit	From	To
Swing arm length	mm	375	460
Swing arm angle	deg	−6	+10
Ride rate	N/mm	17	35
Suspension rate	N/mm	18	40
Ride steer	deg/m	0.0	10.0
Ride camber	deg/m	0.6	4.6
Total roll rate with tire	Nm/deg	350	1150
Roll steer with tire	%	4.0	13.0
Roll camber with tire	deg/deg	−0.7	−0.4
Roll center height	mm	100	250
Lateral force steer	deg/kN	−0.15	0.03
Lateral force camber	deg/kN	−0.45	−0.16
Lateral compliance at wheel center	mm/kN	0.3	3.0
Brake force steer	deg/kN	−0.0	−2.0
Fore/aft compliance	mm/kN	0.07	2.50

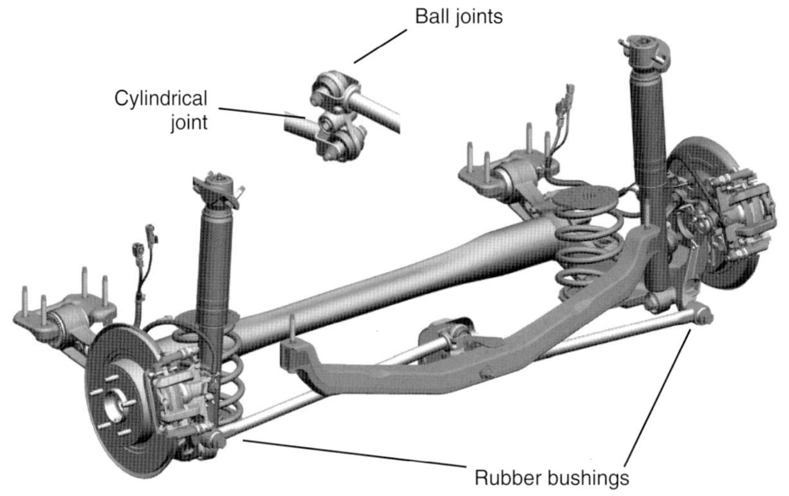

Figure 15. Rear view of a twist-beam with Watt linkage.

characteristics in an optimal way. The compromise lies in moderate, hence safe side force oversteer for the price of relatively harsh impact response and a certain lateral softness.

In 2009, Opel presented the Astra J with a twist-beam axle that was upgraded by two lateral links with a third "differential" link (Harder and Ohligschläger, 2010; Hiemenz and Harder, 2010). This mechanism is called *Watt linkage*. The Watt linkage serves the purpose to maintain the inherent advantages of the twist-beam axle while effectively solving the contradicting requirements for lateral force and steer and fore/aft compliance.

Watt link suspensions for normal front-wheel drive cars have been used before, particularly, in racing together with live axles or with pure torsion beam axles (torsion beam positioned between the wheels). Compared to the historical designs, the attachment for the Watt linkage in the environment of a twist-beam axle has to be reversed. The differential link is attached to the body through an additional subframe and the two lateral links are attached to the axle (Figure 15).

The Watt linkage essentially allows separating the longitudinal support from the lateral support. The lateral forces are reacted by the Watt linkage with minimal lateral displacement. However, if the complete side forces are being compensated by the Watt linkage, then there is still a remaining moment around the vertical axle that will turn the axle into oversteer while cornering. Lateral compliance and lateral force steer are negligible with the Watt linkage. Since the A-bushing is relieved from the support function against side loads and resulting twisting moments, it can be made very compliant in fore/aft direction. This results in an excellent impact behavior and low road noise.

Table 7. Comparison of twist-beam with and without Watt linkage.

	Compound Crank	CC w Watt Link

Lateral force steer	0	+
Lateral force deflection	0	++
Camber stiffness	0	0
Camber compensation	0	0
For/aft compliance	0	+
Side view swing arm angle	0	0
Jounce hysteresis	0	—
Tuning flexibility	0	0
Tolerance sensitivity	0	0
Architecture integration	0	0
Weight	0	—
Development risk	0	—

Table 7 summarizes the advantages of the twist-beam axle with Watt linkage in comparison to the conventional twist-beam.

4.2 Special solutions for four-wheel drive

Fiat Sedici, Suzuki SX4, and Opel Mokka are examples of the increasingly popular "sub-compact sports utility vehicles" that are commonly available with both front-wheel

Table 8. Typical performance attributes of existing bended twist-beam axles.

Performance Attribute	Phys. Unit	From	To
Swing arm length	mm	400	500
Swing arm angle	deg	−7	0
Suspension wheel rate	N/mm	22	30
Suspension ride rate	N/mm	20	24
Ride steer	deg/100 mm	0.00	0.20
Ride camber	deg/100 mm	0.30	0.40
Total roll rate with tire	Nm/deg	600	800
Roll steer with tire	%	13	16
Roll camber with tire	%	−50	−60
Roll center height	mm	200	250
Lateral force steer	deg/kN	0.00	−0.10
Lateral force camber	deg/kN	−0.30	−0.40
Lateral compliance at wheel center	mm/kN	1.00	1.50
Brake force steer	deg/kN	−0.10	−0.20
Fore/aft compliance	mm/kN	1.00	1.50

Figure 16. Four-wheel-drive-capable twist-beam axle. (Reproduced by permission of Adam Opel AG, Germany.)

drive and four-wheel drive. Vehicle size, performance specifications, and cost targets suggest the usage of a twist-beam axle in this segment. Packaging, however, is a severe issue as the crossbeam of the twist-beam normally blocks the space needed for the rear drive module and the propeller shaft.

A bended twist-beam has been found to be a performance-compliant and cost-effective solution that allows the application of a twist-beam axle with both front-wheel drive and four-wheel drive (Figure 16).

In this case, the twisted beam is formed like an arc over the rear drive module. The design does not dictate the static suspension metrics as shown in Table 8. It must be noted, however, that roll steer tuning by rotation of the beam around the y-axis is limited to a few degrees because of the interference of the beam with the driveline.

5 SUMMARY

Starting with the historical designs, the development of the twist-beam axle and its main design parameters that control roll stiffness and roll steer are explained. Chassis designers can take this chapter as a guideline for their own work. Several data samples and additional design hints are offered to help designing such an axle.

The latest developments of twist-beam axles show that some of the intrinsic limitations could be overcome and a modern twist-beam axle is able to have a defined steering behavior and is even able to be used as a driven axle. This will enable the chassis designer to use this lightweight and inexpensive axle for future applications.

RELATED ARTICLES

Volume 4, Chapter 109
Volume 4, Chapter 111
Volume 4, Chapter 114
Volume 4, Chapter 122
Volume 4, Chapter 128

REFERENCES

Harder, M. and Ohligschläger, S. (2010) The New Opel Astra Rear Axle. *Proceedings of the Chassis.Tech Plus Conference*, P 281ff, München, June 8–9.

Heißing, B., Metin Ersoy, M., and Gies, S. (2007) *Fahrwerkhandbuch*, Friedrich Vieweg & Sohn Verlag, Braunschweig.

Hiemenz, R. and Harder, M. (2010) Evolution der Opel Verbundlenkerachse für den neuen Astra. Tag des Fahrwerks, Aachen, October 4.

Hill, A. *et al.* (2013) The Automotive Suspension Systems Report. Stamford, Supplier Business, IHS Global Limited.

FURTHER READING

Arkenbosch, M., Mom, G.P., and Neuwied, J. (1992) *Das Auto und sein Fahrwerk*, Band 1, Motorbuch, Stuttgart.

Bleck, U.N. (2004) Fahrzeugeigenschaften, Fahrdynamik und Fahrkomfort, in *ATZ/AMZ* Sonderausgabe März 2004, Vieweg, Wiesbaden, pp. 76–78.

Bostow, D., Howard, G., and Whitehead, J.P. (2004) Car Suspension and Handling. SAE International, Warrendale: SAE 2004.

Braess, S. (2001) *Handbuch Kraftfahrzeugtechnik*, Vieweg, Wiesbaden.

Heißing, B. (2002) Grundlagen der Fahrdynamik. Seminar, Haus der Technik, Berlin.

Piepereit (2003) Fahrwerk und Fahrsicherheit. Vorlesungsumdruck, FH Osnabrück.

Preukschaeid, A. (1988) *Fahrwerktechnik: Antriebsarten*, Vogel, Würzburg.

Wallentowitz, H. (1998) Quer- und Vertikaldynamik von Fahrzeugen. Vorlesungsumdruck Kraftfahrzeuge 1, IKA Aachen, FKA-Verlag.

Chapter 122

ULSAS—Suspension, A Comparison of Rear Suspension Design

Henning Wallentowitz

RWTH Aachen University, Aachen, Germany

1 INTRODUCTION

Chassis engineering basically refers to the designing of axles and chassis suspension systems and their vehicle integration. A company unique driving performance shall be achieved. Chassis suspension systems, especially those of passenger vehicles, can be categorized by dividing into rigid axles, semirigid axles, and independent suspensions. Figure 1 shows this classification.

Ever since the intensive use of aluminum in vehicle construction, there have been widespread discussions about materials used in the chassis suspension systems. Those discussions are still on-going. Owing to the differentiation between sprung masses (body mass) and unsprung masses (axle parts), the lightweight layout of the unsprung masses with regard to driving comfort, weight, as well as kinematics, is considered in the design of suspension parts. Axles should be as light as possible, hence higher grade steels, lightweight metals, or even synthetics, for example, fiber composites, provide possible solutions, at least on first sight.

Ten years ago, the international steel industry gave an assignment to Lotus engineering to do a survey on lightweight chassis suspension systems. The weight-based competitiveness of steel should be shown for these applied usages. The design proposal, which became known as the *ULSAS-study* (*Ultra light Steel Auto Suspension*) (Lotus, 2000; World Auto Steel, 2012; Stahl-Informations-Zentrum, 2012), focused on rear axles and came up with interesting solutions. This study will be partly reviewed in the following article. For front axles, this volume of the Encyclopedia contains the detailed report "Lightweight front suspension, a comparison" (see Volume 4, Chapter 123) by Mrs Dr. Chang (2006).

2 STARTING SITUATION WITH REAR AXLE SUSPENSION SYSTEMS

The engineers at Lotus started working on the ULSAS study by analyzing existing car chassis suspension systems. They focused on the four most important axles. Figure 2a–d summarizes those axles.

Figure 2a shows the so-called Double Wishbone axle (often known as *Short-Long Arm* suspension in the United States) with a body-connected longitudinal control arm, which is made from a deep-drawn steel plate. The two lower transverse control arms are mounted to a small rear suspension subframe and they are made of forged steel. The front lower transverse control arm carries a suspension strut that is directly held up at the body at its upper end. The knuckle is forged as well and is shaped to follow the contour of the tire until it reaches a ball joint, to which

Encyclopedia of Automotive Engineering, print © 2015 John Wiley & Sons, Ltd.
Edited by David Crolla, David E. Foster, Toshio Kobayashi, and Nick Vaughan.

ISBN: 978-0-470-97402-5
Also published in the *Encyclopedia of Automotive Engineering* (online edition)
DOI: 10.1002/9781118354179.auto006

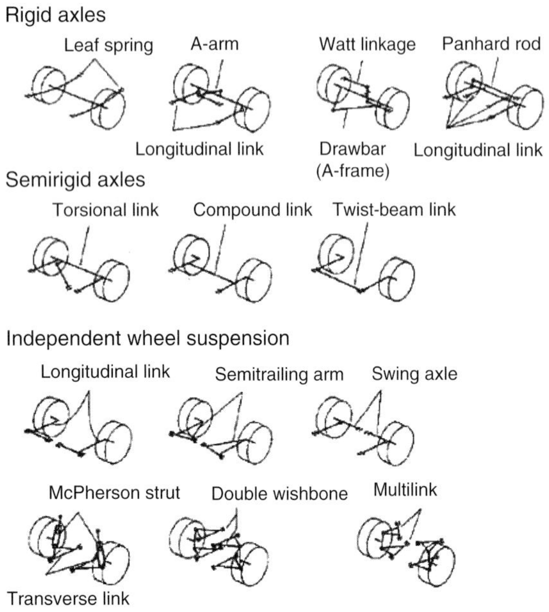

Figure 1. Systematic classification of chassis suspension of passenger vehicles. (Reproduced with permission from ika RWTH Aachen University. © ika Aachen.)

the upper transverse control arm is mounted. This upper control arm is also forged and is directly attached to the body at its other end with an elastomeric bush. According to the ULSAS study, the mass of the axle is 39.8 kg. This nondriven suspension assembly is fitted to the rear of the Honda Accord.

Figure 2b shows a driven rear axle of BMW, the so-called integral axle. The large suspension subframe is made of hydroformed aluminum. Welded consoles, also consisting of aluminum, are the connections to the upper transverse control arms. These control arms are made from forged aluminum. The lower control arm consists of a large box section, which is made up of metal sheet, and is mounted to the suspension subframe with two bearing connections and with one ball joint to the wheel carrier. The so-called integral connecting rod connects the wheel carrier with the large box section transverse control arm (Matschinsky, 1998). The "integral rod" reduces elastic rotational deformations during acceleration and braking. This reduces the so-called wind-up of the wheel carrier that otherwise may lead to wheel hop or "tramping" [reasons are changes in friction (slip) between the tire and the road because of the wind-up of the wheel carrier].

Suspension is done through suspension struts, which are not shown in Figure 2b. Those struts are directly mounted to the wheel carrier. The large suspension subframe enables the connection of the axle in the area of the vehicle's stiff longitudinal chassis beam. This suppresses the transmission of noises into the body, created by the tires or the rear axle differential. The differential is supported in the suspension subframe by elastomeric bushes, as is the suspension subframe in the body. That is why this is called a *double-elastic bearing*. The mass of this BMW integral axle is stated at 46.67 kg in the ULSAS study (aluminum axle).

Axle 2c is a lightweight suspension strut rear axle. It was taken from a Ford Mondeo. Via a tension strut

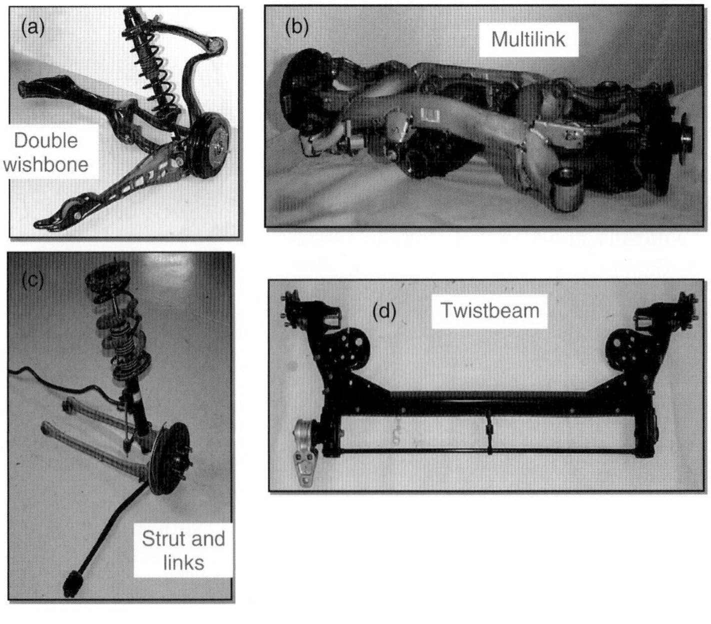

Figure 2. (a–d) Most important axles. (Reproduced from World Auto Steel. © World Auto Steel.)

and connected rubber bearings, the longitudinal forces are transferred from the wheel carrier into thc car body. The transverse forces are transferred to the body through two lower transverse control arms (forged aluminum parts) and the upper suspension strut bearing. This upper suspension strut bearing additionally transmits the vertical forces from spring and damper.

Owing to the upper suspension strut bearing having to brace side forces, the piston in the damper and the piston rod bearing have to handle lateral forces. The induced torque by the vertical forces can be compensated by the inclined function line of the coil spring within the design position of the wheel suspension. This "trick" to compensate torque is used commonly on front suspension McPherson struts, but it can also be applied in rear axles. To achieve a high stabilizer conversion ratio, the stabilizer (or antiroll bar) is mounted to the suspension strut using a rod.

To provide adequate clearance between the tire and the damper strut, the bottom mount of the strut into the hub carrier (often clamp design or welded) is offset laterally inward. To carry the torque between wheel hub and damper, the hub carrier is designed as a cast or forged part to accommodate this. This suspension strut rear axle is designed as a nondriven axle. The ULSAS study quotes a mass of 39.7 kg.

The twist-beam axle that is shown in Figure 2d, and which originates from the Volkswagen Golf, is made from steel plates to achieve the required stiffness. Longitudinal control arms, connected via the twist beam, carry the wheel carrier and the brakes (which are not shown in Figure 2d). The longitudinal arms are mounted directly to the body via rubber bearings. The springs and dampers are then connected to the longitudinal control arms via brackets. As the connections of the brackets, as well as the connections between the longitudinal control arms and the twist beam, are usually achieved by welding, the stiffness is limited through the welding seams. Additionally of disadvantage is the fact that, during cornering, the lateral forces tend to cause the whole axle assembly to turn about a vertical axis, thus the sideslip angles are reduced. The wheels are steering into the turn; this generates an oversteering tendency. The mass of this twist-beam axle is 33.4 kg (without brakes) according to the ULSAS study.

For illustrating the results of the survey, the Lotus engineers used the so-called spider charts. The performance ratio of the four axles in view of certain demand aspects is evaluated. This ratio is then marked on the chart. The higher the ratio, the better is the performance of the axle. Figure 3 shows the spider chart for the four analyzed axles.

The greatest differences can be observed in the categories of cost, manufacturing, and the space demands each individual axle poses to the vehicle structure. The multilink axle is superior to all other axles in respect of "ride and handling" and "refinement (NVH)" performances (NVH, noise–vibration–harshness). The better ride and handling performance is due to the better wheel control afforded by the multilink design, and the better NVH performance is a result of the double-elastic bearing as mentioned earlier. The implemented mass is rated the same with all four axles.

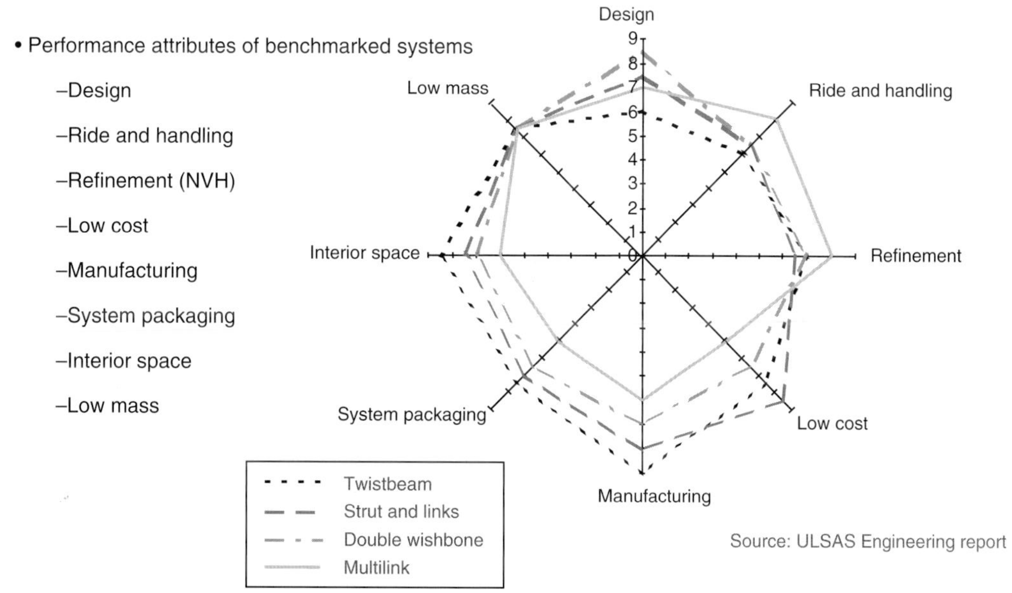

Figure 3. Spider-chart. (Reproduced from World Auto Steel. © World Auto Steel.)

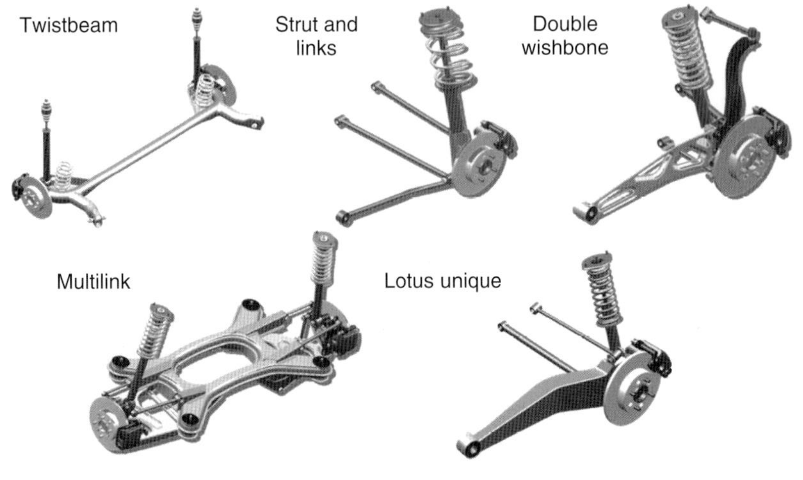

Figure 4. Steel axles as proposed by Lotus engineering. (Reproduced from World Auto Steel. © World Auto Steel.)

3 REAR AXLES DEVELOPED BY LOTUS

Within the development done by Lotus, the engineers proposed rear axles made of just steel. However, different grades of steel were used. The development target was to

1. decrease the mass of the steel-made axles by at least 20% without increasing the costs.
2. achieve the same mass as the aluminum-made axles but using steel instead, and at the same time decrease the costs by at least 20%.

To accomplish these targets, five new axles were developed. They were worked on in design engineering as well as using simulation technology (finite element analysis). Figure 4 shows a compilation of those proposals. In addition to the axles used in practice, there is also another solution shown, called *Lotus Unique*.

Furthermore, they were able to achieve the set goals of reducing the mass, as shown in the comparison (Table 1), illustrating the masses of current axles and the new developments.

Table 1. Mass comparison of the analyzed and the newly developed axles.

	Realized (kg)	ULSAS (kg)	b.	Car
Double wishbone	39.80	32.88	17%	Honda Accord
Multilink	46.67	48.00	3%	BMW 5er
Strut and links	39.70	29.87	25%	Ford Mondeo Taurus
Twistbeam	33.40	27.30	18%	VW Golf
Lotus unique	39.80	26.49	34%	Honda Accord

The differences in cost and the production studies performed by the ULSAS study are not examined any further in this report because of them being rated differently today. However, axle developers are advised to take a closer look into the study, as it is full of interesting suggestions.

3.1 Double wishbone axle

This axle matches the analyzed axle's basic buildup. However, there are differences in some parts. Figure 7 shows these parts; especially the longitudinal chassis beam that is explicitly shown as pressed steel plate made of high grade steel.

The housings for the rubber bearings are tube sections, which are welded; the transverse control arms are also steel tubes with rubber bearings on the end. This enables a connection to the wheel carrier on the one end and to the body on the other end. Whether there is supposed to be a suspension subframe in between the body and the transverse control arm or not is left unanswered in Figure 5. The wheel carrier itself consists of very high grade forged steel. A ball joint connects the upper transverse control arm to the knuckle. The control arm itself consists of only a tube. The suspension strut is directly screwed to the wheel hub and the connection to the body is done conventionally using suspension strut mounts.

The overall rating of this axle was entered into a spider chart as shown in Figure 6. It shows the rating as compared to the other, currently used axles, which were set out as benchmarks. The indications ULSAS D&P refer to are the vehicle classes the axles are intended to be used within. Vehicle class E will be looked at later (Figure 10).

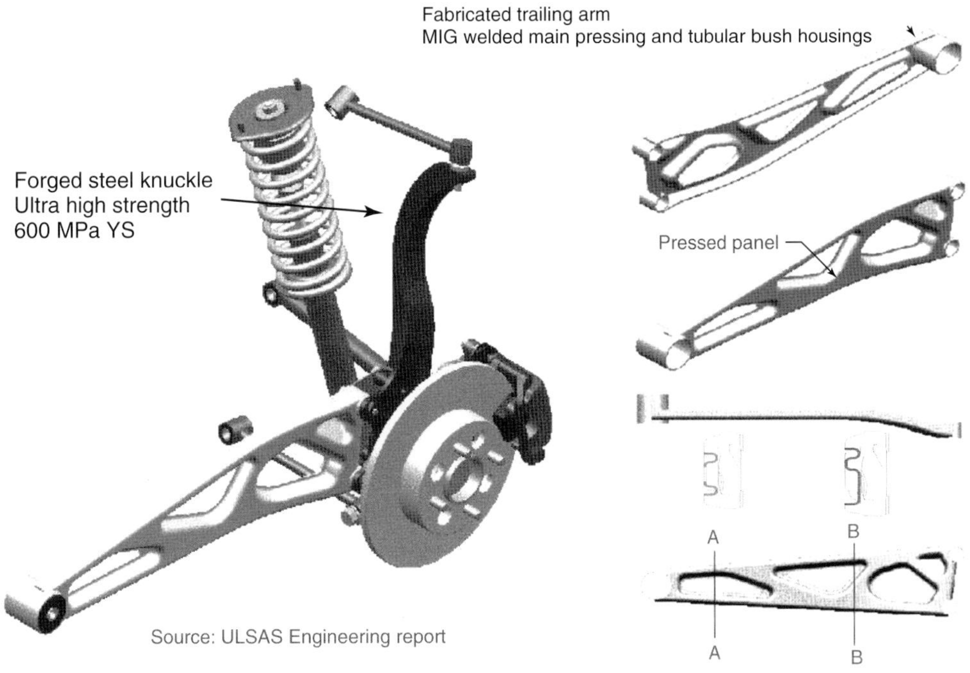

Figure 5. Lightweight construction double wishbone axle. (Reproduced from World Auto Steel. © World Auto Steel.)

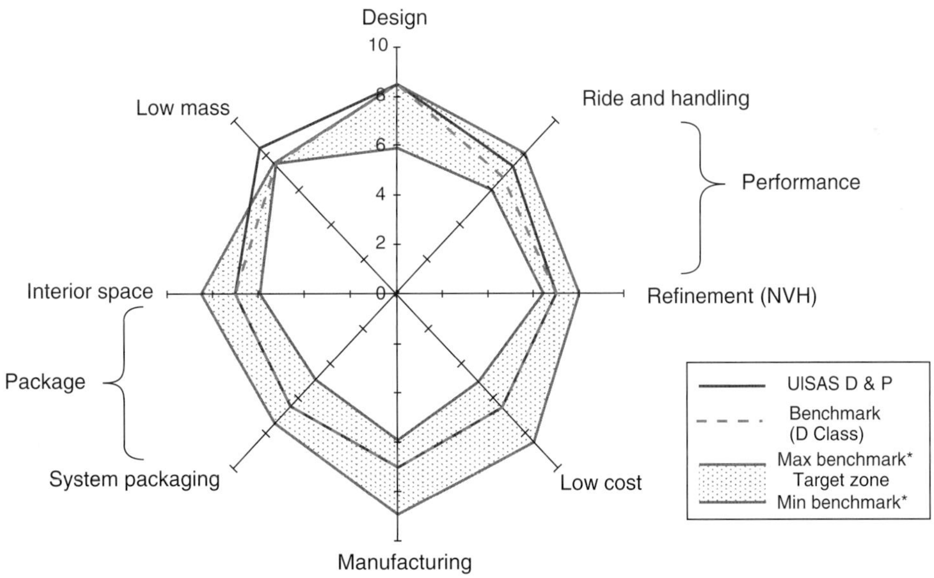

Figure 6. Classification of the double wishbone axle into the benchmark zones as determined before. (Reproduced from World Auto Steel. © World Auto Steel.)

As a result of this classification, it shows that especially the mass could be reduced without creating additional costs. The behavior of the axle (ride, handling, and NVH) competes very well within the competitor class, as well as the other characteristics also staying within the benchmark parameters.

3.2 Multilink rear axle

The multilink rear axle, which in the analyzed integral axle version was equipped with an aluminum suspension subframe and taken from an upscale vehicle class, will also feature a suspension subframe in its steel version.

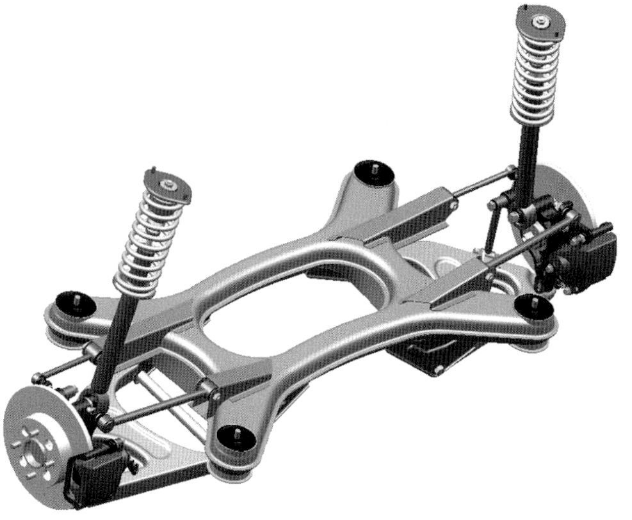

Figure 7. Multilink rear axle. (Reproduced from World Auto Steel. © World Auto Steel.)

Figure 9. View of a multilink rear axle. (Reproduced from World Auto Steel. © World Auto Steel.)

This suspension subframe will, however, not be created by hydroforming but by pressing two sheet metals into frames that are then welded together. The study suggests using 250–500 MPa YS steel. The connectors for the transverse control arms are welded onto the sheet metals of the suspension subframe. These connectors are simply created through angled sheet metal. Figure 7 shows this axle in detail.

Further elements made of steel plates are the lower transverse control arms. They were also designed in a clamshell way. Their production is recommended using 250 MPa YS steel. Figure 8 contains special production instructions for these transverse control arms. Especially important is the flange depleted welding of the two half shells.

The tubular wishbones can be seen clearly in Figure 9. The vehicle-based side has rubber bearings; the wheel-based side has a ball joint. Figure 9 also features the integral link as a connection between the wheel carrier and the lower wishbone.

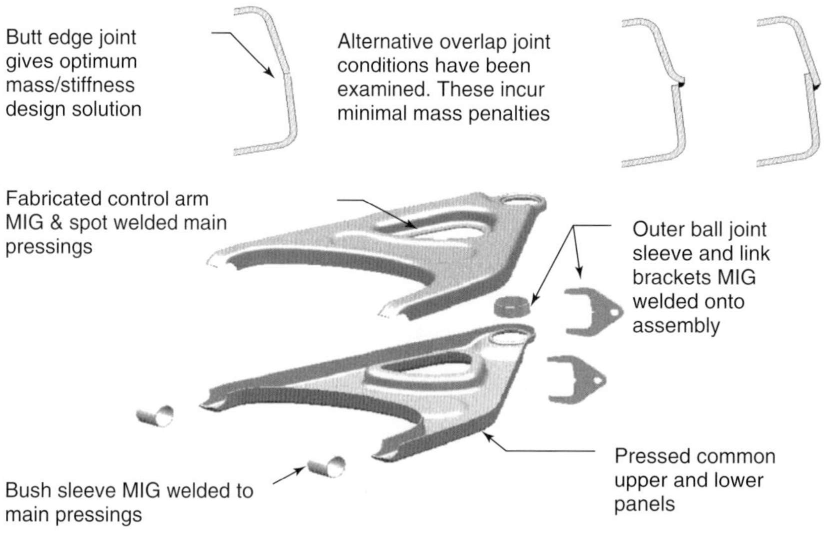
Butt edge joint gives optimum mass/stiffness design solution
Alternative overlap joint conditions have been examined. These incur minimal mass penalties
Fabricated control arm MIG & spot welded main pressings
Outer ball joint sleeve and link brackets MIG welded onto assembly
Pressed common upper and lower panels
Bush sleeve MIG welded to main pressings

Figure 8. Production instructions for the lower transverse control arms of the multilink rear axle. (Reproduced from World Auto Steel. © World Auto Steel.)

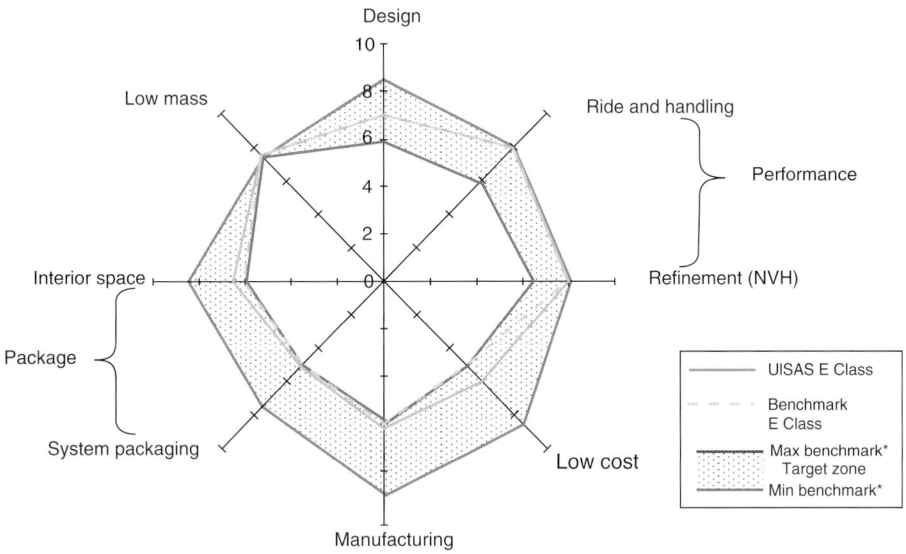

Figure 10. Spider chart of the multilink rear axle. (Reproduced from World Auto Steel. © World Auto Steel.)

The wheel carrier is forged of 750 MPa YS steel; the wishbone tubes consist of 250 MPa YS steel. The engineers at Lotus put the performance data of the axle into a spider chart once again. Figure 10 shows this chart.

It shows that only a minor weight reduction could be achieved, but a significant 30% cost reduction was realized. Performance and packaging match the evaluation criteria of the axles.

3.3 Strut and links rear axle

This axle type was designed for light- and heavyweight vehicles. Figure 11 shows the suggestion for the lighter vehicle. The applied constructions only differentiate in the connection of the tubular longitudinal support beam to the wheel carrier. Instead of connecting to the frontal wishbone, the solution for heavyweight vehicles applies the connection

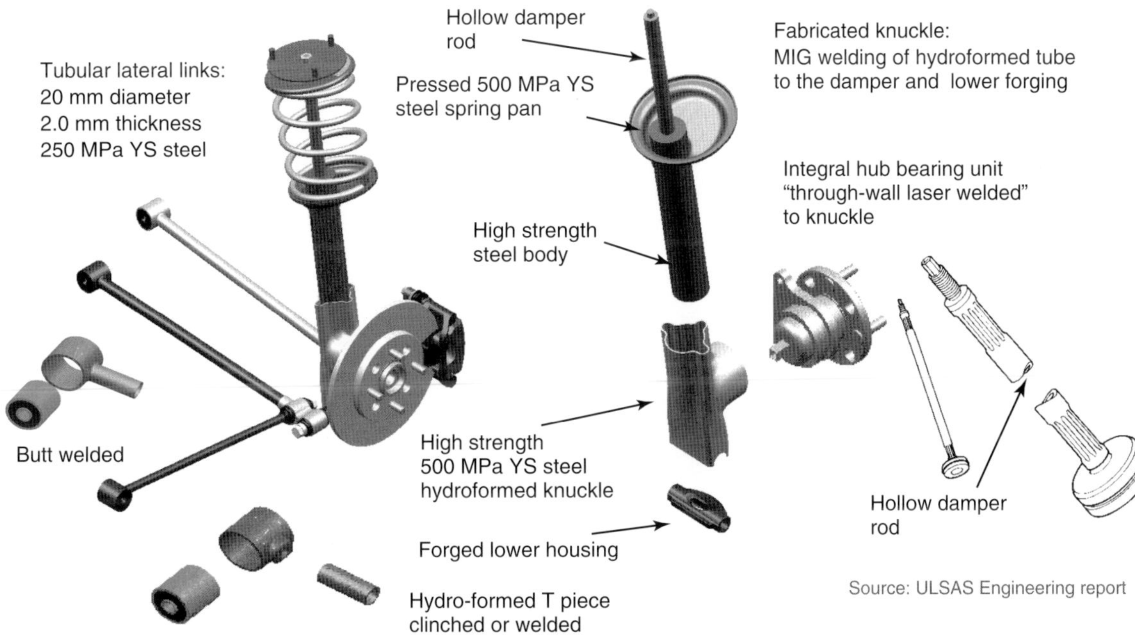

Figure 11. Overview over the suggestions of the strut and links rear axle. (Reproduced from World Auto Steel. © World Auto Steel.)

of the longitudinal support beam to the knuckle. Then, the "forged lower housing" part is altered.

Essential for utilizing the lightweight potential of this axle is the creation of the wheel carrier, which is produced from a hydroformed tube. The hub bearing units, as "finished parts," are welded into the metal sheet knuckle. A special "through penetration welding by laser process" is applied as the laser beam bonds the hydroformed part directly to the lower casing of the wheel hub unit. Accordingly, the steel part, which represents the lower end of the hydroformed wheel carrier, is welded into place as well. This part is welded onto the shock absorber tube. The wheel hub unit also carries the brake caliper mounts. The wishbones are produced as tubes onto which are welded smaller tubular parts that serve as housings for the rubber bearings.

The piston rod for the damper is suggested to be a hollow pipe, which, for stiffness reasons, is manufactured with longitudinal grooves but not in the areas that cross the sealing.

The mass comparison, as given in Table 1, indicates a weight reduction of 25% as compared to contemporary state of technology.

The estimation of performance of the axle for both the light- and heavyweight vehicles is shown in Figure 12. It shows that the mass, but not the costs, could be reduced. Especially ride and handling experience a positive influence with this type of axle construction. Otherwise, the axle is within the benchmark parameters.

3.4 Twist-beam rear axle

With lower and middle class vehicles, which usually have front wheel drive, the rear axle of choice is typically the twist-beam axle. The analysis showed that its welding seams limit the capacity of this axle. Hence, the Lotus engineers put the task on themselves to reduce those seams as much as possible. Figure 13 shows the suggested design and the applied steel grades.

Figure 13 underlines that both the longitudinal control arms and the twist beam are made from one piece of pipe. To achieve the torsion softness of the twist beam, the pipe is cut open with a plasma cutter and the edges are bent a little. This enables only the trailing arm toward the bearing, the spring base, and the mount for the wheel bearing to be welded. The bolted wheel bearing construction then enables the connection of the brakes to the axle. The angling of the rubber bearing in plan view, between the axle and the body that enables some compensation of side force oversteering, can be achieved by simply twisting the trailing arm. This trailing arm is produced by hydroforming. The housing in which the rubber bearing is mounted to the trailing arm can be manufactured by rearranging the trailing arm tube.

In addition, the position of the shear center, which together with the bearings of the longitudinal control arms determines the kinematic data of the axle, can be chosen freely (within a certain range) with this type of design. It only matters as to which side the cross profile is cut open,

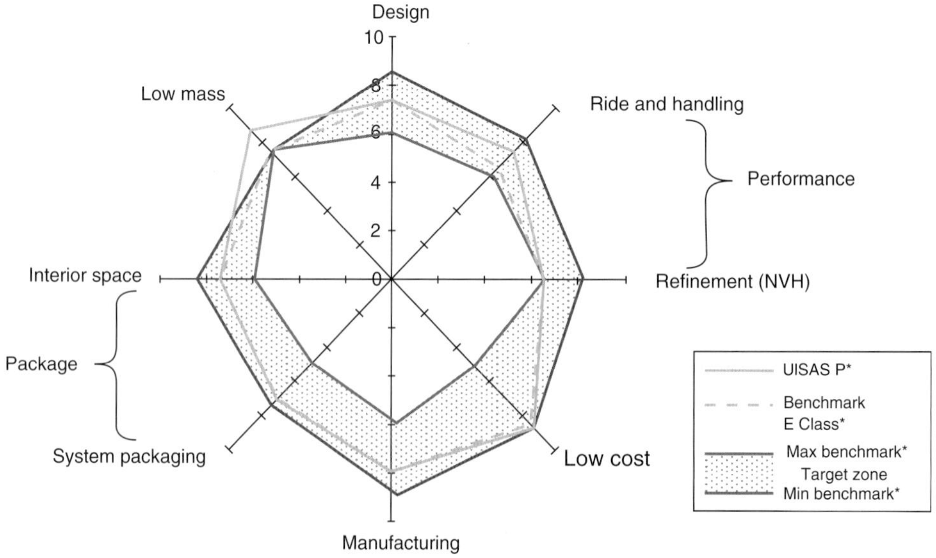

Figure 12. Evaluation of the analyzed strut and links rear axles into the benchmark areas. (Reproduced from World Auto Steel. © World Auto Steel.)

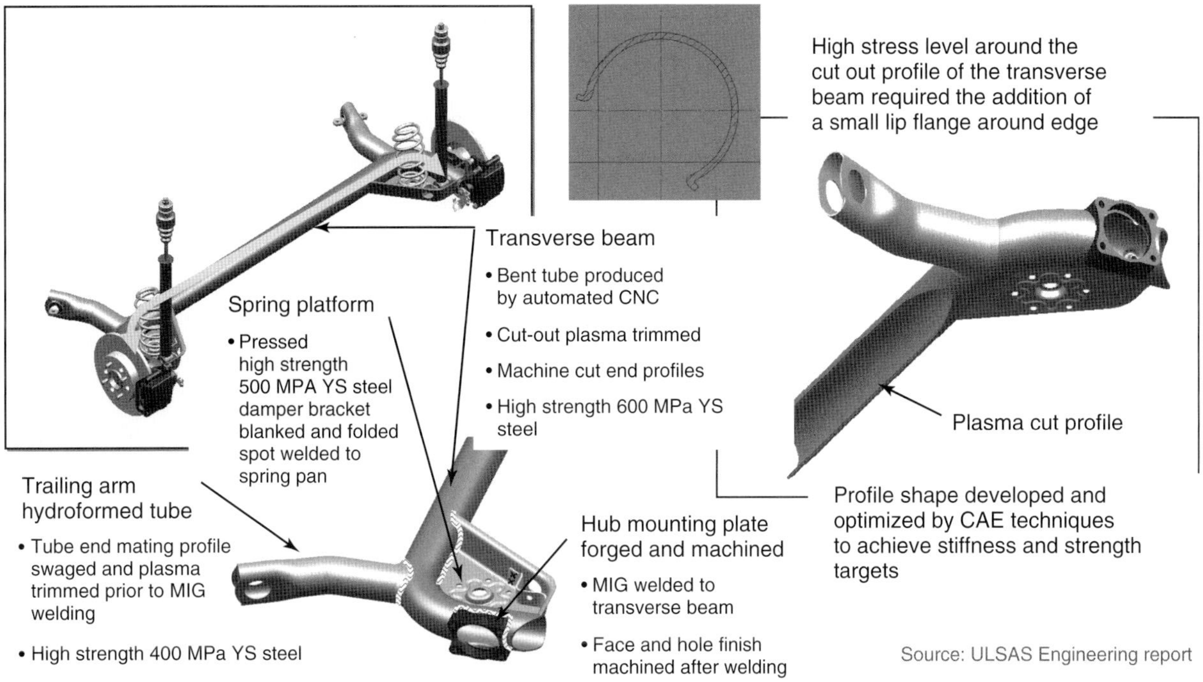

Figure 13. Twist-beam axle as lightweight construction axle. (Reproduced from World Auto Steel. © World Auto Steel.)

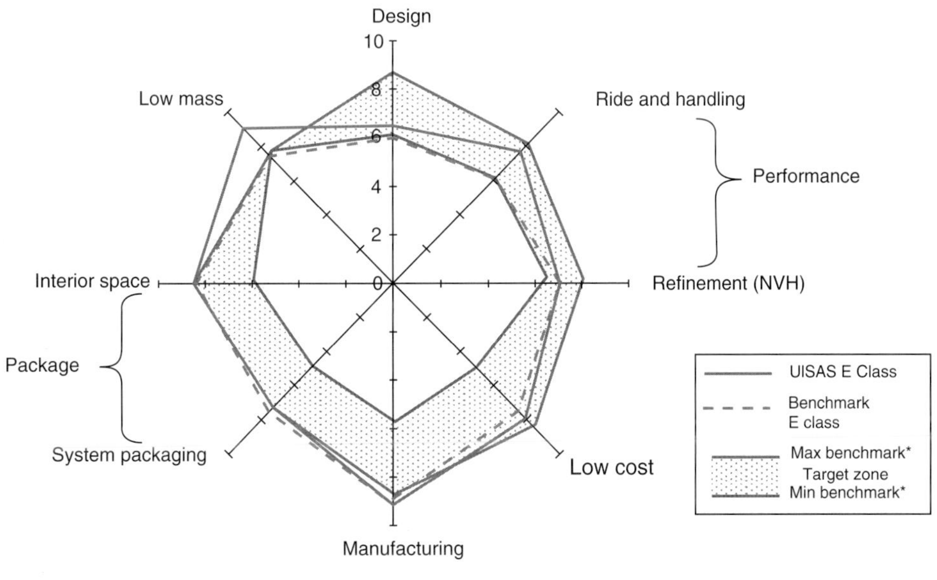

Figure 14. Evaluation of the twist beam axle into the benchmark areas. (Reproduced from World Auto Steel. © World Auto Steel.)

or rather which lengthwise angle is used for welding these longitudinal control arms to the twist-beam tube.

The functionality of the axle is shown in Figure 14.

Especially remarkable are the weight reductions; the cost reductions are less impressive. Design and behavior is even better than the benchmark, whereas production and packaging both meet the benchmark standards.

3.5 Lotus unique rear axle

In addition to the various types of axles considered, the engineers at Lotus also presented their own axle suggestion. This axle can be used as both a driven and a non-driven axle.

A massive body-trailing arm is connected to two tubular wishbones. Thus, the mechanism is statically defined and the spring and damper combination does not have to take any shear forces.

To a certain degree, the different lengths of the wishbones enable an influence on the wheel kinematics (for example, there is no track width alteration in the area of design position of the car). The distinctiveness in the construction of the trailing arms lies in the usage of "Tailor Welded Blanks." Steel plates of different thickness and grade are welded together and are then deep-drawn. Using the strain data available that the specific parts have to withstand enables a strain-related construction. Figure 15 shows these modifications. The trailing arm is welded together from two panels, each of which was manufactured individually. These measures help to achieve the aim of reducing the weight as only as much material as needed is used for production.

These Tailor Welded Blanks can also be used for other axle constructions. By now, there are production techniques that allow rolling metal blanks with varied thickness. The varying stiffness within one part can be achieved by tempering.

The evaluation of the Lotus unique axle by the engineers can be seen in Figure 16. This construction achieved significant reductions in both cost and weight. The packaging demands are evaluated to be adequate; production and assembly is easy. The performance is perceived to be good as well.

The evaluations in Figure 16 show that this axle is positioned in the middle to the upper level of the benchmark. It is therefore advisable to have a closer look at this construction.

4 CONCLUSION

Entering the evaluation figures of all five axles into a spider chart, as done in Figure 17, the advantages of the axles become clearly visible.

This chart exemplifies that the aims to reduce cost and mass were achieved. The multilink rear axle is superior to all others in terms of driving behavior and NVH. Since it is a steel-made axle, it also features a cost advantage in comparison to the axle produced out of aluminum.

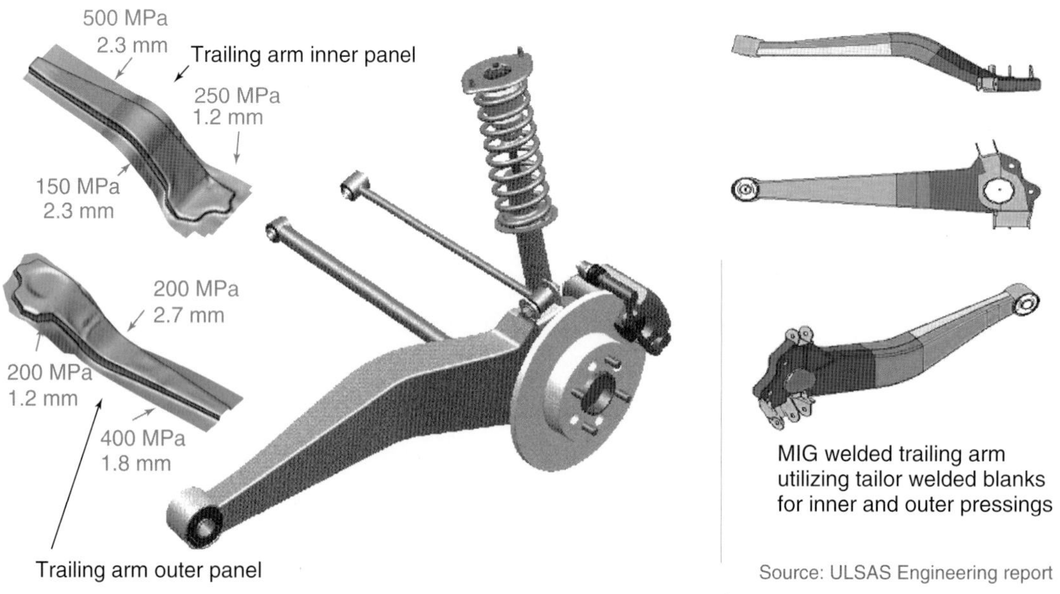

Figure 15. Lotus unique rear axle. (Reproduced from World Auto Steel. © World Auto Steel.)

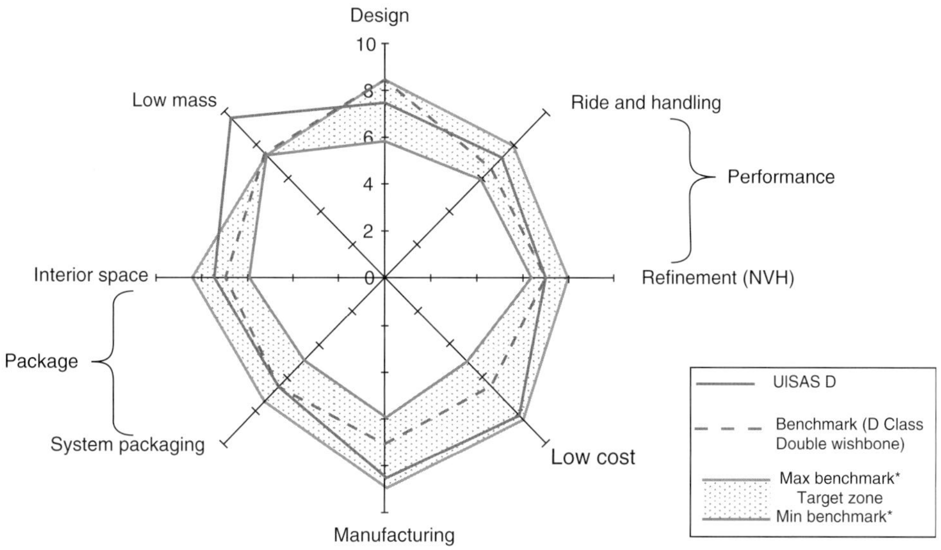

Figure 16. Evaluation of the Lotus unique axle. (Reproduced from World Auto Steel. © World Auto Steel.)

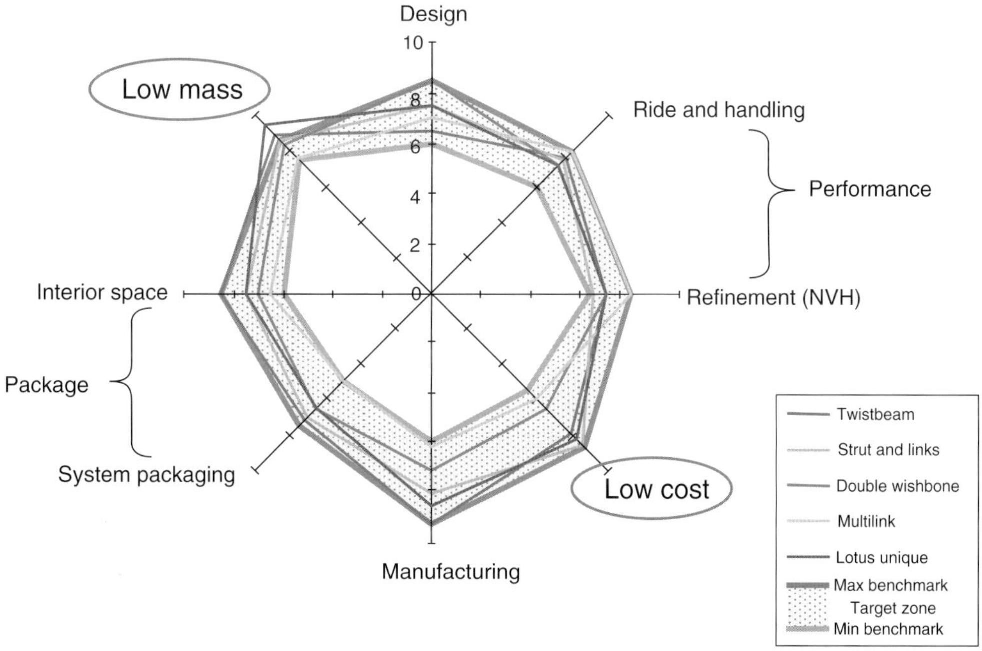

Figure 17. Summary of the different axles. (Reproduced from World Auto Steel. © World Auto Steel.)

Concluding the ULSAS-study, it can be determined that using the suggested steel-axles a mass reduction of up to 34% could be achieved without creating any additional cost. With steel-axles, matching the mass of aluminum-axles there were also no further costs created.

None of the axles compromised the driving performance.

Improved qualities of steel made this progress possible. Since these advancements in steel continued over the last 10 years, a revision of the suggested constructions with contemporary steel qualities and production methods, like flexible rolling and further improved joining technology, is highly suggested. For chassis-developers, there are still some very interesting and innovative ideas left in the ULSAS-study that are still awaiting implementation.

REFERENCES

Chang, M.-Y. (2006) Leichtbau-Vorderachsbauweisen unter Berücksichtigung verschiedener Fahrzeugklassen. Dissertation. RWTH Aachen University, Germany.

Lotus (2000) Ultra Light Steel Auto Suspension, Corus Group.

Matschinsky, W. (1998) Die Radführung der Strassenfahrzeuge Kinematik, Elastokinematik und Konstruktion, Berlin, Germany.

Stahl-Informations-Zentrum (2012) http://www.stahlinfo.de/stahl_im_automobil/ultraleicht_stahlkonzepte/ulsas/ulsas_detailinfo.htm (accessed 12 June).

World Auto Steel (2012) www.worldautosteel.org/project/ulsas (accessed 12 June).

Chapter 123

Lightweight Front Suspensions: A Comparison

Ming-Yen Chang

Modular Management Asia Limited, Taiwan, PR China

1 INTRODUCTION

This contribution differs from purely technical ones, as the identification of the customers' needs and the technical requirements such as lightweight design is combined. Therefore, some tools such as ADAMS/car, QFD (quality function deployment) systems, and finite element modeling (FEM) programs are used to define suspension types that are needed. The "Smart" of the Daimler Company is one of the target vehicles that should be equipped with an as lightweight axle as possible.

2 TARGETSET SETTING AND DEVELOPMENT PROCESS

Figure 1 shows the development process. The first step is the benchmark. From the benchmark, the design requirements could be set. Different front suspension design can be evaluated by the results of the benchmark and also through CAE simulation and QFD. Following those processes, an overall appraisal can be generated.

The weight optimization design can be approached by modification of existing constructions. This chapter will take a transverse leaf spring front suspension as an example. Apart from the functional integration of the components, another possibility for weight reduction is to redesign the components. To meet the requirements, all load cases and misuse events of chassis components must be considered. In order to analyze the stress distribution of components, different front axles are build in MBS-software (e.g., ADAMS/car) and stimulated with different measured hub load events. Computer-aided tools (CAD computer-aided optimization) become a powerful assistant for weight optimization. Considering both, new ideas for suspension systems and optimum design of components, the weight reduction potential is assessed for different front axles of each vehicle class.

3 DATA COLLECTION

The criteria for appraisal are as follows:

- Technical data, kinematic, and compliance data
- Vehicle weight (respectively vehicle mass)
- Layout
- Ride comfort
- Handling
- Brake performance

The vehicles will be evaluated by various methods,

- objective appraisal,
- subjective appraisal, and
- quantitative appraisal.

Encyclopedia of Automotive Engineering, print © 2015 John Wiley & Sons, Ltd.
Edited by David Crolla, David E. Foster, Toshio Kobayashi, and Nick Vaughan.
This article is © 2015 John Wiley & Sons, Ltd. ISBN: 978-0-470-97402-5
Also published in the *Encyclopedia of Automotive Engineering* (online edition)
DOI: 10.1002/9781118354179.auto005

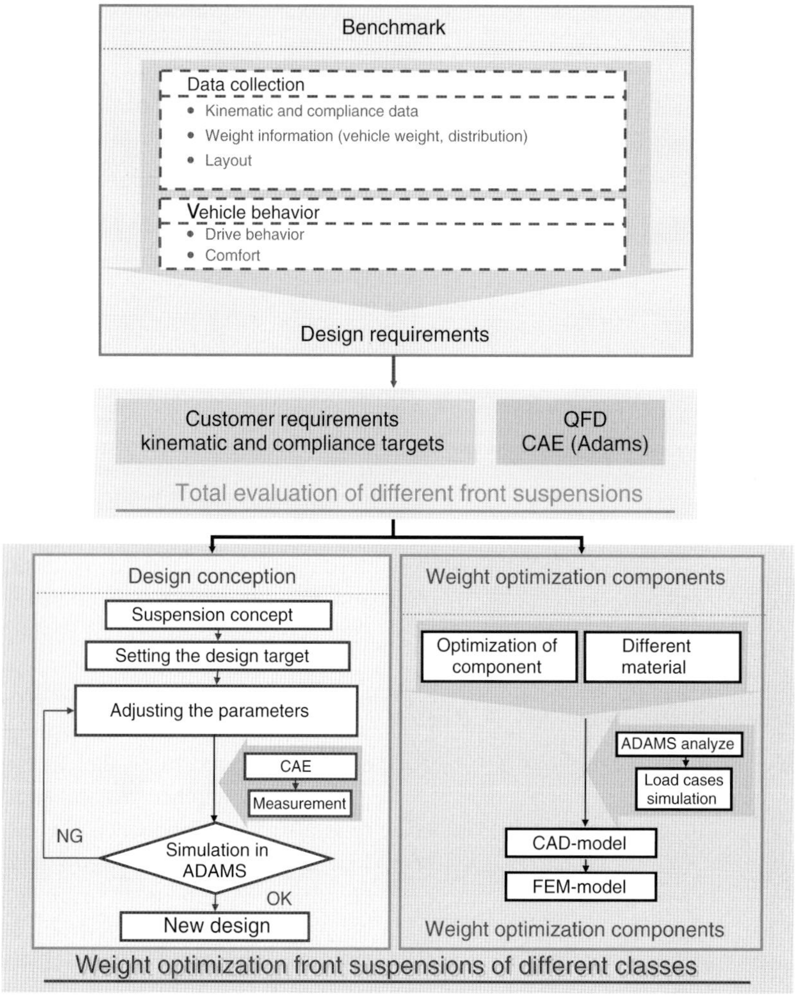

Figure 1. Development process.

Through those comprehensive data, the geometry, kinematic and compliance information, and the basic drive behavior can be roughly estimated. There is a relationship between the subjective evaluations from test drivers and the objective parameters (measured data), which will be applied in the design process.

The suspension designs of different vehicles in the same class are compared. The comparison shows the different priorities, compromises, and preferences, depending on customer expectations of different manufacturers, which will be evaluated subjectively.

The subjective ratings are as follows:

- Potential of the technical development
- Potential of the system integration
- Marketability
- Structural effectiveness.

By test driving, the drive behavior, ride comfort and noise, vibration, and harshness can be evaluated. The ratings include the subjective impressions of the test drivers and also the objective measurement data.

As part of the research, three different vehicle classes and four common front suspension systems are discussed, as shown in Figure 2. The vehicle class is defined as a distinct group of car models, which are characterized by appearance and technical and market orientation features with other competition. Normally, the purchase price and the size of the vehicle in the assignment class are also considered.

3.1 Test driving

As just mentioned, test driving delivers impressions about ride comfort, handling, noise, vibration, and harshness (nvh). Various vibrations of vehicle are caused by both inside and outside disturbances. As the concept of nvh, the

Vehicle classes	Front suspension systems
▪ A~B-segment ▪ C-segment ▪ D-segment	▪ Transverse leaf spring front suspension ▪ McPherson suspension system ▪ Double linkage with strut suspension system ▪ Double wishbone suspension system ▪ Four-linkage suspension system

Figure 2. Different vehicle classes and front suspension systems.

Ride comfort	Driving safety	Braking
- Ride (lightly laded) - Ride (heavily laded) - Front seat - Rear seat - Inner NVH measurement - NVH impression	- Driving safety (lightly laded) - Driving safety (heavily laded) - Vehicle dynamic testing - Handling - Steering - Turning circle - Traction/winter capability - Straight line/wind affection	- Braking deceleration (cold lightly laded) - Braking deceleration 80% V_{max} - Braking deceleration (cold heavily laded) - Braking deceleration (warm heavily laded) - Response characteristic braking

Figure 3. Evaluation factors (Ostmann, 2001). (Data from Ostmann, 2001, Auto Motor und Sport.)

acoustic appearance of a vehicle is characterized. The low frequency mechanical vibration is experienced as comfort by the passengers. The dynamic behavior of the vehicle is its response to the driver's inputs and to the disturbances from the environment during the movement. Good driving behavior gives the possibility of precise route control and, hence, is a key ingredient in the control performance of the overall system. The most important consideration for evaluation of driving behavior is, particularly, the contribution to prevent traffic accidents. Good performance means to minimize the accident risk for drivers and vehicles in traffic.

The driving behavior is evaluated by two appraisal methodologies: subjective evaluation and objective measurement.

Subjective evaluation from experienced drivers is still the most important statement about the quality of vehicle. An absolute scale is not defined; the appraisal is carried out in comparison with other vehicles. Mostly company-defined scales are used. They are very confidential.

The objective measurements for the evaluation of vehicles have either absolute target scales or they are evaluated in comparison with other vehicles. There are intercompany data used, which has been agreed in technical teams.

The subjective and objective evaluation is indicated by quantitative characters. Measurements of certain objective tests and subjective evaluation of the examined parameters is defined according to a predetermined rating scale. To validate the results of the subjective appraisal and objective measurements, data are analyzed.

To give an impression about the working possibilities, evaluation results from "Auto Motor Sports 2001–2003" (a German newspaper) are used. In this report, three parameters are considered: ride comfort, driving safety, and braking. Each parameter includes several evaluation factors (Figure 3).

Figures 4–6 show the maximum point of each of the relevant assessment factors. The percentage of the maximum points on the total score reflects the weighting of the factors. The evaluated cars are shown in Figure 7.

4 DESIGN REQUIREMENTS

On the front axle of vehicle, there are several different requirements to be able to influence and tune the entire spectrum of vehicle characteristics in the most effective way. This chapter discusses the design requirements. The layout, technical data and kinematics characters, weights, and road tests results are shown in the comparison of different classes of vehicles and axle systems.

4.1 Front suspension system

The front suspension systems in common use are as follows:

- Transverse leaf spring front suspension system
- McPherson front suspension system

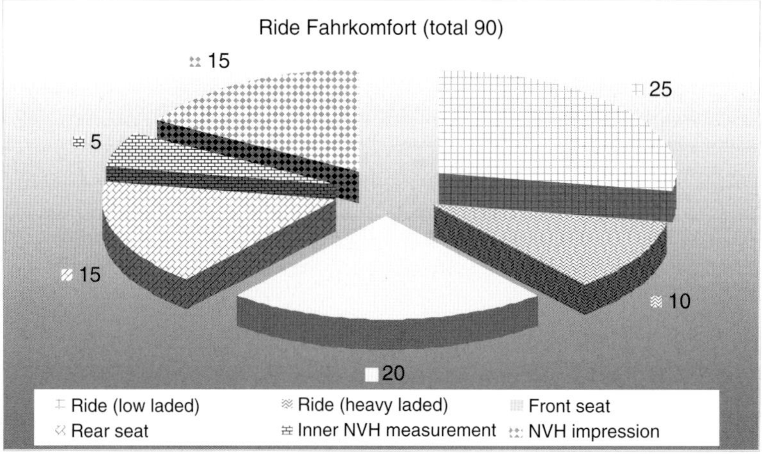

Figure 4. Factors of ride comfort (Ostmann, 2001). (Data from Ostmann, 2001, Auto Motor und Sport.)

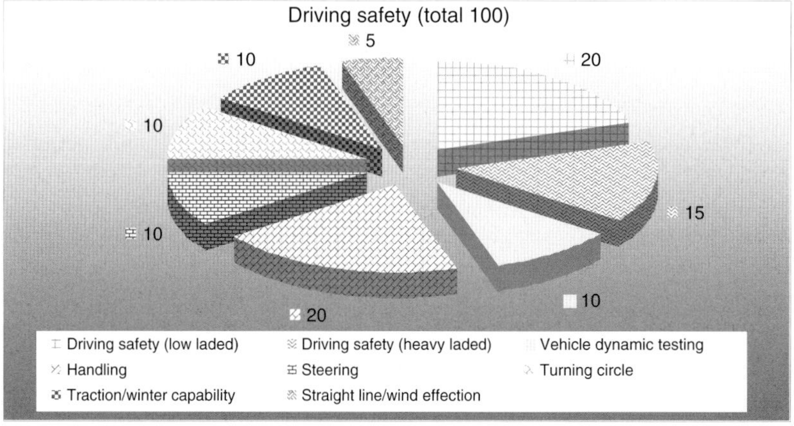

Figure 5. Factors of driving safety (Ostmann, 2001). (Data from Ostmann, 2001, Auto Motor und Sport.)

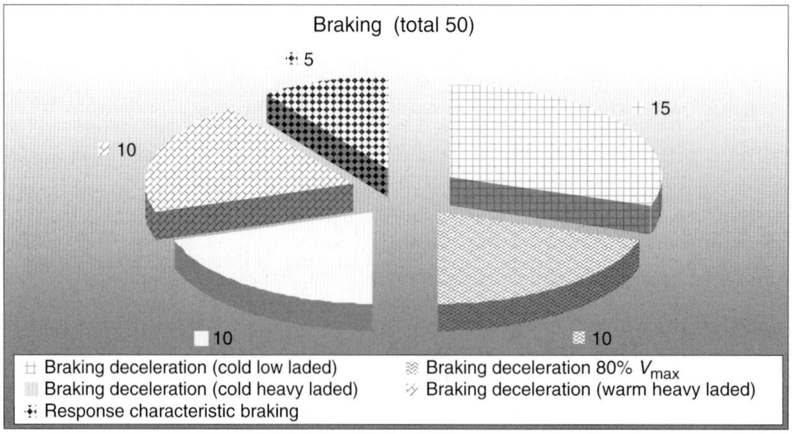

Figure 6. Factors of braking (Ostmann, 2001). (Data from Ostmann, 2001, Auto Motor und Sport.)

A~B segment	C-segment	D-segment
Mercedes A140, Ford Fusion, Nissan Micro, Mazda 2, Honda Jazz, Seat Ibiza, Ford Fiesta, Opel Corsa, Ford Ka, Renault Twingo, Citroen C3, Renault Clio, VW Polo, Fiat Punto, Audi A2, Mini One, Peugeot 206, Toyota Yaris, Opel Astra, VW Golf, Ford Focus, VW Lupo, Skoda Fabia, Seat Arosa, Fiat Seicento, Hynudai Getz, Daewoo Kalos	Honda Accorda, Mazda 6, Toyota Avensis, Audi A3 TDI, Alfa 147, Mercedes C180, BMW 318, Audi A4, Saab 9-3, Peugeot 307, Fiat Stilo, VW Golf TDI, Mini Cooper, Ford Mondeo, Renault Laguna, Mercedes C220, Renault Megnae, Opel Vectra, VW Passat, Toyota Corolla, Opel Omega, Saab 9-5, BMW 325, Citroen C5, Volvo S60	Audi A8, VW Phaeton, Mercedes E320, Mercedes S320, Audi A8, Mercedes S350, BMW 735, Jaguar XJ8, Mercedes S600, BMW 760, BMW 740d, Mercedes S400, BMW 745i, Mercedes S500, Mercedes E430, BMW 540i, VW Passat W8, Lexus LS430, Mercedes S430, BMW 740i, Lexus GS430, Jaguar S-Type, Audi A6

Figure 7. Evaluated cars.

- Double wishbone front suspension system with decomposed lower link
- Double wishbone front suspension system
- Four-link front suspension system.

In the following sections, the various front suspension systems are discussed.

4.1.1 Transverse leaf spring front suspension system

Figure 8 shows a transverse leaf spring front suspension system. It is composed of a laterally disposed leaf spring, damper, lower control arm, and knuckle. The function of the stabilizer (or antisway or antiroll) bar, possibly, could also be fulfilled by the transverse leaf spring. Thus, this arrangement can have a relative simple construction compared to other systems. A more detailed description can be found in the article by Chang (2006).

4.1.2 McPherson front suspension system

With this suspension, the wheel is controlled by a lateral, or track control, arm below the wheel center (usually triangular control arm), a nominally vertical strut and a tie rod. The advantages of this system are low unsprung mass, a strong support, low load distribution, and minimal space requirements. Named after its inventor, the construction was continually developed and is now the standard arrangement for many vehicles up to the middle class (Figure 9).

4.1.3 McPherson front suspension system with two lower links

A system with two separate lower links can be seen in Figure 10. The two ball joints are located immediately next to one another, which can lead to packaging difficulties.

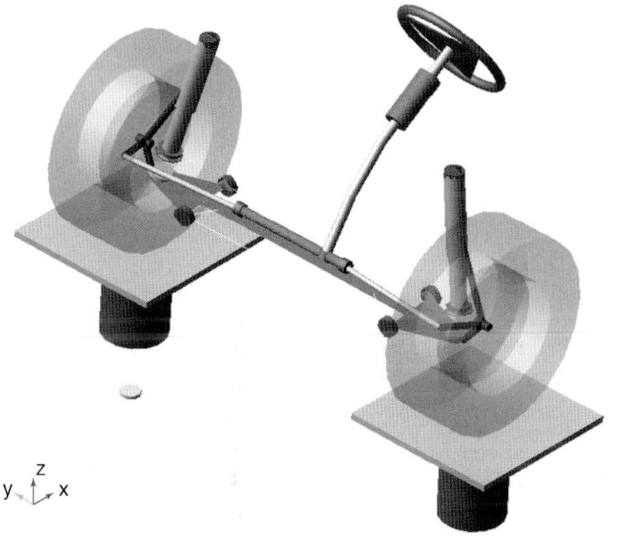

Figure 8. Transverse leaf spring front suspension system (Chang, 2006). (From Chang, 2006. Reproduced by permission of Ming-Yen Chang.)

Figure 9. McPherson front suspension system with single lower arm (triangular control arm).

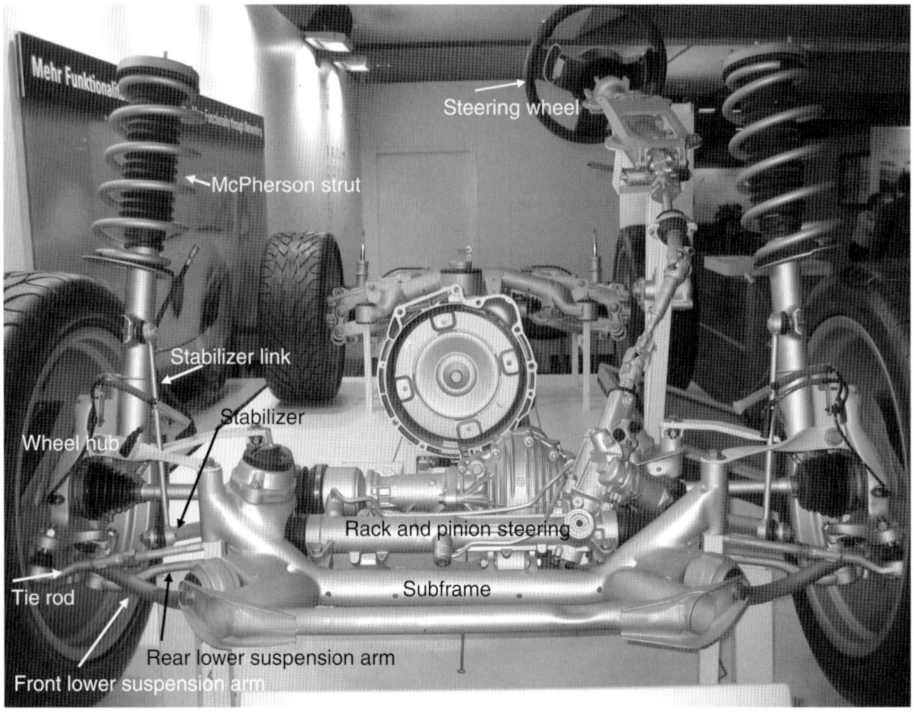

Figure 10. McPherson front suspension system with two lower links. (Reproduced with permission from Henning Wallentowitz. © Henning Wallentowitz.)

This problem can be eliminated using a "double joint" in which the ball joints are mounted directly above one another with their ball studs facing each other.

4.1.4 Double wishbone front suspension system

If a wheel is suspended by two lateral links, one of the links must be above the wheel center and one of the links must be below the wheel center in order to resist all of the forces and moments acting on the wheel (Figure 11a). In case of modern designs, a third lateral link is required to control toe and steering. This configuration is referred to as double lateral link suspension or, more commonly, double wishbone suspension (Figure 11b).

4.1.5 Four-link front suspension system

Figure 12 shows the four-link front suspension system of the Audi A4 and other Audi Ax-cars. In this arrangement, the wheel is guided by four rod-shaped links and the steering tie rod, with the suspension strut reacting the vertical load. The design principle of the A4-axis offers by the independent guiding elements a high potential to optimize the comfort and kinematic and compliance properties.

4.2 Weight

Following a survey of front suspension system from 46 vehicles, the mass reduction potential of the different suspension systems can be demonstrated. The spectrum of different vehicles ranges from compact cars to limousines. The calculated weight is according to the definition shown in Figure 13. For an independent suspension system, the total weight is the sum of the left and right corners.

The mass requirement for the allowable axle loads of different vehicle classes is shown in Figure 14. The red curve is the regression curve, which is calculated from the different masses of the front suspension systems for the permissible axle loads. The boundaries define the demands in Figure 14, which lie below the regression curve. The different vehicle classes are shown by different colors.

4.3 Driving behavior

In this section, the driving evaluation is discussed. On the basis of these assessments, the requirements for the various vehicle classes are determined (Figures 15–17).

The assessments are collected from "Auto Motor Sports 2001–2003." The results included 48 A~B-segment vehicles, 61 B-segment vehicles, and 35 D-segment vehicles.

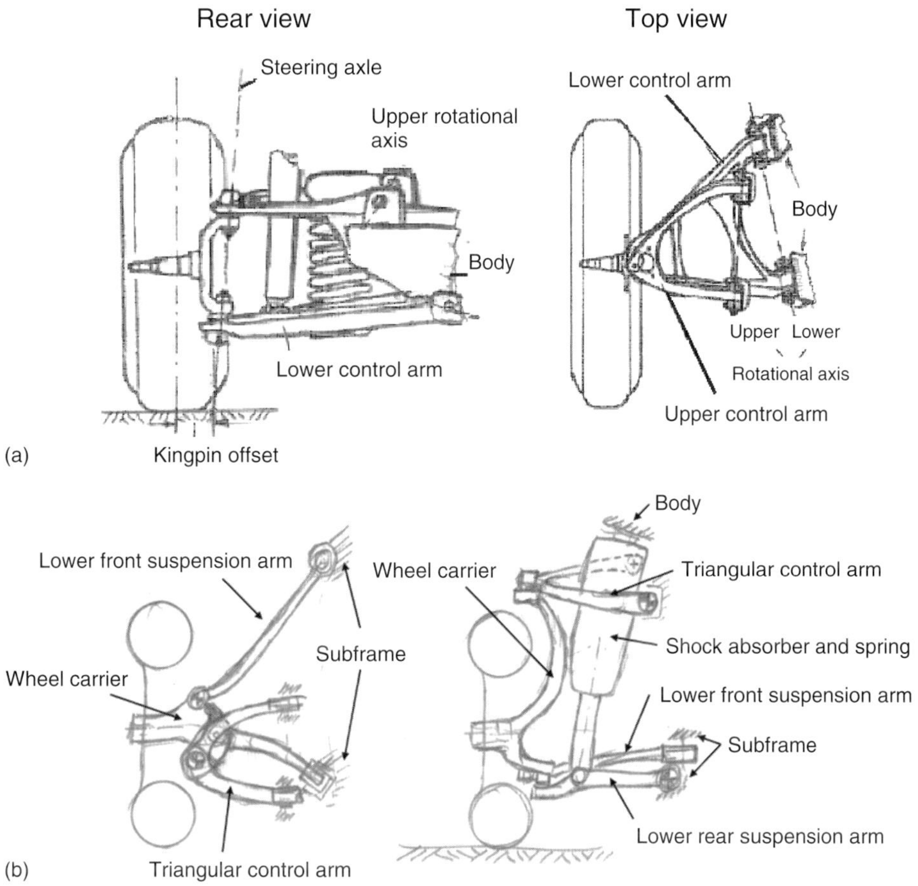

Figure 11. Different double wishbone suspensions. (a) Double wishbone suspension. (*Source*: Reproduced from Heissing, 2011). (Reproduced from Heissing, 2011. With kind permission of Springer Science+Business Media.) (b) Double wishbone suspension, similar to the Honda design. (Reproduced with permission from Henning Wallentowitz. © Henning Wallentowitz.)

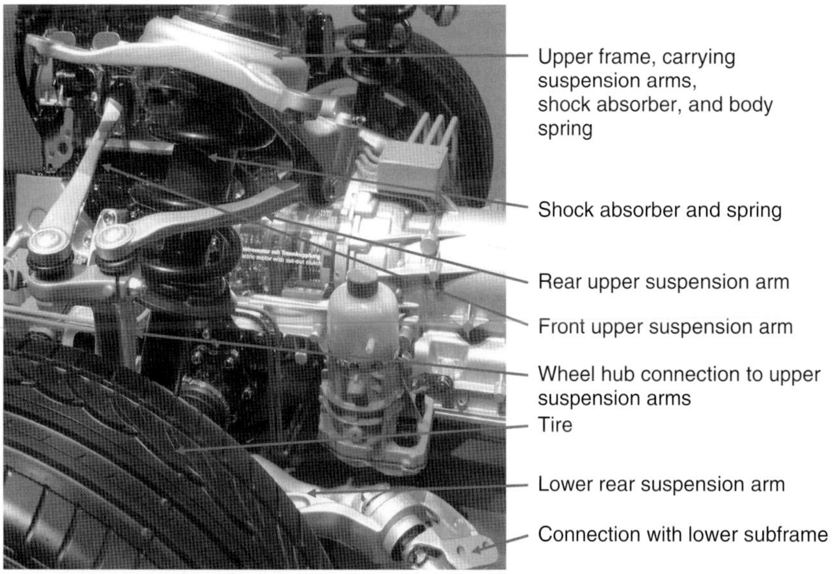

Figure 12. Audi four-link front suspension system. (Reproduced with permission from Henning Wallentowitz. © Henning Wallentowitz.)

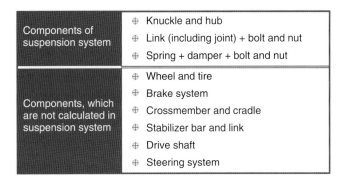

Components of suspension system	⊕ Knuckle and hub ⊕ Link (including joint) + bolt and nut ⊕ Spring + damper + bolt and nut
Components, which are not calculated in suspension system	⊕ Wheel and tire ⊕ Brake system ⊕ Crossmember and cradle ⊕ Stabilizer bar and link ⊕ Drive shaft ⊕ Steering system

Figure 13. Weight definition of suspension system.

There are three disciplines: ride comfort, driving safety, and braking. Each discipline includes various evaluation factors, such as Section 3.1 discusses. The results reveal a distribution of different factors in different vehicle classes (Figure 18). Owing to the different weighting of each factor, the percentage will be used, which is divided itself from the highest rating.

According to the distributions, the requirements of evaluation factors are defined and also the mean values are calculated. The lowest 10% of results are omitted. The rest defines the requirements (the gray area) and the average values (thick line) (Figure 18). All requirements of these three disciplines are clarified and are represented by the so-called spider charts (Figures 19–21) are the requirements of C-segment. Figure 7 shows the evaluated vehicles of different classes.

Each evaluation of different factors is determined by a specific note, whose distribution is shown in Figures 4–6. The spider charts display the percentage of each rating. Figure 22 shows the requirements of different vehicle classes. Except for turning circle, the requirement note is increasing with the vehicle class.

5 ESTIMATION OF DIFFERENT SUSPENSION SYSTEMS

In the following, QFD is introduced, whereby the engineer targets can be set systematically. According to the results of QFD, the potential marketability of various front suspension systems and vehicle class is evaluated.

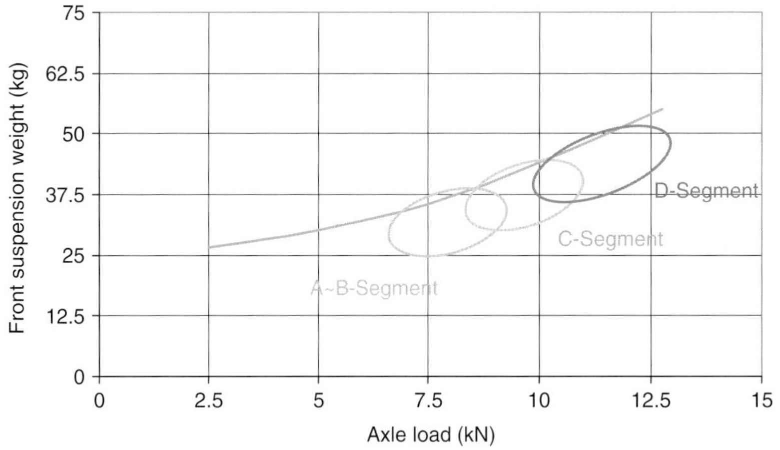

Figure 14. Requirement of the suspension system mass. (Reproduced with permission from Berkefeld, Görich, and Söffge, 1996. © Porsche AG.)

Model		Fiesta	Fusion	A2	Polo	A-Class	A3	Golf	Focus
Cylinder capacity (cc)		1400	1400	1422	1390	1397	1896	1598	1800
Curb	(kg)	1095–1152	1150–1234	1030	1025–1190	1295	1184–1359	1200	1105
GMV	(kg)	1520	1605	1545	1075	1558	1770	1620	1094
Front axle load (GVM)	(kg)		765–840		840		970		
Rear axle load (GVM)	(kg)		765–840		780		870		

Figure 15. Mass and axle load data of A~B-segment.

Model		C-Class	Mondeo	Passat	A4	Accord
Cylinder capacity (cc)		2148	2500	1984	2393	1998
Curb	(kg)	1535	1450	1542–1700	1485	1393–1488
GVM	(kg)	2015	1975	2110	2030	1920
Front axle load (GVM)	(kg)		1070	1090		
Rear axle load (GVM)	(kg)		1015	1080		

Figure 16. Mass and axle load data of C-segment.

Model		A6	E-Class	S-Class	5er-Series	7er-Series
Cylinder capacity (cc)		2496	2685	3724	2494	3600
Curb	(kg)	1560	1665	1810	1575	1935
GVM	(kg)	2110	2200	2340	2010	2440
Front axle load (GVM)	(kg)				985	1165
Rear axle load (GVM)	(kg)				1125	1335

Figure 17. Mass and axle load data of D-segment.

5.1 Quality function deployment

QFD is a matrix and table-driven tool for off-line analysis, communication, and planning system of different procedures; through these procedures, emotional quality experience—"language of the customer"—can be translated into the technical language of engineers, developers, and designers. The procedures also reveal the critical quality characteristics and decision support available in all product development phases.

The QFD methodology is based on the typical QFD analysis principle of the preparation and determination for lists and evaluation forms of characteristic sets (product parameters) and, characteristically, the direct comparison of these two parameters lists, which stretch a correlation matrix out. Through this process, two requirements can be united into one: data sets, more on customer requirements—what should be achieved—the other side more on company needs—how can this be achieved, as illustrated in Figure 23.

5.2 Analysis of design parameters with the help of the QFD

To analyze the potential of various front axles, the relations between the kinematics data and customer requirements are determined. The first step is to discover customer needs and to calculate their weights (the ratings).

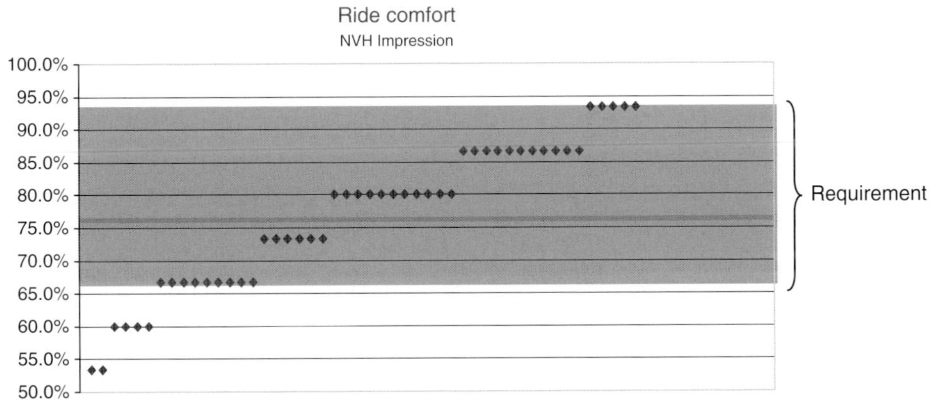

Figure 18. Distribution of NVH impression.

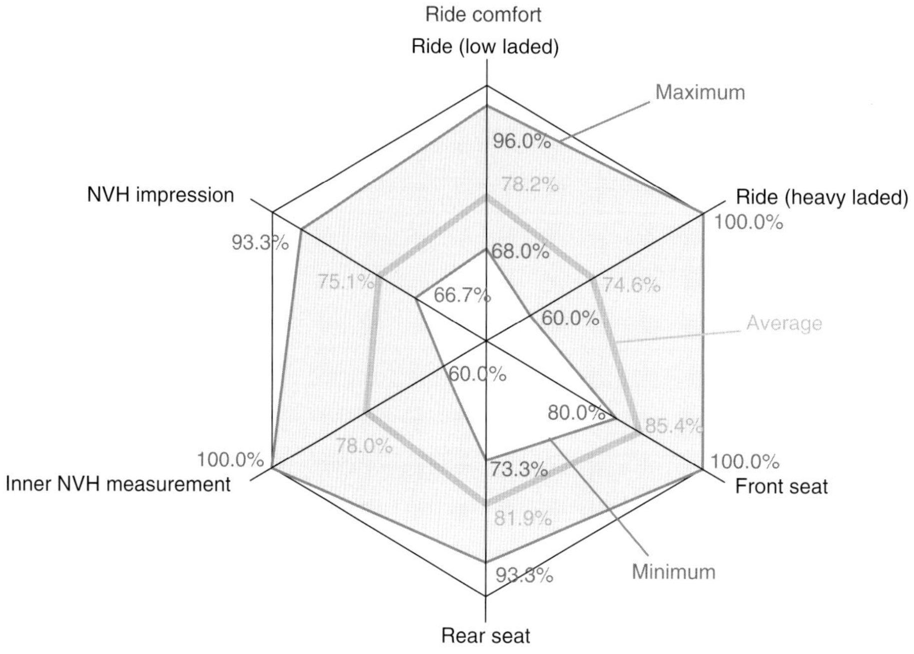

Figure 19. Requirements of ride comfort of C-segment.

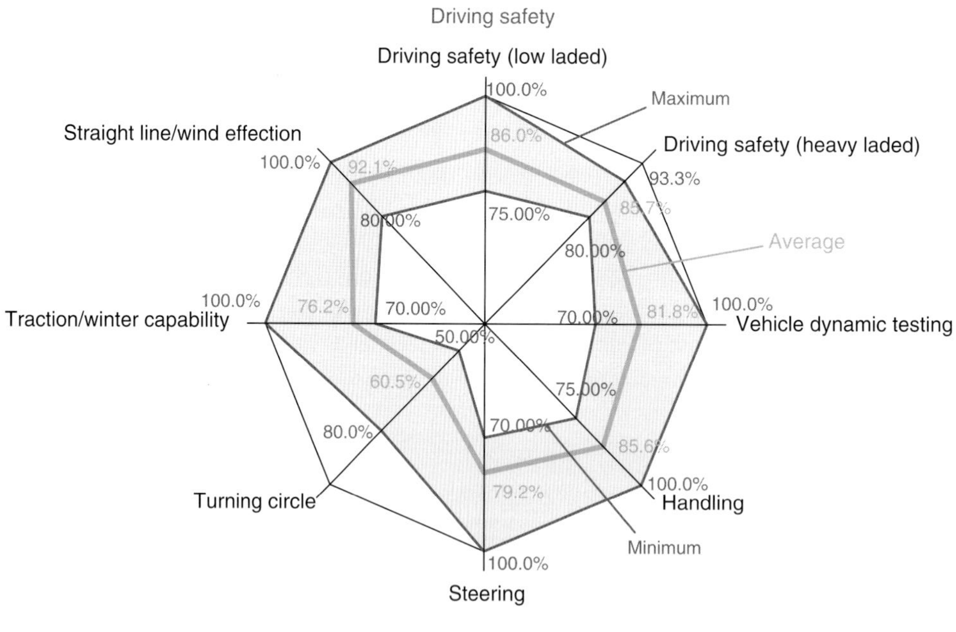

Figure 20. Requirements of driving safety of C-segment.

The customer requirements can be determined by various criteria: ride comfort, driving safety, driving fun, and so on, as the following shows (Figure 24).

The next step is to determine the weightings of each requirement. In this chapter, a comparison matrix is used. It is a relatively complicated method, but it is an objective solution.

Figure 25 shows an example. All requirements are listed in Tables A and B. There are two requirements compared with each other. If A is more important than B, it gets at

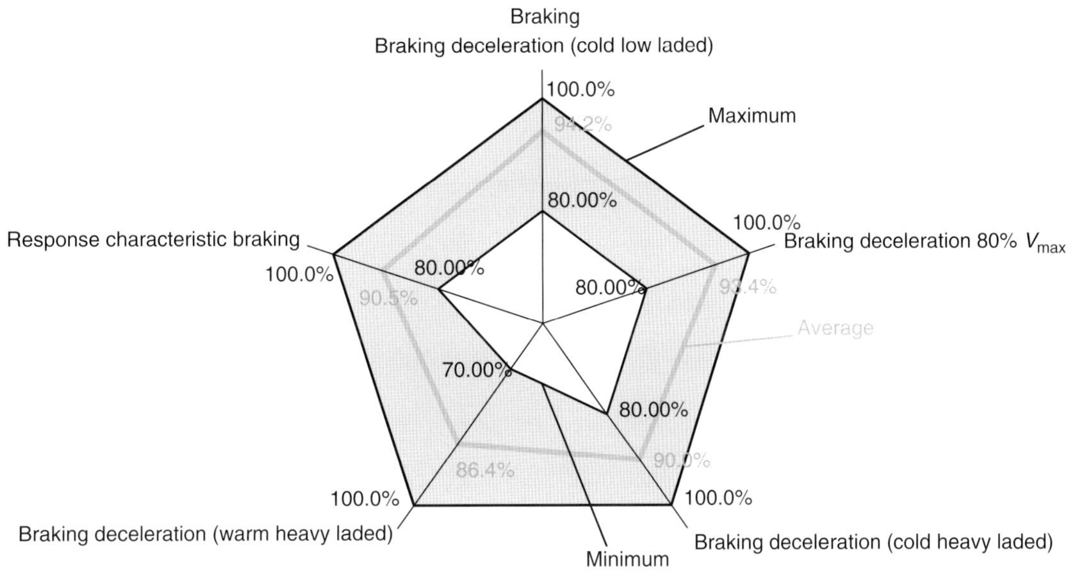

Figure 21. Requirements of braking of C-segment.

Vehicle class		A~B-Segment	C-Segment	D-Segment
Ride comfort				
Ride (low laded)	25	15–24	17–24	19–24
Ride (heavy laded)	10	6–10	6–10	7–10
Front seat	20	13–20	16–20	17–20
Rear seat	15	10–14	11–14	13–15
Inner NVH measurement	5	3–5	3–5	4–5
NVH impression	15	10–14	10–14	11–15
Total	**90**			
Driving safety				
Driving safety (low laded)	20	12–18	15–20	17–20
Driving safety (heavy laded)	15	10–14	12–14	12–15
Vehicle dynamic testing	10	6–10	7–10	8–10
Handling	20	14–20	15–20	17–20
Steering	10	6–10	7–10	8–10
Turning circle	10	5–9	5–8	4–6
Traction/winter capability	10	6–9	7–10	6–10
Straight line/wind affection	5	4–5	4–5	4–5
Total	**100**			
Braking				
Braking deceleration (cold low laded)	15	10–15	12–15	14–15
Braking deceleration 80% V_{max}	10	6–10	8–10	9–10
Braking deceleration (cold heavy laded)	10	6–10	8–10	9–10
Braking deceleration (warm heavy laded)	10	5–10	7–10	9–10
Response characteristic braking	5	4–5	4–5	4–5
Total	**50**			

Figure 22. Requirements of driving behavior for different classes.

upper right half 2 points. If they have the same importance, then it gets 1 point, otherwise there are no points. Similarly, the lower left part is evaluated. Thereby, the absolute and relative weightings are calculated.

The customer opinion is requested by a questionnaire survey. The weightings are the averages of the survey results. Figure 26 shows the weightings of various classes.

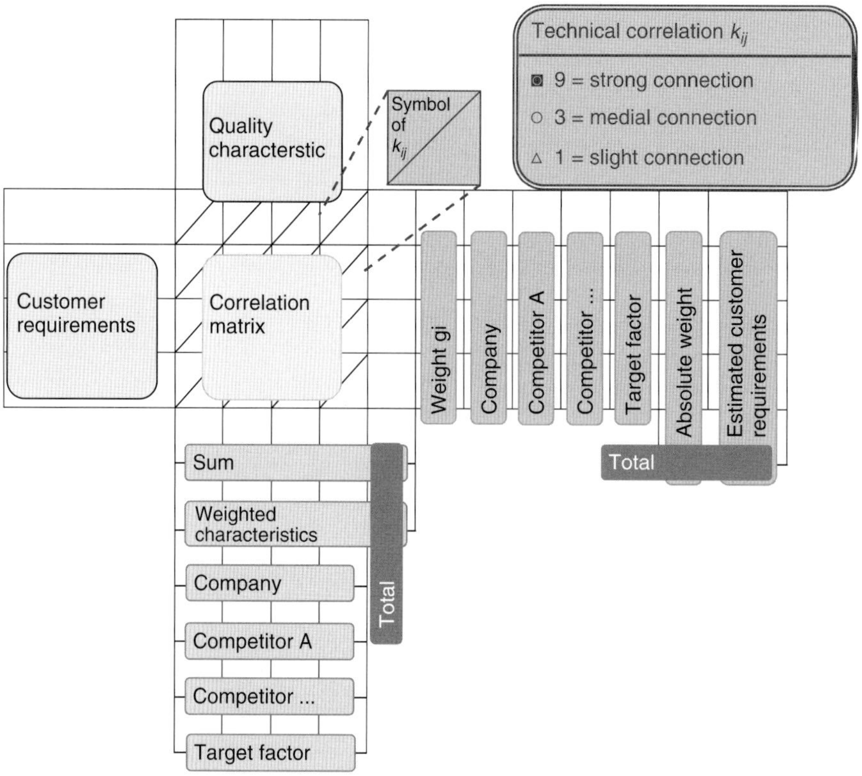

Figure 23. "Quality plan" or "house of quality" to translate the customer requirements into company language (Bors, 1995). (From Bors, 1995. Reproduced by permission of Carl Hanser Verlag.)

Ride comfort	Low disturbance by road unevenness
	Low impact in the steering
	Limited body roll during cornering
	Limited brake and traction influence
	Good comfort by loaded
Driving safety	High driving stability
	Precise steering
	Limited crosswind influence
	High safety by loaded
Driving fun	Feelingful steering
Others	Low cost (fuel emission, maintenance, etc.)

Figure 24. Customer requirements.

5.3 Analysis of design parameters with the help of the ADAMS/car

The next step is to complete the correlation matrix. Analysis of design parameters with the help of the QFD determined according to the rules described in Chapter 5-2 and with the help of ADAMS/car. Thus, the absolute and relative estimations of the suspension characteristics are determined. To objectively determine the correlation matrix, the ADAMS/car simulations are applied. It can also be done by K&C measurement, if the prototype is ready.

By estimation of different suspension characteristics and simulations, the correlation matrix can be generated. Figures 28, 32, and 33 show the QFD worksheet of different classes. The meanings of the symbols are shown in Figure 23.

5.4 Evaluation

The evaluation of the functional properties of suspensions is discussed in this section to learn more about appraisal of front suspension system.

5.4.1 QFD worksheet

Figures 27, 30, and 31 show the technical function characteristics of different suspension systems, which are also the technical language of engineers. With the help of these tables and the described properties, the evaluation of

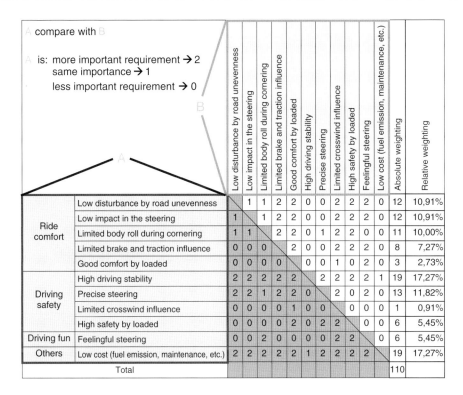

A compare with B

A is: more important requirement → 2
same importance → 1
less important requirement → 0

	A \ B	Low disturbance by road unevenness	Low impact in the steering	Limited body roll during cornering	Limited brake and traction influence	Good comfort by loaded	High driving stability	Precise steering	Limited crosswind influence	High safety by loaded	Feelingful steering	Low cost (fuel emission, maintenance, etc.)	Absolute weighting	Relative weighting
Ride comfort	Low disturbance by road unevenness		1	1	2	2	0	0	2	2	2	0	12	10,91%
	Low impact in the steering	1		1	2	2	0	0	2	2	2	0	12	10,91%
	Limited body roll during cornering	1	1		2	2	0	1	2	2	0	0	11	10,00%
	Limited brake and traction influence	0	0	0		2	0	0	2	2	2	0	8	7,27%
	Good comfort by loaded	0	0	0	0		0	0	1	0	2	0	3	2,73%
Driving safety	High driving stability	2	2	2	2	2		2	2	2	2	1	19	17,27%
	Precise steering	2	2	1	2	2	0		2	0	2	0	13	11,82%
	Limited crosswind influence	0	0	0	0	1	0	0		0	0	0	1	0,91%
	High safety by loaded	0	0	0	0	2	0	2	2		0	0	6	5,45%
Driving fun	Feelingful steering	0	0	2	0	0	0	0	2	2		0	6	5,45%
Others	Low cost (fuel emission, maintenance, etc.)	2	2	2	2	2	1	2	2	2	2		19	17,27%
Total													110	

Figure 25. Comparative table of customer requirements.

	Requirements	A~B-segment	C-segment	D-segment
Ride comfort	Low disturbance by road unevenness	7,27%	8,53%	9,72%
	Low impact in the steering	7,13%	8,46%	9,16%
	Limited body roll during cornering	9,30%	9,79%	10,42%
	Limited brake and traction influence	8,11%	8,32%	9,79%
	Good comfort by loaded	8,74%	7,76%	8,74%
Driving safety	High driving stability	13,78%	13,85%	13,92%
	Precise steering	9,86%	10,98%	11,05%
	Limited crosswind influence	6,85%	7,76%	8,53%
	High safety by loaded	9,79%	9,23%	9,30%
Driving fun	Feelingful steering	6,01%	6,85%	6,36%
Others	Low cost (fuel emission, maintenance, etc.)	13,15%	8,46%	3,01%

Figure 26. Weightings of different classes.

customer requirements and various parameters are obtained (Figure 28). Thus, the targets are set. The sum of the weightings from customer requests and customer appraisals shows the market potential, and the sum of the weightings from the evaluation of various suspension characteristics shows the design advantages (Figure 29). The evaluation is scored from 100-points, where 100 is the best value.

As the A~B-segment, Figures 32 and 33 show the QFD worksheet of C- and D-segments.

5.4.2 *Mass evaluation*

The axle mass is estimated by the sum of all component masses. The component mass data of various vehicle classes are collected. The missing data are replaced by assumed

McPherson		Characters	Evaluation*)
Geometry	Kingpin offset	Determination by kingpin angle and space	5
	Disturbing force level arm	Determination by kingpin angle and offset	3
	Tire load level arm	Determination by kingpin offset, caster offset, caster, and kingpin angle	3
	Longitudinal travel at wheel center	Large longitudinal pole distance behind and below the wheel center	3
	Braking torque compensation angle	Longitudinal polepositiongiven by geometric definition of longitudinal travel at wheel center	3
	Ratio factor	Spring, damper, stabilizer barat knuckle	5
Kinematics	Toe	Change of toe can be tuned by longitudinal and vertical position of tie rod	5
	Camber	Degressive increase in negative camber on the compression wheel travel	3
	Track change	Can be kept low during compression, with strong progressive increase during extenssion	3
	Roll center	Height is tunable, however, the roll center is basically dropped during compression	3
Compliance	Under drive force	Coupling to the kinematics and compliance design under braking forces	4
	Under traction moment	Coupling to the kinematics and compliance design under braking forces	4
	Under brake force	Freely tuned by elastic parts and location of the tie rod	5
	Under lateral force	Freely tuned by elastic parts	5

*) Evaluation: 0 inopportune, 3 medium, 5 opportune

Figure 27. Evaluation of function characters of McPerson suspension system (Berkefeld, Görich, and Söffge, 1996). (Reproduced with permission from Berkefeld, Görich, and Söffge, 1996. © Porsche AG.)

values, which were modified with similar components, for example, rubber bearings. This allows the axle mass to be estimated from different suspensions. The assumptions are compared with the data in Figures 15–17, in order to ensure the accuracy. Figure 34 shows the results.

The evaluation of suspension mass (or the weight, the only different factor is the acceleration of gravity) is calculated by following formula:

$$B = \left(1 - \frac{(W - W_n)}{W_n}\right) \times 100$$

B: Evaluation

W: weight of different suspension system

W_n: weight of lightest suspension system

5.4.3 Overall evaluation

The overall evaluation is done by application intention and the results of the QFD and mass assessment. The market potential is a key position in the design. In this chapter, the mass reduction is also an emphasis, so it also occupies a large part. Of course, the design advantages are also important, but they can be overcome by abundant design experience. Figure 35 shows the weighting distribution.

In the following, you can see the overall evaluation of various front suspensions, as shown in Figures 36, 37 and 38. Although the market potential and the design advantage of the transverse leaf spring suspension is not as good as the McPherson suspension, it is superior to A~B-segment by its potential for mass reduction (this also means weight reduction). The Chapter title "Transverse Leaf Spring Suspension System" explains the design idea and the mass optimization of the transverse leaf spring suspension system.

After the overall evaluation, the four-linkage suspension system is shown as the best (or to name it optimum) for the C- and D-segments (Figures 37 and 38).

6 TRANSVERSE LEAF SPRING SUSPENSION SYSTEM

The axle mass (equivalent the weight) can be reduced by integration of functions and components. Another way is the optimization of the single components. In this chapter, the design concept of transverse leaf spring suspension system and the methodology of mass reduction are introduced.

6.1 Function integration

Figure 8 shows a mass reduction concept of front suspension: a McPherson front suspension comprising track

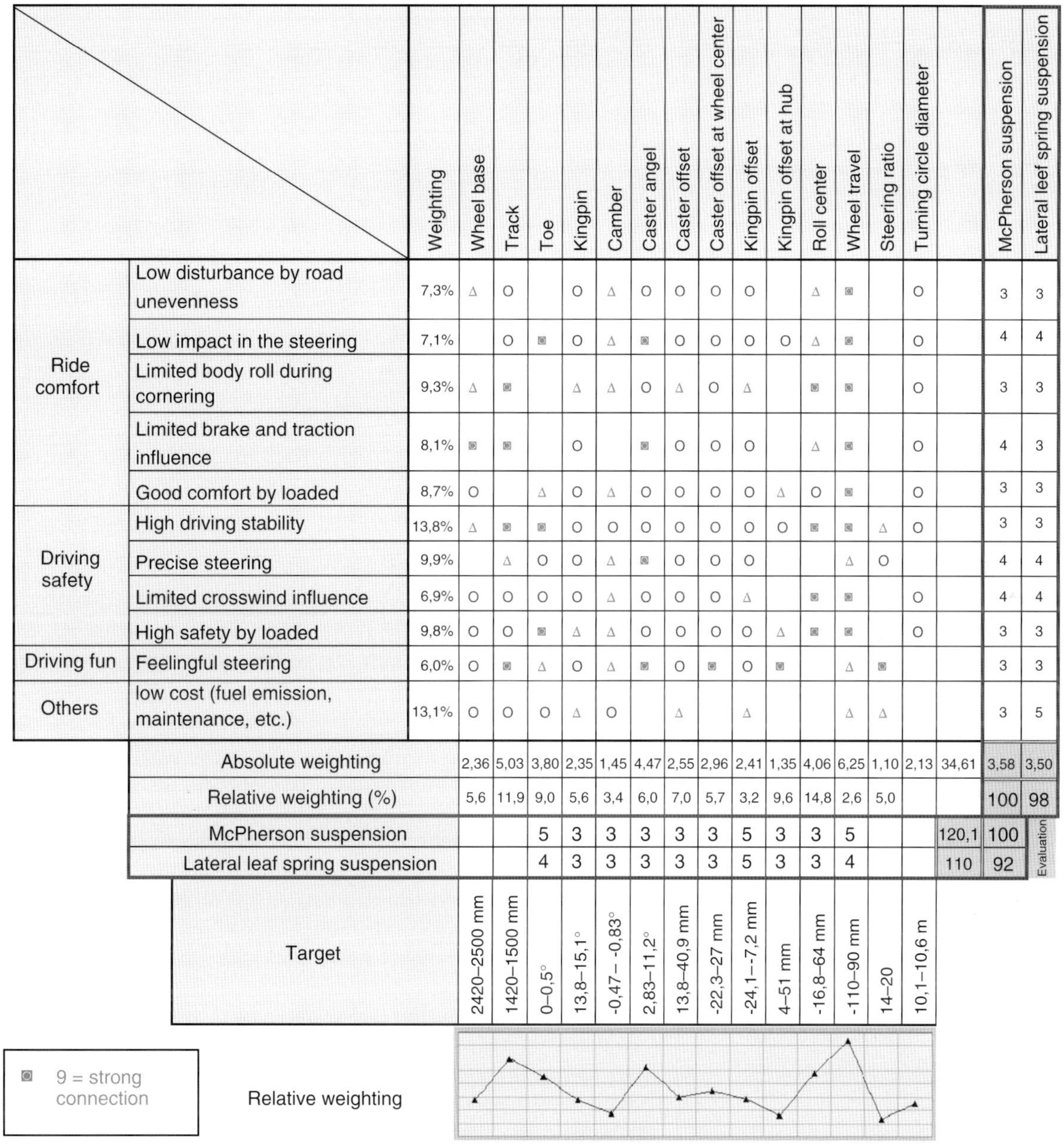

Figure 28. QFD work sheet of A~B-segment.

Figure 29. Evaluation.

control arms, stabilizer system, transverse leaf springs, dampers, knuckle, hub units, and drive shafts. To reduce mass (respectively weight), the suspension arms, the stabilizer system, and the coil spring can be partly replaced by a redesigned transverse leaf spring, as shown in Figure 39.

6.2 Optimization of component

An appropriate optimization of components can decrease the mass efficaciously. However, an applicable testing pattern, inclusive of component, system, and vehicle level and also application of CAE simulation are both crucial.

Double wishbone suspension

		Character	Evaluation*)
Geometry	Kingpin offset	Determination by kingpin angle and space	5
	Disturbing force level arm	Determination by kingpin angle and offset, small kingpin angle is possible	4
	Tire load level arm	Determination by kingpin offset, caster offset, caster, and kingpin angle	4
	Longitudinal travel at wheel center	Medium longitudinal pole? distance behind and below the wheel center	3
	Braking torque compensation angle	Longitudinal pole position given by geometric definition of Longitudinal travel at wheel center	4
	Ratio factor	Spring, damper, Stabilizer baratlateral linkage	3
Kinematics	Toe	Change of toe can be tuned by longitudinal and vertical position of tie rod	5
	Camber	Degressive increase in negative camber on the compression wheel travel (shorter upper arm and longer under arm)	4
	Track change	With shorter upper arm and corresponding spatial location to limited track change	4
	Roll center	Height is tunable, limited change during compression during compression	4
Compliance	Under drive force	Coupling to the kinematics and compliance design under braking forces	4
	Under traction moment	Coupling to the kinematics and compliance design under braking forces	4
	Under brake force	Freely tuned by elastic parts and location of the tie rod	5
	Under lateral force	Freely tuned by elastic parts	5

*) Evaluation: 0 inopportune, 3 medium, 5 opportune

Figure 30. Evaluation of function characters of double wishbone suspension system (Berkefeld, Görich, and Söffge, 1996). (Reproduced with permission from Berkefeld, Görich, and Söffge, 1996. © Porsche AG.)

Multilink suspension

		Character	Evaluation*)
Geometry	Kingpin offset	Determination by the poles of the link plan whose connecting line forms by kingpin axis	5
	Disturbing force level arm	Determination by kingpin angle and offset (kingpin angel is freely selectable within limits)	5
	Tire load level arm	Determination by kingpin offset, caster offset, caster, and kingpin angle	5
	Longitudinal travel at wheel center	Medium longitudinal pole distance behind and below the wheel center	3
	Braking torque compensation angle	Longitudinal pole positiongiven by geometric definition of longitudinal travel at wheel center	4
	Ratio factor	Spring, damper, stabilizer barat knuckle	3
Kinematics	Toe	Change of toe can be tuned by position and length of linkage	5
	Camber	Degressive increase in negative camber on the compression wheel travel (shorter upper arm and longer under arm)	4
	Track change	With shorter upper arm and corresponding spatial location to limit track change	4
	Roll center	Height is tunable, limited change during compression	4
Compliance	Under drive force	Freely tuned by pole position and elastic parts	5
	Under traction moment	Coupling to the kinematics and compliance design under driving forces	4
	Under brake force	Freely tuned by pole position and elastic parts	5
	Under lateral force	Freely tuned by pole position and elastic parts	5

*) Evaluation: 0 inopportune, 3 medium, 5 opportune

Figure 31. Evaluation of function characters of multilink suspension system (Berkefeld, Görich, and Söffge, 1996). (Reproduced with permission from Berkefeld, Görich, and Söffge, 1996. © Porsche AG.)

		Weighting	Wheel Base	Track	Toe	Kingpin	Camber	Caster Angel	Caster offset	Caster offset at wheel center	Kingpin offset	Kingpin offset at hub	Roll center	Wheel travel	Steering ratio	Turning circle diameter		McPherson suspension	Lateral leef spring suspension	Double-linkage with strut suspension	Double wishbone suspension	Modified four-linkage suspension	Four-linkage suspension
Ride comfort	Low disturbance by road unevenness	8,5%	Δ	O		O	Δ	O	O	O	O		Δ	■		O		3	3	4	4	4	5
	Low impact in the steering	8,5%		O	■	O	Δ	■	O	O	O	O	Δ	O		O		4	4	4	4	5	5
	Limited body roll during cornering	9,8%	Δ	■		Δ	Δ	O	Δ	O	Δ		■	■		O		3	3	4	4	4	5
	Limited brake and traction influence	8,3%	■	■		O		■	O	O	O		Δ	■		O		4	3	4	4	5	5
	Good comfort by loaded	7,8%	O	O	Δ	O	Δ	O	O	O	O	Δ	O	■		O		3	3	4	5	4	5
Driving safety	High driving stability	13,8%	Δ	■	■	O	O	O	O	O	O	O	■	■	Δ	O		3	3	4	5	5	4
	Precise steering	11,0%		Δ	O	O	Δ	■	O	O	O			Δ	O			4	4	3	5	5	4
	Limited crosswind influence	7,8%	O	O	O	O	Δ	O	O	O	Δ		■	■		O		4	4	4	5	4	5
	High safety by loaded	9,2%	O	O	■	Δ	Δ	O	O	O	O	Δ	■	■		O		3	3	4	5	5	4
Driving fun	Feelingful steering	6,9%	O	■	Δ	O	Δ	■	O	■	O	■		Δ	■			3	3	4	4	4	5
Others	Low cost (fuel emission, maintenance, etc.)	8,5%	O	O	O	Δ	O		Δ		Δ			Δ	Δ			5	5	4	3	3	3
	Absolute weighting		2,27	5,11	3,80	2,45	1,36	4,83	2,64	3,16	2,48	1,46	4,14	6,39	1,17	2,21	43,486	3,53	3,45	3,89	4,42	4,43	4,50
	Relative weighting (%)		5,2	11,8	8,7	5,6	3,1	11,1	6,1	7,3	5,7	3,4	9,5	14,7	2,7	5,1		78	77	86	98	98	100

	Wheel Base	Track	Toe	Kingpin	Camber	Caster Angel	Caster offset	Caster offset at wheel center	Kingpin offset	Kingpin offset at hub	Roll center	Wheel travel	Steering ratio	Turning circle diameter		Evaluation
McPherson suspension			5	3	3	3	3	3	5	3	3	5			123,5	81
Lateral leef spring suspension			4	3	3	3	3	3	5	3	3	4			113,3	75
Double-linkage with strut suspension			5	4	3	4	4	4	5	3	3	5			136,6	90
Double wishbone suspension			5	4	4	4	4	4	5	4	4	3			143,8	86
Modified four-linkage suspension			5	5	4	5	5	5	5	4	4	3			130,8	95
Four-linkage suspension			5	5	4	5	5	5	5	5	4	4			151,7	100
Target	2500–2700 mm	1490–1520 mm	-0,1–0,6°	13,8–15,1°	-0,47–-0,83°	2,83–11,2°	13,8–40,9 mm	22,3–27 mm	-24,1–-7,2 mm	4–51 mm	-16,8–64 mm	-120–90 mm	14–20	10,5–11 m		

■ 9 = strong connection

Relative weighting

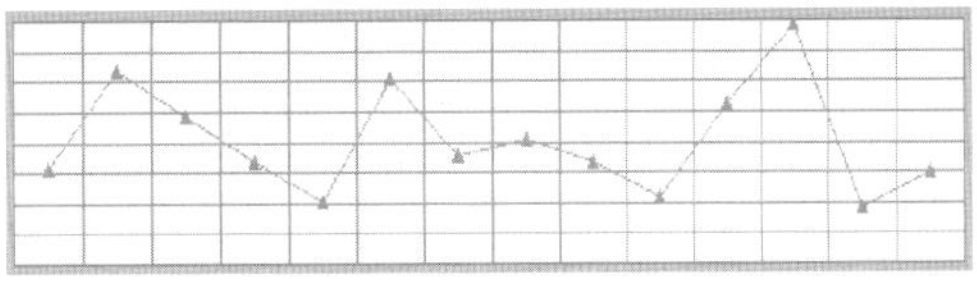

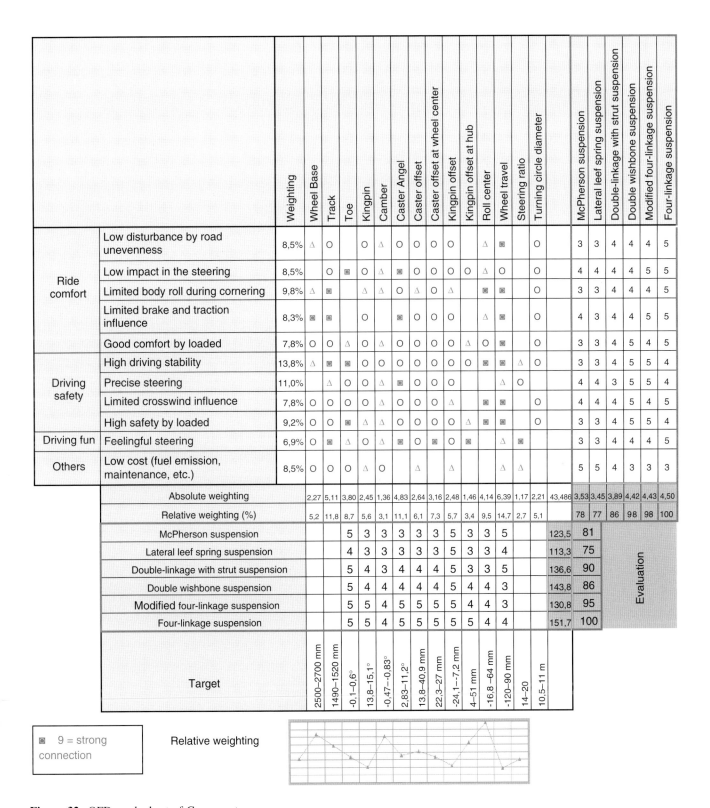

Figure 32. QFD work sheet of C-segment.

		Weighting	Wheel base	Track	Toe	Kingpin	Camber	Caster angel	Caster offset	Caster offset at wheel center	Kingpin offset	Kingpin offset at hub	Roll center	Wheel travel	Steering ratio	Turning circle diameter		McPherson suspension	Lateral leef spring suspension	Double-linkage with strut suspension	Double wishbone suspension	Modified four-linkage suspension	Four-linkage suspension
Ride comfort	Low disturbance by road unevenness	9,7%	Δ	O		O	Δ	O	O	O	O		Δ	■		O		3	3	4	4	4	5
	Low impact in the steering	9,2%		O	■	O	Δ	■	O	O	O	O	Δ	O		O		4	4	4	4	5	5
	Limited body roll during cornering	10,4%	Δ	■		Δ	Δ	O	Δ	O	Δ		■	■		O		3	3	4	4	4	5
	Limited brake and traction influence	9,8%	■	■		O		■	O	O	O		Δ	■		O		4	3	4	4	5	5
	Good comfort by loaded	8,7%	O	O	Δ	O	Δ	O	O	O	O	Δ	O	■		O		3	3	4	5	4	5
Driving safety	High driving stability	13,9%	Δ	■	■	O	O	O	O	O	O	O	■	■	Δ	O		3	3	4	5	5	4
	Precise steering	11,0%		Δ	O	O	Δ	■	O	O	O			Δ	O			4	4	3	5	5	4
	Limited crosswind influence	8,5%	O	O	O	O	Δ	O	O	O	Δ		■	■		O		4	4	4	5	4	5
	High safety by loaded	9,3%	O	O	■	Δ	Δ	O	O	O	O	Δ	■	■		O		3	3	4	5	5	4
Driving fun	Feelingful steering	6,4%	O	■	Δ	O	Δ	■	O	■	O	■		Δ	■			3	3	4	4	4	5
Others	Low cost (fuel emission, maintenance, etc.)	3,0%	O	O	O	Δ	O		Δ		Δ			Δ	Δ			5	5	4	3	3	3
	Absolute weighting		2,30	5,21	3,74	2,55	1,24	5,09	2,73	3,29	2,56	1,45	4,34	6,82	1,07	2,39	44,78	3,45	3,35	3,89	4,49	4,50	4,60
	Relative weighting (%)		5,1	11,6	8,4	5,7	2,8	11,4	6,1	7,4	5,7	3,2	9,7	15,2	2,4	5,3		75	73	85	98	98	100
	McPherson suspension				5	3	3	3	3	3	5	3	3	5			127,7	81	Evaluation				
	Lateral leef spring suspension				4	3	3	3	3	3	5	3	3	4			117,1	75					
	Double-linkage with strut suspension				5	4	3	4	4	4	5	3	3	5			141,3	90					
	Double wishbone suspension				5	4	4	4	4	4	5	4	4	3			134, 7	86					
	Modified four-linkage suspension				5	5	4	5	5	5	5	4	4	3			148,4	95					
	Four-linkage suspension				5	5	4	5	5	5	5	5	4	4			156,7	100					
	Target		2650–2850 mm	1490–1580 mm	-0,1–0,6°	3,6–6°	-0,5–-1°	2,83–11,2°	20–30 mm	-22,3–27 mm	-11–-0 mm	4–51 mm	45–100 mm	-125–130 mm	14–20	11–12 m							

■ 9 = strong connection

Relative weighting

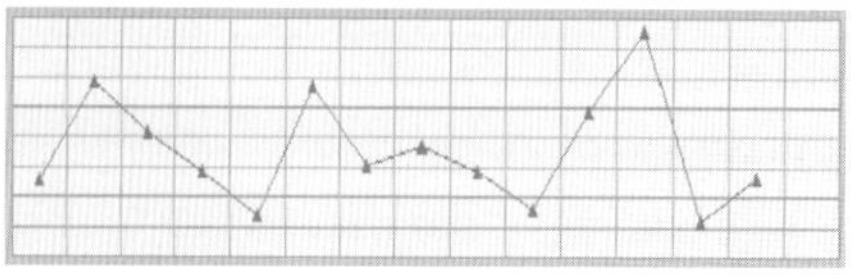

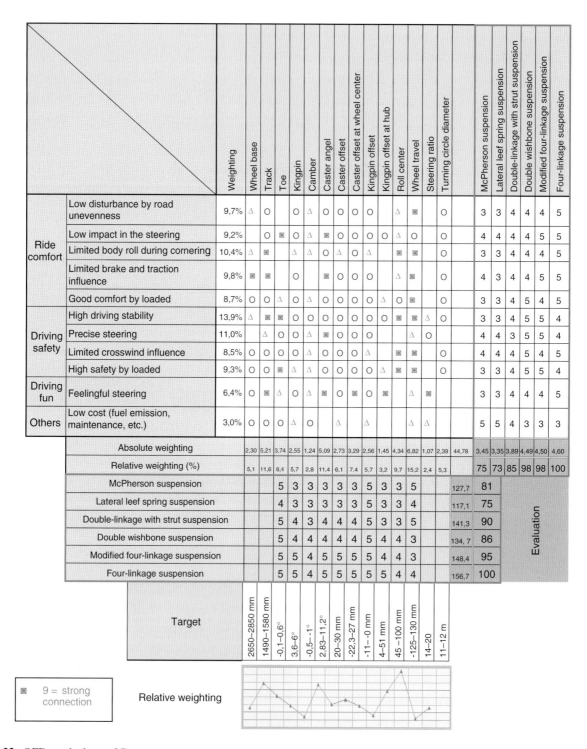

Figure 33. QFD work sheet of D-segment.

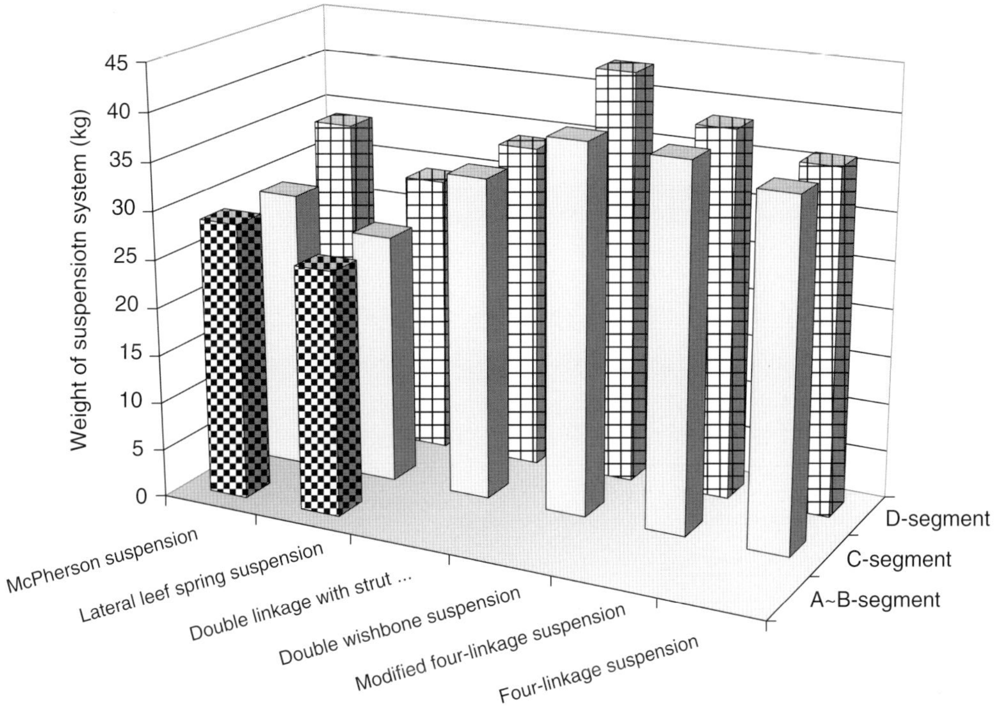

Figure 34. Masses of suspension systems.

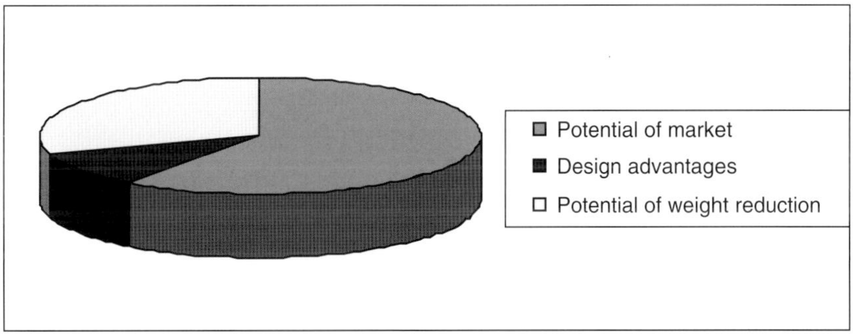

Figure 35. Weighting distribution.

Figure 40 shows the process of component design, which is an iterative process until all targets are met.

The "Smart" is used as a research model. In this new design proposal, the lower control arm, knuckle, stabilizer, and leaf spring are redesigned. The component models are analyzed with FEM software.

After the CAD and FEM simulation, the front suspension mass can be determined. The new transverse leaf spring front suspension is designed not only lighter than the McPherson front axle but also lighter than the original "Smart" front transverse leaf spring (first generation of this car). Figure 41 shows the mass reduction.

7 SUMMARY

The lightweight design has increasingly high priority in the automotive industry. A reduction of suspension component masses (respectively the weights) is not only helpful for energy consumption but it can also improve the driving behavior by reducing unsprung mass.

	Weighting	McPherson suspension	Lateral leef spring suspension
Potential of market	6	100	98
Design advantages	1	100	92
Potential of weight reduction	3	87	100
Total		**96**	**98**

Figure 36. Overall evaluation of A~B-segment.

	Weighting	McPherson suspension	Lateral leef spring suspension	Double-linkage with strut suspension	Double wishbone suspension	Modified four-linkage suspension	Four-linkage suspension
Potential of market	6	78	77	86	98	98	100
Design advantages	1	81	75	90	86	95	100
Potential of weight reduction	3	92	100	79	65	67	72
Total		83	83	85	87	89	92

Figure 37. Overall evaluation of C-segment.

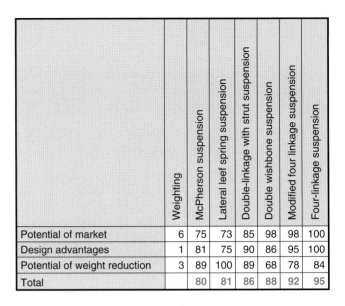

	Weighting	McPherson suspension	Lateral leef spring suspension	Double-linkage with strut suspension	Double wishbone suspension	Modified four linkage suspension	Four-linkage suspension
Potential of market	6	75	73	85	98	98	100
Design advantages	1	81	75	90	86	95	100
Potential of weight reduction	3	89	100	89	68	78	84
Total		80	81	86	88	92	95

Figure 38. Overall evaluation of D-segment.

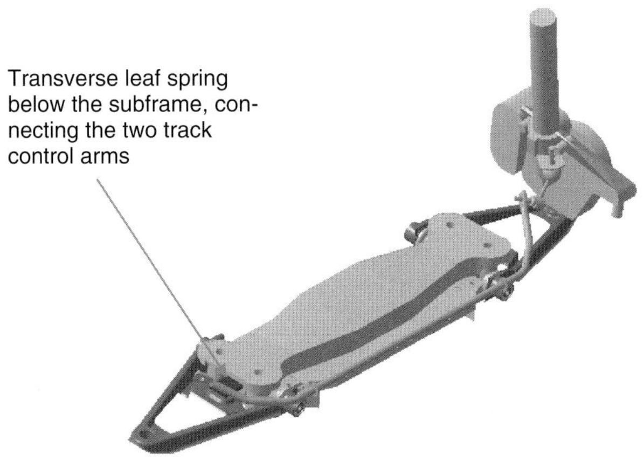
Transverse leaf spring below the subframe, connecting the two track control arms

Figure 39. Transverse leaf spring front suspension, redesigned front suspension system of the Daimler "Smart"-vehicle.

In this chapter, a systematic design process was developed. First, the technical data, masses (weights), technical characteristics, and driving behavior from different vehicles are collected and systematically evaluated. Hence, a comprehensive benchmark was created. From the benchmark data, the requirements of different design parameters have been determined.

Finally, the method of QFD was used to develop the targets systematically. From the results of the QFD, the potentials of various front suspension systems were evaluated for several vehicle classes. Because of the relatively high potential for mass reduction, a lightweight transverse leaf spring front suspension system is selected for further research for the A~B segment.

To decrease mass, the lower control arm, the stabilizer system, and the springs are replaced by a redesigned transverse leaf spring. The design proposal has been incorporated into a model of the vehicle "Smart" and simulated using ADAMS/car. Furthermore, the space has been tested and the optimized devices constructed. The front axle mass has been reduced by more than 20%. Good handling and ride comfort are the key reasons for taking up the four-link suspension as the suspension system for the C- and D-segments.

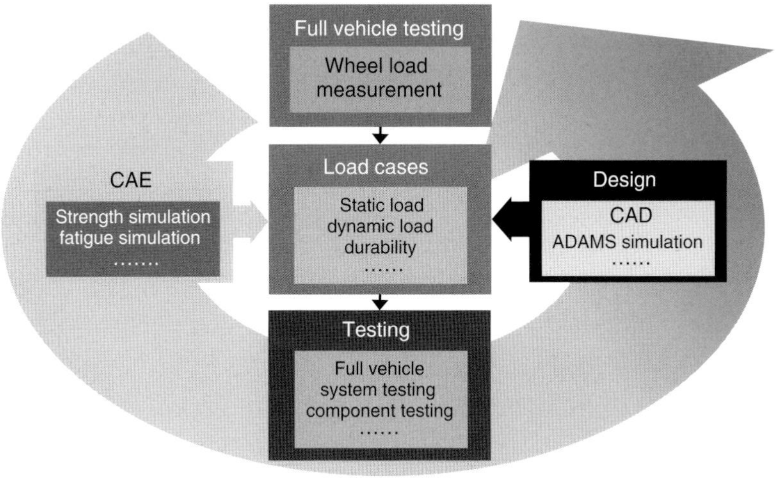

Figure 40. Component design process.

	Smart McPherson	Smart lateral leaf spring	Lateral leaf spring (redesign)
Front suspension weight (without steering and brake)	28,7 kg	25,5 kg	22,5 kg
Lower control arm	3,25 kg	3,25 kg	2,8 kg
Spring	2,5 kg	1,9 kg	1,3 kg
Stabilizer bar	2,7 kg	2,7 kg	1,3 kg
Weight reduction		11,15%	21,60%

Figure 41. Mass reduction result.

RELATED ARTICLES

Volume 4, Chapter 122

REFERENCES

Berkefeld, V., Görich, H.J., and Söffge, F. (1996) Analyse der Achskonzepte für kompakte und leichte Fahrzeuge‘ in *Automobiltechnische Zeitschrift (ATZ)* 98 (1996) 7/8.

Bors, M.E. (1995) *Ergänzung der Konstruktionsmethodik um Quality Function Deployment— ein Beitrag zum qualitätsorientierten Konstruieren*, Carl Hanser Verlag München Wien, Germany.

Chang, M.-Y. (2006) Leichtbau-Vorderachsbauweisen unter Berücksichtigung verschiedener Fahrzeugklassen, PhD-Thesis, RWTH-Aachen University, Germany.

Heissing, B. (2011) Chassis Handbook, Vieweg+Teubner Verlag —Springer Fachmedien Wiesbaden GmbH, Germany.

Ostmann, B. (2001) *Auto, Motor und Sport 2001~2003*, Motor Presse Stuttgart GmbH & Co. KG, Germany.

FURTHER READING

Wallentowitz, H. (2008) Vertikal-/ Querdynamik von Kraftfahrzeugen, Schriftenreihe Automobiltechnik, Institut für Kraftfahrwesen Aachen (ika) RWTH-Aachen: Germany.

Zomotor, A. (1987) *Fahrwerktechnik: Fahrverhalten*, Vogel Buchverlag Würzburg, Germany.

Steering Systems

Chapter 124
Active Front Steering for Passenger Cars

Matthias Wiedmann
AUDI AG, Pfaffenhofen, Germany

1 INTRODUCTION

For manyyears now, power steering has been an indispensable feature of our automobiles and has almost completely replaced mechanical steering. One of the difficulties when selecting power steering settings is deciding on the correct steering ratio, so that the driver enjoys a comfortable driving feel and a sense of safety in all driving situations, from parking to high speed driving. When parking, the driver expects low effort at the steering wheel and easy maneuverability at low speeds, without losing a sense of complete control at high speeds.

This classic conflict of objectives can be solved by a superimposed steering system. A controlled additional angle is superimposed on the steering wheel angle chosen by the driver. In this way, both a variable steering ratio and other functions, such as steering stabilization, can be incorporated.

Encyclopedia of Automotive Engineering, print © 2015 John Wiley & Sons, Ltd.
Edited by David Crolla, David E. Foster, Toshio Kobayashi, and Nick Vaughan.
This article is © 2015 John Wiley & Sons, Ltd. ISBN: 978-0-470-97402-5
Also published in the *Encyclopedia of Automotive Engineering* (online edition)
DOI: 10.1002/9781118354179.auto007

2 HISTORY

The first patent applications for superimposed steering systems were filed in the early 1970s (Pilon, Sattavara, and Schechter, 1972). Practical implementation was unsuccessful then because such complex mechatronics were not available. The patent already contained all the components of a modern superimposed steering system, for instance secure mechanical through-drive with provision for steering angle superimposition. In this case, the superimposition itself was obtained from a double planetary gear set with the ring gear turned by an electric motor through a worm gear drive.

This patent and the further development (Karnopp, 1990) reached series production in 2003 (Köhn *et al.*, 2002), but a year earlier an alternative concept (Musser, 1955) reached series production in which the angle superimposition was obtained by means of a harmonic drive. The functionality will be described in the following chapters.

3 GENERAL FUNCTIONAL PRINCIPLE

All the systems for active front steering in series production are based on superimposing an additional angle on the steering wheel angle chosen by the driver. Other functional principles can be envisaged, for example displacement of the steering system in relation to the vehicle or altering the length of the tie rods, but these technologies have not so far progressed beyond the patent application stage (Mouri, 1992).

Different to pure steer-by-wire systems, superimposed steering systems have a through mechanical drive to the vehicle's road wheels. As the steering system now possesses a further degree of freedom, an additional motor angle δ_M can be superimposed on the driver's chosen steering angle δ_H. The total angle present at the steering system input δ_S can then be calculated with the simple formula

$$\delta_S = \delta_H + \delta_M \quad (1)$$

As torque equilibrium is present in the entire system, the motor must withstand all the torques that occur during vehicle operation. This is done by the high gear ratio (1 : 50) between the electric motor and the gearbox housing. In addition, this design satisfies package and weight requirements.

In the case of vehicle maneuvers with no active contribution from the driver by turning the wheel, for example side wind compensation or vehicle stabilization by electronic stability control (ESC), the driver always has to hold up or compensate for the steering torque that arises from the angle superimposition system.

4 CONSTRUCTION

Three systems differing in their gear-set technology and installed position on the vehicle are currently in use for active front steering. In principle, superimposed steering needs the following components:

(a) superimposition gears;
(b) electric motor;
(c) motor angle sensors; and
(d) a lock to restore the through mechanical connection from the steering wheel to the steering gear in the event of a fault or loss of electric power from the system.

Two types of gear drives have been established on the market:

1. the double planetary gear set as used by BMW/ZFLS[1] and
2. the harmonic drive gear set (in series production at Audi/ZFLS and in a slightly modified form at Lexus/JTEKT[2]).

The main gear-set requirement is a high ratio, so that minimum electric motor torque is needed to withstand steering torque, and the motor can therefore be of small size. In addition, the gear set should (if possible) operate without slack to ensure precise initial steering response. Acoustical considerations also have an increasingly important role to play, when determining the gear set's design ratings, as these steering systems are mostly installed on high quality vehicles that are correspondingly quiet.

In view of the high demands, they must satisfy in respect of control quality and performance, the electric motors are in all cases of the permanent magnet synchronous type (brushless electrically commutated DC motors). One of the main factors to be considered when determining the motor rating is high power density, so that the desired torques are reached with a satisfactorily dynamic action and torque ripple is as low as possible, in order to avoid feedback that can be felt at the steering wheel.

With the aid of a rotary position sensor, the actual position of the rotor in the electric motor is measured as a means of determining the actuating signals and phase currents for the individual motor phases. The closed-loop control circuit thus eliminates deviations between the actual and the desired motor positions.

The locking device is a solenoid that, when its power supply is interrupted, locks the system by means of a preloaded spring. This lock must satisfy the most stringent safety requirements, as it is the system's mechanical fail-safe device. Rapid, reliable locking in all environmental conditions and extremely high failure protection are needed. Among the measures adopted to ensure this are redundant springs to actuate the locking pin.

4.1 Types of system

4.1.1 Hollow-shaft harmonic drive

The actuator is notable for its hollow-shaft concept. The permanent magnet synchronous motor is located concentrically around the input shaft. The other components such as the gear set and motor position sensors are also mounted on a joint shaft. The gearbox technology consists of a harmonic drive gear set with a ratio of 1 : 50. This type of gear set needs very little space and provides a high gear ratio with complete freedom from play (without special measures having to be taken) and excellent torsional rigidity. The gear set is also suitable for high torques, which makes it capable of withstanding misuse situations (parking close to the curbstone and turning the steering wheel with full torque) reliably.

The harmonic drive (Figure 1) operates by means of a flexible thin-race ball bearing (a flex-bearing), which is pressed onto the elliptical motor shaft and thus acquires an elliptical shape as well and acts as the wave generator (WG). This elliptical ball bearing changes the shape of the gearwheel connected to the steering wheel (flexible spline, FS). Thanks to its elliptical shape, the FS (which has 100 teeth) is able to mesh with the circular spline (CS) or ring gear, which has 102 teeth. The output shaft leading

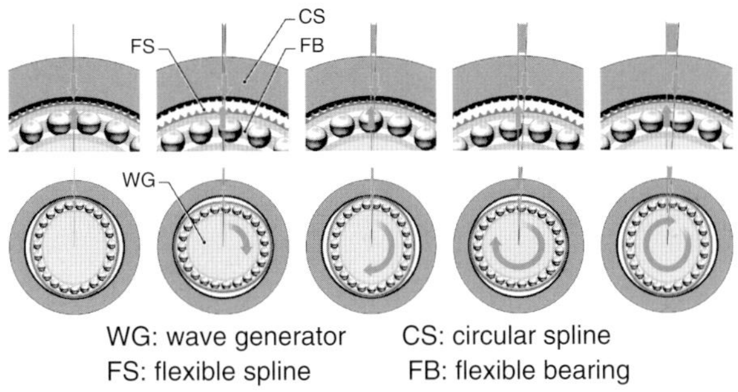

Figure 1. Function of harmonic drive.

to the steering system itself is attached to the circular spline.

When the motor shaft revolves through 360°, the 100 teeth of the FS rotate within the 102 teeth of the circular spline and thus generate a gear ratio of $1:50$ and therefore a superimposition of 7.2° per revolution of the motor. This is added to the angle reaching the circular spline by way of the input shaft and FS.

When the vehicle is parked or if a system fault occurs, the motor shaft is locked by way of the lock ring and locking pin; this inhibits one of the degrees of freedom of the gear set and ensures that the through mechanical drive to the wheels remains operational.

This compact design can be installed in the upper part of the steering column and therefore in the car's interior, which has advantages above all in meeting the environmental requirements that the system must satisfy (temperature and humidity), whereas this might be more difficult if the system is installed in the engine compartment. At the same time, however, acoustical requirements and also questions of crash protection have to be considered when the actuator is located in the passenger compartment.

The superimposition steering system, which is used in the Audi A6 and which is installed in the steering column, is shown in Figure 2. Explanations are given in Schwarz and Dick (2007), whereas another design can be found in Rothmund *et al.* (2006).

4.1.2 Double planetary gear set

A characteristic of this system is that the actuator is mounted directly on the steering system. The superimposition function is performed by a double planetary gear set (Wallbrecher, Schuster, and Herold, 2008). This position on the steering system behind the torque sensor/hydraulic valve has the advantage that effects such as friction and feedback because of the motor's mass moment of inertia can be compensated for by the power steering assistance, thus reducing possibly unwanted feedback at the steering wheel. As an additional reduction gear stage to the electric motor, which is also an electronically commutated, permanent magnet synchronous motor, the planetary gear set has a worm gear. This provides a very high overall reduction ratio and also a degree of self-locking. This self-locking produces a certain degree of redundancy in the locking action. It enhances the safety concept. In addition, the electric motor is prevented from rotating by a locking device (falling pin) when the system is inactive and in the event of a fault.

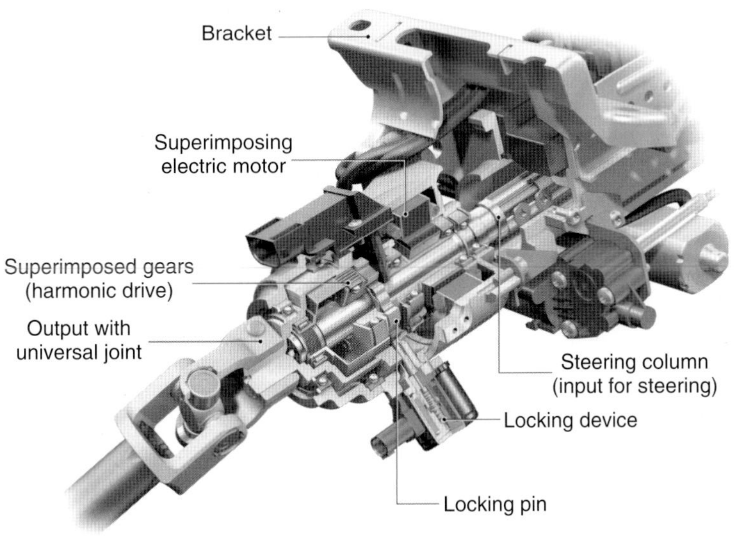

Figure 2. Construction of Audi dynamic steering system.

The actual superimposition is obtained from two planetary gear stages with different ratios. The first of these ($i_1 = 15/12$) has an input shaft (ring gear) with 15 teeth and planet wheels with 12 teeth. The second stage ($i_2 = 13/14$) has an output shaft (ring gear) with 14 teeth and planet wheels with 13 teeth. The two sets of planet wheels are linked by a planet wheel carrier (spider) and therefore always rotate at the same speed. If the electric motor causes the planet wheel carrier to perform a single revolution, the ratio of output to input shaft rotation is 1.35 : 1 ($i_{ges} = \frac{15/12}{13/14}$). As manufacturing tolerances mean that the teeth in these gear sets can never be entirely free from play, the individual planet wheels must make a sprung connection in order to prevent play in the steering. This measure, however, can lead to increased friction in the system.

4.1.3 Harmonic drive with rotating housing

In the same way as the system described in Section 4.1.1, this system is based on a harmonic drive assembly, but unlike that system, the housing also turns with the steering. In addition, there are slight differences in the gear set itself, for instance the circular spline is divided into two rings. One ring has 100 teeth and is connected to the steering wheel and the other has 102 teeth and acts as the output to the steering system. The two rings are linked together by the FS, which also has 100 teeth and is coupled to the electric motor. Here too this is a permanent magnet synchronous motor. In this system, when the electric motor performs one revolution, there is a difference of two teeth between the steering wheel and the steering system. The gear ratio is therefore once again 1 : 50. As the housing also rotates, it is not necessary for components such as the motor and the locking device to be constructed with a hollow shaft. To make the assembly as compact as possible radially, the locking device is integrated into the housing and connected by a pivoting lever. This concept can have the disadvantage of requiring more installation space and also that electric power has to be transmitted by a volute spring.

As this system too is installed ahead of the steering system, additional measures are needed to improve the acoustics and minimize unwanted feedback that would be felt at the steering wheel. A 'Hardy' flexible-joint disk therefore decouples the system.

4.2 Adaptations to the vehicle

Using Audi Dynamic Steering as an example, it can be seen which additional components on the vehicle need to be adapted in order for the superimposed steering to operate safely and conveniently to its full extent.

4.2.1 Steering system

As superimposition increases steering angle speeds at the steering system, especially when performing stabilizing movements, dynamic steering requirements are higher. In the case of hydraulic systems, for example, pumps with increased displacement are needed to ensure that the necessary volumetric flow is reached. In view of the increased energy consumption and therefore the higher hydraulic fluid temperature, electronically regulated pumps are often used, with an electrically energized valve to regulate the flow volume. A power steering oil cooler of larger capacity may also be needed.

Similar requirements apply to vehicles with electromechanical power steering (EPS). If axle loads and therefore the forces acting on the tie rods are high, the electric motor rating and possibly even the supply voltage may have to be modified.

The demands imposed by superimposed steering must also be considered when choosing the steering ratio. In combination with superimposed steering, it is best to adopt the most direct possible basic gear ratio (for example 1 : 13.5), so that assistance from the actuator, especially when parking or driving slowly, is kept low (1 : 12). This improves particularly the acoustics and the amount of feedback felt by the driver in this critical range. However, this approach is subject to limits, as the driver must be capable of supplying the necessary level of passive fail-safe effort in the event of a defect. This applies to the overall steering ratio and also to the sudden jump in the effective ratio if a fault happens.

4.2.2 Acoustic requirements

In particular when parking and driving slowly, the superimposition must not cause any noise that could be regarded as unpleasant. However, a mechatronic system with gears and an electric motor can never be entirely noiseless, and additional measures must therefore be taken. When deciding-g on these measures, a distinction has to be made between structure-borne noise and airborne noise. Airborne noise can be reduced by suitable encapsulation. To prevent the propagation of structure-borne noise, damping elements must be installed. One that has already been mentioned is the "Hardy" flexible disk between the actuator and the steering wheel, but damping elements can also be installed between the actuator and the steering column if the system uses steering-column actuators. A disadvantage of

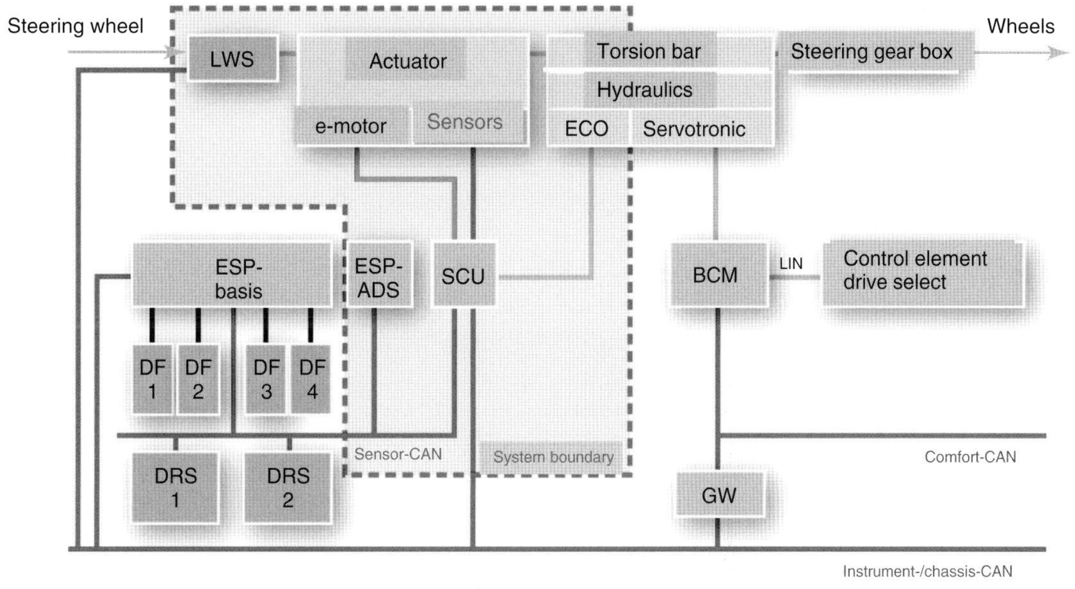

Figure 3. In-car system networking.

these additional measures is that they reduce rigidity. This influences steering response, for instance, and the ability to achieve the required eigenfrequency in the steering column.

It is therefore advisable for the acoustic requirements not to be disregarded when the concept is chosen. Major factors in the selection process are the type of gear set chosen and the gear ratio, though classic conflicts of objective arise here, as the gear ratio also affects the size of the motor and the maximum torque that can be withstood.

4.2.3 Networking adaptations

As superimposed steering effectively communicates with all the vehicle's systems that need steering angle information, networking is accordingly complex. Additional systems supply the input values for the system, for example a driving program switch for the selection of different vehicle dynamics characteristics. If in particular there is also a vehicle stabilization function, this will call for intensive data exchange with the ESC. The ESC itself must satisfy more stringent demands, for example yaw rate-sensing redundancy. If EPS is installed, there will also be a communication with this system, so that further functions such as steering performance limiting can be realized.

Figure 3 shows this signal networking. Networking in such systems is also dealt with in Knoop, Flehming, and Hauler (2008).

5 CUSTOMER FUNCTIONS

5.1 Variable steering ratio

The variable steering ratio is the basic function of a superimposed steering system. This function makes it possible to avoid the compromise that is always present in conventional steering systems between stability at high speeds and avoidance of excessive steering-wheel effort at low speeds. This can be achieved by way of the steering angle (in a similar manner to a variable-ratio steering rack, but with a greater spread of ratios possible) or by way of road speed. The steering ratio can be varied over a wide range between a more direct ratio (for optimal vehicle agility and maneuverability) and a less direct ratio (for high vehicle stability).

The various driving situations can be approximately classified under three headings:

1. Low speed driving (city traffic/parking),
2. Medium speed driving (regular out-of-town roads), and
3. High speed driving (freeways, etc.).

When selecting settings in these three categories, it is important to maintain a harmonious steering ratio characteristic so that the driver does not encounter any unexpected vehicle reactions.

Figure 4 shows the AUDI realized gear ratio function for different vehicle dynamics variants in dependence of the driving velocity.

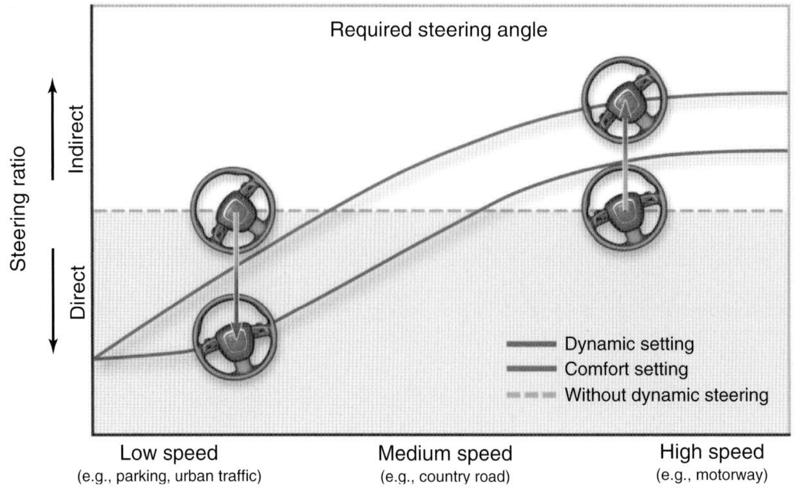

Figure 4. Steering ratio characteristic.

5.1.1 Low speeds

At low speeds, the aim must be to keep driver effort at the steering wheel to a minimum. The steering is normally rated to permit the driver to control the vehicle's direction without having to reposition his or her hands on the steering wheel. Possible factors limiting the degree of directness could be the high degree of superimposition needed, the associated noise, and the driver's ability to adjust to what is felt at the steering wheel. When parking, a superimposed steering system permits the effort required at the steering wheel to be reduced to between one and a half and two turns from lock to lock.

5.1.2 Medium speeds

In this speed range, the vehicle's agility and ease of handling are the decisive criteria. Agility can be described by the increase in the vehicle's yaw rate. The aim in AUDI is to choose a more direct steering ratio than the standard value, one that builds up yaw rate gain plotted against the vehicle's road speed more rapidly (Figure 5). A deeper description can be found in ATZ-Automobiltechnische Zeitschrift (2008).

5.1.3 Road behavior at high speeds

At high speed, a less direct steering ratio is chosen so that the vehicle can be driven in a relaxed manner at high speed. It reacts calmly and less nervously to steering wheel movements. In the BMW superimposed steering even countersteering is installed. At high speed, the drivers steering input is reduced by the electric motor to make the car more stable.

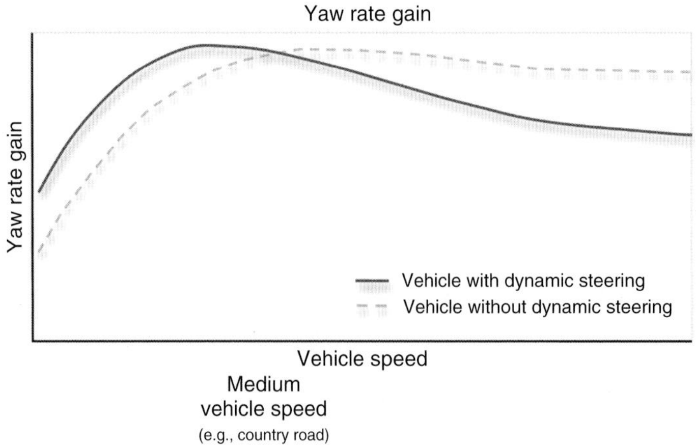

Figure 5. Yaw rate gain plotted against road speed.

5.1.4 Change between characteristics

As the steering ratio can be tuned, the characteristic of the steering gear ratio can be coupled to a driving program switch if this is installed in the vehicle. This gives the driver another means of influencing the steering in addition to the manual torque exerted at the steering wheel. Furthermore, with the aid of a superimposed steering system, it is relatively simple to provide entirely different steering and therefore vehicle characteristics, for example, on a sport model within a model line. This is difficult to achieve with conventional steering systems, as a separate version has to be developed for a limited number of vehicles. The arising costs are prohibiting this. The superimposed steering makes this differentiation possible.

6 STEERING STABILIZATION

As a means of moving the road wheels independently of the steering wheel is now available, it is possible to stabilize the vehicle in critical situations by turning the steering. Until now, vehicle stabilization was only possible with ESC, by applying individual wheel brakes, but as any form of brake application naturally slows the vehicle down, the subjective sense of sportiness is reduced.

At high road speeds, the superimposed steering system demonstrates its advantages most clearly in the form of reaction times that are distinctly faster than any brake application and they are much more comfortable. Steering correction stabilizes the vehicle, whereas the brake application reduces its dynamic performance in an equivalent situation.

At slower speeds or in less critical situations, it may in certain circumstances even be possible to dispense entirely with brake applications. In view of this, a vehicle with steering stabilization maintains its progress more smoothly than one that is only stabilized by means of the brakes. This is especially obvious when friction in the form of tire grip is low. If brake applications are still necessary in such situations, they can be less powerful or will only be needed at a later stage in the critical situation.

Stabilization is effective to a differing degree according to steering performance, which is limited by the actuator. A parameter for this is the steering angle gradient for the stabilizing action. However, the maximum gradient that can be adjusted not only depends on the actuator's performance but also on how it is networked on the vehicle. It is essential to ensure that a superimposition angle that has been called for incorrectly does not lead to a critical safety situation. The latency time of the bus system, for example, also has a decisive part to play.

The decision as to which form of intervention will bring the desired result—brake application, steering stabilization, or a combination of both—is also an important factor for stabilization quality. Comprehensive evaluations of stabilization procedures are reported in Baumgarten *et al.* (2004) and Holle (2003).

6.1 Oversteer

If a vehicle oversteers, an experienced driver will sense its reaction and apply an opposite steering input at the steering wheel. To make this situation easier for less experienced drivers as well, the superimposed steering can apply the opposite steering actively if oversteer occurs, and in this way achieve the optimal front-wheel steering angle. Thus, undesirably high vehicle yaw reaction can be reduced in effect or entirely canceled out. At high vehicle speeds, in particular, the vehicle's fast reaction to the steering angle allows the brake application to take place later and more harmoniously.

Figure 6 demonstrates this behavior of a car. In case of Figure 6a, the brakes are applied only. If the active front wheel steering is added to the stabilization procedure of the car, the brake intervention can be reduced. This is shown in Figure 6b. In addition, it becomes obvious that the total steering effort for the driver is reduced as soon as the superimposed steering is applied.

6.2 Understeer

A vehicle that is understeering is driving more straight ahead at the front wheels, because the maximum force that the tire contact patches on the front axle can transmit has been exceeded (Figure 7). This limits the effect of any action taken through the front steering. The aim must therefore be to provide the driver with access to the maximum adhesion point for as long as possible. When action is taken in this situation, the steering ratio is made less direct so that the steering does not pass the point of maximum friction too quickly.

6.3 μ-Split

μ-Split is the term used to describe a situation in which the amount of grip on one side of the vehicle is noticeably different from the other side. This occurs in the autumn or winter, for example, when one side of the road is covered with wet leaves and the other is dry. When the brakes are applied in this situation, the difference in effective braking force generates a yaw moment that causes the vehicle to turn toward the side with the higher grip. To

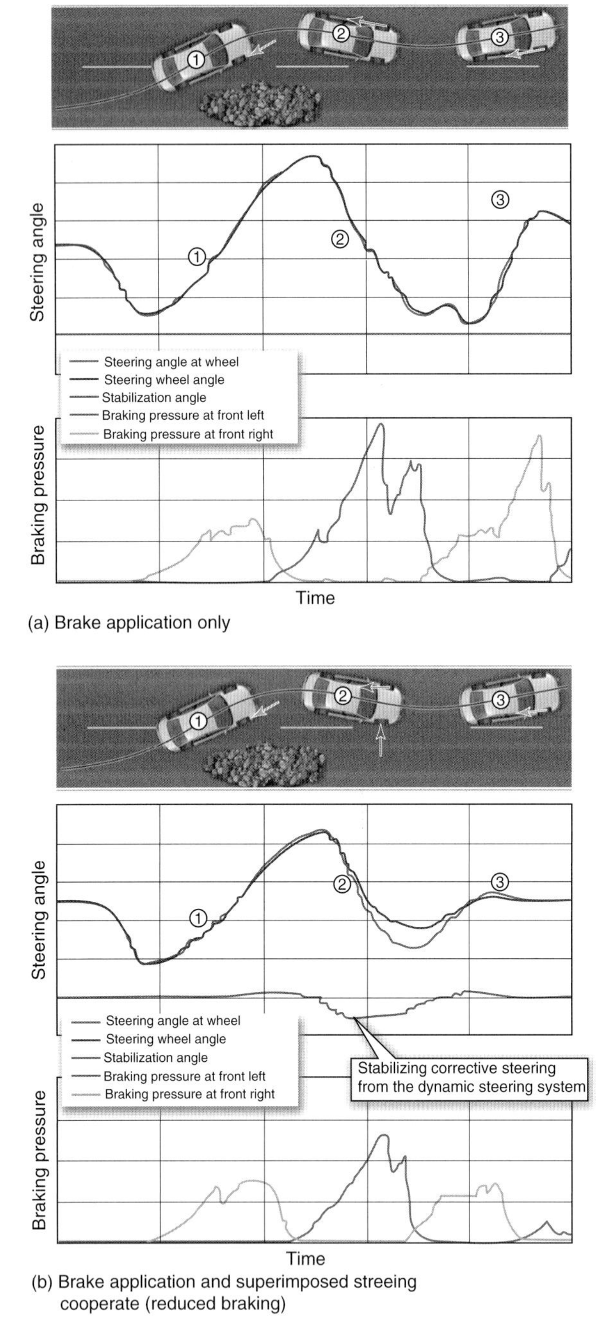

Figure 6. (a,b) Oversteer intervention.

Figure 7. Understeer intervention.

keep the vehicle moving in the chosen direction, the driver has to steer toward where tire grip is lower. To give the driver sufficient time to react, braking pressure is built up relatively slowly.

With superimposed steering, it is possible to select the correction angle at the steering automatically. All the driver has to do is turn the steering wheel in the correct direction, and as the superimposed steering system reacts quickly, braking pressure can be built up more rapidly as well, so that the braking distance can be shortened. The yaw movement will be reduced or even avoided. Figure 8 gives an insight into the brake forces and the steering angles. The brake pressure on the high friction side can be higher than on the low μ side. The front wheel on the high μ side produces side forces to counteract to the yaw moment, which is generated by the different brake forces. The superimposed steering system generates these side forces.

7 SAFETY

Like other mechatronic steering systems, the superimposed steering system is a product with safety relevance. The primary development objective must therefore be to avoid potentially critical situations. These can include

- Preventing reversible or irreversible errors that could be caused by the control unit, the electric motor, or the motor position sensor.
- Monitoring externally computed stabilizing interventions and initiating suitable measures to prevent maximum permissible positioning errors from being exceeded.
- Ensuring that in the event of an error the maximum tolerable gear ratio jump is not exceeded.

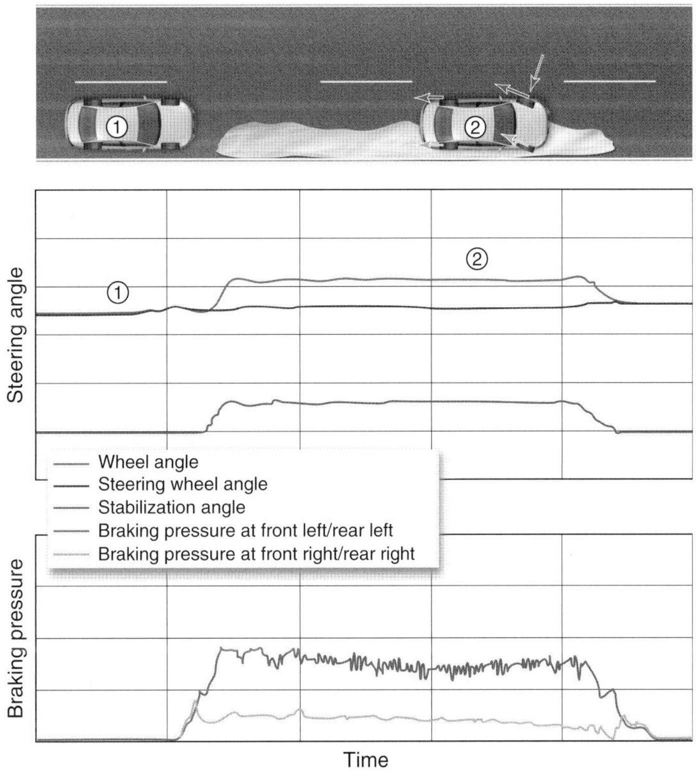

Figure 8. μ-Split braking with and without superimposed steering.

- Preventing any situation in which steering movement is uncontrolled.

Figure 9 shows the three-level steering control unit (SCU) safety concept adopted for the Audi dynamic steering system. All the functionally necessary software modules are integrated into layer 1, including signal plausibility checks and error strategy. All critical paths that could lead to a malfunction are calculated in a diverse manner in layer 2. This ensures that system errors (for example, programming errors) or sporadic RAM defects cannot result in a malfunction. Diverse means that the computer hardware and the used software in layer 2 should be different from layer 1. If possible, different persons should do even the engineering. Layer 3 ensures the correct program sequence, for example, and checks by way of question–answer communication that the set of commands has been carried out correctly (watchdog function).

The difficulty that arises with computing in a diverse manner is achieving the same result in the diverse path with other algorithms as in the main path. Two principal measures enable this:

1. Under error-free conditions, it is possible to achieve the same results with the main function and the diverse function with no relevant time delay. The reason is that the variable steering ratio functions follow feed forward control rather than feedback control.
2. For position control, including sensor evaluation, both paths are monitored in layer 2 by read-back and checking of the motor position signal in relation to its desired angle, as shown in Figure 10.

To ensure high availability, AUDI follows a stepwise degrading of system functions, depending on the error that has occurred:

- Selection of a constant steering ratio (as on vehicles with standard steering) if vehicle velocity information is lacking.
- Inhibition of external stabilizing interventions if reduced performance is to be anticipated, for example, because of vehicle power supply fluctuations.
- If an error is suspected, deactivation of the system, just when the steering wheel angle is zero, in order to prevent the steering wheel from having a constant angle deviation. Other companies are using this constant steering wheel error as a hint of the failed system. The

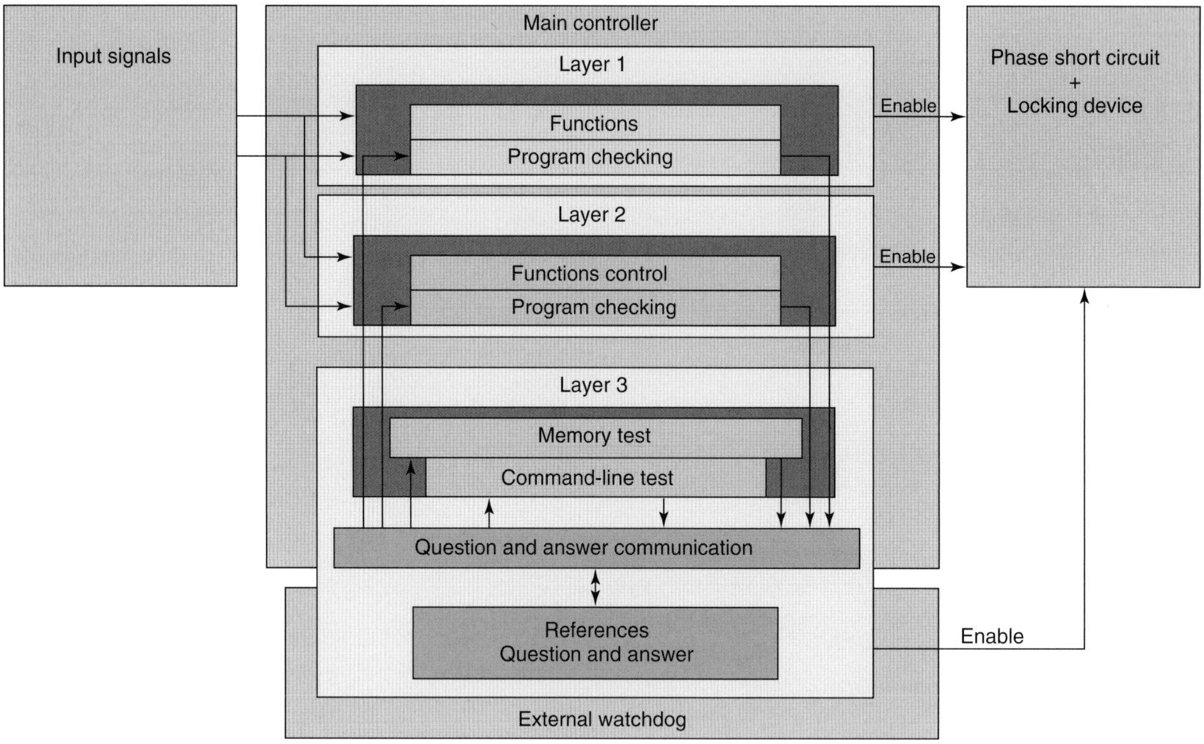

Figure 9. Monitoring levels.

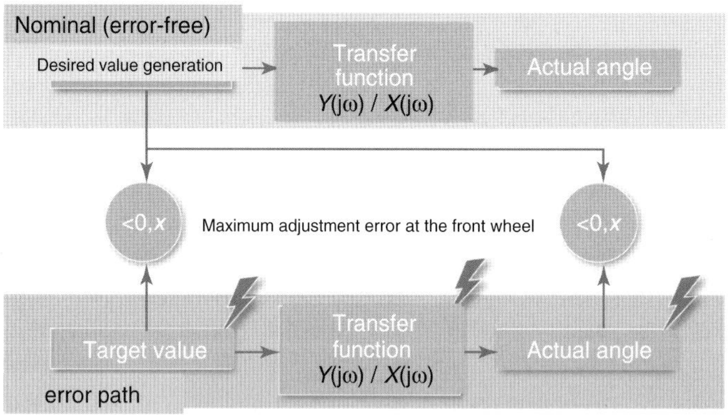

Figure 10. Position control error path.

customer should be advised to go to a workshop and to take the car for repair.

- Immediate, complete deactivation in the event of severe faults.

In the AUDI system, the integrated integral index sensor makes it possible to restart the system without having to take the car to a workshop even if a severe error has occurred, for instance, the loss of electric power. For this purpose, the motor angle is reset in an initializing phase.

The SCU output signals must also satisfy the relevant safety criteria, as other vehicle systems with safety relevance, for example, ESC, react to them. In Jakobi (2011), there are intensive discussions about fallback behaviors of the steering system.

8 SUMMARY

One of the latest mechatronic systems in cars is Active Front Steering. This means that in addition to the driver's steering input an electric motor can steer the front wheels of a car. Driver and the electronical system are normally working in cooperation. The amount of automated steering depends on the car speed, a selected control program, and the driver's steering input.

At present, there are two technical solutions used, the double planetary gear set and the harmonic drive gear set. The systems are described and the different positioning in the steering shaft is mentioned. In addition, the requirements in acoustics and in the oil supply are mentioned. The Active Front Steering is connected to the safety network of the car.

For normal driving, it is possible to use different steering ratio characteristics. The individual driver's wish can be fulfilled. Some examples are explained in this contribution and the basic support for vehicle dynamics is mentioned.

Special requirements for such a system are demanded by the system's safety. The SCU and the used sensors must be diverse. This means that hardware and software must function in different ways. This makes the system more complex. On the other hand, there is so much benefit for the drivers that this effort is justified.

ENDNOTES

1. ZFLS: ZF Lenksysteme GmbH, a Joint Venture of Robert Bosch GmbH and ZF Friedrichshafen AG.
2. JTEKT Corporation Japan (Toyota is one of the shareholders of this company).

REFERENCES

ATZ-Automobiltechnische Zeitschrift (2008) *Dynamiklenkung im AUDI Q5, ATZ Sonderheft Ausgabe Nr.: 2008-02*, Springer Automotive Media, Wiesbaden.

Baumgarten, G., Hofmann, M., Lohninger, R., *et al.* (2004) Die Entwicklung der Stabilisierungsfunktion für die Aktivlenkung, ATZ 106 Heft 9.

Holle, M. (2003) Fahrdynamikoptimierung und Lenkmomentenrückwirkung durch, Überlagerungslenkung, PhD Thesis, RWTH Aachen University, Germany.

Jakobi, F.R. (2011) *Eigensichere Überlagerungslenkung mit elektrischer und mechanischer Rückfallebene, Schriftenreihe des Instituts für Verbrennungsmotoren und Kraftfahrwesen der Universität Stuttgart, Band 57*, Expert Verlag.

Karnopp, D. (1990) Motorbetriebenes Servolenksystem, Deutsches Patent DE 4031316 C2, filed 1990, Robert Bosch GmbH.

Knoop, M., Flehming, F., and Hauler, F. (2008) Improvement of vehicle dynamics by networking of ESP with active steering and torque vectoring. 8th Stuttgart International Symposium, Stuttgart.

Köhn, Ph. *et al.* (2002) Die Aktivlenkung—Das neue fahrdynamische Lenksystem von BMW Aachener Kolloquium Fahrzeug- und Motorentechnik, Aachen.

Mouri, T. (1992) Variable Ratio Steering Gear for a Motor Vehicle, British Patent GB 22 59 062 A, Fuji Yokogyo Kabushiki Kaisha, filed 1992.

Musser, W. (1955) Strain Wave Gearing, US Patent 2906143, United Shoe Machinery Corporation, Felmington N.J., filed.

Pilon, H.M., Sattavara, S.W., and Schechter M.M. (1972) Power Steering Gear Actuator, US Patent 383170. Ford Motor Company, filed.

Rothmund, M., Schwarzhaupt, A., Sulzmann, A., *et al.* (2006) Lenksystem mit Stelleinrichtung und Harmonic-Drive-Getriebe, Deutsches Patent DE 10 2006 055 774 A1, Daimler AG, filed.

Schwarz, R. and Dick, W. (2007) Die neue AUDI Dynamiklenkung, Tagung Reifen, Fahrwerk, Fahrbahn, 11. Internationale Tagung Hannover, 23/24. 10. 2007, Hannover.

Wallbrecher, M., Schuster, M., and Herold, P. (2008) Das neue Lenksystem von BMW—Die Integral Aktivlenkung. Eine Synthese aus Agilität und Souveränität. 17th Aachener Kolloquium Fahrzeug- und Motorentechnik, Aachen.

Chapter 125

New Electrical Power Steering Systems

Mathias Würges
NSK Deutschland GmbH, Ratingen, Germany

1 INTRODUCTION

Following the introduction of the first steering systems with an electromechanical servo unit (electric-power-assisted steering, EPAS) at the end of the 1980s, they have become more and more widespread in recent years. This development is driven by the necessity to economize on energy and thus reduce CO_2 emissions. Depending on vehicle type and driving style, EPAS systems contribute to a reduction in fuel consumption of between 0.3 and 0.5 L/100 km.

The global EPAS market in 2010 already totaled 26 million units, and it is expected to almost double by 2015. This trend stems from the rapidly increasing technological development of electrical and electronic components and safety concepts, which can be used in small-range to top-of-the-range vehicles, and from the expansion of EPAS technologies in high growth markets such as China and Brazil.

Encyclopedia of Automotive Engineering, print
Edited by David Crolla, David E. Foster, Toshio Kobayashi, and Nick Vaughan.
 ISBN: 978-0-470-97402-5
Also published in the *Encyclopedia of Automotive Engineering* (online edition)
DOI: 10.1002/9781118354179.auto008

2 ELECTRIC-POWER-ASSISTED STEERING

Nowadays, there are different EPAS systems on the market, which are used according to the vehicles' boundary conditions and the vehicle manufacturers' technological philosophy. Significant technical factors in the selection of suitable systems are the necessary steering rack force and the steering ratio, that is, the ratio between steering wheel angle and the steering rack stroke or the corresponding front wheel steer angle (Figure 1).

In spite of the different designs of the various EPAS systems, they all share the most important functional requirements:

- Safe operation in all driving situations and a very high level of availability
- Highly dynamic response characteristics in the most varied driving situations
- A sufficient level of steering assist for the driver in the case of intensive actuation forces, for example, parking maneuvers
- Minimal noise during all steering maneuvers. As for this vehicle functions, acoustic feedback is not desirable
- High quality steering characteristics in line with the philosophy of the vehicle brand
- More and more steering functions are being integrated into modern EPAS system, which improve safety or comfort for the driver and can be correspondingly marketed by the vehicle manufacturers.

However, the introduction of EPAS systems—and with it the substitution of hydraulic steering systems—was primarily driven by the reduction in fuel consumption. As it is crucial for the steering assist to respond highly dynamically in every driving situation, the oil pressure

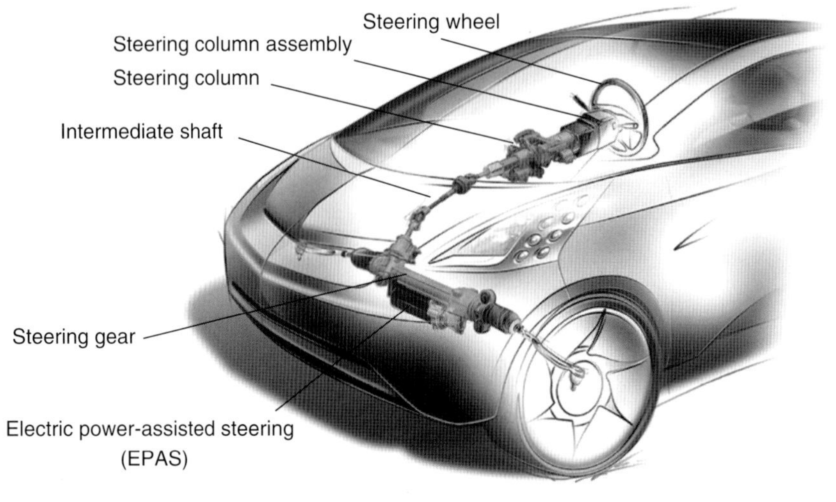

Figure 1. Electric-power-assisted steering (EPAS) system.

in hydraulic systems must be maintained at all times. This results in power dissipation through the continuous operation of the pump and therefore to a high demand for energy. Meanwhile, EPAS systems convert electrical energy from the vehicle power supply, drawing only the amount needed for the length of time required by the respective steering requirement. A further advantage of EPAS is its simple installation and dismantling, because the EPAS actuator system can be connected simply by attaching the power and signal plug connectors. There is no time-consuming handling of hydraulic fluids.

Further advantages of electromechanical steering are its good control capability, the incorporation of the EPAS control unit into the vehicle communication network, the system's highly dynamic properties, and its most temperature-independent characteristics. These properties are also used to introduce new (steering) functions, which increase safety and comfort levels when steering.

The exacting safety demands made of a steering system require the development and validation of extensive safety functions, as well as the application of standardized quality processes. The universal IEC 61508 safety standards for electronic safety-related systems originally applied to the application of safety concepts. However, the new safety standard ISO 26262 specifically for safety-related electrical and electronic systems in road vehicles, which came into force in November 2011, will be the standard for future developments.

2.1 General functions of EPAS systems

The driver applies a manual steering torque to the steering wheel. This is detected by a torque sensor and is transmitted as an analog or digital signal to the electronic control unit (ECU) of the steering system. The ECU calculates the necessary assist torque, considering the driving situation. The drive status is determined using system-internal and system-external information, such as the vehicle speed. The ECU controls the electric motor correspondingly via the power electronics. The steering torque accumulated from the manual torque and assist torque is converted into an actuation force by a pinion on the steering rack and transmitted to the wheel unit via the tie rod (Figure 2).

To minimize loading of the vehicle electrics—and thus the demand for energy—the steering system must operate

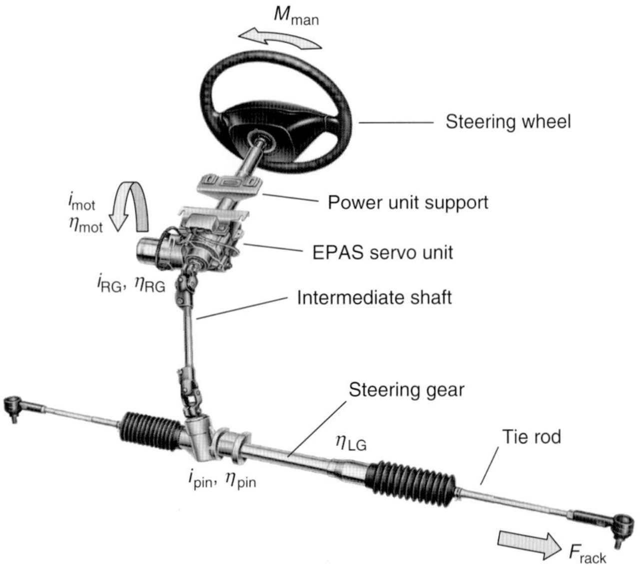

Figure 2. The equilibrium of forces in steering column EPAS.

as efficiently as possible. The system design makes a significant contribution to this with low mechanical friction losses as well as efficient motors and transmission trains. The following equation shows the definitive mechanical influence variables in an example of a steering column electric-power-assisted steering (C EPAS) system

$$F_{\text{rack}} = i_{\text{pin}} \times \eta_{\text{pin}} \times [M_{\text{man}} + (\eta_{\text{mot}} \times M_{\text{mot}}) \times i_{\text{RG}} \times \eta_{\text{RG}}] \times \eta_{\text{LG}}$$

F_{rack}	Actuation force on the tie rod
i_{pin}	Gear ratio of the pinion on the steering column to the steering rack of the steering gear
η_{pin}	Efficiency of the rack-and-pinion gear
M_{man}	Manual steering torque applied by the driver
η_{mot}	Mechanical efficiency of the motor
M_{mot}	Mechanical torque of the motor
i_{RG}	Gear ratio of the reduction gear
η_{RG}	Efficiency of the reduction gear
η_{LG}	Efficiency of the steering gear

This equilibrium of forces clearly illustrates that the manual steering torque is influenced directly by changes to the controlling torque of the motor. Speed-dependent regressive steering assist draws on this principle. At low vehicle speeds, such as when parking, the EPAS motor is activated with a relatively high current (assist current), so that smooth steering is achieved. At high speeds, only small steering movements are generally carried out and the driver requires precise tactile feedback; therefore, a smaller level of steering assist is provided by the EPAS system in this case (Figure 3).

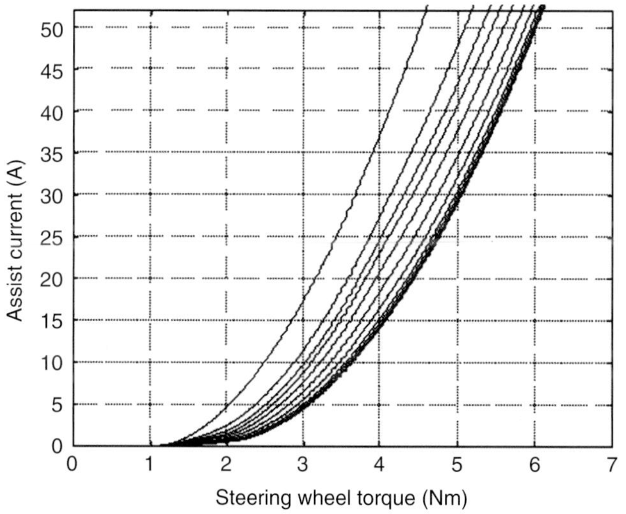

Figure 3. Regressive steering support.

Highly dynamic drives are used in EPAS systems, so that the necessary assist torque can also be provided when completing fast steering movements, for example, when parking or conducting quick evasion maneuvers. Generally, EPAS systems are capable of providing full assist torque, even at angular velocities of 360°/s.

2.2 EPAS technologies

2.2.1 Steering column EPAS

Steering C EPAS is used primarily in small and compact vehicles. Thanks to further enhancements to the technology used for the electric motor, the (power) electronics, the component stiffness, the reduction gears, and the control software; however, modern steering C EPAS is also used in all classes of vehicle, small-range to top-of-the-range models.

In case of steering C EPAS, the motor/control unit and the torque sensors are integrated into the steering column. By means of a reduction gear, the motor transmits a supporting torque to the steering column. As the technology is incorporated into the vehicle interior, this type of EPAS can be configured for relatively moderate ambient conditions.

As the servo unit is integrated directly into the steering column, modern EPAS systems must be small enough to adapt flexibly to the vehicle installation space. Owing to its close proximity to the driver, the servo unit must be very quiet at all steering speeds. The high demands made of the steering feeling are satisfied by the stiff mechanics of the steering column and the intermediate shaft (I-shift), as well as by corresponding control algorithms (Figure 4).

A look at the product portfolio of the globally active Japanese steering manufacturer NSK clearly shows the wide

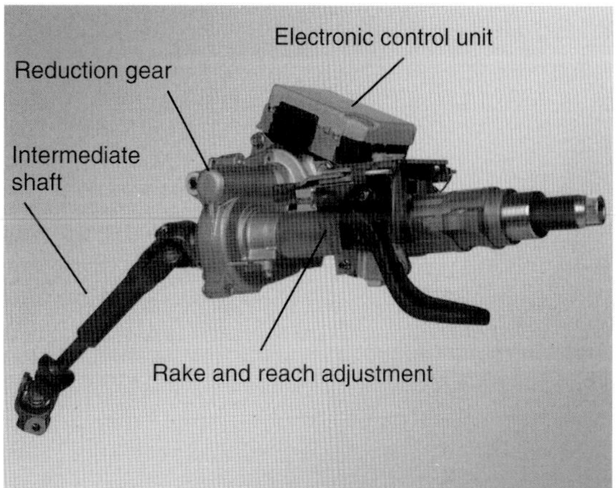

Figure 4. Steering column EPAS. (Reproduced by permission of NSK Deutschland GmbH.)

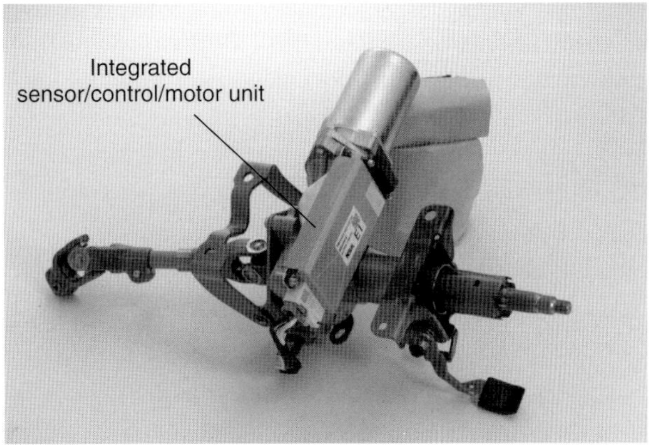

Figure 5. Small package Toyota iQ steering column EPAS. (Reproduced by permission of NSK Deutschland GmbH.)

range of possible applications for steering C EPAS systems. Figure 5 shows the smallest system currently available on the market, which is used in the Toyota iQ. With its steering system for the new Toyota Sienna, NSK sets the standard worldwide for the performance and power density of C EPAS systems. This system provides a steering assist of 12.5 kN rack load and is thus a classic example of an alternative to hydraulic steering assist that can be used for various vehicle classes, from small vehicles to cars with over 3-L engines. At 8.5 kN, the highest performance C EPAS on the European market was also developed by NSK. It is used in models including the Renault Mégane Scénic.

2.2.2 Single-pinion EPAS

In case of the single-pinion EPAS, the servo unit is positioned directly on the steering pinion. Integrating the torque sensor, servo unit and reduction gear into the steering pinion housing results in a compact EPAS system, which, however, has a relatively inflexible layout. Single-pinion EPAS is used in small- and middle-range vehicles.

As they are located in the engine compartment, single-pinion EPAS and the following systems are exposed to higher ambient temperatures than C EPAS systems (up to 135°C) and must be designed accordingly, especially as regards the electronic components. Its positioning also means that the EPAS system is exposed to dirt particles and moisture; therefore, the casing construction must be sealed (Figure 6).

2.2.3 Double-pinion EPAS

As the name suggests, this type of EPAS system features a second pinion. This second unit contains the electric motor, the controls, and the reduction gear as well as the pinion. Because of the principle used, the torque sensor is integrated into the pinion unit on the steering column, similar to the single-pinion EPAS, so that a separate signaling line is routed to the electrical control unit. Compared to the single-pinion concept, double-pinion EPAS offers a greater level of flexibility in the arrangement of the servo unit and thus its placement within the limited space available in the engine compartment. As it is independent of the steering pinion ratio, the gear ratio of the reduction gear can be optimized for applications with somewhat higher controlling torque (Figure 7).

2.2.4 Axially parallel EPAS

Axially parallel drive systems are used in vehicles with medium to high axle loads and are presently enjoying strong growth, as hydraulic systems are now being replaced with electric servo systems in this vehicle class too as EPAS systems becoming increasingly mature.

The unit, which is composed of an electric motor and an ECU, is arranged axially parallel to the steering gear. A pinion on the motor shaft drives a toothed belt, which transfers the torque to the nut of a ball screw drive, whose spindle is on the steering rack. Owing to the effective transmission stages, toothed belt, and ball screw drive, this

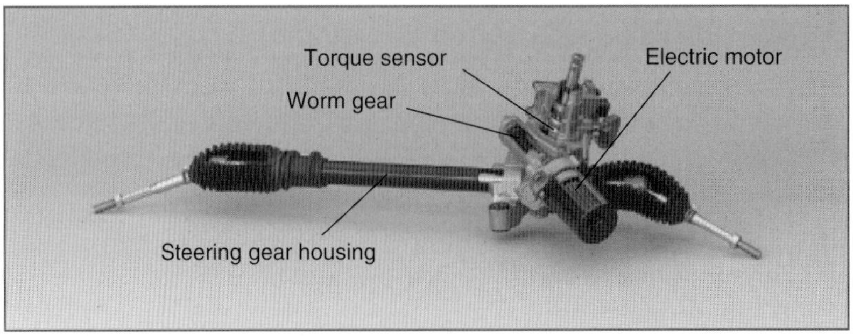

Figure 6. Single-pinion EPAS (with casing open). (Reproduced by permission of NSK Deutschland GmbH.)

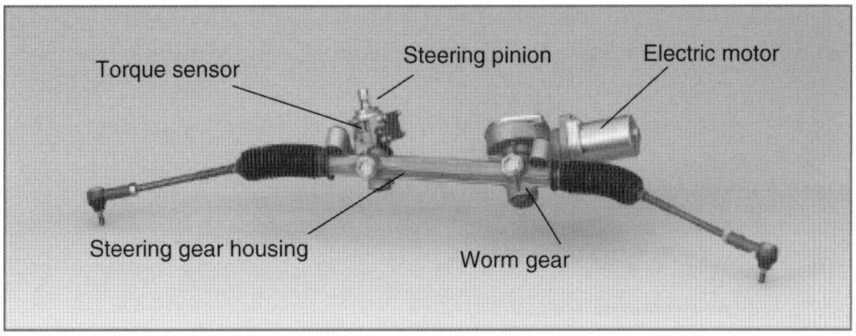

Figure 7. Double-pinion EPAS. (Reproduced by permission of NSK Deutschland GmbH.)

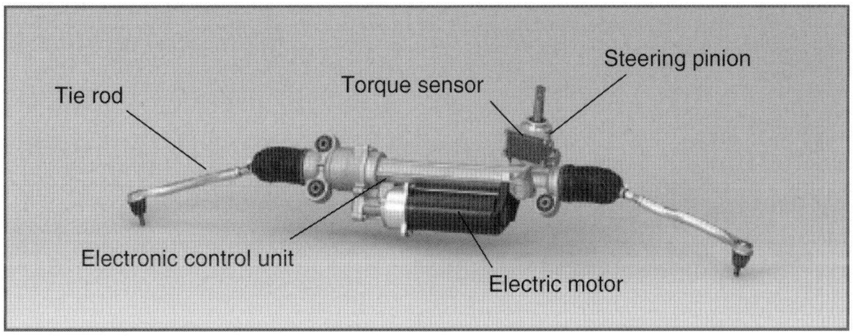

Figure 8. Axially parallel EPAS. (Reproduced by permission of NSK Deutschland GmbH.)

form of EPAS has a high overall efficiency, which enables it to control large steering forces. The torque sensor is in turn integrated into the steering pinion unit and connected with the servo unit via a signaling line (Figure 8).

2.2.5 Rack-concentric EPAS

Rack-concentric EPAS systems are used in vehicles with high axle loads and a correspondingly high actuation force requirement. However, they are still not very widespread today.

With this system, the rotor of the electric motor is seated directly on the ball screw nut. The torque of the electric motor is converted into an actuation—force acting on the steering rack by the ball screw. As the system includes only one transmission stage, the electric motor must provide a very high torque (Figure 9).

3 SYSTEM COMPONENTS

3.1 Mechanics of the steering column in column EPAS

The steering column mechanics have the primary task of absorbing the steering torque applied by the driver and

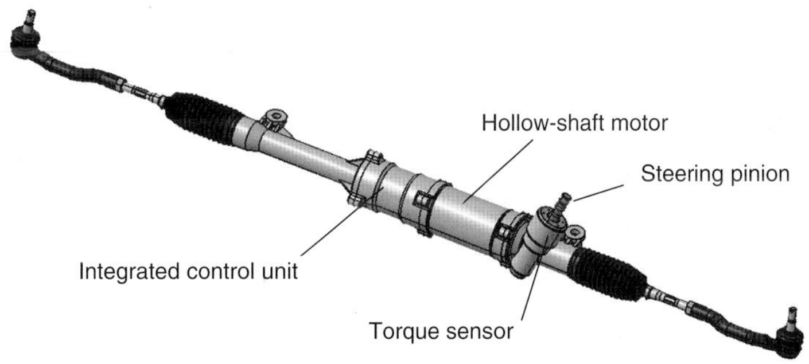

Figure 9. Rack-concentric EPAS.

transferring it to the I-shift with the lowest possible friction losses. As the steering must always be available as a top-priority safety system, the design of the steering mechanics is correspondingly robust and subject to intensive tests. In the case of a crash, the steering column also has the task of absorbing the impact energy by means of a controlled collapse rate should the driver be thrown against the steering wheel.

The steering column has the following mechanical interfaces to the driver/vehicle system (Figure 1):

- Steering wheel (torque input)
- I-shaft (torque transfer)
- Cross car beam (fixing)
- Instrument panel (package interface).

The mechanics of a steering C EPAS system include a divided steering shaft, whose two parts are connected by a torsion bar. Depending on the torsion bar stiffness, typical values are between 2.0 and 25 Nm/°, this design ensures a rotation angle difference of the input and output shafts relative to the manual steering torque. This angular difference is used to detect the steering torque (Section 3.3.1; Torque sensor). The accumulated torque from the manual steering torque and the servo assist torque is transmitted, via an I-shaft, to the steering pinion and translated into the resulting steering rack force.

A distinction is made between the following steering column types based on their adjustability:

- A steering column that cannot be adjusted individually to the driver position is known as a *fixed steering column*.
- For enhanced comfort, the height of some steering columns can be adjusted. In this case, the steering column is released using a lever and set to the right height for the driver's position.
- The third type of steering column can be adjusted for height and reach. This allows the steering wheel position to be moved up or down and in or out.

The adjustable systems feature a divided steering shaft, which is compatible with a clamp system to fix the steering column in position. A special version features electrically adjustment for height and reach. This is mainly found in premium vehicles. Such electrical adjustment systems are made using an electromechanical system separate from the steering actuator system (Sections 3.1.1–3.1.3).

3.1.1 Manual height adjustment

When vehicles are developed, the steering wheel position is designed for a driver of a certain height. Depending on the actual anatomy of the driver and his or her individual seat position, a height-adjustable steering column can increase the level of comfort. Height-adjustable steering columns pivot at their lowest mounting suspension point. The steering column is guided along a slot on the upper fixing, which determines the possible tilt of the steering column (approximately ±20–25 mm is typical). A clamping system guarantees that the steering column is fixed securely in the selected position, even when it is exposed to loads, such as the driver pulling himself or herself up, using the steering wheel. Both friction-based mechanical systems and positive locking are established options.

3.1.2 Manual reach adjustment

A further adaptation of the steering wheel position can be achieved by adjusting it for reach, that is, changing the distance between the driver and the steering wheel. There are two types of reach adjustment, which are used, based on the philosophy of the vehicle manufacturer and the space available.

Type 1: During reach adjustment, only the inner steering column tube is moved, leaving the upper cardan joint of the I-shaft in position.

Type 2: The entire upper steering column is moved, including the upper cardan joint. A sliding mechanism in the I-shaft compensates for this movement in the system.

For both types, reach adjustment is guided by a slot, which defines the possible adjustment (approximately ±20 mm is typical). The steering column is fixed in position using friction clamping or positive locking. The system must be designed to ensure that the forces applied by strong drivers do not accidently alter the steering wheel position. This comfort-enhancing option is generally combined with height adjustment, so that the driver can be optimally adjusting the system.

3.1.3 Steering column with electric height and reach adjustment

Some premium vehicles have electrically adjustable systems. These systems have one electromechanical actuator system for height adjustment and one for reach adjustment, allowing the driver to adjust the steering wheel position within the permitted range by activating the corresponding switch. The mechanism is controlled by a separate system, independently of the EPAS control unit.

These systems are not only used to adjust the steering wheel position when driving, but they also perform additional functions. For example, when the driver exits the vehicle, the steering wheel can be lifted automatically to make getting out easier. After the driver gets in, the steering wheel is automatically returned to the specific position stored for the respective driver.

3.1.4 Passenger safety in the case of a collision

As part of the passive safety system, the steering system includes mechanisms for the protection of the driver in the case of a vehicle collision. If the driver is thrown against the steering wheel, the steering column should absorb energy and thus decrease the forces acting on the driver. Basically, a vehicle collision—in particular a head-on collision—can be divided into two phases, the primary and secondary collisions.

During the primary collision, the vehicle hits the obstacle and the engine compartment is deformed. Because of this, the engine is pushed back, which would move the steering wheel toward the driver. This relative movement between the steering gear and the steering wheel is compensated by a sliding mechanism in the I-shaft—the same as that used for reach adjustment (Figure 11).

The secondary collision refers to the rapid deceleration of the vehicle and its passengers and the reaction of the steering system. The air bag is released and the driver is flung against the steering wheel. The steering system now protects the driver by absorbing largely the impact energy. Defined counterforces are generated here, whose characteristics can be represented in three phases.

As already mentioned in the description of reach adjustment (Section 3.1.2), the steering wheel can also be subject to high loads exerted in the direction of the steering shaft during operation. For example, the steering wheel is often used for support when the driver moves or changes the position of his or her seat. In these cases, the steering column must not collapse, that is, high initial forces acting parallel to the axis of the longitudinal column must be absorbed (typically 1.5–2.5 kN) in order to initiate the first phase of the absorption system. In constructive terms, this breakaway force is represented by crash elements, which activate the collapse of the steering column at a defined force. The crash elements are connecting elements between the carrier bracket and the steering column, which allow movement parallel to the axis above a predefined force threshold. This function can be realized by means of a friction force-oriented design. However, this is generally dependent on the tightening force of the bolts during vehicle assembly. Alternatively, a capsule can be used, whose predetermined breaking points allow the steering column to collapse if the predefined force threshold is exceeded. The properties of this capsule are independent of the vehicle assembly.

In the second phase, the steering column is slid together against a relatively constant force level. One way to absorb the force is to use a design, which generates a defined friction between the internal and external steering column tubes. For instance, ribs can be pressed into the internal steering column tube, which generates almost a constant friction force between the tubes. This phase can be overlaid by additional elements, which allow the counterforce characteristic to be set individually (tuning). This option is particularly interesting in the case of platform applications for steering columns, where it may be necessary to adapt the mechanism to the various models in the design series.

In the third phase of the defined collapse of the steering column, the system reaches its limit, meaning that the residual kinetic energy is absorbed by elasticity in the steering system and in the vehicle. The counterforce rises sharply as a result.

3.1.5 Reduction gear

The reduction gear has the task of converting the fast rotating movements of the electric motor's drive shaft into the rotation of the steering column shaft. As well as the purely functional requirement of converting this power, the EPAS gear has a significant influence on the acoustic and tactile characters of the power steering.

Steering C EPAS systems typically use worm gears as reduction gears (Figure 10). The worm interlocks with the motor shaft and the casing contains roller bearings. The bearing arrangement is elastic, so that the system is constantly subject to a defined preload. The preload, combined with the precise manufacturing process used for the friction partners (the worm and worm wheel), leads to a quiet, uniform transmission of energy. Worm gears are used in steering C EPAS, single-pinion EPAS, and double-pinion EPAS systems.

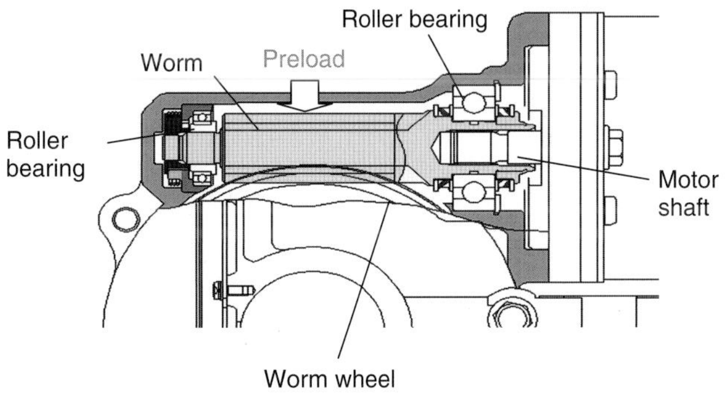

Figure 10. Worm wheel gear in steering column EPAS.

As the reduction gear acts directly in the torque flow of the steering mechanics, any jamming or mechanical damage must not block the steering. A low coefficient of friction in the reduction gear ensures that the driver is also still able to steer the vehicle manually if the steering assist is switched off. As the mechanics are permanently in action, the construction must also prevent any blocking of the motor (Section 5).

3.1.6 Intermediate shaft

The I-shaft uses a positive fitting design to transfer the steering torque from the upper steering column to the pinion of the steering gear.

The dimensions of the I-shaft depend on the steering torque to be transferred and the respective load profile of the steering system. In the case of steering C EPAS, both the manual steering torque applied by the driver and the assist torque of the servo unit must be transferred. The I-shaft of EPAS systems that are integrated into the pinion gear or into the steering gear only need to transfer manual steering torque. On the basis of their application, I-shafts must be designed to meet demanding vehicle-specific or platform-specific load profiles throughout their service life and to cope with the maximum force requirements. As well as safety requirements, stiffness and play tolerance values must be considered in the construction of the I-shaft, depending on the vehicle-specific requirements for steering feeling.

Figure 11 shows an intermediate shaft for an EPAS system. Cardan joints are used at the interfaces to the steering column shaft and to the pinion on the steering gear side, in order to balance the angle and the axis offset between these component parts. An equalizing mechanism compensates for relative movements between the upper and lower cardan joints as they occur, for example during vehicle assembly, in the case of a crash, when adjusting the steering wheel reach, or when driving as a result of elasticity.

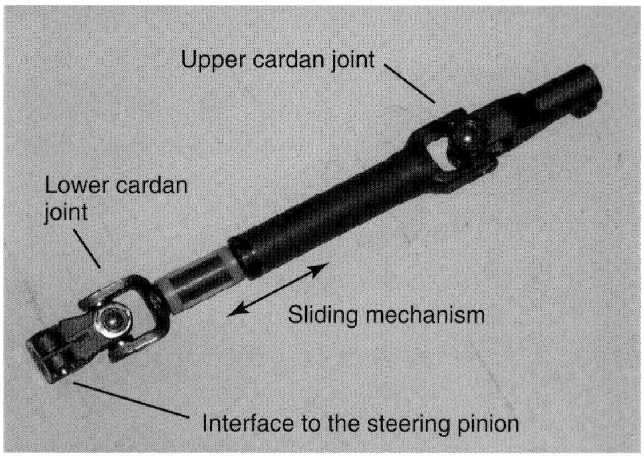

Figure 11. Intermediate shaft. (Reproduced by permission of NSK Deutschland GmbH.)

3.2 Electromechanical system

The main task of the electromechanical system for EPAS is to identify the driver's steering requirements and reliably provide assist torque. Various analyses and secondary functions are integrated into the software to perform this task conveniently and reliably. Additional steering functions are increasingly being incorporated into EPAS systems, which are not perceived directly by the driver, as are functions, which offer more obvious benefits, such as parking assistance. The most important functions are described in Section 4.1.

The electrical components of a modern steering C EPAS system are shown in Figure 12. The torque sensor, ECU, and electric motor are the core components of all forms of EPAS. To identify the driver's steering requirements, the manual steering torque is registered by a torque sensor and transmitted in the form of an electric signal to the ECU. As well as this internal signal from the EPAS system, vehicle data—such as the vehicle's speed or the rotational speed of the combustion engine—can be supplied to the ECU by means of a CAN or FlexRay bus. From this input data,

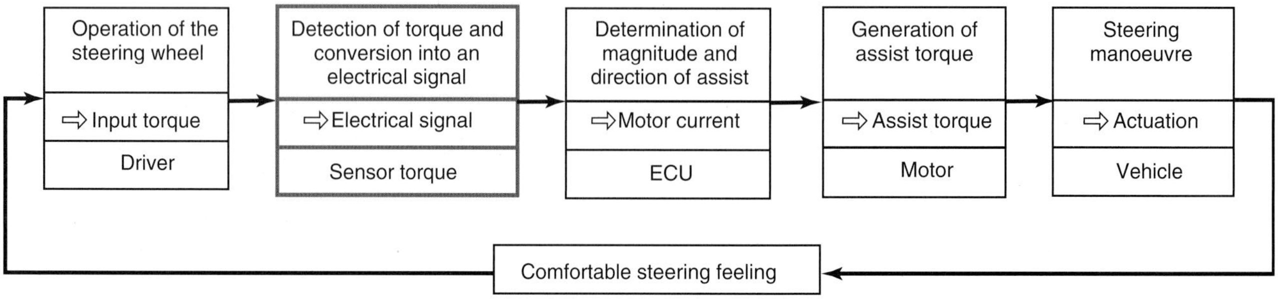

Figure 12. Workflow of the electrical system of an EPAS.

the ECU's processor unit determines the necessary assist torque, and the phase currents are computed for the electric motor and injected into the phases via the power electronics. The engine torque provided is converted by the reduction gear and supplements the driver's manual steering torque.

Knowledge of the absolute position of the steering wheel angle is necessary for different steering functions. The necessary signal can be registered by an external steering angle sensor and made available to the EPAS system via the interface to the vehicle communication. In modern EPAS systems, the absolute steering angle can be calculated using the existing wheel speed signals by means of an integrated electrical hardware system or a complex algorithm. The absolute steering angle position is used not only for different steering functions but also for other vehicle systems such as ESP. As both the steering and the ESP system must satisfy the highest safety demands, the signals are developed according to standardized safety norms (see also Section 6 on functional safety).

The subsystems, which make up the electromechanical system, are described in the following sections.

3.3 Sensor technology

3.3.1 Torque sensor

The steering torque exerted by the driver is a crucial input variable for the calculation of the electrical steering assist. High demands are therefore placed on the torque sensors for EPAS systems. These exceptional requirements are described in the following list:

Exact—In order to be capable of calculating the assist torque required by the driver, a torque sensor with high resolution of about 0.1° at high angular speeds up to 2500–3200 °/s and good signal quality is used.

Safety—The safety of the steering system relies on the magnitude and direction of the steering movement being recorded and the correct signals being transmitted. The construction of the system must safeguard this. Sending incorrect information to the control unit can lead to unwanted steering assist or countersteering. These are major safety-related faults. Such faults must be prevented and/or identified using reliable error detection. Suitable action must be taken to avoid situations, which could lead to accidents.

Reliable—The technology used and the design in question must have a high level of reliability, that is, very low failure rates, even when subject to external influences such as vibrations, high temperatures or fluctuations of temperature, and under the influence of electromagnetic radiation.

Low friction—As modern EPAS mechanics are constructed using a very low friction design, this requirement must also apply for the subsystems, such as the torque sensor. Therefore, low friction systems or systems without contact are used.

Figure 12 shows a simplified control circuit, which illustrates the function and importance of the torque sensor. The sensor acts as the EPAS system's interface to the driver by registering the steering requirements and the steering torque applied and converting these into an electric signal. On the basis of the signal, the necessary assist power or current is calculated in the control unit, considering the driving conditions. The generated actuation force itself initiates a vehicle reaction, which in turn causes a steering action by the driver.

A mechanical system, which generates a physically measurable variable dependent on the torque, forms the basis for the detection of the steering torque. To achieve this, the steering shaft is divided into an input shaft and an output shaft. The mechanical connection for the force transfer is realized via a torsion bar with a defined stiffness. The manual torque input twists the torsion bar, changing the angle between the input and output shafts. This angle is used to calculate the steering torque. Similar arrangements are used in hydraulic power-steering systems.

Figure 13 shows the torque sensor module assembly of a steering C EPAS system. The torque from the input shaft is transferred to the torsion bar via a toothed profile. In order to enable relative movements to be absorbed, for example, in the case of reach adjustment, the input shaft can be displaced axially with respect to the torsion bar. While the assist torque is transmitted to the output shaft directly by the EPAS, the manual torque is transmitted by the torsion bar to the output shaft via a pressed-in pin. A secure mechanical

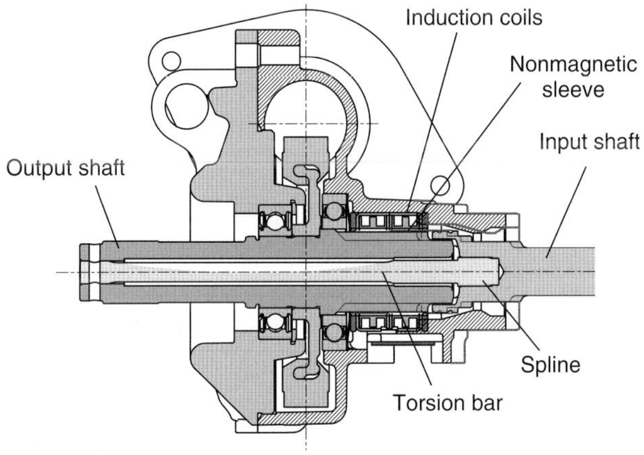

Figure 13. Integrated torque sensor.

layout and intensive validation ensure that the system is extremely reliable. However, even if the torsion bar or its connections fail, a mechanical stop between the input and drive output shaft ensures that the vehicle can be steered manually.

In the system shown, a nonmagnetic sleeve is fixed to the input shaft. The sleeve has windows, which are rotationally offset with constant phase angles. The relative movement of the output shaft's tooth profile in relation to the sleeve changes the magnetic flux in the coils, which causes a change in the coil impedance. This change in the coil impedance is converted by the sensor electronics into an electric signal and transmitted to the ECU. This electric signal is a measure of the required torque assist.

3.4 Motor angle sensor

For the brushless motors used in many systems, as well as for induction generators, the respective motor angle and the motor angular speed are required as a basis for the control algorithm. The motor angle sensor has the task of converting the relative angle position into an electric signal and transmitting it to the ECU. In order to avoid any friction, contact-free principles, such as magnetoresistive (MR) technologies or resolvers, are used in modern EPAS systems.

3.5 Steering angle sensor

The control unit, based on the information supplied by the motor angle sensor, considering its resolution, the mechanical gear ratios, and the mechanical tolerance stack can calculate the relative steering angle. As the motor angle sensors are not multiturn capable, that is, the position can only be determined within one revolution (360°) of the motor shaft, the absolute position of the steering shaft cannot be determined. However, some steering functions require the absolute steering angle, such as the parking assistant.

One way to determine the absolute steering angle without any additional electrical hardware is using an algorithm, which can calculate the straight-line progress using the wheels' rotational speeds. This learning algorithm uses the wheels' rotational speeds determined by the anti-lock brake system (ABS) after the engine is started, in order to index the position of the straight-line progress. By means of the EPAS system's motor angular speed signal, the absolute steering angle can be calculated on the basis of this information. In principle, this process requires a certain driving distance, which is dependent on the steering profile and the road. For most steering functions, determining the absolute steering angle position in this way is sufficient.

However, ESP systems in particular require high precision, rapidly available signals. The so-called true power-on systems provide the absolute steering angle position as soon as the steering system control unit boots up when the engine is started.

The ECU is the distribution center of the electromechanical steering system. All signals arrive here and the suitable assist torques and power flows are computed for the respective driving situation then outputted to the electric motor. The system states and the active processes are monitored by the ECU in a complex safety structure, and appropriate action is taken if a fault is detected (Figure 14).

The input circuits of the signal electronics convert the torque sensor signals from the vehicle power supply and the angle position of the motor into electric signals, which can be processed by the ECU. To generate suitable steering assist, the central processor unit (CPU) uses this information to calculate the current to be injected into the individual motor phases. To activate the motor phases, the CPU sends a control current to the gate driver, which in turn activates the field-effect transistors (FETs) in the power output stage, which is arranged in an H-bridge. As well as the resistive losses, switching losses also occur in the power electronics, which lead to a high level of heating in the component parts. For this reason, the power electronics are placed directly on the housing of the control unit for optimal heat dissipation and the construction ensures that heat is effectively transferred to the steering gear housing.

The monitoring of the CPU is a key component of the safety concept for an EPAS system, because incorrect outputs can lead to fatal faults. The calculated results and/or the computing capability of the processor and the control algorithms must be checked independently. For instance, important calculated results are checked by independent parallel algorithms within the CPU, which have to considerably differentiate from the basic control algorithms. Furthermore, the system diagnostics monitor the software and the electrical hardware of the control unit by running calculations in two independent computer units. Some of the monitoring is completed in the central processor, whose results are checked by a further processor (sub-CPU). If, in the comparison calculations, differences are diagnosed outside of a stipulated tolerance range, which can lead to faults with safety implications, the system goes into safe mode.

Depending on the type of fault identified, the system can revert to replacement actions, which only reduce the level of steering comfort or the degree of steering assist. If a more severe fault is identified, power-assisted steering is switched off and the vehicle must be steered manually. As modern EPAS systems are extremely reliable, these interventions by safety functions are extremely rare occurrences, however.

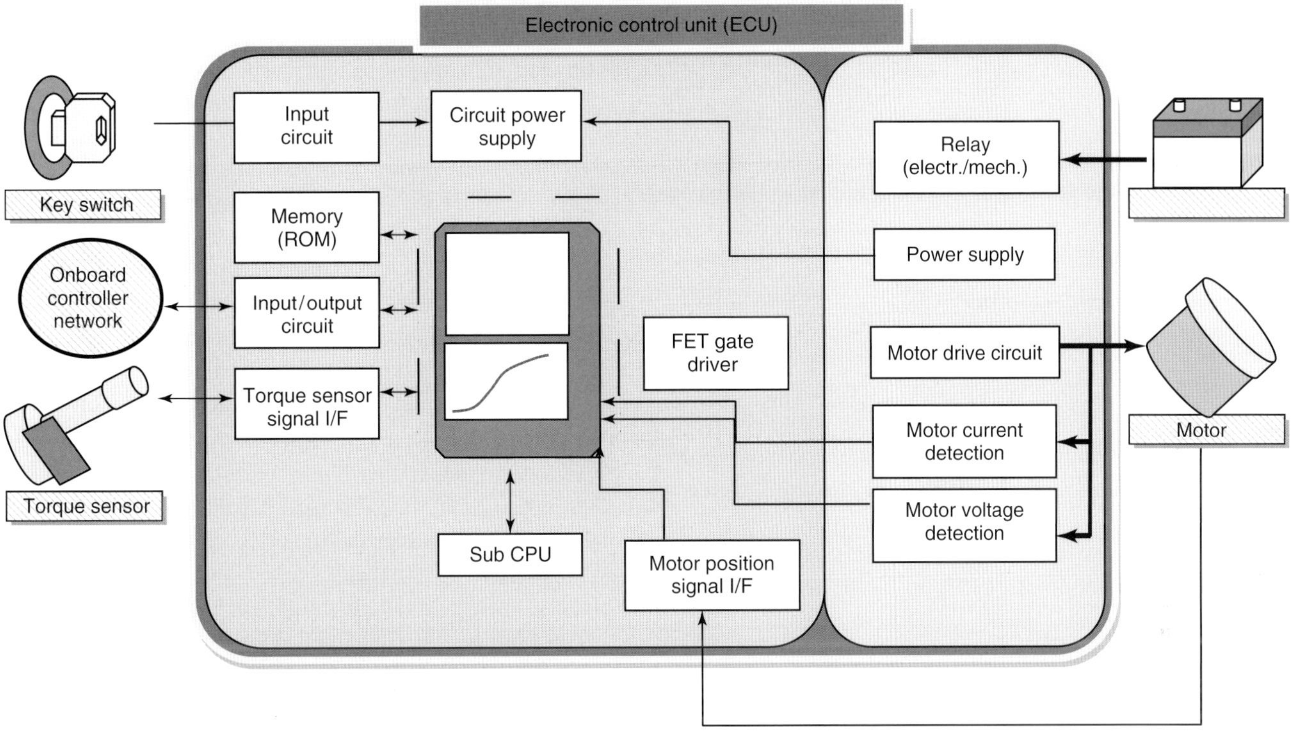

Figure 14. Simplified block diagram of an EPAS electronic control unit.

4 STEERING FUNCTIONS

All functions, which are provided by means of software algorithms within an EPAS system, are referred to as *steering functions*. A distinction can be made between functions, which can be incorporated into the EPAS system and functions, which are supported by the EPAS but are activated by other control units in the vehicle. Generally, the functions, which can be implemented via the EPAS system, serve to support the driver and keep providing him or her with appropriate feedback on the driving situation. However, the driver remains responsible for decision-making regarding the steering. Tactile feedback from the steering system should also place the driver in a position to identify faults in the vehicle, such as a badly balanced wheel or tire pressure loss.

4.1 Basic steering functions

4.1.1 Steering assist

The steering assist provided by the servo motor is the elementary function of an EPAS system. As described in Section 3.3.1, the steering assist is outputted primarily based on the manual steering torque applied by the driver, steering requirements, and the vehicle's speed. The characteristics of the steering assist are extensively coordinated in test runs by vehicle manufacturers and steering suppliers, and these characteristics determine the character of the steering in the respective range of vehicles. The philosophy of the engineers, responsible for coordinating steering in a certain manufacturer's vehicle, reflects the driving feeling that the driver associates with a brand, regardless of the platform in question.

4.1.2 Damping

During operation, the steering system is subject to various influences, which tend to affect the stability of the steering system in a negative manner. In the ideal case of a straight-line drive, no steering intervention and thus no servo assistance would be necessary. In reality, however, acceleration is continuously present from external disturbance variables, such as the stimulation from the road surface or the small steering movements made by the driver, so that these influences do not lead to instabilities in the steering activation, these disturbance variables are detected by the system and dampened in a speed-dependent manner. The character of the damping can be defined by corresponding parameterization in the vehicle coordination phase.

4.1.3 *Friction compensation*

In the electromechanical steering system, friction losses arise at different points, which have a corresponding influence on the steering characteristics. A supplementary assist torque and a corresponding compensation current are calculated to compensate for the friction and added to the basic steering assist.

4.1.4 *Inertia compensation*

Steering a vehicle is a dynamic process, which—considering the transmission ratios typical of EPAS systems—is disturbed by the inertia of the relatively large masses moved. Without further measures, the driver would have to continuously steer against the resulting forces. Inertia compensation means that the driver no longer has to do this and gives him or her the sense that the steering responds immediately and exactly to his or her actions.

4.1.5 *Reduction of the assist power*

If the driver steers into an obstacle, such as a kerb, the wheels are blocked, so that no further steering movement is possible. As long as the driver continues to exert a steering torque, however, the EPAS system will provide a powerful assist torque, which will generate very high power dissipation, leading to heating of the system. In order to reduce this loading and to protect the system, an algorithm is implemented that identifies any blocking and reduces the assist power.

4.2 EPAS steering functions

A great advantage of EPAS systems is the option of going beyond basic steering features by integrating further steering functions via software algorithms without the need for extra hardware. Section 4.2.1 describes functions, which can be realized by the EPAS system using existing information in the vehicle network.

4.2.1 *Active return*

In a high quality steering system, the driver expects the steering wheel to return clearly to the center position depending on the vehicle speed. This means that, when cornering, he or she expects a uniformly increasing counterforce before the steering is reset on the next straight stretch. Furthermore, the neutral position of the steering wheel should be unambiguously reached and maintained as long as the driver exerts no steering torque. On the whole, modern chassis kinematics are no longer designed with this focus in mind as the axle construction considers the option of active return via the EPAS system and the mechanics are optimized with regard to other aspects.

The active return function calculates the necessary active return torque through the EPAS system, considering the steering wheel torque, the vehicle speed, the absolute steering angle, and the steering angle speed. Without steering torque input by the driver, in relation to the vehicle speed, the vehicle automatically goes back to traveling in a straight line after cornering.

4.2.2 *Soft end stop*

Very high rack forces and steering speeds can occur in an EPAS system, depending on the steering requirements. If no other measures are implemented, the steering gear can reach its end stop and be exposed to very high mechanical loading as a result. The mechanics must therefore be designed robustly, which drives up weight and costs.

To optimize the mechanical system, a function can be introduced that can reduce the EPAS assist before it reaches its mechanical end stop, or even oppose the driver's residual manual steering torque. For the realization of this function, knowledge of the absolute position of the steering is necessary, calculated from the absolute angle position of the steering wheel. As well as the mechanical design optimization possibilities offered by this function, the driver also perceives steering toward the end stop as significantly softer, which further enhances comfort.

4.3 Vehicle functions

The technical possibilities of EPAS systems enable different comfort-enhancing functions to be introduced at vehicle level. Other control units in the vehicle manage overall functionality, and the necessary function of the EPAS system is called up via the vehicle communications. However, responding to requirements remains the responsibility of the EPAS system. Many of these additional driver assistance functions lead to an immediately perceptible increase in tactile comfort for the driver, and they can be marketed correspondingly. It is expected that greater use will be made of such assistance functions in the coming years. Some of them are described in the following sections.

4.3.1 *Lane keeping assistance*

If a vehicle changes or leaves its lane following an action such as use of the indicators or an unambiguous steering movement, it is clearly the driver's intention to do so. However, if the vehicle slowly drifts out of its lane, this could be caused by inattentiveness on the part of the driver and could lead to an accident.

Unintentional departure from a lane is generally detected by means of camera systems and feedback about the driver's actions. In order to make the driver aware of this situation, the EPAS system can send tactile feedback to the driver, for example, by making the steering wheel vibrate briefly. However, the driver is still responsible for taking any action to correct the vehicle's direction. This function is also known as *lane departure warning*.

Before active support can be provided to avoid unintentional lane departure, several safety criteria must be met. The driver must be able to override the active steering intervention at all times and the system must always be able to identify whether the driver has his or her hands on the steering wheel by evaluating the manual steering torque. If these prerequisites are met, any active correction by the EPAS system using information from the camera system can be activated.

EPAS can also actively support safety by giving the driver a steering torque recommendation. If oversteering or understeering is registered at vehicle level, or the vehicle skids after braking on μ-split condition, the driver can be encouraged to counteract this by means of short torque pulses to the steering wheel. Here too, the driver must retain ultimate control over the steering, that is, it must also be possible to override these steering pulses.

4.3.2 Parking assistance

Parking assistance helps drivers with parallel and reverse parking. Different types of parking assistance are available, which differ in the degree to which the vehicle's acceleration is automated (accelerating and braking). Steering is always taken over by the vehicle. Autonomous steering places special demands on the steering system in that different steering requirements are communicated by an external signal via the vehicle communication, instead of via the system's internal torque signal. As the driver still has to keep control of the procedure, any possible intervention on the steering wheel is identified by the EPAS system and is responded to appropriately, for example, by terminating the parking procedure.

5 ELECTRIC MOTORS FOR EPAS SYSTEMS

The electric motor in an electromechanical steering system primarily has the task of converting electrical energy into the required mechanical assist torque. Its effect on the character of the EPAS system becomes clear when the most important requirements are considered in more detail:

Performance: From small steering angle speeds to speeds of one steering wheel revolution per second, an EPAS system must be able to make the maximum nominal torque available. Load profiles, which consider the frequency and scales of the assistance provided, are used to design the layout of the motor. For instance, the maximum steering torques only ever have to be made available on a short-term basis and a low level of performance is all that is usually required during operation.

Dynamics: The size and direction of the steering assist changes highly dynamically during steering operations. The motor is therefore required to have the lowest possible rotor inertia and very dynamic response characteristics.

Efficiency: As electric steering uses a comparatively large amount of power, the system must be highly efficient in order to make optimum use of the limited energy available from the vehicle power supply. A high level of efficiency, in combination with a high power density, is also necessary to enable EPAS systems to be positioned in the confined spaces available in the vehicle.

Acoustics: The electric motor for an EPAS system should be as quiet as possible. Any residual noises should be "pleasant," that is, in a relatively low frequency band. Low levels of mechanical friction, the avoidance of resonance, and a good electric design, as well as corresponding motor control, contribute to achieving these objectives.

Tactile properties: A uniformly running rotor with low torque fluctuation results in a high quality steering feeling.

Safety: When considering the functional safety of EPAS systems (Section 6), the focus is on self-steering and blocking in particular. In principle, any blocking of the rotor by a foreign body in the motor can lead to blocked steering. The system is therefore constructed so as to exclude foreign bodies during manufacturing and/or prevent the release of particles during operation. Electric short circuits must also be avoided by constructive means, because these can lead to inadmissible braking torques.

Robust against environmental impacts: EPAS motors are exposed to vibrations and a large ambient temperature range, as well as to different media depending on where they are installed. The motor module assembly must be designed robustly to withstand these influences and provide the required performance reliably throughout the vehicle's service life.

Costs: Costly materials are used in electric motors, such as copper fillings and various magnets. Ongoing improvements to the design and materials utilized for electric motor development reduce motor costs and thus contribute to maintaining the overall costs of EPAS systems at a competitive level.

5.1 Motor types

In EPAS systems, DC motors or brush motors (brushed DC) are used for a mechanical power in the region of 500 W, asynchronous motors (ASMs) are used for 300–500 W, and brushless motors (brushless DC or BLDC) are used up to approximately 900 W.

5.2 Structure of the brushless DC motor

Even though brush motors continue to have their areas of application, BLDC motors are preferred for modern EPAS systems because of their low inertia, long service life, high efficiency, and high power density. The structure of this motor type, which is also designated as a permanent magnet synchronous motor (PMSM) or electronically commutated (EC) motor, is described later.

In order to achieve a high power density, high energy magnets (approximately 360 kJ/m^3) made of neodymium iron boron (NeFeB)—the so-called rare-earth magnets—are used for BLDC motors. During manufacturing, the magnetic powder is pressed, sintered, and then cut into the respective application shape with diamond tools. Ring or segment magnets are used in EPAS motors, which are fixed on the surface of a laminated core. In order to prevent the motor from being dangerously blocked by particles released from the brittle magnetic material, the magnets are bandaged. Rectangular pocket magnets, which are inserted into corresponding pockets in the laminated core, are another magnet construction design. The laminated core with the magnets is pressed onto the motor shaft. This combination forms the rotor, which is supported by roller bearings in the motor housing and in the housing cover.

EPAS motors of the newer generation are generally structured with technology using individual tooth winding. This enables the spatial separation of the phase windings, in line with the safety objective of preventing short circuits between the phases. The running properties of the motor are determined by the combination of the number of slots and the number of poles. 9/6 (9 slots/6 poles), 12/8, and 12/10 motors are frequently used in EPAS systems. As in the case of electrical machines with slotted stators, an undesired cogging torque occurs due to the varying magnetic resistance associated with the change from stator

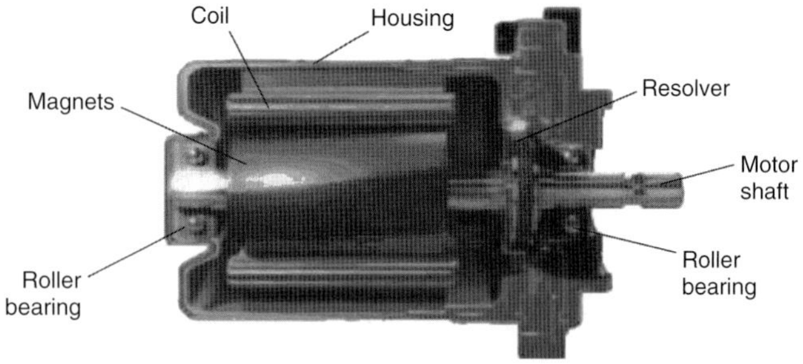

Figure 15. Motor structure of a synchronous motor.

tooth to stator slot. This effect is countered constructively by skewing the rotor or the stator.

BLDC motors in EPAS systems are commutated electronically in relation to their position. Resolvers integrated into the motor module assembly, or MR measurement methods, are used to determine the rotor position. In line with the rotor position, a stator current is injected via the power electronics of the control unit to create the stator magnetic field (Figure 15).

6 FUNCTIONAL SAFETY IN EPAS SYSTEMS

To put it simply, the steering system should always follow the requirements of the driver, that is, the steering torque should be exerted as requested in terms of both magnitude and direction and the requested steering angle should be set. Similarly, it is important that no steering assist is provided when it is not requested and that the system is always steerable.

In order to ensure product safety, the systems are based on standards, which cover the entire life cycle of safety-related electrical and electronic systems. Systematic development in defined processes forms an elementary component of these standards.

The IEC 61508 standard "Functional security of safety-related electrical/electronic/programmable electronic systems" originally dates from 1998 and is used for the development and evaluation of EPAS systems. It covers "all safety-related systems that include electrical, electronic, or programmable electronic components (E/E/PES)." Depending on the risk to persons and the environment, the systems considered are classified into Safety Integrity Levels SIL1 to SIL4. A steering system is classified as SIL3, which corresponds to the highest safety level in the automotive sector and requires the system to be approved by independent third parties.

While IEC 61508 for safety-related electrical/electronic systems covers a range of power systems including EPAS, a specific standard (ISO 262629) has now been developed for automotive electrical/electronic systems on the basis of IEC 61508. This has been officially valid since the end of 2011. This standard will replace IEC 61508 in the development of electrical steering systems, which will affect the product generation process and the product design.

Corresponding to the IEC norm, ISO 26262 also classifies systems using Automotive Safety Integrity Levels from ASIL A to ASIL D. On the basis of an analysis of the system risk, EPAS systems are classified in the highest category, ASIL D. As mechanical aspects are not considered in the standard, fault-free mechanical systems are always assumed in case of the electromechanical systems analysis.

6.1 Identification of the safety hazard

In the risk analysis, the possible safety-related effects of faults are identified in the overall system considered. For an EPAS system, the following safety-related faults are considered:

- *Self-steering*: The servo unit assists in magnitude or direction, which does not correspond to the driver's requirements.
  Self-steering is caused by incorrect torque assist from the electric motor, which is incorrectly activated, for example, by an incorrect signal from the torque sensor, a fault in the electrical circuit or a software error.
- *Steering blocking*: A distinction is made between two scenarios.
  (a) The steering is blocked mechanically. As only the electrical system is considered in IEC and ISO, this case is reduced to a blockage of the electric motor.
  (b) The steering is so stiff that the vehicle can no longer be safely steered. This fault can be caused by a phase short circuit in the motor or in the motor control circuit. In this case, the motor is operated as a generator, so that a corresponding electromagnetic braking torque acts against the steering movement.

6.2 Safety objectives

ISO 26262 requires the application of a redundancy and diagnostics concept, which also guarantees that the system enters safe mode in the case of an error. Measurable target values are defined using special metrics, which are divided into three categories:

- Isometrics give the target value for the maximum acceptable probability of occurrence for safety-related faults. This target value is defined for the occurrence of a fault with reference to the operating time (failure in time, FIT);
- Single point of failure (SPF) refers to errors that lead directly to a safety-related fault;
- Latent faults are undetected faults that, in combination with a further fault, lead to a violation of the safety objective.

In order to ensure that these objectives are met, EPAS systems are developed strictly according to the automotive SPICE model, in combination with further measures from the ISO standard.

6.3 Functional safety in use

The consideration of functional safety includes all components of the electrical system, which have an influence on the actuation force and are capable of causing blocking. The sensor technology, the electrical hardware, the software, and the motor must be considered. Depending on their impact on functional safety, external information is also verified again in the EPAS system.

The top priority is to develop components with a very high level of reliability, that is, to avoid faults in operation. Nevertheless, as there are chances for a fault occurring in operation, a fault identification mechanism must be in place, which analyzes the fault and classifies it according to criticality.

If a fault with a possible impact on safety is identified by the diagnostics, a system response is triggered. Depending on the type of fault, substitute reactions are initiated that, as far as possible, are not perceptible to the driver or are only perceptible to a small degree. In some cases, a gradual reduction of the assist power is necessary to make the vehicle safe. With every fault detection and the corresponding system response, a defining factor is that the driver remains control of the vehicle at all times.

For development test drives and homologation (certification for road use), the results are finally evaluated from the process of developing the electric subsystem and also from the approval tests of the electromechanical overall system before approval.

7 SUMMARY

Modern EPAS systems will continue to become much more widespread in the coming years, with the different systems being used in top-of-the-range vehicles and in small- and middle-range models. While basic functions, reliable operation, and safety were the focus when developing the first generation of EPAS, modern EPAS systems will feature new steering functions, reduced weight, and a more compact design. However, the introduction of the new ISO 26262 standard for functional safety will also lead to further development of the electronic concepts. In order to optimally exhaust the possibilities for further development, it is crucial to develop the system in an integrated way, ensuring compatibility between mechanics, electrical hardware, and software. The global steering manufacturer NSK is one company, which combines a deep knowledge of steering column mechanics with an expansion of the development of steering-specific control units. Such kind of system competence leads to modern safety architectures and steering functions to meet future needs for steering systems.

FURTHER READING

Braess, H.-H. and Seiffert, U. (2001) *Vieweg Handbuch Kraftfahrzeugtechnik* 2. Auflage, Friedrich Vieweg & Sohn Verlagsgesellschaft mbH, Braunschweig/Wiesbaden.

Heißing, B., Ersoy, M. and Gies, S. (2011) *Fahrwerkhandbuch: Grundlagen, Fahrdynamik, Komponenten, Systeme, Mechatronik, Perspektiven (ATZ/MTZ-Fachbuch)*, Vieweg + Teubner, Wiesbaden.

Isermann, R. (2007) *Mechatronische Systeme: Grundlagen*, Springer, Berlin, Heidelberg.

Pfeffer, P. and Harrer, M. (2011) *Lenkungshandbuch (2011)*, Vieweg + Teubner Verlag, Wiesbaden.

Wallentowitz, H. and Reif, K. (2011) *Handbuch Kraftfahrzeugelektronik*, Vieweg + Teubner Verlag, Wiesbaden.

Chapter 126
Steer-by-Wire, Potential, and Challenges

Lutz Eckstein, Lars Hesse, and Michael Klein

RWTH Aachen University, Aachen, Germany

1 INTRODUCTION

Since the invention of the automobile in 1886 by Carl Benz, vehicle technology has rapidly evolved multiplying performance, efficiency, and safety of today's vehicles compared to those of more than 125 years ago. Nevertheless, some technical solutions seem to be immune to technological progress: after a few attempts to steer a vehicle by a crank-like device, the steering wheel commonly, and almost exclusively, became the operating device used to influence the vehicle's direction. The first step to improve the complex vehicle operation back then was made by Alfred Vacheron in a redesigned 1893 Panhard 4 hp driven in the race Paris-Rouen in 1894 (Alexandre, 1894; Dick, 2004). The main reason for the steering wheel's success was the steering gear, which offered an ideal force transmission realizing an effortless set and hold to the vehicle's course.

Encyclopedia of Automotive Engineering, print
Edited by David Crolla, David E. Foster, Toshio Kobayashi, and Nick Vaughan.
 ISBN: 978-0-470-97402-5
Also published in the *Encyclopedia of Automotive Engineering* (online edition)
DOI: 10.1002/9781118354179.auto010

Regarding the steering system, which translates the driver's steering command into a change of the vehicle's direction and—at the same time—provides an adequate feedback in order to support the stability of the driver–vehicle control loop, the predominant solution in street legal passenger and commercial vehicles is still based on a mechanical transfer of forces and torques. Numerous inventions created an evolution from purely mechanical linkages via power-assisted (see Volume 4, Chapter 124; Volume 4, Chapter 125) steering to electromechanical steering systems offering additional but limited functionalities regarding vehicle stabilization and driver assistance.

1.1 Motivation

The motivation for introducing steer-by-wire systems can be attributed to three aspects:

- functionality,
- vehicle architecture, including benefits in production and costs, and
- human factors.

The three-level model for driver assistance systems classifies the vehicle operation into the levels of navigation, guidance, and stabilization. While today's electromechanical steering systems enable functionalities such as speed-dependent power assistance on the level of vehicle stabilization and lane-keeping assistance regarding vehicle guidance, the mechanical part of the system restricts the functionality because of the transfer of torque between steering wheel and steering system. In case of a fixed steering ratio, an operation of the steering actuator yields both a change in steering wheel angle and angle of the steered wheels.

The idea of superimposing an additional angle to the steering wheel angle while having a mechanically safe operating mode was patented for the first time in 1972 (Pilon *et al.*, 1972). However, after another 30 years, Toyoda Machinery Works and Lexus, as well as ZF-Lenksysteme and BMW, have introduced a superimposed steering system into the market (Köhn *et al.*, 2002; NN, 2003). Active front steering introduced by BMW superimposes an electronically commanded steering angle in addition to the steering wheel angle using a planetary gearing. This principle clearly enables additional functionality such as a variable steering ratio and improves driving stability but is still limited by the fact that the driver needs to compensate the change in steering torque resulting from the (see Volume 4, Chapter 125) actuation of the active front steering (Köhn *et al.*, 2002).

The second motivation for steer-by-wire is given by the fact that the mechanical integration of the steering system has a large effect on the vehicle package, as the steering column needs to connect the steering wheel with the steering gear in right- and left-hand-drive vehicles. The geometry of a conventional steering system also limits the vehicle variants that may be derived on a modular basis, as the ergonomically required position and angle of the steering wheel largely changes with the different postures of the driver, for example, in a limousine versus a sports utility vehicle. Introducing steer-by-wire yields a robust package of the steering system and engine compartment and maximum spreading of vehicle variants on one technical platform.

The third motivation also plays a major role in the history of aviation: with increasing aircraft size and performance, the pilot's ability to control the aircraft without significant auxiliary forces and flight control systems rapidly decreased. Moreover, the classical control column in front of the pilot consumed a large amount of valuable space and surfaces in the cockpit that were needed to integrate additional controls and instrumentation. Consequently, the entire cockpit layout has changed with the introduction of fly-by-wire incorporating compact joysticks instead of large column sticks. In addition, in a road vehicle, the cockpit layout can be significantly improved by introducing by-wire controls in terms of human factors, design, active safety, and passive safety (Eckstein, 2001).

1.2 Definition

A steer-by-wire system uses an electronic communication replacing the mechanical linkage between the operating device (e.g., steering wheel) and the steering system, as stated by the term *by-wire*.

Extended definitions of steer-by-wire systems found in literature (Binfet-Kull, 2001) also include other nonmechanical connections between the steering control and the steering actuator(s), for example, hydraulic or pneumatic systems. This broad definition is not adopted here, as these systems significantly differ in terms of technology and functionality.

Consequently, a steer-by-wire system comprises at least one sensor in order to sense the input, an electronic control unit calculating a steering command and one actuator influencing the angle of the steering system.

The exact technical solution depends on the individual requirements concerning operating device, functionality, steering actuation, steering feedback, and system integrity. In order to keep up the steering functionality in case of a system fault, a fallback mode is required that clearly distinguishes pure by-wire systems with an electrical/electronical redundancy from those systems incorporating a hydraulic, pneumatic, or mechanical backup.

2 HISTORY OF STEER-BY-WIRE SYSTEMS

Braess (2001) points out that since the 1950s, the demands on steering systems have increased because of increasingly high engine powers, improved road surfaces, and hence higher vehicle velocities. Consequently, the design of conventional steering systems was largely improved by, for example, introducing hydraulic power-assisted steering, electrohydraulic (see Volume 4, Chapter 125) power-assisted steering, electric power steering (EPS), or superimposed steering. (see Volume 4, Chapter 124 Similarly to the introduction of fly-by-wire systems, the next step would be the introduction of steer-by-wire systems with a for example, mechanical, hydraulic, electric, or integrated fallback (Figure 1).

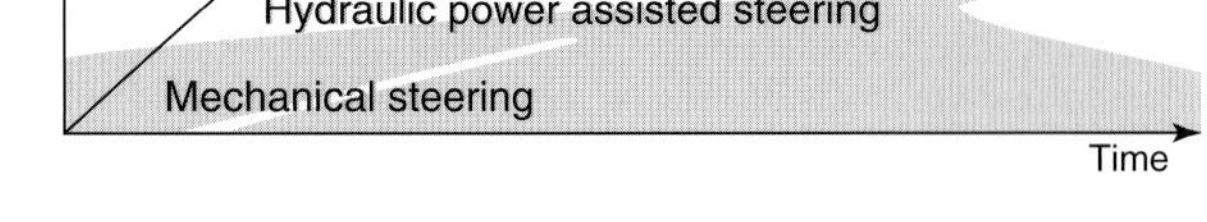

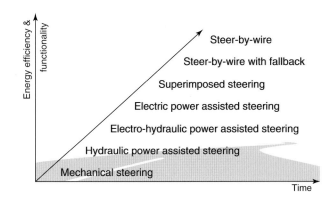

Figure 1. Timeline steering systems.

The first series production of hydraulic power-assisted steering systems started in the Chrysler models New Yorker and Imperial in 1951. The principle itself had already been filed in a patent by F.W. Davis in 1920s and was ready to go into serial production of GM's Cadillac by 1933 (Davis, 1927). However, owing to the world economy crisis, GM planned to sell only 15,000 vehicles, leading to cost-intensive machine tools in production. Finally, the introduction of the system into the market was stopped. During World War II, American and British army forces equipped military vehicles with Bendix–Davis hydraulic power-assisted steering systems and the market for heavy-duty vehicles started to develop. Series production started in 1951; by 1956 already every forth vehicle in the US market was equipped with a hydraulic-assisted power steering system. Right from the beginning, the servohydraulic steering gear has been supplied by a hydraulic pump, which has been directly driven via a belt drive by the combustion engine. Sporadically, steering systems have been equipped with electric motors driving the hydraulic pumps due to packaging reasons. The automotive R&D departments have always been focusing on designing energy-efficient steering systems. At the end of the 1990s, electrohydraulic power steering systems have been introduced into series production. The development of electric motors and pumps brought forth improved degrees of efficiency and the control of the pump rotational speed decreased the total power demand (Pfeffer, 2011; Rixmann, 1962). Following hydraulic and electrohydraulic power-assisted steering systems, EPS systems have decreased energy consumption and broadened the horizon of steering functionality. The patent from Bayle and Ecquevilly (1972) presents an early example for the idea of an electromechanical steering system. This solution shows an electric motor integrated into the steering column. The first vehicle having an EPS system was the Suzuki "Cervo" in 1988. This system developed by Koyo (Japan) has an electric motor as a component integrated in the steering column. As the front axle wheel load was small, the electric motor only needed an input power of 240 W and was connected to the steering shaft via a worm drive with a gear ratio of 1 : 16 (Stoll, 1992). The next innovation in steering design was the superimposed steering system, which has been patented in 1972 (Pilon *et al.*, 1972). In 2002, ZF Lenksysteme and BMW, as well as Toyoda Machinery Works and Lexus, presented a superimposed steering in series production (Köhn *et al.*, 2002; NN, 2003).

Regarding the functional motivation, *automation* was one key driver in research on steer-by-wire systems in close analogy to the aviation industry. Since the 1960s, research activities concentrated on the idea of autonomous driving. At the same time, the Concorde was the first civil airplane to perform, in 1969, its first flight with analog fly-by-wire technology. Three years later, NASA's Vought F-8 Crusader flew with a digital fly-by-wire system (Brockhaus, Alles, and Luckner, 2011; Henke, 2010).

The Eureka project PROMETHEUS (PROgraMme for a European Traffic of Highest Efficiency and Unprecedented Safety) and its successive project MOTIV, initiated by German car manufacturers, bundled the activities of many European companies and Universities starting from 1987 (see Volume 4, Chapter 129). In the same year, the Airbus A320 had the first completely digitalized fly-by-wire system. The electronic system was designed with redundancies. A mechanical pitch elevator and yaw rudder pedals served as mechanical backup (Brockhaus, Alles and Luckner, 2011; Henke, 2010). In addition to the PROMETHEUS project, similar projects have been initiated (e.g., NAHSC, IVI, and AHSRA) (see Volume 4, Chapter 129) (Stiller, 2007; Sun, Bebis, and Miller, 2006).

The conceptual idea of intelligent vehicle highway systems (IVHSs) started in the General Motors Pavilion at the 1939 World's Fair. The automotive future was presented by relaxing drivers in self-driving cars (Fenton, 1994). In recent years, the US Defense Advanced Research Projects Agency (DARPA) organized the Grand and Urban Challenges, which were intended to boost US research activities (see Volume 4, Chapter 129). Broggi *et al.* (2010) presented the VisLab Intercontinental Autonomous Challenge (VIAC), where four electric vehicles aimed to drive autonomously along a 13,000-km trip from Italy to China (see Volume 4, Chapter 129).

Apart from these competitions and races, many car manufacturers investigate the potential of driver assistance systems, which not only support the driver on the longitudinal driving task, such as adaptive cruise control, but also provide steering support. Such a system would offer autonomous driving in defined scenarios, such as traffic on motorways, and constitute an important step toward full autonomous driving.

While there is little official available information on the second key motivation for steer-by-wire, namely benefits regarding vehicle architecture, modularity, and production, the third key driver of research on steer-by-wire, *safety and human factors*, was also inspired by progress in aviation technology. Before the invention of the airbag in 1951 (Linderer, 1951) and its final breakthrough in the Mercedes-Benz W126 in 1980 (Patzelt, Schiesterl, and Seybold, 1971; Kramer, 2009), the steering wheel and column often caused severe injuries and many deaths. Already in 1959, General Motors presented a research vehicle based on a Chevrolet Impala, which had a door-integrated joystick to control the longitudinal and lateral vehicle dynamics (Bidwell and Cataldo, 1958). General Motors designed the vehicle stick displacement to be proportional to lateral acceleration at velocities >10 or 15 mph. As there have been no mechanical connections between stick and the controlled vehicle, the freedom in designing the most suitable steering response characteristics was one principal challenge. The joystick feeds back the car's motion to the driver. According to Bidwell (1958), the vehicle's path stability in critical situations was improved while the control knob acted as an accelerometer and also the sensitivity to wind gusts was decreased. The GM Firebird III was presented at GM's Motorama in the same year. This vehicle concept, which was inspired by the space age, had an integrated single joystick like that of the GM Impala replacing steering wheel, gas, and brake pedal (Bidwell, 1959; Davis 2004).

In recent two decades, numerous research prototypes and concept cars with steer-by-wire systems have been built and as such, only a few examples can be mentioned in this chapter. A concept car, the Saab 9000, having an active joystick with force feedback control was presented in 1991. The sidestick concept, originally developed for a military aircraft, was moveable in lateral direction, whereas a passive spring–damper combination and an electric motor applied the feedback force (Bränneby *et al.*, 1991). The DaimlerChrysler F200 Imagination, presented in 1996 at the Paris Motor Show, was a research vehicle featuring a steer-by-wire system with two hydraulic sidesticks instead of a steering wheel. Eckstein (2000) carried out substantial research on control algorithms and human factors resulting in a Mercedes-Benz SL 500 (R129) prototype with two active joysticks (Figure 2). The control commands for steering, throttle, and brake were based on force transducers, whereas the displacement of the sidesticks provided an active feedback in lateral direction; in longitudinal direction, the joysticks were isometric. In 2002, General Motors presented the GM Hywire, a steer-by-wire concept where steering was achieved by gliding up or down the steering device handgrip creating a feeling similar to that of a conventional steering wheel. In 2003, DaimlerChrysler published a steer-by-wire concept in the F500 Mind on the Tokyo Motor Show.

3 STATE OF THE ART

3.1 Classification of steering systems

The oldest steering system is the turntable steering, which still can be found nowadays in truck trailers. Here, the drawbar connects the rigid axle and the turntable. The articulated-frame steering is another concept in the heavy equipment sector. In this case, the vehicle bends around a joint in the middle. Both designs are not applicable for passenger cars driving at high speeds because of high steering forces, the performance of the steering kinematics, and the required packaging space (Stoll, 1992).

Single-wheel steering and axle-pivot steering are known as *Ackermann steering systems*, which can be distinguished by the way the steering force is generated (Stoll, 1992):

- *Manual Steering System.* The driver has to apply the steering-wheel force. Only the steering gear reduces the force to a manageable level.
- *Power Steering System.* The driver is assisted in the steering task by additional components reducing the required steering wheel torque.

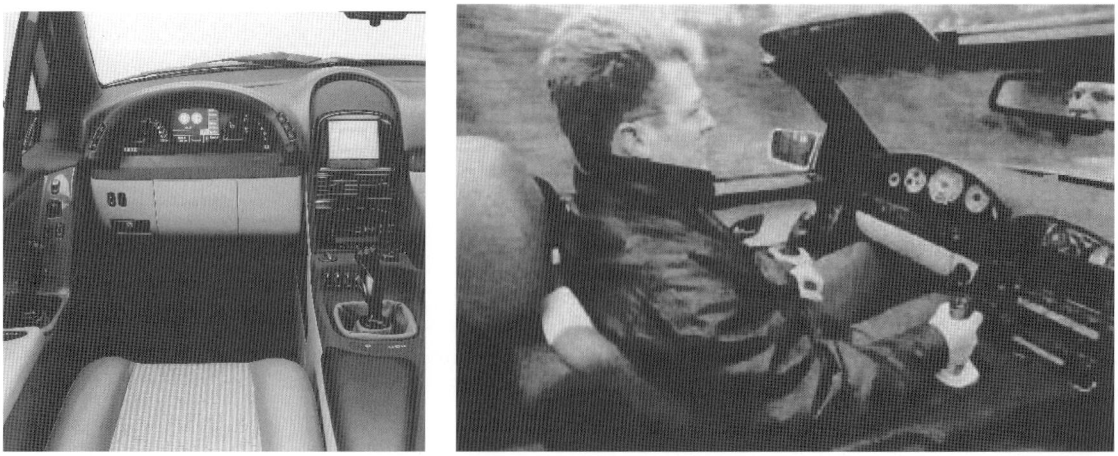

(a) (b)

Figure 2. Steering with joysticks. (a) Cabin of a driving simulator to compare joystick driving with steering wheel driving and (b) joystick driving in a real car.

- *Full-Power Steering Equipment*. The required steering forces are provided solely by one or more energy supplies ECE Reg. 79 (2005).

One potential of steer-by-wire systems is the freedom in designing both the steering torque assistance and the steering ratio. Steering systems can also be classified by the type of force transmission between steering control and steered wheels (Stoll, 1992) as

- mechanical steering systems,
- hydraulic steering systems,
- pneumatic steering systems, and
- electrical steering systems.

Steering systems can also be distinguished by the type of steering gear:

- translational movement (rack-and-pinion steering gear),
- rotational movement (cam-and-roller (Gemmer) or ball-and-nut steering gear), and
- wheel individual steering actuators.

The rack-and-pinion steering gear converts the rotational movement of the steering wheel angle into a translational motion of the tie rod. In contrast, the second type of steering gear transmits the rotational movement of the steering column to a rotation of the pitman arm. Wheel individual steering actuators are often used in specialized on-road and off-road vehicles on the basis of a hydraulic system.

3.2 Steer-by-wire system architectures

In recent decade, many steer-by-wire system architectures have been proposed. This chapter gives an overview of the system architectures without making a claim to be exhaustive. In addition, means to improve system safety are presented.

Steer-by-wire systems do not have a mechanical linkage between the steering control (e.g., steering wheels and joysticks) and the steering system. A basic system configuration comprises a sensor for detecting the driver steering input, an electronic control unit, and an actuator for the operation of the wheels and adequate means for communication between the system elements. Integrating an additional actuator at the control element enables feedback information to the driver, depending, for example, on the steering angle of the wheels or the lateral acceleration of the vehicle. The force feedback actuator thus generates a haptic feedback, which aims at easing vehicle control. The steering actuator is linked to or replaces the conventional steering gear and operates the wheels (Wallentowitz and Reif, 2006). This system is able to provide almost all the functions described in Chapter 4 Potentials in Volume 4, Chapter 126. Splitting the steering tie rod and adding single-wheel steering actuators increases functionality again. However, in case of a fault in the E&E architecture, this extended basic configuration does not achieve the reliability of conventional steering systems. Of course, also the electric power source has to be considered in the respective analysis for system integrity. As steering systems are considered to be safety critical, the system safety has to be guaranteed in all operating conditions and over vehicle lifetime. As a one-channel electronic control unit and mechatronic subsystems do not have a failure occurrence of $<10^{-7}$ per hour (Wallentowitz and Reif, 2006), a system without redundancy would not be sufficient for safe operation. In aviation industry for example, the Joint Aviation Authorities (JAAs) have specified that failures in primary control systems, such as fly-by-wire systems of large airplanes, should have a failure occurrence of $<10^{-9}$ per hour to be considered as improbable (Joint Aviation Authorities Committee, 1989; Reichel, 2004). Thus, in order to make a steer-by-wire system as safe as a conventional steering system, the system architecture has to allow single electric and electronic faults in each of its mechatronic subsystems without leading to loss of control by the driver (Wallentowitz and Reif, 2006).

Hayama *et al.* (2008) show the development of a basic steer-by-wire to a fault-tolerant architecture without considering the electric power source and required sensors. The baseline steer-by-wire architecture (Figure 3) consists of one steering wheel angle sensor, one reaction torque actuator, one steering actuator, one tire angle sensor, two controllers, and one battery. Considering the state

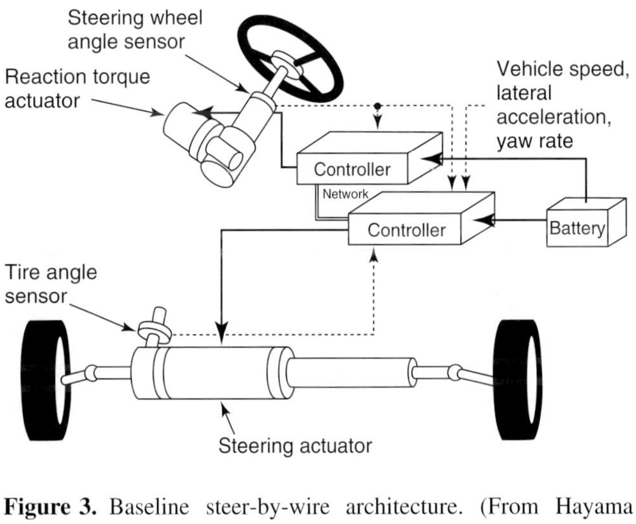

Figure 3. Baseline steer-by-wire architecture. (From Hayama *et al.*, 2008. Copyright © 2008 SAE International. Reprinted with permission.)

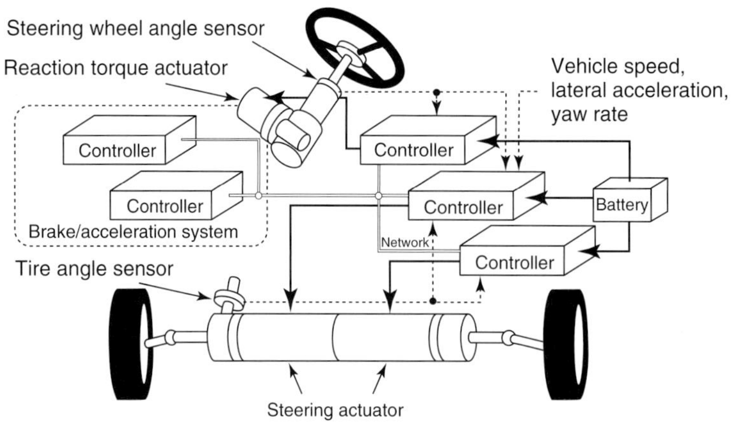

Figure 4. Integrated control steer-by-wire architecture. (From Hayama *et al.*, 2008. Copyright © 2008 SAE International. Reprinted with permission.)

transition of different steer-by-wire architectures, Hayama *et al.* (2008) suggest an integrated steer-by-wire architecture (Figure 4). Compared to the baseline structure, the integrated control architecture has two redundant steering actuators and a controller. In addition, the steering system has an interface to the braking and accelerating system in order to generate a yaw momentum by an appropriate wheel torque distribution, thus supporting the driver's steering intention.

Wallentowitz and Reif (2006) have presented another fault-tolerant architecture that can be used for steer-by-wire systems. The idea is to combine two fail silent units (FSUs) to one fault-tolerant unit (FTU) in order to increase system integrity. (Fail safe means that a system is in a safe state, even in case of a failure; fail silent means that the system does not generate adverse effects, by still being active). The superior fault-tolerant architecture (FTA) consists of one sensor, controller, and actuator FTU. Moreover, the power supply is redundant. The strategy to guarantee system integrity is based on local redundancies for E/E subsystems. Figure 5 shows the E/E architecture of the proposed steer-by-wire architecture. The hand wheel and front axle actuator are designed as FTU with two motors, three sensors on one single shaft, and two electronic control units. Each control unit processes the data of two different sensors and controls one actuator. Two electronic control units representing the central electronic control unit as FTU are connected to the actuator FTUs via a fault-tolerant real-time data bus system. This bus transfers the data of the sensor signals, the actuator set point commands, and the system status.

These two examples show that there is no general solution for a steer-by-wire architecture. The design depends on many factors such as system integration, steering functionality, and vehicle properties but will always comprise some degree of redundancy, which leads to additional costs.

3.3 Current steering systems relevant for steer-by-wire

This chapter gives an overview of steering systems that provide steer-by-wire functionality to some degree and that have already been introduced to the market. So far, no true steer-by-wire system has been presented beyond concept vehicles and prototypes; thus, the presented steering systems only fulfill the extended definition of steer-by-wire, for example, with a hydraulic or electrohydraulic linkage between steering control and steering system.

In the heavy equipment sector, for example, agriculture, construction, and forestry, vehicles with steer-by-wire functionality have already penetrated the market because legal regulations allowed such steering systems for vehicles with low maximum speeds at an early time. The main technical driver for the introduction of hydraulic steering systems without mechanical fallback in this sector have been huge vehicle dimensions resulting in high steering torques also due to heavy-duty tires. Moreover, autonomous driving plays an important role in some off-road applications.

In the agriculture sector, John Deere has presented the ActiveCommand Steering (ACS™) system. The ACS steering system offers a wheel-offset control, a variable steering ratio, and torque and eliminates steering wheel drift. The system consists of a power supply, a feedback unit, two controllers, control valves, and an electric-driven backup pump. In case of a lack of oil, the latter one supplies the oil to the steering system and brakes. The feedback unit applies light feedback in field use, slightly heavier feedback in transport mode and cornering at high speed. The system is able to correct minor tire angles, for example, when forward driving, to reduce necessary driver corrections. Considering the sensor technology, the system comes with a gyroscope to measure the tractor yaw

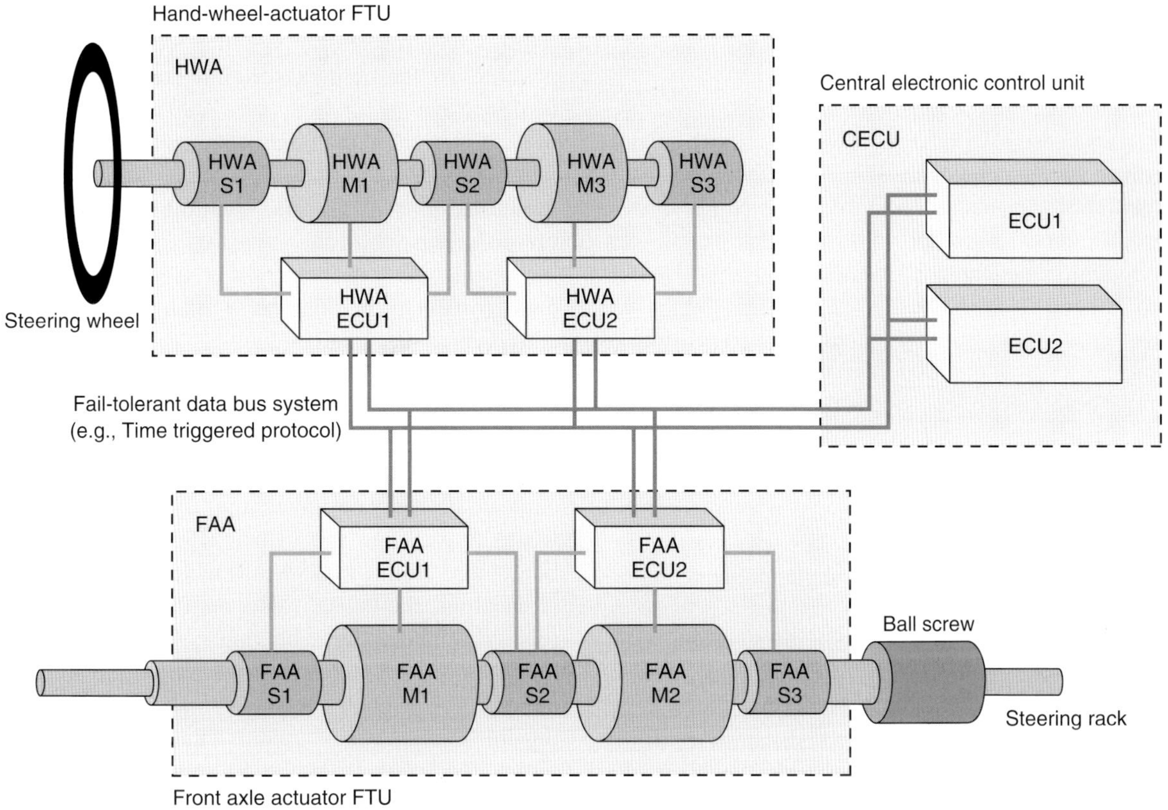

Figure 5. Fault-tolerant steer-by-wire architecture. (Reproduced from Wallentowitz and Reif, 2006. With kind permission of Springer Science+Business Media.)

rate, wheel angle sensors, and steering wheel angle sensors. The ACS is designed to fail operational. Basically, every function is backed up by another component. Thus, the primary controller is replaced by the second one in case of a fault. If the alternator power fails, the battery takes over. The electric-driven backup pump resumes control and supplies hydraulic oil to the system if the engine quits running (NN, 2012).

Other well-known examples are specialized heavy-duty vehicles for transporting, for example, production plants, which comprise several actively steered axles. In these applications, the angle and vertical travel of each wheel is controlled individually and actuated hydraulically.

In aviation technology, fly-by-wire systems are already state of the art. Besides the reduction of weight and volume, fly-by-wire systems offer the possibility of a variable computer-aided flight control. Binfet-Kull (2001) describes, among other advantages, the ability to handle the steering input feeling and the electrical transmission of pilot input signals along short to long distances within a short period of time at minimal energy. In addition, saving time-consuming adjustments of the mechanical steering system reduces assembling and maintenance costs. Binfet-Kull (2001) points out that the demands on steer-by-wire systems compared to fly-by-wire systems differ in major aspects. Examples are the danger potential, the highly trained operation personnel, and the maintenance intervals. First, if a fly-by-wire system fails, this breakdown could directly affect hundreds of human lives. Second, in comparison with normal car drivers, pilots are highly trained personnel who learned in many flight simulator hours how to react if technical faults cause worst-case scenarios. Finally, airplanes are subject to numerous inspection and maintenance intervals, the so-called letter checks, which secure its proper operation.

4 POTENTIAL

Steer-by-wire systems are able to resolve many compromises that occur during the conventional layout process of a steering system. These potentials are widely discussed throughout literature. Binfet-Kull (2001) provides a complete list of the potentials unlocked by steer-by-wire systems:

- Free configuration of steering ratio and steering assistance
- Improvement of active and passive safeties
- Maximum potential of functionality
- Suspension design without regarding steering feedback
- Improved package situation
- Reduction of variants
- Simplification of assembly
- Introduction of new human–machine interfaces (HMIs).

These aspects can be categorized into functionality potentials, package and production potentials, and HMI potentials. These categories are discussed in detail in the following sections.

4.1 Functionality potentials

Steer-by-wire inherently allows a simple implementation of any kind of steering function. Additional hardware is not necessary and communication with other systems can improve overall vehicle functionality beyond the steering system. Several related functions of these categories such as basic steering, vehicle dynamics, advanced driver assistance, and autonomous driving are introduced in the following sections.

4.1.1 Basic steering functions

The basic functionality of a steering system can be divided into two different aspects. On the one hand, the driver needs to set the trajectory of the vehicle. On the other hand, a feedback of the current state of the steering system or the vehicle dynamics can be provided to the driver.

As in a steer-by-wire system no mechanical connection between the steering wheel and the wheels exists, traditional conflicts in the layout process can be resolved. Transmission ratio and operating force can be chosen independently and individually. Thus, the functions of a conventional steering system "variable steering assistance" and "variable steering ratio" are inherently available in a steer-by-wire system. This allows, for example, a freely designed steering feedback, which could inform the driver of a significant change of the operating force when vehicle dynamics limits are about to be reached. As the correlation of steering wheel and wheel movement depends on software only, it can be adapted depending on the actual driving situation or personal preference of the driver.

4.1.2 Vehicle dynamics control

On the basis of a steer-by-wire system, yaw-rate and sideslip angle control can be realized. As no brake forces are utilized, as would be the case in conventional electronic stability control (ESC) systems, control intervention is possible in less critical driving states and is regarded less invasive by drivers. Yih (2005) derives that, in an oversteering situation, control intervention by means of the steering system is more efficient. On the one hand, this is due to the fact that the resulting lever arms from the tire contact patch to the center of gravity are longer, thus influencing the yaw behavior of the vehicle more than a brake force. On the other hand, both tires of the front axle contribute to the stabilization when the invention is realized by the steering system.

Breuer and Bill (2006) show the potential of an active steering system when braking on a μ-split surface. In this situation, the driver of a conventional vehicle is required to compensate the yaw momentum by operating the steering wheel to stabilize the vehicle. The standard driver will likely lose control of the vehicle. To reduce the yaw momentum, the brake force on the μ-high side can be reduced by the antilock braking system (ABS) leading to a longer braking distance. By controlling the steering angle, the yaw momentum can be compensated and maximum possible traction can be used to obtain the minimum braking distance. Results from the EU-project PEIT (powertrain equipped with intelligent technology) (Maisch *et al.*, 2005) show that these advantages can also be transferred to commercial vehicles.

In order to fully unlock the vehicle dynamics potential of a steer-by-wire system, it is necessary to integrate the steering control system into a global chassis control (GCC) system (Semmler and Rieth, 2004; Krüger, Pruckner, and Knobel, 2010). With this holistic approach, all degrees of freedom of the integrated chassis control systems can be used to obtain maximum driving dynamics performance. Consequently, the steer-by-wire system can provide a change in yaw momentum in any driving situation in combination with other chassis systems, not only in critical driving situations, but also during normal driving, for example, in order to compensate crosswind or road disturbances. Installing a more complex steer-by-wire system, for example, with individual wheel steering actuation, it is possible to extend the limits even further, as the friction potential at every wheel can be used in an optimal manner (Eckstein, 2012).

4.1.3 Advanced driver assistance systems (ADAS)

Steer-by-wire systems provide the best possible prerequisites to integrate the steering system into advanced driver assistance system (ADAS). Possible applications include functions such as lane-keeping and lane-change assistance and reverse driving trailer assistance. Another possible feature is (partly) autonomous parking, allowing to smoothly maneuvering the equipped vehicle hands-free into a parking space. Compared to conventional approaches to introduce these functions using an electromechanical steering or a superimposed steering system, steer-by-wire systems allow a realization without steering wheel movement or torques that need to be supported by the driver. Thus, also autonomous driving as demonstrated by Deutschle (2006) can be implemented. Steer-by-wire systems are able to completely fulfill any requirement of autonomous driving as the steering actuation is designed to perform any steering maneuver.

4.2 Package and production potentials

Two main package advantages result from the absence of a mechanical connection between steering wheel and the wheels. First of all, the absence of a steering column allows much better space utilization in the engine compartment as no specific packages have to be considered for right- or left-hand drive. In addition, during a frontal crash, there is less likelihood that the impact will force the steering wheel to intrude into the driver's survival space (Yih, 2005). In case of individually actuated steering of the wheels, the space usually occupied by the lateral connection, for example, by the rack and pinion steering gear is vacant as well.

Secondly, the entire steering mechanism can be designed and installed as a modular unit, thus leading to higher production volumes and easier assembly. Especially for commercial vehicles, the assembly process can be revolutionized with the introduction of steer-by-wire. Today, the chassis and the driver's cab are integrated at a very early stage because of the difficult assembly of the multipart steering column. Steer-by-wire would permit to assemble the chassis and the driver's cab independently, joining them at the very end of the production line. This would save significant space in the production plant and allow a more efficient task sharing.

Thirdly, steer-by-wire would allow an integrated cockpit concept including the steering device. Apart from advantages in production, this concept allows a significantly larger spread of vehicle derivatives especially with respect to the seating position and dashboard height. This is due to the fact that the mechanical steering column determines the angle of the steering wheel, which has a large effect on the driver's posture and is thus an important limiting factor.

4.3 HMI potentials

As, in a steer-by-wire system, operating forces are not directly transferred from the operating element to the wheel, the steering ratio between interface and wheel and the amplification of steering torque can be designed without any restrictions. In literature, this feature is sometimes referred to as *full steer-by-wire functionality*.

The benefit of freely designing the steering transmission ratio is the possibility to reduce the steering effort at low speeds and tight turns or while parking as well as to ensure a precise and stable control at high velocities. In contrast to a superimposed steering system, the spread in steering ratio is not limited by the fact that, in case of a failure, the ratio step must not exceed a value of about two points (e.g., from 17 to 19) in order to maintain sufficient controllability (Freitag *et al*., 2001). In addition, without a direct mechanical connection between the steering wheel and the road wheels, noise, vibration, and harshness (NVH) from the road no longer have a direct path to the driver's hands and arms through the steering wheel (Yih, 2005), thus increasing driving comfort.

Owing to the lack of necessity to apply the complete steering forces manually in case of a disabled power steering system, it is possible to introduce new and innovative steering controls to steer the vehicle. Winner and Heuss (2005) conclude that the use of a steering wheel is caused by the historical development of motor vehicles and does not necessarily need to be continued when steer-by-wire systems are introduced. Among many options, Winner believes that the most likely successful option would be an active joystick, that is, a force actuated stick that is located at both sides of the driver and provides an appropriate steering feedback. As early as 2001, Eckstein shows that this operating concept is suitable for controlling a passenger car (Figure 2). His investigations in a motion-based driving simulator show that novice drivers are able to learn to operate a vehicle with sidesticks just as well as a vehicle with a conventional steering wheel and pedals.

Another option, which arises, is the possibility to allow vehicle control from several positions inside or even remote control from outside a vehicle. This is especially of interest in special-purpose vehicles and heavy-duty trucks and already state of the art.

Winner and Hakuli (2006) introduce a consistent advancement of the steer-by-wire concept. Their paradigm of conduct-by-wire removes the driver from the vehicle stability control loop leaving the responsibility to command the vehicle on the guidance level. The vehicle will be controlled by passing maneuver commands to the system. The actuation of the individual vehicle variables, that is, driving and braking torques as well as steering angles, is determined by a centralized controller and executed autonomously. The responsibility to monitor proper operation of the vehicle remains with the driver. To implement this innovative driving paradigm, a complete x-by-wire architecture of the vehicle is required as well as a very complex sensor concept to capture the environment. The conduct-by-wire approach merges today's ADASs and vehicle stabilization systems into one complex vehicle conducting system.

5 CHALLENGES

Among the powerful potentials that can be realized by deployment of a steer-by-wire system, many challenges originate from it as well. Some of them are discussed in the following sections.

Two main aspects that have to be mentioned are reliability and safety. Besides these technical challenges, perception and acceptance by customers also play an outstanding role for possible success of this technology on the market. Acceptance also depends very much on costs.

In addition to technological and economic challenges, *legal aspects* also need to be addressed. While the relationship among the authorities, the manufacturers, and the driver is not affected in the first place, a clear definition of responsibilities between the vehicle manufacturer and the supplier of a steer-by-wire system is decisive and will also affect the development process and the system layout itself. The legal aspect is discussed first in the following section.

5.1 Legal requirements for steer-by-wire systems

Legal requirements such as the ECE reg. 79 (2005) define rules regarding the approval of steering equipment for vehicles. Revision 2 of the ECE reg. 79 (2005) from 2005 considers the advancements in steering technology and the advantages of steer-by-wire systems compared to steering systems having a mechanical link. Thus, it is nowadays possible to approve steer-by-wire systems having no mechanical connection between wheels and steering control.

According to the definition of ECE reg. 79 (2005), steer-by-wire systems belong to the steering equipment classification of full-power steering equipment. Solely one or more energy supplies provide, that is, the steering forces.

The ECE reg. 79 (2005) provides additional requirements that full-power steering systems need to fulfill. Regarding the system design, the regulation specifies different failure provisions:

- A failure in the transmission that is not purely mechanical has to be indicated to the vehicle driver via predefined warning signals. In this failure mode, a change in the average steering ratio is allowed only if the steering effort is not exceeding a predefined maximum effort value of 300 or 450 N over a period of time of 4 or 6 s, both values depending on the vehicle category.
- In the event of a failure in the energy source of the control transmission, the vehicle having an energy storage level at which the failure was signaled to the driver should be able to drive at least 24 "figure of eight" maneuvers with a loop diameter of 40 m at a velocity of 10 km/h with the same performance level given for an intact system.
- If a failure in the energy transmission occurs, with the exception of the parts not liable to breakage (e.g., steering column), no immediate change in the steering angle is permissible. A vehicle, still able to drive faster than 10 km/h, has to fulfill all test provisions after having finished at least 25 "figure of eight" maneuvers with a 40 m diameter at 10 km/h minimum speed. The energy storage level at the beginning of the test maneuver has to be the same as in the failure of the energy source of the control transmission.

The development process of steer-by-wire systems has to be in accordance with relevant standards, such as the ISO 26262, which address the functional safety of road vehicles. According to the ECE reg. 79 (2005), the manufacturer of complex electronic vehicle control systems has to accomplish specific requirements for documentation, fault strategy, and verification with respect to the safety aspects. Concerning the fault strategy, the design process has to guarantee the safe operation of the vehicle fail-safe procedures, required redundancies or necessary driver warning systems. ECE reg. 79 (2005) gives examples for design provisions for a system failure, which are for example

- fallback to operation using a partial system,
- change over to a separate back-up system, and
- removal of high level functions.

In summary, for approval of such a safety-critical system, the manufacturer has to provide a complete documentation of the transparent design process and on the system means to guarantee its safe operation.

5.2 Technical challenges

From legal requirements, customer demands, and OEM (Original Equipment Manufacturer) design goals, a complex list of technical challenges results. Accordingly, some of the most important challenges include the provision of functional safety, acceptable steering feel, and reducing costs and weight.

5.2.1 Safety

While mechanical systems are considered inherently safe, as a mechanical failure can be detected by inspection and is unlikely to fail in a sudden manner when properly designed and used within its specification boundaries, a mechatronic system can fail without prior signs of damage or wear. Thus, great care has to be taken designing a system that ensures a safe operation as well as acceptable reliability.

To achieve a safe state of a vehicle when a system failure occurs, it is obviously sufficient if the vehicle is stopped at a suitable location, for example, the emergency lane of a highway. To reach this safe state, it is important that any fault does not lead to an instable reaction of the vehicle and that the vehicle remains controllable by the driver during the emergency maneuver. This means that

1. the initial effect of a fault must be limited,
2. the cause of the fault is detected and addressed within a finite time (typically some milliseconds), and
3. the resulting vehicle behavior can be controlled safely by the driver.

Customer requirements by far exceed the demand for a safe system according to the earlier definition. System reliability plays an important role in terms of availability of all system and vehicle functions, as it greatly affects customer acceptance and warranty costs for the OEM. Depending on the likeliness of a fault occurrence and the presumed consequence, the developer needs to decide, which countermeasures need to be taken to reach a system reliability that guarantees sufficient availability.

A comparison to current solutions in the aeronautical industry shows that the general approach is to maintain control over the vehicle with no or only manageable interruptions. This can be achieved by either introducing redundancies that are able to fulfill the original functionality without limitations or time limited, thus providing a fail-operational behavior. Examples for redundancies are duplicated bus systems to ensure data transmission or additional batteries that allow further operation of the system for a limited time in case the main power supply should fail.

Besides the deployment of redundancies, a restricted functionality resulting from a failure of a subsystem or a combination of different faults is often commonly used in aeronautics. For example, it is possible to operate an airplane using only some of the available control surfaces. The plane is still controllable although some maneuvers might not be possible. This state of the system with reduced functionality is referred to as a *degraded state*.

Within the automobile industry, the concept of degraded states is used as well. ESC systems will be deactivated in case of a detected fault, whereas the basic hydraulic brake system is still in operation. Similarly, steering assistance by a power steering system is deactivated when a fault occurs. Both breakdown situations are accepted as a safe state although functionality and controllability are considerably reduced.

Proposals for steer-by-wire system architectures including suggestions for redundancies and degraded states can be found in literature (Binfet-Kull, 2001; Freitag *et al.*, 2001; Heitzer, 2003; Wallentowitz and Reif, 2006; Hayama *et al.*, 2008). To achieve a robust vehicle behavior, fault and event detection is a key factor. In this respect, the analogy with aviation may not be valid, as the required system reaction time in a vehicle driving on the road, for example, to prevent an unintended lane departure, is usually far shorter than those of an airplane.

5.2.2 Steering feel

As for any subjective assessment of a technical system when evaluating steering systems, different approaches and criteria are used depending on brand, vehicle class, or regional preferences. When new technologies are introduced, usually their impact and acceptance are discussed controversially. When, for example, EPS was first implemented in passenger cars, the fact that the steering feel differs from conventional hydraulic power steering systems led to criticism of the innovative system for its synthetic steering feel. As EPS[1] offers several additional innovative functions such as lane keeping, automatical[2] parking, side wind compensation, and even more, it was not surprising that EPS systems significantly penetrated the market. On the one hand, this was certainly due to an improved parameterization of the system functions and a reduction in fuel consumption; on the other hand, it was also successful due to a certain habituation effect by the drivers.

Regarding steer-by-wire systems, it is evident that steering feel needs to be provided completely on a synthetic basis, which may compose of an electromechanical steering feedback actuator as well as passive elements such as springs and dampers. The passive elements can assure the stability of the dynamic behavior of the steering wheel system itself in case of a fault of the feedback actuator. Using all degrees of freedom in designing steering feel dependent on, for example, vehicle speed, lateral acceleration, road friction, or lane keeping, a considerable amount of effort is necessary to achieve an adequate steering feel in all relevant driving states (Koch, 2010). On the other hand, the freedom in designing steering feel must be regarded as a big advantage of a steer-by-wire system, as a specific steering feel can easily be provided for different vehicle models and markets including parameters for individualization and customization.

5.2.3 *Costs and weight*

From an OEM perspective, using the same steer-by-wire system across many vehicle models and markets yields cost savings because of economies of scale and reduction of variants. At least part of these cost savings need to be invested in compensating higher system costs. These can be expected to be higher even compared to a superimposed steering system, as additional components are needed to provide steering feel and redundancy of the steering actuator. The precise effect of this change in technology on costs, weight, and efficiency largely depends on the specific system layout and thus cannot be quantified in general. Aspects such as the number, location and performance of actuators, and the type of driver interface will be most influential. Barthenheier (2002) expects that it will be possible to reduce cost and weight because of the fact that many parts can be removed from the system. Other sources (Fleck, 2003; Grell, 2003) state that no positive effects are to be expected as the effort to achieve a safe system with redundancies will overcompensate the advantages. In the end, the success of *steer-by-wire systems* will largely depend on whether or not the customer has to expect additional costz for the system. This refers to purchase as well as to operating and maintenance costs.

Regarding the development cost for steer-by-wire systems, as for any complex technical innovation, the initial research investment is very high. If steer-by-wire systems are introduced in a small segment such as luxury vehicles, development costs are hardly acceptable (Winner *et al.*, 2004). If, on the other hand, steer-by-wire would be widely introduced in mass production, the risk from an economic and marketing point of view is high if a large number of vehicles need to be recalled in case of quality problems.

5.3 Customer acceptance

When steer-by-wire systems are introduced to the automotive market, they will have to compete with current modern steering systems. As described earlier, steer-by-wire systems do have significant advantages regarding possible steering functions, although the advantages might not be immediately evident for the average driver. The main marketing challenge for the successful introduction of steer-by-wire systems will be to clearly communicate the advantages of the new and innovative system.

Whether this will be possible is not assessable from today's perspective, as customer acceptance is dependent on many influences and is hardly ascertainable in an early stage of the development process as Binfet-Kull (2001) states. Daniels (2003) expects that a major challenge will be skepticism toward the safety of a steering system lacking a mechanical connection between the steering wheel and wheels. The mechanical system has proved its reliability in recent decades and in general customer perception it is regarded as safe, whereas many drivers have already experienced failures of electrical systems—whether in their car or using consumer electronics.

The first steer-by-wire systems thus need to comprise a very convincing safety concept, which is highly effective and reliable. Communication should focus on evident benefits of such a system and ideally avoid the term *by-wire* in order to generate a positive perception of this promising innovation.

6 SUMMARY

While substantial research on steer-by-wire systems has been carried out over the past decades, this innovative approach has not made its way into volume production. Active steering systems superimposing an additional steering angle to the driver's command can be regarded as forerunner of steer-by-wire, as their performance would be sufficient to influence lateral vehicle dynamics under normal driving conditions while holding the steering wheel straight. In contrast to steer-by-wire, the torque provided by the steering actuator needs to be counterbalanced by the driver, and in case of a system malfunction, the mechanical steering column represents a well-known fallback.

While active steering and electric-powered steering systems already enable innovative functionality regarding vehicle stabilization and advanced driver assistance, full steer-by-wire systems offer three areas of potential: firstly, 100% freedom in designing the functional relationship between the operating device and the steered wheels, secondly more flexibility in package and production yielding time and cost savings, and finally maximum design freedom regarding the driver's working place.

On the other hand, the market introduction of steer-by-wire systems faces some major challenges: apart from legal requirements, the main challenge concerns the provision of functional safety while at the same time limiting the costs due to resulting requirements, for example, on redundancies of sensors, communication, and actuators. Finally, customer acceptance is the prerequisite for market success—the benefit of the introduction of steer-by-wire has to be understood immediately providing a unique driving experience.

REFERENCES

Alexandre, H. (1894) Voitures Automobiles. L'ingénieur Civil, September 15, 1894.

Barthenheier, T. (2002) *Steer-by-Wire-Systeme – Stand und Entwicklungsaussichten*. 47. Internationales Wissenschaftliches Kolloquium, X-by-Wire-Workshop, September 2002, Ilmenau.

Bayle, R. and Ecquevilly, Y. (1972) *Servomechanismus*, Patent DE 2237166

Bidwell, J.B. and Cataldo, R.S. (1958) Single Stick Member for Controlling the Operation of Motor Vehicles. US patent 3 022 850.

Bidwell, J.B. (1959) *Vehicles and drivers—1980*. Presentation at the SAE Annual Meeting, Sheraton-Cadillac and Statler Hotels, Detroit, Michigan, January 12–16.

Binfet-Kull, M. (2001) *Entwicklung einer Steer-by-Wire Architektur nach zuverlässigkeits- und sicherheitstechnischen Vorgaben*, Verlag Mainz, Mainz.

Bränneby, P., Palmgren, B., Isaksson, A., *et al.* (1991). Improved Active and Passive Safety by Using Active Lateral Dynamic Control and an Unconventional Steering Unit. *13th International Technical Conference on Experimental Safety Vehicles, Proceedings 1*, Paris, pp. 224–230.

Braess, H.-H. (2001) Lenkung und Lenkverhalten von Personenkraftwagen. Was haben die letzten 50 Jahre gebracht, was kann und muss noch getan werden? *VDI-Berichte*, **1632**, 13–55.

Breuer, B. and Bill, K.H. (2006) *Bremsenhandbuch*, Friedr. Vieweg & Sohn Verlag/GWV Fachverlage GmbH, Wiesbaden.

Brockhaus, R., Alles, W., and Luckner, R. (2011) *Flugregelung*, Springer-Verlag, Berlin Heidelberg.

Broggi, A., Bombini, L., and Cattani, S., *et al.* (2010) *Sensing requirements for a 13,000 km intercontinental autonomous drive*. 2010 IEEE Intelligent Vehicles Symposium, University of California, San Diego, CA, June 21–24, 2010.

Daniels, J. (2003) Hidden wires *European Automotive Design*, **2003**, 29–31.

Davis, F.W. (1927) *Hydraulic Steering Mechanism*, Patent US 1 790 620.

Davis, M.W.R. (2004) *General Motors a Photographic History*, Arcadia Publishing, Charleston SC.

Deutschle, S. (2006) The KONVOI Project—Development and Evaluation of Electronic Truck Platoons on Highways. *Proceedings of 15th Aachen Colloquium "Automobile and Engine Technology" 2006*, Aachen.

Dick, R. (2004) *Mercedes and Auto Racing in the Belle Epoque, 1895–1915*, Mcfarland & Co Inc, Jefferson, NC.

ECE R79 (2005) Regulation No. 79—Rev. 2—Steering Equipment, http://live.unece.org/fileadmin/DAM/trans/main/wp29/wp29regs/r079r2e.pdf (accessed 25 November 2007).

Eckstein, L. (2000) Sidesticks im Kraftfahrzeug – ein alternatives Bedienkonzept oder Spielerei? Ergonomie und Verkehrssicherheit. *GfA Konferenzbeiträge der Herbstkonferenz 2000*, 12.-13. Oktober 2000 an der Technischen Universität München, Herbert Utz Verlag, München.

Eckstein, L. (2001) *Entwicklung und Überprüfung eines Bedienkonzepts und von Algorithmen zum Fahren eines Kraftfahrzeugs mit aktiven Sidesticks*, VDI Verlag GmbH, Düsseldorf.

Eckstein, L. (2012) *Vertical and Lateral Dynamics of Vehicles*, Forschungsgesellschaft Kraftfahrwesen Aachen mbH, Aachen.

Fenton, R.E. (1994) IVHS/AHS: driving into the future *Control Systems, IEEE*, **14** (6), 13–20.

Fleck, R. (2003) Systematic Development of Mechatronic Steering Systems with Steer-by-Wire Functionality. *Proceedings of fahrwerk.tech 2003*, München.

Freitag, R., Moser, M., Hartl, M., *et al.* (2001) Safety Concept Requirements of Steering Systems with Steer-by-Wire Functionality. *Proceedings of "Internationale Tagung Elektronik im Kraftfahrzeug"*, VDI-Berichte Nr. 1646, 2001. VDI Verlag GmbH, Düsseldorf.

Grell, D. (2003) *Innovationslawine in der Autotechnik. C't – Magazin für Computertechnik, 14/2003*, Heise Zeitschriften Verlag GmbH & Co. KG, Hannover.

Hayama, R., Higashi, M., Kawahara, S., *et al.* (2008) Fault-tolerant architecture of yaw moment management with steer-by-wire, active braking and driving-torque distribution integrated control. SAE World Congress & Exhibition, April 2008, Detroit, MI. Session: Safety-Critical Systems (Part 2 of 3).

Heitzer, H. (2003) Entwicklung eines fehlertoleranten Steer-by-Wire Lenksystems. *Tagung "PKW-Lenksysteme"*, Haus der Technik, Essen.

Henke, R. (2010) Vorlesung Flugzeugbau II (SS2010) Version 2.0. Thema 8: Fly-by-Wire.

Joint Aviation Authorities Committee (1989) Joint Aviation Requirements, JAR 25, Large Aeroplanes, ACJ No. 1 to JAR 25.1309, http://www.jaa.nl/publications/crd/jar-25-change13.pdf (accessed 25 November 2007).

Kramer, F. (2009) *Passive Sicherheit von Kraftfahrzeugen*, Vieweg + Teubner Verlag, Wiesbaden.

Krüger, J., Pruckner A., and Knobel C. (2010) Control Allocation for Road Vehicles—A System-Independent Approach for Integrated Vehicle Control. *Proceedings of 19th Aachen Colloquium "Automobile and Engine Technology" 2010*, Aachen.

Koch, T. (2010) Untersuchungen zum Lenkgefühl von Steer-by-Wire Lenksystemen. Dissertation. TU München, München.

Köhn, P., Baumgarten, G., Richter, T., *et al.* (2002) Active Steering—The BMW Approach to Modern Steering Technology. *Proceedings of 11th Aachen Colloquium "Automobile and Engine Technology", Band 2*, Aachen, 7.-9. Okt.

Linderer, W. (1951) *Einrichtung zum Schutze von in Fahrzeugen befindlichen Personen gegen Verletzungen bei Zusammenstößen*, Patent DE 896 312.

Maisch, A., Kandar, T. and Arpad, M. *et al.* (2005) PEIT: powertrain equipped with intelligent technologies. Deliverable D15, Final Report.

NN (2003) *Annual Report 2003 Toyoda Machine Works, LTD*, http://www.jtekt.co.jp/ir/pdf/ar_2003.pdf (accessed 25 November 2007).

NN (2012). *John Deere Homepage—Presentation of Active Command Steering*, http://www.deere.com/wps/dcom/en_US/products/equipment/tractors/row_crop_tractors/8r_8rt_series/8260r/8260r.page (accessed 25 November 2007).

Patzelt, H., Schiesterl, G. and Seybold, A. (1971) *Schutzvorrichtung, insbesondere für die Insassen von Kraftfahrzeugen*, Patent DE 2152902.

Pfeffer, P. (2011) *Lenkungshandbuch*, Vieweg + Teubner Verlag, Wiesbaden.

Pilon, H. M., Park, A., Sattavara, S. W., *et al.* (1972) *Power Steering Gear Actuator* Patent US 3831701.

Reichel, R., Armbruster, M. and Bäuerle, K. (2004) Gemeinsame Systemstrukturen im Flugzeug-und Automotiv-Bereich für sicherheitskritische Steuerfunktionen *5. Braunschweiger Symposium Automatisierungs- und Assistenzsysteme für Transportmittel*, Braunschweig, 200–220.

Rixmann, W. (1962) Die Daimler-Benz Servolenkung *Automobiltechnische Zeitschrift*, Jahrgang 64, Heft 1, 1–9.

Semmler, S.J. and Rieth, P.E. (2004) Global Chassis Control—The Networked Chassis. *Proceedings of 13th Aachen Colloquium "Automobile and Engine Technology" 2004*, Aachen.

Stiller, C. (2007) Autonome Mobile Systeme *Informatik aktuell*, Part 6, 163–170.

Sun, Z., Bebis, G., and Miller, R. (2006) On-road vehicle detection: a review *IEEE Transactions on Pattern Analysis and Machine Intelligence*, **28** (5), 694–711.

Stoll, H. (1992) *Fahrwerktechnik: Lenkanlagen und Hilfskraftlenkungen*, Vogel Verlag, Würzburg.

Wallentowitz, H. and Reif, K. (2006) *Handbuch Kraftfahrzeugelektronik. Grundlagen, Komponenten, Systeme, Anwendungen*, Friedr. Vieweg & Sohn Verlag/GWV Fachverlage GmbH, Wiesbaden.

Winner, H., Isermann, R., Hanselka, H., and Schürr, A. (2004). When Does By-Wire Arrives Brakes and Steering? *Proceedings of AUTOREG 2004*, VDI Berichte 1828, VDI Verlag GmbH, Düsseldorf.

Winner, H. and Heuss, O. (2005) X-by-Wire Betätigungselemente – Überblick und Ausblick. *Darmstädter Kolloquium Mensch & Fahrzeug*, Darmstadt. 08.+09. März, ISBN:3-935089-83-X.

Winner, H. and Hakuli, S. (2006). Conduct-by-Wire – Following a New Paradigm for Driving into the Future. *Proceedings of FISITA World Automotive Congress*, 22.-27. Oktober 2006 in Yokohama, Japan.

Yih, P. (2005) Steer-by-wire: implications for vehicle handling and safety. Dissertation. Stanford University, Stanford.

Handling

Chapter 127

The Potential for Handling Improvements by Global Chassis Control

Thomas Raste and Peter E. Rieth

Continental Corporation, Frankfurt, Germany

1 INTRODUCTION

From a driver's perspective, controlling of a vehicle means controlling the speed and the path curvature (Figure 1). In exceptional circumstances, for example, in emergency evading situations, also the orientation of the vehicle has to be controlled. In a narrower sense, vehicle handling refers to vehicle dynamics such as cornering and swerving and includes the vehicle stability.

In day-to-day use, speed control by braking or accelerating is largely decoupled from handling, because the steering wheel is turned slowly to steer the car. When driving on race tracks or in emergency evading situations, the steering rate reaches very high values. Skilled drivers are able to manage steering rates of 1000 deg/s or more. In those situations, the capability to follow a path is strongly depending on the speed. The most limiting factor is the tire–road friction, which restricts the lateral acceleration depending on the road conditions. Equation 1 shows how lateral (centripetal) acceleration a_y, vehicle speed v, radius R of the path, yaw rate ω, and sideslip ("spin") angle rate $\dot{\beta}$ are related

$$a_y = \frac{v_2}{R}, \quad R = \frac{v}{\omega + \dot{\beta}} \tag{1}$$

2 ACTIVE SYSTEMS

Active systems, which mean systems with auxiliary power and electronic control, are well suited to improve the handling of the vehicle. Especially, active braking has been established as the most effective active safety system. Electronic stability control (ESC) is now becoming mandatory in many countries all over the world. However, also active systems primarily designed for comfort are able to contribute to a better handling of the vehicle (Figure 2).

2.1 Requirements for active systems

Passive components affecting the handling of the vehicle need targets for the design in every stage of the development process (Figure 3). The feedback of extensive evaluations is used "offline" to tune the passive parts. Active systems lay a foundation for a new perspective in this development process. They offer the opportunity to determine their effectiveness "online" during driving based on the design targets. The active systems, therefore, need to be as generic as possible to minimize additional costs when setting up a new configuration (Andreasson, 2007).

Encyclopedia of Automotive Engineering, print © 2015 John Wiley & Sons, Ltd.
Edited by David Crolla, David E. Foster, Toshio Kobayashi, and Nick Vaughan.
This article is © 2015 John Wiley & Sons, Ltd. ISBN: 978-0-470-97402-5
Also published in the *Encyclopedia of Automotive Engineering* (online edition)
DOI: 10.1002/9781118354179.auto022

Figure 1. The driver is controlling the vehicle's speed, path, and orientation. (Reproduced by permission of Continental.)

Figure 2. Vehicle with electronic air suspension and electronic adjustable damper. (Reproduced by permission of Continental.)

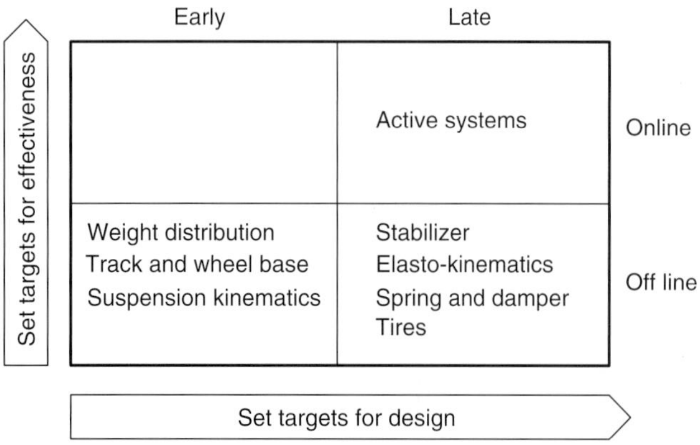

Figure 3. Target setting in the vehicle development process for passive components affecting vehicle handling in comparison to active systems. (Reproduced by permission of Continental.)

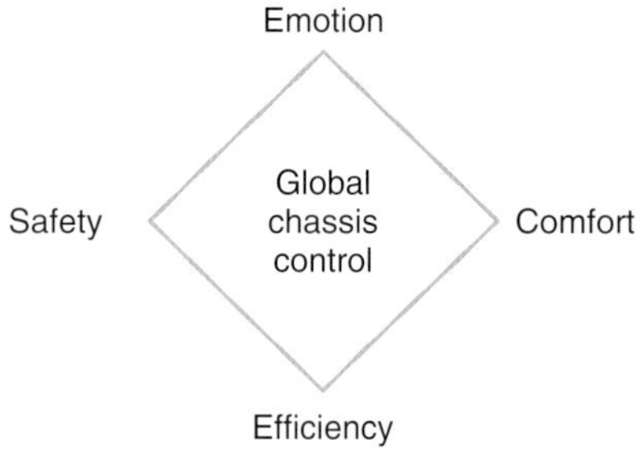

Figure 4. Trade-offs for global chassis control. (Reproduced by permission of Continental.)

The requirements for global chassis control are not unique. Figure 4 illustrates that there is usually a trade-off among emotion, safety, comfort, and efficiency. While safety will never be compromised, the remaining aspects are variable in positioning. There is a trend toward individualization and personalization of vehicle handling functions. The driver can select from different profiles with the push of a button. End consumers can thus buy just one vehicle and yet have the experience of driving different vehicle types. Above and beyond this, for example, for hybrid vehicles to come, a smooth blending of the friction brake and the generator brake for decelerating allows efficient driving with lower CO_2 emissions on a day-to-day basis (Bauer, Raste, and Rieth, 2007).

Active systems are safety-relevant components, which comprise the risk of malfunction. To minimize the risk, the development process has to be according to the safety standard ISO 26262 "Road vehicles—Functional safety." The criticality associated with a function of the system is the result of a hazard analysis and risk assessment and is classified by the Automotive Safety Integrity Level (ASIL). The classification reaches either from QM (quality management, not safety relevant) or from ASIL A (lowest level) to

Table 1. Safety requirements and ASIL of typical active systems.

Active System	Safety Requirement	Typical ASIL
ESC	Avoids dangerous false brake intervention	ASIL D
AFS	Avoids dangerous false steer angle	ASIL D

Reproduced by permission of Continental.

ASIL D (highest level). The ASIL is then inherited by the software and hardware elements that realize the function and defines the safety requirements that must be fulfilled during concept phase, product development, and production and operation of the system. Typical safety requirements and ASIL classification can be found in Table 1.

2.2 Portfolio of active systems

In Figure 5, a portfolio of currently available active systems and their effectiveness in the regions of normal driving and at the friction limit is shown. The effectiveness of the individual standalone systems can be extended significantly by networking with other active systems or surrounding sensor systems (Raste, Bauer, and Rieth, 2008). Currently, the following topics are being further developed:

- specification of the areas in which the vehicle dynamics should be defined by global chassis control;
- composition of the best active system portfolio for a specific vehicle or a family of vehicles;
- partitioning of control functions on a certain electronics architecture with the need for managing complexity.

The path to a consistent, cross-vendor coordination approach for chassis control systems is still far away. However, there is consensus on the objectives: under normal operating range, the controller ensures maximum comfort and driving pleasure. In this case, the vehicle manufacturer determines all degrees of freedom for the individual setting of the vehicle character. In the friction limit range, all available actuators in the system are included in a coordinated manner to reach one target: to support the driver optimally for accident avoidance.

2.3 Potential of active systems

Figure 6 illustrates the contact patch tire forces and velocities of a single wheel during driving and how the wheel contributes to the total yaw moment of the vehicle. The wheel is steered by the angle δ and the actual direction of travel with velocity v_R defines the sideslip angle α. The resultant horizontal force F_R points toward the opposite direction of the contact patch sliding velocity v_G. The sliding velocity components v_{Gx}, v_{Gy}, each related to the longitudinal velocity v_{Rx} of the wheel center, define the tire

Effect plane	Active system	Normal driving range				Friction limit range		
		Ride comfort (z,θ,φ)	Agility (y,ψ)	Operational comfort	Ride safety (y,ψ)	Stability (x,y,ψ,φ)	Stopping distance	Traction
Horizontal	ESC electronic stability control		+	+	+	o	o	o
	ATV active torque vectoring		o	o	+	+		o
	ARK active rear axle kinematics		o	o	+	+	+	
	AFS active front steering		o	o	+	+	+	
	EPS electric power steering			o	+	+	+	
Vertical	EAS electronic air suspension	o		o		+		
	ARS active roll stabilizer	o	o			+		
	EAD electronic adjustable damper	o	o			+	+	+
	ABC active body control	o	o		+	+	+	+

Figure 5. Portfolio of active systems and their effectiveness. The symbol "o" denotes the main improvements of the active system when acting standalone and "+" shows the potential improvements by networking with other active systems or surrounding sensor systems. (Reproduced by permission of Continental.)

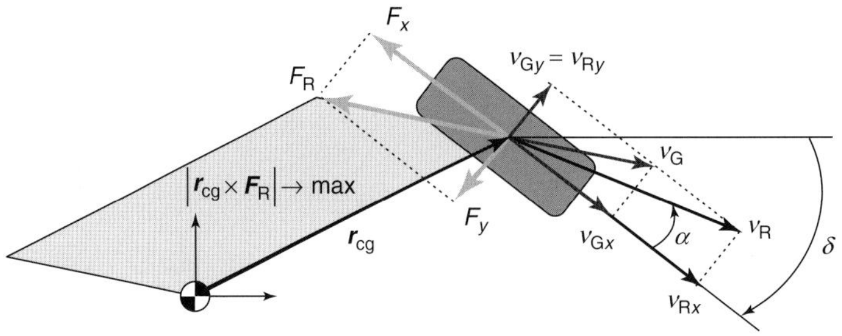

Figure 6. Tire forces, wheel and contact patch velocities, and contribution of the wheel to the yaw moment around the center of gravity of the vehicle. (Reproduced by permission of Continental.)

slip and sideslip angles, which, on the other hand, determine the magnitude of the forces F_x and F_y.

The magnitude of F_R is limited by the friction circle (Kamm's circle). The radius of the friction circle is determined by the product of the tire–road friction coefficient μ and the vertical tire force F_z. The portion of the yaw moment generated by each individual wheel is determined by the scalar product of the total tire force vector F_R and the position vector r_{cg}, denoting the distance between the vehicle center of gravity and the wheel center. The potential of an active system to maximize the scalar product is to increase the resulting area

- by aligning position and force vector orthogonally via steer angle (active rear axle kinematics (ARK), AFS, and electric power steering (EPS));
- by increasing the magnitude of the longitudinal force vector by wheel individual brake or propulsion intervention (ESC and active torque vectoring (ATV));
- by decreasing the magnitude of the lateral force vector by wheel load distribution (EAS, active rear wheel steering (ARS), EAD, and active body control (ABC)).

Figure 7 shows the potential of active brake, steering, and suspension systems selected from Figure 5 to generate additional yaw moments when activated during steady-state driving with constant radius (Schiebahn, Zegelaar, and Hofmann, 2007). It is evident that in the friction limit range, ESC has the highest potential to stabilize an oversteering vehicle. The active steering systems are highly effective to reduce lateral forces within the friction limit range. In case of AFS, this leads to a high turning-out yaw moment, whereas, in the case of ARK, this leads to considerable turning-in yaw moment. Both steering systems show high potential with opposite effectiveness within the normal driving range but only poor authority to increase lateral forces within the friction limit range. The potential of the active suspension system ARS depends as a first approximation on lateral acceleration.

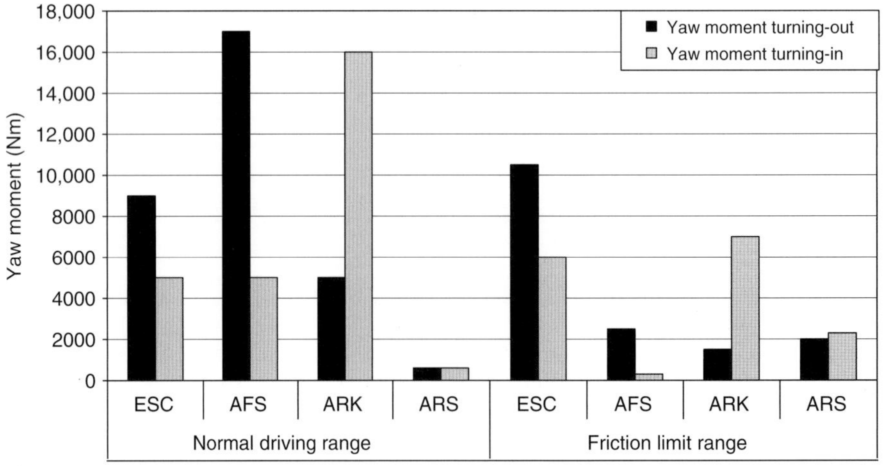

Figure 7. Potential of active brake, steering and suspension systems to generate additional yaw moment to force a vehicle in or out a curve, respectively, when driving in steady state with constant radius. (Reproduced by permission of Continental.)

3 VEHICLE DYNAMICS MODEL

In this section, a mathematical model describing the lateral motion of the vehicle is defined. The model is only valid for lateral accelerations below $0.4g$ and is illustrated in Figure 8.

3.1 Equations of motion

To derive the equations of motion, it is assumed that the center of gravity is on ground level and the steering angle δ_{F} on front axle and δ_{R} rear axle is small. The equations of motion of the lateral and yaw motions are given by

$$mv(\dot{\beta} + \omega) = F_{y\mathrm{F}} + F_{y\mathrm{R}} \tag{2}$$

$$J_z\dot{\omega} = l_{\mathrm{F}}F_{y\mathrm{F}} - l_{\mathrm{R}}F_{y\mathrm{R}} - \frac{b_{\mathrm{F}}}{2}F_{x\mathrm{FL}} + \frac{b_{\mathrm{F}}}{2}F_{x\mathrm{FR}} - \frac{b_{\mathrm{R}}}{2}F_{x\mathrm{RL}} + \frac{b_{\mathrm{R}}}{2}F_{x\mathrm{RR}} \tag{3}$$

where $F_{y\mathrm{F}}$ and $F_{y\mathrm{R}}$ are the combined front and rear lateral tire forces. The model parameters are the vehicle mass m, the moment of inertia around the z-axis J_z, the distances l_{F} and l_{R} from the front and rear axles to the center of gravity, and the front and rear track widths b_{F} and b_{R}. The lateral tire forces are assumed to be a linear function of the sideslip angles of the wheels with the effective cornering stiffness C_{F} and C_{R}

$$F_{y\mathrm{F}} \approx -C_{\mathrm{F}}\alpha_{\mathrm{F}}, \quad \text{where } \alpha_{\mathrm{F}} = -\left(\delta_{\mathrm{F}} - \beta - \frac{l_F}{v}\omega\right) \tag{4}$$

$$F_{y\mathrm{R}} \approx -C_{\mathrm{R}}\alpha_{\mathrm{R}}, \quad \text{where } \alpha_{\mathrm{R}} = -\left(\delta_{\mathrm{R}} - \beta + \frac{l_R}{v}\omega\right) \tag{5}$$

3.2 State-space equation

The state-space equation establishes a relationship between the system's current state and its input, and the future state of the system. The state-space form of the vehicle model is linear and time invariant if the vehicle speed v is considered to be a constant parameter. The system state vector x includes the sideslip angle β and the yaw rate ω. The input vector u contains the two steer angles and the four longitudinal forces illustrated in Figure 8. In addition to these six inputs, two more inputs are introduced, a (virtual) lateral force F_y and a (virtual) yaw moment M_z. These additional inputs are used as virtual control commands for controllability analysis in Section 5.2. The state-space equation of system (A and B) can be derived from Equations 2–5 and is written as

$$\dot{x} = Ax + Bu \tag{6}$$

$$Ax = \begin{bmatrix} -\frac{C_F + C_R}{mv} & -\frac{\mathrm{l}_F C_F - \mathrm{l}_R C_R}{mv^2} - 1 \\ -\frac{\mathrm{l}_F C_F - \mathrm{l}_R C_R}{\mathrm{J_z}} & -\frac{\mathrm{l}_F^2 C_F + \mathrm{l}_R^2 C_R}{\mathrm{J_z}^{\mathrm{v}}} \end{bmatrix} \begin{bmatrix} \beta \\ \omega \end{bmatrix} \tag{7}$$

$$Bu = \begin{bmatrix} \frac{C_F}{mv} & \frac{C_R}{mv} & 0 & 0 & 0 & 0 & \frac{1}{mv} & 0 \\ \frac{l_{\mathrm{F}}C_{\mathrm{F}}}{J_z} & -\frac{l_{\mathrm{R}}C_{\mathrm{R}}}{J_z} & -\frac{b_{\mathrm{F}}}{2J_z} & \frac{b_{\mathrm{F}}}{2J_z} & -\frac{b_{\mathrm{R}}}{2J_z} & \frac{b_{\mathrm{R}}}{2J_z} & 0 & \frac{1}{J_z} \end{bmatrix} \times \begin{bmatrix} \delta_{\mathrm{F}} \\ \delta_{\mathrm{R}} \\ F_{x\mathrm{FL}} \\ F_{x\mathrm{FR}} \\ F_{x\mathrm{RL}} \\ F_{x\mathrm{RR}} \\ F_y \\ M_z \end{bmatrix} \tag{8}$$

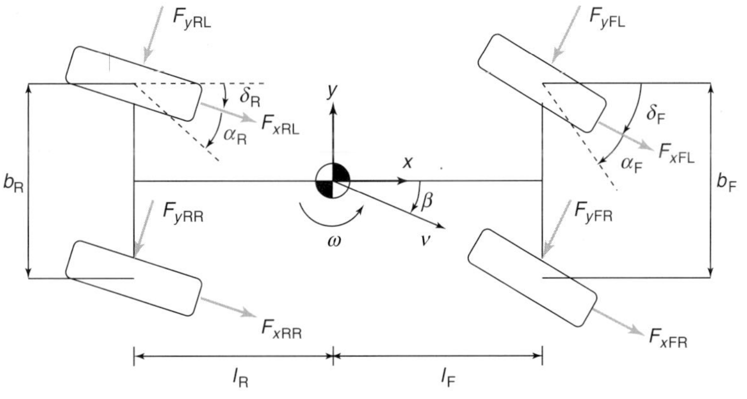

Figure 8. Vehicle model with the tire forces F_x and F_y, the vehicle speed v, the yaw rate ω, and the vehicle sideslip angle β. (Reproduced by permission of Continental.)

The Equations 6–8 can be simplified for stationary operation with $\dot{x} = 0$ and the steer angle inputs only ($F_x = 0$, $F_y = 0$, $M_z = 0$). Under the assumption that the system matrix A is regular (a necessary condition is that $v > 0$), the stationary-state variables (β, ω) can be determined from Equation 6 by inverting A and introducing the parameters wheel base l and understeer gradient K_{us}:

$$x = -A^{-1}Bu \tag{9}$$

$$\beta = K_F\delta_F + K_R\delta_R, \text{ where } K_F = \frac{l_R\left(1 - \frac{ml_Fv^2}{C_Rl_Rl}\right)}{l + K_{us}v^2}$$

$$\text{and } K_R = \frac{l_F\left(1 + \frac{ml_Rv^2}{C_Fl_Fl}\right)}{l + K_{us}v^2} \tag{10}$$

$$\omega = K_\omega(\delta_F - \delta_R), \text{ where } K_\omega = \frac{v}{l + K_{us}v^2}$$

$$\text{and } K_{us} = \frac{m}{l}\left(\frac{l_R}{C_F} - \frac{l_F}{C_R}\right) \tag{11}$$

3.3 Vehicle parameter determination

The moment of inertia J_z can be approximated with the ease to measure total vehicle length L, the wheel base l, and the vehicle mass m. The distances l_F and l_R from the center of gravity to the axles are calculated using the weighted rear axle vehicle load m_R (Equation 12).

$$J_z = 0.1269 \cdot m \cdot l \cdot L, \; l_F = \frac{m_R}{m} \cdot l, \; l_R = l - l_F \tag{12}$$

The tire parameters are determined from a stationary circle vehicle driving test with constant radius R and zero steer angle at rear axle ($\delta_R = 0$). From these tests, two gradients with reference to the lateral acceleration a_y have to be derived from measurement plots. The required gradients are the sideslip angle gradient $d\beta/da_y$ and the steering angle gradient $d\delta_F/da_y$ ($=K_{us}$). The steering reduction ratio i_s between the front steer road wheel angle and the driver's steering wheel angle δ_H is assumed to be constant, that is, $\delta_H = \delta_F i_s$. The steering reduction ratio and the cornering stiffness can be calculated consecutively from

$$i_s = \frac{\delta_H.R}{l}, \quad C_R = -\frac{ml_F}{\frac{d\beta}{da_y}l}, \text{ and } C_F = \frac{ml_RC_R}{K_{us}.lC_R + ml_F} \tag{13}$$

A set of vehicle parameters for a typical sedan car can be found in Table 2.

Table 2. Vehicle model parameters for a generic sedan car.

Symbol	Value	Unit	Description
m	1770	kg	Vehicle mass
J_z	3140	kg m^2	Vehicle yaw moment of inertia
L	4.841	m	Vehicle length
l_F	1.575	m	Distance from center of gravity to front axle
l_R	1.313	m	Distance from center of gravity to rear axle
b_F	1.562	m	Track width of front axle
b_R	1.590	m	Track width of rear axle
C_F	94672	N rad^{-1}	Effective cornering stiffness of front axle
C_R	160880	N rad^{-1}	Effective cornering stiffness of rear axle
i_s	17.6	—	Steering reduction ratio

Reproduced by permission of Continental.

4 HANDLING IMPROVEMENTS

The operational range for handling improvements is separated into two distinct areas, separated by the yaw rate at the friction limit range as illustrated in Figure 9. The "deep slip" range above the critical friction limit yaw rate is the application range of ESC.

Considering the basic idea that "the steering wheel is a yaw rate demand" (Blundell and Harty, 2004), improvements of handling mean that all the following are true:

- The path curvature can quickly and easily be altered with open-loop commands.
- The path curvature can be adjusted by the driver with only small closed-loop content.
- The vehicle response to steering input is predictable for the driver.

4.1 Handling improvements during normal driving

The alteration of the path curvature can easily be achieved by increasing the yaw gain, such that the driver steering input is small. This strategy is only applicable up to a medium speed. The yaw rate's normal driving range decreases significantly with vehicle speed, because the available tire–road friction is saturated at high speed quickly when the steering wheel angle input is too high (Figure 9). For normal tires on dry roads, this dangerous saturation level is reached at a lateral acceleration of 1g. The acceleration levels of 0.7, 0.3, and 0.1g correspond to wet, snowy, and icy roads. The strategy at high speed, therefore, must be to decrease the steady-state yaw gain at higher speed. From Equation 14, it can be seen that the

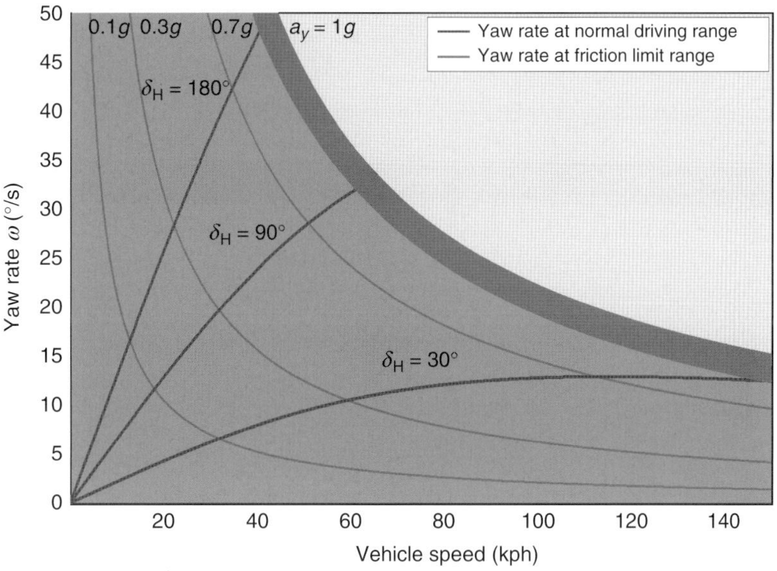

Figure 9. Operational range for handling improvements by global chassis control. (Reproduced by permission of Continental.)

yaw gain can be adjusted either by variation of the steering reduction ratio i_s or using a rear wheel steering. BMW uses both means together in their integral active steering (Herold *et al.*, 2008).

$$\frac{\omega}{\delta_H} = K_\omega\left(\frac{1}{i_s} - \frac{\delta_R}{\delta_H}\right) \qquad (14)$$

The potential for transient handling improvement is demonstrated with a simulated step-steer maneuver. This maneuver demonstrates that altering a path curvature quickly is not in conflict with a good damping behavior, when the controller is designed carefully. The handling controller used for the demonstration of the control potential is acting on the rear axle steer angle. A feedback of the vehicle's state $x = [\beta\ \omega]^T$ with gain matrix K_x improves the closed-loop dynamics of the vehicle, and a feedforward control part K_H is applied for a vanishing steady-state rear axle steer angle (Figure 10). The control law is given by the following equation, with parameters only valid for the given maneuver speed of 150 kph.

$$\delta_R = K_H\delta_H - K_x x, \text{ where } K_H = -0.0239$$
$$\text{and } K_x = (-0.0764 - 0.0896) \qquad (15)$$

The controller increases the response of the vehicle by increasing the yaw moment of approximately 45%. This significant improvement can be seen from the yaw acceleration in Figure 10 and is due to the out-of-phase (= negative) steering of the rear axle at the beginning of the maneuver. This steering input leads to an initially negative lateral force on the rear axle and, therefore, a yaw moment contribution, which enhances the turning-in yaw moment compared to the vehicle with inactive controller. Further advantage of the controller is the reduction of overshoot in yaw rate, sideslip angle, and lateral acceleration. This considerably improved damping behavior provides for an increased safety margin toward the friction limit range. Furthermore, the decreased delay between the steering wheel and the vehicle sideslip angle rate leads to an improved perception of agility by the driver.

4.2 Handling improvements during friction limit driving

At the limit of friction, where safety becomes relevant, the handling controller determines how the vehicle remains stable. All available actuators, that is, those listed in Figure 5, are incorporated and coordinated to reach this goal. The active chassis gives the driver optimal support for avoiding accidents. In the region beyond the limit of friction, the main task of the control system is to prevent the car from heavily skidding such that the car remains on track.

During normal driving, car drivers usually expect a linear yaw response of the vehicle with small phase lag. Most drivers have no experience of loss of linearity caused by saturation of tire forces. If saturation happens at the rear axle, the sideslip angle will increase quickly and, therefore, causes a hazardous driving problem for many drivers. The primary task of the control system should be to keep the vehicle sideslip angle small. An average driver

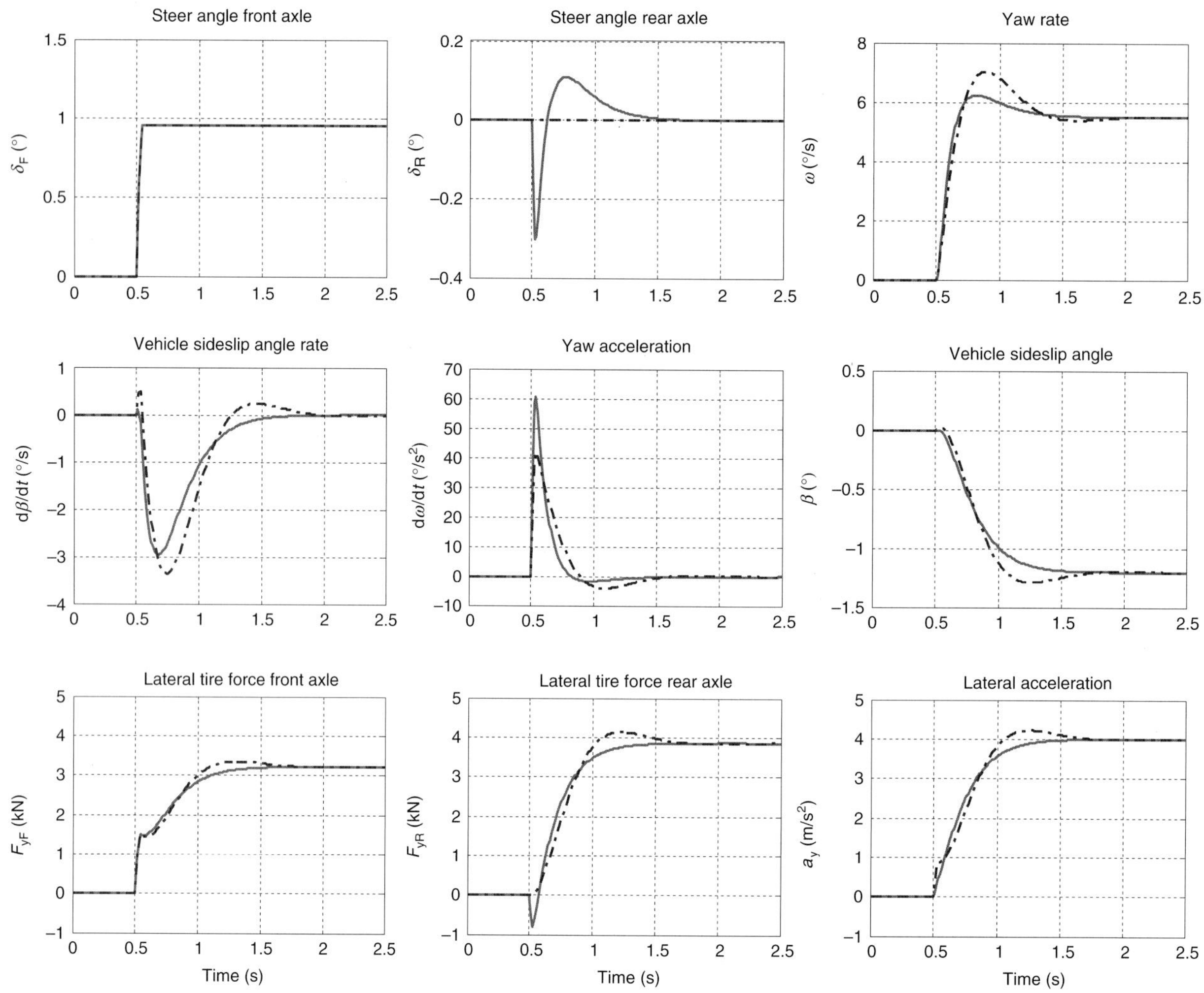

Figure 10. Vehicle states and inputs for a simulated step-steer maneuver with vehicle model Equations 6–8 reaching a steady-state lateral acceleration of 0.4g with a maximum steering wheel angle rate of $\delta_H = 500$ deg/s at a vehicle speed of $v = 150$ kph. The light gray solid line and the dashed dark gray line correspond to the controller active and the controller inactive, respectively. (Reproduced by permission of Continental.)

feels uncomfortable when the magnitude of the sideslip angle exceeds 3°. The state-of-the-art ESC systems limit the sideslip angle indirectly. ESC uses a reference yaw rate limited by the actual acceleration to account for the tire saturation. Additionally, the rate of change of sideslip angle is calculated and also limited. Figure 11 shows the results of a double-lane change vehicle test performed on a low friction surface. With the global chassis control (GCC), the vehicle response is stable and predictable to the driver at that particular speed level. ESC standalone will achieve a similar behavior, but at a much lower speed level. In the data plots of Figure 11, the scaling factor $k = 7$ is used for visualization purposes only. The scaling factor has been introduced to "normalize" the yaw rate for easy comparison of steering angle stimulus and yaw rate response. The unusual delay of the yaw rate response in the ESC standalone case becomes very obvious.

5 GENERIC MOTION CONTROL ARCHITECTURE

5.1 Overall control system

The fundamentals of the integration and coordination of the active systems into a hierarchically structured overall control system are shown in Figure 12. The first subtask in the overall control system is to determine the driver's intention. For that purpose, sensors on brake, steering, and accelerator pedal are interpreted to derive appropriate

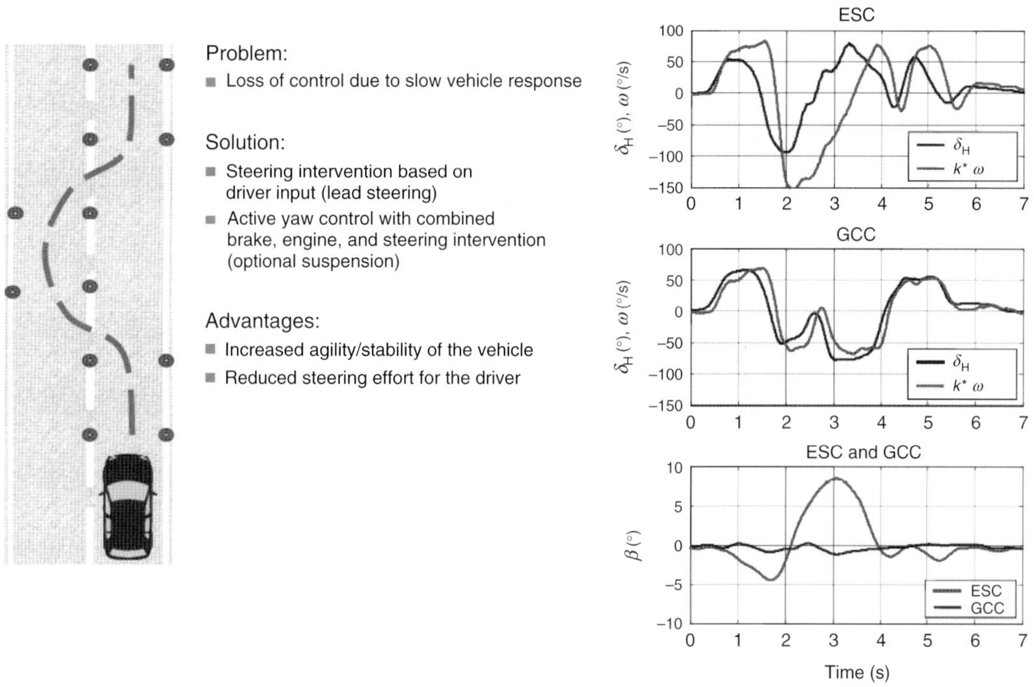

Figure 11. Comparison of vehicle test data from double-lane change (ISO 3888–1) at 70 kph on snow with conventional ESC versus GCC with integrated control of steering at front and rear axles. The light gray solid line and the dark gray line correspond to the GCC controller active and the conventional ESC active, respectively. (Reproduced by permission of Continental.)

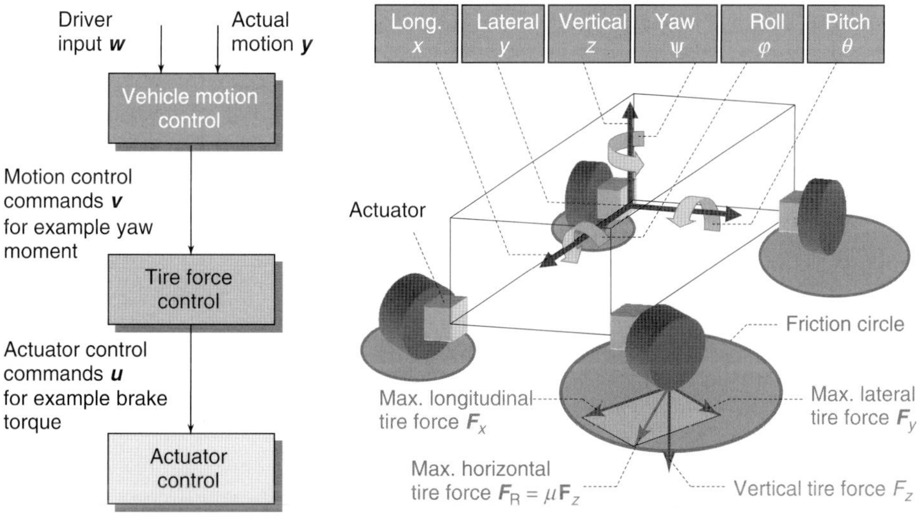

Figure 12. Overall control system. (Reproduced by permission of Continental.)

reference signals. In a second subtask, the driver's intention is compared to the actual vehicle motion, which is measured by inertial and speed sensors. If there are deviations, they are adjusted by calculating target forces and moments to change the actual vehicle's translational and rotational motions according to the driver's intention. The task of the tire force control is to distribute the motion control commands onto the individual wheels in order to change the tire forces in the contact patch area. The tire forces are adjusted by electromechanical or electrohydraulical actuators. A general limitation for the maximum achievable horizontal tire force is the friction circle (Kamm's circle), which depends on the tire–road friction and the load at each wheel.

Tire force control is the task to distribute the motion control commands among the tire forces at each wheel. Mathematically, this is the task to solve an underdetermined, typically constraint system of equations. The ambiguity of the actuator effects is characteristic for vehicles with multiple active systems onboard. For tire force control, a control allocation approach is generally useful, when different combinations of actuator commands can produce the same motion result. When the number of actuators available exceeds the number of degrees of freedom being controlled, the vehicle is called *over-actuated*. A vehicle equipped with ESC falls under this category, because four individual brake actuators control three horizontal degrees of freedom. Control allocation of over-actuated vehicles involves generating an optimal set of actuator control commands while minimizing the control effort and complying with the position and rate constraints of the actuator.

5.2 Selection of actuator configurations

It is a challenging and complex question of how to select an optimal set of actuators. For handling control, a possible answer can be achieved by analyzing the lateral full state (β, ω) controllability property of the vehicle (Raste *et al.*, 2010). The analysis is based on the state-space Equations 6–8. To simplify the analysis, not all longitudinal forces but only the differential forces $\Delta F_{x\mathrm{F}} = F_{x\mathrm{FL}} - F_{x\mathrm{FR}}$ and $\Delta F_{x\mathrm{R}} = F_{x\mathrm{RL}} - F_{x\mathrm{RR}}$ are considered. The analysis requires the controllability Gramian matrix W_c, which can be found as the solution to the Lyapunov matrix Equation 16 (Kailath, 1980).

$$AW_\mathrm{c} + W_\mathrm{c}A^\mathrm{T} = -BB^\mathrm{T} \tag{16}$$

The system state (β, ω) is controllable if the matrix W_c has full rank. If the matrix W_c has at least one eigenvalue equal to 0, then it cannot have full rank and, therefore, the system state is not controllable. Figure 13 shows the normalized smallest eigenvalue λ_min of the controllability Gramian matrix W_c as a function of the vehicle speed. The results can be interpreted as follows:

- Controllability of the full state (β, ω) is given in the whole speed range if front and rear steer angle controls are used, which is approximately equivalent to a virtual control input (F_y, M_z) at low speeds.
- Controllability of the full state (β, ω) with longitudinal differential forces only, which is equivalent to a virtual yaw moment control M_z, is not given at low speeds.

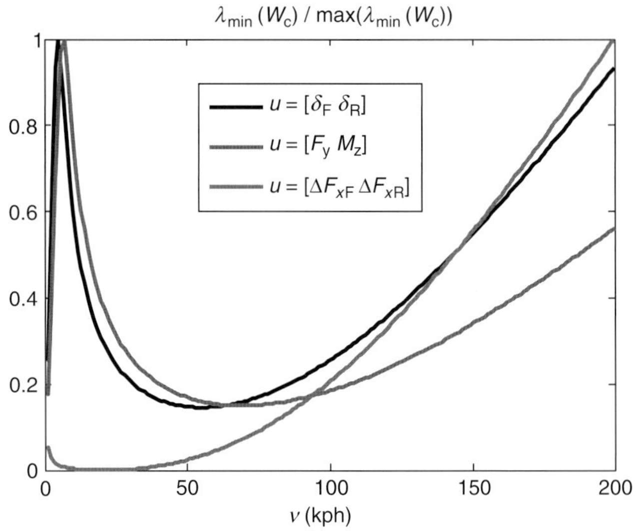

Figure 13. Controllability analysis based on the smallest eigenvalue λ_min of the controllability Gramian matrix W_c for different control input vectors. The black solid line corresponds to steer angle control input. The light gray solid line and the dark gray line correspond to virtual control input and longitudinal differential forces control input, respectively. (Reproduced by permission of Continental.)

- Simultaneous path and orientation controls, that is, full state (β, ω) control over the whole speed range, need both yaw moment actuators (e.g., ESC or ATV) and lateral force actuators (e.g., EPS, AFS, or ARK).

5.3 Compatibility of actuator configurations

The results from the controllability analysis are very useful to design the control system as generic as possible. A potential strategy to make the control compatible with various configurations of actuators is illustrated in Figure 14. The control feature A is originally designed for a rear wheel steering. It is assumed that δ_F is determined by the driver steering input. The task of the converter function is to transform the motion control commands into a target force and moment. These virtual motion control commands F_y and M_z might be used as a common basis for arbitration, for example, when requested by different control features. The control allocation procedure distributes the virtual motion control commands onto real commands for a given actuator configuration. The procedure is successful if the actuators generate a control effect as closely as possible to the virtual control demand.

To verify the feasibility of the strategy, a constraint control allocation library for MATLAB /Simulink has been

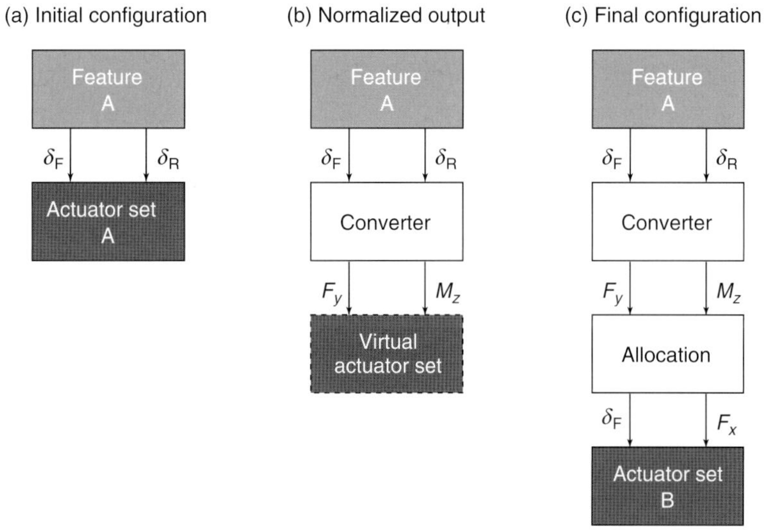

Figure 14. (a–c) Potential strategy to make a control feature compatible with various actuator configurations. (Reproduced by permission of Continental.)

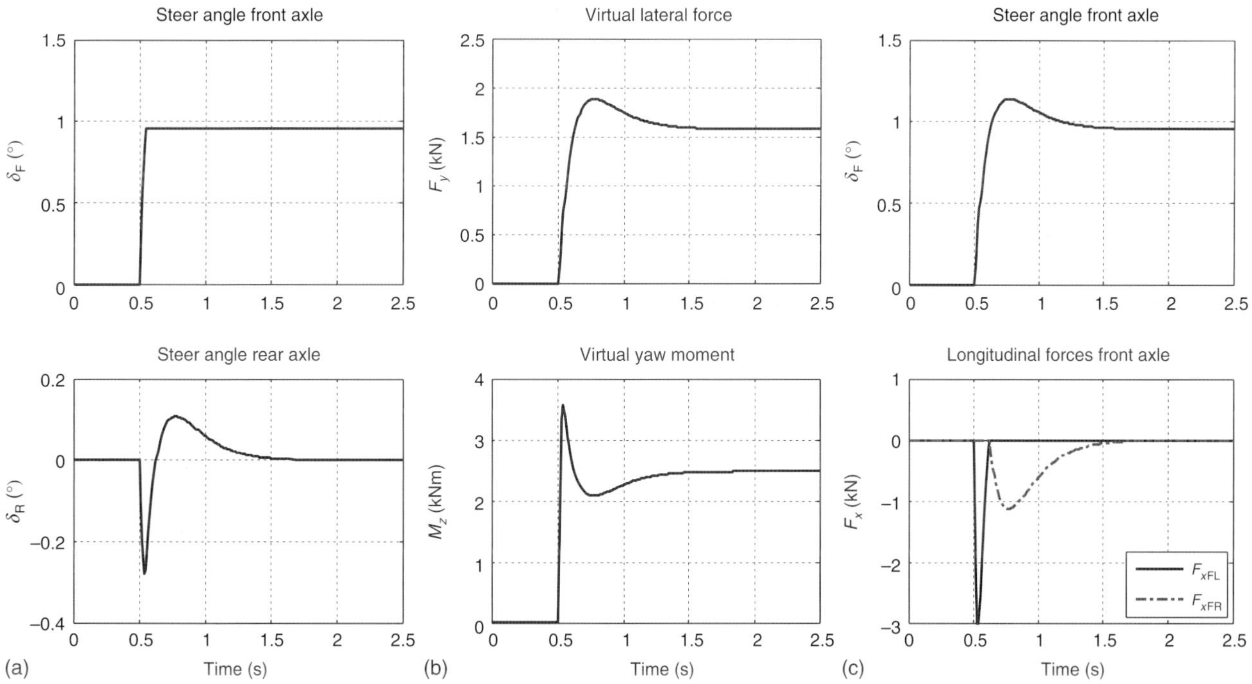

Figure 15. Simulated step-steer maneuver based on vehicle model Equations 6–8 and control feature (Equation 15). The differing input for actuator set A (a), virtual actuator set (b), and actuator set B (c) reproducing identical vehicle state behavior is shown. (Reproduced by permission of Continental.)

applied (Harkegard, 2003). Figure 15 illustrates the results with the same step-steer maneuver, which has been used in the previous section (Figure 10). It is worth to point out that the vehicle state variables β and ω show identical behavior for all three actuator sets. The feature is the transient control with the control law given by Equation 15. The steering signals are converted into the virtual motion control commands shown in Figure 15b. The constraint control allocation is able to consider the position and rate limits of each actuator. In the simulation, a constraint in the positive direction of the force has been used to make the feature available for a brake-based active system. The resulting actuator commands are shown in Figure 15c. The driver input is now slightly modified. Further, simulations have to clarify if the deviation can be expected to be compensated by the driver or if the driver has to be supported by additional active systems, for example, EPS.

6 FUTURE TRENDS OF HANDLING CONTROL

In normal driving and also in hazard situations, the driver is supposed to be the master of control and, therefore, properly the source for reference signals. However, what happens if a driver does not react appropriately? In urban traffic situations, one of the main causes of accidents is driver distraction. The automotive industry has identified the great potential of active systems assisting the driver in such critical situations.

6.1 Emergency brake assist

The emergency brake assist (EBA) shown in Figure 16 intervenes if the driver is inattentive and shows no sign of having recognized the danger of an impending collision. The city version of EBA-City is active at speeds of up to 30 kph. It features an optical sensor that uses infrared beams to monitor the road space in front of the vehicle, up to a distance of about 10 m. Its electronics calculate the distance to the vehicle in front. If there is a risk of collision, EBA initially prepares the brakes, issues a warning to the driver, and then, if time starts to run out, automatically applies the brakes. If the maximum speed differential between the vehicles is no more than 15 kph, a rear-end collision with the vehicle in front can be avoided in most cases. If a collision is unavoidable, automatic emergency braking can significantly reduce the impact velocity and thus the severity of the accident.

6.2 Emergency steer assist

Automotive engineers currently develop systems assisting the driver in hazardous handling situations, when there is no time left for braking (Hartmann, Eckert, and Rieth, 2009). Figure 17 illustrates a typical use case for handling assistance. A driver quickly initiates an evading maneuver to avoid a collision with a suddenly appearing object in front. Although the conventional ESC mitigates the severity of the situation significantly, there is still potential for further improvement. The main benefit comes from surrounding sensors such as radar sensors, which are used to identify an imminent collision. In Figure 17, the object distance information is used to preset an ARK just in time for handling with optimal effectiveness.

The simulation data presented in Figure 18 provides a further insight into the potential of different active systems in combination with predictive handling control. While systems such as adaptive damper or active stabilizer have only limited handling potential, the rear-wheel steering considerably reduces vehicle spinning-out and increases yaw damping when applied early enough. This positive effect on vehicle safety is accompanied by a slightly increased driver-steering effort. The preferred choice of assistance in such a case is to stimulate the haptic channel of the driver via EPS.

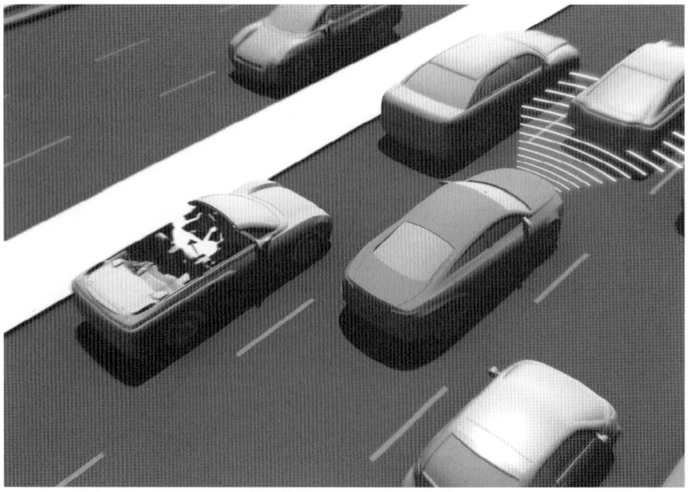

Figure 16. The emergency brake assist for urban areas is already well established. (Reproduced by permission of Continental.)

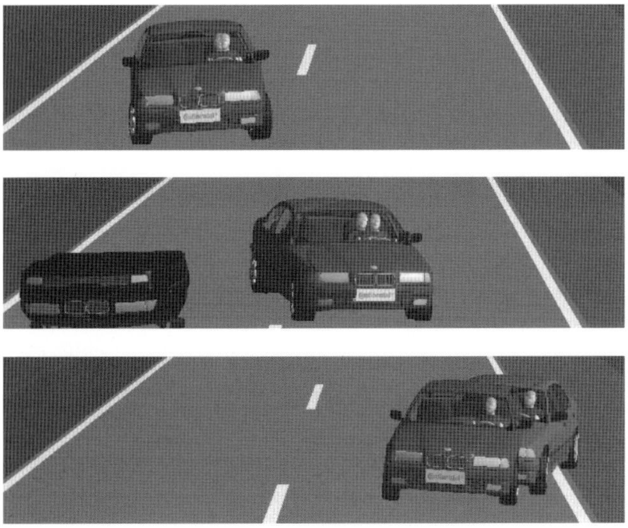

Figure 17. Animation of an emergency evading use case. The vehicle with conventional ESC is skidding off the track and the vehicle with ESC and ARK and predictive handling control remains on track. (Reproduced by permission of Continental.)

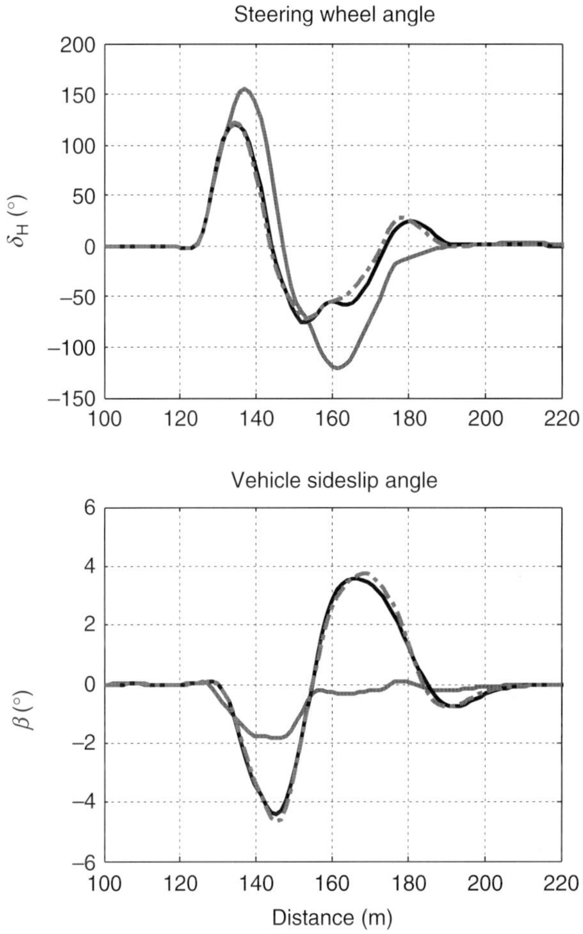

Figure 18. Emergency evading maneuver simulated with a complex nonlinear vehicle and driver model to compare the effectiveness of different active systems when applied with predictive handling control. The light gray solid line corresponds to ESC and ARK, the dark gray dashed line corresponds to ESC and EAD, and the thin black solid line corresponds to ESC and ARS. (Reproduced by permission of Continental.)

6.3 Handling during highly automated driving

Advanced driver assistance systems can offer remedies. On the one hand, they can support the driver in demanding and difficult situations and, on the other hand, develop room for freedom during monotonous driving situations, which are often accompanied by the risk of decreasing attention. Especially, the latter is a potential field of application for highly automated driving.

In the case when the driver is distracted and inattentive, an option of handling assistance could be to carry out the maneuver autonomously. Figure 19 presents a solution based on the procedure shown in Figure 14. The steering maneuver is executed autonomously without the need of steering the front wheels (e.g., when EPS is not available). The control commands are allocated, in the present example, to both friction brakes on the left side and the rear steering. The vehicle behaves in exactly the same way as it would have with steering the front wheels and the controller given by Equation 15 acting at the rear wheels.

Where does the assistance in handling situations go from here? Will drivers understand the meaning of haptic steering feedback in critical situations? What happened if they do not react at all? Autonomous vehicle braking technology has been launched over the last few years with great success. However, a rapid introduction of autonomous vehicle steering is not very likely. The legal situation and the functional safety management raise many questions at the moment. A first step toward a technically feasible solution has been presented by exploiting the opportunities given by the inherent redundancy of multiple active systems.

7 CONCLUSION

Global chassis control delivers significant benefits in normal driving and particularly, in emergency situations. The configuration and the coordinated interaction of the active systems are the key success factors for enhancing the vehicle performance. International standards such as ISO 26262 ensure quality and safety of the overall control system at the highest level. In the near future, the vehicle

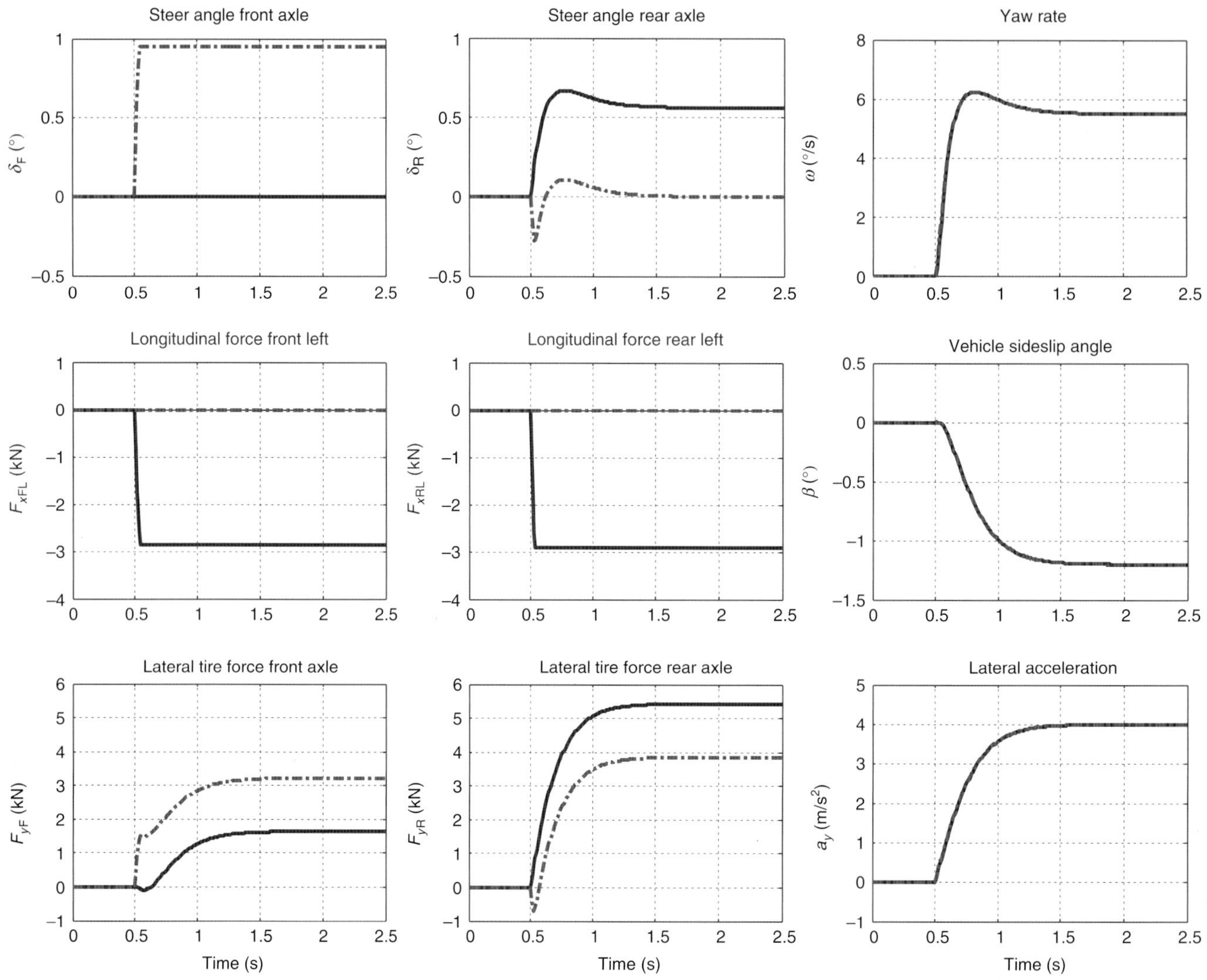

Figure 19. Allocation of control commands in order to substitute a distracted driver. Simulated step-steer maneuver based on vehicle model Equations 6–8 reproducing identical vehicle state behavior for two different input vectors. The light gray solid line corresponds to the maneuver without driver input using longitudinal forces and rear steer angle only. The dark gray dashed line corresponds to the maneuver with driver input on front steer angle and control law given by Equation 15 for the rear steer angle. (Reproduced by permission of Continental.)

is fitted with sensors for monitoring the surroundings, such that predictive motion control interventions become possible.

REFERENCES

Andreasson, J. (2007) On generic road vehicle motion modelling and control. Ph.D. Dissertation. Royal Institute of Technology, Vehicle Dynamics, Stockholm, Sweden.

Bauer, R., Raste, T., and Rieth, P.E. (2007) System integration of hybrid powertrains. *ATZelectronic*, **2** (04), 6–10.

Blundell, M. and Harty, D. (2004) *The Multibody Systems Approach to Vehicle Dynamics*, Elsevier, Burlington.

Harkegard, O. (2003) Backstepping and control allocation with applications to flight control. Ph.D. Dissertation. Department of Electrical Engineering, Linköping University, Linköping, Sweden.

Hartmann, B., Eckert, A., and Rieth, P.E. (2009) Emergency Steer Assist—Assistenzsystem für Ausweichmanöver in Notsituationen. *12. VDI-Tagung Reifen-Fahrwerk-Fahrbahn*, VDI-Berichte No. 2086, pp. 131–148.

Herold, P., Schuster, M., Thalhammer, T., *et al.* (2008) Integral Active Steering. Synthesis of Agility and Sovereignity. *FISITA World Automotive Congress*, Munich.

Kailath, T. (1980) *Linear Systems*, Prentice-Hall, Englewood Cliffs.

Raste, T., Bauer, R., and Rieth, P.E. (2008) Global Chassis Control: Challenges and Benefits within the Networked Chassis. *FISITA World Automotive Congress*, Munich.

Raste, T., Kretschmann, M., Eckert, A., *et al.* (2010) Sideslip Angle based Vehicle Control System to Improve Active Safety. *FISITA World Automotive Congress*, Budapest.

Schiebahn, M., Zegelaar, P., and Hofmann, O. (2007) Theoretical Generation of Additional Yaw Torque. *11. VDI-Tagung Reifen-Fahrwerk-Fahrbahn*, VDI-Berichte No. 2014, pp. 101–119.

Chapter 128

Customer-Oriented Evaluation of Vehicle Handling Characteristics

Adrian Mihailescu, Stephan Poltersdorf, and Lutz Eckstein

RWTH Aachen University, Aachen, Germany

1 INTRODUCTION

This chapter gives an overview on methods that can be used to evaluate the handling quality of passenger cars with respect to the demands of real customers.

1.1 Motivation

The increasing handling quality and driving safety of modern passenger cars makes it difficult for car manufacturers to stand out in this area of vehicle quality. The available development resources must be strictly focused on technical changes that assure improved handling evaluations by customers (and journalists). The vehicle developers, therefore, need efficient methods to evaluate the perceived handling quality in all phases of product development.

Encyclopedia of Automotive Engineering, print © 2015 John Wiley & Sons, Ltd.
Edited by David Crolla, David E. Foster, Toshio Kobayashi, and Nick Vaughan.
This article is © 2015 John Wiley & Sons, Ltd. ISBN: 978-0-470-97402-5
Also published in the *Encyclopedia of Automotive Engineering* (online edition)
DOI: 10.1002/9781118354179.auto021

1.2 Subjective versus objective evaluations

Classical evaluation approaches are subjective evaluations by customers or vehicle dynamics experts. Considering various rules regarding the execution (Section 2) and data analysis (Section 4), this method can generate valuable results.

The reproducibility and validity of subjective evaluations is limited because of several (human) factors, whereas the necessary effort remains high. Furthermore, the necessity for real vehicles conflicts with the increasingly virtual vehicle development process. These problems led to various approaches to support or even replace subjective evaluations by objective methods based on measurements or simulations (described in Section 3). The goal of these approaches is the identification of objective values that significantly correlate with subjective vehicle evaluations, and that can, therefore, be used to predict customer verdicts.

Successful methods for objective evaluations lead to well-defined and testable development targets. The product quality can be evaluated more often and more efficiently, thus reducing the dependence on varying or even biased subjective evaluations.

Over the last decades, a large number of useful objective measures have been identified. Section 5 gives an overview based on an extensive meta-study.

2 SUBJECTIVE VEHICLE HANDLING EVALUATION

In order to gather detailed and reliable handling evaluation based on the perception of individual drivers, several basic rules have to be considered. After a brief introduction of the principles of subjective handling perception, these basic rules are presented in Sections 2.2 and 2.3.

2.1 Principles of subjective handling perception

Each driver stabilizes the vehicle on the chosen trajectory based on various methods of information perception and information processing. The driver's sensations and experiences in the scope of this stabilization determine the subjective evaluation of vehicle handling.

2.1.1 Information perception

The driver perceives information on the current driving situation mainly via three sensory channels: the visual, the vestibular (resulting from the organ of equilibrium), and the haptic (feeling sensors such as hands) channels (Mitschke and Niemann, 1972).

First of all, the high resolution of the visual channel enables the perception of deviations from the desired trajectory, the vehicle orientation, and the steering wheel angle. Velocities and accelerations can also be perceived based on the optical flow. Because of the widespread and precise visual information, this channel is the main information source for trajectory stabilization in the linear handling area (Schimmel, 2010). The equilibrium organs (vestibular system) enable the perception of linear and rotary accelerations. Compared to the visual channel, the vestibular sensation is less accurate but faster. The fastest information is provided by the haptic sense, which enables the indirect perception of vehicle accelerations and of the steering wheel torque across the contact surfaces between driver and vehicle.

2.1.2 Information processing

The processing of the perceived driving-state information can be described by a three-staged model (Rasmussen, 1983): skill-based driver actions constitute the first level and are very rapid and highly automated behavioral patterns not consciously modulated by the driver. On the second level, the driver makes unconscious decisions based on memorized experiences. The speed of information processing at this level is slower than on the first level. If the driver possesses no suitable skills or experiences for the current driving situation, he or she must consciously analyze the driving situation. This level of information processing is the third and the slowest stage of information processing.

With respect to subjective vehicle evaluations, one fact is crucial: because of different levels of skill and experience, different drivers handle identical driving situations using different levels of information processing and differently weighed sensory information. This results in a high variance of driver actions and vehicle evaluations among different drivers. For example, a skilled driver uses the haptic and the vestibular senses much more than an untrained driver, who mainly controls the vehicle based on visual information, which causes differences, for example, in the evaluation of steering feel.

2.2 Execution of subjective vehicle evaluations

Table 1 gives an overview on typical handling characteristics evaluated in subjective evaluation (Heißing and Brandl, 2002; Eckstein, 2011). The table shows the classes of characteristics that can be further detailed into subcharacteristics—especially in the scope of expert

Table 1. Criteria classes of subjective evaluation.

	On-center handling			Corner handling		
Steering feel	Center point feeling	Steering effort	Steering torque level and buildup	Steering torque level and progress	Steering angle demand	Steering response
	Steering response	Self-centering	Steering friction	Stability feel	Steering precision	Feedback of road friction
		Post-pulse oscillation (hands off)		Torque and angle during parking	Steering clearance	Steering returnability
Vehicle behavior	Straight driving stability	Directional stability under braking	Response behavior	Self-steering behavior	Response behavior	Pitch and roll behavior
	Load-change behavior	General (straight) wind sensitivity	Cross wind sensitivity	Load-change behavior	Stability and lane change behavior	Corner exit behavior
	Lateral inclination sensitivity	Trailer stability	Aquaplaning behavior	Corner braking behavior	Aquaplaning behavior	

evaluations. Subjective characteristics that are supported by significantly correlating objective measures (Section 5) are marked in gray.

2.2.1 Choice of drivers

Usually, professional test drivers are used for vehicle evaluations because of their more accurate, differentiated, and reproducible judgment (Schimmel and Heißing, 2009; Pfeffer and Scholz, 2010; Riedel and Arbinger, 1997). In contrast, untrained drivers offer realistic representation of the skill and experience of the real customers. The lower precision and reproducibility of their judgment is sometimes accepted, for example, if questions of driving safety or controllability are investigated (Bubb, 2003). A collective of normal drivers is best generated by random sampling among large heterogeneous groups of possible customers (Breuer, 2009). Bortz (2010) and Bubb (2003) describe how the appropriate size of the driver collective can be calculated based on the questions under investigation, the desired level of significance, the used evaluation scales, and the statistical distribution of data collected in pretests. This chapter can only give rough recommendations based on these sources: tests with professional drivers only need small numbers of drivers (2–5), especially when new vehicle variants are compared to reference variants that have been evaluated by a larger number of drivers. Tests with normal drivers that are used to compare two vehicle variants need more drivers (30–50). If tests with normal drivers shall be used to understand the relationship between subjective evaluations and certain driver characteristics (gender, age, and so on), a much higher number of drivers (e.g., >1000) are needed for highly significant results.

2.2.2 Choice of driving scenario

The vehicles must be evaluated in scenarios that provide realistic representations of real driving situations. The vehicles should be driven "closed loop" on a clearly defined course, so that the task of trajectory stabilization needs to be fulfilled (Kudritzki, 1989).

The shape and velocity profile of the used course should realistically depict the typical driving range with respect to velocity, lateral acceleration, and necessary steering frequencies. This is achieved by choosing a course with a wide range of cornering radiuses (20–500 m). Corners with special characteristics (lateral inclination, increasing curvature, and so on) should also be present with different radiuses (Heißing and Brandl, 2002). The necessary steering frequencies can be influenced by more or less immediate changes in corner direction and by the curvature gradients at the beginning and the end of corners (abrupt vs continuous changes in curvatures).

The maneuver definition must clearly describe in which handling area the vehicle has to be driven (linear area or nonlinear limit area). For limit maneuvers, Kudritzki (1989) and Neukum, Krüger, and Schuller (2001) suggest driver-dependent velocities, as the significance of subjective ratings increases as soon as the evaluating driver approaches his or her individual velocity limit.

2.2.3 Execution of driving tests

The test conditions (tire condition, vehicle load, track temperature, weather, daytime, and so on) must be kept as constant as possible. Evaluations should not last longer than 1 h (Barthenheier, 2004).

All vehicles should be "blind tested," that is, no driver should know which variant he or she is currently driving (Zong, Guo, and Hsin, 2000). Ideally, the vehicle variants should only vary with respect to the evaluated characteristics in order to prevent disturbances caused by the unintended evaluation of characteristics such as design or brand.

Variants should be evaluated together with a reference vehicle that has been evaluated by a large number of drivers before. Thus, relating the evaluation results to the highly significant evaluations of the reference variant can increase the significance of the results.

It is also recommended to repeat the evaluation of some variants, for example, the reference variant, in order to detect whether the scale of individual evaluations drifts off during a test-drive (Dettki, 2005). In order to prevent the so-called transfer effect (the judgment of one variant is influenced by the variants driven before), each driver should drive the variants in a different order (Bubb, 2003).

2.3 Design of subjective evaluation questionnaires

Commonly printed questionnaires with rating scales are used for the evaluation. Sometimes, an interviewer in the car personally asks for the evaluation. Käppler (1993) and Riedel and Arbinger (1997) point out that the low reliability of many subjective evaluations is caused by an inappropriate questionnaire design. Therefore, the evaluation scales and the kind, number, and formulation of the questions should be chosen thoughtfully.

2.3.1 Absolute versus relative evaluations

Absolute vehicle evaluations try to rate the vehicle handling on an absolute and universally valid scale. Different vehicles can afterward be compared and ranked based on their absolute ratings. Relative evaluations do only try to evaluate a vehicle in comparison to a second vehicle. If all vehicle variants are compared to each other relatively, a ranking can be generated based on these relative evaluations.

Even for professional drivers, it is much easier to rate a vehicle in comparison to a reference vehicle than to rate a vehicle on an absolute scale (Mummendey, 2003). If preliminary tests show that reliable and reproducible absolute evaluations are not possible for a given investigation, it is still possible to switch to a more reliable relative evaluation.

2.3.2 *Content and formulation of evaluation questions*

The questions must be adapted to the vocabulary and experience of the drivers, as a prior "teaching" of expert language to normal drivers would not significantly improve the quality of the results (Kudritzki, 2000). The evaluation becomes easier for normal drivers when they are not asked for a technical vehicle rating but for their personal sensations and experiences in the vehicle (see scale example in Figure 2).

The total number of questions should be kept below 10 (Kudritzki, 2000; Heißing and Brandl, 2002).

2.3.3 *Evaluation scales*

Usually, drivers are asked to evaluate the handling with discrete scales. The steps of a scale must be "anchored" by verbal classifiers, that is, by descriptions of the individual scale steps, in order to minimize the room for (mis)interpretations.

Depending on the scope of the investigation, two basic scale types can be used. Open "unipolar" scales are used for absolute evaluations. On these scales, the magnitude of each rating is described on an absolute scale without a reference. Closed "bipolar" scales are used for relative vehicle evaluations. In this case, each vehicle is rated in comparison to a second (reference) vehicle.

When choosing the number of scale steps, a trade-off between reliability losses because of unperceivable scale differences and detail losses because of a too coarse scale has to be found. In general, the standard deviation for the repeated evaluations of identical vehicle variants by identical drivers should stay below 2 scale steps (Käppler, 1993; Bortz, 2010).

Käppler (1993) recommends an effective scale range of 7 steps for unipolar scales and professional drivers. Harrer, Pfeffer, and Johnston (2006) recommend an effective scale range of ±6 steps for bipolar scales and professional drivers.

2.3.4 *Visual design of scales*

According to the western reading direction, the intensity of the scale anchors should increase horizontally from left to right. The full-scale range should be shown visually, continually, and undistorted (Käppler, 1993).

Figures 1 and 2 show two scale variants that have been derived from the explained principles. The first example (Figure 1) shows an absolute scale for professional evaluators and asks for a three-stage evaluation.

The second example (Figure 2) shows a relative, bipolar scale for untrained evaluators.

3 METHODS FOR OBJECTIVE EVALUATION

Various approaches try to identify objective characteristic values that strongly determine the customer rating of a vehicle. Figure 3 provides an overview on these methods.

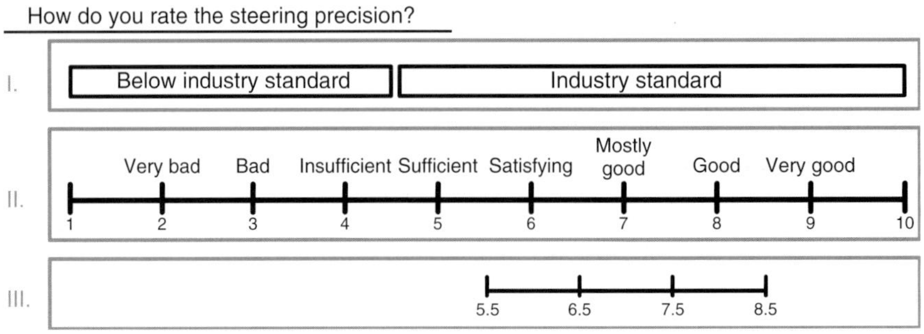

Figure 1. Unipolar scale for professional evaluators.

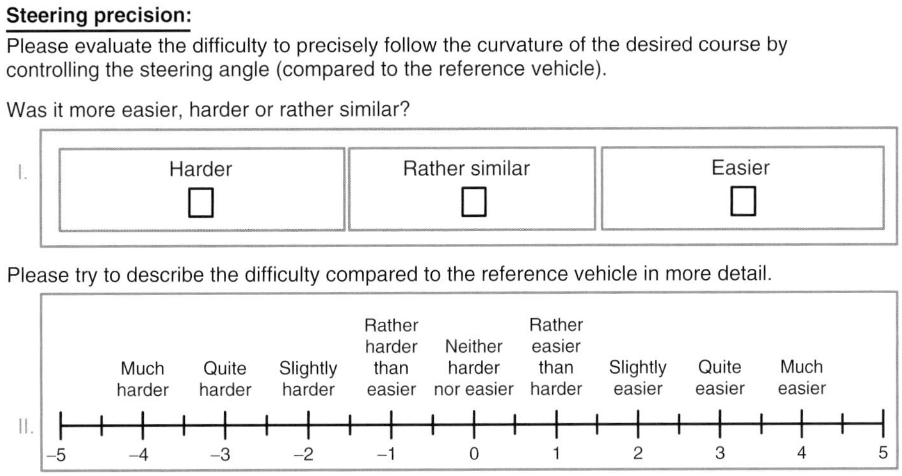

Figure 2. Bipolar scale for untrained evaluators.

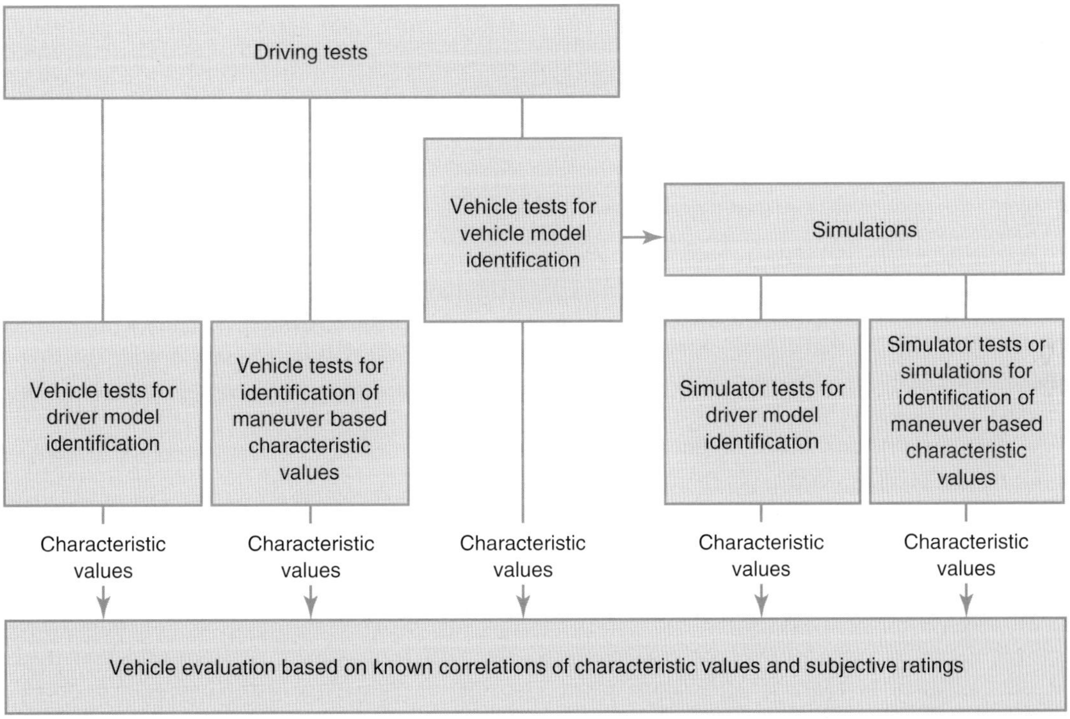

Figure 3. Objective handling evaluation methods.

3.1 Objective evaluation based on characteristic values from driving maneuvers

The classic method of objective vehicle evaluation collects characteristic values from vehicle measurements. Most of these values can also be collected efficiently in vehicle dynamic simulations. Usually, the characteristic values are collected in open-loop maneuvers, as, in this case, a high reproducibility can be achieved because of the reduced driver influence.

It is recommended that the catalog of characteristic values covers all information paths of human drivers. Visual characteristic values (e.g., lateral track offset) should be combined with haptic characteristic values (e.g., steering torque) and vestibular characteristic values (e.g., lateral acceleration) (Wagner, 2003).

Section 5 presents a large number of such characteristic values that correlate significantly with subjective evaluations. Most of these characteristic values are context-sensitive, that is, they only correlate with subjective evaluations from similar driving situations (Jürgensohn, Willumeit, and Irmscher, 1999; Pietsch, Schimmel, and Heißing, 2009a; Sagan, 2003). Therefore, a detailed evaluation of the complete handling behavior can only be achieved by a large variety of maneuvers and characteristic values.

3.2 Objective evaluation based on vehicle models

Vehicle models can be used for objective vehicle evaluations in two ways: one possibility is the determination of characteristic values from the vehicle model parameters themselves. For example, Barthenheier (2004) evaluates the steering behavior of a vehicle based on a simple steering model with the three parameters self-aligning torque, steering damping, and steering friction. So far, only a small number of significant correlations have been identified with this method.

The second and the more promising possibility is the determination of the classic characteristic values (Section 3.1) with the help of vehicle dynamic simulations. Thus, a large number of characteristic values can be gathered with high efficiency and reproducibility.

The models (e.g., extended single-track vehicle model, the dual-track vehicle model, or the steering-system models described by Zschocke, 2009, and Winner *et al.*, 2003) can be parameterized automatically based on test-rig or test-drive measurements (Meyer-Tuve, 2009; Meljnikov, 2003; Pietsch, Schimmel, and Heißing, 2009a; Zschocke, 2009).

3.3 Driver-based characteristic values

A more direct way to predict subjective vehicle evaluations would be the determination of characteristic values based on the reactions of the driver in closed-loop maneuvers. A driver adapts his or her control behavior to the handling properties of a vehicle based on his or her skills and experiences (Wallentowitz, 1979). The easier this adaptation is, the higher his or her evaluation of the vehicle is going to be (Abe and Kano, 2008). This suggests that characteristic values based on the interaction of driver and vehicle highly correlate with subjective vehicle evaluations.

The central problem of all these driver-based objectivity approaches is the high variance among drivers. Using professional drivers and repeated measurements can reduce this problem.

3.3.1 Characteristic values based on driver–vehicle interaction

The results of closed-loop maneuvers can be used to determine characteristic values based on the positions of the control elements, especially the steering angle. For example, the steering angle measured in a double-lane change can be used to evaluate the vehicle stability and controllability in this maneuver. In the end, this approach is a closed-loop, driver-oriented variant of the classical maneuver-based approach.

3.3.2 Characteristic values based on driver models

This approach is based on the assumption that synthetic models can describe the control behavior of the driver. As the driver has to adapt to each vehicle, the parameters of the respective driver model must be vehicle dependent and can be utilized for objective vehicle evaluations (Abe and Kano, 2008; Henze, 2004; Decker, 2008; Dibbern, 1992). Common driver models are based on a combination of a feedforward part and a feedback part. The feedforward part estimates the steering angle necessary to follow the path ahead based on a linear single-track vehicle model. The feedback part compensates the path deviation caused by the inaccuracies of this model with a steering angle controller, for example, a PID controller. The vehicle-dependent parameters of these driver models (i.e., the feedforward and feedback amplification factors) can be automatically identified with the help of the test-drive data, either from a test track or from a driving simulator.

3.3.3 Characteristic values based on driver perception

Schimmel (2010) and Scharpe *et al.* (2012) suggest that the significance of characteristic values increases when the vehicle movement is transformed with respect to the paths of subjective handling perception. For example, characteristic values based on the lateral acceleration of vestibular organs (i.e., the driver head acceleration) or characteristic values based on the lateral pressure between driver and seat are more significant than characteristic values based on the lateral acceleration of the vehicle.

4 EVALUATION OF SUBJECTIVE AND OBJECTIVE DATA

The data gathered in subjective and objective evaluations must be correctly processed and analyzed in order to consider all relevant information and in order to prevent misjudgments. The description of the necessary mathematical methods lies beyond the scope of this chapter. Nevertheless, we would like to emphasize the necessity and the main goals of the data processing.

4.1 Processing of data from subjective evaluation

The subjective data must be tested for four central characteristics:

Objectivity: It must be mathematically proven that the results are independent from the persons conducting the experiments, for example, the interviewer in the car (Lienert and Raatz, 1969).

Reliability: It must be mathematically proven that the data are internally consistent and provide stable results in repeated evaluations (Magnusson, 1975; Riedel and Arbinger, 1997; Redlich, 1994; Bortz, 2010).

Validity: It has to be tested whether a scale/question is really measuring the attribute it is supposed to. It should be possible to distinguish vehicles differing in the analyzed characteristic by comparing the respective answers (Riedel and Arbinger, 1997).

Subjective–subjective correlation: It has to be tested whether there are redundancies in between questions/scales. In this case, removing these redundancies should shorten the questionnaire.

4.2 Processing of data from objective measurements

The objective data from measurements or simulations also have to be mathematically tested for reliability (internal consistency and repeatability) and validity (strong relationship to measured characteristic). An objective–objective correlation can be used to shorten the catalog of characteristic values by removing redundancies (Harrer, 2007). The data preprocessing must ensure the comparability of measurements from different vehicles with different sensor configurations, for example, by the removal of sensor delays and offsets and by the transformation of measurements to default reference points such as the center of gravity of the vehicle (Kobetz, 2004; Pfeffer and Harrer, 2011).

4.3 Correlation and regression analyses

As soon as the subjective and objective data have been processed as described earlier, the correlation and regression analyses are used to analyze the relationship between them. The correlation test determines the statistical significance (i.e., the probability of error) of relationships between objective characteristics and subjective evaluations. The absolute value of the correlation coefficient can be used as a measure for this significance (Bortz, 2010; Cohen and Cohen, 1983).

The regression analysis goes one step further by generating models that can be used to predict subjective ratings based on linear or nonlinear combinations of one or more objective values (Kudritzki, 1989; Dibbern, 1992; Harrer, 2007).

Only significantly correlating characteristics should be used for the vehicle evaluation. A certain amount of remaining probability of error (e.g., 5% for significant or 1% for highly significant correlations) is accepted.

5 OVERVIEW ON SIGNIFICANTLY CORRELATING VEHICLE HANDLING CHARACTERISTICS

Some of the characteristic values found in the literature are statistically proven, some are solely theories, and some are disapproved but often used nevertheless. In order to ensure the customer relevance, the characteristic value has to have been validated by real test-drives and regression analyses. In the further approach of this chapter, only the verified characteristic values of the 250 researched literature sources are considered.

Figure 4 shows the classification structure of the chosen vehicle handling characteristics and the section where it is elucidated.

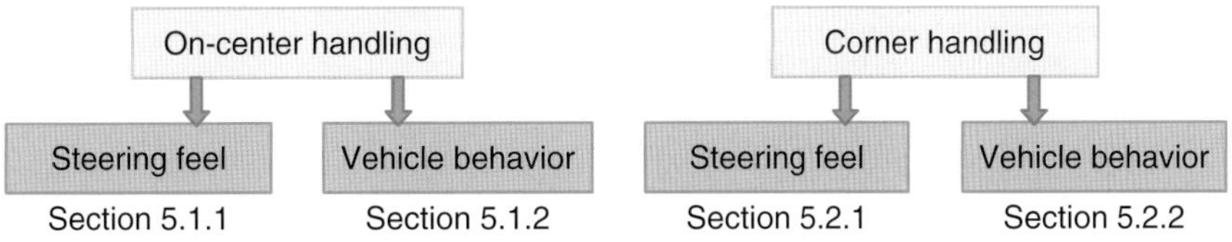

Figure 4. Overview on the handling classification structure.

In the scope of this chapter, we first subdivide the handling into two areas: *on-center handling* refers to the steering behavior on and around straight ahead driving (Farrer, 1993); the handling at higher lateral acceleration is referred to as *corner handling*. In a second classification level, vehicle handling characteristics are categorized according to the vehicle response's effect on the driver: the *steering feel* and the *vehicle behavior* are being considered. The steering feel of the driver is a central handling characteristic and is, therefore, investigated separately for both handling areas. The vehicle behavior section sums up all the literature findings on verified correlations not related to steering feel—for example, yaw behavior or roll behavior.

The follwing sections show the verified correlations and their characteristic values. For each correlation, there is a sign describing the preferable value of the objective characteristic:

⇧ stands for high values;
⇩ stands for low values;
⊙ stands for an optimal value range.

These recommendations are based on investigations done by the referenced sources and not by the authors of this chapter. In some cases, there are further notes added to the correlation. This is marked by a code and is listed in the corresponding section.

5.1 On-center handling

5.1.1 Steering feel

Correlation notes sorted by code (Table 2):

OS1: The evaluation is done between the speeds of 160 and 200 km/h.

$$K_{\mathrm{L}} = \left| \frac{\delta}{\tau v_{\mathrm{res}}^2} \right|_{f_{\mathrm{R,opt}}} \cdot f_{\mathrm{R,opt}} \tag{1}$$

τv^2_{res} is a value describing the intensity of the stochastic crosswind disturbance. The value τ describes the angle between the longitudinal axis of the car and the direction of the resultant wind velocity v^2_{res}, which is build by vector addition of the vehicle speed and the side wind velocity. The quotient is the resonant amplification factor between crosswind and steering reactions. $f_{\mathrm{R,opt}}$ is the resonant frequency of the driver's reaction to the wind disturbance.

OS2: The evaluation is done while driving between the speeds of 160 and 200 km/h.

$$K_{\mathrm{V}} = \frac{\left| \frac{\dot{\psi}}{\tau v_{\mathrm{res}}^2} \right|_{\max}}{\left| \frac{\dot{\psi}_{\tau v_{\mathrm{res}}^2}}{\tau v_{\mathrm{res}}^2} \right|_{\max}} \tag{2}$$

$\dot{\psi}_{\tau v^2_{\mathrm{res}}}$ is the yaw rate with and $\dot{\psi}$ the yaw rate without driver intervention. τv^2_{res} is a value describing the intensity of the stochastic crosswind disturbance.

OS3: The evaluation is done while driving at the speed of 100 km/h. $W_{\delta M_{\mathrm{H}}}$ is the steering work.

$$W_{\delta M_{\mathrm{H}}} = \delta \cdot M_{\mathrm{H}} \tag{3}$$

δ is the steering angle demand and M_{H} the steering torque demand. $W_{\delta M_{\mathrm{H}}}$ values below 5° Nm are preferable.

OS4: The evaluation is done while driving at the speed of 100 km/h.

$$\tilde{\delta}_{\mathrm{H}} = \sqrt{\int_a^b \Phi_{\delta_{\mathrm{H}}}(f)\mathrm{d}f} \tag{4}$$

$\Phi_{\delta\mathrm{H}}(f)$ is the power spectral density of the steering angle. $\Phi_{\delta\mathrm{H}}(f)$ is integrated for frequencies between a and b. $\tilde{\delta}_{\mathrm{H}}$ values below 0.09° for steering frequencies between 0.2 and 1.5 Hz and values below 0.011° for steering frequencies between 1.5 and 3.0 Hz are preferable.

OS5: A delay between 0 and 0.55 ms for a sporty car and between 0.55 and 0.8 ms for a comfortable car is preferable.

OS6: A gradient less than 0.05 Nm/(°)^2 is preferable.

OS7: A quotient between 0.22 and 0.3 Nm/° for a comfortable car and between 0.3 and 0.35 Nm/° for a sporty car is preferable. The evaluation is done at the speed of 100 km/h.

OS8: An amplification factor between 0.22 and 0.28 1/s for a sporty car and between 0.18 and 0.22 1/s for a comfortable car is preferable. The evaluation is done at the speed of 100 km/h.

OS9: This correlation is dependent on the vehicle segment. The evaluation is done at the speed of 120 km/h.

OS10: The evaluation is done at the speed of 80 km/h.

OS11: The evaluation is done at the steering frequencies between 0.4 and 1 Hz.

OS12–14: The evaluation is done at the speed of 120 km/h.

OS15–17: The evaluation is done at the speed of 80 km/h and the steering frequency of 0.5 Hz.

OS18: The evaluation is done at the speed of 120 km/h and the steering frequency of 0.5 Hz. The steering torque is measured during the reduction of the lateral acceleration.

OS19: A height between 0.5 and 1.5 Nm is preferable.

Table 2. Objective criteria on-center handling (steering feel).

On-center handling—Steering feel				
Subjective characteristic	Objective characteristic	Code/note	Reference	Maneuver
Steering effort	Steering angle	⊙	Farrer, 1993	Weave test
	Steering angle rate	⊙		
	Characteristic based on the amplification function between the wind disturbance and steering intervention	*OS1* ⇧	Wagner, 2003	Crosswind
	Characteristic based on the amplification functions between the wind disturbance and the resulting yaw rate with and without driver intervention	*OS2* ⇧		
	Steering work	*OS3* ⇩	Dettki, 2005	Straight driving with lateral inclination or crosswind
	Characteristic based on frequency spectrum of steering correction	*OS4* ⊙		
	Hysteresis area in phase diagram steering angle—lateral acceleration	⇩	Decker, 2008	Weave test
	Steering angle at lateral acceleration of 0 m/s^2	⇧		
	Hysteresis area in phase diagram steering angle/yaw rate	⇩		
Steering torque level and buildup	Steering torque gradient at lateral acceleration of 0 and 1 m/s^2	⊙	Pietsch and Heissing, 2009b	Sine steer
	Ratio of the steering torque/lateral acceleration hysteresis' width and height	⇩	Decker, 2008	Weave test
	Steering torque at 0°/s yaw rate in hysteresis steering torque/yaw rate diagram	⇧		
	Gradients mean value in hysteresis steering torque/yaw rate diagram	⊙		
Steering response	Phase delay between steering angle and yaw rate	*OS5* ⊙	Dettki, 2005	Weave test
	Second derivative of steering torque with respect to the steering angle	*OS6* ⇩		
	Ratio steering torque/steering angle	*OS7* ⊙		
	Yaw rate amplification factor	*OS8* ⊙	Dettki, 2005 Harrer, 2007	
	Gradient of lateral acceleration with respect to steering angle while steering away from center	*OS9* ⊙	Harrer, 2007	
	Steering torque dead band with respect to lateral acceleration	⇩	Farrer, 1993	Weave test or step steer
	Lateral acceleration dead band with respect to steering angle	⇩	Harrer, 2007	
	Lateral acceleration amplification factor	*OS10* ⊙	Harrer, 2007	Single sine
Center point feeling	Ratio steering torque/steering angle	*OS11* ⇧	Farrer, 1993 Schimmel, 2010 Harrer, 2007	Weave test Frequency response Weave test
	Ratio lateral acceleration/steering angle	⇧	Schimmel, 2010	
	Frequency at the phase minimum in the frequency response function between steering angle and steering torque	⇧		Frequency response
	Ratio lateral acceleration/steering angle	⇧		Weave test

(*continued overleaf*)

Table 2. (*Continued*)

On-center handling—Steering feel				
Subjective characteristic	Objective characteristic	Code/note	Reference	Maneuver
Center point feeling	2D rel.: char. 1: ratio steering torque/lateral acceleration; char. 2: ratio yaw velocity/steering angle	*OS12* ⊙	Harrer, 2007	Weave test
	2D rel.: char. 1: ratio steering torque/lateral acceleration; char. 2: ratio lateral acceleration/steering angle	*OS13* ⊙		
	Steering torque at lateral acceleration of 0 m/s²	*OS14* ⊙		
	2D rel.: char 1: ratio yaw velocity/steering angle; char 2: steering friction	*OS15* ⊙		
	2D rel.: char. 1: ratio steering torque/lateral acceleration; char. 2: ratio yaw velocity/steering angle	*OS16* ⊙		
	2D rel.: char. 1: ratio steering torque/lateral acceleration; char. 2: ratio lateral acceleration/steering angle	*OS17* ⊙		
	Steering torque at lateral acceleration of 1 m/s²	*OS18* ⊙		
	Hysteresis height steering angle/steering torque	*OS19* ⊙	Dettki, 2005	
	Hysteresis width steering angle/steering torque	*OS20* ⊙		
Self-centering	Steering torque gradient	⇧	Zschocke, 2009	Weave test
	Ratio steering torque/steering angle	⇧		
Steering friction	2D rel.: char. 1: ratio yaw rate/steering angle; char. 2: phase delay steering torque/steering angle	*OS21* ⊙	Harrer, 2007	Weave test
	2D rel.: char. 1: phase delay yaw rate/steering torque; char. 2: yaw rate response gain	*OS22* ⊙		
	2D rel.: char. 1: phase delay yaw rate/steering torque; char. 2: ratio yaw rate/steering angle	*OS23* ⊙		
	Steering torque rate at lateral acceleration of 0 m/s²	*OS24* ⊙		
	Residual steering angle in lateral acceleration hysteresis	⊙	Zschocke, 2009	

OS20: A width between 1.5 and 5.0° is preferable.
OS21–23: The evaluation is done at the speed of 120 km/h.
OS24: The evaluation is done at the speed of 120 km/h.

5.1.2 Vehicle behavior

Correlation notes sorted by code (Table 3):

OV1: Gradients under 1/90 (°)m²/Ns are preferable. The evaluation is done while driving straight at the speeds of 100, 140, and 180 km/h.

OV2: The evaluation is done while driving straight between the speeds of 160 and 200 km/h.

$$K_G = \left| \frac{\dot{\psi}}{\tau \upsilon_{res}^2} \right|_{max} \tag{5}$$

$\tau \upsilon^2_{res}$ is a value describing the intensity of the stochastic crosswind disturbance.

OV3:

$$\text{RAT} = 11.97 - 0.67 \cdot \dot{\Psi}_{max} - 7.37 \cdot T_{eq} \tag{6}$$

$\dot{\Psi}_{max}$ is the first maximum of the yaw rate. T_{eq} is the equivalent time delay (measured out of separate sine steer maneuver).

OV4: Values below 0.25 1/s are preferable. The maneuver is done at 100 km/h.

OV5: The evaluation is done at the speed of 190 km/h and a constant lateral inclination of 1.5%.

$$\tilde{\dot{\psi}} = \sqrt{\int_a^b \Phi_{\dot{\psi}}(f) \mathrm{d}f} \tag{7}$$

$\Phi_{\dot{\psi}}(f)$ is the power spectral density of the yaw rate. The root of the integral of $\Phi_{\dot{\psi}}(f)$ between the steering frequencies 0.2 and 1.5 Hz shall have values below 0.1°/s.

Table 3. Objective criteria on-center handling (vehicle behavior).

On-center handling—Vehicle behavior				
Subjective characteristic	Objective characteristic	Code/note	Reference	Maneuver
Cross wind sensitivity	Gradient of ratio yaw rate deviation/crosswind pressure	*OV1* ⇩	Dettki, 2005	Crosswind
	Characteristic based on the maximum of the amplification factor between the wind disturbance and the resulting yaw rate	*OV2* ⇩	Wagner, 2003	
	Characteristic based on the first yaw rate maximum and the equivalent time delay (measured out of separate sine steer maneuver)	*OV3* ⇧	Zomotor, Braess, and Rönitz, 1997	
Lateral inclination sensitivity	First derivative of the yaw rate deviation with respect to the roll angle	*OV4* ⇩	Dettki, 2005	Straight driving
	Characteristic based on the frequency spectrum of the yaw rate	*OV5* ⇩	Dettki, 2005	Straight driving with lateral inclination
Straight driving stability	Characteristic based on the steering angle in combination with the time period between two steering corrections	*OV6* ⇧	Engels, 1993	Straight driving
	Characteristic based on the hysteresis between lateral acceleration and steering angle	*OV7* ⇩	Loth, 1997	Sine steer
	Lateral acceleration at zero steering torque	*OV8* ⇧	Zschocke, 2009	Weave test
	Difference between the lateral acceleration amplification factor at 0.2 and at 0.6 Hz	⇧		ISO frequency response test
Response behavior	Steering angle dead-band with respect to yaw rate	⇩	Farrer, 1993	Weave test, transition test
	Time lag between steering angle and yaw rate	⇩		

OV6:

$$SU = 11.67 - 0.47\left(\frac{\bar{\delta}_{\mathrm{Leff}}}{0.3}\right) - 1.59\left(\frac{14.3}{\bar{t}_{\mathrm{Spur}}}\right) \tag{8}$$

$\bar{\delta}_{\mathrm{Leff}}$ is the effective steering angle of a period. $\bar{t}_{\mathrm{Spur}}$ is the time period between two steering corrections. Maneuver is done at 90 and 150 km/h.

OV7:

$$\phi = \arcsin\left(\frac{\delta_{\mathrm{LP}}}{\widehat{\delta}_{\mathrm{L}}}\right) \tag{9}$$

δ_{LP} is half the width of the lateral acceleration/steering angle hysteresis at 0 m/s^2 lateral acceleration. $\widehat{\delta}_{\mathrm{L}}$ is half the width of the hysteresis at maximum lateral acceleration.

OV8: The evaluation is done at the speeds of 100 and 150 km/h.

5.2 Corner handling

5.2.1 Steering feel

Correlation notes sorted by code (Table 4):

CS1–2: The evaluation is done at the speed of 80 km/h, the steering frequency of 0.4 Hz, and the lateral acceleration of 0.4*g*.

CS3–5: A progressive steering torque increase is preferable, that is, low torque at 1 m/s^2, medium torque at 4 m/s^2, and a high torque difference between 0.5 and 1 m/s^2. The evaluation is done at the speeds of 70 and 100 km/h.

CS6–7: The evaluation is done at the speed of 100 km/h and the lateral acceleration of 0.4*g*.

CS8–10: The evaluation is done at the speed of 120 km/h and the steering frequency of 0.5 Hz with a steering amplitude of 10°.

CS11: The evaluation is done up to a lateral acceleration of 0.4*g* with a step steer maneuver at the speed of

Table 4. Objective criteria corner handling (steering feel).

Corner handling—Steering feel				
Subjective characteristic	Objective characteristic	Code/note	Reference	Maneuver
Steering torque level and progress	Ratio peak yaw rate/peak steering torque	*CS1* ⊙	Harrer, 2007	Single sine
	Ratio peak steering torque/peak lateral acceleration	*CS2* ⊙		
	Steering torque at lateral acceleration of $1\,m/s^2$	*CS3* ⇧	Zschocke, 2009	Steering angle ramp
	Steering torque at lateral acceleration of $4\,m/s^2$	*CS4* ⇧		
	Difference between steering torque at lat. acc. of $0.5\,m/s^2$ and $1\,m/s^2$	*CS5* ⇧		
Steering returnability	Stationary residual steering angle (steering release after 2 s cornering at $4\,m/s^2$)	⇩	Zschocke, 2009	Steering return
Steering precision	Steering torque at zero steering angle	⇧	Decker, 2008 Zschocke, 2009	Sine steer/weave
	Hysteresis area in phase diagram steering angle/lateral acceleration	⇩	Decker, 2008	
	Steering angle at lateral acceleration of $0\,m/s^2$	⇩		
	Lateral acceleration at zero steering angle	⇧		
	Phase delay between steering angle and yaw rate at 1 Hz	⇩		
	Characteristic based on the delay time in the lateral acceleration frequency response function	⇩		
	Peak roll rate	*CS6* ⇩	Harrer, 2007	Step steer
	2D rel.: char. 1: peak roll rate; char. 2: ratio of peak roll rate and steady state value of lateral acceleration	*CS7* ⊙		
	2D rel.: char. 1: ratio yaw rate/steering angle; char. 2: steering friction	*CS8* ⊙		Sine steer/weave
	2D rel.: char. 1: ratio yaw rate/steering angle; char. 2: steering angle hysteresis in torque/angle diagram	*CS9* ⊙		
	2D rel.: char. 1: peak value of yaw rate; char. 2: steering angle hysteresis in torque/angle diagram	*CS10* ⊙		
	Ratio lateral acceleration/steering angle	⇧		
	Steering torque hysteresis in torque/angle diagram	⊙		
	Lateral acceleration hysteresis in lateral acceleration/steering angle diagram	⊙		
	Lateral acceleration at 0 Nm steering torque	⊙	Zschocke, 2009 Harrer, 2007	
	Ratio steering torque/steering angle	⇧	Schimmel, 2010	
	Steering torque phase response below 1 Hz	⇧	Zschocke, 2009	
Steering clearance	Phase lead of the steering torque	⇩	Zschocke, 2009	Sine steer/weave
Stability feel	Steering torque hysteresis at 0° steering angle (0.25–0.75 Hz)	⇧	Zschocke, 2009	Sine steer/weave

(*continued overleaf*)

Table 4. *(Continued)*

Corner handling—Steering feel					
Subjective characteristic	Objective characteristic	Code/note		Reference	Maneuver
Steering angle demand	Amplification factor between yaw rate and steering angle		⊙	Decker, 2008 Harrer, 2007 Schimmel, 2010 Zschocke, 2009	Sine steer/step steer/weave/ steady-state cornering
	Amplification factor between lateral acceleration and steering angle		⊙	Decker, 2008 Harrer, 2007 Schimmel, 2010	Sine steer/weave
	Hysteresis area in steering angle/lateral acceleration diagram		⇩	Decker, 2008	Sine steer
	Medial steering ratio		⊙	Zschocke, 2009	Steady-state cornering
Steering response	Amplification factor between yaw rate and steering angle	*CS11*	⊙	Harrer, 2007	Step steer/single sine/weave/ frequency response
	Peak value of lateral acceleration	*CS12*	⊙		Sine steer/weave
	Peak value of yaw rate	*CS13*	⊙		
	Amplification factor between lateral acceleration and steering angle	*CS14*	⊙		Lane change/ single sine/ weave
	Yaw acceleration	*CS15*	⇧	Wolf, 2008	Step steer
	Phase delay between steering angle and yaw angle	*CS16*	⇩		

80 km/h or with a single sine maneuver at the speed of 80 km/h, the steering frequency of 0.2 Hz or with a weave test at the speed of 120 km/h, and the steering frequency of 0.5 Hz with an steering amplitude of 20° or with a frequency response test at the speed of 100 km/h.

CS12–13: The evaluation is done at the speed of 120 km/h.

CS14: The evaluation is done at the speed of 80 or 100 km/h and the lateral acceleration of 0.4*g*.

CS15–16: The evaluation is done at the speed of 80 km/h.

5.2.2 *Vehicle behavior*

Correlation notes sorted by code (Table 5):

CV1: Values around 0.2–0.3 s are optimal.

CV2: The TB2 value is the product between the peak response time $T_{\dot\psi_{\max}}$ of the yaw rate and the maximum vehicle sideslip angle $\beta_{\max}$.

$$\mathrm{TB2} = T_{\dot\psi_{\max}} \cdot \beta_{\max} \tag{10}$$

The TB2 value is not valid for vehicles fitted with rear-wheel steering.

CV3: $\ddot\Psi_{\mathrm{QuMW}}$ is the yaw acceleration mean square value.

$$\ddot\Psi_{\mathrm{QuMW}} = \frac{\left|\sum_{i=0}^{t_e} \frac{\ddot\Psi_i}{|\ddot\Psi_i|}(\ddot\Psi_i)^2\right|}{\sum_{i=0}^{t_e} \frac{\ddot\Psi_i}{|\ddot\Psi_i|}(\ddot\Psi_i)^2} \frac{\sqrt{\left|\sum_{i=0}^{t_e} \frac{\ddot\Psi_i}{|\ddot\Psi_i|}(\ddot\Psi_i)^2\right|}}{100 t_e} \tag{11}$$

t_0 is the point in time of brake application and t_e the point in time of the first driver interaction but has a maximum of 2 s.

CV4:

$$\ddot\Psi_{\mathrm{relAbsMaz}} = \frac{|\ddot\Psi - \ddot\Psi_0|}{\ddot\Psi - \ddot\Psi_0} \cdot \left(|\ddot\Psi - \ddot\Psi_0|\right)_{\max_{t_0-t_e}} \tag{12}$$

$\ddot\Psi_0$ is the reference yaw rate acceleration, t_0 the point in time of brake application, and t_e the point in time of the first driver interaction but has a maximum of 2 s.

CV5:

$$\ddot\Psi_{\mathrm{Diff}} = \left|\left(\ddot\Psi - \ddot\Psi_0\right)_{\max(t_0-t_e)}\right| - \left|\left(\ddot\Psi - \ddot\Psi_0\right)_{\min(t_0-t_e)}\right| \tag{13}$$

$\ddot\Psi_0$ is the reference yaw acceleration and t_0 the point in time of brake application.

Table 5. Objective criteria corner handling (vehicle behavior).

Corner handling—Vehicle behavior					
Subjective characteristic	Objective characteristic	Code/note		Reference	Maneuver
Self-steering behavior	Gradient of steering wheel angle with respect to lateral acceleration at 6 m/s^2		⇩	Zschocke, 2009	Steady-state cornering
Corner braking behavior	Overshoot extent of the lateral acceleration		⊙	Dreyer, 1990	Corner braking/ step steer under braking
	Response time of lateral acceleration and yaw rate	*CV1*	⊙		
	Product of response time of yaw rate and peak vehicle sideslip angle	*CV2*	⇩		
	1 s-value of the yaw angle deviation or of the vehicle sideslip angle		⇩	Zomotor, Braess, and Rönitz, 1997	Corner braking
	Peak vehicle sideslip angle		⇩		
	Characteristic based on the yaw acceleration mean square value	*CV3*	⇩	Schick and Bunz, 2002	
	Characteristic based on the peak relative yaw acceleration	*CV4*	⇩		
	Characteristic based on the difference between the minimum and maximum relative yaw acceleration	*CV5*	⇩		
Pitch and roll behavior	Characteristic based on the wheel travel	*CV6*	⇩	Kawagoe, Suma, and Watanabe, 1997	Cornering under positive longitudinal acceleration
	Characteristic based on the lateral acceleration, roll angle, rate, and acceleration	*CV7*	⇩	Botev, 2008	Lane change
	Roll rate (at 0.8 Hz)		⇩	Seyed-Ghaemi, 2005	Sine steer
	Derivative of the roll angle with respect to lateral acceleration		⇩		Steady-state cornering
	Amplification factor between roll angle and steering angle (0.5 Hz)		⇩	Zschocke, 2009	ISO frequency response
	Overshoot extent of the roll angle at 7 m/s^2		⇩		Step steer
	Characteristic based on steering angle, roll angle, roll rate, and acceleration	*CV8*	⇩	Botev, 2008	Lane change
Response behavior	Yaw Eigen-frequency		⇧	Schimmel, 2010	Frequency response
	Steering torque mean amplitude decrease		⇧		
	Yaw rate peak response time		⊙		Step steer
	Overshoot extent of yaw rate		⊙		
	Amplification factor between yaw rate and steering angle		⊙	Zomotor, Braess, and Rönitz, 1997	Step steer
	Response time of lateral acceleration and yaw rate		⇩		
	Overshoot extent of lateral acceleration and yaw rate		⊙		
	Characteristic based on the product of the yaw rate peak response time and the stationary vehicle sideslip angle	*CV9*	⇩		
	Steering angle at lateral acceleration of 0 m/s^2		⇩	Decker, 2008	Sine steer

(*continued overleaf*)

Table 5. (*Continued*).

Corner handling—Vehicle behavior				
Subjective characteristic	Objective characteristic	Code/note	Reference	Maneuver
Stability and lane change behavior	Vehicle sideslip angle	⇩	Riedel and Arbinger, 1997	Double lane change
	Vehicle sideslip angle rate	⇩	Zschocke, 2009	Steady-state cornering
	Roll rate	⇩	Riedel and Arbinger, 1997	Double lane change
	Time period after which the vehicle sideslip angle exceeds a threshold value	⇩		
	Vehicle sideslip angle relative to the lateral acceleration and to the steering angle of the second steer back	⇩	Botev, 2008	Double lane change
	Phase shift between lateral acceleration and yaw rate (0.5 Hz)	⇩	Huneke *et al.*, 2010	Sine steer
	Characteristic based on the unbalance of the steering angle history of the second lane change and the time delay between steering angle and lateral acceleration	*CV10* ⇩	Dibbern, 1992	Double lane change
Load-change behavior	Characteristic based on the mean yaw rate deviation divided by a reaction time (0.75 s) and the mean yaw acceleration	⊙	Zomotor, Braess, and Rönitz, 1997	Load-change deceleration
	Characteristic based on the yaw rate and reference yaw rate 1.5 s after an accelerator pedal kick	*CV11* ⊙	Schweers and Röth, 1995	Load-change acceleration

CV6:

$$\mathrm{RMI} = \frac{1}{2}[(z_{\mathrm{i,f}} + z_{\mathrm{o,f}}) - (z_{\mathrm{i,r}} + z_{\mathrm{o,r}})] \tag{14}$$

z is the vertical wheel travel with the suffixes i (inner), o (outer), f (front), and r (rear). A negative RMI value is preferable.

CV7:

$$\mathrm{WI} = \frac{\ddot{\varphi}}{a_y} \cdot h_{\mathrm{k}} \cdot \frac{1}{\pi} + \frac{\dot{\varphi}}{a_y} \cdot h_1 + \frac{\varphi}{a_y} \cdot h_1 \cdot \pi \tag{15}$$

h_{k} is the height distance between the driver's head and the roll axle, a_y is the lateral acceleration, φ is the roll rate, and h_1 equals the length of 1 m.

CV8:

$$\mathrm{AWP} = \frac{\ddot{\varphi}}{\delta_{\mathrm{H}}} \cdot h_{\mathrm{k}} \cdot \frac{1}{\pi} + \frac{\dot{\varphi}}{\delta_{\mathrm{H}}} \cdot h_1 + \frac{\varphi}{\delta_{\mathrm{H}}} \cdot h_1 \cdot \pi \tag{16}$$

h_{k} is the height distance between the driver's head and the roll axle, δ_{H} is the maximal steering angle, φ is the roll rate, and h_1 equals the length of 1 m.

CV9: The TB value is the product between the peak response time $T_{\dot{\psi}_{\max}}$ and the stationary vehicle sideslip angle β_{stat}.

$$\mathrm{TB} = T_{\dot{\psi}_{\max}} \cdot \beta_{\mathrm{stat}} \tag{17}$$

The TB value is not valid for vehicles fitted with rear-wheel steering.

CV10: $\Delta\delta_{\max,2}$ is the dissymmetry of the steering angle history of the second lane change and $T_{0,(\delta,a_y)}$ the time delay between steering angle and lateral acceleration.

$$\mathrm{KD} = \Delta\delta_{\max,2} + 2.5 \cdot T_{0,(\delta,a_y)} \tag{18}$$

CV11: $\dot{\Psi}$ is the yaw rate and $\dot{\Psi}_{\mathrm{ref}}$ the reference yaw rate.

$$\frac{\dot{\Psi}}{\dot{\Psi}_{\mathrm{ref}}}(t_0 + 1.5\,\mathrm{s}) \tag{19}$$

t_0 is the time of the accelerator pedal kick.

6 SUMMARY

This chapter describes how the handling quality of passenger cars can be measured by subjective and objective evaluations. Beginning with an introduction of the basic principles of the subjective vehicle handling perception, practical advice for the planning, conduct, and analysis of subjective evaluations with groups of selected customers or professional drivers was given.

The compliance with the presented rules can significantly increase the informative value of these evaluations. Objective evaluation methods based on characteristic values promise a more efficient way to assess vehicle handling and facilitate evaluations in virtual simulation environments. Therefore, this chapter has provided an overview on possible methods of objective evaluation. The necessary analysis of subjective or objective data and the identification of the relationship among both by means of correlation and regression analyses were described. The correlation coefficient has proven to be a reasonable measure for the statistical significance of the relationship. Finally, on the basis of an extensive meta-study, an overview on customer-relevant vehicle handling measures was given. In order to ensure customer relevance, only characteristic values that have been validated by real test-drives and regression analyses were considered. The identified measures have been summarized for practical use in the form of Tables 2–5.

REFERENCES

Abe, M. and Kano, Y. (2008) A study on vehicle handling evaluation by model based driver steering behavior. FISITA 2008-03-022.

Barthenheier, T. (2004) Potenzial einer fahrertyp- und fahrsituationsabhängigen Lenkradmomentgestaltung. *Fortschritt-Berichte VDI Reihe*, **12**, 584.

Bortz, J. (2010) *Statistik für Human- und Sozialwissenschaftler*, 7th edn, Springer Verlag, Heidelberg.

Botev, S. (2008) Digitale Gesamtfahrzeugabstimmung für Ride und Handling. *Fortschritt-Berichte VDI Reihe*, **12**, 684.

Breuer, J. (2009) Bewertungsverfahren von Fahrerassistenzsystemen, in *Handbuch Fahrassistenzsysteme* (eds H. Winner, S. Hakuli, G. Wolf), Springer Verlag, Heidelberg, pp. 55–68.

Bubb, H. (2003) Fahrversuche mit Probanden - Nutzwert und Risiko. *Darmstädter Kolloquium Mensch & Fahrzeug*, TU Darmstadt.

Cohen, J. and Cohen, P. (1983) *Applied Multiple Regression/Correlation Analysis for the Behavioral Sciences*, 2nd edn, Erlbaum, Hillsdale, NJ.

Decker, M. (2008) Zur Beurteilung der Querdynamik von Personenkraftwagen. Thesis. TU Munich.

Dettki, F. (2005) Methoden zur objektiven Bewertung des Geradeauslaufs von Personenkraftwagen. Thesis. Universität Stuttgart.

Dibbern, K. (1992) Ermittlung eines Kennwertes für den ISO-Fahrspurwechsel in Versuch und Simulation. *Fortschritt-Berichte VDI Reihe*, **12**, 164.

Dreyer, A. (1990) Untersuchung des Fahrverhaltens von PKW bei kombinierten Lenk- und Bremseingaben. Thesis. RWTH Aachen University.

Eckstein, L. (2011) Vertical- and lateral dynamics of vehicles. Lecture Automotive Engineering II, RWTH Aachen University.

Engels, A. (1993) Geradeauslaufkriterien für Pkw und deren Bewertung. Thesis. TU Braunschweig.

Farrer, D.G. (1993) An objective measurement technique for the quantification of on-centre handling quality. SAE Paper 930827.

Harrer, M. (2007) Steering feel—objective assessment of passenger cars—analysis of steering feel and vehicle handling. Thesis. University of Bath.

Harrer, M., Pfeffer, P.E., and Johnston, N.D. (2006) Steering feel—objective assessment of passenger car analysis of steering feel and vehicle handling. FISITA 2006V165.

Heißing, B. and Brandl, H.J. (2002) *Subjektive Beurteilung des Fahrverhaltens*, Vogel Fachbuch Verlag, Würzburg.

Henze, R. (2004) Beurteilung von Fahrzeugen mit Hilfe eines Fahrermodells. Dissertation. Technische Universität Braunschweig, Schriftenreihe des Instituts für Fahrzeugtechnick TU Braunschweig, 7.

Huneke, M., Pascali, L., Strecker, F., *et al.* (2010) Identifikation von Fahrdynamikmodellen zur Fahrverhaltensbewertung und deren Anwendung in der Erprobung und Simulation bei der Fahrdynamikentwicklung. *15. VDI-Tagung Erprobung und Simulation in der Fahrzeugentwicklung, VDI-Berichte* 2016.

Jürgensohn, T., Willumeit, H.-P., and Irmscher, M. (1999) Fahrermodelle als Hilfsmittel zur Objektivierung von subjektiven Bewertungen der Fahrbarkeit. *Fortschritt-Berichte VDI Reihe*, **22**, 1.

Käppler, W.-D. (1993) Beitrag zur Vorhersage von Einschätzungen des Fahrverhaltens. *Fortschritt-Berichte VDI Reihe*, **12**, 198.

Kawagoe, K., Suma, K., and Watanabe, M. (1997) Evaluation and improvement of vehicle roll behavior. SAE Paper 970093.

Kobetz, C. (2004) Modellbasierte Fahrdynamikanalyse durch ein an Fahrmanövern parameteridentifiziertes querdynamisches Simulationsmodell. Thesis. TU Wien.

Kudritzki, D. (1989) Zum Einfluss querdynamischer Bewegungsgrössen auf die Beurteilung des Fahrverhaltens. *Fortschritt-Berichte VDI Reihe*, **12**, 132.

Kudritzki, D. (2000) Möglichkeiten zur Objektivierung subjektiver Beurteilungen des Fahrzeugverhaltens, in *Subjektive Fahreindrücke sichtbar machen: Korrelation zwischen CAE-Berechnung, Versuch und Messung von Versuchsfahrzeugen und –komponenten* (ed. K. Becker), Expert-Verlag, Düsseldorf, pp. 11–26.

Lienert, G. and Raatz, U. (1969) *Testaufbau und Testanalyse*, 3rd edn, Verlag Julius Beltz, Weinheim.

Loth, S. (1997) Fahrdynamische Einflußgrößen beim Geradeauslauf von Pkw. Thesis. TU Braunschweig.

Magnusson, D. (1975) *Testtheorie*, 2nd edn, Deuticke, Wien.

Meljnikov, D. (2003) Entwicklung von Modellen zur Bewertung des Fahrverhaltens von Kraftfahrzeugen. Thesis. Universität Stuttgart.

Meyer-Tuve, H. (2009) Modellbasiertes Analysetool zur Bewertung der Fahrzeugquerdynamik anhand von objektiven Bewegungsgrößen. Thesis. TU Munich.

Mitschke, M. and Niemann, K. (1972) *Die Regeltaetigkeit des Autofahrers bei Kursabweichungen Deutsche Kraftfahrtforschung und Strassenverkehrstechnik*, VDI-Verlag, Düsseldorf, 221.

Mummendey, H.D. (2003) *Die Fragebogen Methode*, Hogrefe-Verlag, Göttingen.

Neukum, A., Krüger, H.-P., and Schuller, J. (2001) *Der Fahrer als Messinstrument für fahrdynamische Eigenschaften?VDI-Berichte Nr. 1613: Der Fahrer im 21. Jahrhundert*, VDI-Verlag, Düsseldorf.

Pfeffer, P.E. and Harrer, M. (2011) Lenkgefühl: Die Kunst der Beschreibung, in *Lenkungshandbuch: Lenksysteme, Lenkgefühl, Fahrdynamik von Kraftfahrzeugen* (ed. M. Harrer), Vieweg+Teubner, Heidelberg.

Pfeffer, P.E. and Scholz, H. (2010) Present-day cars—subjective evaluation of steering feel. *Chassis.tech plus 2010*, Munich.

Pietsch, R., Schimmel, C., and Heißing, B. (2009) Objective assessment of handling performance. C*hassis.tech 2009*, Munich.

Pietsch, R. and Heißing, B. (2009) Modellbasierte Beurteilung des Lenkgefühls, in *Subjektive Fahreindrücke sichtbar machen IV: Korrelation zwischen objektiver Messung und subjektiver Beurteilung in der Fahrzeugentwicklung* (ed. K. Becker), Expert-Verlag, Essen.

Rasmussen, J. (1983) Skills, rules and knowledge; signals, signs and symbols, and other distinctions in human performance models. *IEEE Transactions on Systems, Man and Cybernetics*, **SMC-13** (3), 257–266.

Redlich, P. (1994) Objektive und subjektive Beurteilung aktiver Vierradlenkstrategien. Thesis. RWTH Aachen University.

Riedel, A. and Arbinger, R. (1997) Subjektive und objektive Beurteilung des Fahrverhaltens von PKW. *FAT-Schriftenreihe*, 139.

Sagan, E. (2003) *Zur Beurteilung von Fahreigenschaften in fahrdynamischen Testverfahren*Reifen*, Fahrwerk, Fahrbahn 2003*.

Scharpe, B., Prokop, G., Golloch, D., and Schick, B. (2012) Design in the Loop—bridging the gap between subjective perception and objective evaluation to achieve desired handling characteristics. C*hassis.tech plus 2010*, Munich.

Schick, B. and Bunz, D. (2002) Fahrzeuggierstabilität beim Kurvenbremsen aus hohen Geschwindigkeiten. *brems.tech 2002*, Munich.

Schimmel, C. and Heißing, B. (2009) Fahrerbasierte Objektivierung subjektiver Fahreindrücke, in *Subjektive Fahreindrücke sichtbar machen IV: Korrelation zwischen objektiver Messung und subjektiver Beurteilung in der Fahrzeugentwicklung* (ed. K. Becker), Expert-Verlag, Essen.

Schimmel, C. (2010) Entwicklung eines fahrerbasierten Werkzeugs zur Objektivierung subjektiver Fahreindrücke. Thesis. TU Munich.

Schweers, T.F. and Röth, H. (1995) *Entwicklung eines objektiven Testverfahrens für Pkw mit Antriebsschlupf-Regelung Reifen, Fahrwerk, Fahrbahn 1995*, VDI-Verlag, Düsseldorf.

Seyed-Ghaemi, A. (2005) *Bewertungskriterien zur objektiven Festlegung des Wankverhaltens aus Fahrdynamikmessungen*, Thesis, Landshut.

Wagner, A. (2003) Ein Verfahren zur Vorhersage und Bewertung der Fahrerreaktion bei Seitenwind. Thesis. Universität Stuttgart.

Wallentowitz, H. (1979) Fahrer-Fahrzeug-Seitenwind. Thesis. TU Braunschweig.

Winner, H., Barthenheier, T., Fecher, N., and Luh, S. (2003) Fahrversuche mit Probanden zur Funktionsbewertung von aktuellen und zukünftigen Fahrerassistenzsystemen. *Fortschritt-Berichte VDI Reihe*, **12**, 557.

Wolf, H.J. (2008) Untersuchung des Lenkgefühls unter besonderer Berücksichtigung ergonomischer Erkenntnisse und Methoden. Thesis. TU Munich.

Zomotor, A., Braess, H.-H., and Rönitz, R. (1997) Verfahren und Kriterien zur Bewertung des Fahrverhaltens von Personenkraftwagen—Ein Rückblick auf die letzten 20 Jahre. *ATZ Automobiltechnische Zeitschrift*, (12), 1998 – 03.

Zong, C., Guo, K., and Hsin, G. (2000) Research on closed-loop comprehensive evaluation method of vehicle handling and stability. SAE Paper 2000-01-0694.

Zschocke, A.K. (2009) Ein Beitrag zur objektiven und subjektiven Evaluierung des Lenkkomforts von Kraftfahrzeugen. *IPEK Forschungsberichte Band 34*. Thesis. Universität Karlsruhe.

Chapter 129
Automated Driving

Adrian Zlocki
Institut für Kraftfahrzeuge (ika), Aachen, Germany

1 INTRODUCTION

First assistance functions for automobiles were introduced shortly after the automobile itself had been invented, for example, the introduction of the automatic engine starter in 1913. In addition, the idea of automated driving vehicles was already developed in the early twentieth century. One of the first steps toward this idea was the introduction of the first cruise control system in the Chrysler Imperial as the so-called auto pilot in 1958 (Rowsome, 1958). This mechanical system was able to keep a constant speed, set by the driver, and therefore was able to take over the longitudinal vehicle control on a motorway. With the development of microelectronics and the introduction of computer-controlled systems in the last quarter of the twentieth century, the possibilities to realize automated driving vehicles became available. On the basis of these possibilities, initiatives were started such as PROMETHEUS (PROgraMme for a European Traffic of Highest Efficiency and Unprecedented Safety), the American National Automated Highway System Consortium (NAHSC), the Intelligent Vehicle Initiative (IVI), and the Japanese Advanced Cruise-Assist Highway System Research Association (AHSRA) in the end of the 1980s. At present, automated driving basically can be divided into three main categories:

- Advanced driver assistance systems (ADAS): ADAS partly take over the driving task and support the driver in complex and monotonous situations.
- Infrastructure-based automated driving vehicles: These self-driving vehicles operate in a dedicated infrastructure such as dedicated lanes or closed areas automated.
- Fully automated driving vehicles: This type of vehicles are operated driverless.

2 REQUIREMENTS FOR AUTOMATED DRIVING VEHICLES

Automated driving vehicles need to fulfill technological and legal requirements. The necessary technology for automated driving vehicles needs to provide sophisticated information about the surrounding environment and the current vehicle status. This information is processed to determine the situation awareness and provide the correct action or even reaction by the actuators of the vehicle.

In contrast to conventional controlled vehicles, automated driving vehicles need to perceive the surrounding environment, process the information, and react to them fully automated (Figure 1). This reaction is targeted to the normal car operation and the supervision and avoidance of dangerous situations.

While the driver depends on his own recognition for information perception, for automated systems, vehicle and

Encyclopedia of Automotive Engineering, print © 2015 John Wiley & Sons, Ltd.
Edited by David Crolla, David E. Foster, Toshio Kobayashi, and Nick Vaughan.
This article is © 2015 John Wiley & Sons, Ltd. ISBN: 978-0-470-97402-5
Also published in the *Encyclopedia of Automotive Engineering* (online edition)
DOI: 10.1002/9781118354179.auto023

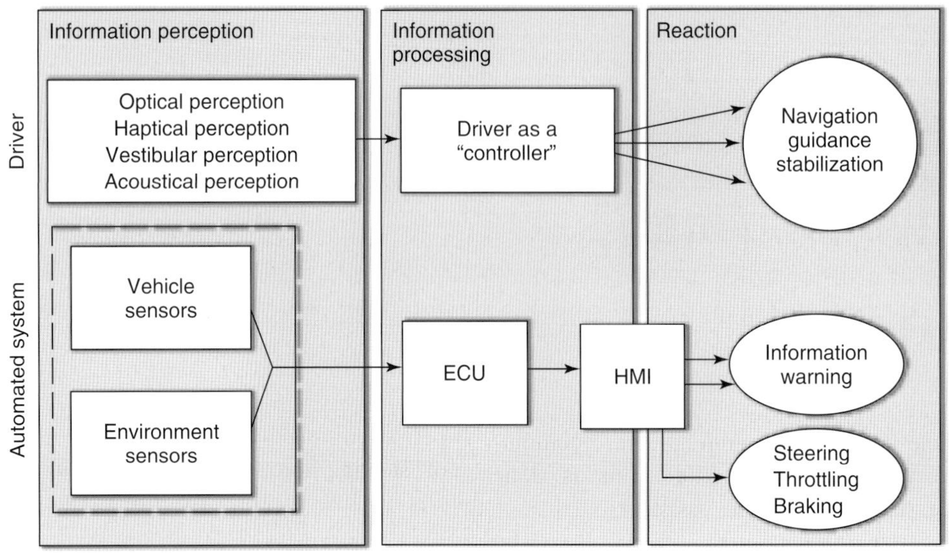

Figure 1. Vehicle control comparison between driver and automated system.

environment sensors are necessary for the perception. These sensors detect the driving condition (speed, acceleration, yaw rate, etc.) and the surrounding traffic situation (other traffic participants, objects in the driving path, current position, etc.). Information processing takes place in an electronic computer unit (ECU). The driver activates the system by means of an appropriate human–machine interface (HMI). Actuators such as the steering system, the accelerator, and the brake actuator or even the automatic gear shifter transfer the reaction of the system into changes of the vehicle's motion.

2.1 Sensor technology

Sensor technology is one of the keys for automated vehicles. The sensors need to be able to detect the vehicle's surroundings and to measure the dynamic status of the other vehicles in relation to the own vehicle. The frontal area of the vehicle is divided into a near field and a far field. The near field is the area up to 50 m around the vehicle. In this area, a wide observation angle is substantial. The far field goes up to 200 m around the vehicle. The range of the sensors and the evaluation of the relative velocities to others are important factors for the far field. Furthermore, position sensors and the communication with the infrastructure and with other traffic participants can provide additional information.

Different principles for environmental detection are available. The most important ones are radar (Radio Detection and Ranging), laser (Light Amplification by Stimulated Emission of Radiation), and image processing by camera picture analysis. Ultrasonic sensors are also necessary for very short ranges.

Radar sensors operate with electromagnetic waves, using frequencies in the centimeter or micrometer distance for object detection. Frequencies for automotive applications of radar sensor are in the 24 GHz and the 76–77 GHz bands.

The frequency-modulated continuous wave-radar (FMCW-radar) represents the most common used radar principle. The measurement of the distance to the relevant target of the FMCW-radar is based on the phase shift between the transmitted and the received signal with a periodic phase. In case the transmitter and the receiver are moved with a relative velocity, a frequency shift occurs between the transmitted and the received frequencies. This shift is due to the Doppler effect. The frequency increases in case the target is approached. The frequency difference depends on the relative velocity, which can be determined based on this effect.

Lidar sensors, which are based on laser technology, use the reflection of transmitted electromagnetic waves with lengths from 0.78 to 1 μm and thus in the infrared range, invisible to the human eye. In automotive technology, two methods are common, the transit time method and the laser-Doppler-shift method, which is based on short laser impulses being emitted by the laser. The Doppler-shift method is used to measure additionally the relative velocity. The frequency shift can be generated by the superposition of the reflected light impulse and a reference impulse of the same laser source. Another type of laser sensor is the laserscanner. Laserscanners use mechanically rotating mirrors. Owing to this, these sensors can scan a wide area and therefore they have high opening angles up to 270°. The own vehicle limits the complete 360° view, unless the sensor is mounted on the top of the vehicle.

Image processing is also used as a sensor. The environment is recorded by means of an optical device. Afterward the image data are converted into electrical signals by means of semiconductor image sensors (charge-coupled device or complementary metal oxide semiconductor). The digital image is analyzed by means of object detection algorithms. A very promising application for image processing systems is used by sensor fusion with other sensor systems, as it is used for collision mitigation and collision avoidance systems.

Vehicle-to-vehicle and vehicle-to-infrastructure communications provide additional information. This is done for short distance (e.g., intersection assistance) or even long distance (μ-factor information of the road surface or traffic sign information). Various communication technologies with different frequency bands are available today (e.g., Wireless LAN IEEE 802.11, HiperLAN, CALM etc.). A frequency range from 5.855 to 5.925 GHz is assigned for vehicle-to-vehicle communication in Europe.

Sensors for vehicle status detection, for example, position, friction coefficient of the road, and rain, have become of high importance for modern assistance systems. In particular, digital map data are used for many different ADAS as they contain attributes about the road infrastructure and legal information such as, that is, velocity limitations.

2.2 Actuator technology

In order to convert the signals of the ECU into the desired vehicle reaction, the vehicle needs to be equipped with actuators for steering, braking, and acceleration. Apart from these main groups of actuators, additional actuators (e.g., curve light and pre-crash) are available for ADAS.

At present, steering support is provided by means of hydraulic power steering (HPS). Additional support is provided by the adaptive HPS, the so-called Servotronic. The Servotronic steering system is able to reduce the steering support with rising vehicle velocity and therefore can offer an easy moving steering wheel for parking in the low velocity range as well as a comfortable steering behavior at high velocity.

Compared to the HPS, the electrohydraulic power steering (EHPS) consists of an electronic powered hydraulic pump, which provides the steering force. The control is managed demand controlled.

The electric power steering (EPS), also called *Servoelectric*, provides the steering force to the servo hydraulic power assistance with the help of an electric actuator. The electric motor and the related transmission exemplarily are fixed with the steering column, with the steering gear pinion or with the gear rack. The EPS provides the possibility of a steering boost by any dimension.

Additional steering angle can be applied by the active front steering. The additional angle is applied by means of a planetary gearbox, integrated into the steering system. This superposes the driver's steering intention with any desired angle. The mechanical connection between the steering wheel and the wheels leads to a direct response in lateral dynamic driving conditions.

By-wire steering systems decouple the steering-wheel input from the front wheels. The concept is based on performing steering maneuvers without mechanical transfer of steering torque. There is no mechanical connection between the steering wheel and the front wheels existent. In order to achieve necessary safety requirements, complete system redundancy is necessary. The vehicle has to remain steerable, even in the case of a system error.

The necessary brake interference for ADAS and automated driving vehicles can be realized on the basis of

- Brake booster:
  The brake booster is able to regulate a desired target brake pressure hydraulically and can therefore brake the vehicle without usage of the brake pedal by the driver.
- Electrohydraulic brake (EHB) system:
  The EHB system can divide the brake force up to each single wheel. The advantage of EHB system is the improved operating time because the control unit recognizes the demand for more braking by observing the pedals or their operating rate and acts before a fully hydraulic brake could build up pressure.
- Electromechanical brake (EMB) system:
  The EMB system is a rearrangement of the brake system to a completely electronically operated brake with dry function for the brake disks. This brake system offers subfunctions for automatic driving (e.g., automatic parking or automatic emergency braking).
- Hybrid brake system:
  The hybrid brake system represents a combination of a hydraulic and an EHB at the front axle and a dry electrical brake at the rear axle of the vehicle. An advantage of the hybrid brake system is the simplified package at the rear axle by the omission of hydraulic components.
- Brake-by-wire system:
  In brake-by-wire systems, the desired deceleration by the driver is electronically passed on from the pedal to the brake system. The by-wire-brake system works electrically without a hydraulic medium. This allows a simple electronic interface for driver assistance systems

as well as the realization of additional comfort functions and package advantages. Redundancies are absolutely necessary. A high quality battery and energy management is necessary because the brake system does not work on hydraulic basis.

Automated acceleration can be provided by means of electronic acceleration pedals (e-gas). These pedals are equipped with electric motors, pedal sensors, and electronic throttle-position controllers. The pedal sensor transforms the accelerator pedal position into an electronic signal, which is transferred into an appropriate position of the throttle. Modern ADAS pedals provide force feedback functionality, for example, for giving drivers feedback in critical driving situations or urging them to reduce driving velocity.

2.3 Legal aspects

From the today's point of view, automated vehicles are facing different legal issues. In a very first but important step, it must be understood that legal issues will usually strongly depend on the nature of access rights to the roads: as far as automation is to be used on roads with unrestricted access, the legal issues mentioned in the following are strongly relevant. In case, however, that the testing of the automation is done on test grounds with only restricted access by the public (e.g., enclosed working areas or test tracks), the risks for life, health, and property of employees as well as visitors and even trespassers will need consideration. It potentially raises the issue of liability in case of an accident. Apart from this, the limiting issues in case of higher automation levels generally are the national regulatory law on the conduct of the driver, product liability, road traffic liabilities, and vehicle type approval (Gasser, 2010a, b).

The most important starting point to identify those legal issues, which are relevant in case of automation on publicly accessible roads, is to clearly describe the level of automation provided. The ADAS, which are already available today, show quite easy how high the level of legally permissible automation already is today. In terms of regulatory law of conduct of the driver, this automation level is possible, as the driver always remains in the position to take over control immediately and at any point. The driver therefore necessarily observes the surrounding traffic, which belongs to his tasks from the legal situation, found in the terms of regulatory law on driver's conduct. As far as systems actively intervene into the task of driving, it is their overrideability that ensures compatibility with regulatory law (such as the Vienna Convention on international level). This will usually be an issue in case of systems, addressing near accident situations (Gasser, 2010a).

Apart from this, product liability takes an important role in the implementation of the automation for road transport. As far as automation is meant in the way that the vehicle remains under control of the driver and does not reach the level of autonomy, the risks, combined with the estimation of foreseeable use, in terms of product liability remain fundamental. In technical terms, any higher level of automation will require precautional technical measures in terms of reliability, in order to avoid running into product liability. Currently, a nationally focused project group has investigated the legal issues combined with the increasing amount of automation in road traffic, lead by BASt (Federal Highway Research Institute) in Germany. The report is expected in 2011 and it will cover the issues of automation in traffic in general. The report will not be restricted to interventions into the task of driving. These are rather well researched already and discussed with respect to the international law (Vienna Convention on Road Traffic) (Gasser, 2010a).

For near-accident situations, the most important aspects discussed so far are again regulatory law, highlighted in the discussion on the meaning of the Vienna Convention on Road Traffic for intervening systems as well as—most important again—product liability.

Art. 8 (1) of the Vienna Convention from 1968 on Road Traffic postulates that

> *"Every moving vehicle or combination of vehicles shall have a driver."*

Consequently, Art. 8 (5) VC constitutes the driver's obligation to be able to control his vehicle permanently:

> *"Every driver shall at all times be able to control his vehicle or to guide his animals."*

Art. 13 (1) VC substantiates this obligation with regard to speed and distance between vehicles; Art. 13 (1) VC says (in extracts):

> *"Every driver of a vehicle shall in all circumstances have his vehicle under control so as to be able to exercise due and proper care and to be at all times in a position to perform all manoeuvres required of him [. . .]."*

National road traffic regulations such as the German Road Traffic Regulations reflect this basic idea of permanent controllability.

Interventions of automated systems in the vehicle guidance, which do not comply with the driver's will and which cannot be corrected and overridden are therefore considered as incompatible with controllability in terms of the Vienna Convention (Seiniger *et al.*, 2011).

13	Uniform provisions concerning the approval of vehicles of categories M, N, and O with regard to braking
13-H	Uniform provisions concerning the approval of passenger cars with regard to braking
79	Uniform provisions concerning the approval of vehicles with regard to steering equipment

Figure 2. Relevant regulations concerning the type approval of added functionality. (Reproduced from Seiniger *et al.*, 2011. © The interactIVe Consortium.)

The second crucial aspect concerning near-accident interventions into driving is product liability. Liability claims arising from damages caused by a defective product may be based on three distinct liability systems: product liability (based on the Product Liability Directive 85/374/EEC), contract (contractual liability), and/or tort (extracontractual liability) in EU Member States (Seiniger *et al.*, 2011).

With regard to the liability deriving from the above-mentioned sources of law, a product should comply with the state of the art in science and technology—in order to be able to prove that this state of the art was adhered to during the design, the construction, and the production processes and with that in order to reduce product liability risks, relevant systems of rules such as the RESPONSE 3 Code of Practice (Knapp *et al.*, 2009), and technical standards such as the FDIS/ISO 26262 (Sauler and Kriso, 2009) should be observed. From a product liability point of view, it is recommendable to design near-accident interventions in a way allowing the driver to override automated braking and/or steering interventions any time the driver wishes to do so.

On EU-level, Product Safety Law is based on the General Product Safety Directive (GPSD) 2001/95/EC. Owing to its character as a directive, the GPSD had to be transposed into national law by the individual EU Member States. Art. 2 GPSD defines terms such as "product," "safe product," "dangerous product," "recall," and "withdrawal" for the purposes of the GPSD (Seiniger *et al.*, 2011).

A very technically oriented legal aspect to be assessed in detail is that of type approval. The mandate to approve vehicles for traffic belongs to the government of each country. However, European countries accept requirements defined by the United Nations Economic Commission for Europe's World Forum for the Harmonization of Vehicle Regulations (UN ECE WP.29) because of the transposition of the EU directives 2007/46/EC, 2002/24/EC, and 2003/37/EC.

There are two different types of vehicle regulations: the 1958 agreement system, which requires vehicles to be certified by an independent technical service (Europe, Japan, rest of the world), and the 1998 agreement (the United States, China, most of the 1958 states), which requires the vehicle manufacturers to certify their vehicles themselves.

The 1958 agreement with its ECE regulations covers most of the world with the exception of the United States and China. It is considered as the most important set of vehicle regulations.

Intervening systems and automated systems act on the vehicle brakes, throttle, and steering systems. The following regulations are of relevance, concerning the type approval of the added functionality (Figure 2).

3 ADVANCED DRIVER ASSISTANCE SYSTEMS

ADAS compensate the known weaknesses of human drivers and support them in their driving tasks. The motivation for the introduction of ADAS and the research of automated and automated vehicles are manifold. The four most important objectives are as follows:

- Increase of driving comfort: Comfort systems relieve the driver from annoying and monotonous tasks in order to ease the drive. Those systems already have a positive effect on traffic safety, as they are for instance keeping the distance to other cars and they avoid quick changes in speed.
- Increase of vehicle safety: ADAS support the driver by increasing the active safety of a vehicle. Active safety can be separated into perception safety (driver support in order to perceive relevant traffic information of the vehicle to other traffic participants), control and interaction safety (driver support to interact with vehicle control elements and perception of vehicle status information), driving safety (support the driver in critical driving situations), and condition safety of

the vehicle (support the driver during the drive, that is, by comfortable climate conditions).

- Improvement of traffic efficiency: ADAS support the improvement of street capacity. Thus, traffic jams can be prevented or are dissolved faster. In addition, the vehicles, which are approaching the traffic jam, can be redirected automatically.
- Reduction of environmental impact: ADAS support the reduction of fuel consumption and noise emissions, for example, assistance systems can give suggestions for gear changes, acceleration maneuvers, or early reduction of vehicle velocity depending on the situation and traffic. In addition, for the optimization for complex drive train structures (hybrid engines), the operating strategy can be implemented.

Support systems or automated systems require the interaction with the environment and an action of the driver, for at least the activation of the system. ADAS, which take over part of the driving task, are part of the closed control loop.

The driver support is divided into one of the three levels of the driving task (Donges, 1982). ADAS are classified, based on their functions and their type of driver support, into the levels navigation, guidance, and stabilization (Figure 3).

On the navigation level, the driver decides on his route inside an existing road network. While driving, the navigation includes the perception of necessary information to maintain the route. If needed, an adjustment of the route due to changed boundary conditions can be pursued. The time demand for the driver on the navigation level is not critical and amounts to over 10 s.

On the guidance level, the driver adapts his driving style to the perceived course of the road and the surrounding traffic. The guidance level comprises subtasks such as lane keeping, following, overtaking, and reaction to road signs. The tasks of this level can be divided into lateral and longitudinal guidance and are realized between 1 and 10 s.

The stabilization level is characterized by changes of the chosen driving strategy in vehicle-related control variables by the driver (steering movement, accelerator, brake, and gear choice). There is a permanent comparison between set and current value of speed and lane position. "Stabilization" for the driver means the avoidance of unsupervised momentum of the vehicle. The stabilization level is highly time critical with a time demand below 1 s. On the basis of this, time criticality automatic intervening systems are used for the stabilization level. The driver does not carry out the necessary action for the vehicle stabilization on his own. Systems on the stabilization level are mostly equipped with a possibility to activate or deactivate the HMI. In some cases, there is no visible HMI during driving (e.g., ABS and ESP).

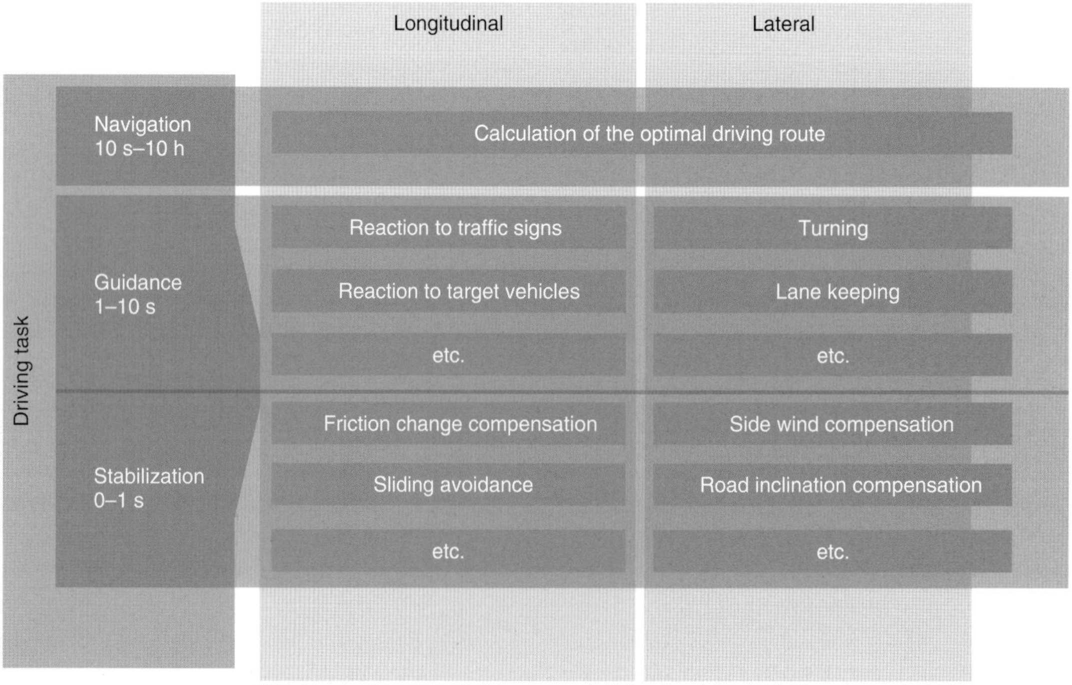

Figure 3. Classification of ADAS in three levels of the driving task.

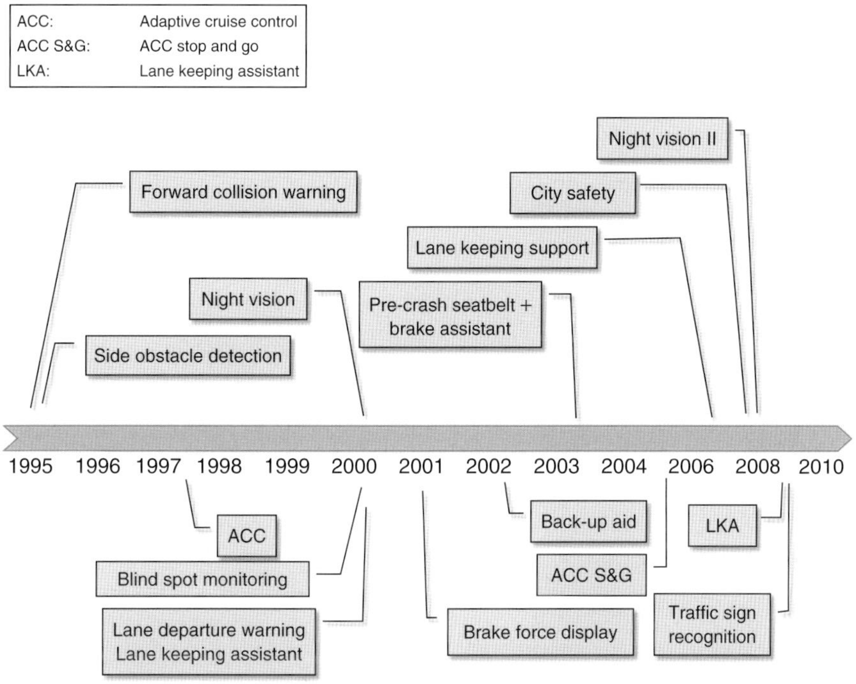

Figure 4. Market introduction of ADAS.

3.1 Overview of ADAS

One of the first ADAS introduced in the market was a collision warning system for heavy goods vehicles in 1995, which was based on a 24 GHz radar sensor for object detection and forward collision warning system. Figure 4 shows the market introduction of different ADAS on guidance and navigation level.

Currently, various different ADAS are available in premium vehicles. They are also being introduced in middle class vehicles. Examples are systems for parking support, driving at nighttime, longitudinal, and lateral vehicle control.

One of the first systems introduced to provide continuous support was adaptive cruise control (ACC). ACC is an enhancement of the cruise control, which was one of the focus research fields of the PROMETHEUS project. The system particularly supports drivers on motorways and country roads by keeping a safe distance and controlling the velocity with respect to vehicles in front. In case the vehicle approaches a slower vehicle, the system intervenes by means of reducing the throttle first and then operates the brake system in order to keep a safe distance to the preceding vehicle. Therefore, the ACC uses its distance sensor in order to detect the range and range rate of the vehicle in front. The desired values regarding the distance are mostly dependent on the velocity and the drivers can adjust the distance by defining a time gap to the vehicle in front. The acceleration and the deceleration behaviors of the vehicle are controlled depending on the driving status by an ECU. The driver is able to turn the function off or override it at any time by pushing the accelerator or the brake pedal. In case no target vehicle is present, the ACC is used as a conventional cruise control.

The market introduction of ACC for different vehicle manufacturers in Europe is given in Figure 5 as an example.

3.2 Research on ADAS

In addition to available ADAS, research on future systems is ongoing in different research areas. The most important ones are as follows:

- Safety (e.g., collision mitigation and avoidance)
- Environmental protection (decarbonization and noise reduction)
- Integrated HMIs and integrated interaction strategies
- New mobility concepts
- Automation

With regard to automated driving vehicles, research on platooning provides the next step to automation. A platoon is described by at least two vehicles, which are electronically connected without any mechanical

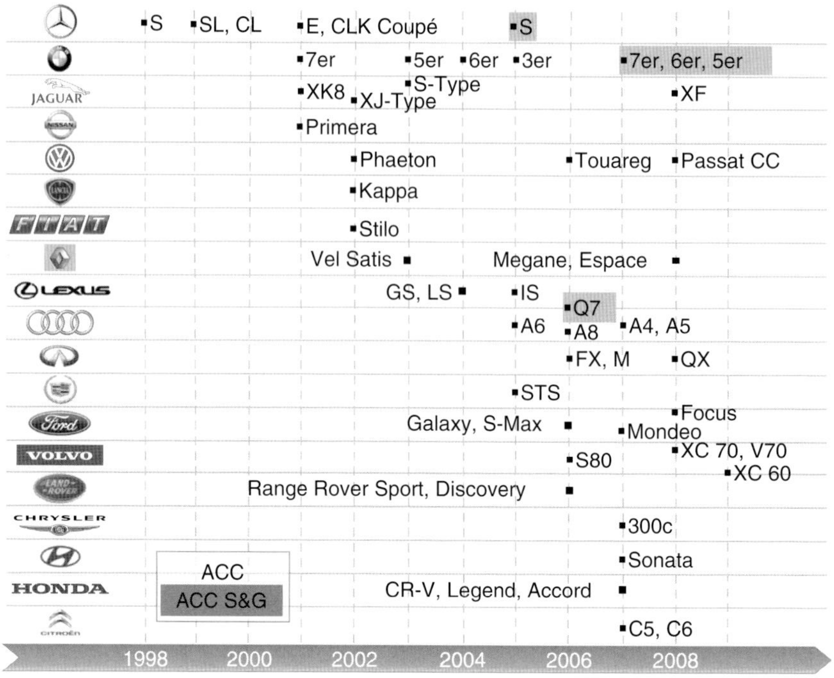

Figure 5. Market introduction of ACC in Europe.

connection between the vehicles. Work in this area was undertaken by VW called *CONVOY Driving* in the PROMETHEUS project (Ioannou 1997), by the PATH (Partners for Advanced Transit and Highways) project of the Institute of Transportation Studies (ITSs) of the University of California in Berkeley together with the California Department of Transportation (Caltrans) in the Unites States in 1996 and the ARTS project (Advanced Road Transportation System), which was funded by the Japanese Ministry of Construction. The ARTS project was merged to the ITS (Transportation System) project in 1995.

The lateral vehicle control of the automated vehicles in these projects is based on magnetic studs, which are integrated into the infrastructure in the form of a dedicated guide way every approximately 3 m. The nails provide the trajectory of the vehicles. Longitudinal vehicle guidance in the platoon is provided by distance and velocity detection in each single vehicle. This information is transmitted via vehicle-to-vehicle communication or vehicle-to-infrastructure communication.

The aim of PATH was to develop long-term strategies in order to cope with the immense traffic in California. Between 7 and 10 of August 1997, a vehicle platoon consisting of eight Buick LeSabres vehicles was demonstrated. The distance between each longitudinal and lateral controlled vehicle was 6.5 m at a velocity of up to 96 km/h. The distance accuracy was measured in the range 10 cm at constant driving and 20 cm at acceleration and deceleration maneuvers (PATH Program, 1997). However, even today, there is no impact on the road traffic in the United States.

In contrast to the PATH approach (only lateral control by means of magnetic nails, longitudinal control provided only to following vehicles based on vehicle sensors), the point-follower approach (infrastructure/vehicle sensors, no target vehicle necessary) demands additional communication between the road infrastructure and the vehicles. This

Figure 6. Automated driving bus using a guide wire in the ground of the driving way. (Reproduced by permission of MAN.)

communication can be provided by LCX (Leakage CoaXial Cable) cables at the surface of the road. The LCX cable provides data transmission and acts as an antenna. Field tests were conducted on a test tracks in Germany and Japan (Gehring, 2000). Figure 6 shows field tests with a MAN bus demonstrator using a guide wire in the ground of the driving way in 1977 (Zeit, 1980).

In Europe, truck platooning was performed for the first time in the European research projects PROMOTE CHAUFFEUR 1 and 2 of the fifth European framework program. The following distance between the three used demonstrator trucks was between 6 and 16 m at a velocity of 80 km/h. On the basis of the reduced air resistance, fuel in the range between 15% and 20% was reduced compared to conventional driving.

The control approach was called *tow-bar* principle, which was realized by means of vehicle-to-vehicle communication in the 5.8 GHz band and an infrared camera system. The distance to the target vehicle as well as the velocity and the acceleration of each vehicle was transmitted to all following vehicles in the platoon. The infrared camera determined the relative lateral position between the vehicles.

Next to the "tow-bar" principle, the so-called CHAUFFEUR assistant was introduced in PROMOTE CHAUFFEUR 2. The assistant was a combination of an ACC system and the lateral control in order to support the driver of the first vehicle in the platoon.

The final demonstration of all functions was done on the IVECO test track in Balocco, Italy, on 7 May 2003 (Bonnet *et al.*, 2003).

Similar to the PROMOTE CHAUFFEUR project, the national funded German research initiative KONVOI performed research on truck platooning. Next to technical research questions, covering the steering hardware and the function development for automated longitudinal and lateral control impacts on surrounding traffic, the strategies to build the platoon, acceptance studies, and impacts on truck drivers were investigated in detail. In addition, legal aspects for an introduction of truck platoons were analyzed. The demonstration of a truck platoon with four tractor–semitrailer combinations took place on the German motorway in 2009 (Figure 7) (Deutschle, 2008).

SARTRE is a European project, cofunded by the seventh framework program of the European Commission, that targets to develop strategies and technologies to allow vehicle platoons to operate on normal public highways with significant environmental, safety, and comfort benefits.

In the SARTRE platooning system, a human driver like in KONVOI drives the lead vehicle manually. The other vehicles follow the trajectory of the lead vehicle. Thus, the lead vehicle driver has a huge responsibility. In order to improve the safety in a platoon, different ADAS systems are installed in the lead truck to support the driver.

In order to improve the comfort and support the drivers to find suitable platoons, a back office unit calculates the meeting point for the potential following of the platoon and provides navigation instructions to the drivers (Deutschle, 2008; SATRE project, 2011).

4 INFRASTRUCTURE-BASED AUTOMATED DRIVING VEHICLES

Operation of automated vehicles in dedicated infrastructure provides several advantages compared to fully automated driving vehicles. Dedicated infrastructure can separate automated vehicles from ordinary traffic participants, which solves integration issues. Furthermore, dedicated infrastructure can be equipped with sensors, for example, LCX cables for platooning. Basically two different scenarios are possible for automated vehicles in dedicated infrastructures:

- Guide ways to separate automated vehicles from other traffic participants
- Guide ways to support automated vehicles or ADAS

Different vehicle types are possible to be operated in these scenarios:

- Personal rapit transit (PRT)
- High tech busses
- Vehicle platoons
- Automated vehicles

Infrastructure-based automated driving vehicles provide new mobility concepts especially in the field of public transport, that is, PRT and high tech busses.

PRT is a fully automated transport system using small driverless vehicles (Figure 8). The vehicles navigate automatically along a network of dedicated guide ways. PRT

Figure 7. KONVOI truck platoon demonstration on the German motorway in 2009.

Figure 8. (a,b) Examples for PRT systems. (a: Reproduced from Bly and Lowson, 2009. © European Commission, b: Reproduced by permission of 2getthere B.V. (http://www.new.2getthere.eu).)

offers a driverless taxi, providing on-demand. By this, a nonstop transport from the origin station to any other destination station selected by the passenger on the whole dedicated guide way network will be possible. Therefore, PRT can be described as a horizontal lift or elevator. All PRT systems under current development are electric powered and thus (locally) emissions free. PRT systems aim to have empty vehicles waiting at stations for arriving passengers, so that there is little or no waiting time. Because the electric vehicles are quiet and emit no exhaust pollutants, they can be routed through buildings.

One system installed at the airport in Heathrow is "ULTra," developed by Advanced Transport Systems of Thornbury, the United Kingdom. The four-seater vehicles have four rubber-tired wheels and the size of a small car. They are powered by electric motors, running from lead-acid batteries. ULTra vehicles are driving on a 2 m wide concrete or metal track. Batteries are recharged in the stations. The maximum driving velocity is 40 km/h, and the vehicles can climb a 10% gradient and negotiate 5 m radii. They can go reverse and steer backward, offering great flexibility in maneuvering (Bly and Lowson, 2009). Another PRT system in operation is located in Masdar, Abu Dhabi (2getthere, 2011).

High tech buses in separated lanes are also already in operation. An example for such a BRT (bus rapid transit) system is the Phileas program in Eindhoven/Veldhoven. Phileas is a new concept for comfortable passenger transport on high frequency dedicated bus lanes (Figure 9). The guide way is fitted with magnetic markers for electronic lane assistance and precision docking. Phileas has hybrid electric propulsion, a large transport capacity, and precision docking, which makes it possible for passengers to quickly enter and exit the vehicle. Thus, the stop times are short and the average speed can be kept as high as possible. In comparison with tram or metro systems, the investment and maintenance costs for the infrastructure are low. Overhead wires and rails are not needed. Phileas combines the advantages of tram and metro systems with the flexibility and low costs of a bus system (Phileas, 2011).

Figure 9. Phileas high tech bus on a dedicated guide way (Phileas, 2011). (Reproduced by permission of VDL Bus & Coach bv.)

In industrial applications, different types of automated and driverless vehicles are already utilized since many years. These are vehicles, which transport goods on dedicated driving routes in factories or work yards.

In 2001, the FOX GmbH in Germany to transport large amounts of euro pallets on a factory yard has modified two commercial vehicles. The vehicles were equipped with additional sensors for automated driving. The driving route of 190 m is marked by transponders in the floor. These transponders are detected by the vehicles with an accuracy of about 5 mm. Therefore, the lateral control of the vehicles on the driving route is possible within a 2 cm range at a velocity range of about 5 km/h. A laser scanner detects obstacles in front of the vehicles. In addition, the vehicles have sensors integrated in the front bumpers. The bumpers are made of soft foam. In case of contact between an obstacle and the bumpers, the sensors activate the braking system and the vehicles are able to halt within a range of 0.4 m after contact. The vehicle position and movement is transferred to a control room by means of communication technology. The vehicles are driving automatically, but a driver can override the overall system at any time (Götting KG, Abt. FOX, 2008).

5 FULLY AUTOMATED DRIVING VEHICLES

Automated driving vehicles are under research since the 1970s. Different research activities demonstrated automated driving according to the latest state-of-the-art technology available at the time of the activity. In recent years, sensor technology for environment detection and electronically controlled actuators became available in production vehicles. Vehicle manufacturers use their knowledge on the necessary technology and provide demonstrator vehicles for automated driving in order to underline the public perception of the company image.

As legal issues still do not allow automated vehicles to be driven on public roads, first systems in low velocity ranges are available at dedicated infrastructures such as company depots, work yards, or for demonstration purpose within inner city limits, separated from conventional traffic.

In order to stimulate recent research on automated vehicles and automated driving by universities and research institutions, different challenges have been initiated by different stakeholders (e.g., military and governments). On the basis of the experience of the research work and the availability of sensor systems for path planning and obstacle detection, automated driving vehicles demonstrators have been set up to fulfill manifold purposes, such as creating development platforms, organizing driver training, or defining new markets. Furthermore, new mobility concepts are being investigated by means of automated driving vehicles.

5.1 Research activities

In the last decade of the past century, different research activities on automated vehicle guidance were conducted with the results of automated driving demonstrations. Within these activities, prototype vehicles were built up and tested. Some tests of the prototypes were carried out as journeys on public roads in real traffic:

- VAMP: test on a journey of approximately 1600 km from Munich (Germany) to Odense (Denmark) in 1995 (Maurer *et al.*, 1996);
- NavLab 5: test on a journey of approximately 4587 km from Pittsburgh (USA) to San Diego (USA) in 1995 (Bertozze, Broggi, Fascioli, 2000);
- ARGO: test on a journey of approximately 2000 km in the MilleMiglia tour (Italy) in 1998 (Bertozze, Broggi, Fascioli, 2000);
- Google Cars: test drives of approximately 1600 km without human intervention and more than 230,000 km in total (Markoff, 2010).

In the recent past, challenges were announced by different stakeholders. The American DARPA (Defense Advanced Research Projects Agency) initiated the DARPA Grand Challenges and the DARPA URBAN Challenge in order to demonstrate the state of the art for automated driven vehicles. In Europe, the Grand Cooperative Driving Challenge (GCDC) was initiated to combine different European research activities in cooperative and automated driving.

5.1.1 VaMoRs/VAMP

One of the first automated driving vehicles was called *VaMoRs* (Versuchsfahrzeug für autonome Mobilität und Rechnersehen) of the German Bundeswehr University of Munich. The second vehicle from Daimler was called *VITA*.

Between 1985 and 2004, a transporter of the type Mercedes DB 508 was converted in order to control electronically the lateral and longitudinal dynamics of the vehicle. The core of the research activity was machine vision, which was performed by means of camera systems. The vision system detected objects and lane markings in a distance of up to 120 m in front of the vehicle. By means of this collision avoidance, automated distance control, lateral control, and lane change maneuvers were implemented. Next to machine vision, additional sensors of a digital map and satellite position system were used in order to detect intersections and turnings.

The achieved know-how of the VaMoRs project was used to build up the vehicles VAMP (VaMoRs Passenger Car) and VITS-II. These Mercedes Benz 500 SEL vehicles were equipped with a radar distance sensor additionally to the machine vision system.

After thousands of test kilometers automated driving for testing purpose, a demonstration was conducted from Munich (Germany) to Odense (Denmark) in 1995. The test route had a length of approximately 1600 km. A total of 95% of the test track was driven in automated mode at velocities of up to 180 km/h. In total, approximately 400 lane changes were performed during the test drive (Maurer *et al.*, 1996).

5.1.2 No hands across America

In 2005, the "No Hands Across America" tour was performed by the Robotics Institute of the Carnegie Mellon University, Delco Electronics, and AssistWare Technology. The prototyped vehicle called *NavLab 5* drove approximately 4587 km from Pittsburgh to San Diego (USA).

The vehicle was equipped a windshield mounted camera, a GPS receiver, and a radar sensor for obstacle detection, which were used for lateral control of the vehicle. Longitudinal vehicle control was performed manually by the driver.

NavLab 5 was able to drive 98.2% of the journey (4503 km of a total of 4587 km) with automated lateral control. The system proved to be robust with respect to real road and traffic conditions. The major problems encountered were due to rain, low sun reflection, shadows of overpasses, construction zones, road, and road markings deterioration (Bertozze, Broggi, Fascioli, 2000).

5.1.3 The ARGO project and the VisLab intercontinental autonomous challenge

From 1996–2001, the University of Parma worked on the ARGO project, which had the goal to enable automatic lane following on motorways for a modified Lancia Thema. In 1998, a demonstration of a 2000 km long journey on the motorways of northern Italy with an average speed of 90 km/h took place. A total of 94% of the time the car was in fully automated mode, with a longest automatic stretch of 54 km. The vehicle was equipped with only two black-and-white cameras and used stereoscopic vision algorithms to perceive its environment.

On the basis of the results of the ARGO project, the idea for the intercontinental autonomous challenge was born. In total, 13,000 km was driven from Parma (Italy) to Shanghai (China) with four automated controller electric vehicles (two vehicles traveling and two vehicles as backups). The challenge was performed from 26 July 2010 to 28 October 2010 including two different continents with changing geographical morphology, traffic conditions, weather, infrastructures, and so on.

Figure 10. Vehicles of the VisLab Intercontinental Autonomous Challenge. (Reproduced by permission of Vislab.)

Each test vehicle was equipped with seven cameras, four laserscanners, GPS receivers, and vehicle-to-vehicle communication. The vehicles are shown in Figure 10.

During the journey, all vehicle data and all perception data were logged. The amount of data added up to about 50 TB and provided a data set for a variety of different driving situations, weather conditions, and infrastructure conditions.

In difficult situations, for example, bad weather or heavy and chaotic traffic conditions, driver had to take over control and switch to manual mode because not sufficient information from the perception system had been available (Broggi, 2011).

5.1.4 DARPA grand challenges

The DARPA Grand Challenge took place in the years 2004 and 2005. The task of the Grand Challenges was to cover a track of about 150 and 132 miles through the desert in fully automated mode with no human intervention. The route was announced shortly before the start of the challenges so that the competing teams could not tune their automated vehicles according to the track.

In the first challenge of 2004, no vehicle was able to cover the route. The best team was only able to cover around 5% of the track. In 2005, four vehicles reached the destination within the given time limit of 10 h.

The winner of the Grand Challenge 2005 was a VW Touareg called *Stanley* of the Stanford University. The vehicle was equipped with four laser scanners, 24 GHz radar sensors, mono- and stereo image processing systems, and a GPS system. The vehicle movement was measured by means of an internal measurement unit. The sensor data and the control algorithms were processed by seven Pentium M processors with 1.6 GHz calculation speed each (Thrun, 2006).

5.1.5 *DARPA urban challenge*

The DARPA Urban Challenge was conducted in 2007 on an old military area close to Victorville in the state of California in the United States. The main differences to the Grand Challenges were the involvement of other traffic participants and the urban scenario with road regulations and urban infrastructure. Especially, the behavior of the other traffic participants, which were vehicles driven by stuntmen, needed to be considered. Pedestrians and bicycles were not part of the Urban Challenge. The route was shortened to 60 miles, which needed to be covered in <6 h.

The winner of the DARPA Urban Challenge was the team Tartan Racing of the Carnegie Mellon University with the vehicle called *Boss* (Urmson *et al.*, 2008).

5.1.6 *GCDC in Europe*

In contrast to the DARPA Grand Challenges and the Urban Challenge, the European GCDC has no military background. It was initiated by TNO, a governmental research organization of the Netherlands and High Tech Automotive Systems (HTASs). The goal was to create a liaison between research groups on the topic of cooperative and automated driving. The first GCDC was held at Helmond in the Netherlands in 2011. It took place on the highway A270, which was closed for public traffic while the challenge was performed.

The focus of the first GCDC was on longitudinal vehicle control for platooning by means of vehicle-to-vehicle communication for cooperation. Different communication and sensor hardware and different types of vehicles needed to be part of the control strategy. In order to provide boundary conditions, the communication protocol, the signals to be communicated, safety measures, the wireless communication, and a mandatory message set needed to be used by all participants. Each participant could choose the vehicle, in-car architecture, the environmental sensors, the control strategy, and the usage of the communicated information freely.

In total, 11 teams competed in the four given scenarios (platoon stability, joining a platoon, joining a platoon at a traffic light, and merge on intersections). The teams were grouped into 2×2 platoons on parallel lanes. The organizers provided the lead vehicles, which were defining the driving velocity of the platoon. Points were provided for different criteria such as string stability and smoothness. The first GCDC was won by team AnnieWay of the Karlsruhe Institute of Technology in Germany (Grand Cooperative Driving Challenge, 2011).

5.1.7 *Google cars*

The Internet Company Google also works on automated vehicle in terms of using artificial intelligence software for driverless vehicle control. The vehicle combines information from the company's Google Street View service with sensor signals from video cameras inside the vehicle, a LIDAR sensor on top of the vehicle, radar sensors on the front of the vehicle, and a position sensor attached to one of the rear wheels. The system drives at the speed limit, which is stored on its digital map and maintains its distance from other vehicles using the environmental sensors. The driver can override the automated driving function at anytime (Markoff, 2010). In 2010, seven test vehicles have driven 1600 km without human intervention and more than 230,000 km with only occasional human intervention. Next to the technical feasibility demonstration, Google lobbied for two changes of state laws that made the state of Nevada (USA) the first state, where driverless vehicles can be legally operated on public roads. The first bill is an amendment to an electric vehicle bill that provides for the licensing and testing of automated vehicles. The second bill will provide an exemption from the ban on distracted driving to permit occupants to send text messages while sitting behind the wheel (Markoff, 2011).

5.2 Examples for automated vehicle demonstrators and research platforms

In recent years, car manufactures, computer companies, and universities established a variety of different research platforms. On the basis of the experience of the DAPRA challenges, research on automated driving is proceeding in manifold ways.

At the University of Braunschweig, Germany, automated driving in the city's inner ring road of Braunschweig is demonstrated in the so-called *Stadtpilot* project. The goal is to drive fully automated in the traffic flow and to behave according to traffic rules (Saust *et al.*, 2011).

The Volkswagen Group Electronics Research Laboratory and Stanford University are working on the Audi TTS Pikes Peak automated vehicle. The vehicle is a modified Audi TTS Coupé Quattro called *Shelly*. The goal is to improve vehicle automation and explore capabilities of current and future driver assistance systems by automated drive up the legendary 12.42 miles Pikes Peak Hill Climb route in Colorado, the United States. The focus of the research work is on path planning and vehicle stability. The route to Pikes Peak consists of paved and graveled roads and can contain weather changes at all time. Therefore, vehicle guidance of the Audi TTS Pikes Peak is working entirely on differential GPS and vehicle state information such as speed and acceleration measured by wheel-speed sensors, an accelerometer, and gyroscopes, but without additional sensors for environmental detection. The control algorithm for path planning and vehicle stability compares the sensor data to a digital map of the route to determine possible deviations and necessary maneuvers. The resulting action for longitudinal and lateral vehicle control is performed by the vehicles production hardware, which already exists in Audi production vehicles (Blackman, 2010).

The VW Golf 53 + 1 is an automated driving robot, which was developed to perform reproducible driving maneuvers in the high dynamic range on test tracks (Kompaß, 2008). Reproducible driving maneuvers allow vehicle development by means of a driver-independent analysis of vehicle dynamics (function or system), driver behavior (impact of the system), and boundary conditions (environment). The driving robot is able to take over the function of the driver with sufficient accuracy and reproducibility. Longitudinal and lateral vehicle dynamics are controlled by the robot and therefore provide automated driving on a predefined course. The driving trajectory is determined by means of a differential GPS. The transmission of the additional correction signal allows a positioning in the centimeter range. Traffic cones mark the driving track, which are detected by a laser scanner in the front of the VW Golf. The vehicle ECU fuses the GPS and the laser scanner data and calculates the driving trajectory on the track. Within the driving path, the ideal racing line is calculated for lateral vehicle dynamic control. These calculations result in desired values for the maximum driving velocity and the vehicle acceleration. The vehicle is equipped with a brake booster, which allows high dynamic braking maneuvers (Kompaß, 2008).

The BMW "Track Trainer" was designed by BMW for race driver training (Kompaß, 2008). The ideal racing line of the racetrack is calculated off-line. The calculation considers position data from GPS as well as vehicle parameters from test runs of professional drivers. An inertial measurement unit compensates the GPS position in order to detect inaccuracies of the positioning. The vehicle ECU controls longitudinal and lateral vehicle dynamics and keeps the vehicle automated on the ideal racing line of the course. The driver experiences the vehicle behavior for some test rounds on the racing course and therefore learns how to stay on the ideal racing line. In a second phase, the driver takes over vehicle control and is supported by the vehicle (haptic and acoustic advises) in case he leaves the ideal racing line. All vehicle data are logged during the test drive. Afterward the driving style can be analyzed in detail and recommendations can be given by the system (Kompaß, 2008).

5.3 Cybercars

Owing to the usage of vehicles in highly populated urban environment, severe problems with respect to pollution, noise, and safety arise. A new mobility concept in terms of Cybercars uses the advantages of automobiles combined with automation technology. The idea of Cybercars was developed in the European funded research project "Cybermove" (Cybermove project, 2004). Cybercars are slow-moving automated driving vehicles for existing road infrastructure. The design of the vehicles allows easy and clean transportation within inner cities. Different persons, depending on the vehicle size, can share the electric vehicles. With existing technology, the velocity is limited to 30 km/h and therefore sufficient for inner urban usage. Some of these vehicles can also allow traditional manual driving in order to run among normal traffic. In these cases, the vehicles are called *dual-mode vehicles* and their automated capabilities allow them to be used in platoons, for example, in order to collect the cars.

Several companies and research organizations have been involved in recent years in the development of Cybercars. The first systems have been put in operation in the Netherlands at the end of 1997 and have been running successfully 24 h a day. Several other systems have been implemented in European research projects on this topic, see Cybermove project (2004), Cybercars project (2006), and CityMobil project (2011). The focus of the research is on the development of tools and systems to enable the driverless vehicles to perform the necessary driving maneuvers in a cooperative manner, that is, in cooperation with each other and also in cooperation with conventional driven vehicles (Vlacic and Parent, 2009).

6 SUMMARY

A first step toward automated driving is the introduction of ADAS into the market. Currently, the development of new assistance functions and the improvement of existing functions are ongoing. Especially, progress in sensor and actuator technology allows a shift from pure comfort function toward active safety and continuous driver support. In the recent past for longitudinal and lateral support in all velocity ranges from parking to city driving up to high speed driving became available.

Research activities in the field of automated driving demonstrate technical solutions for fully automated systems. At present, these are being utilized in niche markets. However, in the evolution of today's ADAS toward automated driving systems, the legal aspects and the assessment of the systems need to be considered. High market penetration of ADAS in the future will be the foundation toward the fully automated driving vehicle.

RELATED ARTICLES

Volume 4, Chapter 117
Volume 4, Chapter 124
Volume 4, Chapter 126
Volume 4, Chapter 130
Volume 5, Chapter 149
Volume 5, Chapter 150
Volume 5, Chapter 151
Volume 5, Chapter 170
Volume 5, Chapter 179
Volume 6, Chapter 197
Volume 6, Chapter 199
Volume 6, Chapter 206
Volume 6, Chapter 207
Volume 6, Chapter 208

REFERENCES

2getthere (2011) http://www.new.2getthere.eu (accessed 18 October 2013).

Bertozze, M., Broggi, A., and Fascioli, A. (2000) Vision-based Intelligent Vehicles: State of the Art and Perspectives. *Robotics and Autonomous Systems* **32**, 1–16.

Blackman, C. (2010) Stanford's robotic Audi to brave Pikes Peak without a drive. Stanford Report, issue no. 3 (February).

Bly, P. and Lowson, M. (2009) Outline description of the Heathrow Pilot PRT scheme. Deliverable D1.2.2.2, CityMobil project.

Bonnet, C., Schulze, M., and Dieckmann, T. (2003) Promote Chauffeur 2 - Final Report. Deliverable D24, Promote Chauffeur 2 project.

Broggi, A. (2011) The VisLab Intercontinental Autonomous Challenge. *Keynote Speech at CityMobil Conference*, La Rochelle, France, CityMobil project.

CityMobil project (2011) http://www.citymobil-project.eu/ (accessed 18 October 2013).

Cybermove project (2004) http://www.cybermove.org (accessed 18 October 2013).

Cybercars project (2006) http://www.cybercars.org/ (accessed 18 October 2013).

Gasser, T.M. (2010a) 'Die "Projektgruppe Automatisierung": Rechtsfolgen zunehmender Fahrzeugautomatisierung' 11. Braunschweiger Symposium AAET, Proceedings, Intelligente Transport- und Verkehrssysteme und –dienste Niedersachsen e.V., 978-3-937655-23-9, Braunschweig.

Gasser, T.M. (2010b) Legal Aspects of Driver Assistance Systems' 19. Aachener Kolloquium Fahrzeug und Motorentechnik 2010, RWTH Aachen University, Aachen.

Deutschle, S. (2008) Das KONVOI Projekt - Entwicklung und Untersuchung des Einsatzes von Lkw-Konvois, Aachener Kolloquium Fahrzeug und Motorentechnik 2008, Aachen.

Donges, E. (1982) Aspekte der aktiven Sicherheit bei der Führung von Personenkraftwagen *Automobil-Industrie*, **27** (2), 183–190.

Grand Cooperative Driving Challenge (2011) http://www.gcdc.net (accessed 18 October 2013).

Gehring, O. (2000) Automatische Längs- und Querführung einer Lastkraftwagenkolonne. Dissertation. Institut A für Mechanik der Universität Stuttgart, Stuttgart

Götting KG, Abt. FOX (2008) automatisierte Serienfahrzeuge, Infobroschüre der Abteilung FOX, Götting KG, Lehrte.

Ioannou, P. (1997) *Automated Highway Systems*, Plenum Press, New York. ISBN: 0-306-45469-6

Knapp, A., Neumann, M., Brockmann, M., *et al.* (2009) RESPONSE 3: Code of practice for the design and evaluation of ADAS; version 5, http://www.acea.be/images/uploads/files/20090831_Code_of_Practice_ADAS.pdf (accessed 18 October 2013).

Kompaß, K. (2008) *Fahrerassistenzsysteme der Zukunft - auf dem Weg zum autonomen PKW, Forschung für das Auto von Morgen*, Springer Verlag, Berlin.

Markoff, J. (2010) Google cars drive themselves, in traffic *The New York Times*, 9.10.2010,

Markoff, J. (2011) Google lobbies Nevada to allow self-driving cars, The New York Times, 11.05.2011

Maurer, M., Behringer, R., Fürst, S., *et al.* (1996) A Compact Vision System for Road Vehicle Guidance. Proceedings of ICRP 1996, IEEE.

PATH Program (1997) Vehicle Platooning and Automated Highways, PATH Fact Sheets, University of California, Berkeley.

Phileas (2011) http://www.apts-phileas.com (accessed 18 October 2013).

Rowsome, F. (1958) What it's like to drive an auto-pilot car. Popular Science Monthly, USA, April.

SATRE project (2011) www.sartre-project.eu (accessed 18 October 2013).

Sauler, J. and Kriso, S. (2009) ISO 26262 – Die zukünftige Norm zur funktionalen Sicherheit von Straßenfahrzeugen, http://www.elektronikpraxis.vogel.de/themen/elektronikmanagement/projektqualitaetsmanagement/articles/242243/ (accessed 18 October 2013).

Saust, F., Wille, J.M., Lichte, B., and Maurer, M. (2011) Autonomous vehicle guidance on braunschweig's inner ring road within the stadtpilot project. Intelligent Vehicles Symposium (IV), IEEE.

Seiniger, P., Westhoff, D., Fahrenkrog, F., and Zlocki A. (2011) Legal Aspects, Deliverable D7.3, interactIVe project, www.interactive-ip.eu/ (accessed 18 October 2013).

Thrun, S. (2006) Towards self-driving cars, 7. Braunschweiger Symposium AAET, 22./23.2.2006, DLR Braunschweig, Gesamtzentrum für Verkehr Braunschweig e.V.

Urmson, C., Anhalt, J., and Bagnell, D. (2008) Autonomous driving in urban environments: boss and the urban challenge *Journal of Field Robotics*, **25** (8), 425–466. DOI: 10.1002/rob.20255, Wiley Periodicals, Inc

Vlacic, L. and Parent, M. (2009) Cybercars2 - final report V2.0. Cybercars2 project.

Zeit (1980) Immer in der Spur, newspaper article, 09.05.1980

Chassis Control Systems

Chapter 130

Global Chassis Control in Passenger Cars

Hideo Inoue[1], Takashi Yonekawa[2], Masayuki Soga[1], Kenji Nishikawa[1], Eiichi Ono[3], and Makoto Yamakado[4]

[1]*Toyota Motor Corporation, Toyota, Japan*
[2]*Toyota Motor Corporation, Susono, Japan*
[3]*Toyota Central R&D Labs. Inc., Nagakute, Japan*
[4]*Hitachi, Ltd., Ibaraki, Japan*

1 INTRODUCTION

Full-scale global chassis control systems are being developed by various companies for use in passenger cars. This development started in Europe in 1983 with the Daimler-Benz 4Matic four-wheel drive (4WD) system, which used online calculation of vehicle states. Calculation results from a bicycle model and yaw rate measurement results from the wheel speeds were compared to determine whether to engage the front axle to the drive train using electronically controlled clutches. This system used longitudinal slip and lateral vehicle dynamics for control for the first time (German patent DE 35 05 455 CS, 1985). Other early examples include the integrated active suspension, active rear steer (ARS) system, traction control system (TCS), and antilock brake system (ABS) in the Toyota Soarer in 1991 (Tanaka *et al.*, 1992) and the introduction of an in-vehicle local area network (LAN) with the 4WDi-Four and ARS system in the Toyota Crown Majesta in 1992. In addition, in 1992, BMW offered dynamic stability control (DSC), the first electronic stability control (ESC) in its 850i sports car. This system controlled the throttle and the ignition timing after comparing the vehicle behavior to a bicycle model. Mercedes and Toyota as ESC released additional brake intervention controls in 1995. These were followed in 1997 by the release of the Toyota Hybrid System (THS) in the Prius, which featured the first example of technology that combined braking performance with the recovery of regenerative energy. Subsequently, various companies are now developing global control systems that integrate powertrain, steering, and active braking systems. These include the integrated chassis management (ICM) system developed by BMW, the vehicle dynamics management (VDM) system developed by Bosch, the vehicle dynamics integrated management (VDIM) system developed by Toyota, and so on. Such systems are beginning to be considered as fundamental technology for enhancing safety and environmental performance as well as driving enjoyment. In addition, advances in vehicle environment recognition technology have led to the development of driver assistance systems such as precrash safety (PCS) and the like. This chapter describes an outline of integrated vehicle systems, focusing on global chassis control. It also

Encyclopedia of Automotive Engineering, print © 2015 John Wiley & Sons, Ltd.
Edited by David Crolla, David E. Foster, Toshio Kobayashi, and Nick Vaughan.
This article is © 2015 John Wiley & Sons, Ltd. ISBN: 978-0-470-97402-5
Also published in the *Encyclopedia of Automotive Engineering* (online edition)
DOI: 10.1002/9781118354179.auto030

presents an overview of the integration of driver assistance systems and global vehicle dynamics controls to enhance safety technology and discusses the value of the regenerative energy recovery and integrated vehicle dynamics control functions of hybrid electric vehicle (HEV) systems in enhancing environmental technology.

2 HIERARCHY OF INTEGRATED VEHICLE SYSTEMS

The process of driving a vehicle consists of three related factors: the driver, the vehicle, and the traffic environment. The component technologies that make up integrated vehicle control must be structured simply and rationally while considering the requirements to enhance the overall value of the vehicle in terms of safety, driving enjoyment, and environmental performance. Figure 1 shows the hierarchical structure of an integrated vehicle system. The system consists of the following five parts.

1. VDM
2. driver assistance management
3. energy management
4. human–machine interface (HMI) management
5. occupant protection management.

Currently, the most important of these parts are VDM and driver assistance management. It is likely that further advances will occur in the field of sensing, such as cooperative communication among vehicle environment recognition systems using autonomous sensors such as radar and camera, navigation systems, and roadside infrastructure. Such advances are particularly likely for integrated safety systems and VDM is a key technology supporting this progress.

2.1 Integrated safety

This section briefly discusses the concept of integrated safety. Figure 2 shows the trends in active and passive safety technology. Advances in the field of passive safety technology include optimized body structures, enhanced restraint systems such as airbags, and measures for various different types of accident formats (i.e., compatibility in crashes with different vehicle types, pedestrian safety, and the like). However, active safety has grown in importance in recent years. The aim of active safety systems such as ESC, VDIM, and PCS is to help reduce the number of accidents that occur. Systems that coordinate with roadside infrastructure using communication technology may also be developed in the future. On the basis of these trends, it is likely that active and passive safety systems will be introduced together as integrated packages rather than as separate technologies. The aim of integrated safety management is to create a simpler overall system that helps to seamlessly operate each function together (Figure 3). For

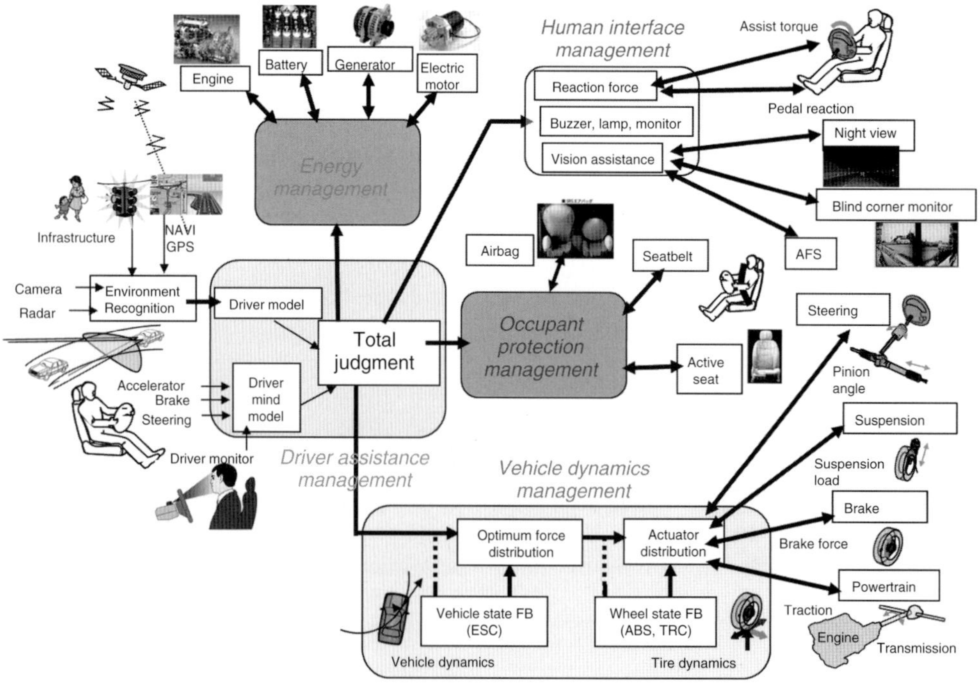

Figure 1. Hierarchical structure of integrated vehicle control.

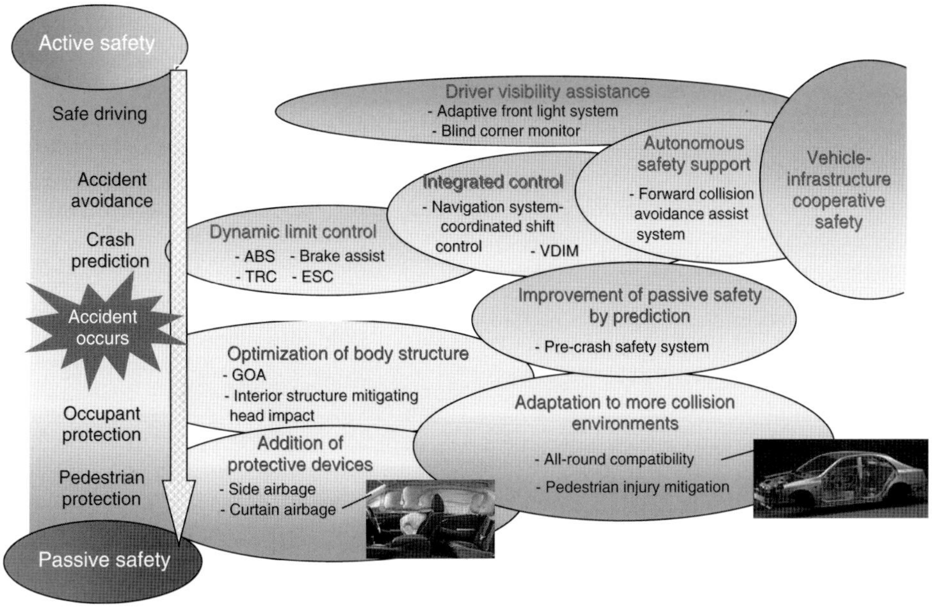

Figure 2. Trends in passive and active safety technology.

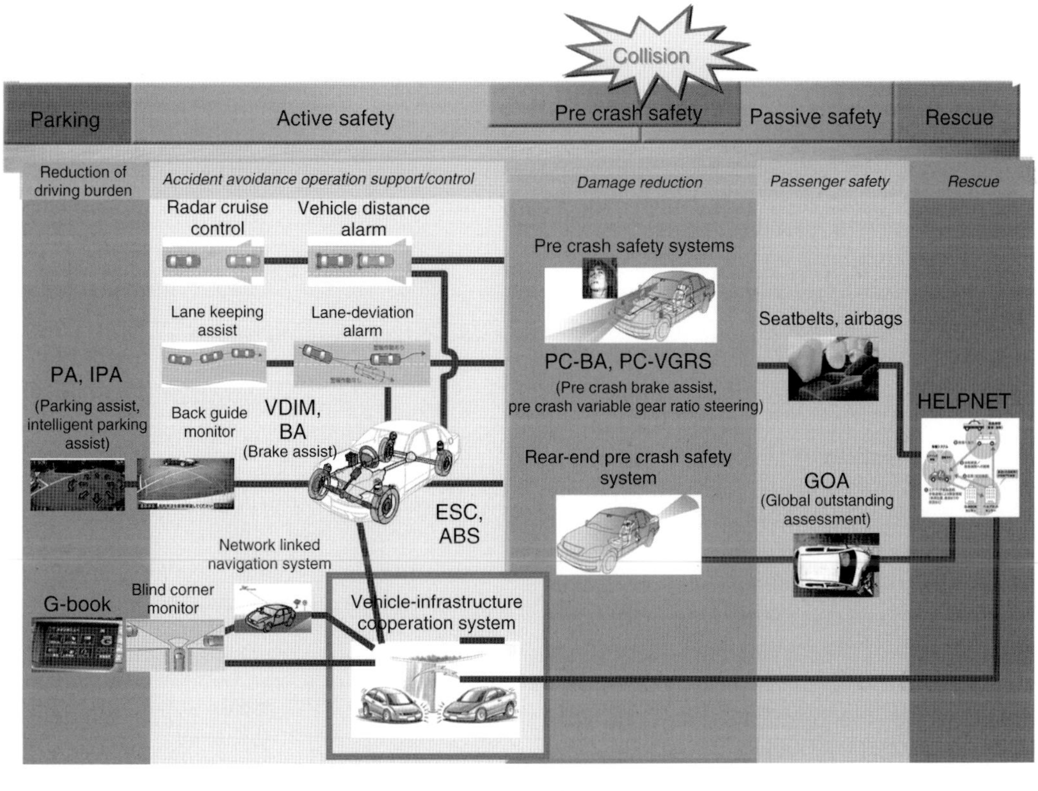

Figure 3. Integrated safety management.

this reason, VDM will play a fundamental role acting as the muscles and nerve system of the vehicle.

3 EVOLUTION OF CHASSIS CONTROL DEVICES

The development of more advanced chassis control devices that act as the muscles of the car to control the tires is an essential part of enhancing vehicle dynamic performance. This requires expanding the degree of freedom of control over the forces generated at each tire contact point from the longitudinal direction to the lateral and vertical directions. This can only be achieved by smooth and highly responsive control in all directions. The following sections describe recent development trends in chassis control devices and examples of their application.

3.1 Brake control

The development of brake control systems has accelerated because of the adoption of ABS and electronic brake distribution (EBD) systems as standard equipment. In combination with advances in ESC technology, the major effect of integrated brake control systems in helping to reduce accidents has been confirmed in the real world. As a result, the United States and other countries have begun to mandate their usage and such systems are likely to become standard equipment in the future.

The popularization of HEVs in recent years has also contributed to the development of electronically controlled brake (ECB) systems (Figure 4) with the aims of achieving linear hydraulic brake control and improving response (Nakamura, 2002). This is a brake-by-wire system in which the hydraulic brake pressure at each wheel is isolated from the brake pedal and braking control is performed by a high pressure source through linear solenoids. This system facilitates braking force coordination during regenerative braking in an HEV. As its smooth and highly responsive controllability can also be utilized as part of ESC systems, ECB systems are also spreading to other vehicle types in addition to HEVs.

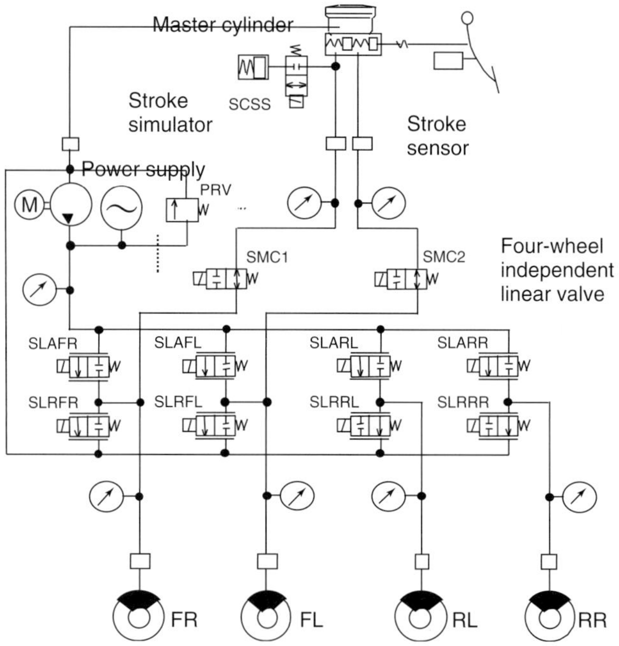

Figure 4. ECB structure.

3.2 Drive system control

This field has seen progress in active controls such as systems that transfer torque to the front and rear wheels in a 4WD vehicle, as well as limited slip mechanisms in center, front, and rear differentials. In recent years, systems for distributing driving force to the left and right wheels using a speed-increasing mechanisms installed inside the differential have been developed. The characteristics of these systems are being actively used to improve cornering performance.

In contrast, the electrification of driving forces has started in HEVs and electronically controlled 4WD vehicles to enable smother and more responsive driving force control. The practical application of motors capable of driving each wheel independently in the future may also help to further enhance dynamic performance.

3.3 Steering control

Electric power steering (EPS) systems are being rapidly adopted to improve fuel efficiency, as well as for use in HEVs and the like. In addition to simply replacing hydraulic power steering, the role of EPS is spreading to new functions that make active use of its capability to freely vary the assistance force.

For example, it has already been introduced in some vehicles to perform the following functions in coordination with ESC. EPS can assist the driver's steering effort to facilitate countersteering when the rear wheels slip or to prevent understeering when the front wheels slip. It can also be used as an actuator in functions that help the driver keep the vehicle in its lane by assisting the driver's steering effort. Alternatively, EPS can also function as an actuator in automatic driving systems such as Intelligent Parking Assist.

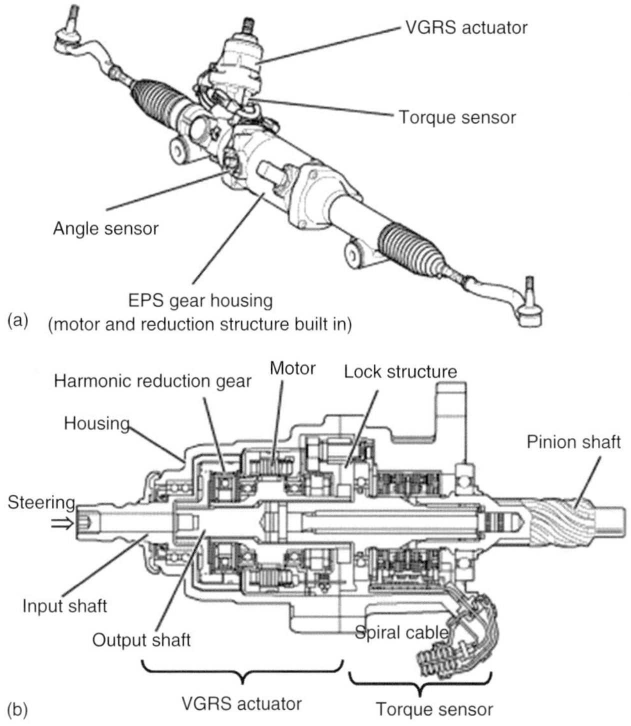

Figure 5. (a,b) VGRS actuator.

In addition, variable gear ratio steering (VGRS) systems have been developed that vary the steering response characteristics of the vehicle yaw angular velocity (Figure 5). Conventionally, the steering characteristics of the yaw angular velocity were determined by the specifications of the suspension. VGRS is capable of varying these characteristics to a constant optimum level in accordance with the driving environment. Furthermore, ESC devices are starting to be used, which utilize the capability of VGRS to control the turn angle of the front wheels independently of the steering wheel angle.

ARS has also been adopted on some vehicles, and its enhancement of dynamic performance has been verified. In recent years, vehicles have been released that combine ARS with front-wheel active steering systems to further improve dynamic performance (Katayama, 2007; Kojo, 2002; Ono, 2007).

3.4 Suspension control

The suspension of a vehicle consists of springs, shock absorbers, and link mechanisms. However, there is a long history of control technology applied to suspensions to improve both ride comfort and vehicle stability. Examples include the semiactive suspensions introduced in the 1980s for controlling shock absorber damping force. Improvements have continued since then. Recent years have seen the development of electronically controlled active stabilizer suspension systems that actively reduce vehicle roll (Figure 6). Progress is also being made toward the development of electronically controlled fully active suspensions. As a result, the use of actuators in suspension control is spreading, and these are expected to help improve vehicle performance through the active variation of vehicle attitude and vertical load.

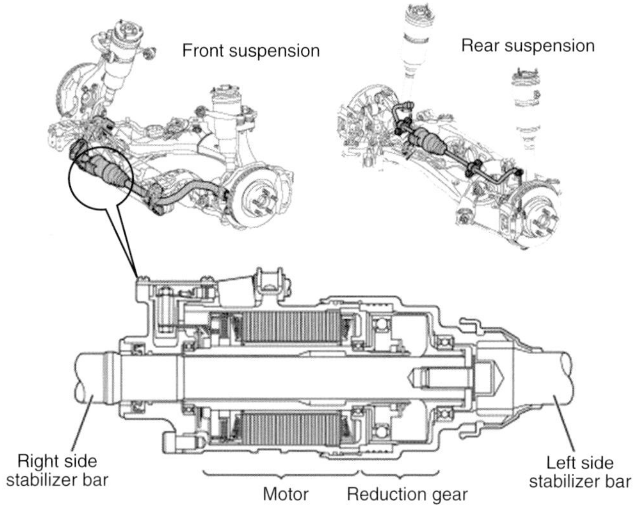

Figure 6. Active stabilizer actuator.

3.5 Evolution of vehicle environment recognition technology

One current trend is vehicle environment recognition technology that supports the cognitive processes of the driver. The development of environment recognition sensors such as radar and cameras is advancing rapidly.

Forward recognition technology using radar is already used by cruise control systems. PCS systems have been developed that help the driver to avoid accidents through coordinated control with the brakes (Tsuchida, 2007). A more accurate system that combines information from radar with camera images (Figure 7) and a system that coordinates information with a camera that detects the orientation of the driver's face have also been developed.

Lane Keeping Assist, Intelligent Parking Assist, and other driving assistance systems that use image-recognition technology have already been commercialized. However, the number of systems in this field is likely to grow further as researchers study ways of utilizing information provided by roadside infrastructure.

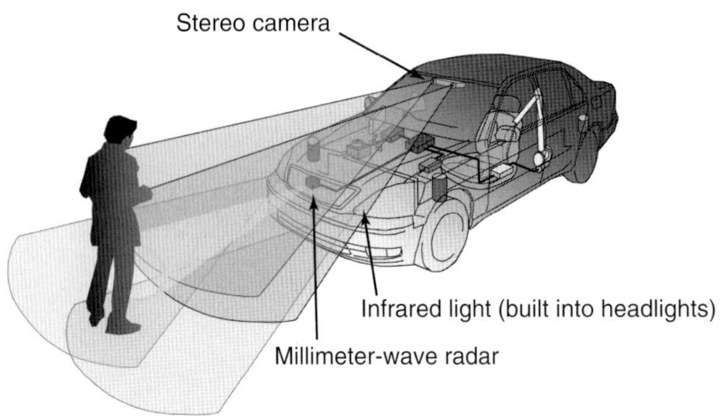

Figure 7. Forward recognition system using radar and stereo camera.

4 EVOLUTION OF CONTROL USING VDIM

4.1 Concept of VDIM

Since the 1980s, various attempts have been made to enhance vehicle dynamic performance using active chassis control. In 1986, the Daimler-Benz 4Matic system was the first to use lateral dynamics for control purposes. Direct yaw moment control systems with active braking such as ESC enable good performance in the critical limit region (Shibahata *et al.*, 1993; Koibuchi *et al.*, 1996; Van Zanten, 1996). The aim of the next generation of vehicle dynamic control systems is to provide seamless vehicle maneuverability and stability at all times through the integrated control of driving forces to all four wheels. Figure 8 shows the concept of Toyota's VDIM system. This illustration is called the *ball in a bowl* concept (Hattori *et al.*, 2002). The ball corresponds to the state of the vehicle, which is maintained within a bowl constructed by the control. The inside of the bowl is the stable region, and the outside is the unstable region. In conventional systems, the walls of the bowl are constructed from independent functions such as ABS, ESC, and TCS, which form sheer boundaries before an emergency occurs.

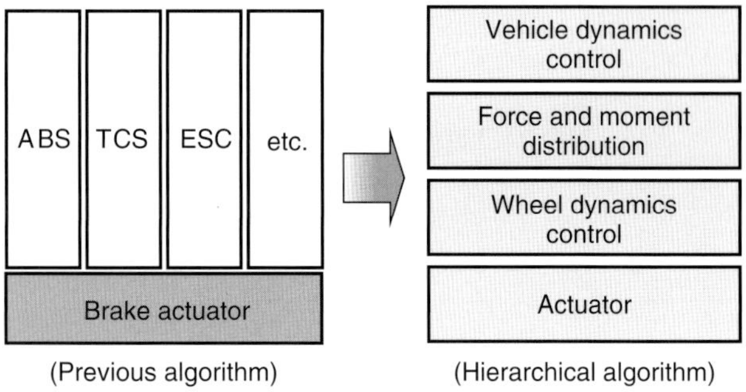

Figure 9. Hierarchical control algorithm.

As a result, although these functions are capable of stabilizing vehicle motion, the motion may be discontinuous in some cases. In contrast, VDIM realizes smoother behavior because the conventional control systems are restructured to form a continuous and smooth wall.

As vehicle control systems are becoming more diversified, the algorithm is required to perform cooperative control of many systems, such as the drive train, braking, and steering, easily. Accordingly, the compatibility of the algorithm with various system configurations is important. The hierarchical control system structure shown in Figure 9 has been adopted for VDIM (Hattori, 2002; Fukatani, 2005).

The first layer (vehicle dynamics control) calculates the target forces and moments of the vehicle to achieve the desirable vehicle motion corresponding to the driver's pedal input and steering wheel angle. There are several examples of research for the first layer. In the critical limit region, the determined target resultant force and moment also satisfy robust stability conditions to avoid vehicle spin (Ono *et al.*, 1998). However, in the moderate region, Yamakado *et al.* (2010) have proposed a target longitudinal acceleration/deceleration model determined by predicted lateral jerk to improve driving enjoyment. The target resultant force and moment of the vehicle motion are distributed to the target tire forces of each wheel based on the friction circle of each wheel in the second layer (force and moment

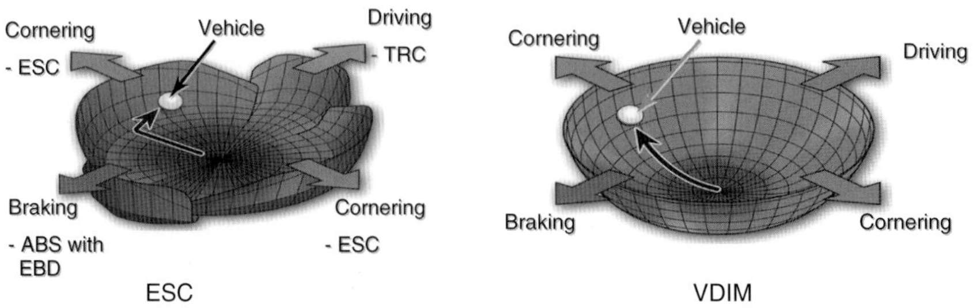

Figure 8. Evolution of control using VDIM.

distribution). The third layer (wheel dynamics control) controls each wheel motion to achieve the target tire force. There are redundant degrees of freedom in the second layer. The vehicle dynamics performance in the critical limit region depends on the force and moment distribution algorithm, which uses these redundant degrees of freedom.

The motion of a vehicle in the three degrees of freedom (longitudinal, lateral, and yaw) is controlled by the steering and traction or braking forces from the four tires. If each tire can be individually steered and operated for traction or braking, the control system has redundancy. Vehicles move using the friction between the tires and the ground. The frictional forces at the tires have limits dependent on the conditions of the road surface. These limits are called the *friction circle*, and a tire cannot exert force on the road surface in excess of the friction circle. To extend the limits of the performance of vehicle dynamics, it is necessary to ensure that the forces exerted by all tires work efficiently in cooperation with each other. The problem of integrated control of vehicle motion then becomes how to best use the redundant degrees of freedom. As the friction circle has nonlinear constraints because of the limitations on the frictional forces at each wheel, the distribution of the longitudinal and lateral forces and the yaw moment (vehicle forces and moments) to each tire force becomes a nonlinear problem.

Ono *et al.* (2009) proposed a vehicle dynamics integrated control algorithm using an online nonlinear optimization method for four-wheel-distributed steering and four-wheel-distributed traction and braking systems. The proposed distribution algorithm calculates the magnitude and direction of the tire forces to satisfy constraints corresponding to the target resultant forces and moments of vehicle motion and also to minimize the maximum μ rate (=tire force/friction circle) of each tire. This research demonstrated the convexity of this problem and guaranteed the global optimality of the convergent solution of the recursive algorithm. This implies that the theoretical limitation performance of vehicle dynamics integrated control can be reached.

Comparing it with general quadratic programming can show the efficiency of the proposed algorithm, which calculates the theoretical limitations of vehicle forces and moments. The following minimization problem of the sum of squares of the μ rate may be considered as a benchmark. This is an extension of the problem described by Mokhimar and Abe (2003). In this simulation, the generated vehicle longitudinal forces are compared for straight-line braking on a split road with different coefficients of friction μ (μ =1.0, 0.2).

Figure 10 shows the tire forces of a vehicle controlled by the proposed method and a vehicle controlled by quadratic programming. Both of the controls achieve the reference braking force within a moderate area when the reference braking force is 7000 N. However, unlike the vehicle controlled by quadratic programming, the vehicle controlled by the proposed method can also achieve the reference braking force in the critical region when the reference braking force is 10,000 N.

4.2 Configuration of VDIM

Figure 11 shows the overall configuration of the VDIM system. Using in-vehicle sensors to detect the yaw rate and steering angle, VDIM collects together various items of data for optimally controlling the brakes and front steering to stabilize the vehicle attitude. It achieves a steer-by-wire function using an active front steering (AFS) system with two actuators (VGRS and EPS) to control the steering angle of the front wheels and the reaction torque of steering. It also acts as a wide-ranging safety control function in coordination with the ECB system.

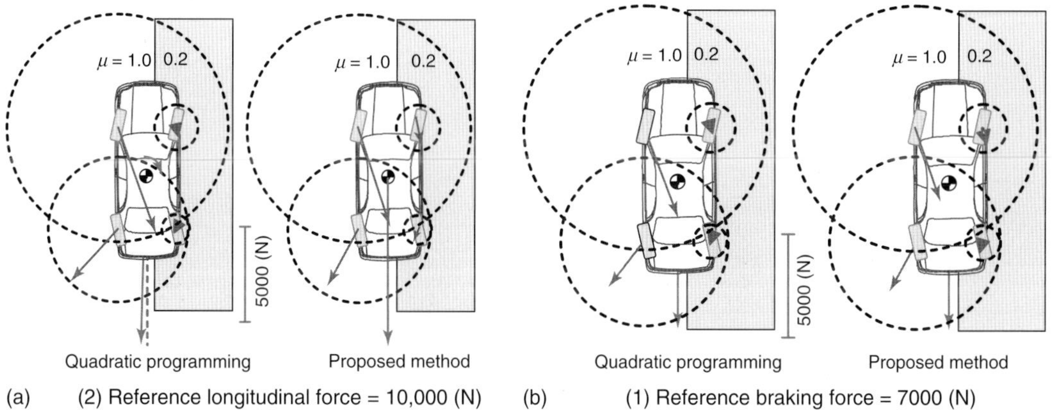

Figure 10. (a,b) Straight-line braking on split μ road.

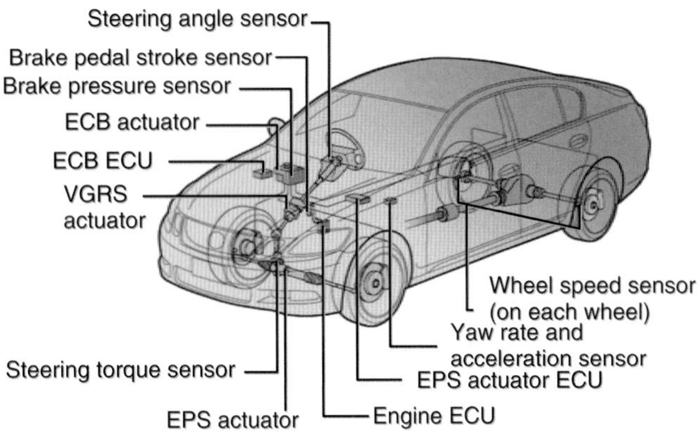

Figure 11. Configuration of VDIM system.

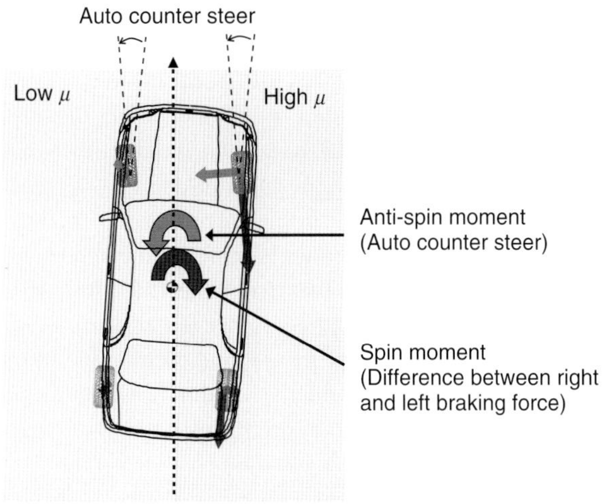

Figure 13. Corrective steering control during braking on split μ road.

Figure 12 shows the layout of functions for controlling vehicle dynamics and behavior using VDIM. Creating this layout diagram clarifies the role of each function individually and in combination with other functions. VDIM enables true integration of vehicle control by emphasizing the development of each function and its actions.

4.3 Performance of VDIM

Figure 13 shows the outline of control when driving on a road with different coefficients of friction (μ) under the left and right wheels. As shown in the figure, when the driver brakes, spin moment is generated in accordance with the difference between the left and right braking forces, resulting in deflection of the vehicle. On this type of road surface, a vehicle dynamics control system that uses just longitudinal forces cannot achieve a high degree of stability and braking simultaneously with driving performance. In contrast, this can be achieved by an AFS system that is capable of controlling the vehicle in the lateral direction.

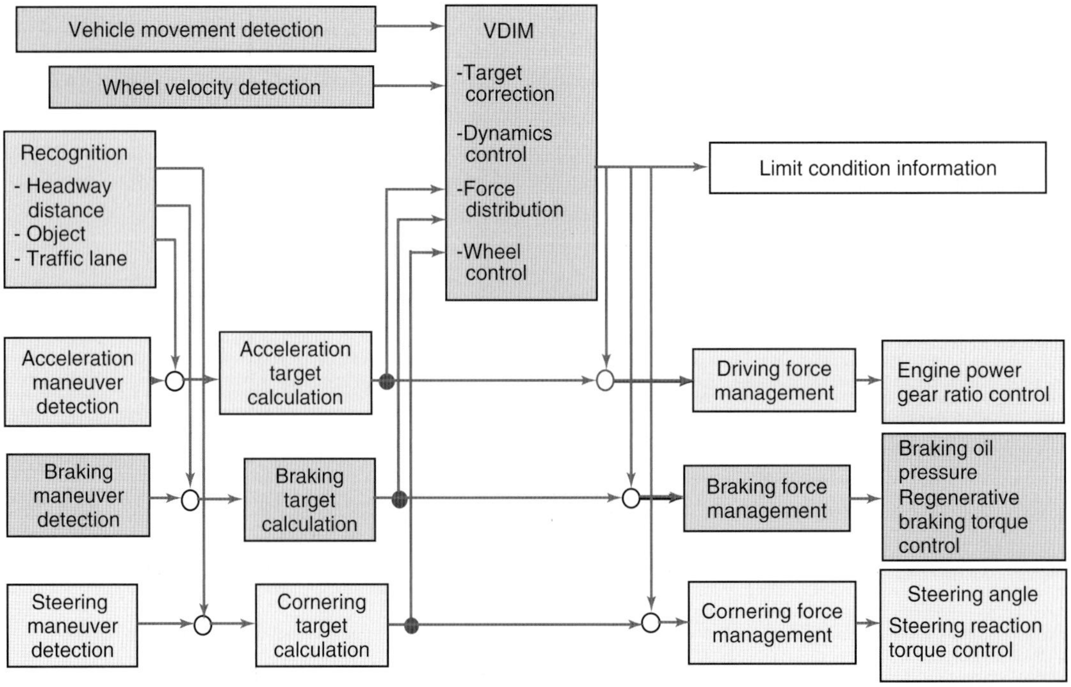

Figure 12. Layout of VDIM software functions.

Figure 14 shows the test results of straight-line braking on a split μ road. It shows the normalized steering angle and peak yaw rate after braking (state without control = 1). With the active steering control, the peak yaw rate was reduced by approximately 50% compared to the rate without the control. In addition, the AFS control requires less corrective steering effort from the driver and generates a lower peak yaw rate on the vehicle than conventional ABS.

AFS has a large effect on side slipping of the front wheels. In particular, for a vehicle braking on a road with different coefficients of friction under the left and right wheels, AFS is capable of maintaining both vehicle stability and driving force. It also helps to suppress vehicle spin behavior when the driver performs an evasive maneuver (Figure 15). AFS is more effective at suppressing spin as it can generate moment by controlling the slip angle of the front wheels in addition to the conventional control that generates moment to reduce spin using braking force.

Figure 16 shows the results of a test comparing the effects of VDIM and ESC. In the test, the vehicle was steered mechanically through a slalom using constant steering operations while accelerating on an artificial low friction surface ($\mu = \sim 0.3$) simulating a snowy road. At the limit regions, the slip angle with VDIM was half that of ESC (+TCD), which shows that VDIM is capable of greatly improving the vehicle stability.

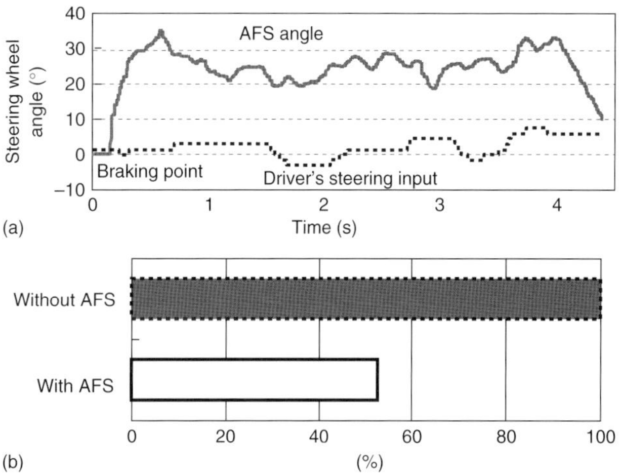

Figure 14. Effect of AFS on braking on split μ road. (a) Time domain data example and (b) normalized peak yaw rate.

4.4 Further evolution in control using VDIM

4.4.1 Enhancement of collision-avoidance performance using environment recognition technology

Toyota developed a collision-avoidance support system in 2006 that aimed to help reduce accidents by assisting evasive maneuvers by the driver. This system uses a forward monitoring function consisting of a millimeter wave radar and stereo camera to judge the risk of a collision with an object and assists the driver to avoid the collision by varying the steering gear ratio and braking. This system consists of a block that detects objects and judges the collision risk, a block that determines the evasive maneuver by the driver, and a block that controls vehicle behavior (Figure 17). The system operates as follows (Figure 18).

1. If a high collision risk is judged, the system warns the driver to take evasive action.
2. The system reduces the VGRS steering gear ratio to assist the driver's evasive steering maneuver.
3. If a high collision risk is judged, the system begins automatic braking. When the driver performs an evasive steering maneuver, deceleration during the maneuver is achieved by reducing braking force with a gradual gradient. In this event, VDIM controls the steering and brakes appropriately in accordance with the vehicle state.

Figure 19 shows the results of a double lane change test with and without system operation. The system reduces the driver's steering wheel angle and steering velocity to enhance the object evasion capability of the vehicle.

4.4.2 G-vectoring

VDIM is basically a feedback control that aims to restore the ball to the bottom of the bowl in the ball in a bowl concept. In contrast, G-vectoring control (GVC) is a feed-forward control mechanism that enables the ball to start rolling toward the edge of the bowl in coordination with driver's maneuvers. In 2008, Yamakado and Abe identified an original trade-off strategy between longitudinal traction and cornering force using jerk information to observe an expert driver's voluntary braking and turning actions (Yamakado and Abe, 2008). This strategy was used to develop GVC, which is basically a mechanism for achieving automatic longitudinal acceleration control in accordance with vehicle lateral jerk caused by the driver's steering maneuvers (Yamakado *et al.*, 2010). With GVC, the direction of the resultant acceleration (G) changes seamlessly (i.e., by vectoring) in the same way as it does when an expert driver is behind the steering wheel. In this way, ideal vehicle motion can be achieved. The following

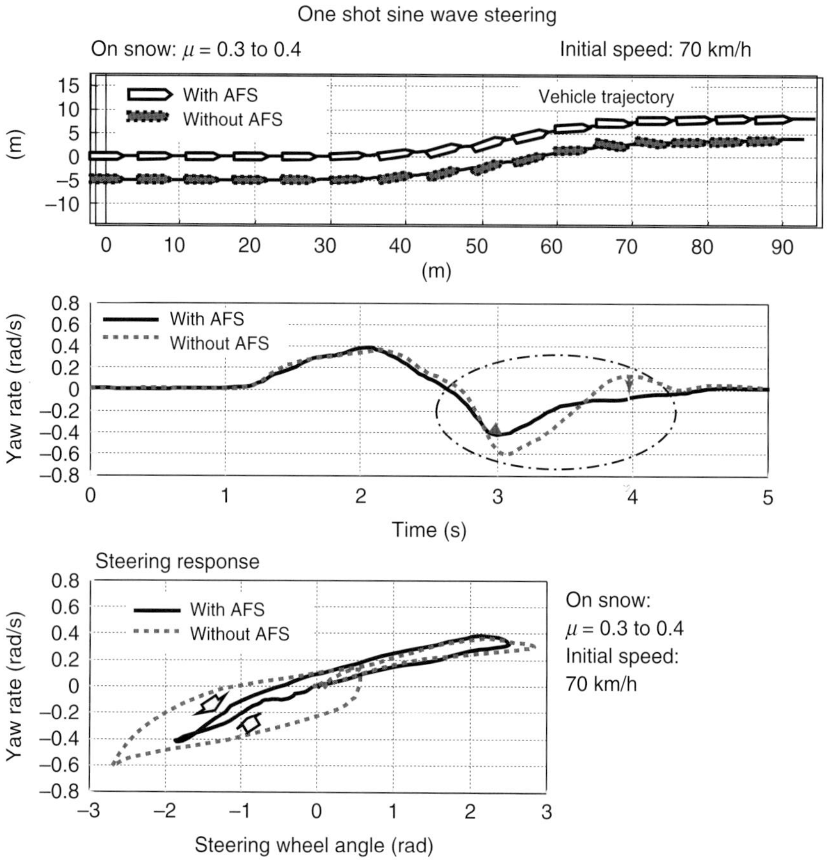

Figure 15. Comparison of vehicle behavior when changing lanes.

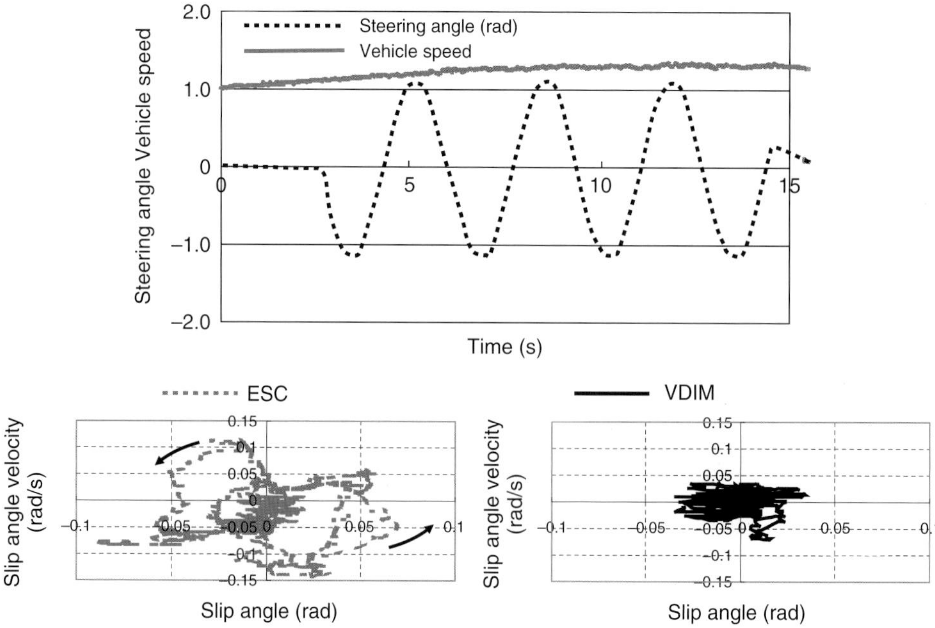

Figure 16. Vehicle stability with ESC and VDIM.

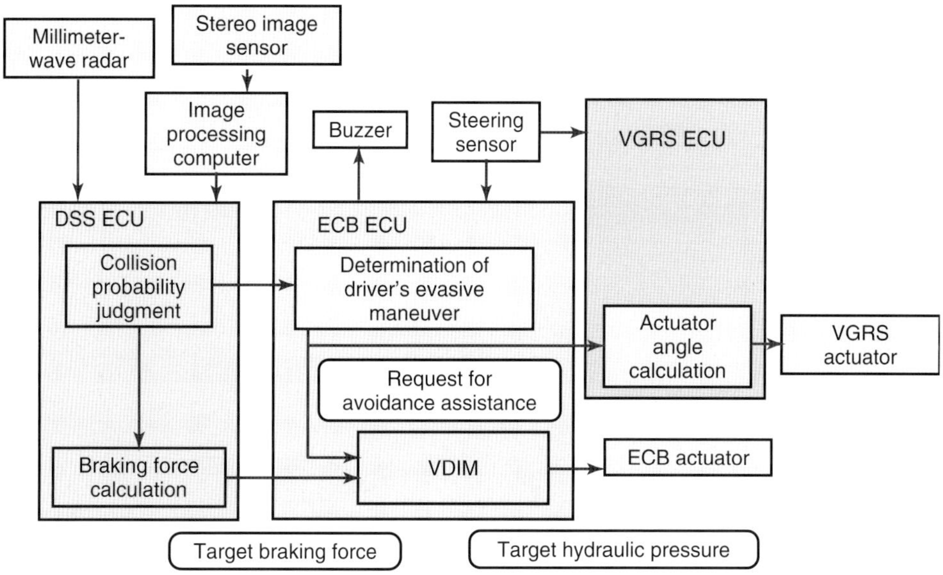

Figure 17. Functional configuration of object avoidance assist system.

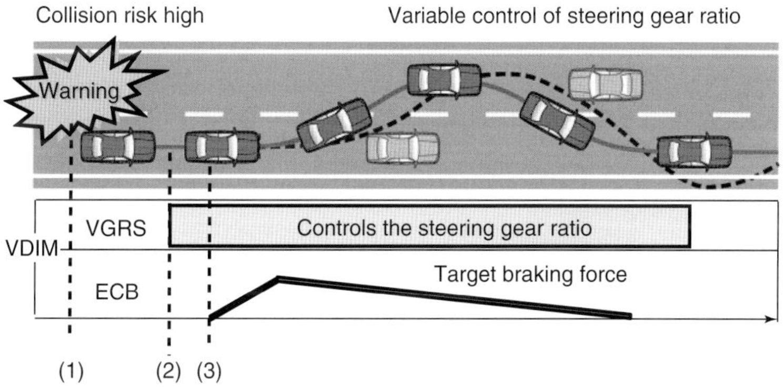

Figure 18. Control procedure of object avoidance assist.

equation was proposed as the fundamental equation for GVC.

$$G_{xt} = -sgn(G_y \cdot \dot{G}_y)\frac{C_{xy}}{1 + Ts}|\dot{G}_y| \quad (1)$$

where G_{xt} is the longitudinal acceleration command, C_{xy} the gain, and G_y the lateral jerk.

Figure 20 illustrates the GVC concept. When the vehicle starts turning a corner, it starts braking simultaneously as lateral jerk increases (vehicle positions 1–3). After that, the braking stops during steady-state cornering (vehicle positions 4 and 5) because the lateral jerk becomes zero. The vehicle begins to accelerate when it begins to return to straight-ahead driving (vehicle positions 6 and 7). If a bowl were fixed to the vehicle, a ball in the bowl would move smoothly along the level curve as shown at the top of the figure (ball positions 3–5) because of the change in inertial force caused by the acceleration of the vehicle. Considering the shift in wheel vertical load between the front and rear wheels (which is caused by the acceleration and deceleration of the vehicle), the handling of the vehicle when entering a corner and its stability when leaving a corner will be improved.

Figures 21–23 show the measurement results obtained for cases with and without GVC. These results show that applying GVC makes it quite possible to emulate expert driving.

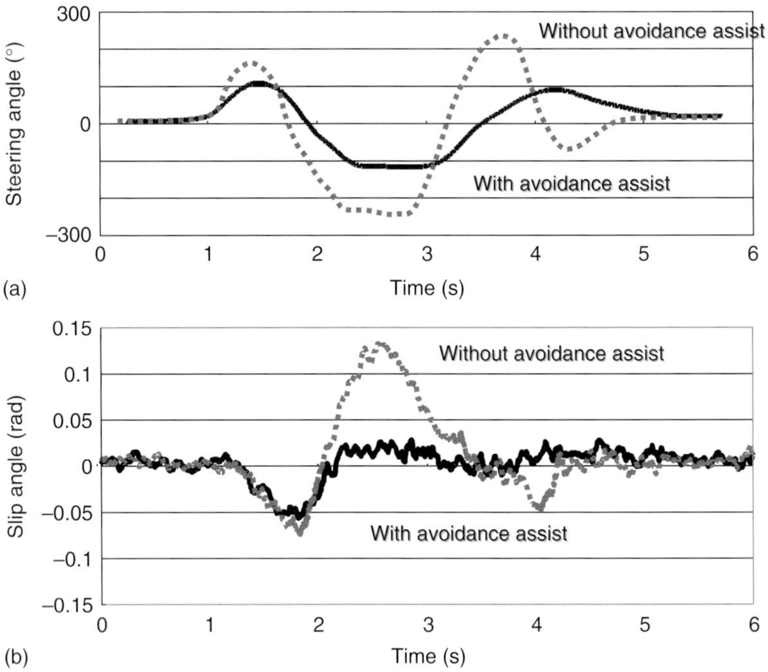

Figure 19. (a,b) Steering angle comparison when avoiding object.

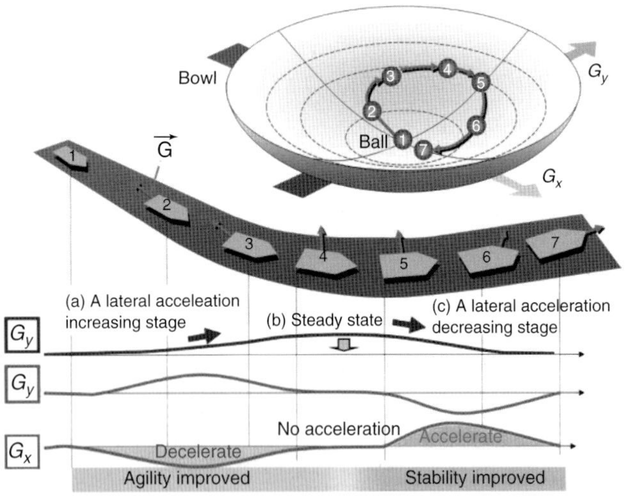

Figure 20. G-vectoring control concept.

5 METHODS OF DESIGNING INTEGRATED CONTROLS

The previous sections have described how integrated controls have a large potential to enhance vehicle dynamic performance. However, the integration of controls increases the complexity of the systems, substantially increases the scale of development and the work hours required, and makes it more difficult to secure reliability. Consequently, the original goal of improving performance also becomes more difficult to achieve. For this reason, the key concepts for designing integrated control are to create a hierarchy and to mask and abstract information.

5.1 Creating a hierarchy of function and application levels

When constructing a control, it is important to consider the failsafe conditions and other operations in the event of an abnormality, in addition to approaching the control from the standpoint of the normal targeted performance and operation. This is particularly important for integrated controls. The control has to be constructed on a layered hierarchical basis, categorizing the functions to be integrated and the functions that operate independently and autonomously. In creating the hierarchy, it is convenient to envision the relationship between the actions of a person's hands and feet and the reflexes of the brain and spinal column. Actions that require fast reactions and actions with a fixed pattern are achieved by spinal column reflex and the situation is reported to the brain. The brain directs overall actions by sending commands to the hands and feet. VDIM was created based on this concept (Figure 24). The roles are determined in sequence from the control devices equivalent to the lower order muscles, and the configuration is

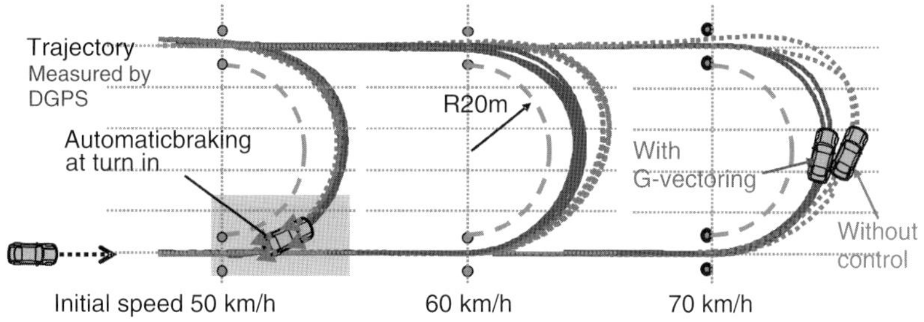

Figure 21. Trajectory evaluation.

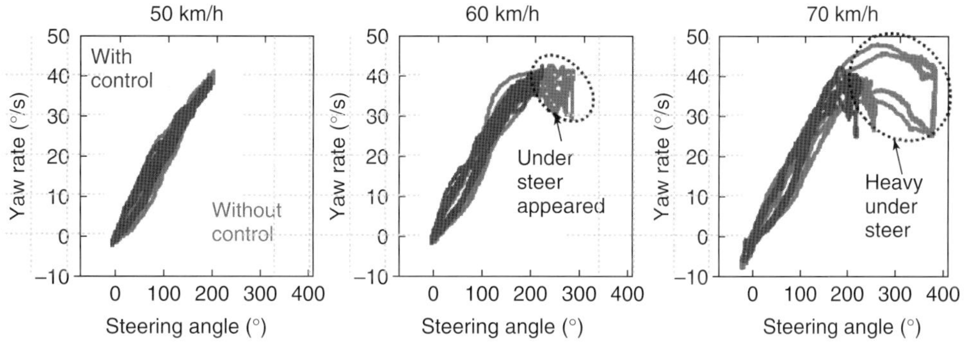

Figure 22. Steering angle versus yaw rate.

designed as much as possible to allow independent action and the selection of functional operation.

5.2 Masking and abstracting information

When constructing various functions in a hierarchical structure, another key point is the masking and abstracting of information when collecting information and transmitting commands. From this standpoint, it is simple to envision the relationship between a team manager and the team members in a corporate organization. Normally, the team members perform work based on instructions from the manager. However, the manager does not have a detailed grasp of each specific aspect of the work of the team members and the manager does not give specific detailed instructions about that work. Therefore, if the manager is ill or in another abnormal situation, the team members can operate autonomously to a certain level. In addition, the manager above the team manager is capable of running the organization without knowing the last detail of the work of the team members. Therefore, this manager can produce results that would not be possible individually. A simple and highly reliable system can be constructed by making use of this type of information collection and command transmission system.

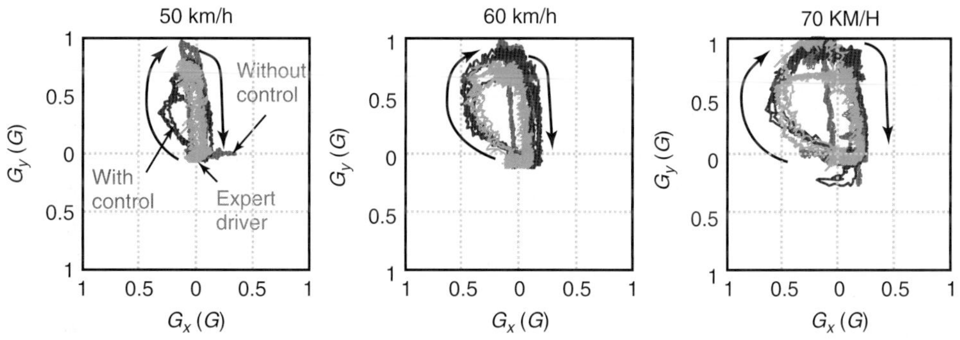

Figure 23. "g–g" diagram.

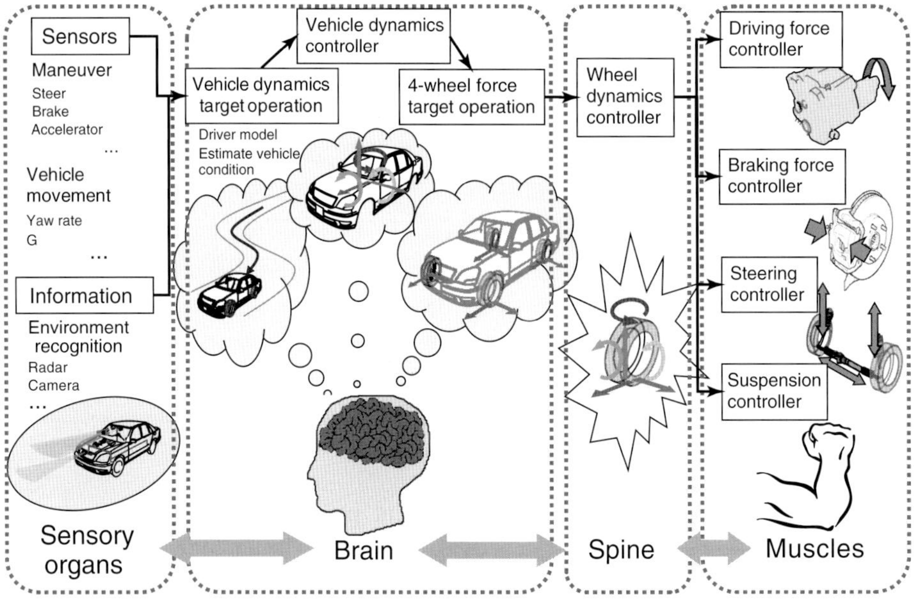

Figure 24. Hierarchy of VDIM and outline of components.

If the upper level system relies on detailed information from the lower level system, it becomes difficult to change the system configuration or add new functions. For example, in the case of brake control, the upper level system can simply control the total braking force even when in combination with a motor or another device. This is carried out by masking internal operations based on an abstraction and normalization concept. This concept positions braking force and braking G above hydraulic pressure and hydraulic pressure above the solenoid current of the actuator. Furthermore, if a theoretically different brake actuator is added into the system, there is only very little impact on the upper level system.

5.3 Creating packaging level hierarchies

The adoption of software platforms and operating systems (OS) is advancing to absorb differences in inputs and outputs, as well as in the communication between hardware and software, and to create freedom in application package locations for integration into ECUs. Conventional software structures are constructed differently based on the approach of each automaker and software supplier. However, standardization efforts are under way to achieve integration. Future development will also have to consider these trends.

5.4 AUTOSAR activities

AUTOSAR (automotive open system architecture) is an enabling technology for integrating systems in a vehicle. As mentioned in Section 5.3, AUTOSAR defines the basic software architecture, which consists of a hardware abstraction layer (HAL), system/communication services, and a runtime environment (RTE). RTE is an embodiment of a virtual functional bus (VFB) that enables the integration of application software and the physical allocation of applications into each ECU (Figure 25).

The system/communication services provide standard functions such as communication and OS. HAL absorbs the differences between microcontrollers, sensors, and the like. Figure 26 shows the basic AUTOSAR architecture.

AUTOSAR is being advanced by a development partnership, in which many global automotive companies are participating in AUTOSAR activities to develop AUTOSAR specifications as a worldwide de-facto standard (Figure 27).

6 FUTURE TRENDS OF INTEGRATED SYSTEMS

6.1 Expansion of active safety and driver assistance

The previous sections have highlighted the significant contribution of integrated control technology in active safety systems. The systems are likely to grow even more important in the future. Normally, the driving process consists of cognition, judgment, and action phases. An error in even one of these driving processes may result in an

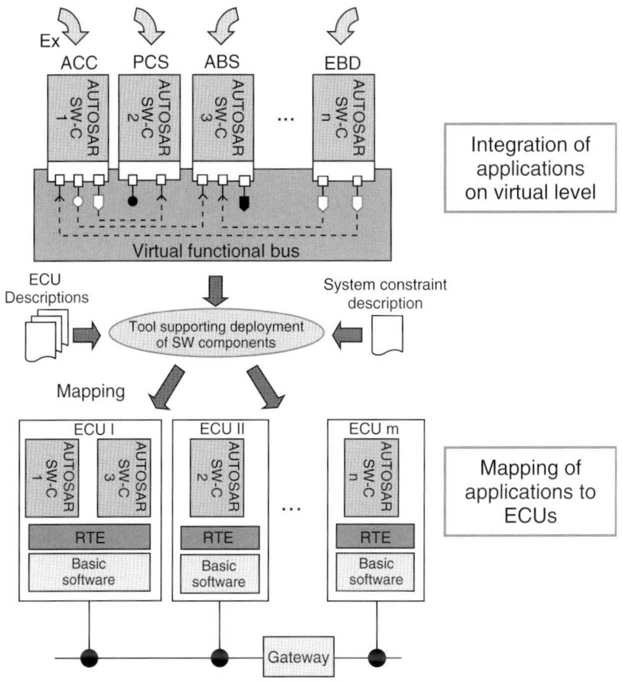

Figure 25. Integration of applications.

accident. Consequently, active safety technology is being developed to support each phase (Figure 28).

Figure 29 shows the technological areas of the various control systems that have already been commercialized. The direction of brake technology has already changed from brake assist (BA) type systems to PCS and other automatic braking systems. However, there are still many undeveloped areas on the horizontal axis, which is likely to be the direction of development in the future.

Although the physical limits are determined by the performance of the tires, brakes, and the like, technologies such as ESC have been developed that support driver operations at these limits. These technologies are now also being integrated with steering controls. In the future, it is likely that development will continue toward the commercialization of vehicle dynamic control that can stretch the possibilities at the physical limits. This development will be based on research such as the verification of theoretical limits when four-wheel independent steering and four-wheel independent braking and traction systems are combined with force control technology through suspension control.

In addition, technology to assist the cognition and judgment phases is likely to become more sophisticated as recognition technology advances. Coordination between dynamics control to enable automatic evasive maneuvering in the lateral direction will probably progress. In preparation for these developments, vehicle dynamics control must have the capability to freely control vehicle behavior. Integrated control technology will play a major role in accomplishing this aim.

6.2 Future trends

Figure 31 shows a matrix depicting the concept for total vehicle system integration. In addition to the integration of energy within the vehicle, further integrated controls are being considered that incorporate the vehicle, driver, and the traffic environment in the same way as active safety. Possible methods of helping to improve the environment

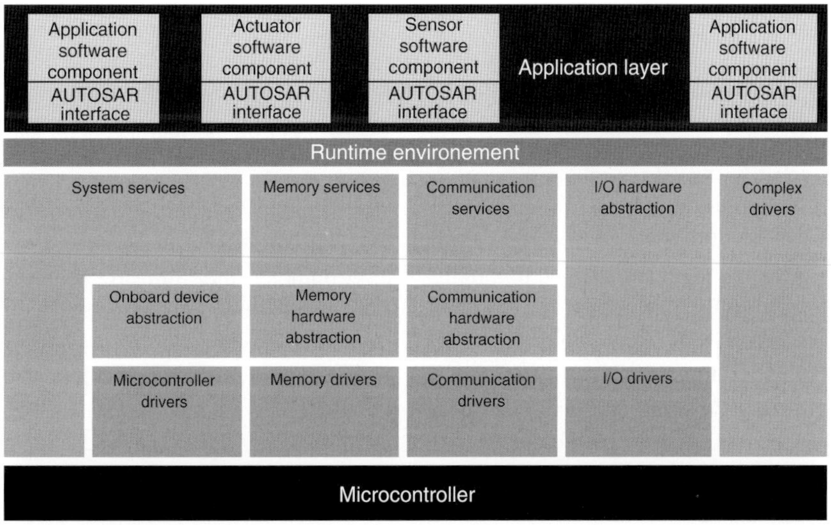

Figure 26. Basic AUTOSAR architecture.

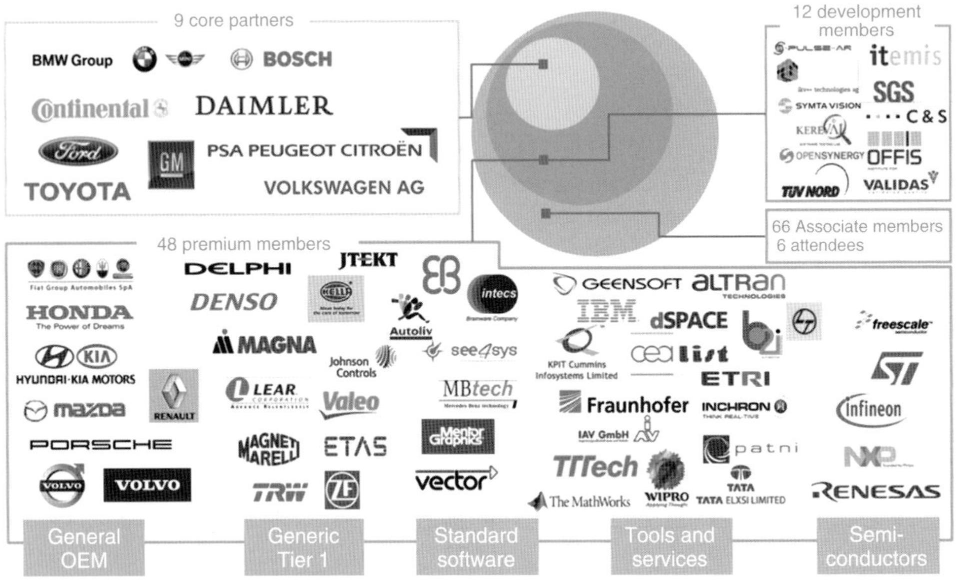

Figure 27. AUTOSAR partnership and members (2010).

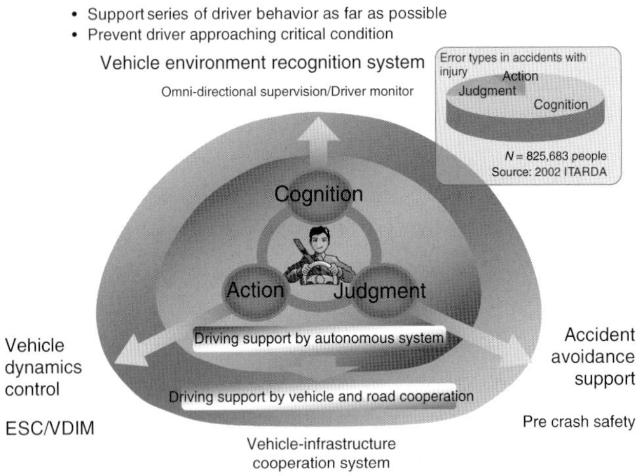

Figure 28. Future trends in active safety technology.

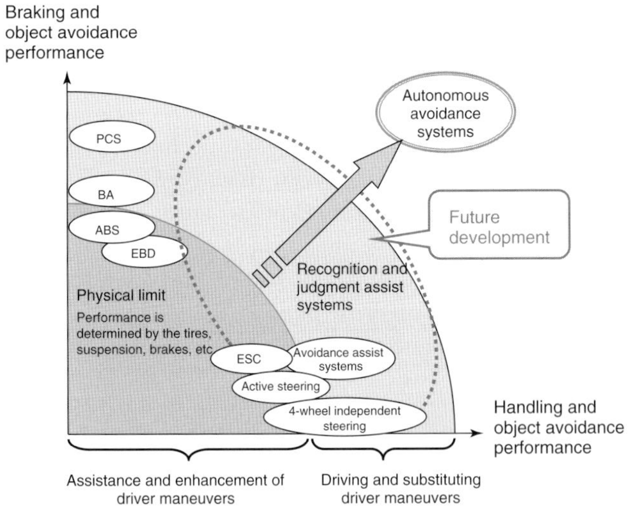

Figure 29. Current control systems and future development areas.

include the shift lever indictor and CO_2 reduction control using ACC. Integrated functions based on traffic signal controls and ITS (information technology services) that also factor in the traffic environment may be developed that help to alleviate congestion.

From the standpoint of active safety, the evolution from integrated vehicle dynamics controls to integrated active safety controls will focus on the development of driver monitoring functions in PCS systems. In the future, these technologies will develop into systems that provide assistance to individual drivers as appropriate in coordination with the driver and integrated safety systems that are coordinated with the traffic environment and infrastructure.

The matrix in Figure 30 is also related to the achievement of sustainable mobility in terms of vehicle safety, the environment, and comfort. In addition to the autonomous and infrastructure-coordinated driving environment detection technologies shown in Figure 31, an area of growing importance will be individual applications designed in accordance with navigation system and traffic control ITS information and the state and personal characteristics of the driver. Therefore, an integrated HMI that incorporates individual information, instructions, alerts, and warnings will be the key for creating an integrated control that helps to enhance safety, the environment, and comfort.

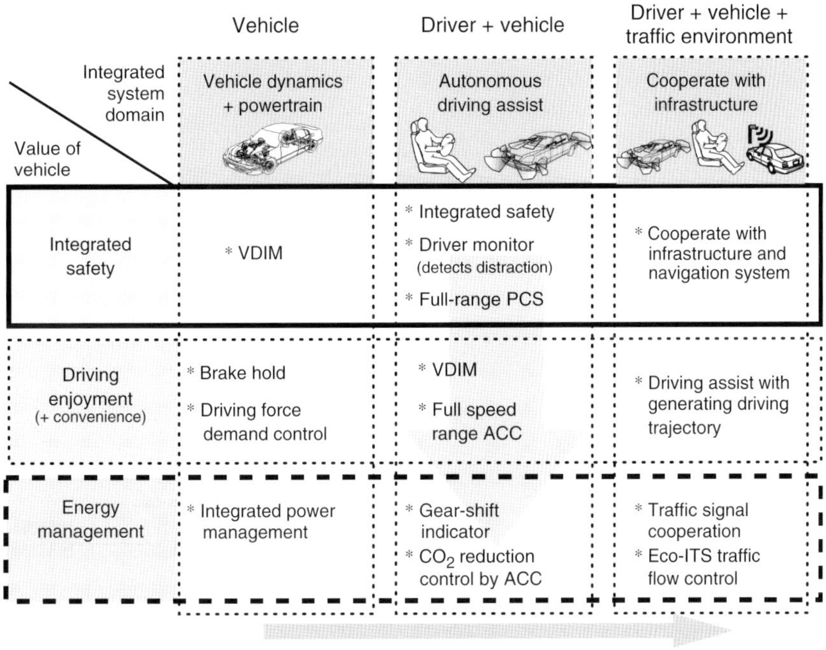

Figure 30. Total integrated vehicle system concept.

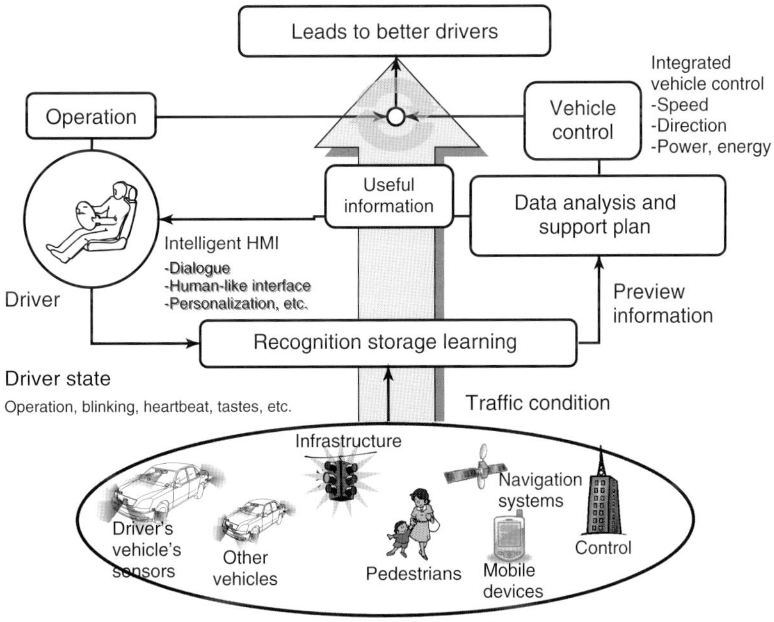

Figure 31. Integrated driver assistance concept.

7 CONCLUSION

Dynamic control technology for controls related to the suspension, steering, braking, driving forces, and the like is advancing relentlessly. At the same time, cognition and judgment functions equivalent to the eyes and brain are also being rapidly developed. However, the number of people hurt or killed in traffic accidents remains at a high level.

Therefore, this technology has to be consolidated and applied properly to fulfill the responsibility of automakers to develop vehicles that do not cause accidents. Control systems are required that are highly reliable, flexible, and have the potential for widespread use. This chapter has described the trends and configurations of these systems. Comfort and driving enjoyment are essential parts of a vehicle and these must not be sacrificed. For this reason, the development of these systems will continue while enhancing basic vehicle performance.

RELATED ARTICLES

Volume 4, Chapter 125
Volume 4, Chapter 126
Volume 4, Chapter 129
Volume 4, Chapter 132
Volume 5, Chapter 143
Volume 5, Chapter 152
Volume 5, Chapter 158
Volume 5, Chapter 161
Volume 5, Chapter 162
Volume 5, Chapter 163
Volume 5, Chapter 165
Volume 5, Chapter 170
Volume 5, Chapter 179
Volume 6, Chapter 204
Volume 6, Chapter 207
Volume 6, Chapter 208
Volume 6, Chapter 209

REFERENCES

Fukatani, K. (2005) Vehicle dynamics integrated control system: overall vehicle modeling and control, Modelling and control of the car understood thoroughly; Workshop teaching materials, published by Nihon Machinery Society, 43–50.

German patent DE 35 05 455 CS (1985) Vorrichtung zur automatischen Zu- oder Abschaltung von Antriebselementen eines Kraftfahrzeuges.

Hattori, Y. (2002) Force and Moment Control with Nonlinear Optimum Distribution for Vehicle Dynamics. *AVEC*, 02024.

Hattori, Y., Koibuchi, K., and Yokoyama, T. (2002) Force and Moment Control with Nonlinear Optimum Distribution for Vehicle Dynamics. *JSAE Transactions*, **35** (3), 215-221.7.

Katayama, K. (2007) Development of four-wheel active steer system. *Nissan Technical Review*, **60**, 5–9.

Koibuchi, K., Yamamoto, M., and Fukada, Y. (1996) Vehicle stability control in limit cornering by active brake. SAE Paper 960487.

Kojo, T. (2002) Development of Front Steering Control System. *AVEC*, 149.

Mokhimar, O. and Abe, M. (2003) Effects of an optimum cooperative chassis control from the view points of tire workload. *Proceedings of Society of Automotive Engineers of Japan Annual Congress*, **33**(03), 15–20.

Nakamura, E. (2002) Development of electronically controlled brake system for hybrid vehicle. SAE World Congress 2002-01-0300.

Ono, E. (2007) Improvement in Safety and Comfort by Active Four Wheel Steering. *Proceedings of the Society of Automotive Engineers of Japan*, 11-07.

Ono, E., Hosoe, S., Tuan, H., *et al.* (1998) Bifurcation in vehicle dynamics and robust front wheel steering control. *IEEE Transactions on Control Systems Technology*, **6**(3), 412–420.

Ono, E., Hattori, Y., Aizawa, H., *et al.* (2009) Clarification and achievement of theoretical limitation in vehicle dynamics integrated management. *Journal of Environment and Engineering*, **4**(1), 89–100.

Shibahata, Y., Shimada, K., Tomari, T., *et al.* (1993) Improvement of vehicle maneuverability by direct yaw moment control. *Vehicle System Dynamics*, **22**, 465–481.

Tanaka, H., Inoue, H., and Iwata, H. (1992) Development of a vehicle integrated control system. 24th FISITA Paper Award.

Tsuchida, J. (2007) The advanced sensor fusion algorithm for pre-crash safety system. SAE International World Congress 2007-01-0402.

Van Zanten, A.T. (1996) Control Aspects of Bosch-VDC. *AVEC 96*, pp. 573-608.

Yamakado, M. and Abe, M. (2008) An experimentally confirmed driver longitudinal acceleration control model combined with vehicle lateral motion. *Vehicle System Dynamics*, **46**, 129–149.

Yamakado, M., Takahashi, J., Saito, S., *et al.* (2010) Improvement in vehicle agility and stability by G-vectoring control. *Vehicle System Dynamics*, **48**, 231–254.

Chapter 131
Chassis Control Systems

Gerhard Klapper and Ralf Leiter

TRW Automotive, Koblenz, Germany

1 INTRODUCTION

The appreciation of vehicles and the differentiation between the manufacturers and models is mainly gained from the increased use of more and more complicated control systems, including their electronic components. This trend will rapidly accelerate in the near future, as indicated by the top models from various manufacturers already on the market. In some companies, there is still the "world car" idea; but in the meantime, nearly everybody has also learned that different regions require different coordination of the chassis. Moreover, the greatest dream is still to achieve this using software coding.

In general, the target of the chassis development can be described as monitoring the vertical wheel forces in a way that the necessary longitudinal and lateral forces can be transferred to the road with simultaneous maximum comfort and safety for passengers and minimal fuel consumption. The classic chassis presents many compromises in design and performance; the use of electronics permits design shifts to improve the performance even further.

Legal requirements and environmental protection can be achieved by bringing microprocessors into the vehicle. The exhaust laws of today can only be met by regulating ignition timing and injection quantity in dependence of the catalyst sensor. This was followed by safety systems such as antilock braking system (ABS) and the airbag, which brought additional microprocessors into the vehicle. Over time, components became so reliable that now complex control strategies can be implemented with simple control systems.

This chapter defines the classic components as the chassis system:

steering,
suspension and damping, and
chassis and brake control.

These components cover the controls for longitudinal and lateral guidance of the vehicle, while simultaneously controlling the vertical movement.

2 STEERING

The steering of a vehicle presents the most sensitive components in the control loop of driver-vehicle. The driver has decisive influence over the behavior of the machine to be operated. A vehicle is moving dependent on the feel and the individual capability of the vehicle's driver. Next to the general vehicle design, steering is the most important feedback to the impressions and control information reported to the driver.

Encyclopedia of Automotive Engineering, print © 2015 John Wiley & Sons, Ltd.
Edited by David Crolla, David E. Foster, Toshio Kobayashi, and Nick Vaughan.
This article is © 2015 John Wiley & Sons, Ltd. ISBN: 978-0-470-97402-5
Also published in the *Encyclopedia of Automotive Engineering* (online edition)
DOI: 10.1002/9781118354179.auto032

The driver receives feedback about the street, the environment, and the vehicle behavior from the respective steering design that he can immediately incorporate into his behavior to operate the vehicle safely. This is a variety of visual (e.g., deviations from the wanted position on the lane), acoustic (e.g., warning strip noises from the tires), and dynamic information (e.g., the torque on the steering wheel, the yaw angle velocity, or also the lateral acceleration) that the driver must simultaneously receive and process. The driver can derive and implement the necessary control interventions with regard to steering the vehicle, releasing the acceleration pedal or even pushing the brake pedal. Since the drivers themselves function as dynamic controllers, it is important that they do not make the entire system unstable. Their basic dynamic behavior is a second order system, which can become unstable.

Today's steering systems do not only ensure the sensors and feedback of important information for driving a vehicle, but depending on execution, they also provide much additional assistance for the driver. This applies to the typical simple servo-support (auxiliary power units to relieve the steering forces the driver must use) until the steering movement support that is mainly dependent on the speed (additional steering angle), in some situations even from the yaw rate sensor.

The latest step forward in steering has been realized by electrical power steering systems. Even those systems can be developed further. The different types of electrical power steering are making this obvious.

2.1 Electrically powered steering

Especially for smaller vehicles with less axle loads, electric drives for power steering are becoming reality. With regard to positioning and space required, the electric drive is superior to the hydraulic drive. This is also true with respect to fuel consumption.

Electric drives reduce the range of their components to:

- cable harness,
- drive motor,
- clutch,
- transmission, transmission housing,
- steering wheel torque sensor,
- control unit or partial integration in present control unit, and
- mechanically functioning steering gear.

In addition, advantages of the electric assistant steering systems are:

- reduction in the number of parts (no pump, pump drive, oil container, and hydraulic lines),
- less weight,
- less space required,
- independent of vehicle motor, operation as possible with combustion engine shut off,
- less loss in driving power,
- better adaptation to the power requirements, reducing gasoline consumption,
- steering force regulation as needed: speed or load dependence is possible, and
- no need for inconvenient hydraulic maintenance.

There are also disadvantages that limit the complete introduction of the electric power steering in the automotive industry. They are evidenced by:

- limited performance, dependent on available energy sources (12-V electrical system);
- considerable protection needed from system failure due to use of electronic components;
- increased development expenses due to software and hardware adaptation work; and
- ever-increasing manufacturing costs (compared to hydraulic power steering systems).

Potential installation positions for the electric steering assist are more numerous than with hydraulic systems. Figure 1 gives an impression of the different possibilities. Electric motors can also be arranged:

- directly under the steering wheel on the steering column;
- like all hydraulic systems on the steering gear and on the rack-and-pinion; or
- for example, using a belt on the recirculation ball gear.
- in parallel to thes steering rack.

Electric power steering systems can also be preassembled modularly, as they are independent of hydraulic lines, which enable verifiability during production and can significantly reduce the production costs again.

The various manufacturers of such systems on the market use a variety of different drive and sensor concepts. Unlike with hydraulic systems, the steering angle and also torque applied are recorded by sensors and converted to electric signals. The control unit evaluates these signals and converts them into corresponding commands for the drive motor. This now works on the mechanical steering gear with the torque associated with the command.

(a) Electro-hydraulic steering

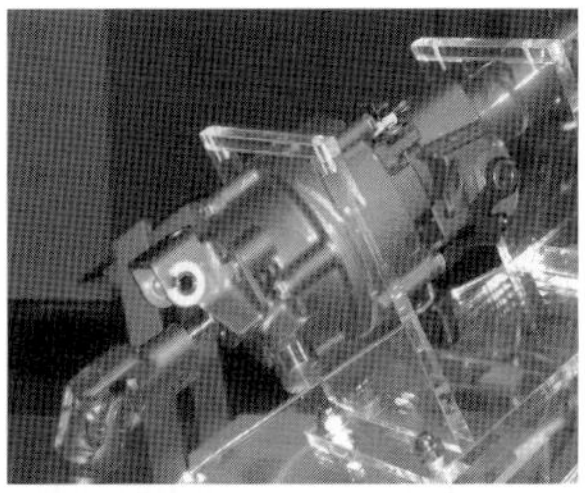

(b) Electric actuator in line with the upper part of the steering column

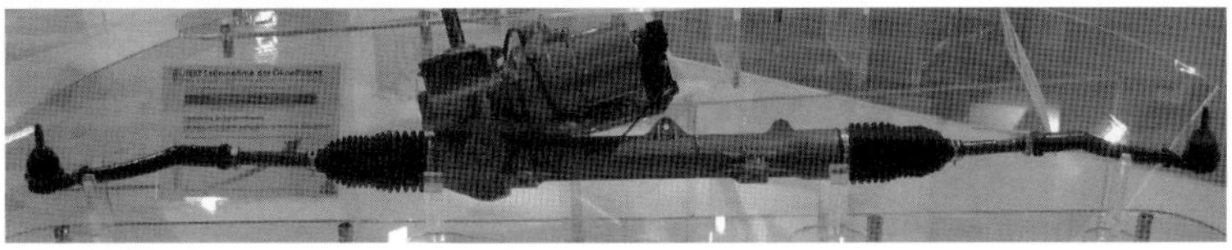

(c) Electric motor at the pinion

(d) Electric actuator in line with the rack

(e) Electric motor rectangular to the upper part of the steering column

Figure 1. (a–e) Potential installation positions for the electrically powered power steering. (Reproduced with permission from Henning Wallentowitz. © H. Wallentowitz.)

2.1.1 Column drive (drive arranged on the steering column)

In this application (Figure 2), the electric motor is directly fastened to the steering column. The unit can be pre-manufactured separately from the vehicle assembly and delivered as a module to the assembly line.

The additionally required electronic control unit can either be integrated into the module or be positioned separately from the servo unit in the vehicle. Depending on the vehicle concept, the control unit can also be integrated in other control units. Synergies with this procedure are evident in the reduction of electronic components. Reductions in costs are the result.

Figure 2. Steering tube drive. (Reproduced by permission of TRW.)

The steering tube or the steering spindle between the steering wheel and the steering gear are powered.

To record the necessary input sizes, a torque sensor, or in individual cases, a rotary sensor and possibly according to requirements also speed sensors are integrated into the system.

2.1.2 Rack-and-pinion drive (with concentrically arranged motor)

The application in this case requires a rack-and-pinion steering with side pickup. Accordingly, other potential rack-and-pinion variants are conceivable, however, with more effort for the realization. Placing a coaxially working motor requires corresponding installation space in the middle region of the rack-and-pinion mechanics. Figure 3 shows one possibility. Often, the engine/gearbox unit requires this space.

2.1.3 Belt drive

In order to save space and still be able to directly grasp the rack, this variant (Figure 4) is available for electrical application of the assistance. The overall design of the unit provides significantly more clearance. In individual cases, the correct variant of those possible must be determined by an installation study, due to difficult installation dimensions in the front of the vehicle. The corresponding suggestion is therefore to evaluate cost/effort.

As is typical for the introduction of electronic components in safety-relevant vehicle systems, the demand for quality controls and safety tests increases considerably.

This is supposed to prevent critical driving conditions and drive situations, caused by malfunctions in the electronic components.

Such malfunctions can occur with:

- Too high current load at the motor
- Motor blockade due to seizing components such as bearing or rotor body

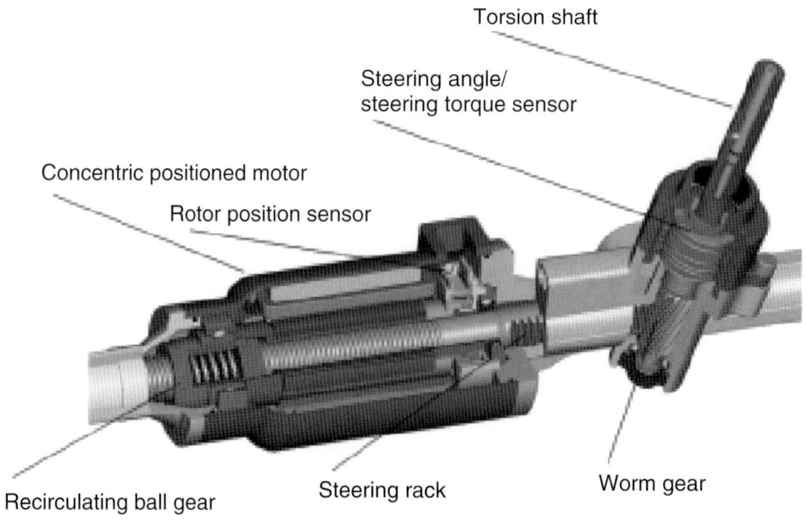

Figure 3. Steering gear drive sectional view. (Reproduced by permission of TRW.)

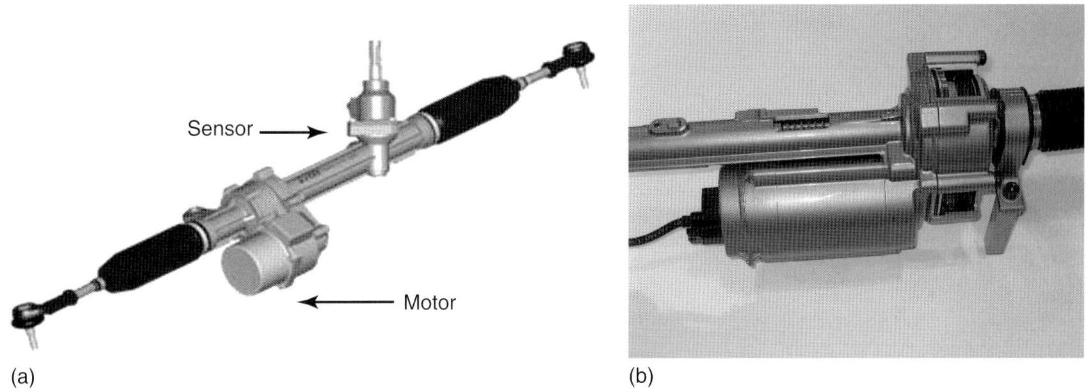

Figure 4. (a, b) Belt drive. (a: Reproduced by permission of TRW, b: Reproduced by permission of Henning Wallentowitz.)

- Manipulation of electronics or the electrical system
- Sensors that give incorrect signals

To avoid all these fault possibilities, vehicle manufacturers exclusively use high quality materials and components. In addition, in production, up to 100% checks can be used (production of safety components).

In addition, redundant systems are required and introduced, which can balance the failure of parallel working systems on one hand, and on the other hand find use for control purposes of main systems and switch off the system if there are deviations in the signals.

Furthermore:

- The current values are technically limited;
- Voltages and sensors are permanently checked; and
- Torques are monitored.
- As further safety equipment, plug connectors that cannot be manipulated and so-called *electric couplers* are used.

Self-diagnostic programs constantly monitor all-important functions. This occurs by comparing the actual announced values with the specified target values or target value windows.

Fault detection shuts down the auxiliary systems and simultaneously notifies about faults using the system lights in the dashboard. The driver also recognizes the fault function by the increased steering forces. These behave exactly like with a strictly mechanical steering system without power steering.

Since the 1980/1990s, there have been cars with rear wheel steering in the market (mechanically powered (Abe, Ogura, and Sato, 1988), or different hydraulically powered systems (Donges, 1989; Kuroki and Irie, 1991). Now, there are systems coming to the market with electrically powered rear wheel steering systems. They may be considered in the next paragraph.

2.2 Active rear steering

Four-wheel steering is a known method to improve the maneuverability at low speeds (opposite impact to reduce the curve radius) and to enable better vehicle stabilization at high speeds. The simplest algorithms adjust the steering angle of the rear axle linear to the steering wheel angle. Advanced algorithms support the development of the yaw moment, dependent on speed, first with opposite turning, but then control back to the same direction, releasing the yaw angle and increasing driving stability (Abe, Ogura, and Sato, 1988).

The active control of slip angles at the rear wheels allows a targeted increase of the cornering forces. This can influence the steering of a vehicle (e.g., development of sideslip angle in the vehicle's center of gravity). Side wind influences can also be automatically compensated, without the driver needing to intervene.

Four-wheel steering can make the vehicle significantly more agile, but it can also change lanes without (in reality with reduced) a yaw movement occurring. That opens new possibilities to the algorithms of automatic accident prevention. The first electronically operated system has been introduced by BMW in the beginning of the 1990s (Donges, 1989; Wallentowitz, Donges, and Wimberger, 1994).

2.3 Superimposed active steering

Another modern steering system, which is in the market since 2003 (Köhn *et al.*, 2003), is the superimposed steering system. AUDI followed with an own solution in 2007. This

Figure 5. Electric superimposed steering. (Reproduced by permission of EBM-Papst.)

is the first step to combine steering commands generated by the driver with steering movements coming from an electronic.

The superimposed steering makes it possible to realize a steering intervention on the front axle independent of the driver, without separating the mechanical coupling between steering wheel and front axle. An additional gear, an electric motor, sensors, and a control unit, expands the normal steering system. Figure 5 shows the additional gearbox in such a system. This system now enables the continuous change of the steering gear ratio—dependent on the driving situation detected. The effective steering angle at the wheels can be larger or smaller than the driver adjusted with the steering wheel. This just depends on whether the auxiliary system is steering in the same direction as the driver, or whether it is steering in the opposite direction. Thus, in city traffic, there is less steering required by the driver; at higher speeds, the vehicle can be better stabilized, thanks to an automatic and limited "counter-steering" of the system.

If the electric motor is not actuated, there is a direct mechanical connection between the steering wheel and the front wheels, as with conventional steering. Thus, such "intelligent" steering fulfills the existing legal requirements.

2.4 Steer by wire

In the last level of the current development, the driver is mechanically and hydraulically completely separated from the power units of the steering systems (like with the control of engines and automated transmission gear boxes). Steering angles are accepted as input parameters by the driver using the steering wheel or in the future, the so-called *joystick*, but are converted into electronic control commands for the actuators, moving the wheels.

Currently, law still requires permanent connection either mechanically or hydraulically (then limitation) vehicle speed to be able to react as needed if the electronics fail. This is called *fully redundant safety measures*. However, the entire vehicle industry is working on other "clever" systems to avoid this expense in the future and to be able to change the laws accordingly.

A complete separation of the input interfaces of the power aggregates would lead to:

- Enormous savings in materials,
- Reduction of parts, and
- Crash behavior improvement

The environment can also profit from this, as no poisonous hydraulic fluids would have to be disposed of. Rubber seals are not used in the large circumference; expensive manufactured tubes and hose lines are no longer needed; as well as a part of the mechanics, using a lot of energy during manufacturing, will not be necessary any longer.

The steering reaction can from this time forward be coordinated to all potential input parameters of the variety of sensors already present in today's vehicles (steering angle and steering torque sensors, yaw rates information, or camera images). Even safety-promoting interventions of the electronics into the handling (already realized in partial areas) are conceivable.

The potential configurations with redundant mechanical backup systems could look as shown in Figure 6.

In the basic idea, the "steer by wire" will not have any mechanical connection between the driver and the front wheel. Obviously, it is extremely important that in this case several redundant electronic systems must efficiently, and above all, safely prevent the steering from failing.

Figure 7 gives an impression of such an idea. Nissan is just now suggesting a similar system, but in addition to the electronic backup, there will be still a mechanical backup with a clutch, which is normally open (Kubota).

Ultimately, the vehicle manufacturers and their development partners have not yet reached series introduction of

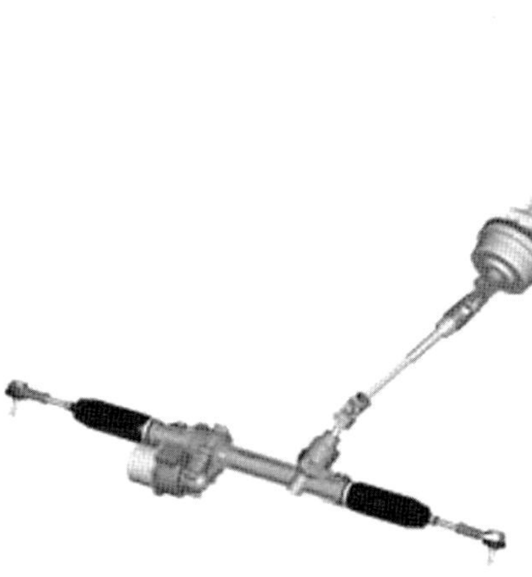

Figure 6. Fully electronic steering with mechanical backup. (Reproduced by permission of TRW.)

these systems. The cheaper and always mechanically redundant hydraulic power steering used in series today is still the main percentage of power steering systems used on the market. Electric/electronic steering systems suffer the enormous price pressure under which the vehicles must be built today. An additional hurdle is coming from the numerous electronic components, which are installed mainly for comfort reasons. This stresses to the limit of the currently used 12-V electrical systems of the vehicles. Fundamental changes of the electrical infrastructure of today's vehicles are currently being controversially discussed from this viewpoint (Amsel, 2003; Wallentowitz and Amsel, 2003).

3 SUSPENSION AND DAMPING

Advanced suspension systems are realized by air springs and by hydraulically actuated conventional steel springs. This chapter describes some of the systems and the possibilities for advancements.

3.1 Air spring

The trend for more comfort in the passenger compartment and less vibration in the vehicle's suspension, independent of load, for continuous control of the height level as well as level adaptations to entry and loading assists, air suspensions in passenger cars, after a first application in the 1960s, became common equipment in passenger cars of the upper class in the 1990s. The suspensions of the 1960s are not comparable to the modern systems.

In semitrucks, where, thanks to the pneumatic brake system, the air energy is already available, air suspension for street transportation vehicles has become routine. As the air compressor and the air dryer are consuming a lot of energy, even here changes are wanted.

Today, the compressors still mostly work in open systems. The air is taken from the environment, compressed and either pressed directly into the air spring or is kept available in a reservoir. Excess air (when lowering the vehicle structure) is returned to the environment. The vehicle structure can only be raised relatively slowly and with limited frequency (load of battery and compressor). Closed systems found a solution where the air is pumped

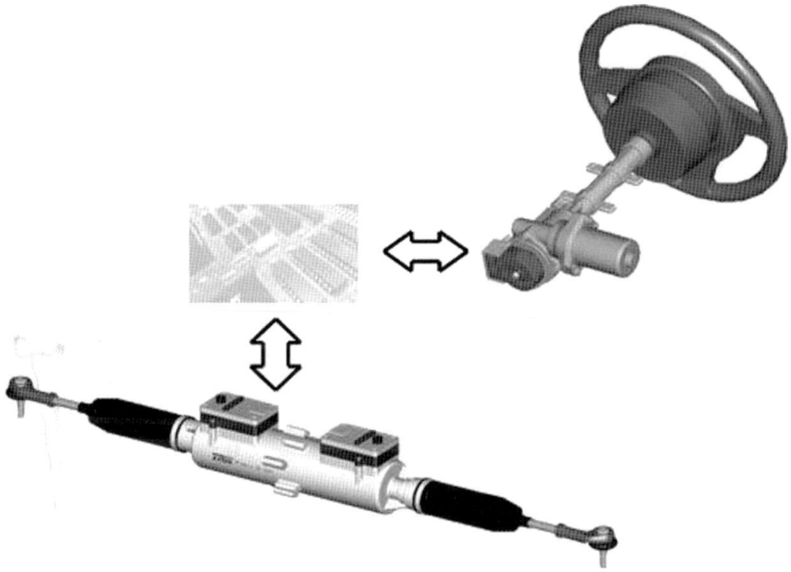

Figure 7. Fully electronic steering with electronic backup. (Reproduced by permission of TRW.)

Figure 8. Air spring with variable effective area. (Reproduced by permission of Meritor.)

back and forth between the spring and reservoir at high pressure. This clearly permits faster-level changes with less needed energy. This technology now also allows lowering the chassis while driving (lower air resistance) or raising the vehicles body (more clearance with poor street surface or driving out in the country).

An interesting variant is changing the effective area of the air springs with an additional interior air bellow. This allows the spring stiffness to increase threefold in a few milliseconds or to decrease without needing much air volume. Figure 8 shows this spring in soft condition (a) and in stiff position (b). There is only compressed air necessary to blow up the small air spring, which changes the rolling piston diameter (Lloyd, 2010).

This "active air spring" enables various operating modes, adapted to the vehicle driving conditions. The main advantage can be created during cornering. The wheel on the outer side of the curve not compress as much with a stiff air spring; the car body remains more horizontal. During acceleration or braking, the anti-dive and anti-pitch can be organized by these air springs. Unfortunately, these types of air springs are not commonly used in the market. For keeping the car horizontal, today active stabilizers are used. Their functionality is described in the next paragraph.

3.2 Active stabilizer

Conventional mechanical chassis are always a compromise between comfort and driving safety. When driving in curves, the vehicles roll around the rolling axle. The roll angle depends on the lateral acceleration. The softer the stiffness of the body springs and the larger the distance between center of gravity and rolling axle, the more roll angle can be built up. This soft roll stiffness can even cause roll over (elk test) if there are quick directional changes and simultaneous minimal roll damping. To decrease the tilt angle, the stabilizer bar rigidity must be increased.

The stabilizer is an additional connection between chassis and structure. The roll moment when driving in curves leads to the wheel load differences becoming larger with increased lateral acceleration (but the sum of wheel load of an axle remains the same). The wheel on the outside driving around a curve compresses and the wheel at the inner side rebounds. Its coupling rod rotates the stabilizer, and the forces at the bearing point create a torque, which counteracts the roll movement of the car's body.

The rigidity of the stabilizer has considerable influence on the roll stiffness of the vehicle. With the coordination of front and rear stabilizers, the vehicle behavior is changed between understeering and oversteering. Additional influential parameters for the driving behavior are the stiffness of the springs between chassis and body (as already mentioned above) and the position of the roll centers on the front and the rear axle. The specific roll center height results from the design of the used axle. The roll centers may be above the road, on the road, or even below. They are (mostly) virtual points.

As the stabilizer represents an additional spring, which is also active for one-sided deflection, named copying, active stabilizers have been invented. They reduce copying and they can in addition reduce the roll angle of the vehicle during cornering. Active stabilizers are realized with electric or hydraulic actuators in two or more stages. Yaw rate, lateral acceleration, steering angle, and speed are

Figure 9. BMW with active chassis control (Dynamic Drive) (Jurr *et al.*, 2001; Konik *et al.*, 2000). (Reproduced by permission of BMW AG.)

considered as incoming parameters. Two pressure sensors, six valves, and one electronic controller provide the "right" driving behavior, which is adjusted using hydraulic pressure from a motor-powered tandem pump. Figure 9 gives an impression about the installation of the system in a car.

Hydraulic motors can even actively intervene in the chassis regulation with a very quick electric tandem pump unit. However, the pump power must be approximately as great as that of the electric hydraulic power steering. This leads to short-term electrical system load >100 A.

After switching on, the control units run a self-test and perform a functional self-test on the connected valves and sensors. Body control does not occur with standing vehicle. At approximately 20 km/h, the systems are active. Cornering is detected by the lateral acceleration signal, steering wheel angle, and speed. With up to 180-bar hydraulic pressure, counter-torque of 600 Nm (front axle) and until 800 Nm (rear axle) is generated on the stabilizer. At higher speeds, understeering drive behavior is preferred; thus, a higher roll torque amount is required on the front axle. To increase agility, a higher roll torque is temporarily set on the rear axle.

Simpler active stabilizers, especially with all-terrain vehicles, can only decouple the two stabilizer ends on driver command and permit an increased linking of the axles, which significantly improves the traction out in the country. Logically, this occurs at speeds below 50 km/h. When driving straight-ahead, the comfort is improved by decoupling (minimizing the copying tendency), but for normal cornering, these stabilizers work like rigid coupled ones.

3.3 Hydraulic chassis

The hydro-pneumatic suspension (a gas suspension with interconnected oil column) was already used in 1953 in passenger cars from the Citroen Company. Every wheel is equipped with a suspension/damping element, which contains nitrogen in the upper half of the hydro-pneumatic accumulator. The gas (constant mass of the gas) compresses and releases during the wheel movements and takes care of the reaction forces with respect to the wheel loads. A rubber membrane separates the gas from the hydraulic oil in the lower half of the strut.

During compression, the piston presses the oil through the pressure valve, arranged in the cylinder. During rebound, the gas presses the oil down through the harder adjusted rebound valve. In principle, the strut is an upside-down one-tube damper. The design level position in the vehicle is set using a lever-operated valve by adding or releasing oil, independent of load.

In the mid 1990s, Citroen added an active hydraulic stabilization system to its hydraulic suspension. This system was able to improve the driving comfort, but all in all, hydro-pneumatic suspension was replaced by air spring suspension. Only special solutions are still on the market, such as the system mentioned in the following paragraph.

3.4 ABC chassis

The active body control of Daimler AG is an electric-hydraulic chassis to meet the goal to always keep the

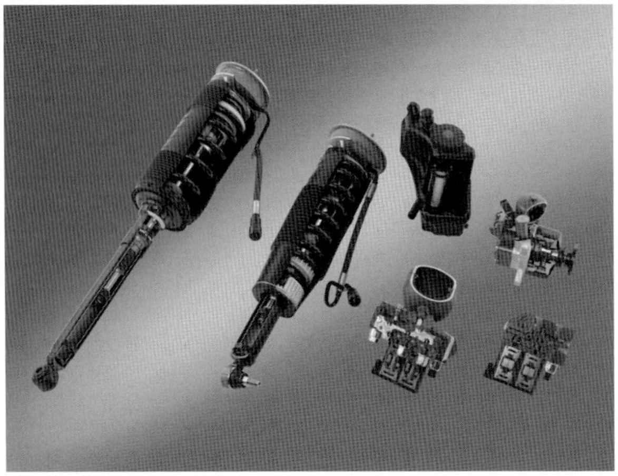

Figure 10. ABC chassis components (Wiesinger). (Reproduced by permission of Daimler.)

vehicle structure at the same height (Wiesinger). In addition to steering angle, wheel speed, yaw moment, and longitudinal and lateral acceleration, the height of the vehicle body is sensed against the wheels. The control unit "observes" the body behavior, and it can almost completely compensate the vertical body movement, the roll, and pitch with assistance of hydraulic cylinders on the support of the coil springs. The base points of the steel springs are hydraulically controlled with frequencies until 5 Hz—the shock absorbers are responsible for the wheel damping.

Electric-magnetic stop and control valves permit the individual control of each strut. With this system, individual wheel loads can be specifically increased or decreased, depending on what the drive situation requires. It is also easily possible to use wheel load control to optimize the brake and side forces in the tire contact area. Acceleration sensors on the vehicle body support the control strategy. Roll torque can be distributed to front and rear axle. A car with ABC suspension does not have a stabilizer bar.

Figure 10 shows the chassis components of such a suspension.

Right before opening the driver's door, the vehicle is lifted to entry height. When driving in curves, the outer springs are pressed together. To reduce roll, the spring base point is lifted until the compression stroke of the spring has been compensated. While braking, this base point tracking occurs on the front springs that are compressed. On the rear axle, the steel spring releases and the base point shifts downward. During acceleration, the base points are shifted just opposite. This compensates pitch and bounce movements.

Level sensors record bumps and correct for these. In a speed range between 65 and 140 km/h, the vehicle body is continuously lowered to 15 mm. Another 40-mm clearance is available with the push of a button. At 80 km/h, the crosswind stabilization is available, which can, however, be canceled again by quick and strong steering movements.

The advantages of such a system are: the car body remains horizontal, even during cornering, accelerating, or braking; mechanical stabilizers are no longer required; a radial piston pump provides the system with up to 210-bar oil pressure; and individual pressure sensors monitor the control movements. One disadvantage is the additional energy consumption to operate the system.

Pre-scan system is just now in series, which allows the ABC system to work with foresight by laser scanning the lane in front of the vehicle, but such systems are still under development.

3.5 Adaptive damping

In 1987, BMW was the first manufacturer in Europe to introduce adaptive damping. The previous systems could only select between several fixed damping characteristics (sport, normal, and comfort), but the BMW EDC has already worked with continuous adjustment. Today's systems also do this. Control programs that use the various areas of the shock absorber parameters are used in every case. In addition to automatic control, the driver can tune the system toward sportier or more comfortable operation as desired using a switch. However, in any way, the system reacts continuously. The damper characteristics are adapted to the driving situation as much as possible.

The former situation with the fixed settings of the damper can be explained by the so-called *conflict diagram* (Figure 11). Every combination of body spring stiffness and damper characteristic delivers driving comfort and driving safety results above the limiting curve. There is no simultaneous optimization of driving comfort and driving safety possible. To overcome this limiting curve was only possible with adaptive dampers.

Three acceleration sensors (two on the front axle, one on the rear axle), the steering wheel angle sensor, and the wheel speeds are the input parameters for the system, which then can adapt the chassis to the driving situation using the four adjustable dampers.

Usually, forces are calculated from the signals. Different simultaneous active control loops try to determine the optimum damping for the respective driving situation. There are vertical, lateral, and longitudinal control algorithms installed. The algorithm with the hardest setting determines the damping characteristics to be set on the shock absorbers. If possible, the soft setting is used until another setting is required.

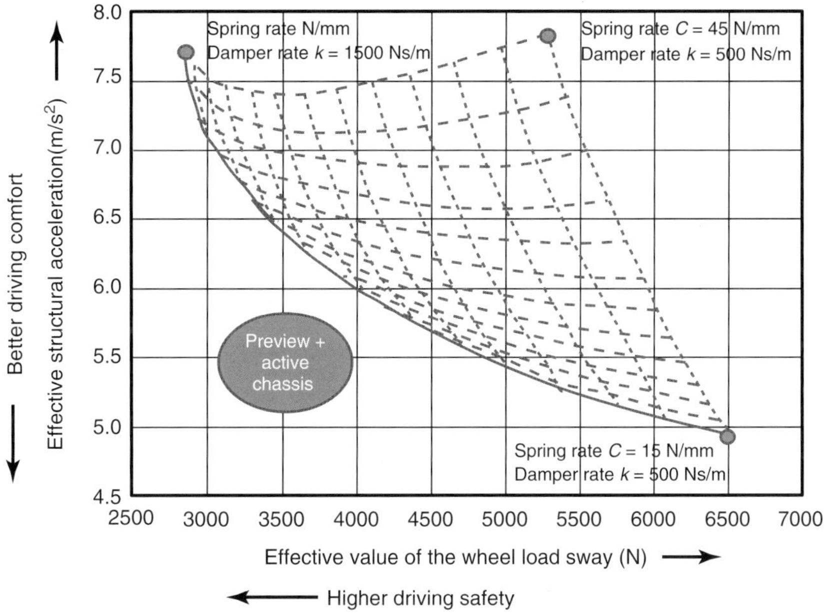

Figure 11. Conflict diagram passive wheel suspension.

The vertical dynamics controller (also called *hub controller*) reacts to the movements of the vehicle in vertical direction. Input parameters are wheel and body accelerations and their time integrals, that means wheel and body speeds. The wheel vibrations are between 10 and 16 Hz. When driving on uneven roads, irregularities occur in the wheel speeds, which can be used to estimate the quality of the street surface. The vertical body speed (the body vibrates approximately with 1 Hz) is determined with consideration of the driving speed, the frequency range of the street surface, the load, and the size of the excitation. Every axle is first individually controlled; a downstream parallel controller then provides for parallel movements.

The longitudinal controller reacts to accelerations and brake processes. The wheel signals (two per cable directly from electronic stability program (ESP), two additional ones via controller area network (CAN), and the vehicle speed) are the input parameters to calculate the static and the dynamic portion of the longitudinal acceleration.

On the basis of this, most of the dampers for both axles are adjusted the same to better support the vehicle.

The lateral dynamic controller detects yaw moments very early on using driving speed and steering angle. The dampers, which are adjusted harder (separate regulation of front and rear axle), already permit better vehicle support during cornering.

The vehicle roll with one-sided compression and rebound of the wheels while driving straight is called *copying*. It is reinforced by the axle stabilizer, which transfers the compression of one wheel (as a result of unevenness, the vehicle structure is lifted on this side) to the other wheel on the axle and also allows this wheel to compress. This increases the roll angle of the body. This is uncomfortable. The vehicle body reacts to this one-sided ground unevenness when in the comfort setting. The right and left body accelerations that are measured above the front axle detect this copying. All dampers are set to be harder and work against this unwanted body movement.

The diminishing mechanical damping characteristics are compensated by a tolerance adaptation. Thus, even defect dampers can be identified. If there are systems faults, an average damping is set that would correspond with a passive chassis, or if there is a total failure, the dampers automatically go to the hard settings.

For comfort reasons, the goal is to drive as long as possible with "soft" dampers and only switch to "hard" for safety reasons. This takes energy from the body and the wheel movements. In Figure 12, a cross section of a damper piston with an adaptive control mechanism can be seen.

Whether the damper rate is set using magnetically influenced rheological fluid or by a proportional valve is only a secondary issue as seen from a technical view.

The recuperation of the energy, now converted into heat in the dampers and radiated into the ambient air, is currently discussed as source for electricity generation. Whether it will be used in series is yet to be seen.

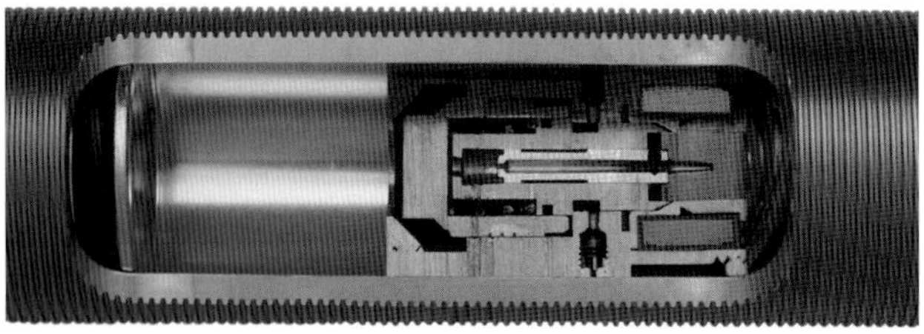

Figure 12. Adaptive damper, cross section of the piston. (Reproduced by permission of Meritor.)

3.6 Tire pressure monitoring

Air is a structural component of today's tires. It provides both lateral rigidity (cornering stiffness) and vertical stiffness (spring affect) of the contact surfaces between tire and road. If there is air loss, the tires lose a substantial part of this rigidity and they deform more than the design allows.

If there is not enough air in the tires, they deform more while rolling. This causes friction between the reinforcement materials (textures) in the tire and the rubber. Heat is generated in the tires. This heat leads to degeneration of the tire materials until pyrolysis—finally leading to total failure of the tire.

The sidewall reinforcements of modern tires are so large that visual monitoring cannot detect a pressure loss of 20%. Even daily pressure monitoring has its limits with strong temperature fluctuations between day and night. The pressure difference can easily amount to 0.4 bar (from 21 to 0°C). While the majority of drivers (82%) claim to check tire pressure at least every 3 months, tests have shown that only 13% actually do. The Ford-Firestone case made this clear. Even today, many accidents caused by tire damage can be led back to tire pressure that is too low. This was found out by checking the other tires of a car after an accident. In addition to lack of maintenance, more than 30% of the pressure gauges at the gas stations indicate the pressure too high. After the cost for fuel, tires are in position two on the maintenance cost of a car.

The longitudinal and the lateral dynamic behavior of the car depends on an intact interface between the tires and the road. If this is disturbed, these vehicle dynamics cannot reach their full capacity. Therefore, the demand is made more and more to combine the tire pressure monitoring (already legally required in the United States of America and Europe) with the brake regulating system, to adapt the controller of longitudinal and lateral dynamics to the pressure loss. This adaptation already starts with the simple determination of the real vehicle speed, which like wheel speed measurement is dependent on the wheel circumference. An additional potential function would be the reduction in the maximum speed due to pressure loss.

3.7 Tire pressure adjustment

With the availability of compressors fed by the electrical system (for instance, for air suspension) and the minimal probability of perception of indicator lamps, it is logical that an active tire pressure control will be useful in chassis control. These technologies are already available for military vehicles and for fire engines today.

The road condition identification of the ESP supply the input parameters for air pressure selection in the stages mud, sand, snow, country road, expressway, and escape. Obviously, the driver can also preselect the pressure. By entering a tire model and the automatic wheel pressure detection (e.g., using the air suspension pressure), the correct pressure is automatically selected. The controlling occurs per axle until permanent refilling for acute loss of pressure. This requires a permanent monitoring. The normal cycle of the pressure measurement every 15 min is reduced to 15 s in case of pressure loss detection. The maximum compressor power is used to stabilize the tires and thus the vehicle. If the system cannot balance the pressure, the driver is warned after 2 min.

4 CHASSIS CONTROL BY USING VEHICLE DYNAMICS THEORY

A wide field of chassis control was opened, when the microcomputers became more powerful at the beginning of the 1980s. Until that time, the ABS (antilock devices for brakes) electronics used specific hardware, which limited the flexibility of the control tasks.

With the expanded electronics, it was possible to use the bicycle model and to evaluate the car behavior as a result of speed, vehicle mass, wheelbase, center of gravity position, and above all the cornering stiffness of the front and the rear axle. Figure 13 shows the yaw amplification factor as function of vehicle speed (Wallentowitz, 2007).

Understeering vehicles have a declining characteristic and are defined with the so-called *characteristic velocity*. With the equation in Figure 13, it is possible to evaluate the vehicle characteristic by using the data. In case the yaw rate, the vehicle speed, and the steering wheel angle are measured, the real relationship between these data can be compared to the theoretical ones (evaluated by online simulation of the equation). In case the real relationship is below the characteristic line, the car is more understeering, in case this point is above, the car tends toward oversteering.

Yaw amplification factor:

$$\left(\frac{\dot{\psi}}{\delta}\right)_{\text{stat.}} = \frac{v}{l + \frac{m}{l}\cdot\left(\frac{l_h}{c_{\text{lat,f}}} - \frac{l_f}{c_{\text{lat,r}}}\right)\cdot v^2}$$

$$\left(\frac{\dot{\psi}}{\delta}\right)_{\text{stat.}} = \frac{v}{l + \frac{d\delta}{da_y}\cdot v^2}$$

$$\left(\frac{\dot{\psi}}{\delta}\right)_{\text{stat.}} = \frac{v}{l\cdot\left(1 + \frac{v^2}{v_{\text{char}}{}^2}\right)}$$

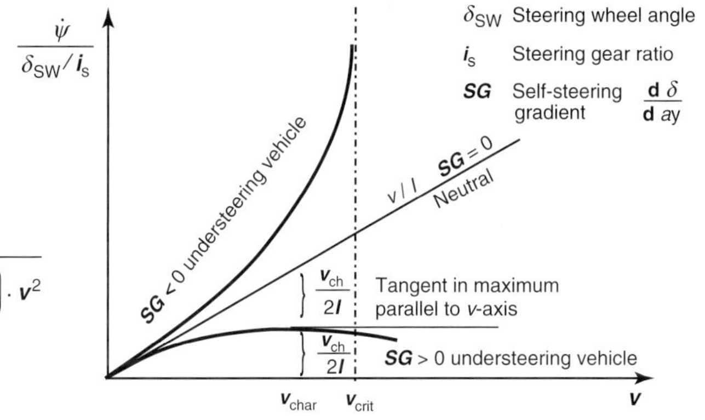

Figure 13. Yaw amplification factor as a function of vehicle speed. (Reproduced by permission of Henning Wallentowitz.)

These theoretical background was used first in the Daimler-Benz All-Wheel-Drive 4Matic—starting in 1986 and in BMW dynamic stability control (DSC) system (dynamic stability program) in 1992 (Debes *et al*., 1997). The next step was the introduction of ESP in the market, which has been done by BOSCH in 1996 (van Zanten, 2000). Other companies followed. There is also the name TRC used.

The comparison of these data derived measures for switching on the all-wheel-drive or controlling the engine. The other systems used at that time only monitored the longitudinal dynamics. Thus, the ASR (traction control system while starting to drive) was realized over a motor interface, to quickly decrease the excessive motor torque at the source. In addition, braking was used to control the spinning wheels, to reach traction with the wheel that is not spinning. The drag torque control was possible due to the active throttle in modern cars.

4.1 Electronic stability program

Since 1996, the ESP is in the market. In addition to the DSC function (Debes *et al*., 1997), it uses brake applications to steer the car in the direction the driver intents to steer to. Under oversteering identification (comp. Figure 13), one front wheel is applied, under understeer, one of the rear wheels is braking, to keep the right driving direction. The values of yaw rate and sideslip angle in the center of gravity are the decisive information. To identify the actual sideslip angle, the measurement of vehicle speed and yaw rate is necessary. The sideslip angle itself is computed during driving.

In addition to the basic functions, electronic brake variator (EBV) (electronic brake force distribution), brake assist system (BAS) (electronic brake assistant), ABS (antilock system), dynamic traction control (DTC) (traction control while driving), ASC (automatic stability control), and DSC,

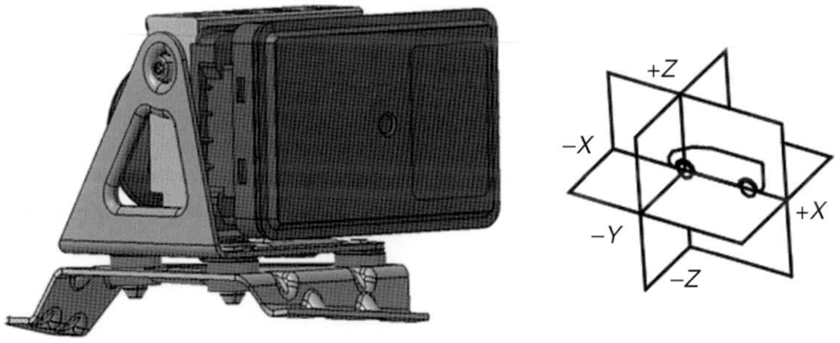

Figure 14. ESP with integrated sensors for yaw rate and longitudinal and lateral acceleration. (Reproduced by permission of TRW.)

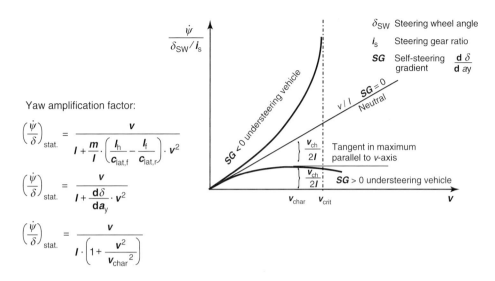

modern ESP systems also have CBC functions (curve brake control—with consideration of the varying wheel loads in the curves and active braking of inner curve wheels), electronically controlled deceleration (ECD) (electronic braking due to other systems such as wheel sensors or electrical parking brake), HDC (driving down mountains), AAS (automatic trailer stabilization), and FLR (the control of driving performance—reduction for brakes that are already overloaded).

Even the brake lining-wear sensors are now evaluated by the ESP, in passenger cars still with one or two steps, in semitrucks already continuously. The originally separate electric parking brake is integrated into the ESP—the control unit must now execute the complete dynamic management (including the brake and warning light control) and the standstill management (including auto-hold) for the vehicle (Figure 14).

Internal temperature models of the brake disks are required for the fading brake support (FBS) (brake fading support) as well as for HTR (tension algorithms of the parking brake at high temperatures).

4.1.1 Networking of ESP and electrical steering

One function of ESP is steering of the vehicle by using the brakes. The driver enters the steering wheel and defines the direction. Single brakes are applied to follow this direction. Now, an automatically generated steering angle of the front wheels can be as effective as the brake intervention, but much more comfortable. And the combination of both actions can be even more effective.

If the vehicle oversteers, the combined system with an earlier steering intervention can possibly avoid the brake intervention. If the vehicle understeers, the steering wheel torque drops and the ESP begins with engine and brake interventions at the rear axle. An active feedback by steering already before interventions from ESP can help the driver avoid the situation.

If the combined system detects braking with different friction coefficients on the right and left lane side, it can automatically take on the necessary steering angle correction, without waiting for the driver. Driving and braking of a vehicle with a flat tire requires more skill to brake than the normal driver has. At the same time, a steering wheel turn is required. If the tire pressure monitor signals which tire is affected, the steering can compensate for the difference torque. By this additional steering input, the stopping procedure remains balanced. This allows quick stopping with high driving stability. This performance will be further improved with an additional networking with the lane assistant (usually realized optical via video cameras). This can realize an automated stabilization of the vehicle in the lane—when networked with the radar system even in the foresighted safe free lane. Fully automatic rescue support will thus be possible in the future.

4.1.2 Networking of ESP and transmission

The introduction of gears, for which not only the distribution of the drive torque between front and rear axle is possible but also the distribution between right and left drive wheel (torque vectoring), opens completely new possibilities for vehicle stabilization. In addition to steering and braking interventions, one-sided drive forces can now also influence the yaw moment of the vehicle. The necessary information signals (torque distribution between front and rear, friction utilization of the different wheels and the wheel speeds) are available from the ESP electronics. Thus, torque vectoring can be a subroutine of ESP.

4.2 EV (electric vehicle) and HEV (hybrid-electric vehicle) and electric brakes

The next level of complexity is achieved with hybrid and electric vehicles (EVs) and electric brakes. At commercial semitrucks, it is typical that in case of drive stability control, all disturbing systems (such as retarders, automatic transmissions, and exhaust brakes in engines) are switched inactive to have only strict brake control of the friction brakes. Now, from the development of the (not yet introduced) electric brakes, it is common knowledge that differences in torque caused by tolerances in spindles, gears, and electric motors are already noticeable with light braking. The car is slightly pulled sideward. Different cable lengths and two different energy reservoirs (X-split of the batteries) reinforce these symptoms as well as different temperature left/right.

If the manufacturing tolerances could be caught up with a final calibration, then there are also influences during driving such as temperatures (one-sided sun, batteries, cable, and motors) and uneven chassis and tire wear. The necessary corrections of offset and tolerances are done by long-term corrections (per ignition run and per brake procedure).

If two drive motors are installed on one axle, the same correction principle can and must also be used for drive torque and generator torque. The necessary prepared signals (wheel speed, yaw rate, lateral acceleration, longitudinal acceleration, steering wheel angle, and steering torque) are already available in the ESP as control unit for the longitudinal and lateral coordination, as well as for the necessary compensation algorithms. It is to be expected that the complete compensation of tolerances in drive and brake systems must be performed in the ESP.

4.3 Wheel hub electric motors

Ferdinand Porsche already used wheel hub motors more than 100 years ago in his Lohner vehicles. However, afterward, the drive concepts were characterized by a centralization of the torque generation. The idea of wheel hub motors did not return until EV and hybrid-electric vehicle (HEV). The unsuspended masses are significantly larger by adding the electric motors to the wheels. This is the same problem as with the increasing combustion engine power: larger wheels were the consequence. The 20-inch wheels mainly serve to accommodate larger and heavier brakes. This decreases the axle natural frequency and the vehicle body structure is more excited, which leads to decrease in comfort. At the same time, the wheel load fluctuations increase—which finally leads to semi-active or active chassis due to the better controllability.

The use of decentralized motors now allows a vehicle stabilization even using the drive. In doing so, targeted one-sided drive torques are created to move the vehicle in the direction the steering wheel angle sensor is indicating.

All technology steps listed above can each only improve individual compromises of the conventional chassis design. Their combination, however, cannot be used functionally independently of each other. Using the steering wheel, drivers specify the driving direction and with the accelerator pedal the vehicle speed is determined via the drive train. The necessary forces in the vehicle center of gravity (the mass is estimated first or derived from the air suspension pressures or wheel deflection suspension) deliver the measured lateral and longitudinal accelerations. From there, the necessary forces are calculated at the four-wheel contact points (all three force vectors). These can now (depending on vehicle equipment) be influenced by various systems. The used one is determined by the energy requirement of each individual function. The system with the least energy requirements is selected (potentially, the vehicle can be steered more simply with the wheel individual motors than with a steering intervention). In addition, those systems with the best efficiency are advantageous. A comfort calculation is superimposed to keep roll, yawing, and vertical movements within limits. As superior hierarchy element, the safety functions control the system behavior. This determines whether the maneuver can be driven this way, or whether a different focal point must be set, such as relations to other traffic participants (avoiding accidents).

5 CENTRAL COMPUTER COMPARED TO DISTRIBUTED INTELLIGENCE

First, electronics for single systems have been installed in the vehicles. The next step was to combine functions. A motor control unit can simply be expanded to the function of a speed control system by reading in additional sensors. The typical "butter soft" shifting of the automatic transmission today already requires a more complex interaction of the control algorithms of the engine and the automatic transmission.

In the start of the 1990s, some engineers tried to put a large central computer into the vehicle. That was also supposed to take on further control tasks, such as transmission control or the ABS.

The main arguments for suggesting this were that distributed intelligences waste a large part of the performance for communication. In addition, it was intended to reduce costs. A large processor is cheaper than two smaller ones. In addition, redundancies are necessary inevitably the central unit only needs one (e.g., hardware: housing, power supply, EEPROM and sensor inputs; and in the software: operating system and diagnostics). While in distributed systems. Theoretically, synergies take effect with the summary of several functionalities.

It was also already tried to implement the idea of the central control unit by summarizing comfort control units from several suppliers. Inconsistencies regarding the resource administration, the product responsibility, the protection of each supplier's knowledge, the cost and coordinate of change management during series production, and the allocation of errors in the warrantee processing brought the specially founded company to a quick end.

The reality went a different direction. In any case, increased functionality means more sensors and actuators, thus, more cables and a higher complexity. If the *one*, central control unit is in a central site, all cables must be routed there. The electric mirror adjustment might be an example: the button is in the left door and the motors are in the left and in the right door. With one central computer, this positioning resulted in many meters of cable to the central control unit and back again. Cables are weight, space, fault sources, and not least costs.

An additional argument against the centralization is the complexity. Every control task demands that the associated sensors are scanned in a fixed pattern and/or processed at very specific times. The sensor requirements of an ABS or an engine controller are significantly different. If sporadic tasks are added, such as controlling an exterior mirror, this places considerable demands on the processor; the corresponding processor performance must be available for the potential concurrency. The controllability of so many layers of requirements is much more complex in one program. The standard PC operating system is certainly an example for such complex programs. The complexity is not controlled so well, that everything is always available at critical moments.

Like with any software, in the chassis systems, there remains a residual error probability of approximately 2500 ppm at the start of series production (aviation is at approximately 250 ppm). This means in a million rows of C-Code, there are 2500 errors that are equally distributed to light, average and critical errors. Half of the errors are caused by incorrect specification, thus, surprising nonfunctionality or failures that were not planned. The probability of errors is approximately 10 times higher than the probability of hardware failures.

To be able to control the entire complexity, to not have to route all cables to one site, to switch off systems in case of error, to compensate for malfunctions of individual systems, every task requires its own control unit. The knowledge of the supplier and its protection lies multiplied in the software. The allocation of legal and financial responsibility is easily possible with single systems.

Finally, it should be noted that the current specifications of the automobile manufacturer already now exceed the capabilities of the microprocessors coming on the market in 2 years. This is true with regard to the memory capacity as well as the loop-time. All demanded algorithms can no longer be processed by the time slot specified by the driving dynamics. The electronic engineers like to compare software to the ideal gas: the software immediately fills all available storage and consumes every available calculation time, without generating noticeable advantages to the end customer.

This makes all safety consideration unanswerable when central computer systems will be in the vehicle. In the future, there will not be any "computer center" in the car.

6 COMMUNICATION OF DISTRIBUTED INTELLIGENCES

To avoid the disadvantage of multiple sensors or actuators with distributed intelligences, data must be exchanged. Let us take the example of cruise control: The engine control unit has obtained the speed signal from the speedometer. The ABS constantly determines the wheel speeds. What could be easier than for the cruise control to use the ABS speed signal? Logically, the speedometer signal should also be generated from the ABS speeds. Doing this generates the advantages for the automotive manufacturer that the transmission pickup can be omitted and the costs could be reduced.

To use data multiple times, a means for transport is required. There should be a communication system that is inexpensive and secure. The vehicle manufacturers and the large suppliers had already recognized the necessity of such a system at an early stage and began doing research in the 1970s.

6.1 CAN bus

In the early 1980s, Bosch began testing the bus systems on hand to see whether they were suitable for vehicles. After none of the existing systems satisfied the requirements of Bosch, they began specifying their own system in 1983. Daimler soon joined and Intel was gained as a hardware supplier. In early 1986, CAN was born.

CAN is a serial multi-master system. Every device connected to the bus determines itself when and how many messages are placed on the bus. The Ethernet network in the office networking is also one such serial multi-master system. However, only one message can be transferred to the bus at a time using the serial system. With the Ethernet, everything is regulated by "trial and error". If two messages collide with each other, both senders must try again using a specified schematic. With a higher busload—many participants try to transmit something at the same time—it can happen that no messages can be transmitted.

CAN is avoiding this with a nondestructive prioritized arbitration. In addition, a series of fault-detection mechanisms are integrated that ensure that transmission faults can be detected securely and generally can be avoided because of repetition of the message.

To simplify the handling, large parts of this standard were realized in hardware. Today, the processors used in the automotive industry are available with up to three integrated CAN interfaces. Every bus participant must be able to transmit its messages at the right time, complete and error-free.

At the start of networking, this was still simple: if the automatic transmission and the engine control unit communicate in 20-ms rhythm, there is never any danger that a message would be lost.

Later, the ABS was added to enable starting assistance together with the engine control unit and to be able to realize ESP functions with additional control units. However, these safety functions are only a small part of the control units in the modern vehicle. Mirror adjustment, window lifter, and electric seat adjuster are raising the need for communication exceedingly. The amount of information is limited by the bus width and the cycle speed of the bus. With a serial bus, the calculation is simple. For 500-kBit/s bus frequency (the standard frequency for high speed CAN) and 20-ms cycle time for the critical signals, the following results:

$$500\ \mathrm{kBit/s} \times 0.02\ \mathrm{s} = 10,000\ \mathrm{Bit}$$

This information rate could be transferred. Those are approximately 76 messages with each 130 bit message length (8 byte user data + overhead)

6.1.1 Structuring

The messages that must be transferred to the bus can be subdivided into important, less important, urgent, and less urgent messages. The messages of safety relevant systems are naturally extremely important, most also urgent. However, non-safety relevant controls can also require a high data transfer speed. The critical systems have an uninterrupted need to communicate—there are closed control loops that expect and send data in fixed time intervals—whose information environment is deterministic. Lost messages are not acceptable.

The vehicle manufacturers first routed two busses in the vehicle, so that critical messages are separated from the less critical messages:

- A high speed bus for the critical messages that are characterized by all messages are routed in cycles on the bus and their busload never exceeds 60% (e.g., drive train bus).
- A low speed bus (comfort bus) that must handle the rest of the communication needs.

To administer the comfort bus, generally, a network management is used to ensure that every participant receives its messages in acceptable times on the bus. Here, the bandwidth of the busses in critical conditions (all simultaneous messages are to be released) is not enough.

The bandwidth is a cost factor. By executing the comfort bus as a low speed bus, the system costs are reduced. Today, in Europe, the CAN is the most widely used bus system in vehicles. However, there were also alternatives, as, for example, VAN in France or in the USA Standard J1850 for Class B Network.

The CAN has a constant cost share with its safety, realized in hardware. If the requirements for cost and safety are lower, the local interconnect network (LIN) bus will be in favor, which halves the costs for a bus node.

6.1.2 Wireless

For networking the vehicle with the environment, for example, for cell phone, personal digital assistant (PDA) connection, and so on, bluetooth (e.g.) is already being used. This type of data transfer is unsuitable for safety systems, because of how it is influenced by electric and magnetic fields.

6.1.3 New requirements

In applications that are currently still covered by CAN, there are new requirements coming up. Driving stability systems that connect active suspension systems with electric steering and with ESP, engine control and adaptive cruise control, to be able to perform active corrective maneuvers, require closed control circuits that cannot be managed by an event-controlled CAN.

The main problem is that the time allocation of the messages cannot be guaranteed due to an accumulation of events or faults. (Highly prioritized messages receive priority and defective messages are repeated.) The variety of participating control units bursts the bandwidth that is available. High transmission rates from up to 10 Mbit/s are required.

The CAN is no longer able to cover the safety requirements set by the X-by-Wire functions (brake-by-wire, steer-by-wire). In some systems, there will be no mechanical fallbacks any more. In the currently used EHB systems (electric-hydraulic brake, the brand name at Daimler is SBC), some brake pressure can still be generated in emergencies, using the pedal. However, it is obvious that the development has reached a limit, which now requires new solutions to expand the technology.

6.2 Flexray

The bus system that is step by step being implemented in the automobile is called *Flexray*. It is a communication system as it has a fixed communication protocol and fixed hardware. It permits a deterministic, collision-free data transfer with guaranteed latency and jitter period with a scalable synchronous and asynchronous data transfer. The latency is the time between an event and the (delayed) reaction following. Jitter is the first derivative of the delay; the back and forth and the cycle frequencies during transfer of digital signals are designated by it (Figure 15).

Minimal latency and jitter times save signal filter times at the recipient. The support of redundant transfer channels, as well as the error tolerance and time-controlled actions implemented in the hardware, permit a quick fault detection and short reaction times.

Instead of arbitration such as with CAN, the Flexray has a fault tolerant synchronous time basis as global time basis of all systems in the vehicle. Owing to the support of Star Technology, partial switch-offs can be realized.

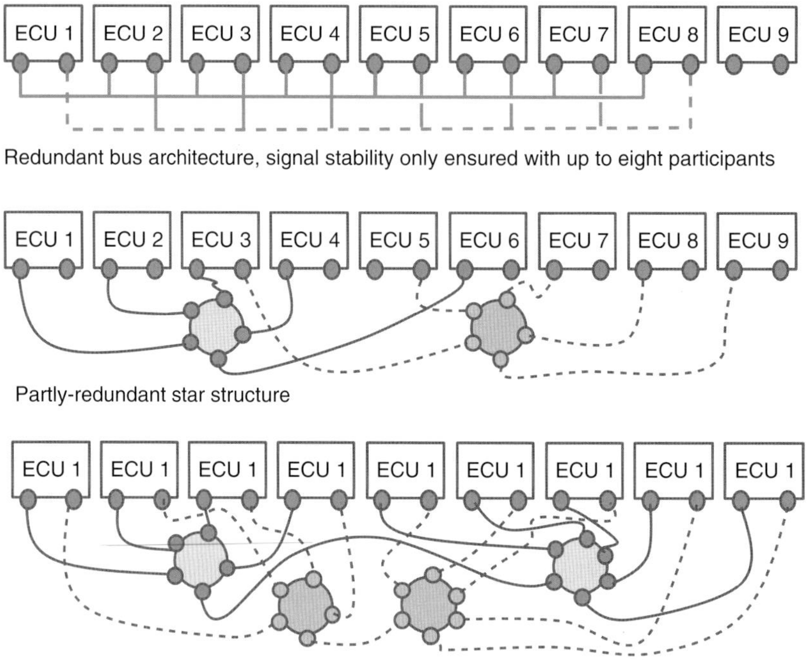

Figure 15. Flexray networking structures. (Reproduced by permission of R. Leiter.)

The Flexray can offer approximately 20 times the bandwidth of the CAN (until 10 Mbit/s) and it permits short cycle times (starting with approximately 1 ms). The consequences are doubling the costs for the bus nodes and the necessity that these bus systems must already be designed during design of the E/E architecture of the vehicle, including all signals and transfer time points. The automobile manufacturer is the only one who can do this. He must rebuild the necessary expertise (including the detailed knowledge of the individual system). Subsequent changes (e.g., adding new bus participants) are extraordinarily costly and time-consuming.

6.3 Gateway

The accumulation of control units forces structuring. As described above, as no bus fulfills all requirements, the bus systems are fit into the vehicle corresponding to the tasks. This means that a vehicle today can easily have five different busses installed; thus, the vehicle has an IT—infrastructure that is not inferior to that of the average production operation.

To make information that is required by many control units available to all systems, for example, the setting of the ignition key or the vehicle speed, that, for example, switches away the television receiver during driving, the information must be transferred to the various bus systems. Gateways take on this task.

6.4 Networking

Without the network, many of the currently available functions, such as vehicle stability control or adaptive cruise control, would not be possible. The new developments build on progressing networking in the vehicle. The comfort for the driver is increased; active safety and driver support systems can be realized with it.

6.5 Hierarchies in the control systems

The limited options of individual control units require cooperation that must be regulated. This will lead to a grouping in function areas—exactly like the supplier and automotive manufacturer are set up in their departments.

Thus, in the lower level of the driver's domain, there are the longitudinal and lateral dynamics, the vertical dynamics, and the drive and energy recovery. In the comfort domain, we have the air-conditioning/heating and the operating forces; and in the communication domain, the geophysical position, the site-related information, the information from other vehicles and local hosts, the driver's intention detection, driver feedback, and information for or about the driver (e.g., recording health status) (Figure 16).

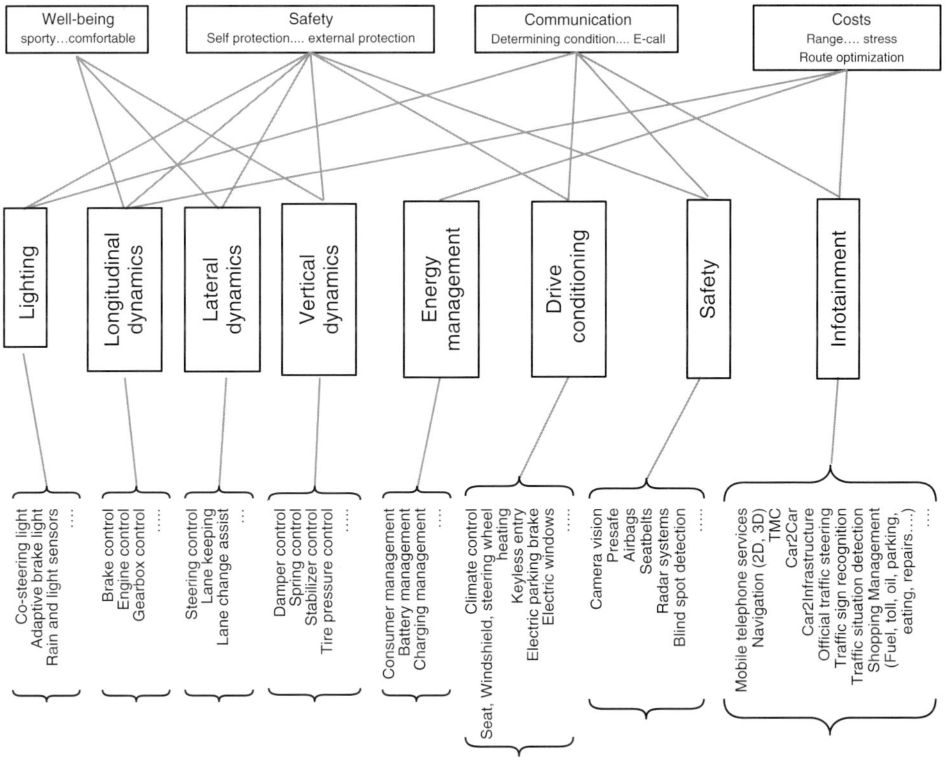

Figure 16. Domain structure. (Reproduced by permission of R. Leiter.)

Each domain has its own controller in which the tasks are joined. In addition, the vehicle manufacturer can place the characterizing master computer that sets the standard specialties specific to the vehicle, and in the case of fault, provides alternative strategies.

6.6 Shortened development times, tight resources

The new technology cannot be introduced without secondary effects. The classic development systems were tested in the vehicles long before the start of series production. Then, high quality and safety standards of the automotive manufacturer were brought into series.

Currently, the manufacturers' desire to innovate is keen and fast. Not just *one* new system is integrated in a new vehicle.

The shortened times to market introduction now mercilessly show their negative sides. All these new systems are specified early on, but naturally due to overall short resources, are only finished at the time when the final integration tests are supposed to be performed. Problems that no one thought of do not appear until then.

An example: the electric parking brake contains a logic that does not operate the brake in dynamic condition when the accelerator pedal is pressed while driving (preventing the driving against the brake). On the CAN, the electric parking brake algorithm only shows the respective brake condition once it has been reached. The cruise control should be abandoned by operating the parking brake (dynamic delay). As the cruise control pays attention to a brake-information, that was not visible at the start of the action, it tries to counteract this force at the start of braking and increases the engine power. The parking brake sees the gas pedal information that the cruise control uses to increase the motor power and it does not execute the command to decelerate. Consequence: the networking function does not work.

The reasons for this problem have been small misunderstandings of the specifications. Since such things do not show up until a time when the actual finished system should run endurance tests, the test times cannot be adhered to, with known and unknown consequences.

6.7 Integration and simulation

The way out from the dilemma is simulation and integration tests on simulation basis. To do this, HIL (hardware in the loop) and SIL (software in the loop) systems are combined. Available control units are connected to vehicle simulation (HIL). In these vehicle simulations, the high level algorithms of the new systems are incorporated as software simulations, so that the overall vehicle can be tested long before the first prototype.

These techniques are available, yet not developed so widely that they could also detect the last problems. Unfortunately, the work never reflects the current development status of the communication participants, but rather the specification that always left interpretation gaps. The human language is not able to sufficiently define the IT: there are communication difficulties between the vehicle dynamics specialists and the "software workers" that are to implement the ideas of the vehicle dynamics specialists.

6.8 Black box and the operating system

With the high degree of networking, it is no longer possible that the automotive manufacturers purchase black boxes from their system suppliers and also shift the responsibility on them.

To be able to successfully integrate all systems, the vehicle manufacturer must have intimate knowledge of the subsystems and provide sufficient engineering resources. The vehicle manufacturer becomes the system-integrator, even if several functions are calculated in one control until. He can only accept the responsibilities connected with this, if he knows the processes in the control unit sufficiently well. The "game rules" must be followed. Accepting the prescriptions of operating systems such as AUTOSAR can do this. Industry started with OSEK, to define software standards. Currently, AUTOSAR 4.0 is used. If the suppliers used to have their own operating systems (and the responsibility for them), today they have to follow the strategy of the automotive manufacturers. The search for errors is very difficult in such complex systems, where due to networking, an overlapping of influence fields happens. Thus, there is also a possibility for faults.

6.9 Risk factor driver

The variety of system functions is supposed to support the driver and increase both his comfort and his safety. However, every automotive manufacturer has its own operating philosophy, also to keep its brand identity.

Most drivers do not take the time to study the operating instructions, and they expect self-explanatory system functions. The same system can function differently with a different manufacturer; even within a fleet of a manufacturer, there could be function differences due to different generations of systems (e.g., autonomous cruise control systems of generation 1 functioned between 30 and 180 km/h, the same technology in generation 2 now works until standstill).

There are many opportunities for faults by the driver, when a foreign vehicle must be controlled.

Once the driver is accustomed to the design of a safety function, he must be able to count on it. If, in a critical driving situation, partial systems fail, that can be critical. The system may overextend most drivers.

7 SUMMARY

The complexity of chassis control in cars is exploding.

Networking is one of the basic technologies of the innovation explosion.

The diagnostic systems and the diagnostics are not keeping up.

REFERENCES

Abe, M., Ogura, M., and Sato, T. (1988) Mechanism for Steering Front and Rear Wheels of Four-Wheel Vehicle. Patent: US 4792007 A.

Amsel, C. (2003) Schutzkonzepte und Designregeln für den Kurzschluss 42 V/14 V im dualen Bordnetz. Dissertation RWTH Aachen, Schriftenreihe Automobiltechnik.

Debes, M., Herb, E., Müller, R., *et al.* (1997) Dynamische Stabilitäts Control *DSC* der Baureihe 7 von *BMW. Automobiltechnische Zeitschrift 99*, **3**, 4.

Donges, E. (1989) Funktion und Sicherheitskonzept der Aktiven Hinterachskinematik von BMW. Tagung "Allradlenkung bei Personenwagen" Haus der Technik, Essen.

Jurr, R., Behnsen, S., Bruns, H., *et al.* (2001) *Das aktive Wankstabilisierungssystem Dynamic Drive*, Springer Verlag, Wiesbaden.

Köhn, P., Wachinger, M., Fleck, R., *et al.* (2003) *Aufbau und Funktion der Aktivlenkung von BMW*, Tagung Fahrwerktechnik des Haus der Technik, München, Juni.

Konik, D., Bartz, R., Bärnthold, F. *et al.* (2000) Dynamic Drive – Das neue aktive Wankstabilisierungssystem der BMW Group; 9. Aachener Kolloquium Fahrzeug und Motorentechnik.

Kubota, Y. (2013) Nissan to install electronic "steer-by-wire" in Infiniti cars, http://www.reuters.com/article/2012/10/17/us-nissan-technology-idUSBRE89G03A20121017 (accessed 4 September 2013)

Kuroki, J. and Irie, N. (1991) HICAS: Nissan Vierradlenkungsechnologie, Fortschritte der Fahrzeugtechnik, Band Nr. 7, in *Allradlenksysteme bei Personenwagen* (ed. H. Wallentowitz) (Hrsg), Vieweg Verlag, Wiesbaden.

Lloyd, J.M. (2010) Cross-linked variable piston air suspension. US Patent: US 2010/0230912 A1; September 6.

Wallentowitz, H. (2007) Lecture Vehicle Dynamics 2. "Vertical and lateral dynamics of vehicles" Aachen.

Wallentowitz, H. and Amsel, C. (eds) (2003) *42 V – Power Nets*, Springer.

Wallentowitz, H., Donges, E., and Wimberger, J. (1994) Die Aktive Hinterachskinematik (AHK) des BMW 850 Ci, 850 Csi. *Automobiltechnische Zeitschrift*, Springer Verlag, Wiesbaden, **96**(11), 674–689.

Wiesinger, J. : Active Body Control (ABC), http://www.kfztech.de/kfztechnik/fahrwerk/federung/abc_aktive_body_control.htm; (accessed 14 April 2013)

van Zanten, A. (2000) Bosch ESP systems: 5 years of experience. SAE Technical Paper 2000-01-1633.

Chapter 132

Torque Vectoring for Drivetrain Systems

Claus Granzow and Matthias Arzner
ZF Friedrichshafen AG, Friedrichshafen, Germany

1 TORQUE VECTORING

The term *torque vectoring* describes the active generation of a vehicle yaw moment by a directed distribution of input torques over the left and right sides of the vehicle. Torque vectoring can be used to influence the degree of yaw of the vehicle and in this way to actively control the driving dynamics (active yaw control).

Encyclopedia of Automotive Engineering, print © 2015 John Wiley & Sons, Ltd.
Edited by David Crolla, David E. Foster, Toshio Kobayashi, and Nick Vaughan.
This article is © 2015 John Wiley & Sons, Ltd. ISBN: 978-0-470-97402-5
Also published in the *Encyclopedia of Automotive Engineering* (online edition)
DOI: 10.1002/9781118354179.auto029

2 TORQUE DISTRIBUTION IN THE DRIVELINE

To avoid stress on the driveline and tire wear during cornering, a transmission is needed that provides free speed compensation. These compensating or differential transmissions create a fixed torque distribution over the two outputs, while making different rotational speeds possible.

Relating to one axle, that is, in the transverse vehicle direction, the differential distributes the torque 50%/50% on the two wheel sides. In the longitudinal vehicle direction of a multi-axle-driven vehicle, differentials are also used for the necessary speed compensation. Not only equal distribution, but also asymmetric torque distributions are possible here (e.g., 33%/67%, 38%/62%). They are selected depending on the traction potential of the axles and the desired handling behavior.

Although differentials are unavoidable for passenger cars, a permanently fixed torque distribution has disadvantages in many driving situations. Widely different road-tire friction coefficients of the wheels on one or different axles can cause individual wheels to build up high slip that limits the torque transfer capability. The wheel with the lowest friction coefficient determines the total transferable torque of the driveline. This means that free speed compensation impairs the traction behavior on different road surfaces. For instance, if one wheel is on ice, it builds up a high level of tire slip, whereas all the other wheels on asphalt cannot transfer any more torque than the wheel rotating on the ice. This gives the vehicle a poor start-up performance.

This can be remedied by limited slip differentials (LSDs) that can prevent the compensating effect in critical driving situations. Principally, a distinction can be made between three types of LSDs:

- Shiftable black–white differential locks that can lock the speed compensation by 100% (automatically or manually).
- Self-locking differentials either in torque-sensing or in speed-sensing design. The first type locks the speed compensation as a function of the torque to be transmitted, whereas the locking rate of the second type represents a function of the differential speed.
- LSD that can be activated externally. These systems are usually electronically controlled and feature a hydraulic or electromechanical actuator. Usually, the locking rate is continuously variable between 0% and 100% and is determined by a strategy depending on the driving situation.

2.1 Definition of locking value

The locking value S is a characteristic value that defines the degree to which the compensating effect of the differential is prevented. The definition is as follows:

$$S = \frac{\text{Locking torque } T_B}{\text{Input torque } T} = \frac{T_{\text{right}} - T_{\text{left}}}{T_{\text{right}} + T_{\text{left}}} \quad (1)$$

The possible value range is between 0% and 100%. An ideal open differential has a locking value of $S = 0\%$, whereas a fully blocked differential has a locking value of $S = 100\%$. Normal open differentials have locking values of approximately $S = 5–10\%$. Torque-sensing limited slip differentials have a constant locking value.

Limited slip differentials can be used mainly to increase traction by compensating any slippage that occurs. The drive torque is largely transmitted to the vehicle side or vehicle axle that still has potential for transmission. The originally fixed torque distribution between the differential outputs (e.g., 50%/50%) is superimposed by the differential locking torque. As a basic principle, a limited slip differential only enables torque transmission from the side that is rotating faster to the side that is rotating more slowly. When a vehicle drives round a curve with an even friction coefficient, actuating the limited slip differential tends to lead to understeering. The limited slip differential can be used to stabilize the driving dynamics.

Torque-vectoring systems expand the function of controllable limited slip differentials. The directed asymmetric torque distribution over one axle can be used to apply a yaw moment around the vehicle vertical axis. Beside the differential locking effect, a torque-vectoring system must also be capable of transmitting torques to the shaft that is rotating faster and therefore increasing the difference in speed of rotation between the left and right vehicle sides. During cornering, it is possible to affect an inward turning; this means a more agile vehicle handling.

2.2 Definition of vectoring torque

The difference between the two-wheel torques (in the direction of travel) is called the *vectoring torque of an axle*:

$$T_{\text{TV}} = T_{\text{right}} - T_{\text{left}} \quad (2)$$

Positive vectoring torques lead to a counterclockwise yaw moment of the vehicle around the vertical axis. The maximum vectoring torque depends on the vehicle moment of inertia around the vehicle vertical axis as well as the required operating range of the torque-vectoring system. Typically, affecting traction at low speeds requires higher values of vectoring torques than affecting driving dynamics.

2.3 Hang-on clutches and torque splitters

As an alternative to differentials with superimposition units (limited slip differentials or torque-vectoring units), there is also the option of achieving speed compensation using a slipping multidisk clutch. Here, there is no differential, because the required torque is completely transmitted via a multidisk clutch operated during slipping. A basic distinction is made between clutches for torque distribution in longitudinal and transverse configurations.

Clutches that distribute the torque between different axles replace an interaxle differential and are called *hang-on clutches*. A permanent all-wheel-drive system with hang-on clutch requires a specially designed clutch as well as a suitable actuating system and operating strategy. The torque flow to the hang-on or secondary axle can be freely selected within the range of the speed conditions according to aspects of traction and driving dynamics. The strategy must also achieve the speed compensation in case of cornering.

The differential relating to one axle as speed compensation in the transverse direction can be functionally replaced by a twin clutch system. Here, the two wheel drive torques are set by the respective clutches and used individually for influencing the traction and driving dynamics. Twin clutch systems are also known as *torque splitters* because they are only able to individually distribute the input torque to the two drives. In contrast, torque-vectoring systems with the above-mentioned superimposition units can generate wheel differential torques and therefore yaw moments independently of loads.

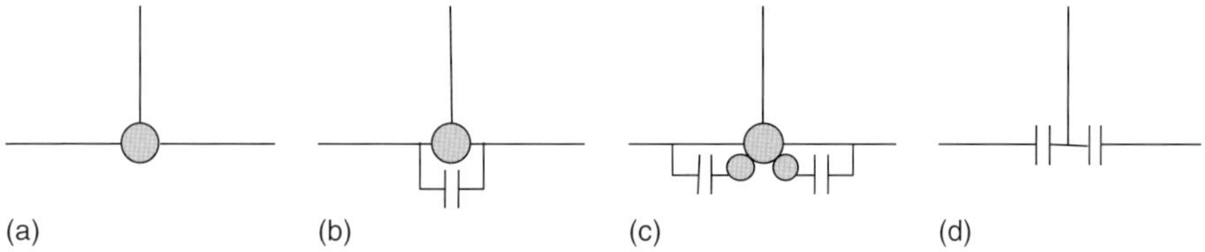

Figure 1. Diagram of the basic elements of transverse torque splitting. (a) Open differential with fixed torque distribution (50%:50%), full speed compensation. (b) Differential with lock and basic distribution (50%:50%), active compensation of differential speeds, torque transfer from faster to slower sides, mainly used to increase traction. (c) Differential with superimposition unit and basic distribution (50%:50%), additional torque can be superimposed, torque transfer to faster side also possible, active yaw rate control possible. (d) Twin clutch system without basic distribution, individual distribution of the input torque possible, no generation of differential torques in coast situation, also covers active longitudinal distribution (hang-on system).

2.4 Speed error

Another essential design criterion apart from the maximum vectoring torque is the speed error. The speed error is the maximum speed difference of the two output shafts in percentage until which torque transmission to the faster rotating output shaft is possible. At the vehicle level, this is the minimum curve radius at which agility-increasing vectoring activities are still possible. While stabilizing or traction-increasing interventions are possible under any speed conditions, agility-increasing interventions are not possible below a minimum curve radius.

Very large speed errors collectively cause increased friction power on the clutches. That is why, it is important when determining the speed error to weigh up the smallest curve radius that is still relevant for torque-vectoring interventions and what collective friction power can be collectively covered by the clutches.

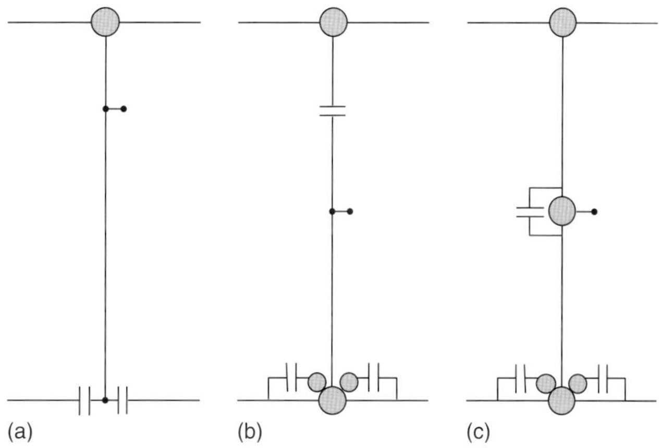

Figure 2. Diagram of driveline examples with active transverse torque distribution. (a) Twin clutch on rear axle, front axle as primary axle (e.g., Honda SH-AWC). (b) Rear axle torque vectoring and hang-on front axle (e.g., BMW X6). (c) Rear axle torque vectoring and Torsen center differential (e.g., Audi Sportdifferential).

2.5 Torque distribution strategies/designs

Figure 1 shows the basic elements of transverse torque distribution. Together with the various options for longitudinal torque distribution, this can be used to arrange various drivelines. Figure 2 shows some driveline configurations available on the market.

3 INFLUENCING VEHICLE HANDLING WITH TORQUE VECTORING

The free distribution of torques between the four vehicle wheels can be used to actively influence the vehicle handling. The way this works is comparable to the steering principle of a tracked vehicle. When the driver of a tracked vehicle wishes to change the travel direction, different track speeds are applied to the inner and outer tracks. The different speeds cause the vehicle to turn around its vertical axis.

Torque-vectoring systems in a passenger car driveline influence the direction of travel in a similar way. Depending on the system characteristics, this can occur depending on or independently of the current input torque of the vehicle. Simple torque splitters can only influence the ride direction when there is input torque. To utilize the effect, the driver must consciously select trailing throttle or acceleration mode.

Torque-vectoring systems with superimposition function can generate the required differential torque without limitation by the input torque. They offer a much wider scope for influencing the driving dynamics. As they work independently of the input torque, they can also be seen as components purely for influencing vehicle dynamics, comparable with steering systems.

3.1 Yaw moment by torque vectoring

The vectoring torque, also called *wheel difference torque*, occurs due to different longitudinal forces on the two wheels of a vehicle axle. In the simplest case, these are two longitudinal wheel forces with identical magnitude and preceding plus or minus signs. However, in most cases, the differential torque is superimposed by symmetrical drive or braking forces.

According to the track width of the vehicle, the wheel difference force leads to a free torque around the vehicle's vertical axis. This torque influences the yawing motion of the vehicle and can be used to change the direction of travel.

The complex friction conditions on the vehicle tires can influence the effect of torque vectoring especially during cornering. In respect to the traction limit given by the circle of forces, the lateral wheel force can be weakened by superimposed longitudinal wheel force. Superimposed vectoring torque reduces the amount of lateral wheel force in the same way as additional drive or braking torque. In any case, the additional longitudinal force causes the amount of lateral wheel force and generates an additional influence on the vehicles yaw behavior.

Depending on the place of installation of the torque-vectoring device at front or rear axle and the sign of the vectoring torque, this effect can reinforce or weaken the effect on the vehicle yaw moment.

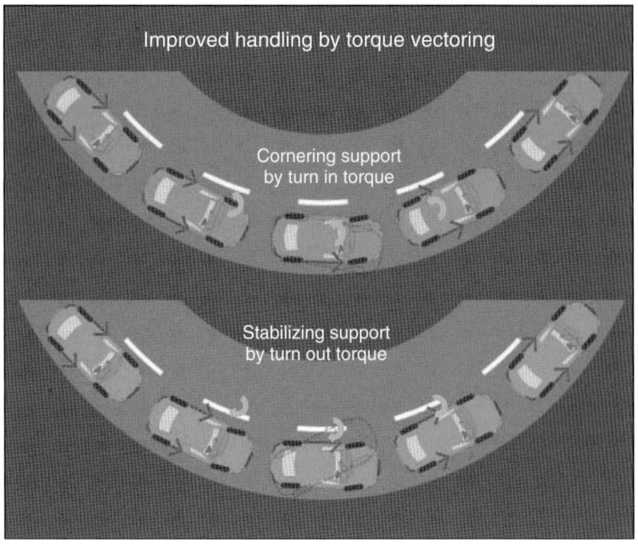

Figure 3. Influencing the driving dynamics with the torque-vectoring function. Agility-increasing and stabilizing yaw moment.

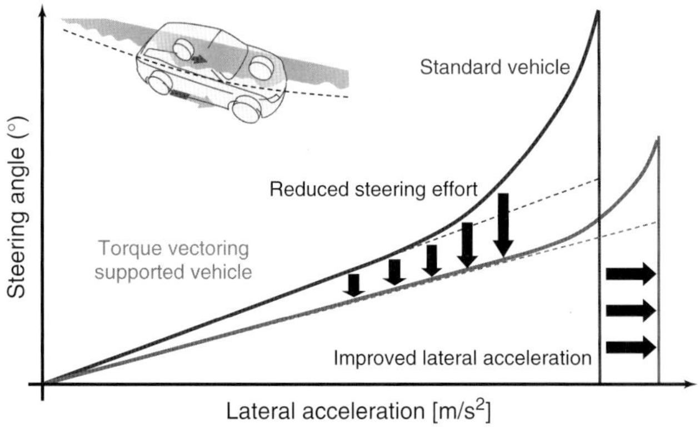

Figure 4. Altering the self-steering response by superimposing the torque-vectoring function.

3.2 Inward- and outward-turning differential torque

The yawing motion of the vehicle around its vertical axis can have a stabilizing effect or can increase the dynamics. In principle, a difference is made between an inward-turning and an outward-turning torque.

Inward-turning torque supports vehicle cornering by reinforcing the yawing motion around the vertical axis. The inward-turning torque reduces the curve radius. The outward-turning torque slows the vehicle yawing motion and increases the curve radius. This is why the outward-turning torque is used to stabilize the vehicle. Figure 3 describes the relation between direction of yaw torque and vehicle behavior during cornering.

3.3 Influencing self-steering using torque vectoring

For reasons of safety and controllability, modern vehicles built to today's standards are designed to have a general understeering behavior at the driving limits. This means that with increasing lateral acceleration, the steering angle needed to further reduce the curve radius increases disproportionately. The maximum possible cornering forces of the front tires limit the minimum curve radius. The vehicle stability limit is indicated to the driver by a stable vehicle with slipping front tires. The relation in terms of driving dynamics between the steering angle and lateral acceleration is also called the *vehicle self-steering effect* (Figure 4).

This understeering tendency of the vehicle is intensified when a braking or input torque is additionally applied to the steered axle. Especially in the case of front-wheel or all-wheel-drive vehicles, this is called *load understeering*.

In combination with a suitable, dynamic steering system, a torque-vectoring system offers the option of freely influencing the self-steering effect. In most cases, it is used to reduce or compensate for the understeering tendency described. If, due to overloading or high cornering speed, the front axle reaches the limits of its cornering force, the yawing motion of the vehicle can be increased by the application of a free yaw moment around the vertical axis. The vehicle continues to follow the driver's steering command and behaves in a sporty and neutral way.

In particular, drivers often find the responsive reaction of the vehicle to their steering commands and the very late or even complete absence of understeering a very positive characteristic.

In general, the torque-vectoring unit is actuated by means of a driving dynamics control. In a similar way to a brake-based stability system (ESC), this monitors the driving condition of the vehicle by comparing a measured vehicle yawing rate with an ideal yawing rate calculated from the steering angle and the vehicle speed. The great advantage of torque-vectoring-based regulation of driving dynamics is the system's continuous mode of operation. It corrects the direction of travel gently and smoothly without any speed-reducing braking interventions.

3.4 Driving dynamics limits of torque-vectoring systems

As with any driven wheel, the torque transmission capability is limited by the friction coefficient and the vertical tire force. The vertical tire force is particularly important because it can change dramatically during cornering.

Constant cornering produces a positive wheel load distribution on the two wheels on the outside of the curve. The force potential gained here can be used to supply the outer wheels with additional input torque. In this way, the wheel load distribution toward the outside of the curve supports the generation of an inward-turning yaw moment.

In the opposite case, the load is reduced on the two inner wheels of the vehicle during high lateral acceleration. The accompanying reduction in wheel force potential limits the force buildup of a stabilizing yaw moment. Therefore, a torque-vectoring system is particularly suited to creating an inward-turning yaw moment.

4 MECHANICAL CONCEPTS OF MODERN TORQUE-VECTORING AXLE DRIVES

Generally, a torque-vectoring system is understood as an axle drive that can generate a positive or negative wheel differential torque irrespective of the drive situation.

The basic power distribution is always provided by a differential compensating transmission. The differential decouples the two wheel speeds and distributes the input torque between both wheels, ideally 50% per wheel. The function can be performed by a conventional bevel gear differential. For reasons of space saving and effectiveness, a spur gear differential is used in some applications.

The torque-vectoring function itself is achieved with the help of so-called *superimposition units*. They make it possible to apply alternative power flows inside the transmission and enable controlled power distribution between the two wheels.

In general, torque shifting is required bidirectionally between both wheels of an axle (inward- and outward-turning redistribution), so two separate superimposition units are installed in a transmission. Here, each unit takes care of one redistribution direction. Depending on the transmission concept, the units can be installed symmetrically on the left and right of the basic axle transmission (see ZF and Magna concept as described in Sections 5 and 7), or arranged nested on one side of the axle drive (Mitsubishi).

A single superimposition unit always consists of a friction clutch and a transmission stage. While the friction clutch provides the actuation moment necessary for torque distribution, the transmission stage ensures the speed error required for initiating and intensifying the torque-vectoring effect.

5 ZF VECTOR DRIVE© CONCEPT

The ZF Vector Drive© axle drive consists of a conventional bevel gear differential and two symmetrically arranged superimposition units. The superimposition unit consists of a planetary gearset, which itself consists of two sun gears, one planet carrier, and three planetary gears. The inner sun gear is connected to the differential cage, whereas the outer sun gear is connected to the relevant side shaft. A multidisk brake acting between the planet carrier and the transmission housing acts as a torque modulation element (Figures 5 and 6).

With the help of a roughly 11% speed error in the planetary gearset, the application of a braking torque achieves a power flow from the axle drive via the planetary gearset to the individual wheel.

The ZF Vector Drive© transmission was used for the first time in volume production in 2008 in the BMW X6.

6 THE MITSUBISHI CONCEPT

The classic, asymmetric torque-vectoring transmission design is called the *Mitsubishi concept*. In the current version, it features a planetary-designed axle drive. The

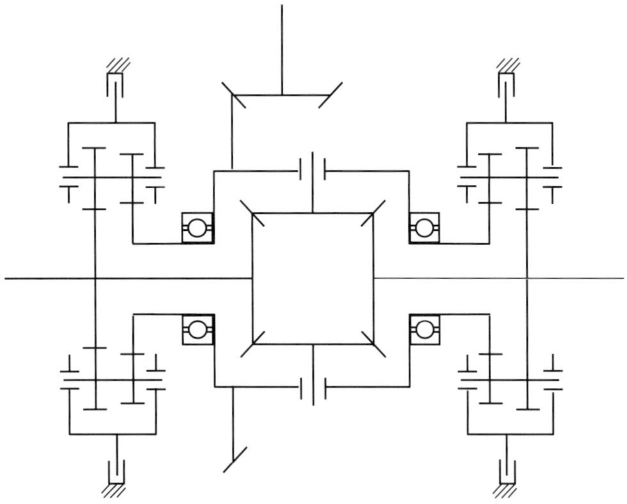

Figure 5. Schematic diagram of the ZF Vector Drive© system.

Figure 6. Sectional drawing of the ZF Vector Drive© system.

advantages of this design are the low self-locking value, the narrow construction shape, and the direct action on the opposite output shaft.

Added to the axle drive is a nested superimposition unit installed on the right side. The input and output of the superimposition unit are each connected to the two side shafts of the axle drive. The superimposition unit itself consists of two multidisk clutches and two transmission stages allocated to the individual clutches (Figures 7 and 8).

The difference is that the transmission stage on the inner clutch generates increased slippage compared to the wheel differential speed, and the transmission stage on the outer clutch provides a reduced clutch slippage compared to the wheel differential speed. Depending on which clutch is

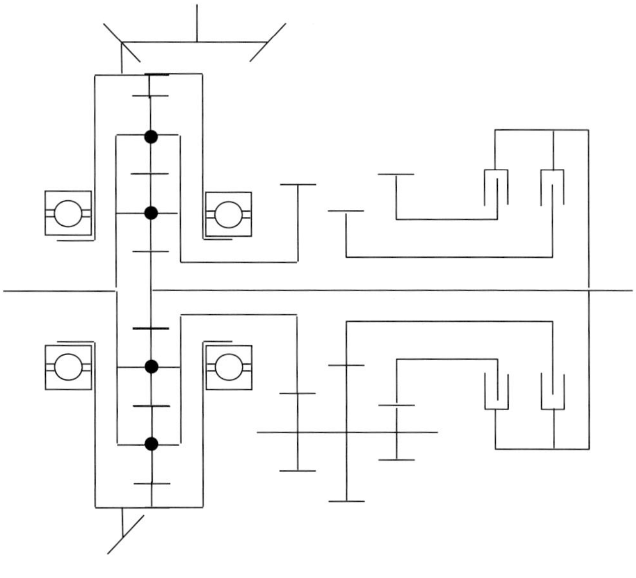

Figure 7. Schematic diagram of the Mitsubishi torque-vectoring system.

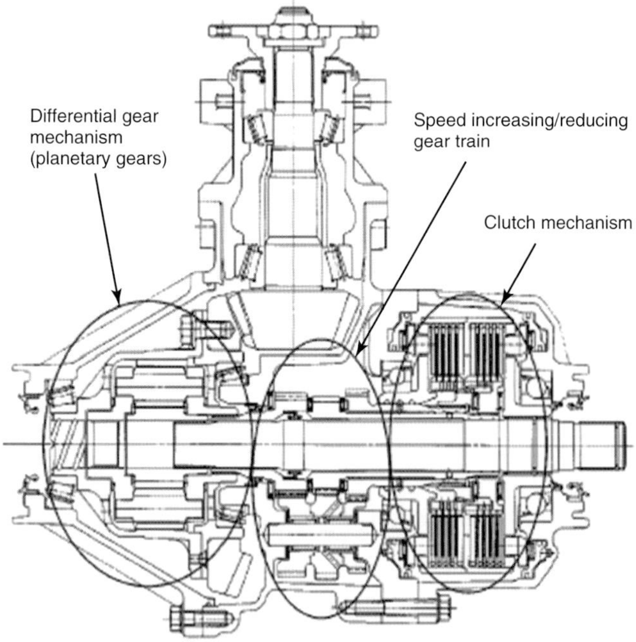

Figure 8. Sectional drawing of the Mitsubishi torque-vectoring system. (Photo: Mitsubishi.)

actuated, the different differential speeds of the two clutches can create a specific power flow directly between the two rear wheels.

The first time a torque-vectoring transmission was used in volume production was by Mitsubishi in 1996 in its Lancer Evolution 4 model. That makes Mitsubishi along with Honda one of the pioneers in the application of this technology. The planetary basic differential design presented here has been in use in the Lancer model since the Lancer Evolution 8 generation.

7 THE MAGNA CONCEPT

The Magna concept also has a symmetrical design, consisting of a bevel-gear-based drive unit and two individual superimposition units attached outboard. The superimposition units each feature two individual, single-stage transmissions connected by a multiplate clutch. Each of these transmission stages consists of a sun gear and a ring gear, which has a radial displacement in respect to the sun gear. The sun gear of the inner transmission stage is connected to the differential cage, and the sun gear of the outer transmission stage is connected to the output shaft. To generate the required speed error within the superimposition unit, the ratios of the two stages differ slightly by about 10%.

Friction clutches are used to control the torque distribution. The two wet-running clutches are each installed between the two ring gears of each superimposition unit and also run with a radial offset to the drive units' output axis (Figures 9 and 10).

When the clutch is actuated, the positive speed error creates a power flow from the axle drive to the wheel via the first transmission stage, the clutch package, and the second transmission stage.

The Magna torque-vectoring transmission was introduced for the first time in 2009 in the Audi S4 under the name "Sportdifferential", and then made available as an optional extra in further Audi models.

8 TORQUE SPLITTER AND LIMITED SLIP DIFFERENTIALS

The two types torque splitter and limited slip differential are special forms of torque-vectoring axle drives. They also enable splitting of the input torque between the wheels, although with the above-described limitations (Figure 11).

The torque splitter does not use a conventional differential-based axle drive. The power or torque splitting is achieved directly by two control elements that control individual wheels. The control elements usually consist of friction clutches that are connected either directly or with the help of planetary gearsets. In both cases, a sophisticated ratio selection must ensure that both clutches have a differential speed in order to generate a positive power flow to the individual wheel.

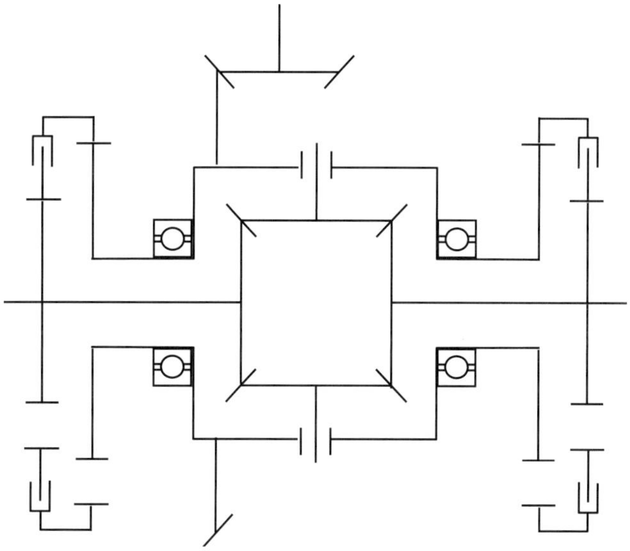

Figure 9. Schematic diagram of the Magna torque-vectoring system. (Photo: Mitsubishi.)

Usually, an electronically controlled limited slip differential consists of a bevel gear differential and an additional multidisk clutch. The clutch is installed between the differential cage and one of the side shafts. It enables a direct flow of power between the differential cage and the side shaft.

9 ACTUATING SYSTEMS

A further distinction between different torque-vectoring transmissions is the selected concept of actuating system. The whole range of mechatronic designs can be used to actuate the multidisk clutches installed. Apart from classic hydraulic actuation, electromechanical and electromagnetic actuations are also used for products in volume production.

Each of these technologies comes with advantages and disadvantages. Electrohydraulic actuation features a high power density and enables central energy generation for both clutches. Compared to separate actuators for each clutch hydraulic actuation can have favorable effects in terms of installation space and weight, but have to be weighed up against disadvantages in efficiency and dynamics at low temperatures.

Electromechanical actuation is attractive due to its integrated position sensing with very precise controllability and high actuating dynamics. Another advantage is the temperature resistance especially under cold conditions. Viscosity

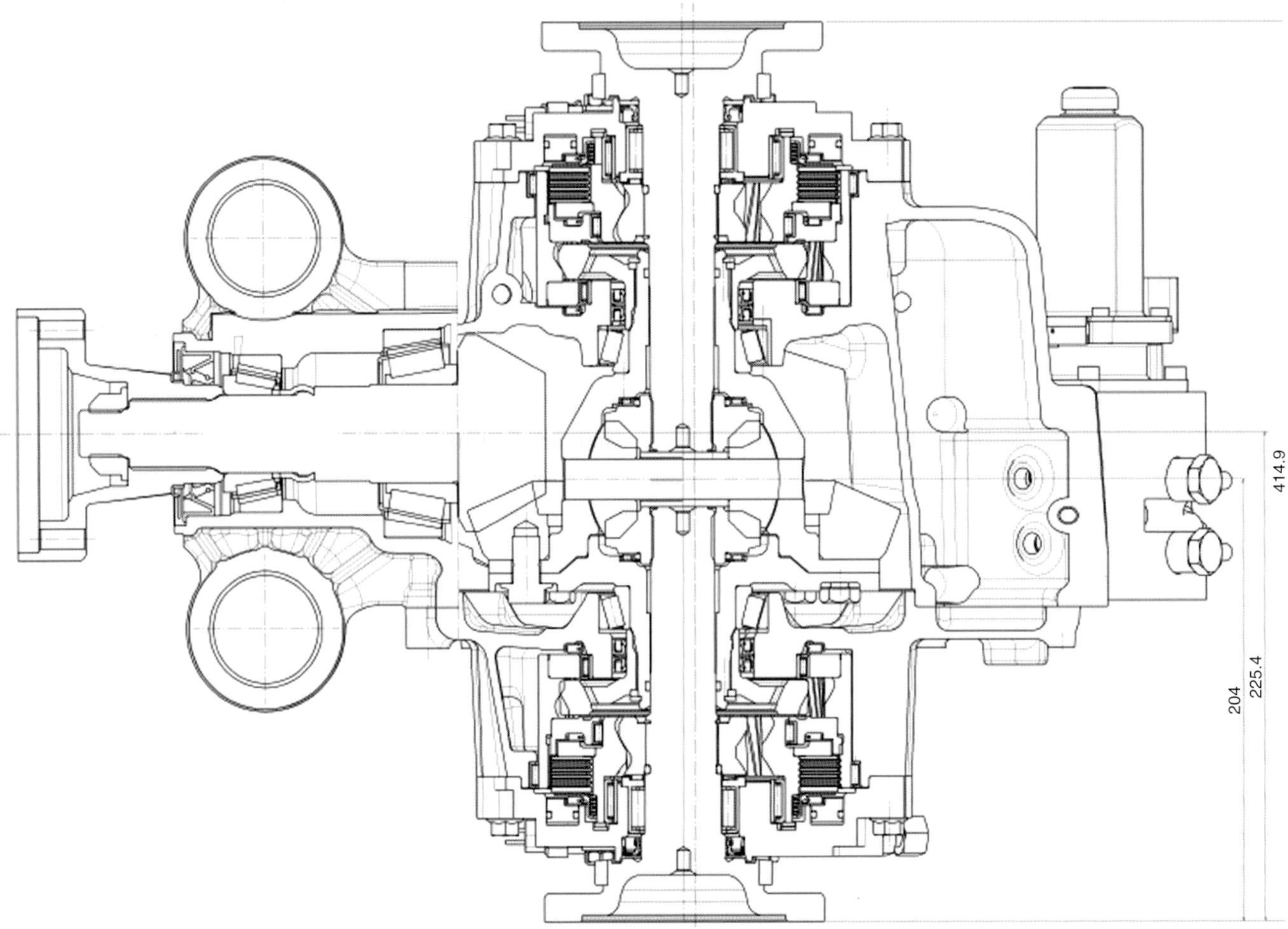

Figure 10. Sectional drawing of the Magna torque-vectoring system. (Reproduced from Sackl, W. and Sankar, M. (2006) 'Simulation and Definition of an Active Yaw Control Device', presented at *7th All Wheel Drive Congress*, Graz, Austria.)

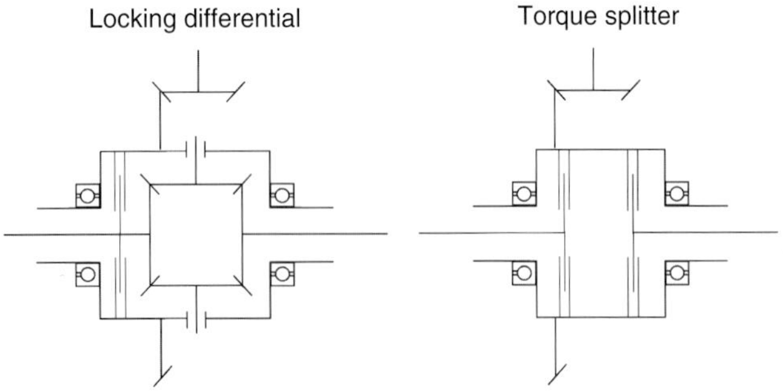

Figure 11. Schematic diagram of a limited slip differential and of a torque splitter. (Reproduced from Sackl and Sankar, 2006 © W. Sackl and M. Sankar)

effects only have minor influence on the actuation function. Compromises must be made in the package and weight because the actuator mechanics cannot be installed freely on the drive unit.

Both electrohydraulic and electromechanical actuation use electric motors to generate the control power. When high dynamic handling is required, the motors provide for short-term higher energy takeoff from the vehicle main power system.

The advantage of electromagnetic systems is their better package and high dynamics. The concentric arrangement of the actuator coils around the side output shafts allows for compact, space-saving installation. The disadvantages of direct magnet actuation are higher weight and lower energy density. To generate high clutch torques, large and heavy coils must be supplied with high currents. These disadvantages can be avoided by using electromagnets to actuate only a pre-control clutch. The pre-control clutch turns a ball-ramp mechanism, using the rotary motion from the driveline. Clutches based on this principle are a common and very cost-effective solution for many all-wheel applications.

10 FURTHER DETAILS OF A TORQUE-VECTORING SYSTEM USING THE EXAMPLE OF ZF VECTOR DRIVE©

10.1 Torque Bias Effect

The direct influence of the self-locking coefficient on the control accuracy of a modern torque-vectoring drive unit is called *torque bias effect*. This effect occurs when the drive unit is included in the power flow for generating the differential torque (see ZF and Magna concept as described in Sections 5 and 7).

Depending on the type of load (trailing throttle, acceleration, or coast) and the required vectoring torque, a residual torque occurs within the drive unit. This residual torque relates to the differential itself and is not transferred by the superimposition unit.

Owing to the locking value S of an open bevel gear differential, the drive unit generates an outward-turning differential torque. The sum of the outward-turning differential torque and the vectoring torque by the superimposition unit determines the overall torque split between the two output shafts.

Depending on the value of the self-locking coefficient, the differential torque in the (very important) case of inward turning is weakened, and in the (less important) case of outward turning, it is strengthened. A self-locking coefficient that is as low as possible and remains constant over the entire lifetime is key to the constant and efficient functioning of a torque-vectoring system.

Figure 12 shows the torque flow of the ZF Vector Drive© system during accelerated cornering in connection with an inward-turning differential torque.

10.2 Actuation

Active control of self-steering properties and of the driving-dynamics-related vehicle response through highly dynamic transversal distribution of wheel torques requires an actuator that can perform very quick and precise actuations. Electromechanical actuating systems are compact and very efficient, with low power consumption. The torque of the E-motor is boosted by a spur gearset before it is transformed into an axial force via a ball ramp. The axial force is used to pressurize the multiplate disc and thus for the transfer of the specifically desired torque (Figure 13). The geometry of the ball ramp can be designed in such a manner that ensures the fastest-possible free travel with optimal resolution and controllability of the operating range. The adjustment time from 0% to 90% of the maximum possible differential torque of 1800 Nm is just 80–100 ms. Figure 14 gives an example of control accuracy measurement with a step-type vectoring- torque request.

10.2.1 Electric motor

In addition to the requirements in terms of dynamics and precision, the E-motor must fulfill the requirements relating to safety and the environmental impact. The safety concept for a de-energized (failure) condition requires that the torque-vectoring moment is immediately reduced and that the transmission behaves like an ordinary rear drive unit

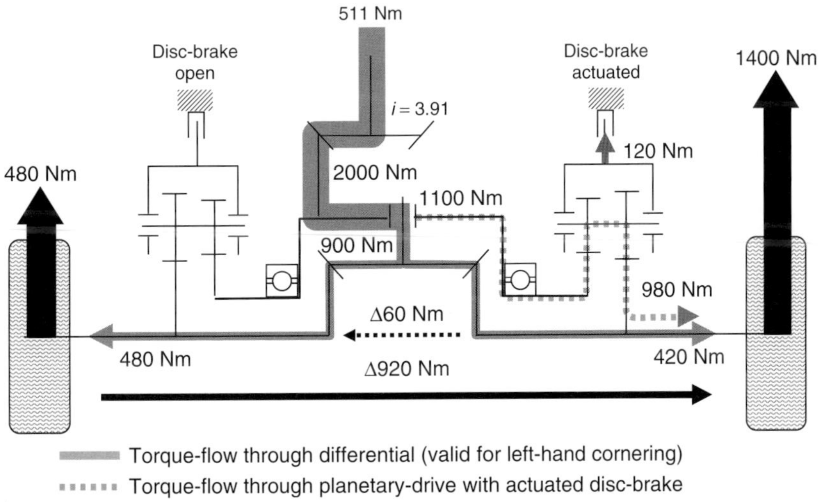

Figure 12. Torque flow and torque bias effect in the ZF Vector Drive © system.

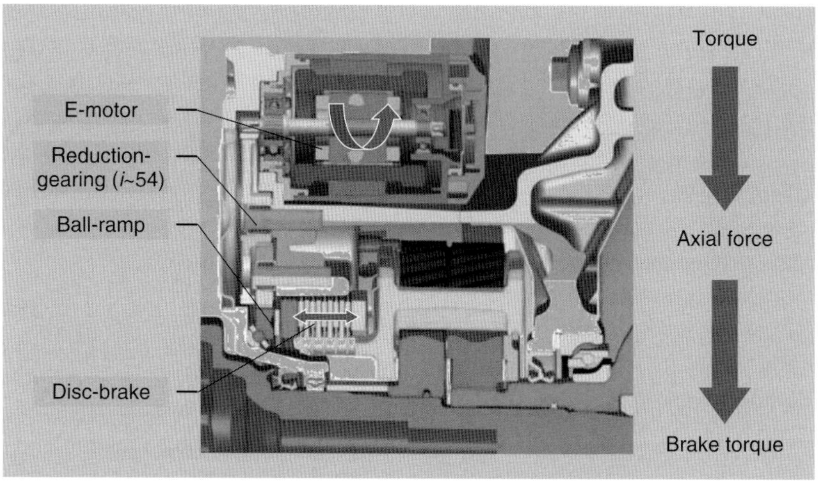

Figure 13. Actuation concept of the ZF Vector Drive © system.

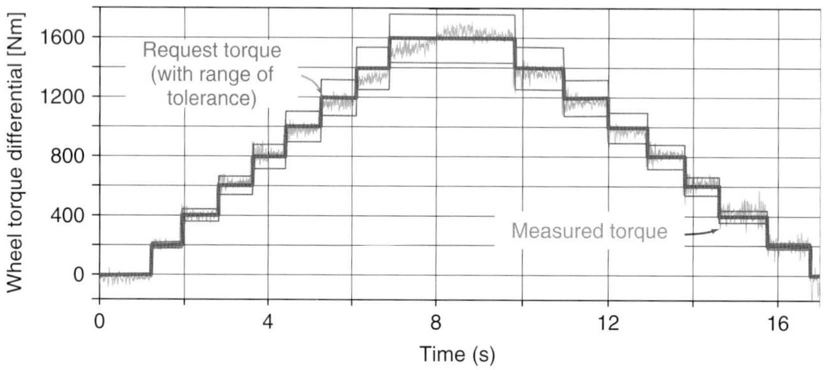

Figure 14. Control accuracy sequence of the ZF Vector Drive © system.

with open differential (fail-safe). As regards the E-motor, this means that the automatic opening of the multiplate disc must not be hindered by a locking moment. Use of the E-motor in the area of the differential implies extraordinary strains in terms of leak tightness, vibration, and temperature.

All these requirements are optimally fulfilled by asynchronous E-motors. These E-motors do not need a brush, nor do they need permanent magnets. They are thus free from any locking moment and ideally support de-energized opening in the event of a failure. Moreover, they are wear-free and provide a high level of ruggedness. Operation is possible with the E-motor open toward the drive unit, which permits free oil flow between motor and space of drive unit. The oil of the drive unit thus also lubricates the bearings of the rotor shaft, and separate venting of the E-motor is not necessary any more.

10.3 Electronics and safety concept

A dual-controller unit with integrated power electronics controls the drive. The vectoring torque calculated by the driving dynamics function is processed further in respect to present operating conditions. This means the controlled electric motors apply the multidisk brakes to generate an accurate vectoring torque.

Angle-of-rotation sensors sense the motor position and permit precise and prompt actuation of the multidisk brake. The software includes compensation functions for temperature, speed, and aging effects.

Owing to the intervention in the driving dynamics and stability of the vehicle, the safety concept was classified according to SIL 3, which corresponds with an error response equivalent to that of an active steering angle intervention. The safety concept implemented in the control unit constantly monitors all units and ensures an immediate reaction to any sudden faults, so that the vehicle response is never compromised.

A dual-controller control unit implements the safety concept on the basis of mutual monitoring through redundancy functions. If the results from the two controllers differ, the vectoring torque will immediately be reduced to establish a safe condition of the system. This ensures that an error will not result in a safety-critical driving condition, which might endanger the driver and which he might not be able to handle. To meet the high system requirements, a precise tuning of the safety functions especially in respect to the units' mechanical characteristics is necessary.

10.4 Performance

Modern driving dynamics functions are based on vehicle models, and use sensor information such as steering angle, lateral acceleration, yaw rate, and vehicle speed to calculate the necessary vectoring torque highly accurately. In general, the overall control consists of two main control elements. The pre-control element sets the optimal and reproducible wheel differential torque for any driving situation. At the same time, the yaw rate control element corrects any deviations between the target and actual yaw rates, ensuring the required driving dynamics are maintained. This ensures a high degree of agility, controllability, and safety even at the stability limits of the vehicle. High setting accuracy and dynamics of the torque-vectoring system is essential to provide these benefits (Figures 14 and 15).

10.5 Efficiency factor and drag torques

As part of current efforts to reduce energy consumption and carbon dioxide emissions, efficiency, including reducing losses that occur within the axle drive, plays an increasingly important role. Every torque-vectoring system creates additional losses due to its components, gearing, and bearings. However, the levels of loss are different depending on the torque-vectoring design and the driving situation. The goal is to keep the extra consumption as low as possible in the driving cycle that is relevant to the driver.

Figure 16 shows the proportion of drag power caused by the two TV units of the ZF Vector Drive© system examined. The driving situation this is based on is straight-ahead driving at constant speed without actuation, such as occurs on freeways. Compared to the total losses from axle drives, the losses here are relatively low. The reason for this is that in this driving situation, the planetary gearset rotates as a block so that there is no gearing roll-off. Furthermore, a careful choice of oil and lining material means the fast-running multidisk clutches have a very low drag torque.

10.6 System networking with other driving dynamics control systems

Modern vehicles can feature a large number of other control systems apart from torque-vectoring systems, which make it possible to actively influence the driving dynamics. Active steering systems offer an additional superimposition steering angle that can be used to influence driving dynamics. Active rear axle steering systems can be used for active sideslip-angle control. Active roll stabilization impacts the self-steering response. Furthermore, braking interventions on selected wheels can also be used in addition to ESC to influence driving dynamics. What all these systems have in common is that they can electronically control the movement of the vehicle around the vertical

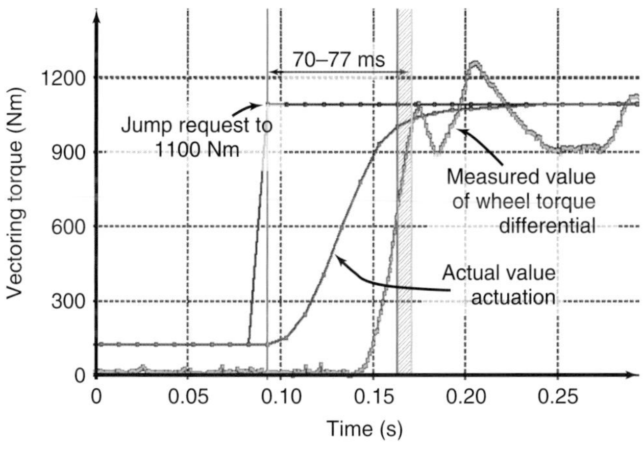

Figure 15. Dynamics characteristic of the ZF Vector Drive © system.

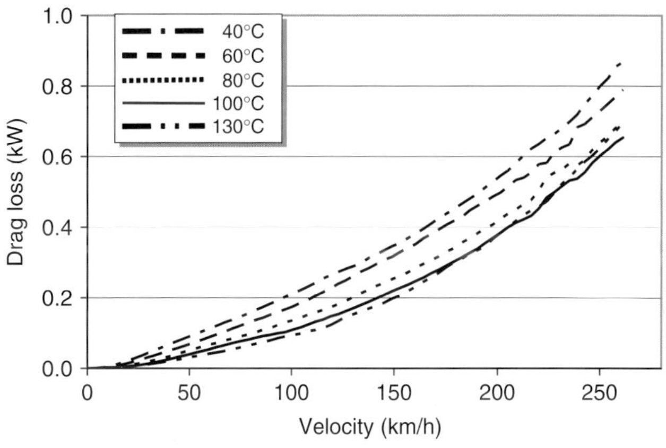

Figure 16. Drag power contribution of the two torque-vectoring units in the ZF Vector Drive ©. The figure shows the measured drag power at constant speed and straight-ahead driving without actuation.

axis. Super ordinate strategies link up the individual driving dynamics systems and help optimally coordinate interventions. This not only increases functionality by using interventions that supplement or support each other but also ensures maximum availability and increased driving safety.

11 DISCUSSION AND CONCLUSION

Torque vectoring for drivetrain systems represents a powerful vehicle technology to improve vehicle performance. The systems allow the intelligent distribution of propulsion torque to influence traction and vehicle dynamics. Different concepts, approaches, and designs have been developed and are nowadays available on the customer market. Several car manufactures use the technology of torque-vectoring to further improve the precision and agility of their modern and sporty vehicles.

The microcontroller-based control strategy of modern torque-vectoring systems allows a defined tuning of the vehicle behavior. The possibility of asymmetric torque distribution generates a yaw moment on the vertical axis and provides a strategy-based input for the intended vehicle behavior. Depending on the application and the intention of the manufacturer, different priorities in performance gain can be chosen during the development of the vehicle.

Beside the dynamic improvement, the optimization of the traction behavior still plays a crucial role, as this is one of the most important motivations for all-wheel-drive vehicles. Torque vectoring can also support the demand for increased traction and off-road performance by the lateral distribution of propulsion torque.

The technical effort to implement a torque-vectoring system at a modern vehicle unfortunately increases the weight and cost of the car. Additional gearsets, clutches, and actuators incorporate losses, which have negative effects on the efficiency and emissions. The losses are quite different, depending on the realized concept but nevertheless, they can restrain the technology from a wide spreading on the present car market.

RELATED ARTICLES

Volume 3, Chapter 88
Volume 3, Chapter 99
Volume 3, Chapter 100
Volume 3, Chapter 101
Volume 3, Chapter 102
Volume 3, Chapter 104
Volume 4, Chapter 127
Volume 4, Chapter 130

Modeling and Simulation

Chapter 133

Chassis Modeling and Optimization by the Advanced Method ABE

Ingo Albers

ZF Friedrichshafen AG, Dielingen, Germany

1 INTRODUCTION

Wheel suspensions have the important task to serve as a connection between the road surface and the chassis, guiding the wheel while driving. The suspension has to fulfill many tasks: cushioning, damping, steering, force transmission, powering, and braking.

The way in which a wheel is guided significantly determines the kinematics of the system and hence which type of movement the wheel is capable of doing. One differentiates between inflexible (or rigid) and elastic kinematics. With the inflexible kinematics, it is given that all connecting components such as suspension arms and joints are ideally stiff and the joints only show one or more pure rotation degrees of freedom toward the desired directions. In reality, the elements of wheel suspensions are seldom considered to be ideally stiff. Especially, elastomer bushings have a considerable effect on the kinematics of the wheel suspensions. In addition, structural elasticity of the involved elements is to be considered to precisely analyze the kinematics. The kinematics of a wheel under the influence of a real element rigidity and the so-called soft elements such as rubber bushings, overload springs, or bump stops is usually called *elasto-kinematics*. The elasto-kinematics can be differentiated drastically from the idealized stiff-kinematics, still it is highly suggested that a clear analysis and synthesis of the stiff-kinematics is performed before looking at the real elasto-kinematics.

To make the kinematics of wheel suspensions tangible and especially comparable, numerous values on axle kinematics have been developed in the past, which have established themselves in the daily life of chassis engineers. This applies to both single values, valid for the so-called design position, and complete characteristic data. These trends are identified and evaluated for compression and rebound and steering or action of forces – or torque conditions (braking, powering, cornering, and vertical loads).

Nowadays, within the automotive industry, usually a first model in 3D-CAD and a model in the multibody system (MBS) are defined when designing the layout of an axle. For a fast prelayout without MBS knowledge, for university purposes, for Formula Student suspensions, or in all other cases in which an MBS model is not feasible (e.g., due to high fees), a simple program can be self-created to calculate the stiff-kinematics. To do so, the mathematical approach by Matschinsky (1992) is used, which is complementary to a program family called *ABE* (axle calculation in Excel) (Albers, 2003, 2009).

Encyclopedia of Automotive Engineering, print © 2015 John Wiley & Sons, Ltd.
Edited by David Crolla, David E. Foster, Toshio Kobayashi, and Nick Vaughan.
This article is © 2015 John Wiley & Sons, Ltd. ISBN: 978-0-470-97402-5
Also published in the *Encyclopedia of Automotive Engineering* (online edition)
DOI: 10.1002/9781118354179.auto037

2 THEORETICAL BASICS ON INDEPENDENT WHEEL SUSPENSIONS

2.1 Necessary degrees of freedom of a wheel suspension

Fundamentally, a stiff body has six degrees of freedom in space and can perform a translational movement in *X*-, *Y*-, or *Z*-direction or a rotational movement around every axis of a Cartesian *X*-*Y*-*Z* coordinate system.

As road roughness has to be compensated and high accelerations of the vehicle body have to be avoided, an independent suspension basically has to have a vertical movement option. The degree of freedom is, thus, $F = 1$. This degree of freedom does not necessarily have to be a parallel movement, but can generally also be a combined stroke, cross, and tilt movement. With such a combination of translational and rotational moving parameters, all parameters are in direct relation to each other, the so-called kinematic constrained motion. A body suspension spring is responsible for saving and releasing the energy. Simultaneously, a damper is dissipating energy and reducing vibrations. The suspension system adjusts to the statically defined weight balance of the vehicle. The spring element represents a highly elastic suspension arm; the degree of freedom $F = 1$ is, thus, appropriate only if a force is applied (Albers, 2003).

If a body is stiffly hung to six suspension arms in a room, which illustrates a wheel with a wheel bearing solidly attached to the vehicle. To maintain the degree of freedom of $F = 1$, one of the six suspension arms has to be removed and one gets the basic type of an independent suspension, the so-called common space 5 rod wheel guidance. The calculation program described here is, thus, only applicable to independent suspension systems, which can be illustrated or be derived from the "common 5 rod suspension." This, for example, also includes the double wishbone axle as well as the common McPherson axle at the front (Albers, 2003).

The guidance of the suspension and definition of the degrees of freedom takes place with the components shown in the following.

2.2 Components of wheel suspensions

A wheel suspension can be considered as a kinematic chain. The wheel hub is the connecting rod. Suspension arms serve as intermediate components of the chain. Joints are the smallest elements of such a chain. The total degree of freedom results from the degrees of freedom of the bodies to be guided, all joints as well as the confining degrees of freedom of the suspension arms. In the following, the different types of joints are shown in Figure 1. How many degrees of freedom is required by the respective joint constrain will be explained. A small "f" identifies the joint degrees of freedom (Albers, 2003).

2.3 Characteristic values of axle kinematics

In the automotive literature, there are numerous different vehicle suspension characteristic values. Matschinsky (1992) describes them in great detail in his PhD thesis, thus only a small overview is given in Tables 1 and 2.

2.4 Velocity status of the wheel carrier

Looking at the respective definitions of the kinematic values shown (Matschinsky, 1992; Albers, 2003), it becomes evident that they can be calculated through either geometrical positions or translational as well as rotational velocity conditions of either the tire contact point or the wheel carrier. The complete velocity status consists of the translational velocity vector $\vec{\mathbf{v}}_M$ and the angular velocity vector

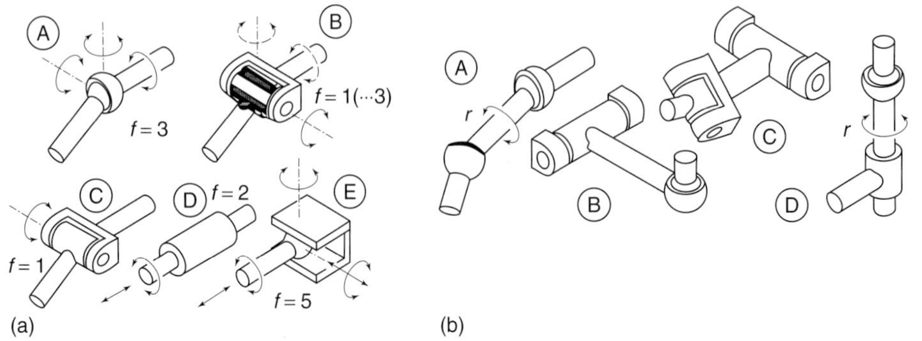

Figure 1. Overview of (a) joint types and (b) steering types. (Reproduced with permission from Matschinsky, 1992. © W. Matschinsky.)

Table 1. Overview of usual characteristics during compression and rebound.

Characteristics During Compression and Rebound	
Basic characteristics	Roll center
Spring travel (mm)	Roll center height (vertical movement) (mm)
Wheel travel (mm)	Roll center position (roll movement) (mm)
Spring ratio (spring travel/wheel travel) (−)	Steering axle/castor axle
Track width (mm)	Castor angle (deg)
Wheel position characteristics	Castor trail (mm)
Toe in (deg)	Castor offset (mm)
Camber angle (deg)	Steering axle/king pin angle
Antidive/antisquat	King pin inclination angle (deg)
Real brake supporting angle (deg)	King pin offset (mm)
Optimal brake supporting angle (deg)	Disturbance force lever (DFL)
Brake pitch compensation (%)	DFL braking (mm)
Real acceleration support angle (deg)	DFL traction (mm)
Optimal acceleration support angle (deg)	Other characteristics
Drive pitch compensation (%)	Suspension oblique angle (deg)
Acceleration balance (%)	Brake support angle (deg)

Table 2. Overview of usual characteristics during steering.

Characteristics During Steering	
Basic characteristics	Steering axle/castor axle
Tie rod travel (mm)	Castor angle (deg)
Wheel steering angle (deg)	Castor trail (mm)
Track width (mm)	Castor offset (mm)
Wheel position characteristics	Castor angle (deg)
Camber angle (deg)	Steering axle/king pin angle
Wheel travel (mm)	Castor offset (mm)
Steering characteristics	Steering axle/king pin angle
Wheel steering angle inside (deg)	King pin inclination angle (deg)
Wheel steering angle outside (deg)	King pin offset (mm)
Averaged wheel steering angle (deg)	Disturbance force lever
Toe difference angle (deg)	Disturbance force lever braking (mm)
Ackermann angle (outside) (deg)	Disturbance force lever traction (mm)
Ackermann percentage (%)	

$\vec{\omega}_K$. The translational velocity is only valid in the reference point, which, reasonably, is the center of the wheel (M). Within a solid body, the angular velocity vector is identical, thus it is defined for the wheel carrier body (K) (Albers, 2009).

The wheel carrier including the tire contact point, which is rigidly coupled with the wheel carrier over the wheel, is abstracted by a wheel carrier level, which, at the beginning, is defined by the tire contact point (A), the wheel center (M), as well as a once defined auxiliary point (H) (Figure 2). The auxiliary point is conveniently defined as being part of the wheel axle; in addition, the lines MA and MH are perpendicular to each other. The distance between H and M is simply the wheel radius, hence points A, M, and H construct an orthogonal, equally sided triangle in space (Albers, 2009).

This plane is being incrementally moved in spring and steering activity. In doing so, the kinematic velocity is being analyzed. Through summation by integration of the incremental speeds, the position is also always known.

As an example for a position-orientated calculation, the toe and camber calculation is shown. The track width is a value, which is trivially determined from the positions of the tire contact points. For the identification of the geometrical constraints of points A, M, and H, it is advisable to show a wheel in a random position and then examine it in two layers. It is very important to determine the respective distances correctly in their respective depiction in the layer. Shown in Figure 2 is a wheel with a positive camber and a positive wheel steer angle. A positive wheel steer angle at the left wheel means negative toe-in (Albers, 2009).

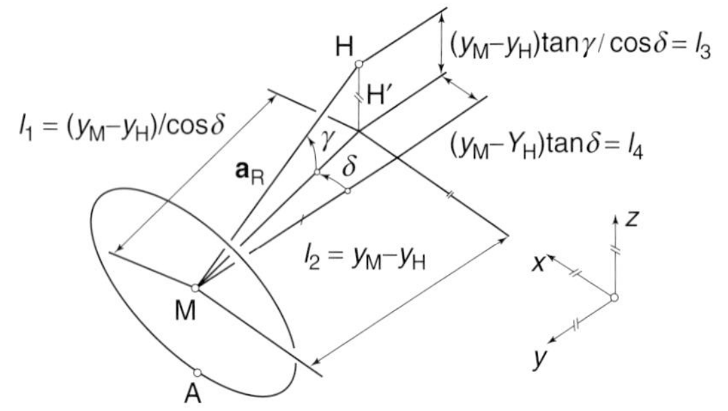

Figure 2. Wheel and wheel carrier in optional, steering situation. (Reproduced from Matschinsky, 1998. With kind permission from Springer Science+Business Media.)

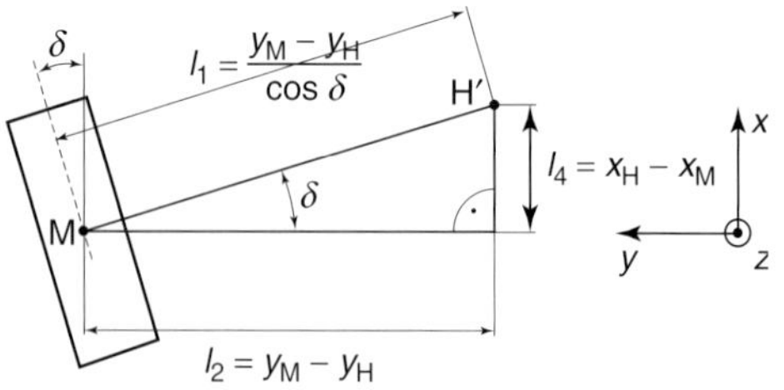

Figure 3. View from above in *X*-*Y*-level.

Viewed from above, the picture is shown in the *X*-*Y*-level, cut at the wheel center M (Figure 3).

The lines named in Figure 2 can easily be found here again and can be depicted geometrically through known values. "H′" is the vertical projection of the point H in the shown level. The respective sections are trigonometrically further explained in the figures.

In the following, the view from the rear upon the wheel axle a_R is illustrated as a cut through the wheel center M and the help point H (Figure 4). Hence, the cut level shown is the *Y*-*Z*-level, turned by the steering angle δ.

With the help of these geometrical correlations, camber- and toe-in with respect to the front wheel steering angle can be determined directly. For instance, when the coordinates of the points M and H of the wheel axle a_R are known, one can calculate the steering angle δ as well as the camber angle γ. From Figures 3 and 4, one gets the following correlation for the steering angle:

$$\delta = \arcsin \frac{x_H - x_M}{\sqrt{(x_H - x_M)^2 + (y_M - y_H)^2}} \tag{1}$$

The toe-in angle for a left wheel is, according to the preceding definition (DIN 70000), obtainable by switching the algebraic sign. The camber angle can be obtained

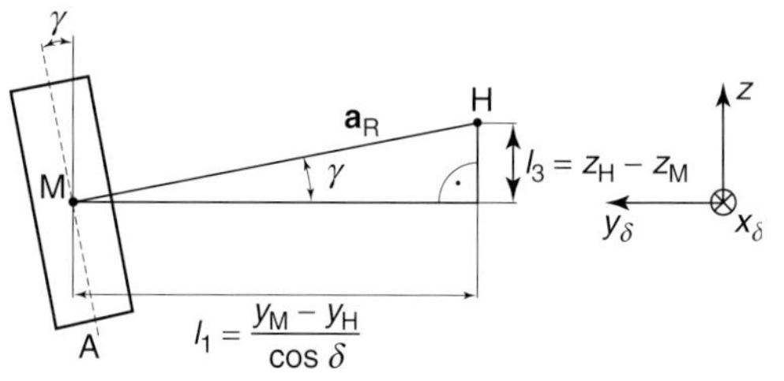

Figure 4. View from behind on the *Y*-*Z*-level of the left wheel.

likewise through the following equation:

$$\gamma = \arctan \frac{z_H - z_M}{\sqrt{(x_H - x_M)^2 + (y_M - y_H)^2}} \tag{2}$$

In this case, the current coordinates of the wheel center point as well as of the help point H on the wheel axle are needed (Albers, 2009).

2.5 The ABE-core: determining the velocity situation

Now, it is possible to establish the core of the calculation program. Considering a solid body, such as a wheel carrier of a wheel suspension in the case given, its velocity situation is explicitly described by specifying the velocity $\vec{\mathbf{v}}_M$ in the wheel center point and $\vec{\boldsymbol{\omega}}_K$ of the wheel carrier; hence, it consists of six components:

$$\vec{\mathbf{v}} = (\vec{\mathbf{v}}_M, \vec{\boldsymbol{\omega}}_K)^T = (\mathbf{v}_{Mx}, \mathbf{v}_{My}, \mathbf{v}_{Mz}, \boldsymbol{\omega}_{Kx}, \boldsymbol{\omega}_{Ky}, \boldsymbol{\omega}_{Kz})^T \tag{3}$$

Thus, for a clear calculation, six independent equations are needed as well. The translational velocity of any random point P_i in a body can be calculated directly if the distance between P_i to its point of reference (in this case the wheel center point M) is known as the *respective distance vector* $\mathbf{r}_i$. The current velocity situation $\mathbf{v}_i$ of a random point P_i is calculated through

$$\vec{\mathbf{v}}_i = \vec{\mathbf{v}}_M + \vec{\boldsymbol{\omega}}_K \times \vec{\mathbf{r}}_i \tag{4}$$

Figure 5 shows the necessary components.

With the calculation algorithm given, it is assumed that the connection rods of the suspension are stiff. Thus, the wheel-carrier- and the vehicle-based velocity can be correlated to each other. The velocity in the direction of the connection rod has to be identical on both sides. This is illustrated with a simple scheme (Figure 6; Albers, 2009).

One can clearly see that there are different velocities $\vec{\mathbf{v}}_i$ and $\vec{\mathbf{v}}'_i$ effective on the wheel-carrier-based point i and

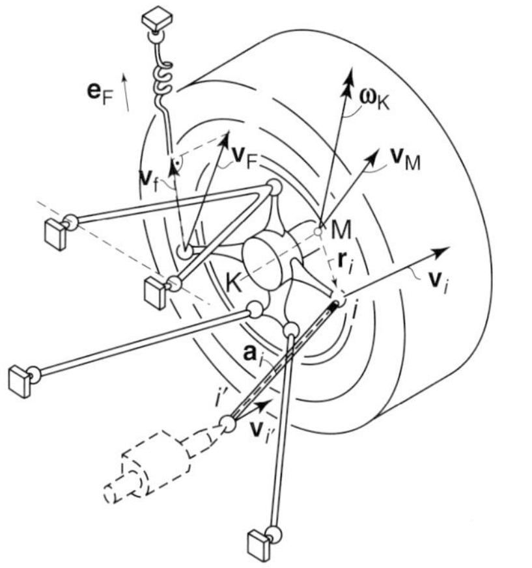

Figure 5. Common velocity situation of a wheel carrier. (Reproduced with permission from Matschinsky, 1992. © W. Matschinsky.)

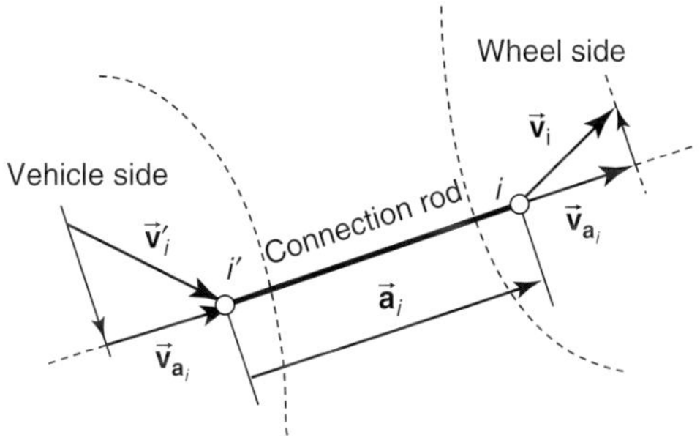

Figure 6. Identical velocity along a connection rod vector. (Reproduced with permission from Albers, 2009. © Ingo Albers.)

on the vehicle-based point i' of a stiff connecting rod, the velocity in the direction of the rod ($\vec{\mathbf{v}}_{\mathbf{a}_i}$), however, has to be identical. The scalar velocity in the direction of the rod results in the common case from the scalar product from velocity vector and steering rod vector, which points from the vehicle to the wheel carrier:

$$\vec{\mathbf{v}}_i \cdot \vec{\mathbf{a}}_i = \vec{\mathbf{v}}'_i \cdot \vec{\mathbf{a}}_i \tag{5}$$

This equation is called the *connection rod requirement*. The connection rod vector $\vec{\mathbf{a}}_i$ is, just like the location vector $\vec{\mathbf{r}}_i$, known from the original coordinates of the wheel suspension. With Equations 4 and 5, the fundamental equation for this discussion is as follows:

$$(\vec{\mathbf{v}}_M + \vec{\boldsymbol{\omega}}_K \times \vec{\mathbf{r}}_i) \cdot \vec{\mathbf{a}}_i = \vec{\mathbf{v}}'_i \cdot \vec{\mathbf{a}}_i \tag{6}$$

which puts the velocity situation of the wheel carrier in correlation to the wheel-carrier- and vehicle-based joints' velocities at the ends of the examined connection rod. This fundamental equation is called *common velocity equation* (CVE) and is an elementary core of the calculation program. The CVE is being used in many different forms and adjusted according to the specific needs (Albers, 2009).

The CVE, formulated for three different basic suspension variants, is as follows:

- Common five-rod suspension with spring element on the wheel carrier (5LR)
- Common five-rod suspension with spring element at a connection rod (5LL)
- McPherson strut (McP)

As an example, the CVE is developed and drafted for a five-rod suspension with connection of the spring element to the connection rod (5LL). For both the other basic types (5LR and McP), refer to the study by Albers (2009).

The CVE originates in its basic form (Matschinsky, 1992) but is further developed to adjust to other applications. Hence, the CVE is formulated in such a way that the spring element incrementally varies its length or the tie rod is incrementally shifted from the vehicle side. The two following moving patterns can be observed:

- *Spring action*: Change in length of the spring elements.
- *Steering action*: Moving of the vehicle-based tie rod joint (equals the connection to a rack and pinion gear box).

In between these two moving actions, on spring movements, the difference is not only that a different spring rod is affected than that on the steering (tie rod) but especially that the length of the spring rod is changed through the stepwise speed increment, whereas upon steering, the length of the tie rod is constant. Here, the steering speed increment is externally impacted, that is, from the vehicle-based side (from a translational operating steering gear box) to the wheel suspension system (Albers, 2009).

For both these common types of movements of the wheel suspension, which can perform independently from each other, the CVE will be developed in such a way that it can be used in a calculation algorithm.

2.5.1 *Derivation of the ABE-core of a wheel suspension system*

With a common five-rod suspension, as shown in Figure 5, the CVE can be freely set up for a spring movement with the indices 1 to 5 for the first five connection rods. The entire derivation originates from the study by Albers (2009).

The necessary sixth equation generally evolves from contemplating the spring element. This spring element is considered to be stiff in each incremental move and completes the linear equation system with the sixth equation. This spring equation has to be compiled independently from the type of movement.

In case of a spring movement, the spring element is moved out of its design position with the spring speed $\mathbf{v}_\mathrm{f}$. Principally, this spring speed increment $\mathbf{v}_\mathrm{f}$ is responsible for a change in position of the wheel suspension, as it induces the velocity $\vec{\mathbf{v}}$ of the wheel carrier. This again results in a velocity $\mathbf{v}_i$ on all wheel-carrier-based joints, which can be calculated with Equation 4. The wheel suspension system moves, that is, it experiences a change in position. The tie rod does not move in this case, it is said to be a "blocked steering."

A similar consideration is valid for the steering process at which the strut compression velocity $\mathbf{v}_\mathrm{f}$ is put to zero. This activity is called a *blocked suspension*. The tie rod is moved at $\mathbf{v}_\mathrm{L}$ and it induces a movement of the wheel carrier and, thus, the entire wheel suspension.

Generally speaking, at first, a scalar strut compression velocity $\mathbf{v}_\mathrm{f}$ is defined, which points in the direction of the unit vector of the spring and is positive during compression from the design position. Put into the CVE according to Equation 4, the result is

$$\mathbf{v}_\mathrm{f} = (\vec{\mathbf{v}}_\mathrm{M} + \vec{\boldsymbol{\omega}}_\mathrm{K} \times \vec{\mathbf{r}}_\mathrm{F}) \cdot \vec{\mathbf{e}}_\mathrm{F} \tag{7}$$

with $\vec{\mathbf{r}}_\mathrm{F}$ (distance vector from the wheel center to the spring function line) and $\vec{\mathbf{e}}_\mathrm{F}$ (unit vector of the spring element). The connection rod vector of the spring is named with $\vec{\mathbf{a}}_6$ and points, as defined earlier, from the vehicle structure to the wheel carrier and to the pivot joint of one of the connecting rods, and thus contrarily to $\mathbf{v}_\mathrm{f}$.

The case that the spring element is mounted directly to a wheel carrier can be expressed more easily. Here, a more complicated case with the connection of the spring to the connection rod is to be shown. It can easily be imagined that in such a case the connection point between the spring element and the connection rod (suspended rod) does not instantaneously follow the velocity situation of the wheel carrier but receives a coupled movement via the suspended rod. It is, thus, important for the development of the spring equation to include the information on where the spring element is mounted. For the derivation of the ABE-core, the following steering indices are defined: tie rod (1), spring element (6), suspended rod (5), as well as the remaining three connection rods (2, 3, and 4). Figure 7 shows the velocity plan of the suspended connection.

It is possible to apply the intercept theorem here to determine the relation of the values of the connection rod

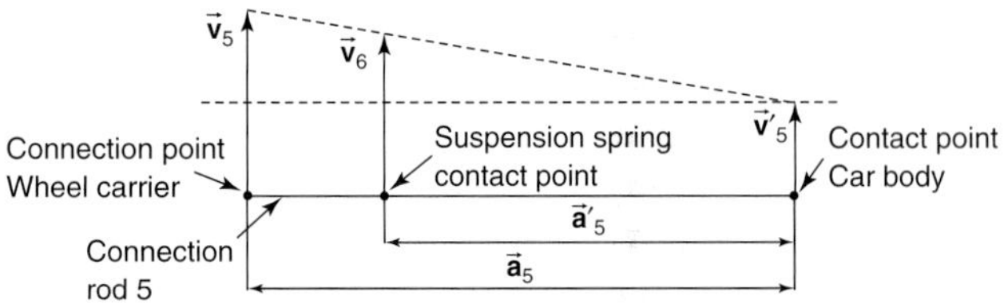

Figure 7. Application of the intercept theorem with a spring, acting to the connection rod. (Reproduced with permission from Albers, 2003. © Ingo Albers.)

vector $\vec{\mathbf{a}}_5$ as well as $\vec{\mathbf{a}}'_5$. The vector $\vec{\mathbf{a}}'_5$ is the connecting vector from the vehicle-based joint to the point of action of the spring at point 6. It is $\mathbf{a}_5 = |\vec{\mathbf{a}}_5|$ as well as $\mathbf{a}'_5 = |\vec{\mathbf{a}}'_5|$. The distance relation is viewed according to amount and converted to

$$\mathbf{v}_6 = \mathbf{v}'_5 + \frac{\mathbf{a}'_5}{\mathbf{a}_5}.(\mathbf{v}_5 - \mathbf{v}'_5) \tag{8}$$

With an introduced leverage $\mu_\mathrm{HBV} = \frac{\mathbf{a}'_5}{\mathbf{a}_5}$, it comes out to

$$\mathbf{v}_6 = \mathbf{v}'_5 + \mu_\mathrm{HBV}.(\mathbf{v}_5 - \mathbf{v}'_5) \tag{9}$$

As the spring compression speed $\mathbf{v}_\mathrm{f}$ is divergent to the connection rod vector, it is

$$\vec{\mathbf{v}}_6 \cdot \vec{\mathbf{a}}_6 = -\mathbf{v}_\mathrm{f} \cdot |\vec{\mathbf{a}}_6| \tag{10}$$

Fundamentally, for a spring element it is, according to the connection rod requirement equation 5

$$\vec{\mathbf{v}}_6 \cdot \vec{\mathbf{a}}_6 = \vec{\mathbf{v}}'_6 \cdot \vec{\mathbf{a}}_6 \tag{11}$$

In combination with Equations 9 and 10, this results in

$$[\vec{\mathbf{v}}'_5 + \mu_\mathrm{HBV} \cdot (\vec{\mathbf{v}}_5 - \vec{\mathbf{v}}'_5)] \cdot \vec{\mathbf{a}}_6 = -\mathbf{v}_\mathrm{f} \cdot |\vec{\mathbf{a}}_6| \tag{12}$$

These are as usual the common requirements of the superposition of translational and rotational speeds $\vec{\mathbf{v}}_5 = \vec{\mathbf{v}}_\mathrm{M} + \vec{\boldsymbol{\omega}}_\mathrm{K} \times \vec{\mathbf{r}}_5$. With this information, the spring equation can be formulated step by step for the ABE-core from Equation 12:

$$\Rightarrow [\vec{\mathbf{v}}'_5 + \mu_\mathrm{HBV} \cdot (\vec{\mathbf{v}}_\mathrm{M} + \vec{\boldsymbol{\omega}}_\mathrm{K} \times \vec{r}_5 - \vec{\mathbf{v}}'_5)] \cdot \vec{\mathbf{a}}_6 = -\mathbf{v}_\mathrm{f} \cdot |\vec{\mathbf{a}}_6|$$

$$\Longleftrightarrow (\vec{\mathbf{v}}_\mathrm{M} + \vec{\boldsymbol{\omega}}_\mathrm{K} \times \vec{\mathbf{r}}_5) \cdot \vec{\mathbf{a}}_6 = -\frac{1}{\mu_\mathrm{HBV}} \cdot [\mathbf{v}_\mathrm{f} \cdot |\vec{\mathbf{a}}_6| + \vec{\mathbf{v}}'_5 \cdot \vec{\mathbf{a}}_6 \cdot (1 - \mu_\mathrm{HBV}) \tag{13}$$

This equation is, component based, completed to

$$\begin{aligned}&[\mathbf{v}_{Mx} + \boldsymbol{\omega}_{Ky} \cdot (z_5 - z_M) - \boldsymbol{\omega}_{Kz} \cdot (y_5 - y_M)] \cdot (x_6 - x_6')\\&+ [\mathbf{v}_{My} + \boldsymbol{\omega}_{kz} \cdot (x_5 - x_M) - \boldsymbol{\omega}_{Kx} \cdot (z_5 - z_M)].(y_6 - y_6')\\&+ [\mathbf{v}_{Mz} + \boldsymbol{\omega}_{Kx} \cdot (y_5 - y_M) - \boldsymbol{\omega}_{ky} \cdot (x_5 - x_M)] \cdot (z_6 - z_6')\\&= -\frac{1}{\mu_{HBV}} \cdot \mathbf{v}_f \cdot \sqrt{(x_6 - x_6')^2 + (y_6 - y_6')^2 + (z_6 - z_6')^2}\\&= -\frac{1 - \mu_{HBV}}{\mu_{HBV}} \cdot [\mathbf{v}'_{5x} \cdot (x_6 - x_6') + \mathbf{v}'_{5y} \cdot (y_6 - y_6')\\&+ \mathbf{v}'_{5z} \cdot (z_6 - z_6')]\end{aligned} \tag{14}$$

The CVE from Equation 6 represents, after completely writing the cross product with the indices $i = 1 \ldots 5$, the first five equations of the ABE-core

$$\begin{aligned}&[\mathbf{v}_{Mx} + \boldsymbol{\omega}_{Ky} \cdot (z_i - z_M) - \boldsymbol{\omega}_{kz} \cdot (y_i - y_M)] \cdot (x_i - x_i')\\&+ [\mathbf{v}_{My} + \boldsymbol{\omega}_{Kz} \cdot (x_i - x_M) - \boldsymbol{\omega}_{Kx} \cdot (z_i - z_M) \cdot (y_i - y_i')]\\&+ [\mathbf{v}_{Mz} + \boldsymbol{\omega}_{Kx} \cdot (y_i - y_M) - \boldsymbol{\omega}_{Ky} \cdot (x_i - x_M)] \cdot (z_i - z_i')\\&= \mathbf{v}_i' = \mathbf{v}'_{ix} \cdot (x_i - x_i') + \mathbf{v}'_{iy} \cdot (y_i - y_i') + \mathbf{v}'_{iz} \cdot (z_i - z_i')\end{aligned} \tag{15}$$

with $\mathbf{v}_i'$ as external, vehicle-based joint moving speed (Albers, 2009).

This equation is, however, only valid when there are no internal connection rod extensions within the connection rods with the indices 1 to 5. To include those in the common matrix as well, the CVE is used once more:

$$\vec{\mathbf{v}}_i \cdot \vec{\mathbf{a}}_i = (\vec{\mathbf{v}}_M + \vec{\boldsymbol{\omega}}_K \times \vec{\mathbf{r}}_i) \cdot \vec{\mathbf{a}}_i = \vec{\mathbf{v}}_i' \cdot \vec{\mathbf{a}}_i \tag{16}$$

The vehicle-based speed scalar product $\vec{\mathbf{v}}_i' * \vec{\mathbf{a}}$ is replaced by the connection rod change speed $\mathbf{v}_{LL}$ multiplied with the value of the connection rod vector $\vec{\mathbf{a}}_i$. As the connection rod change speed $\mathbf{v}_{LL}$ is contrary to the connection rod vector, it is

$$\vec{\mathbf{v}}_i \cdot \vec{\mathbf{a}}_i = -\mathbf{v}_{LL} \cdot |\vec{\mathbf{a}}_i| \tag{17}$$

From Equation 16, thus, it follows that

$$(\vec{\mathbf{v}}_M + \vec{\boldsymbol{\omega}}_K \times \vec{\mathbf{r}}_i) \cdot \vec{\mathbf{a}}_i = -\mathbf{v}_{LL} \cdot |\vec{\mathbf{a}}_i| \tag{18}$$

Solving the vector product, like in Equation 15, calculating the value of the connection rod vector $\vec{\mathbf{a}}_i'$ and performing the scalar multiplication, one gets the equations for the connection rods 1 to 5 in component form:

$$\begin{aligned}&[\mathbf{v}_{Mx} + \boldsymbol{\omega}_{Ky} \cdot (z_i - z_M) - \boldsymbol{\omega}_{kz} \cdot (y_i - y_M)] \cdot (x_i - x_i')\\&+ [\mathbf{v}_{My} + \boldsymbol{\omega}_{kz} \cdot (x_i - x_M) - \boldsymbol{\omega}_{Kx} \cdot (z_i - z_M)] \cdot (y_i - y_i')\\&+ [\mathbf{v}_{Mz} + \boldsymbol{\omega}_{Kx} \cdot (y_i - y_M) - \boldsymbol{\omega}_{Ky} \cdot (x_i - x_M)] \cdot (z_i - z_i')\\&= \mathbf{v}_{LLi} \cdot \sqrt{(x_i - x_i')^2 + (y_i - y_i')^2 + (z_i - z_i')^2}\end{aligned} \tag{19}$$

The left-hand sides of Equations 15 and 19 are identical, whereas they differ significantly on the right-hand sides. Equation 15 takes an external joint movement into consideration (point movement, short PM), whereas Equation 19 provides for an internal connection rod length change (LC). As both movements can only be performed separately, the common equation for the first five connection rod indices can be shown by superposition:

$$\begin{aligned}&[\mathbf{v}_{Mx} + \boldsymbol{\omega}_{Ky} \cdot (z_i - z_M) - \boldsymbol{\omega}_{Kz} \cdot (y_i - y_M)].(x_i - x_i')\\&+ [\mathbf{v}_{My} + \boldsymbol{\omega}_{kz} \cdot (x_i - x_M) - \boldsymbol{\omega}_{Kx} \cdot (z_i - z_M)] \cdot (y_i - y_i')\\&+ [\mathbf{v}_{Mz} + \boldsymbol{\omega}_{Kx} \cdot (y_i - y_M) - \boldsymbol{\omega}_{Ky} \cdot (x_i - x_M)] \cdot (z_i - z_i')\\&= [\mathbf{v}_i'] + [\mathbf{v}_{LLi} \cdot \sqrt{(x_i - x_i')^2 + (y_i - y_i')^2 + (z_i - z_i')^2}\,]\end{aligned} \tag{20}$$

In this and in the spring equation 14, it is to be seen that next to the searched velocity components $\vec{\mathbf{x}} = (\mathbf{v}_{Mx}, \mathbf{v}_{My}, \mathbf{v}_{Mz}, \boldsymbol{\omega}_{Kx}, \boldsymbol{\omega}_{Ky}, \boldsymbol{\omega}_{Kz})^T$, the velocities $\mathbf{v}_i'$ of the vehicle-based pivot joints as well as the spring compression velocity $\mathbf{v}_f$ and the connection rod length change velocity $\mathbf{v}_{LLi}$ are also unknown.

Equations 15 and 14 are separated into a matrix to create a linear equation system. It is shown to be

$$\mathbf{A} \cdot \vec{\mathbf{x}} = \vec{\mathbf{b}} \tag{21}$$

with the square matrix $\mathbf{A} = \mathbf{A}_{(m \times n)}$ as well as the vector $\mathbf{b} = \mathbf{b}_{(m \times 1)}$ with $m, n = 1 \ldots 6$.

$$\begin{pmatrix} \mathbf{a}_{11} & \cdots & \mathbf{a}_{1n} \\ \cdots & \cdots & \cdots \\ \mathbf{a}_{m1} & \cdots & \mathbf{a}_{mn} \end{pmatrix} \cdot (\mathbf{v}_{Mx}, \mathbf{v}_{My}, \mathbf{v}_{Mz}, \boldsymbol{\omega}_{kx}, \boldsymbol{\omega}_{kz})^T = \begin{pmatrix} \mathbf{b}_1 \\ \cdots \\ \mathbf{b}_m \end{pmatrix} \tag{22}$$

To do so, the solution vector components from Equations 15 and 14 have to be isolated. For lines 1 to 6 of matrix **A**, the following six row entries are to be valid:

$$\begin{aligned}
\mathbf{a}_{i1} &= (x_i - x_i') \\
\mathbf{a}_{i2} &= (y_i - y_i') \\
\mathbf{a}_{i3} &= (z_i - z_i') \\
\mathbf{a}_{i4} &= (y_i - y_M)\cdot(z_i - z_i') - (z_i - z_M)\cdot(y_i - y_i') \\
\mathbf{a}_{i5} &= (z_i - z_M)\cdot(x_i - x_i') - (x_i - x_M)\cdot(z_i - z_i') \\
\mathbf{a}_{i6} &= (x_i - x_M)\cdot(y_i - y_i') - (y_i - y_M)\cdot(x_i - x_i') \qquad (23)
\end{aligned}$$

The first five entries of vector **b** are the result of Equation 15 for the indices 1 … 5:

$$\mathbf{b}_i = [\mathbf{v}_i'] + [\mathbf{v}_{\mathrm{LL}i}\cdot\sqrt{(x_i - x_i')^2 + (y_i - y_i')^2 + (z_i - z_i')^2}] \qquad (24)$$

The sixth vector entry originates from Equation 14 and is to be enlarged by the term $\mathbf{v}_{\mathrm{LL}}$ through superposition

$$\begin{aligned}
\mathbf{b}_6 = &-\frac{1}{\mu_{\mathrm{HBV}}}\cdot\mathbf{v}_{\mathrm{f}}\cdot\sqrt{(x_6 - x_6')^2 + (y_6 - y_6')^2 + (z_6 - z_6')^2} \\
&-\frac{1-\mu_{\mathrm{HBV}}}{\mu_{\mathrm{HBV}}}\cdot[\mathbf{v}_{5\mathrm{x}}'\cdot(x_6 - x_6') \\
&+\mathbf{v}_{5\mathrm{y}}'\cdot(y_6 - y_6') + \mathbf{v}_{5\mathrm{z}}'\cdot(z_6 - z_6')] \\
&+\mathbf{v}_{\mathrm{LL}6}\cdot\sqrt{(x_6 - x_6')^2 + (y_6 - y_6')^2 + (z_6 - z_6')^2} \qquad (25)
\end{aligned}$$

To generate a solvable linear equation system with six variables of this equation system, further boundary conditions, depending on the type of movement, have to be set.

Generally speaking, the body can be seen as stationary with every suspension or steering action. This is a true analog to many real suspension test benches.

For the conventional spring movement, all vehicle-based velocities $\mathbf{v}_i'$ are set to zero. In addition, there are no connection rod length changes except at the spring element; thus, all $\mathbf{v}_{\mathrm{LL}i}$ are set to zero. The spring compression velocity is a quotient, originating from the total spring travel and the number of calculation steps and is, thus, exactly defined. The matrix **A** is a pure geometrical matrix of the reviewed wheel suspension and stays unchanged. Vector **b** simplifies noticeably to

$$\begin{aligned}
\vec{\mathbf{b}} &= (0,0,0,0,0,0,\mathbf{b}_6)^{\mathrm{T}} \text{ with} \\
\mathbf{b}_6 = &-\frac{1}{\mu_{\mathrm{HBV}}}\cdot\mathbf{v}_{\mathrm{f}}\cdot\sqrt{(x_6 - x_6')^2 + (y_6 - y_6')\,(z_6 - z_6')^2} \\
&-\frac{1-\mu_{\mathrm{HBV}}}{\mu_{\mathrm{HBV}}}\cdot[\mathbf{v}_{5x}'\cdot(x_6 - x_6') + \mathbf{v}_{5y}'\cdot(y_6 - y_6') \\
&+\mathbf{v}_{5z}'\cdot(z_6 - z_6')] \qquad (26)
\end{aligned}$$

Matrix **A** and vector **b** are, thus, complete and definitely covered.

For a conventional steering movement, all connection rod length changes $\mathbf{v}_{\mathrm{LL}i}$ are set to zero. The spring compression velocity $\mathbf{v}_{\mathrm{f}}$ is set to zero as well. The tie rod bar of the wheel suspension is moved in *y*-direction, this is where the steering gear connects. The steering velocity $\mathbf{v}_1'$ (index 1 for the tie rod) is the quotient from the total traverse path related to the number of calculation steps. All other $\mathbf{v}_{\mathrm{i}}'$ are set to zero. This reduces vector **b** to an indicator of the steering velocity.

$$\vec{\mathbf{b}} = (\mathbf{v}_1', 0, 0, 0, 0, 0)^{\mathrm{T}} \qquad (27)$$

Matrix **A** and vector **b** are, thus, complete and definitely covered for this case as well.

Next to the conventional analysis of the spring movements or steering actions, this calculation core can also be used for a tolerance or sensitivity analysis. It can be investigated how sensitive the system reacts to tolerances in the body-based pivot joints, for example, through production tolerances in body fabrication or differently long suspension arms. To do so, the spring compression velocity $\mathbf{v}_1$ as well as the steering velocity $\mathbf{v}_1'$ is initially set to zero.

For the analysis of different connection rod lengths, the connection rod length $\mathbf{v}_{\mathrm{LL}i}$ changes with the respective index as the set value for the calculation. For the tolerance analysis of the body pivot joints, the respective entry for the vehicle-based speeds $\mathbf{v}_{\mathrm{i}}'$ is chosen and defined. The tolerance fault can then be iteratively calculated through these new reference inputs.

2.6 Solution and embedding the ABE-core

With all elements of matrix **A** and vector **b** complete after Section 2.5, the solution vector is calculated in the ABE-core. In the programming used, the calculation via the inverted matrix turned out to be the fastest alternative. To do so, the initial system $\mathbf{A}^*\vec{\mathbf{x}} = \vec{\mathbf{b}}$ is multiplied with the inverted matrix on the right-hand side:

$$\begin{aligned}
\mathbf{A}^{-1}\cdot\mathbf{A}\cdot\vec{\mathbf{x}} &= \mathbf{A}^{-1}\cdot\vec{\mathbf{b}} \\
\Longleftrightarrow \vec{\mathbf{x}} &= \mathbf{A}^{-1}\cdot\vec{\mathbf{b}} \qquad (28)
\end{aligned}$$

It is possible to calculate the inverted matrix $\mathbf{A}^{-1}$ only when matrix **A** is a regular and not a singular matrix. A squared matrix with the dimension *n* is considered regular only when the rank of the matrix is equal to the dimension *n* [BRO05]. Only then the matrix has *n* linear and independent equations. If the determinant det(**A**) unequal to zero, then

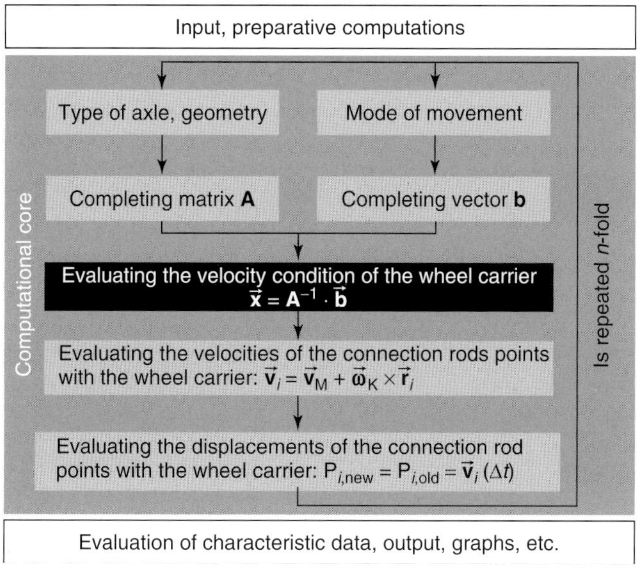

Figure 8. Embedding of the ABE-core.

the matrix is considered to be regular. That is why the determinant is checked before inverting the matrix.

The central ABE-core is centrally embedded to the total flow. It is very flexibly designed and programmed in its basic form as well as constantly called upon by different program modules. It is especially essential to program this calculation block efficiently in view of calculation speed (Figure 8).

This core can, for example, be programmed in Microsoft Excel using a VBA programming language and designed with a respective interface for input and output of data.

3 REVIEW

Chassis modeling and optimization is very often done by commercially available programs, which do not facilitate understanding of the theoretical relationship between input and output. This chapter describes a methodology that avoids these disadvantages and starts with the theoretical background of chassis design to formulate equations and explain the movements of suspensions under vertical and rotational inputs.

The equations, that both explain and describe the wheel movement, are finally programmed in Excel and the kinematic characteristics can be found. Changing the coordinates of the connection points within the suspension linkage, the new chassis characteristics can be easily found rapidly. This design tool, therefore, enables the optimization of chassis systems, even before any hardware has been built. In addition, the engineers can identify the relationship between their design changes and the new chassis behavior.

The technical background of all these procedures are described by Matschinsky (2000) and the computational transfer into a simple and fast simulation program is the merit of Albers (2009).

REFERENCES

Albers, T. (2003) Erstellung eines Berechnungstools zur starrkinematischen Analyse von Einzel-radaufhängungen. Diploma thesis. RWTH Aachen, Aachen, Germany.

Albers, I. (2009) Auslegungs- und Optimierungswerkzeuge für die effziente Fahwerkentwicklung. PhD thesis. RWTH Aachen, Aachen, Germany.

Matschinsky, W. (2000) *Road Vehicle Suspensions*, Professional Engineering Pub, Wiley-Blackwell, UK.

Matschinsky, W. (1998) *Radführungen der Strassenfahrzeuge*, Springer Verlag, Berlin, Germany.

Matschinsky, W. (1992) Bestimmung Mechanischer Kenngrössen von Radaufhängungen. PhD thesis. Universität Hannover, Hannover, Germany.

Chapter 134

System Simulation in DSH*plus*, What Applications are Possible?

Christian von Grabe[1], Olivier Reinertz[1], René von Dombrowski[2], and Hubertus Murrenhoff[1]

[1]*RWTH Aachen University, Aachen, Germany*
[2]*FLUIDON Gesellschaft für Fluidtechnik mbH, Aachen, Germany*

1 INTRODUCTION

The rising demands on technical systems in automotive and general applications in mechanical engineering result in products with an elevated degree of complexity and a wide range of integrated functionalities. To fulfill the requirements regarding system performance and reliability, modern technical systems must combine subsystems of different technical disciplines such as pneumatics, hydraulics, electronics, and mechanics as well as informatics and control techniques.

These very different technical subsystems must interact precisely in order to realize the required functionality of the entire system. In contrast to developments in the past, the engineers involved in the development process of such systems can no longer rely on estimated formulas or simple static design techniques. The use of simulation techniques is therefore indispensable in today's development processes. Furthermore, the used system simulation tools are required to provide functionalities for the coupling between different technical disciplines to coordinate the interactions between the implemented subsystems.

To meet the requirements, several software tools such as DSH*plus*, AMEsim, and SimulationX are available to aid the engineer during the whole development process.

2 SYSTEM SIMULATION

2.1 Simulation procedure

In the following, the general modeling procedure and the basic calculation methods of one-dimensional simulation tools will be illustrated, using the example of DSH*plus*. System simulation tools provide a detailed description of the real system, which is to be investigated, using sophisticated models of the component and transfer properties in the form of mathematical equations.

To ease the creation and compilation of those equations, most simulation suites provide a graphical user interface, which allows the synthesis of the system from single functional component units. These are available in libraries that contain hydraulic and pneumatic as well as electrical, control engineering, and mechanical elements. The functional units usually constitute a technical component such as a valve or a cylinder, but can as well be used to synthesize complex technical systems. Behind every functional unit,

Encyclopedia of Automotive Engineering, print © 2015 John Wiley & Sons, Ltd.
Edited by David Crolla, David E. Foster, Toshio Kobayashi, and Nick Vaughan.
This article is © 2015 John Wiley & Sons, Ltd. ISBN: 978-0-470-97402-5
Also published in the *Encyclopedia of Automotive Engineering* (online edition)
DOI: 10.1002/9781118354179.auto034

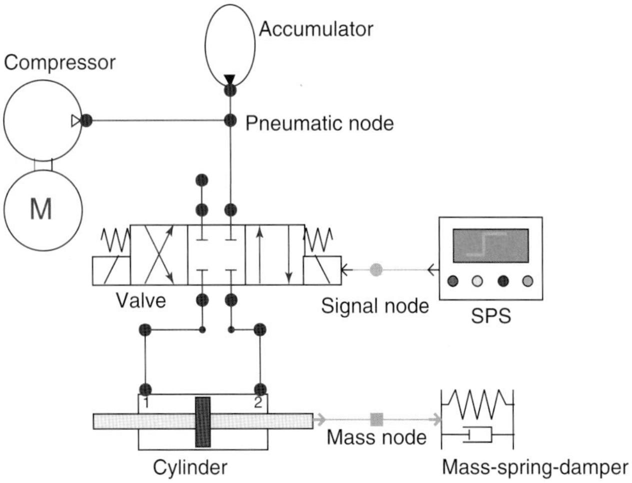

Figure 1. Simple pneumatic system.

such as a high pressure injection pump or a piston pump, for example, lies a mathematical description of the component. The components are thereby available with different levels of detail according to the needs of the simulation focus. To remain on the example of pumps, a pump, for example, can be modeled by just providing a constant volume flow to the system or in very detailed way considering all geometric circumstances and physical effects. In the second case, the volume flow of the pump will show all volume flow pulsations resulting from the number of pistons, for example. This is not necessarily needful for the simulation of entire systems but it might be of great importance for the investigation of noise problems.

If the provided libraries are not adequate to model the desired component functionalities, most simulation environment offer the possibility to create user-specific function units in a common computer language. In case of DSH*plus*, the programming language *C++* is utilized to create components from scratch. This enables the user to implement even the most complex physical models, as long as they are mathematically conveyable and the computation time is economically acceptable. The downside to the use of *C++* is the requirement for the user to be familiar with at least basic programming techniques.

In the further modeling process, different functional units can be connected by the user via nodes in the graphical user interface depending on their physical interaction with other functional units. A simple simulation model of a pneumatic cylinder drive consisting of different functional units and component connections is shown in Figure 1.

On the basis of the connections of the functional units, the mathematical equations of the physical system are automatically generated by the simulation tool. The dynamic behavior of the technical system is represented in a differential equation of the *n*th order or *n* differential equations of first order. The notation in *n*-differential equations of first order is far more suitable for the numerical integration because numerical solution methods utilize the same form of equation.

With a given set of parameters for the system components and starting values for the state variables, the system of equations can be treated as an initial value problem of common differential equations (FLUIDON, 2007):

$$\dot{\underline{y}} = \underline{f}(t, \underline{y})$$
$$\underline{y}(t_0) = \underline{y_0} \tag{1}$$

with

$$\underline{y} = \begin{pmatrix} y_t\,(t) \\ y_2(t) \\ \cdots \\ y_n\,(t) \end{pmatrix} \tag{2}$$

and

$$\underline{f} = \begin{pmatrix} f_t\left(t, y_1, y_2, \ldots, y_n\right) \\ f_2(t, y_1, y_2, \ldots, y_n) \\ \cdots \\ f_n(t, y_1, y_2, \ldots, y_n) \end{pmatrix} \tag{3}$$

Solving the set of differential equations and integrating the state variables over time provides values to the state variables at a later time step.

2.2 Calculation of pneumatic and hydraulic values

The node-oriented model structure and a concentrated calculation of the state variables within the connecting nodes are the basic principles of all system simulation tools, which therefore are also called *lumped parameters simulation tools*. Figure 2 visualizes the calculation core of such lumped parameter approaches for the example of a pneumatic system. Within the nodes, the pressure derivatives are calculated from the balances of the entering and leaving mass flows. These pressure derivatives are then integrated to pressures, which are transferred to the components. Mass flows, however, are calculated in the components from the difference of the pressures applied to their connections. To define the characteristics of the pressure buildup, for example, in a pneumatic connection node, the starting values for the volume, temperature, and pressure must be specified by the user. In case of volume changing components, as for example cylinders, the volume and its derivative are transferred to the nodes to allow their consideration in the pressure buildup equation.

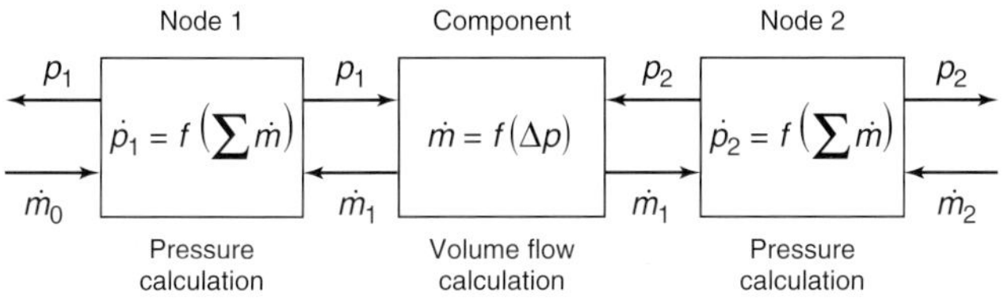

Figure 2. Calculation method of simulations with lumped parameters.

The correlation of volume flow and mass flow is described by the density ρ, which can be calculated by ideal gas equation (Equation 4) using the temperature T, the pressure p, and the gas constant R.

$$\rho = \frac{p}{T \cdot R} \tag{4}$$

The change in pressure of a pneumatic system is caused by mass flows, heat transfer, and volume change. The pressure change is described by the first law of thermodynamics for open systems and results in Equation 5:

$$\dot{p} = \dot{p}_{\dot{m}} + \dot{p}_{\text{th}} + \dot{p}_{\dot{V}} \tag{5}$$

The energy exchange resulting from the mass flows in between the components and the nodes is considered in Equation 6 by the variable $\dot{p}_{\dot{m}}$:

$$\dot{p}_{\dot{m}} = \frac{\kappa}{V} \cdot R \cdot \left(\sum \left(\dot{m}_{\text{in}} \cdot T_{\text{in}}\right) + \sum \left(\dot{m}_{\text{out}} \cdot T_{\text{out}}\right)\right) \tag{6}$$

Furthermore, the heat flow from and to the environment is represented by $\dot{p}_{\text{th}}$ in Equation 7. κ is the polytropic exponent. For air, this can be an adiabatic process and then κ would be near 1.4. In an isothermic process, the κ is 1.0. The heat transmission coefficient α describes the ratio of heat flux per unit area to the temperature difference from component wall with the contact area A to the air inside the balanced system

$$\dot{p}_{\text{th}} = \frac{\kappa - 1}{V} \cdot [\alpha \cdot (T_W - T) \cdot A] \tag{7}$$

The performed work due to the volume change at the balanced system is considered by $\dot{p}_{\dot{V}}$ in Equation 8:

$$\dot{p}_{\dot{V}} = -\frac{\kappa}{V} \cdot p \cdot \dot{V} \tag{8}$$

The temperature is acquired by means of the ideal gas equation as shown in Equation 9:

$$T = \frac{p \cdot V}{m \cdot R} \tag{9}$$

The time derivative of the ideal gas equation delivers the change in temperature over time:

$$\dot{T} = \frac{\dot{p}}{p} \cdot T + \frac{\dot{V}}{V} \cdot T - \dot{m} \cdot \frac{RT^2}{pV} \tag{10}$$

To minimize the computation time of the simulation, the pressure change $\dot{p}$ is calculated first. In a second step, the preceding results are used to determine the temperature change $\dot{T}$.

2.3 Numerical integration

The system of differential equations representing the modeled system describes the derivative of all state variables in all operating points. To solve the system of differential equations of the entire simulation model, DSH*plus* provides different numeric integrators, which allow an explicit as well as an implicit solving on a time-discrete basis.

The explicit solver only uses values, which are already calculated and therefore available from former time steps. The implicit solver in contrast uses the solution of the current time step as well.

The integral interval $[t_0, t_n]$ is divided into time increments.

$$t_0 < t_1 < t_2 < \cdots < t_n$$

with the local time increment of $h_i = t_{i+1} - t_i$ for $i = 0 \ldots (n-1)$. An approximation for $y(t_i)$ is obtained by Equation 11 with $i = 0 \ldots (n-1)$.

$$\underline{y}(t_{i+1}) = \underline{y}(t_i) + \int_{t_i}^{t_{i+1}} f(t, y(t))\mathrm{d}t \tag{11}$$

Furthermore, modern system simulation tools offer solvers, which work with an automatic step size control instead of a constant step size integration. The advantage of this multistep methods is a small time increment in high dynamic operation modes and a bigger time step, when no or only small changes in the model behavior are present. Multistep methods therefore reduce the computation time. As a consequence of the numerical integration, a discretization error has to be encountered for. To reduce this error, DHS*plus* offers an automatic step size control in order to achieve optimal results for highly nonlinear very stiff systems. The step size control adapts the step size in such manner that a specific error tolerance is met.

2.4 Typical applications

Typical applications of system simulation tools such as DSH*plus* within the automotive engineering are all fluid-containing systems, such as functional systems, safety systems, or comfort systems. As an example, comfort systems such as air springs or semiactive dampers are virtually analyzed and optimized with the help of system simulation tools. Further, common simulation applications are power steering or braking units as well as safety systems such as ABS (antilock brake system), ESP (electronic stability program), or active roll systems with their related control strategies.

Beneath simulation applications on the entire system level, it is also common practice to look at the single component as a system. Applications on the component level are fuel injection pumps, hydraulic or pneumatic lines, fuel injectors, hydro bearings, or torque converters, for example.

3 SIMULATION EXAMPLES

To illustrate the wide range of possible applications for DSH*plus*, a rather unusual example of a modeling approach for pneumatic seat components is described. The example shows the whole process including data acquisition for component parameterization, the buildup of the simulation model in DSH*plus*, and the comparison of the simulation results with measurements. The example not only reveals the advantages of simulation tools in the design process but also displays the flexibility of modern system simulation tools.

3.1 Example 1: pneumatic seat adjustment

Pneumatically actuated comfort systems for the dynamic adjustment of the shape of automotive seats are characterized by their compactness and consist of miniature pneumatic components. These components are a compressor, a number of valves—depending on the application—inflatable plastic cushions, as well as the necessary connectors and hoses. Figure 3 depicts the functions and the pneumatic components built into an automotive seat.

Commonly, the components mentioned earlier will be delivered to the seat manufacturer by different suppliers. Therefore, a fast, exact, and energy-efficient method is necessary to compare components of different producers or to characterize or assess single components regarding their usability for the automotive seat. The pneumatic elements are low priced components according to their usage in the automotive sector, which is well known as a *mass market*. Their functionality and fabrication result in a large variation of their pneumatic characteristics. This variation of the components' characteristics leads to a strong variation of the system behavior. Therefore, an integral approach for the characterization and simulation to determine the system behavior is of great importance. DSH*plus* is utilized to evaluate the applicability of the components regarding the specified behavior of the complete system. As pneumatic cushions are quite rare system components in typical pneumatic applications, there is no library containing similar components available in DSH*plus*. Furthermore, because of the early development stage of the components, there are no specific datasheets available to parameterize the components. Hence, the parameterization of such a model by estimates would result in an inaccurate description of the system behavior. To ensure accuracy of the model,

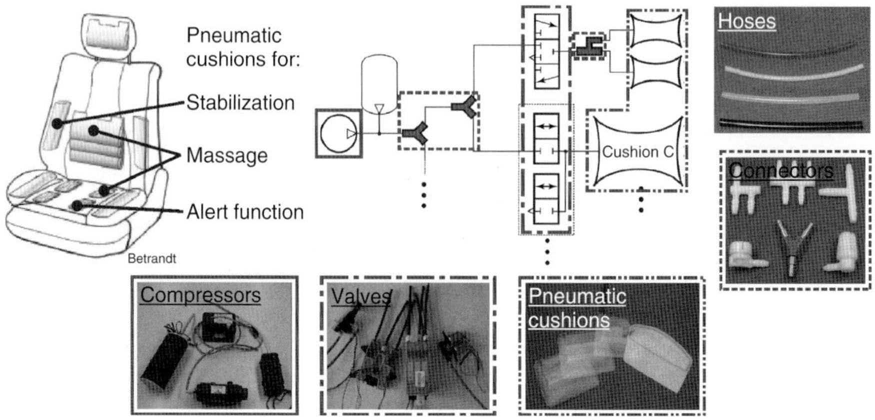

Figure 3. Pneumatically actuated comfort system of an automotive seat with its functions and its builtin components.

components are calibrated and the results are used to reproduce the exact component behavior in the system simulation in DSH*plus*.

3.1.1 Valves

Figures 4 and 5 depict the mass flow characteristics derived from measurements and the mentioned parameters of an exemplary valve for the flow path $1 \rightarrow 2$ and $2 \rightarrow 3$, respectively.

The result of one gauging procedure is the mass flow characteristic over pressure p_1. Regarding the flow through the valve, p_1 represents the static pressure in front of the valve (input pressure). A set of characteristics results from the variation of the pressure p_2 downstream of the valve. According to the interesting pressure range from p_{Amb} to $1.5\,\text{bar}_{\text{Abs}}$, p_2 was varied in 0.1 bar steps. The set is completed by changing the flow direction from "normal" flow to "reverse" flow.

In a second step, the outlet metering edge was measured (Figure 5). In addition, the switching delay t_{SD} was determined (Siebertz *et al.*, 2010). This is an important parameter, which characterizes the valve's opening dynamics. A second parameter, the pressure loss coefficient $K_{\text{V},\Delta p}$, describes the valve's leakage behavior. Both parameters are implemented in the simulation model described in the following sections.

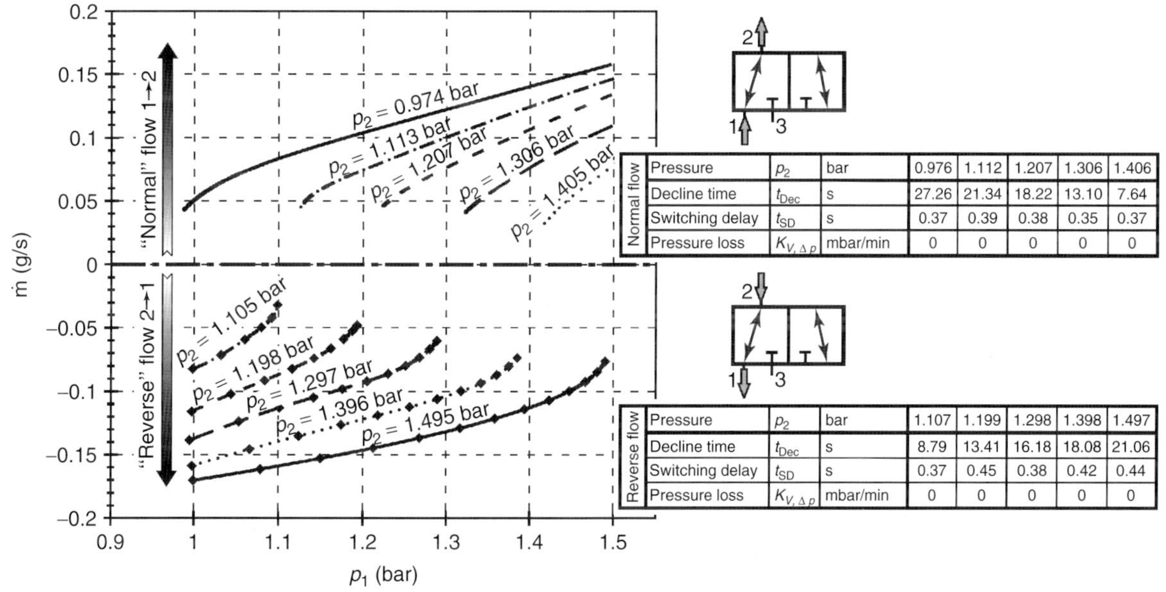

Normal flow							
Pressure	p_2	bar	0.976	1.112	1.207	1.306	1.406
Decline time	t_{Dec}	s	27.26	21.34	18.22	13.10	7.64
Switching delay	t_{SD}	s	0.37	0.39	0.38	0.35	0.37
Pressure loss	$K_{V,\Delta p}$	mbar/min	0	0	0	0	0

Reverse flow							
Pressure	p_2	bar	1.107	1.199	1.298	1.398	1.497
Decline time	t_{Dec}	s	8.79	13.41	16.18	18.08	21.06
Switching delay	t_{SD}	s	0.37	0.45	0.38	0.42	0.44
Pressure loss	$K_{V,\Delta p}$	mbar/min	0	0	0	0	0

Figure 4. Characteristics and parameters of the flow path $1 \rightarrow 2$ of an exemplary valve.

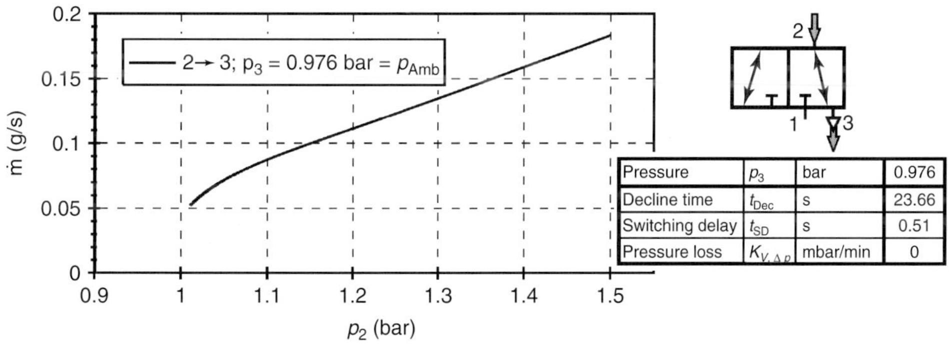

Pressure	p_3	bar	0.976
Decline time	t_{Dec}	s	23.66
Switching delay	t_{SD}	s	0.51
Pressure loss	$K_{V,\Delta p}$	mbar/min	0

Figure 5. Characteristics and parameters of the flow path $2 \rightarrow 3$ of an exemplary valve.

3.1.2 Compressor

The result of the compressor gauging is its conveying characteristics and a few other parameters, as depicted in Figure 6. The characteristics over the relative pressure difference of two exemplary compressors are shown. According to the compressor's performance, the pressurization of the measurement volume is only possible up to the compressor's maximum pressure $\Delta p_{\mathrm{Max}} = p_{\mathrm{Max}} - p_{\mathrm{Amb}}$. The leakage rate $K_{\mathrm{C},\Delta p}$ is determined in order to specify the compressor's quality. Driving the compressor's motor with 12 V, the power consumption of the compressor is quantified by the averaged current value I_{Avg}. While the compressor's characteristic is used for simulation, the other parameters can be used as quality attributes of the compressor, too.

3.1.3 Cushions

The volume as a function of the internal pressure characterizes the seat cushions' pneumatic behavior. It is determined by scaling the cushion while filling it with water, as water can be deemed to be incompressible in the investigated pressure range and its specific weight is well known. To allow the measurement of pressure-dependent characteristics, the water reservoir, connected to the cushion, is pressurized by an entering airflow, whereas the internal pressure of the seat cushion is determined by a pressure sensor at its inlet. Figure 7 shows the behavior of three different cushions, measured without applying a load. It becomes obvious that—after a filling period with minimal pressure rise—a linear relationship of pressure and volume exists. The results show steady-state volumes, whereas the strain of the cushions' plastics leads to a time dependence, comparable to a first-order lag element. The knowledge of the system behavior is necessary to enable a correct implementation of the cushions in DSH*plus*.

3.1.4 Measured data processing

The behavior of the measured compressors and valves has to be transferred into characteristic fields for their use in the one-dimensional simulation. These characteristic fields have—in the case of valves—two input parameters (the inlet and outlet pressures p_1 and p_2) and one output (the mass flow $\dot{m}$). The simulation software DSH*plus* requires tables, providing an output value for each possible combination of input sampling points (FLUIDON, 2007). This

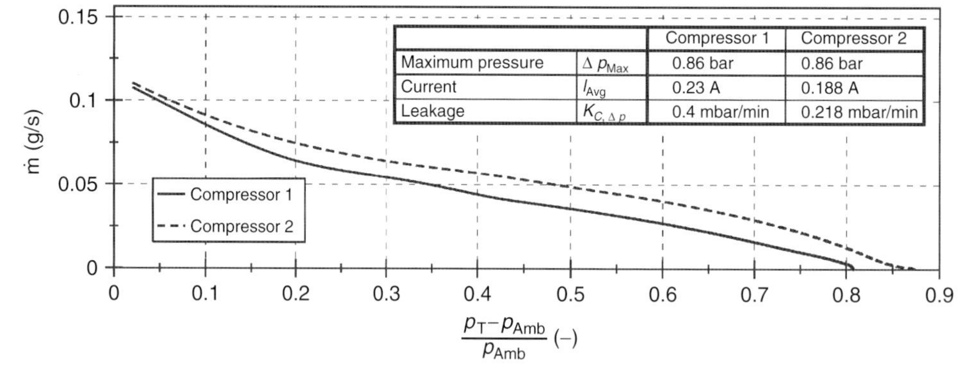

		Compressor 1	Compressor 2
Maximum pressure	Δp_{Max}	0.86 bar	0.86 bar
Current	I_{Avg}	0.23 A	0.188 A
Leakage	$K_{C,\Delta p}$	0.4 mbar/min	0.218 mbar/min

Figure 6. Characteristics and parameters of two exemplary compressors.

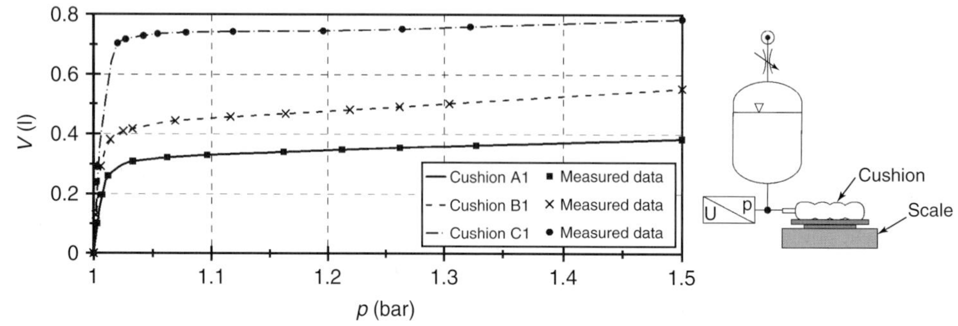

Figure 7. Cushion volume as a function of its internal pressure.

means, in case of the above-mentioned gauging method, that the sampling points of the inlet pressure have to be identical for each measurement, whereas the outlet pressure is varied. The measured inlet pressure is influenced by the mass flow across the measured object. Thus, at different points, the measurements will vary. Common sampling points have to be generated by a spline interpolation algorithm, interpolating from the measured points at defined pressure levels. Here, the "radar-spline" method can be applied to generate additional curves between the measured ones and thus can be used to avoid the above-stated problems, occurring during linear and common spline interpolations. This method considers zero mass flow in case of pressure equality, represented by an additional diagonal in the characteristic diagram. Its endpoints serve as center points for the radar-splines, defined by one center point and crossing points with the present curves. Finally, additional curves between the measured ones can be achieved by interpolation of sampling points generated by radar-splines. In this way, a characteristic diagram with a more continuous behavior (Figure 8) is generated. The level of discontinuity and the difference from the above-described physical law decrease with a higher number of data points in the characteristic map. As the number of data points to be generated with the radar-spline method is unlimited, the model behavior can be optimized by a compromise between model quality and data volume. The data import into DSH*plus* is realized either by a provided characteristic map creator for two-dimensional characteristic maps (one input parameter results in one output parameter) or by an ASCII-text file. The wide compatibility of almost any data acquisition software to ASCII-text files allows the import of measured data from almost any source to DSH*plus*. Furthermore, this method allows the import of three-dimensional characteristic maps by simply appending a two-dimensional characteristic map for every coordinate in the third dimension to the text file.

The deflating valve can be described by a one-dimensional characteristic (Figure 5), as the ambient pressure is considered to be constant ($p_3 = p_{\mathrm{Amb}} =$ constant). The measured compressor mass flow as a function of its outlet pressure can also be imported as a one-dimensional characteristic without further calculations (Figure 6). The inflatable seat cushions are described by their pressure-dependent volume. The dynamical strain behavior is considered by summation of a proportional and a first-order lag element.

3.1.5 Simulation model

The validation of the component models in DSH*plus* can be achieved by simulating and measuring the time variant pressure in a system, consisting of elements measured beforehand. The system includes a compressor, a triple valve (three single 3/2-way valves in one housing), three cushions of two different types, and several hoses. In the DSH*plus* model, each component is represented by a corresponding component model, parameterized by measured characteristics, whereas the volume of the hoses is contained in the depicted volume nodes. Compressor and valves of the simulation model are driven by the measured control signals of the test bench to achieve identical conditions for simulation and measurement. Figure 9 shows the test bench and the corresponding simulation model.

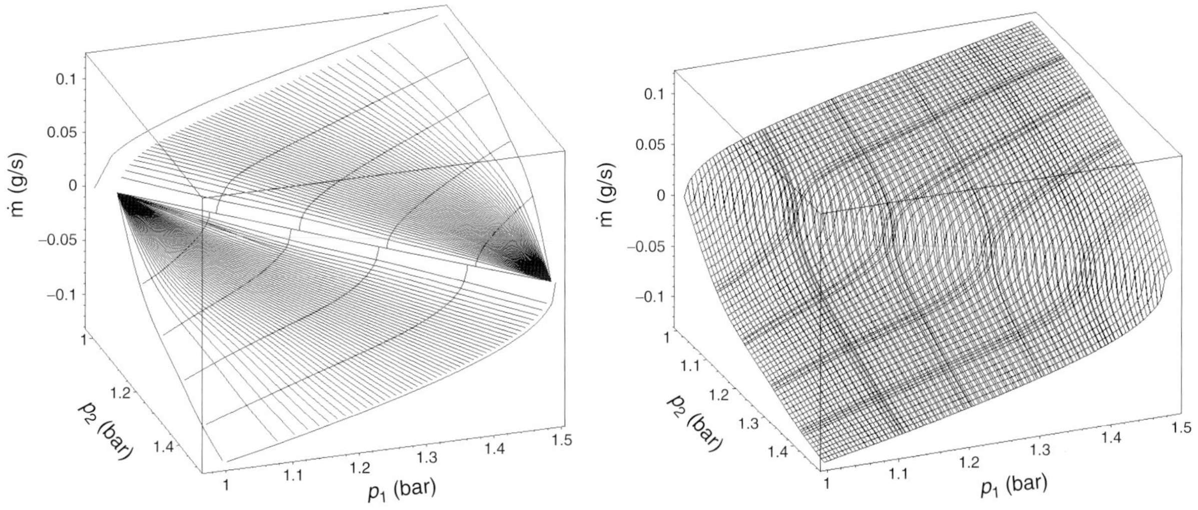

Figure 8. Characteristics map of the valve using the radar-splines method.

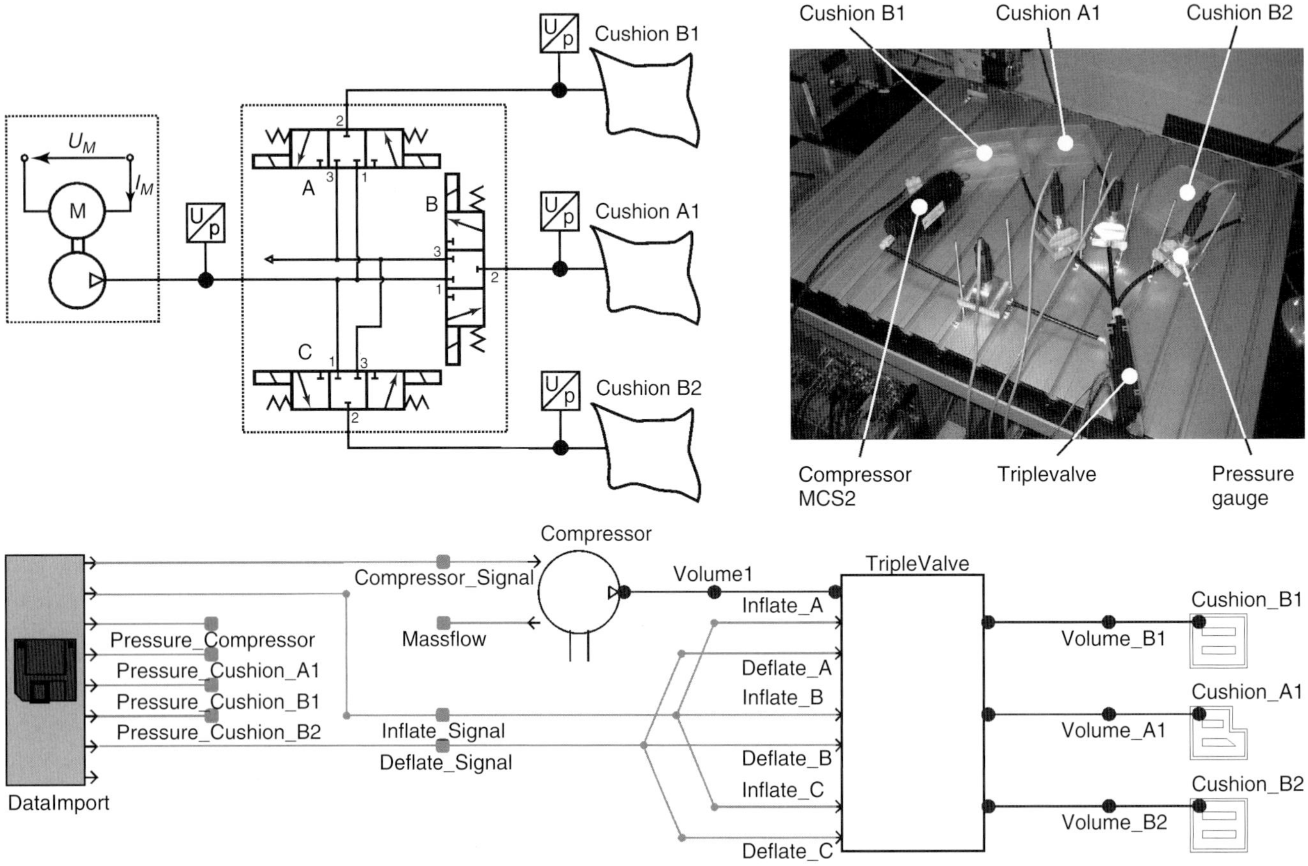

Figure 9. Test bench and corresponding simulation model.

The very good correlation between simulation results and measurements shown in Figure 10 emphasizes the high quality of the gauging method and the modeling. In combination with the fast gauging method for all system components, the simulation model enables the user to optimize the contour adjustment of automotive seats in a very short time. This is a big improvement compared to the state of the art method consisting of trial and error bench testing of hardware components.

The one-dimensional simulation enables the user to conduct parameter studies in a very fast and comfortable way, because no additional pneumatic components are required.

3.1.6 Summary of example 1

As an example of a complex pneumatic system, a pneumatically actuated comfort system for the contour adjustment of automotive seats was presented. A measurement method for the characterization of the system's elements can be used for quality inspection tasks, fault detection, and quantification, as well as the extraction of element-specific maps and parameters. These maps and parameters can be used for parameterization of a simulation model. Pneumatic valves, compressors, and cushions were investigated in order to determine their theoretical model behavior. The measurement method is not limited to pneumatic miniature elements and can be used for common elements of industrial pneumatics, likewise. The good correlation between measured pressure characteristics and its simulated values during the validation phase emphasizes the high quality of the simulation model and the need for highly accurate data to parameterize the components. Therefore, the simulation model can be used for a preestimation of the system's behavior or an investigation of the parameters' sensitivity.

3.2 Example 2: simulation of tubes and hoses

In automotive engineering, tubes and hoses are used most of the time to connect different pneumatic and hydraulic subsystems and components, such as the fuel system, the air conditioning, brake systems, exhaust systems, and power steering unit. These components typically consist of highly

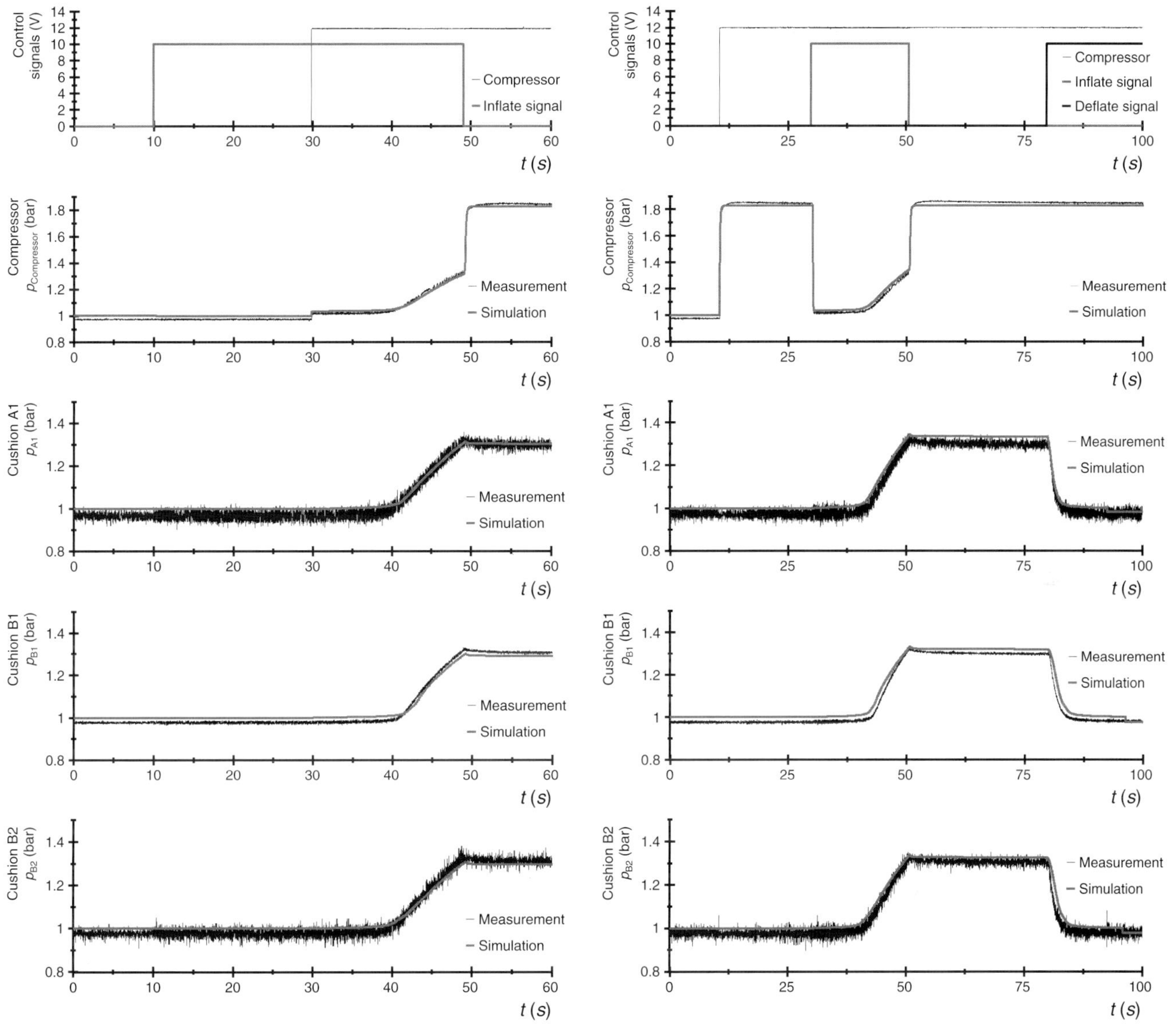

Figure 10. Validation of the simulation model.

integrated components, which must be are arranged on a very limited installation space. These limited installation conditions require flexible pathways and good mechanical features. Tube and hose connections fulfill those requirements and represent very flexible connection elements.

Figure 11 shows a typical expandable automotive line of average complexity, used in power steering units.

Owing to excitation from different sources, such as pumps, valves, and vibrations caused by the engine or other ancillary units, unwanted pulsations in the tubes or hoses can arise. These pulsations can lead to mechanical vibrations, which could cause unwanted noise emission in the form of structure-borne sound. The vibrations can cause a disturbance in the passenger cabin, which deteriorates the comfort of the vehicle. Furthermore, pressure peaks caused by resonance can result in high component strains especially in high pressure connections. These effects are summarized within the term *NVH* (*noise, vibration, and harshness*), which is also used to describe the study and modification of noise and vibration characteristics of vehicles.

Well known as a source of noise within the interior of a vehicle is the hydraulically assisted power steering system. The hydraulic pump of the steering system initiates a pressure ripple into the hydraulic circuit in which the pressure ripples propagate throughout the system as fluid-borne noise. To minimize the noise emitted by the power steering system, different noise reduction techniques are used. The most convenient way to reduce the noise emission

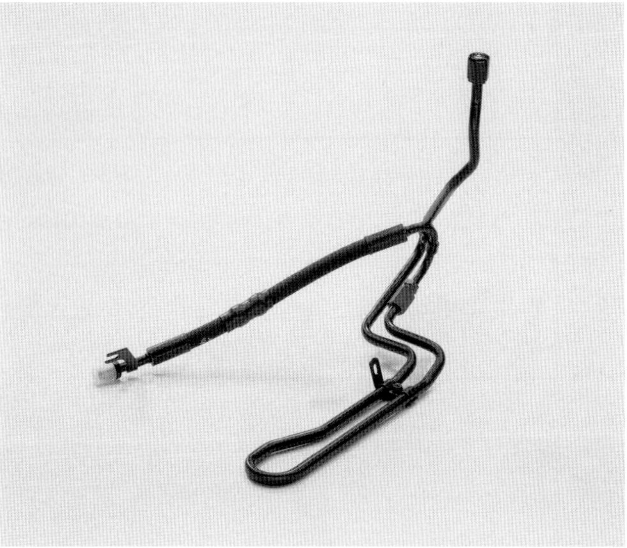

Figure 11. Typical expandable automotive line with average complexity (Johanning, Baum, and Wurmann, 2009). (Reproduced by permission of Heiko Baum.)

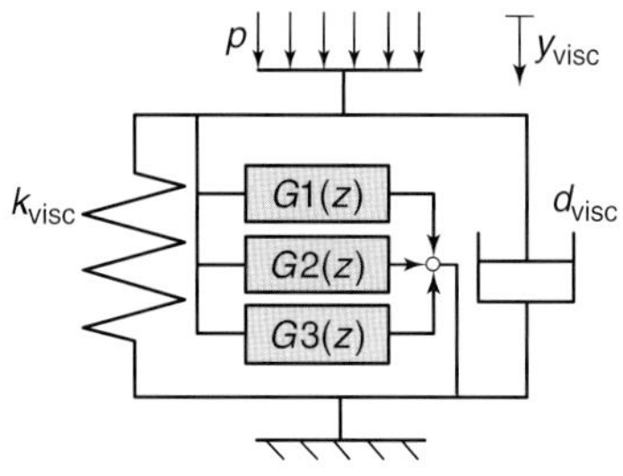

Characteristic equation in the *x*-*t*-plane

$$C^{+}: \frac{1}{\sqrt{\rho E_{Fl}}}\frac{dp}{dt}+\frac{d\upsilon}{dt}+\frac{1}{p}\frac{\partial p_f}{\partial x}+\frac{2c}{Rd_{visc}}(G(z)+(p-k_{visc}y_{visc}))=0$$

$$C^{-}: \frac{1}{\sqrt{\rho E_{Fl}}}\frac{dp}{dt}+\frac{d\upsilon}{dt}+\frac{1}{p}\frac{\partial p_f}{\partial x}+\frac{2c}{Rd_{visc}}(G(z)+(p-k_{visc}y_{visc}))=0$$

Figure 12. Implementation of parametric resonance into the characteristic equation (Baum and Johanning, 2009). (Reproduced by permission of Heiko Baum.)

is the installation of a tuning cable with a certain length to induce a destructive interference and therefore attenuate the amplitude of the pressure ripple (Visteon, 2005).

In addition to the destructive interference, structural damping can be used for noise reduction. The length of elastic hose elements in the hose assembly is increased, so that the energy from the pressure ripples is absorbed by the expansion of the hose wall. Furthermore, the hose reduces the wave speed in the fluid (Visteon, 2005).

If excessive noise emission and vibrations occur in the final product, the reinstatement work usually takes a lot of effort and time. To compensate or suppress the propagation of these pulsations throughout the connections and components, the mechanical features of the connections must be considered in an early stage of the development process.

In the past, a combination of different methodologies using analytical (Beater, 1999) and experimental approaches were used. However, all of these methodologies show significant disadvantages. The experimental methodology using a trial-and-error approach on a physical model is very time consuming and therefore costly. The analytical methodology uses a set of equations in a closed form and is therefore often restricted to simplified geometry and material properties. Furthermore, the increasing complexity of, for example, the power steering system and the typically nonlinear system behavior complicate the analytical description. The use of a finite element approach allows an accurate description of the isolated tubes and hoses but struggles with the simulation of the whole system because of the complexity of component interactions, which must be regarded.

A methodology without the above-mentioned disadvantages facilitates a time-based modeling approach provided by modern system simulation tools, such as DSH*plus*. In contrast to finite element method, the time-based approach allows the accurate and fast evaluation of the NVH characteristics of the system, including system vibration, airborne noise, and fluid-flow characteristics because the parameters at the boundaries are calculated by physically modeled system components such as the pump and valve of a power steering unit. This even enables the simulation of time-dependent operating conditions such as driving maneuvers.

This makes the system simulation the first choice to perceive and circumvent problems regarding NVH. Furthermore, changes can be applied and tested in an early development stage, which leads to a significant cost saving and reduction of time to market. In addition, system simulation tools can be used to troubleshoot systems in which NVH problems occur. The system simulation allows to clearly identify the source of vibration in a depicted and parameterized system. Furthermore, the system simulation enables the user to develop and test adequate countermeasures in a short period of time.

To obtain accurate results, all relevant physical aspects of the connection element must be implemented and properly parameterized. In the modeling process, this poses the greatest challenge. For example, steel tubing allows a simple phenomenological approach to describe the viscoelastic properties of the tube walls by a simple spring damper model (Müller, 2002), whereas the damping of hoses is essentially influenced by the temperature- and pressure-dependent viscoelastic properties of the material used for the hoses. Furthermore, measurements showed that typical hoses used in power steering units allow volumetric expansion rates of 15–35 ccm/m at 23°C, which is substantially higher than steel tubes.

The viscoelastic properties have a significant influence on the damping and are adjustable by the hose material and the winding characteristics. In a simple spring damper model of the hose wall, the interaction of the characteristic frequency of the connection element and the damping of the connection wall is neglected. The representation of these specific characteristics requires a more complex model. Therefore, a parametric resonance is implemented into the model Figure 12.

The parameterization of the phenomenological models is done by measurements on a test bench, which allows the characterization of a specific material property of a hose.

On the basis of the characterization of different hoses, the characterization for frequently used fittings and hose elements, as shown in Figure 13, is available in the DSH*plus* component library.

By combining these basic elements, almost any desired elastic hose configuration can be modeled. Furthermore, the combination of basic elements allows the integration of resonators such as volume, Helmholtz, and cavity resonator, as well as throttles or orifices. Furthermore, a combination with a tuner cable is configurable and allows the optimization of a whole hose assembly.

3.2.1 Measured data processing

The foundation of realistic simulation results is the correct parameterization of the system components in the system

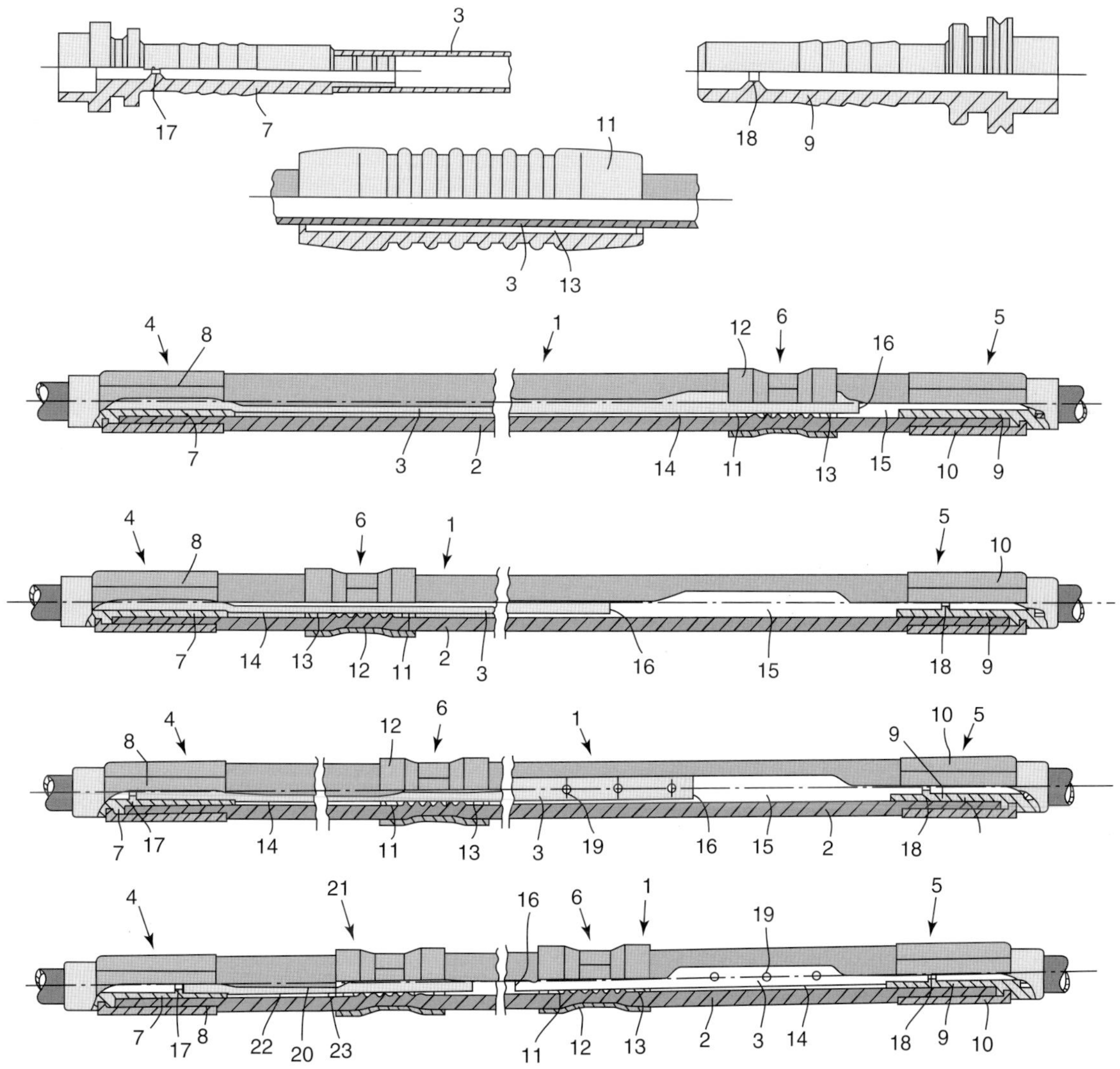

Figure 13. Examples for fittings and flexible hose elements. (Eaton, 1992. Reproduced by permission of Eaton Aeroquip GmbH.)

simulation. To correctly parameterize connections in the simulation, their dynamic properties are required. A well-established approach to determine the dynamic properties from experimental data is the quadrupole analysis, which allows the determination of the wave propagation in tubes and hoses in a wide frequency spectrum. On the basis of the measured data, the transmission behavior of the connection can reliably be reproduced in the simulation tool.

The quadrupole analysis is based on a process, originated in the field of electrical engineering, and is used to characterize an electrical network with two pairs of terminals connected together internally by the electrical network, which is to be investigated. This allows the mathematical description of any linear circuit detached from the physical buildup, provided that the circuit does not contain an independent source and satisfies the port condition. The port condition requires that the same current must leave and enter each pair of the terminals.

By the transferability of electrical effects to fluid power effects, the current corresponds to the volume flow and the potential corresponds to the pressure.

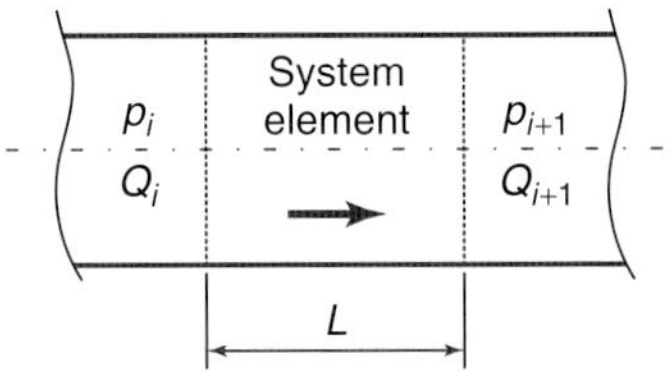

Figure 14. Line segment.

Therefore, a linear fluid technical system element can be described similarly to an electrical system element by four parameters. The transfer matrix consists of the elements T_{ij}, which correlate the pressure p_i and volume flow Q_i of the input side with the pressure p_{i+1} and volume flow Q_{i+1} of the output side of a line segment (Figure 14).

$$\begin{pmatrix} p_i \\ Q_i \end{pmatrix} = \begin{bmatrix} T_{11} & T_{12} \\ T_{21} & T_{22} \end{bmatrix} \begin{pmatrix} p_{i+1} \\ Q_{i+1} \end{pmatrix} \tag{12}$$

The vector of state variables in the frequency domain of $\hat{p}_1$ and $\hat{Q}_1$ on the input side results from the multiplication

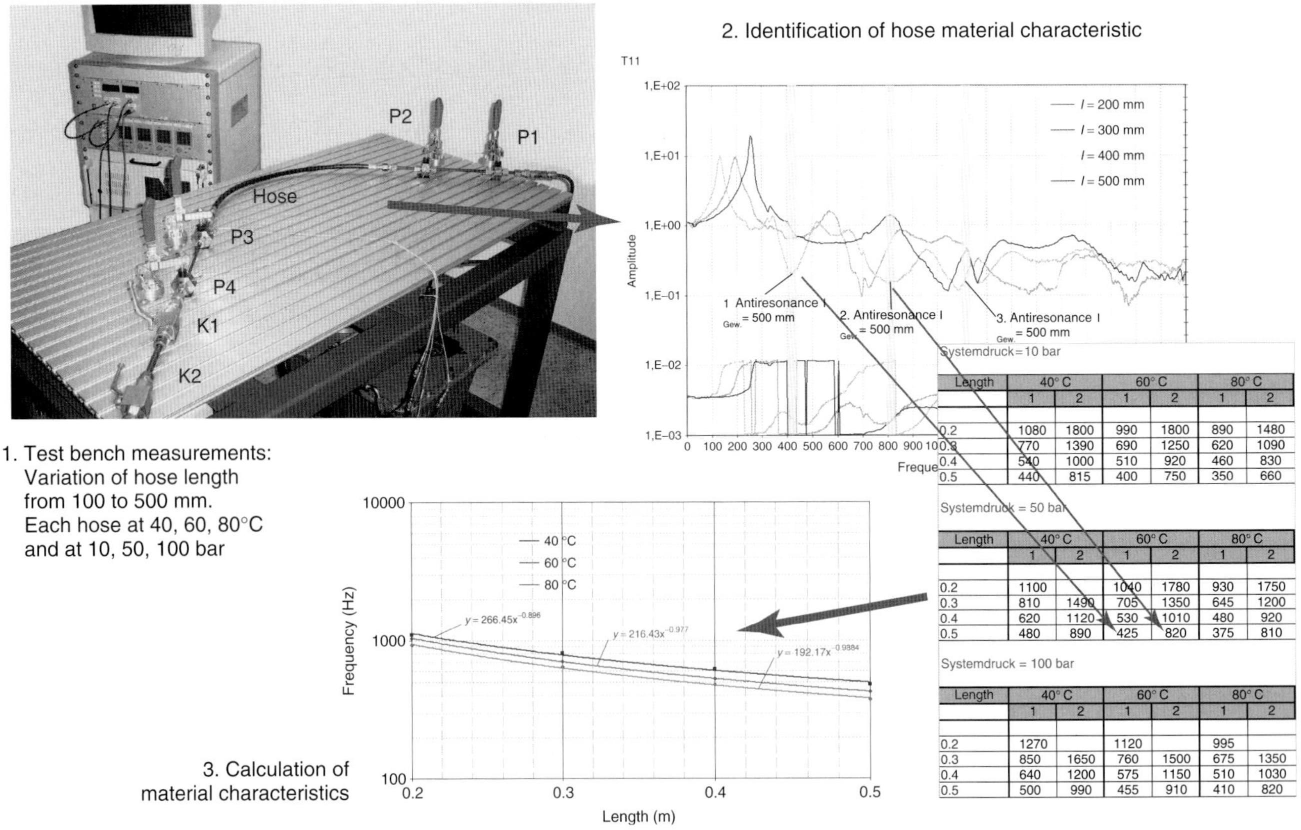

Length	40° C		60° C		80° C	
	1	2	1	2	1	2
0.2	1080	1800	990	1800	890	1480
0.3	770	1390	690	1250	620	1090
0.4	540	1000	510	920	460	830
0.5	440	815	400	750	350	660

Length	40° C		60° C		80° C	
	1	2	1	2	1	2
0.2	1100		1040	1780	930	1750
0.3	810	1490	705	1350	645	1200
0.4	620	1120	530	1010	480	920
0.5	480	890	425	820	375	810

Length	40° C		60° C		80° C	
	1	2	1	2	1	2
0.2	1270		1120		995	
0.3	850	1650	760	1500	675	1350
0.4	640	1200	575	1150	510	1030
0.5	500	990	455	910	410	820

Figure 15. Test bench for identification of hose material characteristics (Baum and Johanning, 2009). (Reproduced by permission of Heiko Baum.)

of the transfer matrix T and the state variables $\hat{p}_2$ and $\hat{Q}_2$ on the output side.

To determine the transfer matrix T, the pressure pulsations can be measured directly with high resolution using pressure sensors. An indirect measurement method is necessary to determine the volume flow because volume flow rate sensors either have an influence on the pressure propagation or lack the required dynamic properties. Therefore, the volume flow rate is usually determined in a reference tube, whose characteristics are well known, so that the flow rate can be determined indirectly from the measured pressure signal.

3.2.2 Test bench

The transfer matrix can be determined from the Fourier transformation of the measured pressure signals of any given tube or hose. Figure 15 shows the test bench with a mounted hose segment.

The measured tube characteristics allow the system simulation to consider the connection properties. Hence, any combination of these elements can be simulated. The model validation of a line assembly is conducted on the real test bench, which is also available as fully parameterized component in DSH*plus*, depicted in Figure 16. Therefore, the validation process of newly implemented components is narrowed down to comparing the simulated line characteristics to the measured line characteristics.

The following example shows the optimization process of pulsation propagation in DSH*plus*.

At the beginning of every development, process stands the customer specifications regarding the dynamic system behavior of the assembly of expandable hoses and tubes. The goal of the simulation process is to find an adequate assembly of hoses and tubes, which fulfill the targeted pressure losses, frequency response, or amplitude damping.

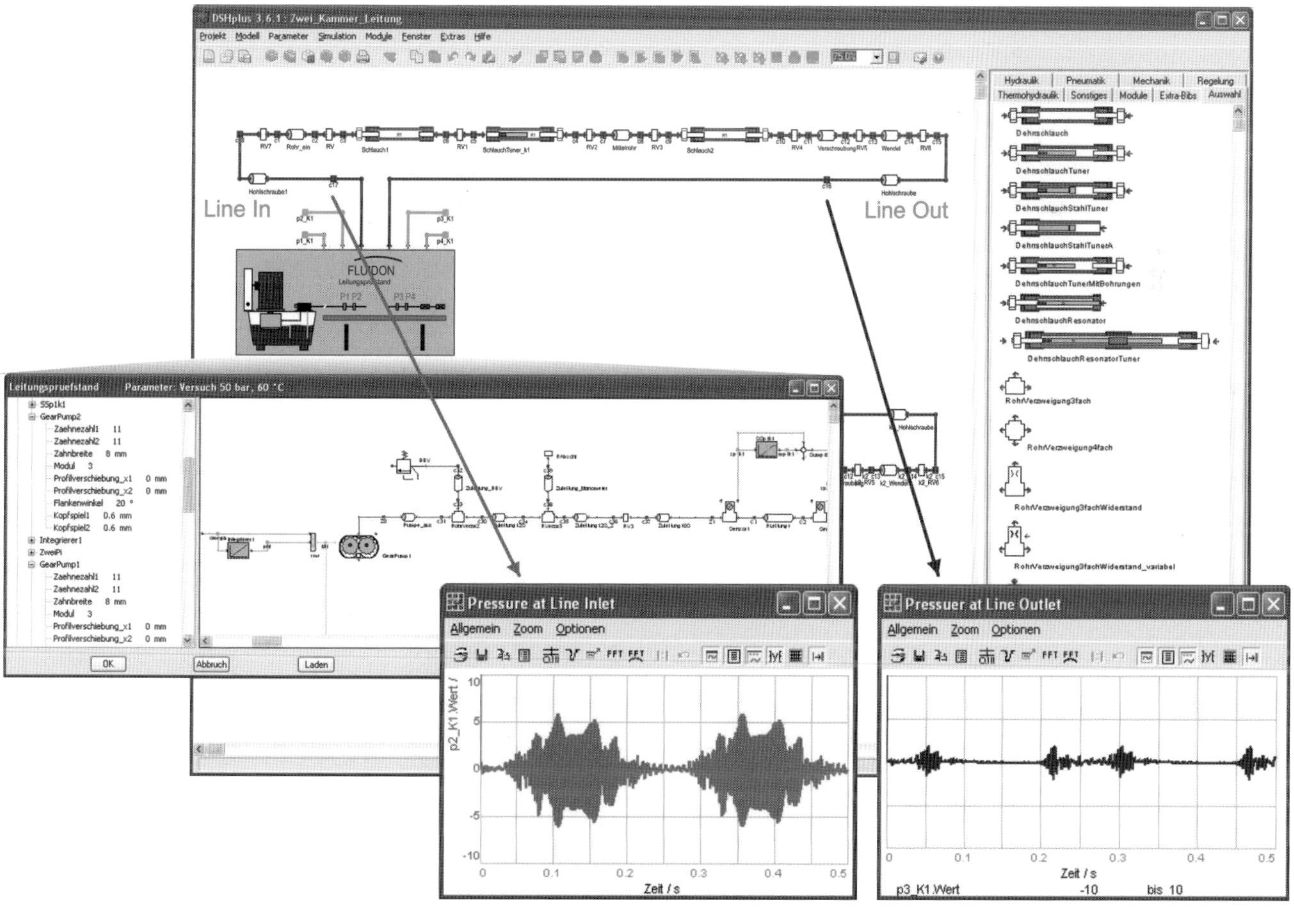

Figure 16. Flexible hose element library and simulation model of test bench with installed line (Baum and Johanning, 2009). (Reproduced by permission of Heiko Baum.)

For the final design, only little tuning on the test bench or the vehicle is required. This allows a significant time saving compared to a typical trial and error procedure.

Figure 17 depicts the procedure of the optimization process. Beginning with the CAD data of the packaging space and connection alignment, a first connection design is created, outgoing from developments in former projects.

On the basis of the geometrical data of the CAD model, the assembly is recreated in DSH*plus* from the component library and parameterized by the geometric dimensions of the design. Only the expandable hose sections are parameterized by a data set that describes the elastic properties of the hose material. If the data set of the specific material is not available in the material database, the material properties must be measured once before the design process and are then available for future projects from the material database.

Integrating the design into the virtual hose test bench, which is equilibrated with the real test bench, offers a simple way to validate the simulation model.

Furthermore, the flow conditions of the desired application on the ends of the hose assembly must be set.

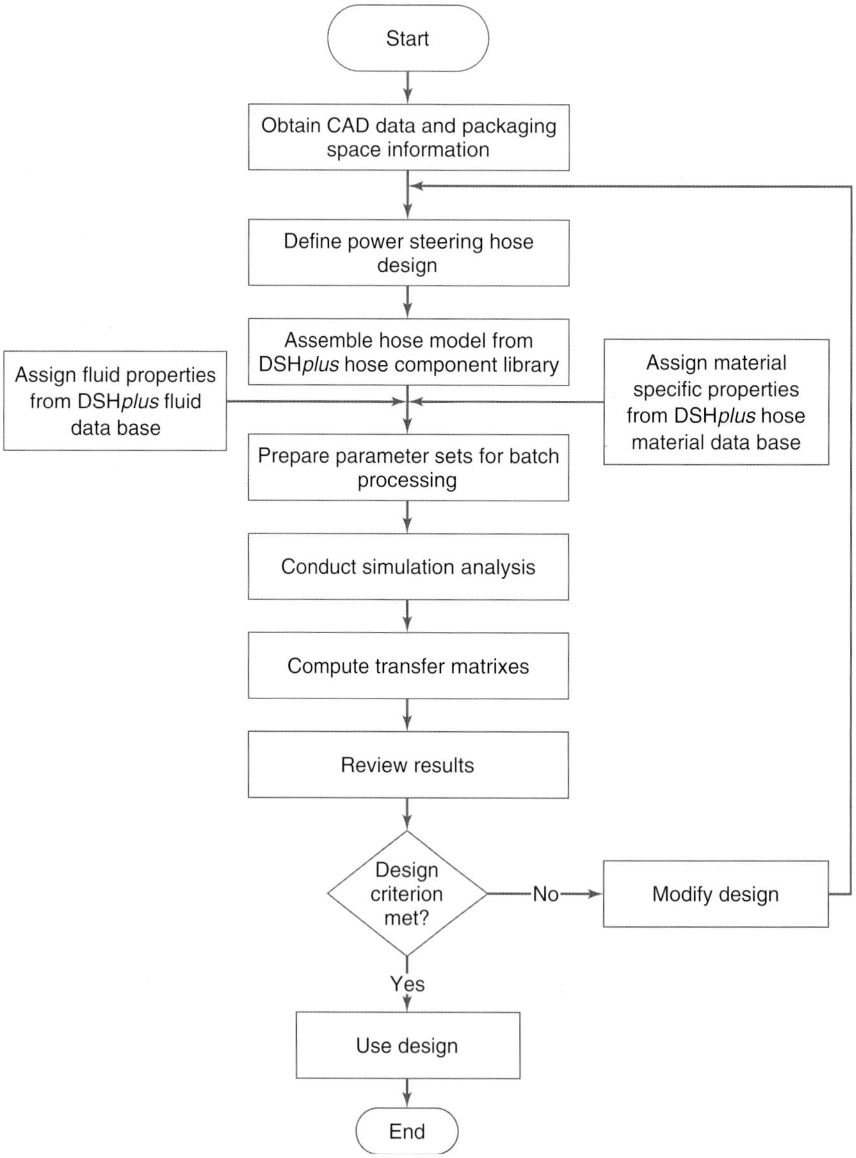

Figure 17. Development methodology to optimize automotive hose assemblies (Baum and Johanning, 2009). (Reproduced by permission of Heiko Baum.)

This can be achieved either by modeling the necessary system components such as valves or pumps or by a characteristics diagram of the connected components. If the parameterization of the hose assembly is finalized, an automated design variation under consideration of the prescribed packaging space and design requirements can be initiated. The quadrupole analysis of DSH*plus* allows the graphical representation of the findings.

Furthermore, DSH*plus* allows a parameter-driven design optimization. The user specifies quality criteria over a certain frequency spectrum, which is then used to drive the parameter variation. Furthermore, the influence of different geometric design parameters can be investigated by a design of experiment (DOE) analysis.

If the developed geometry satisfies the design requirements, a prototype can be build up; otherwise, the user should return to the beginning of the design process and start again with a new starting design.

3.2.3 Application of the tube and hose model

Using the example of a two-chamber automotive line with PTFE tuner and a single-chamber automotive line with steel-flex-tuner, the suitability of DSH*plus* as simulation tool is presented. The two connections are existing stock parts and have been modeled in DSH*plus* to use in a design parameter study. The design parameter study allows the automated variation of geometric lengths and diameters of the tuners, hoses, and tube segments as well as the position of the tuner elements.

To validate the simulation results, the flow characteristics of the connections have been measured on the test bench.

Figure 18 shows the transmission behavior of the two-chamber automotive line. The transmission behavior of the initial configuration A shows a distinct rise in the amplitude ratio in the frequency range 500–700 Hz. The optimization process of configuration B results in an improved damping

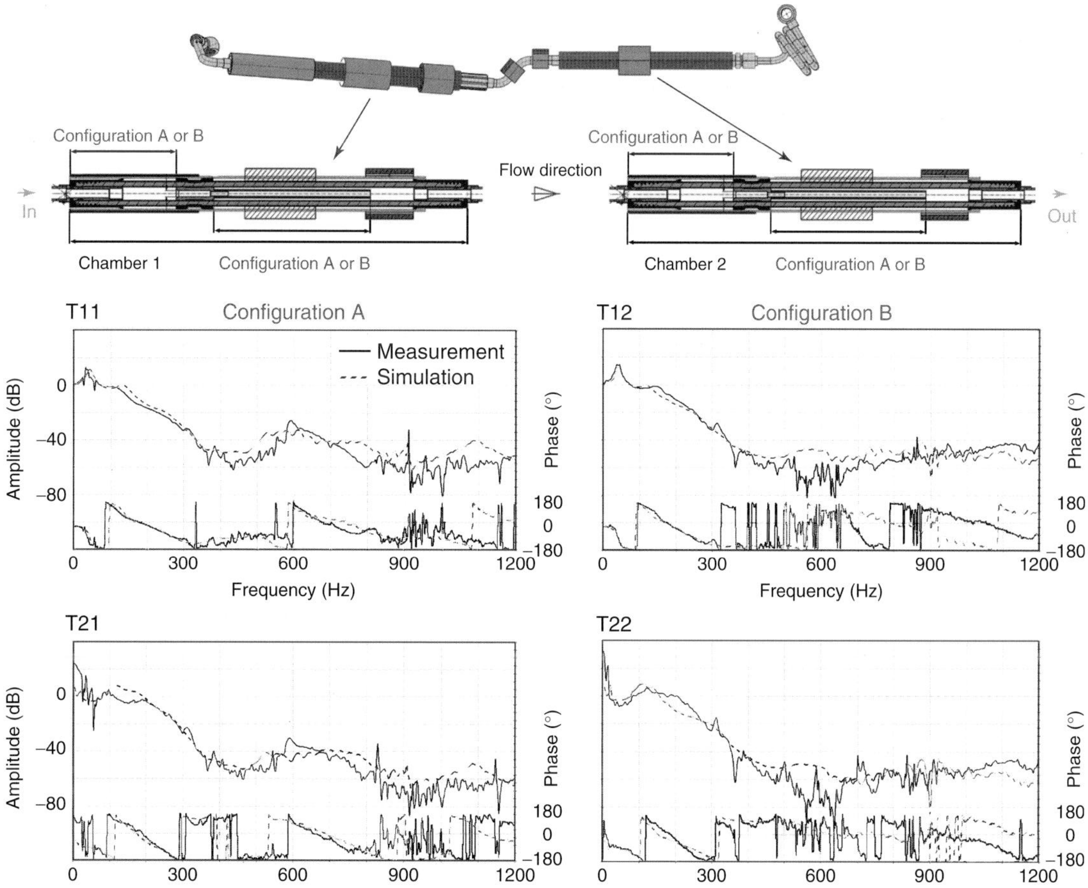

Figure 18. Measuring and simulation of two-chamber automotive line (Baum and Johanning, 2009). (Reproduced by permission of Heiko Baum.)

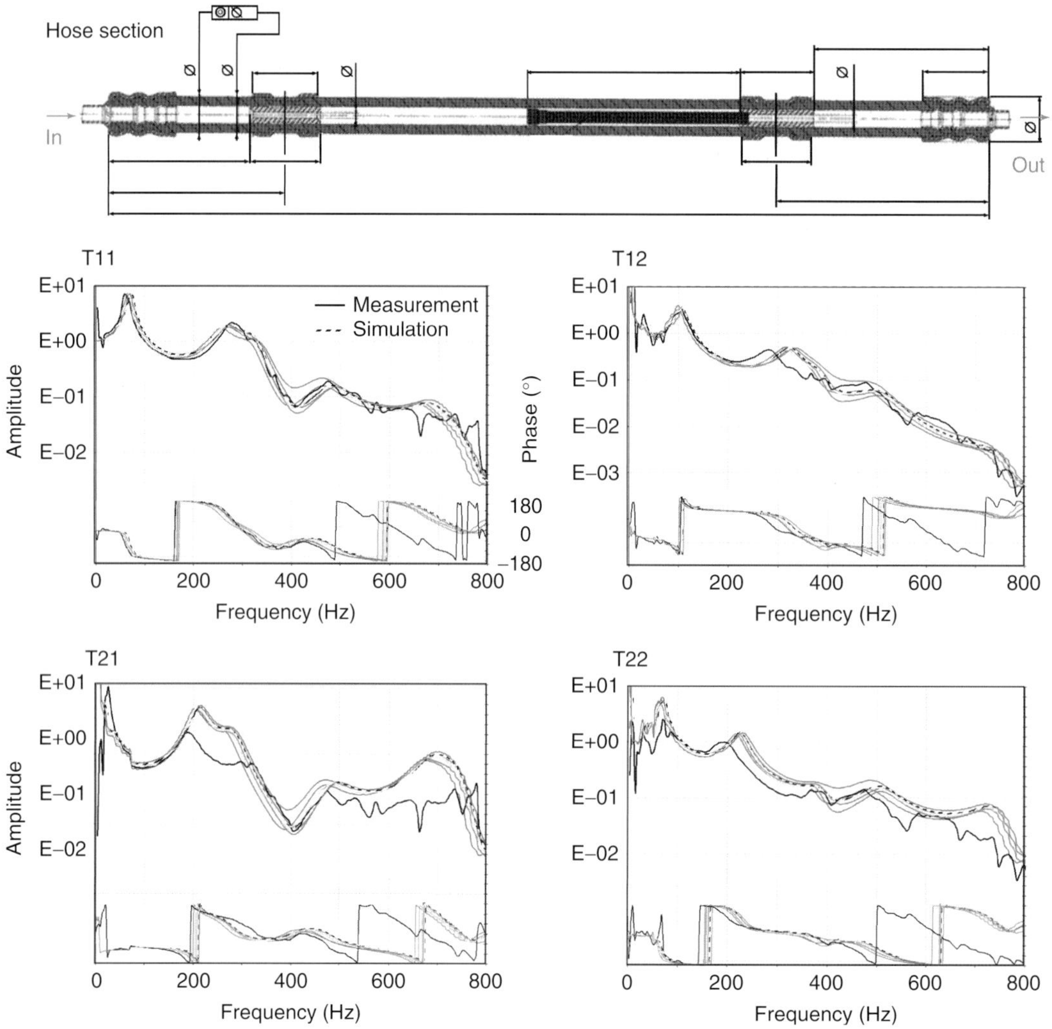

Figure 19. Robustness analysis of a one-chamber line with steel tuner (Baum and Johanning, 2009).

in the frequency range, owing to a variation of the throttle position and length of the tuner cable.

Figure 19 depicts the measurement and simulation of a flexible one-chamber-line with steel tuner.

In the first simulation of the one-chamber line, the exact geometric dimension of the existing line was modeled. On the basis of the performed simulation, a parameter variation of the relevant geometric dimensions, as depicted in Figure 19, is conducted. The limits for the parameter variation of the geometric dimensions were deduced from the tolerance specification of the line. This analysis shows the variation in the dynamic properties of the line within the accepted tolerance span of manufacturing. The simulated variations envelope the curve of the simulation using the exact geometric dimensions. The parameter variation enables the identification of sensitive geometric dimensions. On the basis of the simulation results in an early development stage, appropriate countermeasurements can be considered.

3.2.4 Summary of example 2

The use of system simulation tools accompanying the development process of expandable automotive lines in the early stages allows a significant cost and time saving because protracted trial-and-error studies can be circumvented.

DSH*plus*, among others, has all necessary software modules to aid the designer during the whole design process. The component library for automotive lines allows the user to start with a virtual prototype for a variety of different application fields not only limited to the automotive sector. On the basis of the virtual prototype, the design and optimization process regarding the specified design requirements can be performed. In order to do this, DSH*plus* offers a comfortable way to perform optimization and robustness – as well as sensitivity studies for the line design. Furthermore, DSH*plus* allows the evaluation of different parameters utilizing methods of DOE. This accelerates the overall design process and significantly reduces the number of time-consuming field tests.

4 SUMMARY

Modern system simulation tools are an indispensible utility in the modern design process, applicable to a wide variety of fields in engineering. System simulation tools such as DSH*plus* provide the user with a certain level of planning security in an early design stage. Furthermore, modern simulation tools provide a possibility to test different design concepts with relatively little effort in time and money. Later in the design process, system simulation tools offer valuable optimization tools to optimize an initial design under consideration of the specified design requirements. The resulting design is optimally attuned to the specifications in a short time span compared to conventional methods of trial and error. In addition, the provided tools allow an elaborate comparison and assessment of different designs.

Furthermore, system simulation can be used to illustrate and evaluate the system behavior of interconnected components. Early system evaluation significantly reduces failure during first startup. Furthermore, the depiction of whole systems allows optimal choice of components before realization. Hence, a cost optimal solution can be found for the manufacturer and for the customer.

RELATED ARTICLES

Volume 2, Chapter 43
Volume 5, Chapter 144

REFERENCES

Baum, H. and Johanning, H.-P. (2009) *Simulation of flexible hose lines for power steering and active chassis application*. Aachener Kolloquium Fahrzeug- und Motorentechnik 2009.

Beater, B. (1999) *Entwurf hydraulischer Maschinen: Modellbildung, Stabilitätsanalyse und Simulation hydrostatischer Antriebe und Steuerungen (VDI-Buch)*, vol. 1999, Springer-Verlag, Berlin Heidelberg.

Eaton Aeroquip GmbH (1992) Expandable Hose that Reduces the Hammering Produced in Hydraulic Systems by Pumps. United States Patent 5094271, Publication Date: 1992-03-10.

FLUIDON GmbH (2007) User Manual DSH*plus* 3.7, FLUIDON GmbH, Aachen, Germany.

Johanning, H.-P., Baum, H., and Wurmann, G. (2009) Simulation von Dehnschlauchleitungen für Lenkung und Fahrwerk. ATZ-Automobiltechnische Zeitschrift, 2009-06.

Müller, B. (2002) Einsatz der Simulation zur Pulsations- und Geräuschminderung hydraulischer Anlagen. Dissertation RWTH Aachen.

Siebertz, K., Reinertz, O., Fritz, S., and Murrenhoff, H. (2010) Rapid Gauging Method and Generic Modelling Approach for Pneumatic Seat Components. *7th International Fluid Power Conference*, 22–24 March, Aachen, 2010.

Visteon Global Technologies, Inc. (2005) Method of Power Steering Hose Assembly Design and Analysis. United States Patent 6917907, Publication Date: 2005-07-12.

Chapter 135

Comparison of the Modeling Techniques for Chassis Applications. An Advice for the User

Frank Heßeler, Alexander Katriniok, Matthias Reiter, Jan Maschuw, and Dirk Abel

RWTH Aachen University, Aachen, Germany

1 INTRODUCTION

Progress in simulation technology leads to a wider application of simulation software for the reduction of testing costs in many engineering disciplines. This trend is also true for vehicle design at all development levels. The level of detail of the models is sufficient for simulation of entire vehicles as well as for detailed simulations of separate mechanical or electrical components. Each engineering discipline applies its own software tools, developed specifically for the purposes of application. An updated and complete overview of single programs is thus not easy to provide and lies beyond the scope of this chapter.

In order to meet the increasing requirements to safety and comfort in a vehicle, pure mechanical systems are complemented or replaced by mechatronic systems. This tendency can be also observed in chassis systems. Typical examples of such systems are antilock braking system (ABS) and electronic stability control (ESC), which are currently part of a standard equipment of each modern vehicle. Moreover, there is an increase in the development and marketing of comfort features, such as adaptive damping or electric power steering. Such systems can be strongly connected with safety critical systems, such as ESC. To exemplify this statement, electric power steering can be named, which receives steering directions from the ESC and thus also implements a safety critical task.

In addition to the pure mechanical construction of the chassis components, the development of mechatronic systems receives a steadily growing attention of the automobile manufacturers, as mechatronic systems can help to meet both safety and comfort requirements. However, interconnection of mechanical and electrical components and integration of the control software make the development even more complex. In this area too, there are a number of modeling technologies for simulating a vehicle. Modern functional developments are based on the methods of Rapid Control Prototyping (RCP). With these methods, it is possible to test the developed control functions during early stages of the development cycle. In addition to established signal-oriented modeling methods, new physically based, object-oriented modeling languages, such as Modelica, are applied.

The purpose of this chapter is to provide an overview of various modeling approaches in the area of vehicle simulation for the application in the functional design and to show how these models can be used in the development

Encyclopedia of Automotive Engineering, print © 2015 John Wiley & Sons, Ltd.
Edited by David Crolla, David E. Foster, Toshio Kobayashi, and Nick Vaughan.
This article is © 2015 John Wiley & Sons, Ltd. ISBN: 978-0-470-97402-5
Also published in the *Encyclopedia of Automotive Engineering* (online edition)
DOI: 10.1002/9781118354179.auto035

process of RCP. In addition to the explanation of the different modeling approaches, two different ways to model these differential equations on a digital computer will be introduced: (1) the established signal-oriented methods, as they are implemented in Simulink and (2) an object-oriented method using the modeling language Modelica.

The chapter is organized as follows: in Section 2, the principles of RCP as a method of modern functional development are elucidated, and the resulting necessity of having various model types is demonstrated. Section 3 gives an overview of popular modeling approaches for the dynamical description of the longitudinal, lateral, and vertical dynamics of a vehicle. An attempt is made to provide the reader with the instructions on various applications to facilitate the choice of an appropriate model. This chapter provides only a general overview and description of the methods and refers the interested reader to the related literature for details. Section 4 presents a comparison between signal-oriented and object-oriented methods for differential equations. On the basis of these methods, the widely used software Simulink by the Mathworks and Dymola are described. This chapter concludes with a summary.

2 RAPID CONTROL PROTOTYPING

Examination of the developments in the automobile industry during the past decade shows that the number of mechatronic systems in a vehicle grows steadily and will continue to grow in the future. The same is also true for chassis control. This trend is particularly encouraged by the increase in safety and comfort that is possible with the use of mechatronic systems. ABS and ESC are the most typical examples of such systems. To a great extent, these functions are realized by software that is integrated as an embedded system into a vehicle. A further trend is a stronger interconnection of separate functions in the vehicle, which makes the functional development of new systems even more difficult and time-consuming.

To guarantee goal-oriented and efficient development of the control software in these complex systems, the concept of RCP has been established for controller development. It combines the advantages of the classic method of V-modeling with the possibility of early testing at separate development stages in a standardized development environment. This testing helps to eliminate the disadvantages of the classic modeling in which certain design errors are only detected rather late in the development cycle. The basic prerequisite here is the use of a continuous tool chain to allow the designer to concentrate on the core competencies. Figure 1 depicts the V-model of a controller design for the graphical visualization of the single steps and their succession in the development.

According to the V-model, the design comprises the following steps: requirement analysis, specification, rough and detailed design, simulation, component test, and system test. In the classic development cycle, these steps are performed one after another, and the tests are performed at the end. In the RCP concept, the tests can be performed much earlier in the development cycle using the corresponding software tools. The development steps need not occur in the above-mentioned succession and can be performed repeatedly, rather quickly, with the assistance

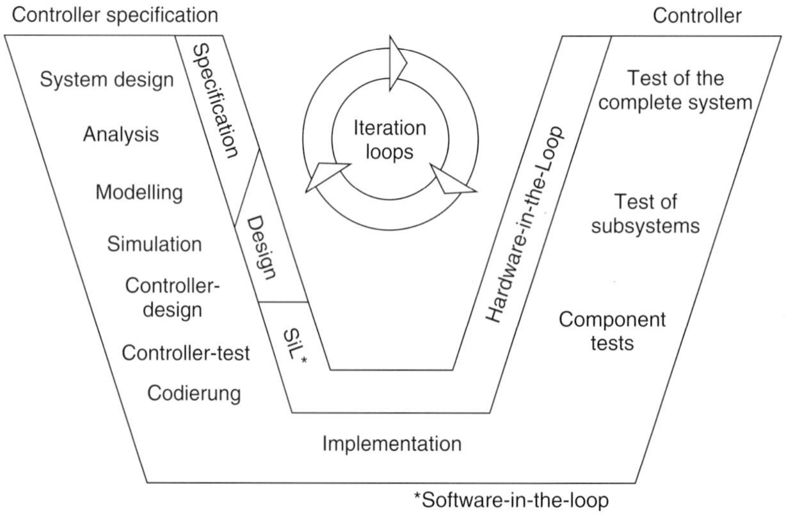

Figure 1. V-model. (Reproduced from *Rapid Control Prototyping*, 2005, D. Abel. With kind permission of Springer Science+Business Media.)

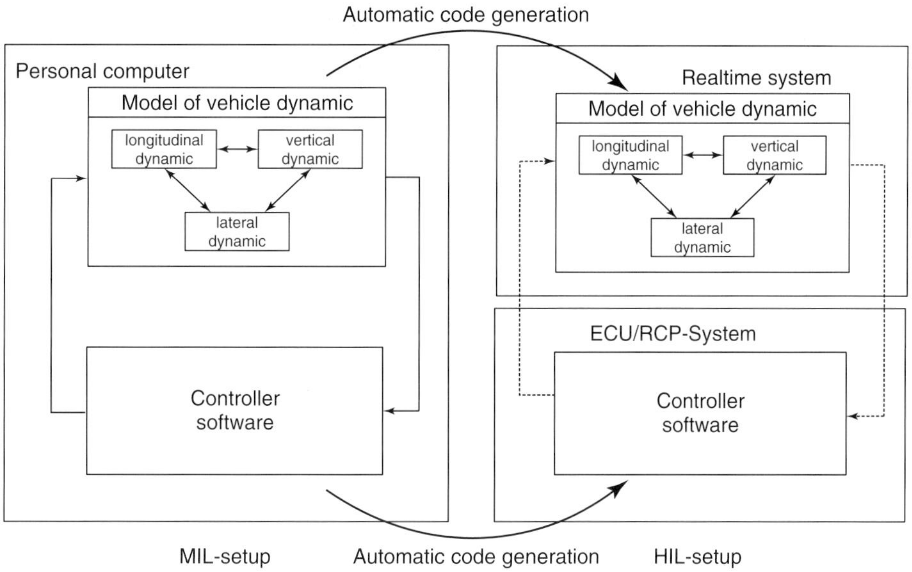

Figure 2. MIL-test.

of the software tools, in order to even more quickly detect and eliminate design and implementation errors.

Essential features of the methods of RCP development are Model-in-the-loop (MIL) and Hardware-in-the-loop (HIL) tests. They help perform tests during various stages of the development cycle, for example, for the purpose of also implementing horizontal iteration loops. Thus, it allows an early creation of controller software in the process of development by using automatic code generation, as well as testing with the available real components or with models of yet unavailable real components.

In Figure 2, two scenarios are compared: the MIL and the HIL tests. The MIL test can be performed during early stages of the development cycle, as the development environment can be, for example, Matlab/Simulink by The Mathworks, in which the controller development also often takes place. The primary goal of a MIL test is the development of the controller's concept and validation of control quality. Therefore, it is important that the MIL models describe the real vehicle behavior as good as possible.

In contrast, in a HIL test, both components (system model and controller model) must be executed in real time. Using automatic code generation, it is possible to transfer the models to a real-time simulation platform. This step is one of the central elements of RCP because the complex manual software design can be automated. It allows making quick changes and tests directly in a real-time system by the press of a button. The primary goal of HIL tests is performing system tests in which the designed controller structure runs on the target platform and is tested there in combination with all other components.

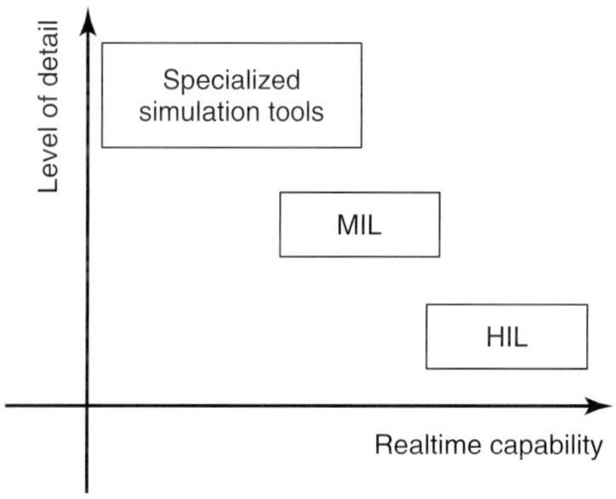

Figure 3. Comparison between different types of models.

Simulation models of the controlled system play a crucial role in all these testing strategies. On the basis of their application, the MIL and HIL models differ in their level of detail and in real-time capability.

Figure 3 shows the relation between real-time capability and level of detail for various model types. Special simulation tools such as ADAMS by MSC Software or Simpack by SIMPACK AG, that can be used to design the mechanical components of a chassis, possess the highest level of detail and also require the highest amount of computation time. Generally, these models are not suitable for simulation of an entire vehicle because of their computation time, and thus are seldom used for controller development.

Concerning MIL models, they can simulate an entire vehicle with all its essential components at a high level of detail. Examples of these models are CarMaker by IPG or CarSim by Mechanical Simulation Corporation. Depending on the selected level of detail, these models are capable of providing real-time performance on very powerful computers. Because the simulation of an entire vehicle they provide is of a high quality, these models are used for MIL tests, which are primarily intended for validation of the control functions. The MIL test need not be run in a real-time environment and can be performed directly on a personal computer. As the controller development is implemented mostly in Matlab/Simulink, the MIL test is also often performed as a co-simulation in Simulink. Most software producers offer corresponding interfaces for Simulink.

Another model type is a HIL model. The main function of the HIL models is real-time performance that, however, reduces the level of detail and thus the accuracy of the models. The model accuracy must be sufficient for the interaction of all designed features and the corresponding diagnosis functions. The control quality does not have the primary importance here, as it has already been tested during the MIL tests. The real benchmark is always to be found in a real vehicle where the quality of the target results of the described process can be judged on the basis of real tests.

Thus, both detailed and real-time capable system models are equally important for modern controller development. In addition to the application of the models for validation of the controller software, these models can also be used in the development of the control concept. A model-based controller design usually requires a linear control model that can be applied either for control design or even as a component of the control concept itself. Thus, the application of simulation models gains further importance, simultaneously making it more difficult to choose a suitable model for a certain purpose. Section 3 deals with various models with different levels of detail for a number of applications. Each model is illustrated with an example of its application to facilitate the reader's choice of a right model.

3 OVERVIEW OF MODELING TECHNIQUES FOR DIFFERENT CHASSIS APPLICATIONS

3.1 Aim of modeling

Developing a chassis application requires a mathematical model that describes the vehicle's dynamic behavior in an adequate manner with respect to the considered application. In particular, the relevant system inputs (e.g., the steering angle) and outputs (e.g., the yaw rate) have to be identified. Subsequently, the dynamic system behavior has to be described in terms of differential equations. The level of detail, that is, the number of differential equations and the number of employed inputs and outputs, inherently depends on the chassis application. From the perspective of control engineering, a process model is required for the purposes of *analysis*, *controller synthesis*, and *validation*, as subsequently explained. As far as process *analysis* is concerned, a plant model is used to analyze the open-loop, that is, the uncontrolled, process behavior. In the linear case, the process' eigenvalues are crucial for the plant's stability, its dynamic time constants, and the damping ratio. For the purpose of *controller synthesis*, a simplified plant model is commonly employed to design the controller in such a way that the closed-loop system fulfills the desired requirements such as stability, offset-free tracking of reference values, the ability to suppress external disturbances, or further performance requirements. Before the controller is applied to the real plant, it is *validated* in numerical simulations. Simplified (linear) plant models are used for controller synthesis, whereas plant models with much higher complexity are employed for validation purposes.

In order to illustrate the above-mentioned issues, the process of modeling is subsequently illustrated for several chassis applications having different requirements as far as model complexity is concerned. In this context, applications of longitudinal, lateral, and vertical vehicle dynamics are taken into account.

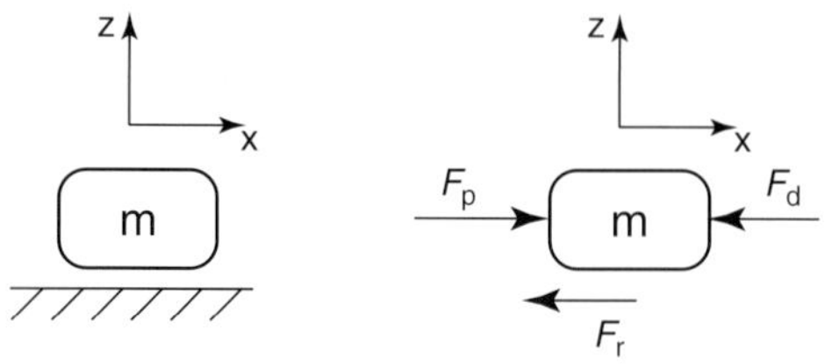

Figure 4. Free body diagram.

3.2 Longitudinal vehicle dynamics

In this section, model approximations for applications in which primarily the longitudinal dynamics are of interest will be presented. In particular, linear model approximations, that are often used for the application of adaptive cruise control (ACC), and nonlinear effects, that are needed, for example, to model problems addressed by ABS, will be discussed. In the following sections, the chassis will be modeled as point mass that is subject to the forces applied from the surrounding environment. A free body diagram is given in Figure 4.

F_p, F_d, and F_r represent the propulsion force, the drag force, and the rolling resistance. According to (Mitschke and Wallentowitz, 2004), the main influencing forces and consequently the longitudinal vehicle dynamics on flat terrain can be described by:

$$m\dot{v}_x = \underbrace{\frac{T_e\left(n_e, \alpha_{th}\right) \bullet \eta \bullet i_g}{R}}_{F_p} - \underbrace{c_w A \frac{\rho}{2} v_x^2}_{F_d} - F_r \qquad (1)$$

Thereby T_e, n_e, and α_{th} denote the engine torque, the engine rotational speed, and the throttle input, respectively. The gear ratio, efficiency, and static tire radius are given by i_g, η, and R while the drag coefficient, the reference area, and the air density are given by c_d, A, and ρ. The above equation holds for engine operation; for brake operation, the term F_p changes sign and has to be replaced by an actuator dynamics that is controlled by a brake pedal position α_{br}.

3.2.1 Linear approximations

To differentiate between absolute values and differences with respect to the operating point, we introduce the perturbation $\tilde{v}_x = (v_x - v_{x,0})$, and $\tilde{\alpha}_{th} = (\alpha_{th} - \alpha_{th,0})$. If we neglect the velocity dependence of the rolling resistance and assume that the engine rotational speed can be related to the velocity as $n_e = v_x \bullet i_g/(2\pi r)$ with an effective tire radius r, the Taylor-series expansion of Equation 1 results in the following approximation of the velocity dynamics:

$$m\dot{\tilde{v}}_x + \left(c_d A\rho \bullet v_{x,0} - \frac{\eta \bullet i_g^2}{R \bullet 2\pi r} \bullet \left.\frac{\partial T_e}{\partial n_e}\right|_0 \right) \tilde{v}_x = \frac{\eta i_g}{R} \bullet \left.\frac{\partial T_e}{\partial \alpha_{th}}\right|_0 \tilde{\alpha}_{th} \qquad (2)$$

Equation 2 exhibits a first-order lag behavior that can be summarized by

$$\tau \bullet \dot{\tilde{v}}_x + \tilde{v}_x = K_v \bullet \tilde{\alpha}_{th} \qquad (3)$$

where the time constant τ and the static gain K_v depend on the above-mentioned vehicle parameters and the operating point. Likewise, the dynamics during braking can be approximated by a linear model. Therefore, the partial derivative of the braking force with respect to the brake pedal position α_{br} has to be included instead of the derivatives of the engine torque.

The above model assumes that the engine torque or a braking pressure (or force) are used as system input that can directly be controlled. However, this is often not the case, and further dynamics have to be included. As a result, another model that is frequently used in literature describes the longitudinal acceleration denoted by a_x. Very often, lower level controls are applied to linearize the resulting dynamics such that this can be approximated again by a first-order lag element as

$$\tau \bullet \dot{a}_x + a_x = a_{x,ref} \qquad (4)$$

where τ denotes the resulting time constant. Concerning the approximation of drivetrain dynamics and underlying linearizing controls further reference is given to (Lu and Hedrick, 2004; Rajamani, 2006) and (Ha, Tugcu, and Boustany, 1989) respectively.

3.2.2 Nonlinear effects

The linear models introduced by Equations 2 and 4 stem from a Taylor-series expansion that is only valid for small deviations from the operating point. Typical driving conditions do normally involve a wide interval of operating points (e.g., different throttle or braking commands at different velocities) and hence cannot be expressed by only one of the models mentioned above. Often, this problem can be dealt with by switching between several different linear models that best describe the current operating point. It should be noted that this is only possible as long as operating conditions change slowly; otherwise, it might be better to use true nonlinear models to describe the drivetrain dynamics.

Besides different operating conditions, another source of nonlinearity arises from the force transmitted between tire and road. For the above-mentioned models, we were basically assuming that any torque applied to the wheels (through engine or brake commands) is transmitted to the road and results in a net force accelerating the mass m. But, the angular speed of the wheels ω and the longitudinal vehicle speed v_x are coupled by a nonlinear friction-slip characteristic that limits the transmittable force between tire and road. If the rolling resistance is neglected, the transmittable force can be expressed by

$$F_x = \mu \bullet F_z \qquad (5)$$

where μ and the tire slip λ are related by a (generally) nonlinear static mapping (friction-slip characteristic). The longitudinal tire slip is defined by

$$\lambda = \frac{\omega \bullet r - v_x}{v_x} \text{ (braking)}, \quad \lambda = \frac{\omega \bullet r - v_x}{\omega \bullet r} \text{ (acceleration)} \qquad (6)$$

For low slip values, a (quasi-linear) ascent of friction (or force) goes along with an increasing tire slip. If the slip exceeds a critical value, friction and hence the transmittable force do not further increase but decrease. Effective force-slip characteristics of a Pacejka tire model are shown in Figure 6a for the longitudinal direction discussed here and in Figure 6b for the lateral direction. A more detailed view of (nonlinear) tire models is given in Section 3.3.2 focusing on the lateral direction. Owing to the decreasing force (beyond the critical slip value), the

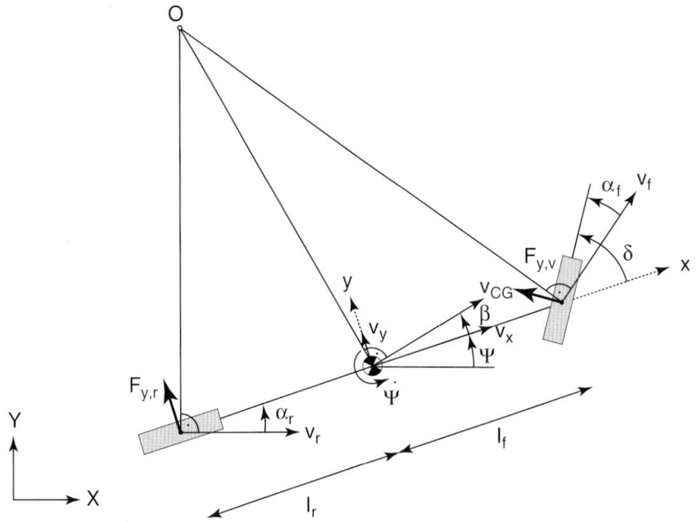

Figure 5. Free body diagram of single track vehicle model.

slip further increases until the wheel spins (during acceleration) or skids (during braking). Using ABS, this problem can be reduced and the maximum force can be achieved during the braking maneuver. As a result, the transmittable force and hence the longitudinal acceleration are limited by a maximum value. For critical maneuvers with high acceleration values, this effect has to be taken into account. As an example, a simplified extension of the linear model in Equation 4 would be an additional saturation of the effective acceleration to account for the related effects.

3.3 Lateral vehicle dynamics

3.3.1 Single-track model

Applications that aim at guiding the vehicle in the lateral direction require a plant model that is basically different from those that have been illustrated in the previous section. When considering advanced driver assistance systems (ADAS) such as lane keeping systems or even autonomous driving systems that are only capable of changing the vehicle's steering angle, a single-track model (see Figure 5) according to (Mitschke and Wallentowitz, 2004) is commonly employed to describe horizontal vehicle dynamics.

The single-track model combines the two wheels of an axle to a single one, while the center of gravity (CG) is placed on the road surface, that is, $h_{CG} = 0\text{m}$. In Equations 7–9 all rolling resistances as well as aerodynamic drag are neglected. Furthermore, it is assumed that no longitudinal forces are applied at the wheel. In particular, Equations 7 and 8 describe the translational degrees of freedom in the longitudinal and lateral directions with respect to the vehicle reference frame. Furthermore, yaw dynamics are modeled by Equation 9.

$$m\dot{v}_x = m\dot{\psi}v_y - F_{y,f}\sin(\delta) \tag{7}$$

$$m\dot{v}_y = -m\dot{\psi}v_x + F_{y,f}\cos(\delta) + F_{y,r} \tag{8}$$

$$J_z\ddot{\psi} = F_{y,f}\cos(\delta)l_f - F_{y,r}l_r \tag{9}$$

In this context, v_x and v_y denote the longitudinal and lateral velocities at CG with respect to the vehicle reference frame, $\dot{\psi}$ the yaw rate, δ the wheel steering angle at the front axle, m the vehicle mass, J_z the mass moment of inertia with respect to the vertical axis, and l_f and l_r the front and the rear wheel bases. When considering a lane keeping or path following system, additional differential equations are required to model the relative motion between the vehicle and the desired path, see (Keßler et al., 2007). As this section primarily focuses on the issue of modeling lateral vehicle dynamics, these equations are omitted for reasons of clarity. In order to determine the tire force $F_{y,f}$ at the front and $F_{y,r}$ at the rear axle, a tire model is required. In the following section, a modeling approach for these tire forces is discussed depending on the magnitude of the considered lateral accelerations.

3.3.2 Tire models

If small lateral accelerations are considered, that is, $|a_y| \leq 0.4g$ on a dry surface, a linear tire model is sufficient to model the resulting tire forces, see (Mitschke and Wallentowitz, 2004). In detail, the side force $F_{y,i}$ at each wheel

$$F_{y,i} = c_{\alpha,i} \bullet \alpha_i, i \in \{f, r\} \tag{10}$$

is assumed to depend linearly on the tire sideslip angles α_i where $c_{\alpha,i}$ describes the nominal tire cornering stiffness for pure cornering. In this context, the vertical tire load is assumed to be constant. The tire sideslip angles define the difference angle between the velocity vector at the wheel and the longitudinal wheel axis, that is,

$$\alpha_f = \delta - \arctan\left(\frac{v_y + l_f\dot{\psi}}{v_x}\right) \tag{11}$$

$$\alpha_r = -\arctan\left(\frac{v_y - l_r\dot{\psi}}{v_x}\right) \tag{12}$$

When considering larger lateral accelerations, nonlinear tire behavior has to be taken into account, see (Katriniok and Abel, 2011). In literature, the Pacejka Magic Formula tire model (Pacejka and Bakker, 1992) is frequently employed to model the steady-state lateral tire force at pure cornering, i.e. only lateral forces are transmitted at the wheel, as a nonlinear function of the tire sideslip angle

$$F_{y,i} = D_i \sin\left[C_i \arctan\left\{B_i\alpha_i - E_i\left(B_i\alpha_i - \arctan\left(B_i\alpha_i\right)\right)\right\}\right], \quad i \in \{f, r\} \tag{13}$$

In general, this model can also be used to determine the longitudinal tire force as well as the aligning torque. An example of the Pacejka tire model for modeling the side force is shown in Figure 6b.

It can be seen that there are basically three regions in the tire model: one "stable" region of static friction (i.e., region II) and two "unstable" regions of sliding friction (i.e., regions I and III). In particular, the maximum absolute tire force that is physically feasible is limited to $F_{y,i,PMF,max}$ for a tire sideslip angle of $\alpha_{i,PMF,max}$ respectively $\alpha_{i,PMF,min}$. If the absolute sideslip angle increases even more, the tire begins to slide over the road surface such that the resulting absolute tire force is decreased. When comparing both tire models, it can be noticed that the cornering stiffness

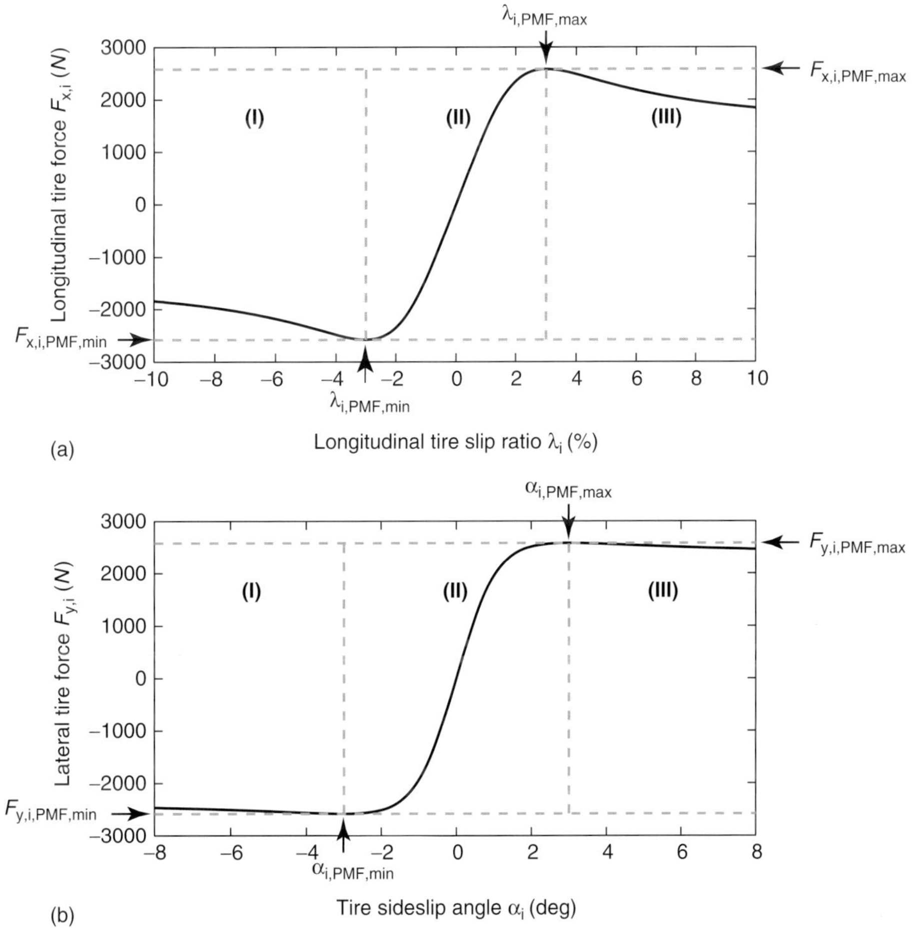

Figure 6. Pacejka Magic Formula tire model.

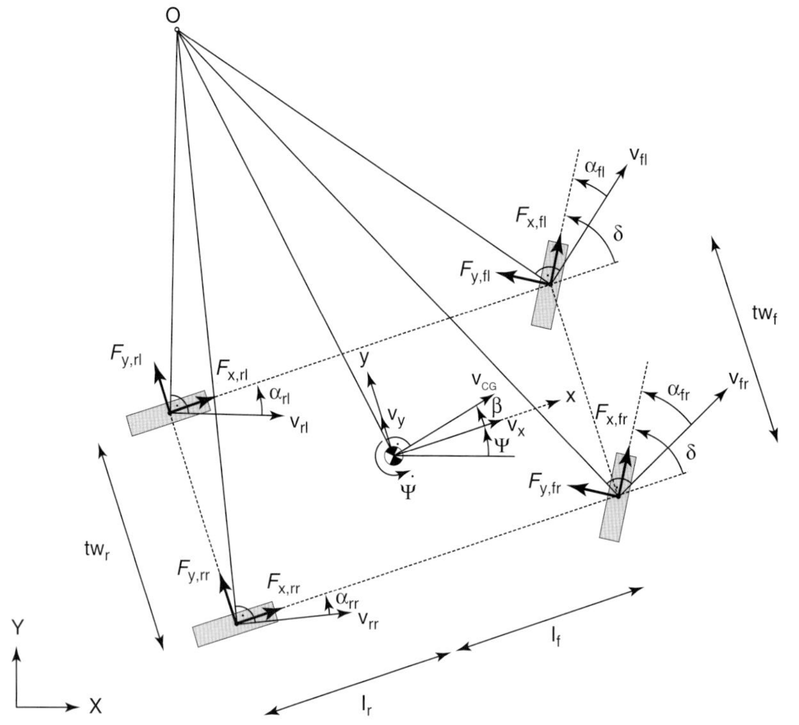

Figure 7. Free body diagram of the two-track vehicle model.

$c_{\alpha,i}$ of the linear tire model corresponds to $dF_{y,i}/d\alpha_i$ for $\alpha_i = 0 rad$ with respect to the Pacejka Magic Formula tire model. The same principle, which is illustrated in Figure 6a, holds for the transmission of longitudinal tire forces that basically depend on the longitudinal tire slip λ_i. If the considered application requires that combined slip, that is, transmission of longitudinal and lateral forces at the same time has to be taken into account, there are several approaches to include this issue into the Pacejka tire model, see (Johansson and Gäfvert, 2004). Furthermore, load transfer can also be incorporated into the tire model. Especially, if the chassis application operates at the vehicle handling limits, the application has to be aware of the maximum feasible tire forces that inherently change with the vertical tire load.

3.3.3 Two-track model

So far, this section has mainly focused on chassis applications that are only able to change the vehicle's steering angle. If the simultaneous transmission of longitudinal and lateral tire forces at each wheel has to be considered, for example, for an ESC or torque vectoring, the single-track model is not sufficient to describe horizontal vehicle dynamics in the required level of detail. Therefore, the two-track model that is illustrated in Figure 7 is commonly employed.

In contrast to the single-track model, all the four wheels are considered such that Equations 14–16 contain the forces $F_{x,i}$ (longitudinal) and $F_{y,i}$ (lateral) at each wheel $i \in \{fl, fr, rl, rr\}$. These forces can again be determined by employing one of the tire models described in the previous section. Furthermore, tw_f and tw_r denote, respectively, the front and the rear track widths.

$$m(\dot{v}_x - v_y\dot{\psi}) = \left(F_{x,fl} + F_{x,fr}\right)\cos(\delta) - \left(F_{y,fl} + F_{y,fr}\right) \sin(\delta) + F_{x,rl} + F_{x,rr} \tag{14}$$

$$m(\dot{v}_y + v_x\dot{\psi}) = (F_{y,fl} + F_{y,fr})\cos(\delta) + (F_{x,rl} + F_{x,fr})\sin(\delta) + F_{y,fl} + F_{y,rr} \tag{15}$$

$$J_z\ddot{\psi} = l_f[(F_{y,fl} + F_{y,fr})\cos(\delta) + (F_{x,fl} + F_{x,fr}) \sin(\delta)] - l_r(F_{y,rl} + F_{y,rr}) + \frac{tw_f}{2}[(F_{x,fr} - F_{x,fl})\cos(\delta) + (F_{y,fl} - F_{y,fr}) \sin(\delta)] + \frac{tw_r}{2}\left(F_{x,rr} - F_{x,rl}\right) \tag{16}$$

3.4 Vertical vehicle dynamics

Finally, this section illustrates the issue of determining a sufficient dynamic plant model considering vertical vehicle dynamics. In particular, this section focuses on the application of an active suspension, that is, the vertical movement of the vehicle body (the sprung mass) that results from an uneven pavement should be suppressed. For this purpose, the vehicle and its suspension can be reduced to a quarter car that consists of an unsprung mass m_u—the tire—connected to the road through a spring (i.e., the stiffness of the tire) and a sprung mass m_s—the vehicle body—connected to the unsprung mass through a spring c_s and a damper d_s (i.e., the suspension), see Figure 8. In order to model an active suspension, a force actuator is added between the sprung and the unsprung mass that is able to apply a force F_a to suppress external disturbances of the road.

In this context, z_s denotes the vertical position of the sprung mass, z_u the position of the unsprung mass and z_r the road elevation. According to (Mitschke and Wallentowitz, 2004), the dynamic behavior of the plant can be described with respect to the equilibrium point by the following differential equations

$$m_s \ddot{z}_s = c_s \cdot (z_u - z_s) + d_s \cdot (\dot{z}_u - \dot{z}_s) + F_a \quad (17)$$

$$m_u \ddot{z}_u = c_s \cdot (z_s - z_u) + d_s \cdot (\dot{z}_s - \dot{z}_u) - F_a + c_u \cdot (z_r - z_u) \quad (18)$$

In Equations 17 and 18, the springs and dampers are modeled as linear elements. If nonlinear effects have to be taken into account, the spring stiffnesses c_u and c_s and the damping ratio d_s have to be modeled as nonlinear functions of the vertical deflection and the vertical velocity. If the consideration of linear elements is sufficient, it inherently depends on the considered chassis application and has to be determined during the development process. If, for example, a controller for suppressing external disturbances of the road by applying the force F_a has been developed using a linear model and tested with satisfying performance on the real plant, nonlinear effects need not be taken into account for controller synthesis. Nevertheless, it might be useful to consider nonlinear effects in the validation model that is used to validate the controller. Finally, it has to be stated that the presented suspension model is well suited for a general analysis of its dynamic behavior, that is, its damping or eigenfrequency. Furthermore, vertical vehicle dynamics can be combined with longitudinal and lateral vehicle dynamics when the presented plant model is combined with single- and two-track models, see (Mitschke and Wallentowitz, 2004).

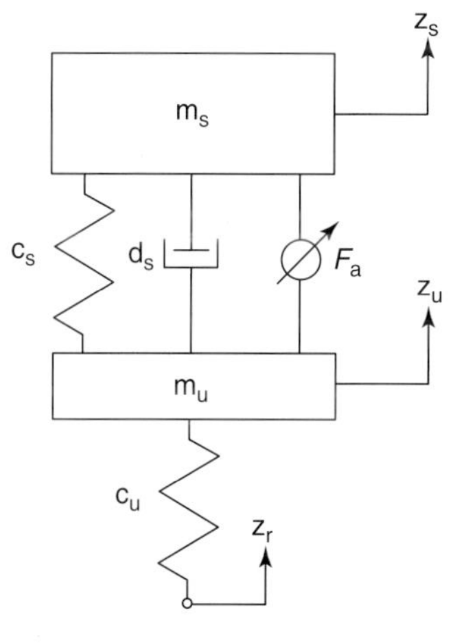

Figure 8. Free body diagram of a quarter car model.

4 USAGE OF DIFFERENT MODELING TECHNIQUES IN COMMON SOFTWARE TOOLS

4.1 Signal-oriented versus object-oriented modeling

Two main modeling concepts are common in many modern simulation tools for simulating dynamic systems: block-oriented (or signal-oriented) and object-oriented modeling. The main differences are found in the way the equations governing the behavior of the system are described and implemented.

A very easy-to-understand method is the block- or signal-oriented approach. For signal-oriented models, cause–effect relationships are modeled directly. A signal-oriented model consists of blocks that are interconnected by signals. The direction of the signals defines causality. Blocks have fixed inputs and outputs, signals only "flow" from block outputs to block inputs; therefore, there will be no feedback from a downstream block to an upstream block if no explicit feedback path is established. In order to be able to create a signal-oriented model, knowledge of the basic governing equations is needed. If an equation contains derivatives, it should be solved for the highest derivative before implemented in order to generate a well-defined block-oriented model. Thus, the lower derivatives can be established by integration of the higher derivatives, preserving causality. For example, if a moving mass is to be considered, its equations of motion should be solved for the acceleration. The velocity is then calculated by integration of the acceleration and the position is calculated by the integration of the velocity.

In Figure 9, an oscillator consisting of two point-shaped masses interconnected by a spring shall serve as an example.

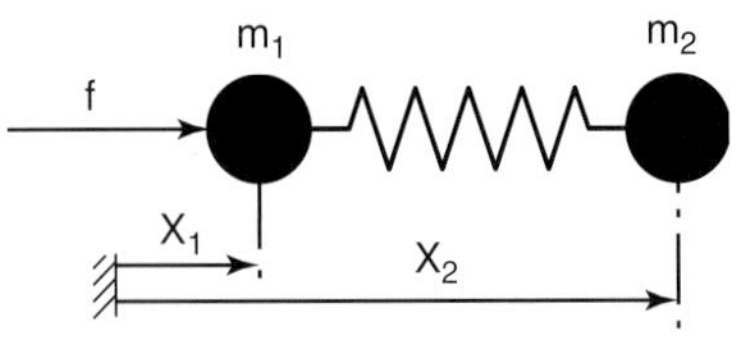

Figure 9. Example: Simple two-mass oscillator.

In order to generate a block model, first the equations of motion of the individual masses are derived:

$$m_1 \bullet \ddot{x}_1 = f_{\text{Spring}} + f$$
$$m_2 \bullet \ddot{x}_2 = -f_{\text{Spring}} \quad (19)$$

Using the equation for the force created by the spring $f = C \bullet (x_2 - x_1)$ and solving for the highest derivates of each state, two differential equations are found:

$$\ddot{x}_1 = \frac{1}{m_1} \bullet \left[f + C \bullet (x_2 - x_1)\right]$$
$$\ddot{x}_2 = \frac{1}{m_2} \bullet \left[-C \bullet (x_2 - x_1)\right] \quad (20)$$

On this basis, the equations can be implemented using standard blocks, for example, using Simulink. The resulting block diagram can be seen in Figure 10.

As one can see, the structure of block models does not necessarily correspond to the equivalent physical system (Fritzson, 2003); therefore, modeling of complex physical systems can become quite difficult. One main reason for this is the above-mentioned absence of implicit feedback due to the use of unidirectional signals. As of Newton's third law, according to which every action is accompanied by an equal and opposite reaction, almost all real physical systems incorporate a feedback loop that has to be accounted for, which in block-oriented modeling language leads to additional signal paths. Hence, although there is only one physical connection between the two bodies, there are two signal paths - one into and one out of the blocks representing the body. Still, a signal-oriented approach can be a good choice for a physical model. Especially for relatively simple systems, signal-oriented models are easy to understand. Most importantly, however, solving the underlying equations is very straightforward and can be done using a relatively simple ordinary differential equation (ODE) solver. This can especially be useful if a real-time capable model is needed, that is, if the computational effort for simulation needs to be predictable. This is the case here because ODE solvers do not necessarily need to use iterations. Apart from the use as a modeling formulation for physical systems, block models are widely used in control engineering. They are very useful for the description of digital control algorithms, which essentially are calculation instructions and therefore have built-in causality. Especially, as solving the equations is straightforward and well defined, it is possible—under certain restrictions—to automatically generate code from a block diagram that can be implemented on a microcontroller. Therefore, the block-oriented description approach can also be seen as an intermediate step between control engineering theory and controller programming.

Another modeling approach is object-oriented modeling. Here, the very universal object-oriented modeling language "Modelica" shall serve for illustration. A thorough introduction to object-oriented modeling with Modelica can be found in (Tummescheit, 2002). In Modelica,

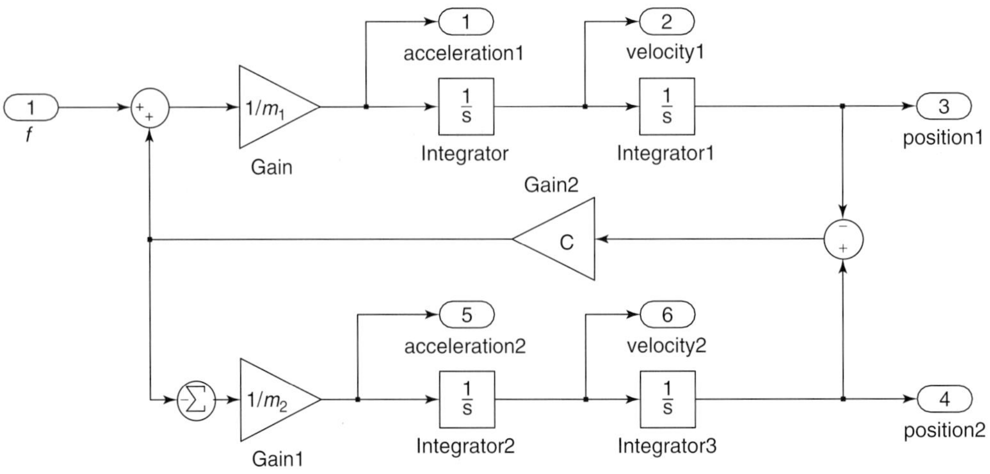

Figure 10. Implementation of two-mass oscillator in Simulink.

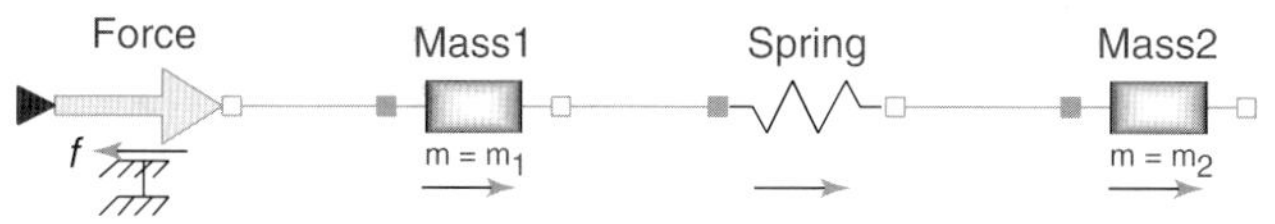

Figure 11. Implementation of two-mass oscillator in Dymola.

physical systems are described using so-called objects that can be interconnected. For example, the two-mass oscillator mentioned earlier can be described as a connection between two objects of type "mass" and one object of type "spring." Figure 11 displays the resulting model as implemented in Dymola, a commercial simulation environment for simulation of Modelica models. All objects can have states, parameters, and equations that contribute to the overall model. They also have connectors that define interfaces to other objects. For example, in this case, each mass object has a parameter containing its mass as well as state variables containing its speed and location. The spring is described by an equation that relates its deflection to the force it exerts on its connected objects.

Unlike signals, which have a fixed direction, interconnectors establish additional relationships between connected elements, but do not yet determine the direction of causality. The connector equations account for both cause and effect, so it is not necessary to explicitly incorporate the resulting opposite reaction. Also, conservation of energy is automatically accounted for, greatly simplifying the task of creating models that are physically correct.

An important aspect of object-oriented programming is the concept of classes and inheritance. A class is a description of a certain type of object. For example, the two masses in the example are both of the class "point mass," both having the same structure. So the actual objects are instances of that class which only has to be defined once, leading to high reusability. In addition, class properties can also be inherited by other classes. For example, going back to the case of the quarter-car model, a "tire" class could be created. A tire would contain an unsprung mass of the type mass as well as a spring. Therefore, the class "tire" automatically inherits the states of the tire (velocity, position) and the equation for the spring, but for the end user can be displayed as a tire. It is also possible to incorporate interchangeable components. For example, it would be possible to give the user a choice of springs to simulate the behavior of different types of tires.

Especially when the system to be modeled becomes more complex, high level languages such as Modelica can be advantageous. However, it should be noted that a Modelica model describes physical systems but not the simulation process. Therefore, additional software is required to simulate the model using the equations specified in Modelica. Modelica models can incorporate differential, algebraic, and discrete equations (Tummescheit, 2002). Their solution can require high computational effort and iterative solution methods. Thus, the models may not always be suitable for real-time computation.

4.2 Modeling example in Matlab/Simulink, and Dymola

4.2.1 Implementation in Simulink

In this section, the implementation of a plant model to be used for numerical simulations will be illustrated using Matlab/Simulink. For this purpose, the application of an active suspension, introduced in Section 3.4, is considered again. According to Section 3.4, the dynamic behavior of the plant is described by the following differential equations

$$m_s \ddot{z}_s = c_s \cdot (z_u - z_s) + d_s \cdot (\dot{z}_u - \dot{z}_s) + F_a \tag{21}$$

$$m_u \ddot{z}_u = c_s \cdot (z_s - z_u) + d_s \cdot (\dot{z}_s - \dot{z}_u) - F_a + c_u \cdot (z_r - z_u) \tag{22}$$

One possibility to implement the plant model (Equations 21 and 22) in Matlab/Simulink is to use integrator, gain, and sum blocks that are part of the Simulink library. Figure 12 shows a possible realization in Matlab/Simulink.

Equation 21, describing the vertical movement of the sprung mass, is modeled in the upper part of the Simulink model; the coupling of the two masses through the spring and the damper in the middle and the vertical movement of the unsprung mass, defined by (22), is modeled at the bottom. Considering the sprung mass, the acceleration $\ddot{z}_s$ is computed by dividing the sum of all forces acting on m_s by the sprung mass m_s. Subsequently, the velocity $\dot{z}_s$ and the position z_s of the sprung mass are determined using two integrators. As the spring and damper forces are proportional to the relative distance and velocity of the two masses and the road, these forces are modeled using the sum as well as gain blocks. The system inputs, that is, the road's elevation z_r as well as the actuator force F_a, and the considered system outputs, that is, the vertical velocity and position of both masses, can be recognized as sources and sinks.

In order to obtain a clearly arranged implementation, Simulink allows for creating subsystems. Thus, the plant can be reduced to its inputs and outputs as illustrated in Figure 13.

Another possibility to implement the quartercar in Matlab/Simulink in a clearly arranged way is to use the *state space*-block, which belongs to the Simulink block

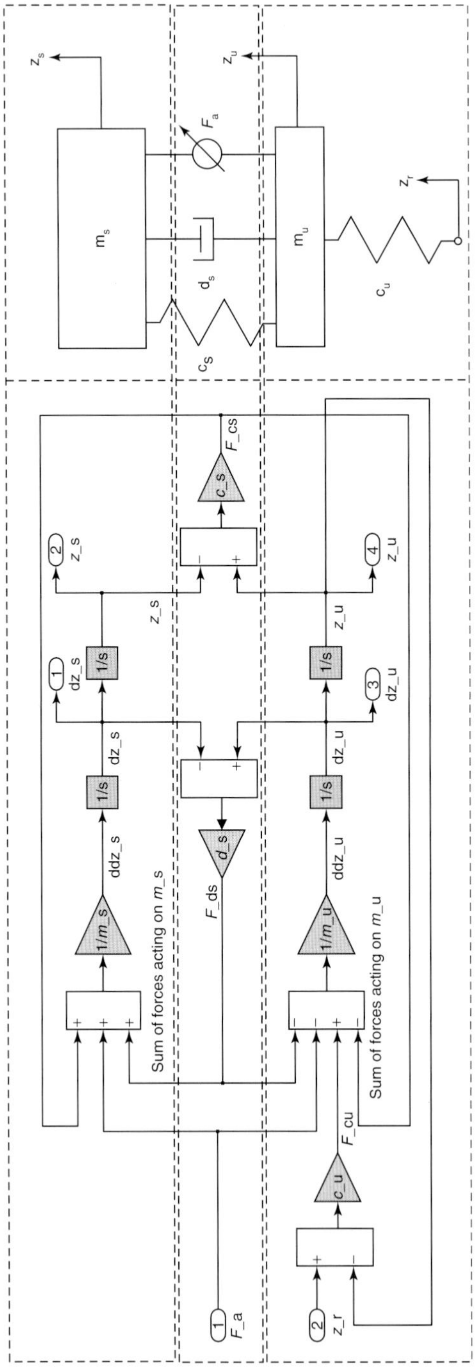

Figure 12. Simulink implementation of the quartercar example using single blocks.

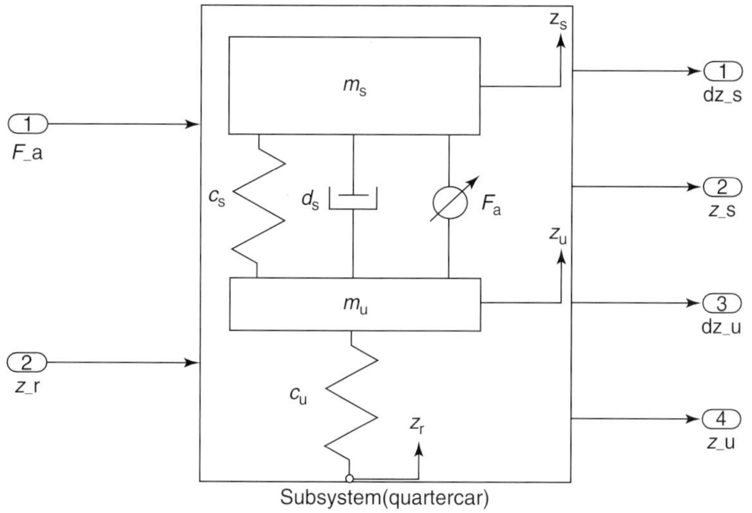

Figure 13. Simulink implementation of the quartercar example using subsystems.

library and supports linear time-invariant systems having the structure

$$\dot{x} = \mathbf{A}\mathbf{x} + \mathbf{B}\mathbf{u} \tag{23}$$

$$y = \mathbf{C}\mathbf{x} + \mathbf{D}\mathbf{u} \tag{24}$$

Using $\mathbf{x} = [\dot{z}_s\ z_s\ \dot{z}_u\ z_u]^T$ as state vector, the actuator force $u = F_a$ as control input, the road elevation $z = z_r$ as system disturbance, and $\mathbf{y} = [\dot{z}_s\ z_s\ \dot{z}_u\ z_u]^T$ as system output, Equations 21 and 22 can be rewritten as

$$\begin{bmatrix} \ddot{z}_s \\ \dot{z}_s \\ \ddot{z}_u \\ \dot{z}_u \end{bmatrix} = \begin{bmatrix} -\frac{d_s}{m_s} & -\frac{c_s}{m_s} & \frac{d_s}{m_s} & \frac{c_s}{m_s} \\ 1 & 0 & 0 & 0 \\ \frac{d_s}{m_u} & \frac{c_s}{m_u} & -\frac{d_s}{m_u} & -\frac{c_s+c_u}{m_u} \\ 0 & 0 & 1 & 0 \end{bmatrix} \begin{bmatrix} \dot{z}_s \\ z_s \\ \dot{z}_u \\ z_u \end{bmatrix} + \begin{bmatrix} 1 \\ 0 \\ -1 \\ 0 \end{bmatrix} F_a + \begin{bmatrix} 0 \\ 0 \\ \frac{c_u}{m_u} \\ 0 \end{bmatrix} z_r \tag{25}$$

$$\mathbf{y} = \begin{bmatrix} 1 & 0 & 0 & 0 \\ 0 & 1 & 0 & 0 \\ 0 & 0 & 1 & 0 \\ 0 & 0 & 0 & 1 \end{bmatrix} \begin{bmatrix} \dot{z}_s \\ z_s \\ \dot{z}_u \\ z_u \end{bmatrix} \tag{26}$$

As described earlier, only linear dynamic systems having the structure as in (Equation 23) can be modeled using the state space block. Thus, z_r has to be defined as an input such that Equation 25 can be written as

$$\begin{bmatrix} \ddot{z}_s \\ \dot{z}_s \\ \ddot{z}_u \\ \dot{z}_u \end{bmatrix} = \begin{bmatrix} -\frac{d_s}{m_s} & -\frac{c_s}{m_s} & \frac{d_s}{m_s} & \frac{c_s}{m_s} \\ 1 & 0 & 0 & 0 \\ \frac{d_s}{m_u} & \frac{c_s}{m_u} & -\frac{d_s}{m_u} & -\frac{c_s+c_u}{m_u} \\ 0 & 0 & 1 & 0 \end{bmatrix} \begin{bmatrix} \dot{z}_s \\ z_s \\ \dot{z}_u \\ z_u \end{bmatrix} + \begin{bmatrix} 1 & 0 \\ 0 & 0 \\ -1 & \frac{c_u}{m_u} \\ 0 & 0 \end{bmatrix} \begin{bmatrix} F_a \\ z_r \end{bmatrix} \tag{27}$$

With this modification, Equations 25 and 26 can be implemented in Simulink as illustrated in Figure 14.

The overview that has been given earlier is just a compendium of the possible modeling techniques that can be employed in Simulink. The particular implementation depends on the complexity of the considered plant. If, for example, a model of an entire car is implemented, it is reasonable to implement, for example, the vehicle

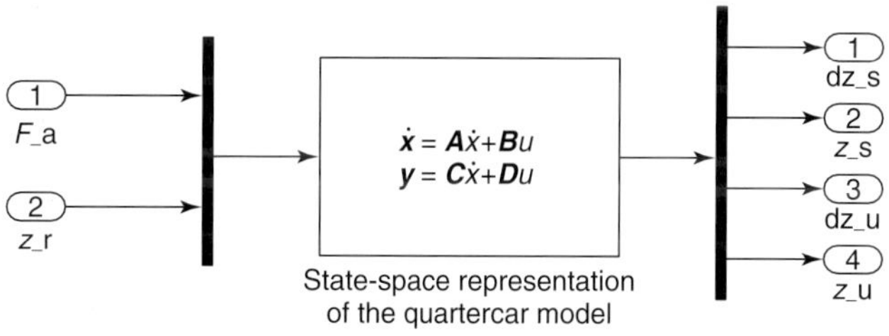

Figure 14. Simulink implementation of the quartercar using a state space representation.

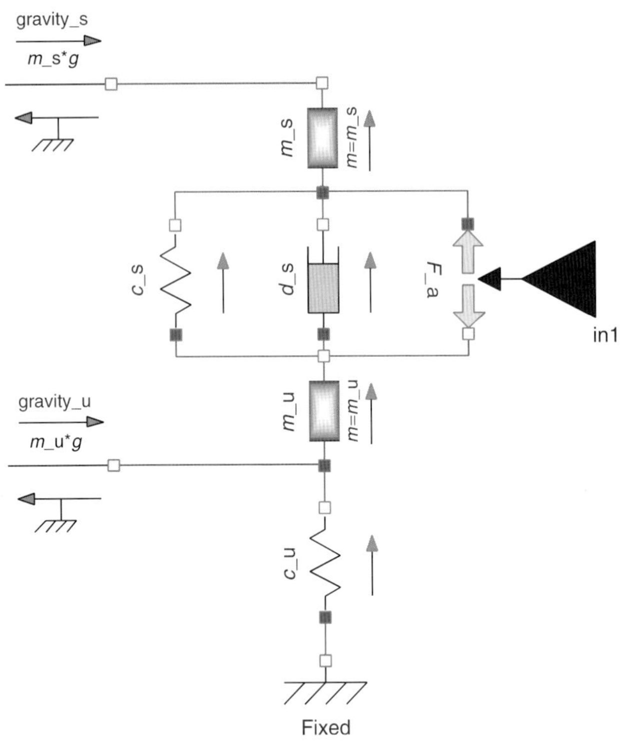

Figure 15. Dymola implementation of the quartercar example.

body, the suspension, and the power train as interconnected subsystems.

As mentioned in Section 3.1, plant models are required and employed for different purposes. For controller design, the model that has been described earlier has an appropriate complexity to be used for a first analysis of the dynamic plant behavior as well as controller design. For the purpose of controller validation, plant models of higher complexity that consider nonlinear effects are commonly employed. In this regard, it is reasonable to run co-simulations of Simulink and third-party software tools. As far as vehicle dynamic control applications are concerned, IPG CarMaker and CarSim^{TM}are two powerful tools that provide several vehicle models as well as validated tire datasets to be used for validation purposes. In this context, both software packages allow for MIL as well as HIL testing procedures.

4.2.2 Implementation in Dymola

The chapter concludes with an implementation of the quartercar model in Dymola (Figure 15). Using blocks that are part of the free Modelica Standard Library Modelica Association (1998–2008), the model can be assembled using the predefined components for the masses, springs, and the damper as well as the interface elements "force" and "force2." The interface element "force"—used here to account for gravity—applies an additional force to its connected element, whereas "force2" creates a generic force between two elements and is used here to introduce the actuator force acting between sprung and unsprung mass. As one can see, the structure of the model is very close to the structure of the physical system it describes, making it more accessible. This is mainly due to the above-mentioned use of connectors as opposed to unidirectional signals. Still, it is also possible to incorporate model parts that have fixed causality. For example, a control algorithm calculating the desired actuator force can be implemented using the signal input "in1."

RELATED ARTICLES

Volume 3, Chapter 106
Volume 4, Chapter 130
Volume 4, Chapter 131
Volume 4, Chapter 132
Volume 4, Chapter 133
Volume 4, Chapter 134
Volume 5, Chapter 144
Volume 5, Chapter 151

REFERENCES

Abel, D. (2005) *Rapid Control Prototyping*, Springer.

Fritzson, P. (2003) *Principles of Object-Oriented Modeling and Simulation with Modelica 2.1*, Wiley-IEEE Press.

Ha, J., Tugcu, A.K. and Boustany, N.M. (1989) Feedback linearizing control of vehicle longitudinal acceleration. *IEEE Transactions on Automatic Control*, **34**, 689–698.

Johansson, B. and Gäfvert, M. (2004) Untripped SUV rollover detection and prevention. IEEE Conference on Decision and Control, pp. 5461–5466.

Katriniok, A. and Abel, D. (2011) LTV-MPC approach for lateral vehicle guidance by front steering at the limits of vehicle dynamics. In IEEE Conference on Decision and Control and European Control Conference, pp. 6828–6833.

Keßler, G.C., Maschuw, J.P., Zambou, N. and Bollig, A. (2007) Concept for the generation of reference variables and model-based predictive control for the lateral guidance of heavy-duty vehicle platoons. *Automatisierungstechnik*, **55** (6), 298–305.

Lu, X.Y. and Hedrick, J.K. (2004) Practical string stability for longitudinal control of automated vehicles. *Vehicle Systems Dynamics Supplement*, **41**, 577–586.

Mitschke, M. and Wallentowitz, H. (2004) *Dynamik der Kraftfahrzeuge*, Springer, Heidelberg.

Modelica Association (1998–2008) *Modelica Standard Library*, Modelica Association.

Pacejka, H. and Bakker, E. (1992) The magic formular tire model. *Vehicle System Dynamics: International Journal of Vehicle Mechanics and Mobility*, **21** (1), 1–18.

Rajamani, R. (2006) *Vehicle Dynamics and Control*, Mechanical Engineering Series Edition, Springer, New York.

Tummescheit, H. (2002) *Design and Implementation of ObjectOriented Model Libraries using Modelica*. Department of Automatic Control, Lund Institute of Technology.

PART 6
Electrical and Electronic Systems

Background

Chapter 136

Historical Overview of Electronics and Automobiles: Breakthroughs and Innovation by Electronics and Electrical Technology

Mitsuharu Kato
DENSO Corporation, Kariya, Japan

1 INTRODUCTION

This chapter describes the relationship between automobiles and electronics. Figure 1 shows an outline of this history.

1.1 Vehicles (Casey, 2008)

The fledgling automotive technology that sprung up in Germany in the 1880s took shape as an industrial product in the form of the Model T Ford released in 1908, which went on to sell a total of 15 million units. According to Ford's internal records, the Model T was the first vehicle to be mass produced on a conveyor belt. This was state-of-the-art production line technology for the time and its basic premise is still in use today. In contrast, however, cars have completely changed. In 1908, cars were made up entirely of mechanical parts. Soon after, light bulbs began to be used in headlamps, and starter motors were introduced. Subsequently, automotive technology remained based on mechanical systems, supplemented by a small range of traditional electrical technologies, such as light bulbs, the starter, battery charger, battery, and spark plugs. At the time, the term *electronics* referred to vacuum tubes. By today's standards, the production of 15 million vehicles is an astounding feat and demonstrates the impressive evolution of mechanical engineering since the industrial revolution.

1.2 Semiconductors (Shockley, 1950)

The first practical semiconductor devices were invented in 1947, 40 years after the Model T. William Shockley was responsible for confirming the characteristics of diodes and amplification. Although already known based on solid-state physics and quantum mechanics theory, the practical verification of these characteristics was a major turning point. Soon after, p-type and n-type semiconductors were combined to form a rectifying diode, and the first transistor with an amplification function was developed. These new devices were achieved as a result of the advances in solid-state physics and quantum mechanics research that occurred in the twentieth century. In 1958, the integrated circuit (IC)

Encyclopedia of Automotive Engineering, print © 2015 John Wiley & Sons, Ltd.
Edited by David Crolla, David E. Foster, Toshio Kobayashi, and Nick Vaughan.
This article is © 2015 John Wiley & Sons, Ltd. ISBN: 978-0-470-97402-5
Also published in the *Encyclopedia of Automotive Engineering* (online edition)
DOI: 10.1002/9781118354179.auto212

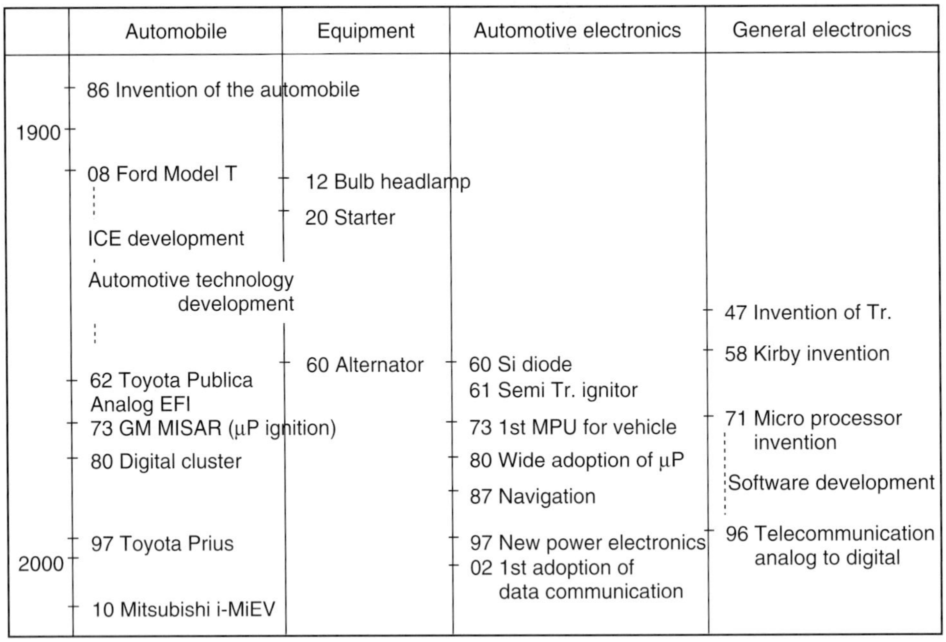

Automobile
Equipment
Automotive electronics
General electronics
1900
2000
86 Invention of the automobile
08 Ford Model T
ICE development
Automotive technology development
62 Toyota Publica
Analog EFI
73 GM MISAR (μP ignition)
80 Digital cluster
97 Toyota Prius
10 Mitsubishi i-MiEV
12 Bulb headlamp
20 Starter
60 Alternator
60 Si diode
61 Semi Tr. ignitor
73 1st MPU for vehicle
80 Wide adoption of μP
87 Navigation
97 New power electronics
02 1st adoption of data communication
47 Invention of Tr.
58 Kirby invention
71 Micro processor invention
Software development
96 Telecommunication analog to digital

Figure 1. Automobiles and electronics (history). (Reproduced with permission from Kato, 2010. © Denso Corporation.)

that used multiple transistors was invented and patented by Jack Kilby. This invention heralded the launch of analog and digital ICs. The semiconductor device invented by Shockley in 1947 was the bipolar transistor, which used the amplification effect of a p-n junction. This was followed by the development of analog and digital ICs that used bipolar devices, which combined pnp and npn transistors. Operational amplifiers consisting of these devices in combination represented a major breakthrough as the basic components of analog ICs. In comparison to bipolar transistors, the practical application of metal-oxide-semiconductor (MOS) transistors took time due to the instability of the MOS interface. However, once these characteristics were stabilized, the simplicity of the MOS structure enabled large-scale integration (LSI) devices to enter mainstream use. Since then, the number of transistors mounted on a single IC has increased exponentially. Figure 2 shows the upward trend of device integration. As described by Moore's law, integration has roughly quadrupled every three years. This has been accomplished by the development of increasingly fine semiconductor manufacturing technologies and innovations in device structures.

1.3 Fusion of megatechnologies in the twentieth century

The vehicle is an example of a megatechnology, that is, a product in which many technologies intertwine to create a new easy-to-use technology that has the potential of helping realize the dreams of its users (Kato, 2010). As vehicles became more and more widespread, exhaust emissions created issues such as pollution, typified by the infamous smog of Los Angeles. The emissions regulations that were introduced in 1980 could not be met by conventional technology, and depended on the development of electronics, another megatechnology of the twentieth century. Figure 3 shows the principle configuration of the system developed at the time to meet these regulations.

The configuration shown in the figure is required to ensure complete combustion by optimizing the air/fuel (A/F) ratio and ignition timing, providing feedback control through O_2 sensors, and the like. This system is not entirely electronics-based. It also utilized the evolution of materials technologies such as three-way catalysts and the feedback system shown in the figure to greatly reduce emissions. It is no coincidence that the accelerated global spread of the automobile was sparked by the practical application of semiconductor technologies and the collaboration between automotive mechanical and electronics engineering to meet such regulations. This system may only represent a tiny step forward by today's standards, but it would have been unfeasible without the contribution of electronics. In this way, the negative effects on the global environment caused by the mass production and popularization of cars from 1910 were halted by the advent of electronic controls. These controls that overcame the limits of mechanical technology became feasible through the 1970s up until

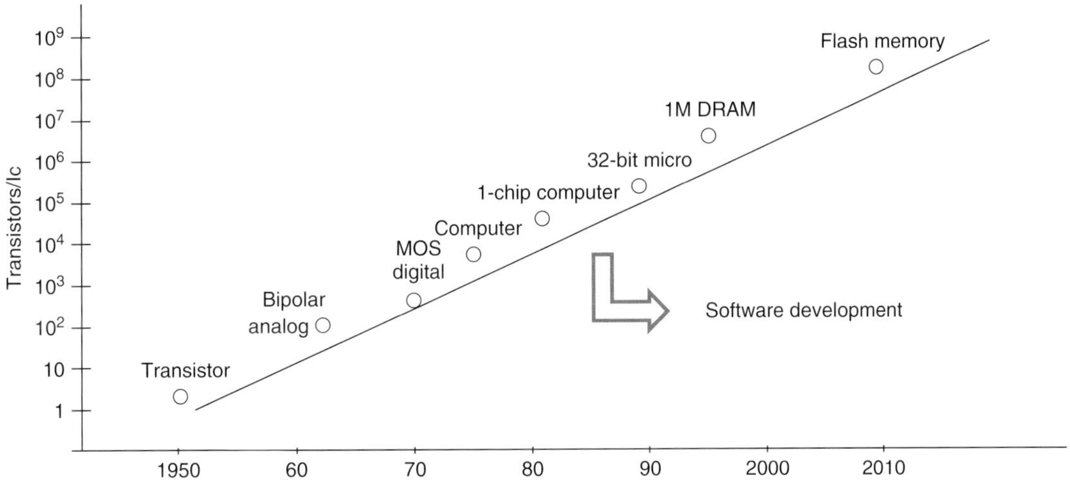

Figure 2. Integration trends.

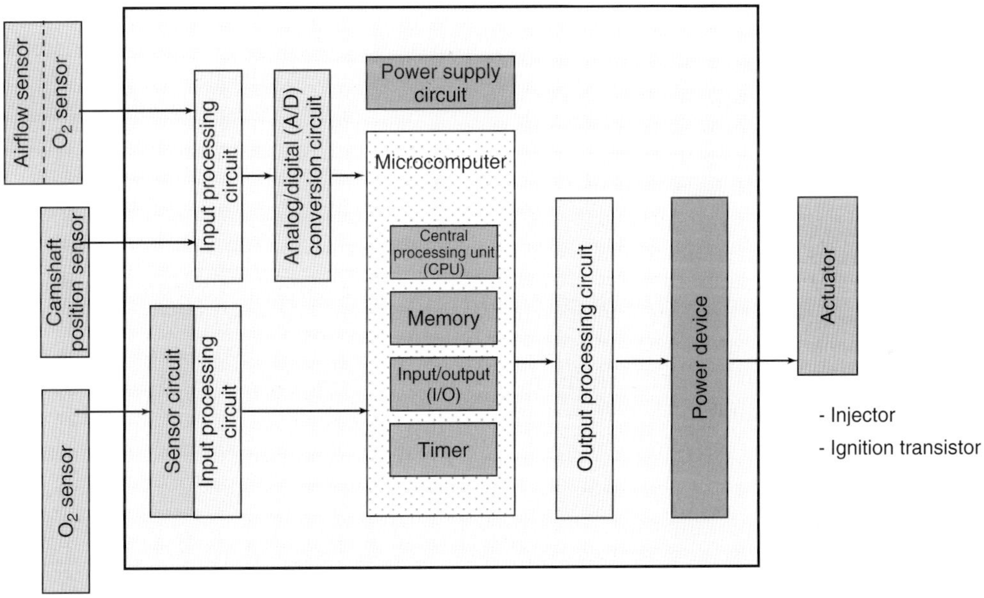

Figure 3. Basic ECU configuration.

around 1980. Modern electronics can be seen as a product of this era, in which solutions were presented to meet such regulatory requirements. This was the trigger for full-scale application of electronics to automobiles, and it inspired the development of various vehicle controls and products. In this era, the megatechnologies of automobiles and electronics finally came together.

2 ADOPTION: INITIAL PERIOD

The invention of semiconductor diodes enabled the development of solid-state rectifiers, which was a significant breakthrough because it allowed the adoption of more efficient alternating current (AC) generators in place of direct current (DC) generators. Figure 4a shows a rectifying diode. The principle of the rectifying function has not changed in the last 50 years. As shown in Figure 4b, the first transistors were used for ignition. Conventionally, a mechanical contact was used to shut off the current of the ignition coil. However, the high voltage generated by the coil would cause an arc, which limited the size of the generated voltage. Transistor ignition was capable of shutting off the current without using a contact. It enabled the generation of higher voltages and resulted in the development

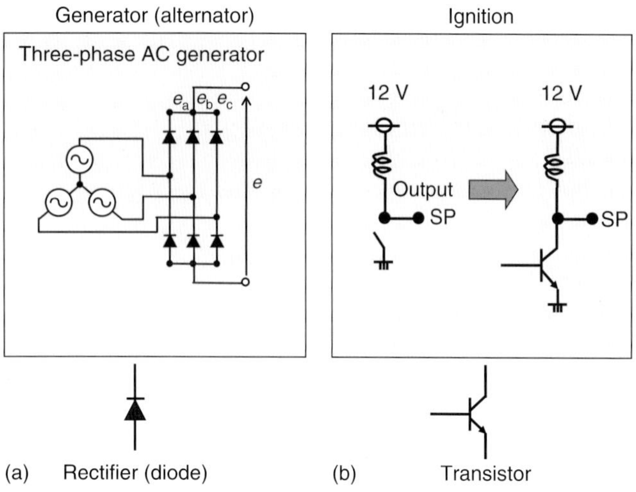

Figure 4. (a, b) Initial period of diode and transistor development (beginning of 1960s).

of gasoline engine ignition systems with high combustion efficiency.

2.1 Adoption to electronic fuel injection (EFI) (Manger, 1998)

The popularization of electronic circuits through analog circuits and analog computers, which combine operational amplifiers and make use of their functions, started in household appliances. However, when this trend began in the 1970s, such circuits were quickly adopted in vehicles as well. The first full-scale adoption of electronic circuits was in A/F ratio control systems that measured the amount of intake air in a gasoline engine with an airflow sensor and determined the optimum amount of fuel to be injected. This had been conventionally performed by mechanical systems that used the principle of atomization by a carburetor. In these systems, an amount of fuel commensurate to the airflow was mixed in the manifold and dispatched to the cylinders. In contrast, electronic control enabled the cylinder to combust the correct A/F ratio for the intake air volume. A/F ratio control was a major breakthrough because it allowed the calculation and control of engine conditions beneficial to ignition and torque output. The configuration of this system is shown above in Figure 3. Even today, the basic engine control unit (ECU) configuration of sensors, control circuits, actuator drivers, and a power source has not changed. Sensor signals include inputs such as the crankshaft, airflow, air temperature signals, and the like. This system controls parameters, such as the time that power is applied to the injectors, to determine the amount of fuel injection. Initially, this control used analog control ICs. Digital control was introduced after the invention of the microcomputer.

2.2 ECU development (Numazawa, 1998)

The development of a practical microcomputer opened the floodgates for the application of electronics to cars. Figure 5 shows the history of this application. Meter and body system microcomputers started to be adopted from around 1980. The first utilization in meter systems was in the form of digital meters that used fluorescent display tubes. Subsequently, higher performance systems started to be used as liquid crystal meters became more available. One particular use of high performance microcomputers was for

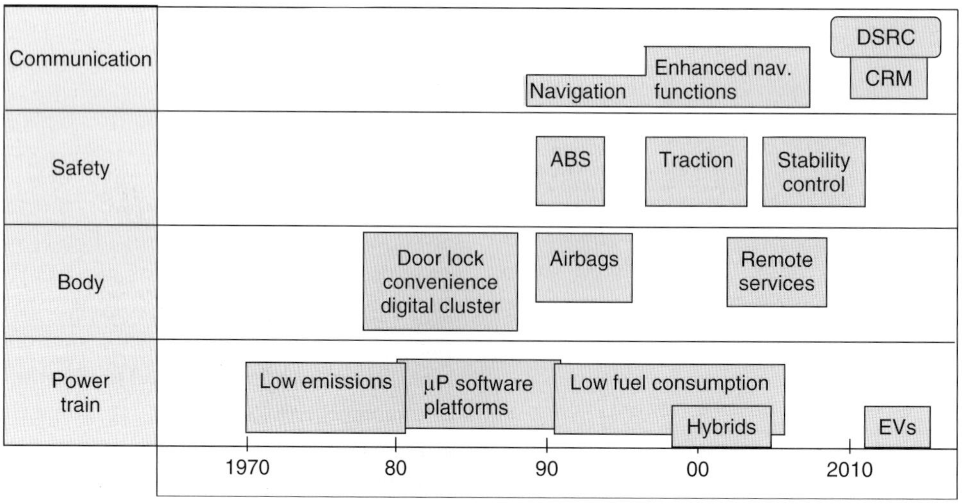

Figure 5. History of automotive electronics.

information display. Later in the 1980s, microcomputers were adopted in safety controls. This started with seat belt interlock functions, and then spread to airbags and antilock brake systems (ABSs). The development of these systems at this time all depended on microcomputer technology.

3 MICROCOMPUTERS

Intel developed the first 4-bit microprocessor in 1971. The first automotive application was in GM's MISAR ignition control system in 1973, which used a 10-bit microprocessor. In Japan, 12-bit microcomputers were used in fuel injection systems. Although the microcomputers invented by Intel are now available in 4-, 8-, 16-, and 32-bit configurations, the first microprocessor was only available in anomalous 10- and 12-bit configurations. This was because the performance of a general purpose microcomputer was insufficient. These anomalous bit lengths were adopted to increase calculation speeds through the use of integrated instruction and data sets. Although use of these 10- and 12-bit microcomputers was limited, the same basic configuration is still present. In particular, the free running timer and compare register configuration has become a basic circuit block for real-time control. A significant breakthrough was the use of special timers to compensate for inconsistencies between the microcomputer calculation speed and the speed required for control. The use of such circuit blocks to reduce the load of the microcomputer remains unchanged today.

3.1 Microcomputer configuration

The first microcomputers required various externally attached parts, such as ROM, RAM, I/O, and the like, centered on a CPU. A watchdog timer was also required to prevent microcomputer overrun. Modern microcomputers are now realized on a single chip with in-built ROM, RAM, I/O, and program monitoring functions.

3.2 Various types of microcomputer

In addition to fields that depend on control speed, microcomputers have also been developed to meet needs for programmability and diversity. One dominant example is the 1-bit microcontroller (see the image and configuration in Figure 6). The 1-bit microcontroller was cutting-edge technology when the cost of 4- and 8-bit microcomputers was too high for use with single functions and developers were consolidating peripheral parts onto a single chip. Although the 1-bit configuration limited their coefficient of performance, these microcontrollers could be used to control event chains in single bit increments. As a result, these were ideal for monitoring and controlling door

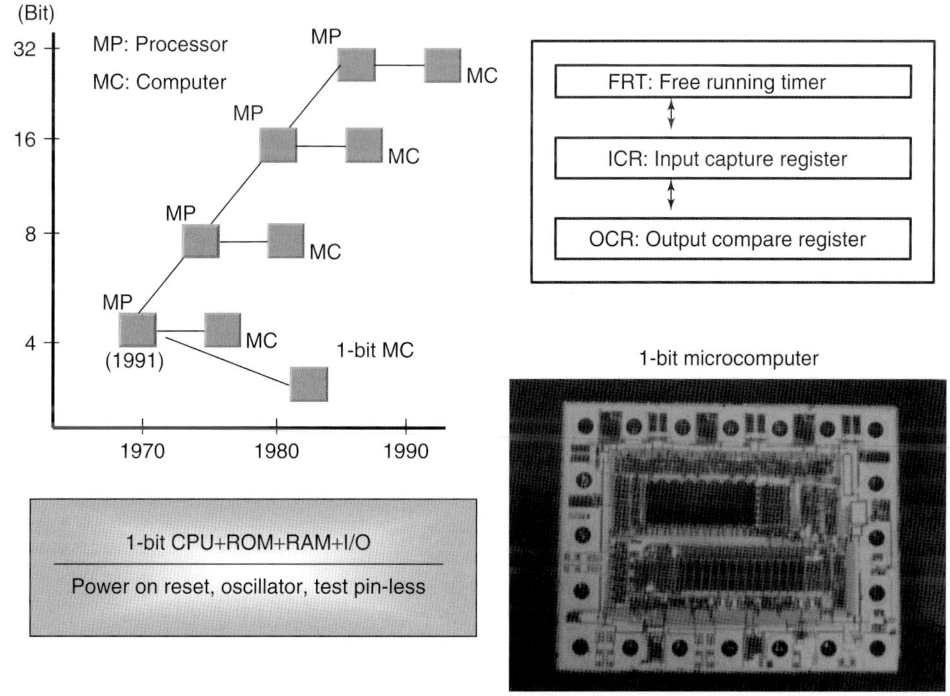

Figure 6. Microprocessor history.

opening/closing and switch on/off functions. 1-bit microcontrollers consolidated all peripheral parts onto a single chip and have evolved into today's embedded microcomputers. Mainstream microcomputers have adopted 16- and 32-bit configurations. Innovations unique to the automotive industry such as easy-to-program architectures and free running timers mean that the automobile has created a separate field of microcomputers.

4 ECU CONFIGURATION AND HARDWARE

Although the configuration of ECU hardware has not changed from that shown in Figure 3, the shape of functional parts has changed greatly in accordance with the cyclical changes of semiconductors. Figure 7a shows that early ECUs used printed circuit boards with through holes, which were used to attach parts such as capacitors, resistors, and ICs packaged with leads. The ECU in Figure 7a uses three analog ICs. Subsequently, parts become more compact and devices were mounted directly onto the surface of the circuit board. Excluding connectors, these boards had no through holes (Figure 7b). This figure shows an ECU with a 32-bit microcomputer. The vast increase in parts mounted on the circuit board is easily noticeable. As the performance of microcomputers increased, vehicles began to use the latest microcomputer technology.

4.1 Features of on-board ECUs (Kato, 2010)

The defining feature of automotive ECUs is their harsh operating temperature range, which may reach as high as 125°C in the engine compartment and fall as low as −40°C in winter. This is vastly different from the operating temperature range of household appliances. Figure 8 lists the features of automotive ECUs. To operate in these conditions, although ECUs in the occupant compartment use conventional printed circuit boards, ECUs in the engine compartment tend to use ceramics. Actuators have been integrated onto electronic circuits from the start, and devices used with ceramic circuit boards have adopted a bare chip configuration. In particular, systems such as generators and ignition devices that have always been installed in the engine compartment adopt special packaging. One of the characteristics of automotive ECUs is the high recovery rate of defective parts. As shown in Figure 8, if a defect occurs in the market, a system has been developed that returns the defective ECUs to the manufacturer through the dealer. This enables the location of the defect to be identified and helps understand issues, particularly

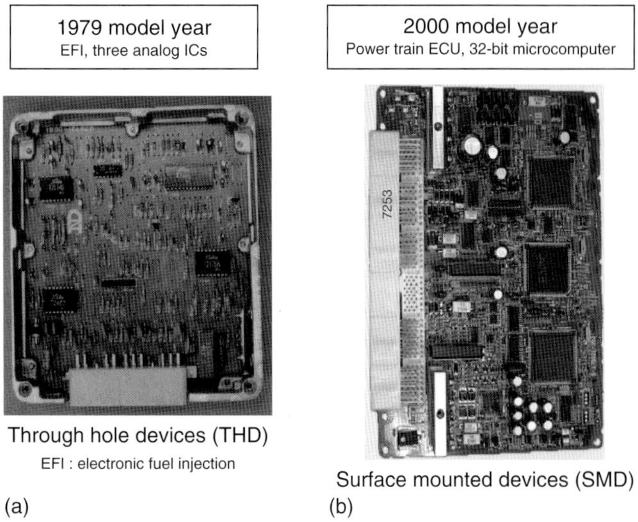

Figure 7. (a, b) Photographs of ECUs.

with semiconductors. Lessons from these steps have been reflected in improved semiconductor structures and manufacturing methods. As a result, this system has also been applied to household appliances as well as automotive semiconductors, thereby greatly helping improve the quality of the semiconductor industry as a whole.

4.2 Electromagnetic compatibility (EMC) and noise of on-board ECUs

One of the consequences of the proliferation of automotive electronics is the increase in the number of noise sources (Kato, 2010; CISPR 12; CISPR 25; ISO7637-1; ISO7637-2; ISO11451; ISO11452). As indicated in Figure 9, typical noise sources include radio frequency noise from the ignition system, surges, and electrostatic discharges. Power wiring voltages vary from 4.5 V used by 12-V systems when the starter is in operation to 17 V. Signal wiring may generate overvoltages of 200 V or more when switching parts with embedded inductance. Therefore, during ECU design, diodes and the like must be provided to protect electronic components from malfunction or breakage due to voltages transmitted through power or signal wiring. In recent years, the amount of high frequency noise emitted from ECUs has increased as microcomputer operation has speeded up. This is a potential cause of audiovisual (AV) system malfunction. Another major issue is ECU malfunction due to a reduction in signal levels caused by greater semiconductor integration. These issues are referred to as *electromagnetic interference (EMI)* and *electromagnetic compatibility (EMC)*, respectively. Measures for EMI include reducing the strength of electromagnetic

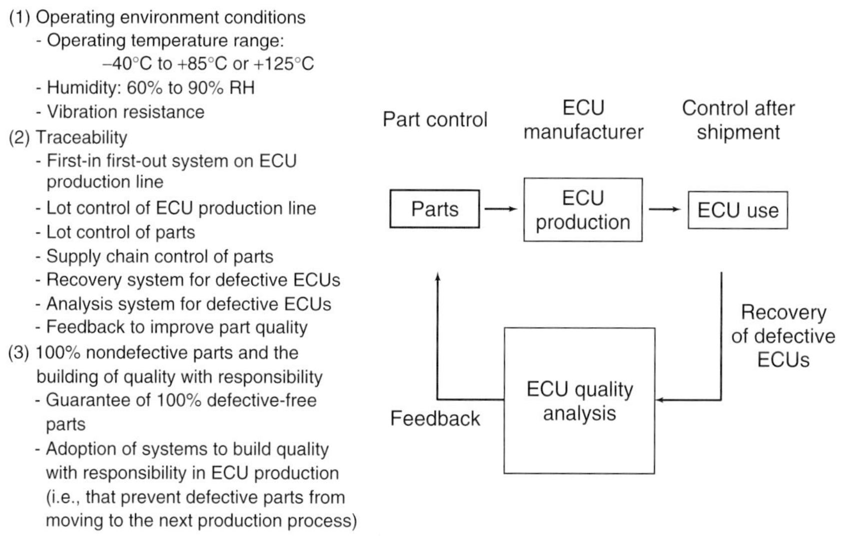
(1) Operating environment conditions
- Operating temperature range:
−40°C to +85°C or +125°C
- Humidity: 60% to 90% RH
- Vibration resistance
(2) Traceability
- First-in first-out system on ECU production line
- Lot control of ECU production line
- Lot control of parts
- Supply chain control of parts
- Recovery system for defective ECUs
- Analysis system for defective ECUs
- Feedback to improve part quality
(3) 100% nondefective parts and the building of quality with responsibility
- Guarantee of 100% defective-free parts
- Adoption of systems to build quality with responsibility in ECU production (i.e., that prevent defective parts from moving to the next production process)
Part control
ECU manufacturer
Control after shipment
Parts
ECU production
ECU use
Recovery of defective ECUs
ECU quality analysis
Feedback

Figure 8. Features of automotive electronics.

Type		Source	Effects
Ignition system radio noise		• Spark plugs • Distributor	• Obstruction of communication with general devices (regulatory compliance) • Electronic devices: Malfunction
Surge	Low frequency	• Alternator • Ignition coil • Relays • Solenoids, motors	• Electronic devices: Malfunction • Electronic devices: Breakage
	High frequency	• Relay contacts • Motors	
Electrostatic discharge		• Between occupants and vehicle • Static electricity from workers on assembly line	• Electronic devices: Malfunction • Electronic devices: Breakage

Figure 9. Sources and effects of ECU electrical and radio noise.

wave emissions as a part of electronic circuit design. EMC is a potential case of electronic circuit malfunction, and robust design measures are required to prevent electromagnetic interference. Various international test standards have been established with respect to these issues. Figure 10 shows some relevant examples. EMI and EMC will play an increasingly significant role when improving the performance of electronic products in the future.

5 SEMICONDUCTOR SENSORS (INA, 1988)

Semiconductors have another critical role in electronic systems as sensors. A glance at early electronic fuel injection (EFI) control shows that the airflow sensor played a key part. Rudimentary airflow sensors measure the angle of a damper to gauge the airflow. These evolved into sensors that measured the Karman vortex and then into the indirect measurement of intake air by monitoring the manifold pressure. The use of pressure sensors has greatly reduced the size of these systems. A dedicated semiconductor pressure sensor was developed to measure the manifold pressure, which signaled the start of the groundbreaking technique of semiconductor micromachining. The semiconductor sensor shown in Figure 11 uses the Piezo resistance effect. This provided the stimulus for the development of various sensors using different semiconductor properties (Figure 12). Semiconductors are now used in pressure, acceleration (G), photo, and magnetic sensors. Silicon plays a key role as the base material for micromachining. Angular acceleration sensors have

Evaluation items			International standards	Use scenarios
Immunity	Static electricity		ISO 10605	On contact
	RF immunity		ISO 11452-2, -3, -4	Operation in strong electrical fields
Emissions	Transient characteristics	Load dump	ISO 7637-2	Positive voltage surge when battery terminals are disconnected
		Field decay	ISO 7637-2	Negative voltage from inductive load of alternator
	Radiant emissions		ICISPR 25	Other devices in the occupant compartment (radio waves to AM/FM radio, TV)
	Conductive emissions		ICISPR 25	Other devices in the occupant compartment (radio waves to AM/FM radio, TV)

Figure 10. EMI and EMC evaluations for ECUs.

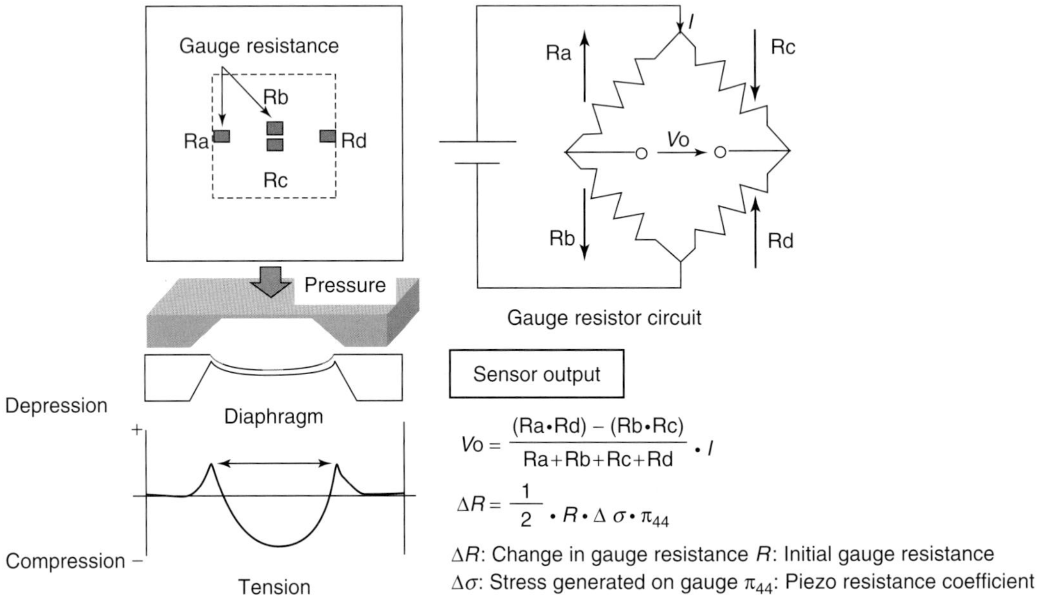

Figure 11. Pressure sensor. (Reproduced with permission from Kato, 2010. © Denso Corporation.)

also been practically adopted. Magnetic sensor devices that mount thin film magnetic resonance (MR) devices on an IC are a crucial part of proximity sensors that identify minute vector changes in magnetic fields. The feasibility of control systems depends on the enhancement of sensor performance as well as the microcomputer processing capability. For example, engine control systems are now capable of fine controls using linear sensors that monitor the oxygen concentration in emissions. Vehicle attitude control uses *G* and yaw rate sensors. The detection and control of states using sensors involves calculation by a microcomputer, and feedback of the results to actuator outputs.

6 SOFTWARE STORAGE PROGRAM MEMORY

The configuration of program memory has switched from mask ROM to EPROM, EEPROM, and electronically rewriteable flash memory. The usability of program memory has changed substantially. Writeable technology for program memory using masks required the details of the program to be determined a few weeks before the IC was packaged. Therefore, technology was developed to push the program writing step further back in the semiconductor manufacturing process to minimize the program creation lead time. The program still required four to six weeks,

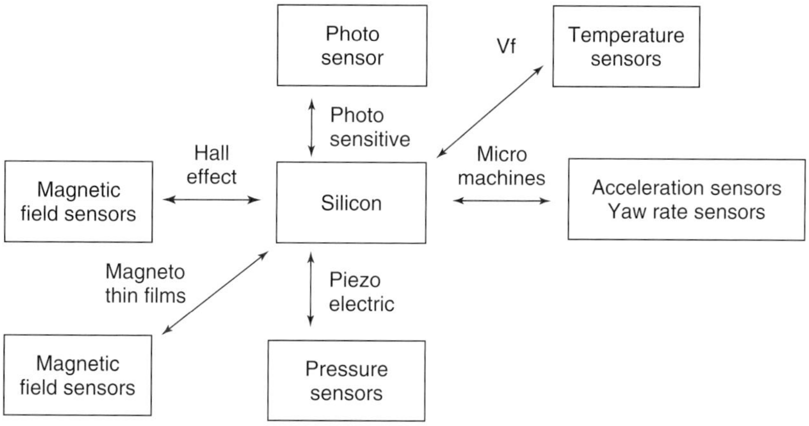

Figure 12. Semiconductor sensors.

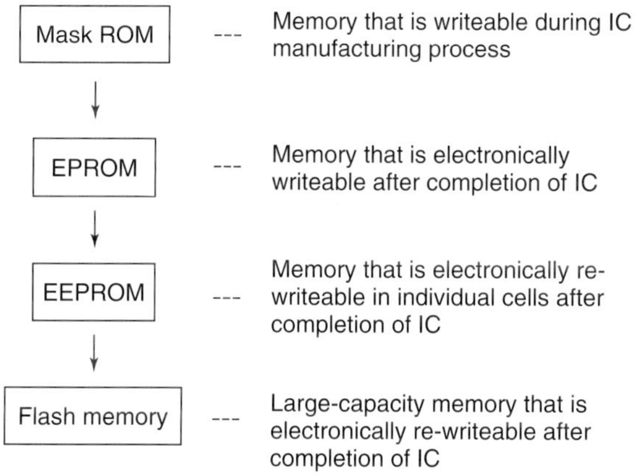

Figure 13. Program memory.

which became a major issue once software control became the mainstream control method. For this reason, the appearance of EPROM technology was a major breakthrough as it allowed the contents of the ROM to be electronically written. Electronic programming could then be set as the final step of ECU manufacture. Semiconductor technology continued to evolve and perfect the basic technologies for writing programs into memory electronically. Flash memory is a good example of this. It can be used to update ECU programs through a service connector while the ECU is mounted in the vehicle. This has helped further accelerate the pace of vehicle control development (Figure 13).

6.1 Importance of software technology

High performance ECUs that use microcomputers have given rise to new system controls, which have in turn spurred the development and introduction of even higher performance microcomputers. The scale of software has increased as a result of more sophisticated microcomputers and cheaper memory, making system controls more complicated. Consequently, the quality of software has become more important. Software program languages have evolved from assembly language to advanced languages such as C and C++. Programs now consist of between several tens of thousand to several hundreds of thousand lines of code. Advanced languages are a prerequisite for software creation productivity. Software developers are facing a growing burden, which means that software architecture is playing a more central role and a critical issue has become how to improve reusability rates. A capability to build architectures with individual software modules that have independent characteristics is the most important current issue for improving both productivity and quality. From a systems standpoint, detailed studies of control specifications are more important than ever, and software developers are driven by the need to build fault-tolerant systems (Figure 14).

7 ECU–ECU NETWORKS AND FUSION OF ECUs (KATO, 2010)

The number of wire harnesses (W/H) has increased in accordance with the spread of electronic systems. As a result, W/H weight and complexity is on the increase. This increase in W/H is the direct consequence of the importance of linking ECUs. Multiplex communication has also become more important as the number of W/H has exceeded 1500. Historically, ECU communication began with the controller area network (CAN) used by engine systems and spread with the adoption of J-1850 body control systems. More recently, low, medium, and high speed CAN systems have been adopted, and FlexRay has

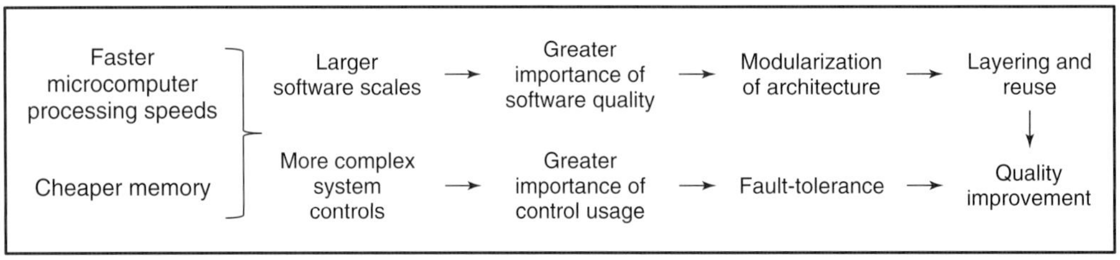

Figure 14. Software trends.

emerged as a feasible direction for sensor systems. Standard multimedia systems include Media Oriented Systems Transport (MOST). Although such systems are capable of eliminating several hundred W/H, increasing system complexity is causing the number of W/H to increase. ECU fusion is one basic method of suppressing system complexity and W/H increase. This aims to consolidate similar ECUs, thereby reducing their number as well as the connecting W/H. ECUs can be categorized into five main types: powertrain systems, chassis control systems, body systems, entertainment systems, and systems related to external information. However, the best way to consolidate each ECU type remains the key issue. Possible bases for combining ECUs include standard functions, differences in optional functions, and differences in development cycles. Furthermore, the growing popularity of hybrid vehicles is driving even greater ECU complexity by making control systems more complicated, further diversifying the information that must be communicated between ECUs, increasing the driving assistance functions that can be performed by connecting with external networks, and the like. The development of fault-tolerant systems is becoming even more crucial.

8 THE INFORMATION AGE (KATO, 2010)

In the 1990s, the application of electronic systems became further oriented toward information. Although the first navigation system was actually developed in 1987, the full-scale adoption of navigation systems began at the end of the 1990s. This was due to the spread of accurate position detection using GPS technology, the development of digital maps on compact discs (CDs), and the improvement in microcomputer performance. Figure 15 shows these trends. In addition, display technology was enhanced by the development of the liquid crystal display (LCD). As a result, it became possible to show other information rather than just maps on a multipurpose screen, such as information captured by rearview cameras.

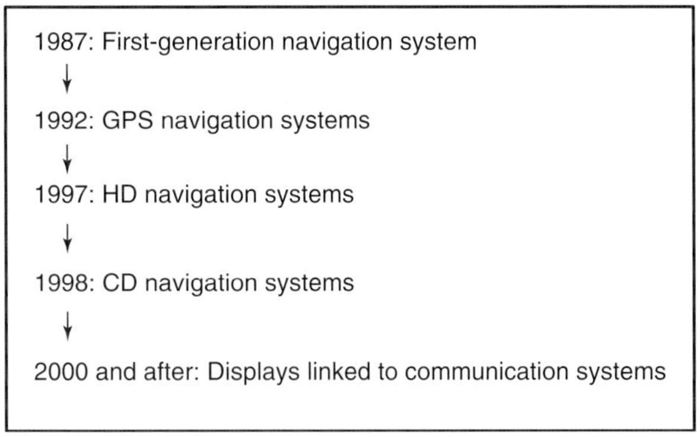

Figure 15. Information device trends.

The importance of multipurpose displays has increased further. The digitalization of communication has significantly speeded up rates of information transmission. As a result, communication technologies have become an indispensable aspect of vehicles in the 2000s. In addition to destination information as typified by navigation systems, emergency notification systems such as OnStar from GM and eCall from Mercedes have also been developed. Toyota's G-Book, Nissan's CARWINGS, and Honda's Internavi are other specific examples of enhanced communication systems. These developments highlight the growing significance of information communication in twenty-first century. Further progress will see the permanent connection of vehicles to communication systems and the adoption of more innovative communication applications.

8.1 The age of ITS

Improving the efficiency of communication, information, and traffic flow is a key part of modern vehicle development. Helping alleviate the congestion and negative environmental impacts caused by the popularization of the automobile and reduce the number of traffic accidents require comprehensive solutions. In the future, it is possible that automated vehicles will work as an ideal solution. Major short-term issues include reducing accidents by alleviating driver burden and congestion avoidance. The navigation system plays an indispensable role in addressing these points. The identification of the vehicle's current location by GPS and vehicle speed sensors, and detailed maps that include traffic information can help guide the driver to target destinations, minimize congestion, and reduce energy consumption. Navigation software exceeds half-a-million lines of code, the most of all automotive systems. This field is likely to become even more important in the future to help improve usability and save energy, as well as to provide detailed road information and safety assistance.

8.2 Expansion of automotive communication technologies

Communication technologies will play an even more important role in the future. Radio waves, which were originally used for AM and FM radio systems, have also been adopted for remote entry systems and tire pressure monitoring systems (TPMSs), as well as for external communication functions such as the Vehicle Information and Communication System (VICS), electronic toll collection (ETC), and mobile telephones. Figure 16 shows the history and status of radio wave usage in Japan. Effective radio wave technologies have already evolved from AM modulation, FM modulation, time division multiple access (TDMA), and code division multiple access (CDMA) protocols. In the future, the long-term evolution (LTE) protocol is expected to be adopted. These advances have been made feasible by the greater integration and processing speed of semiconductor devices. In addition to helping boost the performance of microcomputers, the evolution of semiconductor technology has also contributed greatly to the advancement of communication technology, that is, the effective use of finite frequency resources. In the future, it is expected that 1-GHz broadband technology will become even more sophisticated and spread to vehicles by improving communication quality.

9 THE FUTURE OF COLLABORATION

Until now, control system ECUs and electronics have been able to overcome the limitations of mechanical technology. In turn, this has encouraged the evolution of mechanical systems, a synergistic effect that has accelerated the evolution of automotive controls. Figure 17 illustrates these trends. Taking ABS as an example, microcomputers spurred the evolution of hydraulic systems. Compact and high performance oil pumps were then integrated with electronic circuits, which have been further integrated with actuators. These systems have diverged from household applications where devices are mounted on printed circuit boards to form new derived technology. Packaging technology has advanced to the stage where devices can be mounted on ceramic boards with excellent heat resistance. Although this is only a single example, the growing complexity of controls is likely to require the development of even more

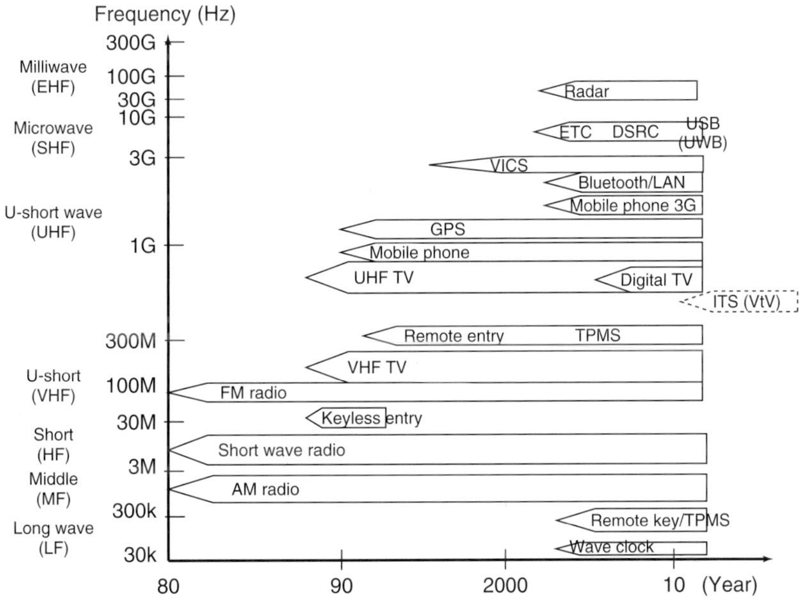

Figure 16. Wave frequencies and automotive applications. (Reproduced with permission from Kato, 2010. © Denso Corporation.)

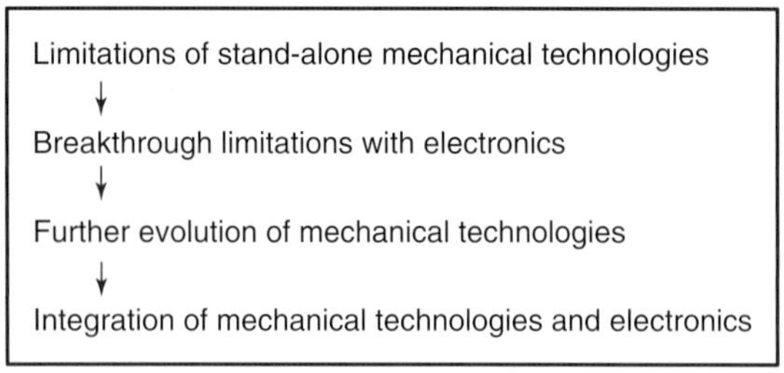

Figure 17. Collaboration trends (1).

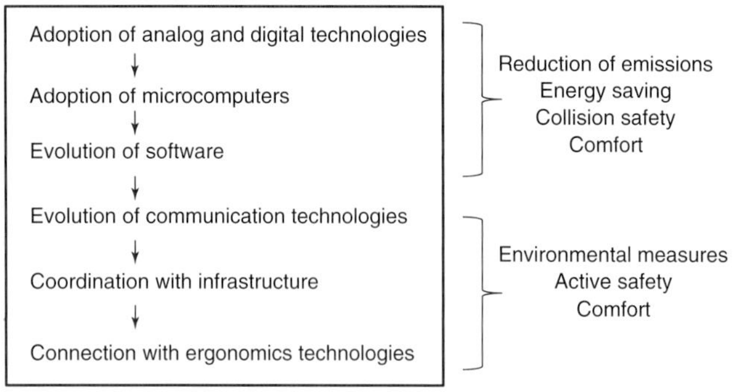

Figure 18. Collaboration trends (2).

unique technology. One possibility is the electronic correction and reduction of mechanical tolerances that exist in individual torque converters to improve the efficiency of automatic transmissions. It should be possible in the future to electronically adjust individual mechanical tolerances. This will require the integration of mechanical products and electronic circuits, which is dependent on the development of new packaging methods. Furthermore, although technology applications started with analog and digital technology, microcomputers have taken center stage and larger scale software has evolved. It is likely that wireless communication technology will become more advanced and find greater applications in the fields of environmentally friendly technology, safety, comfort, and the like. At the same time, this is likely to trigger further discussion about communication reliability. The introduction of CO_2 regulations as vehicles become even more widespread, safety concerns, and growing demands for active safety systems will usher in an age where image processing and recognition technologies, and technologies that incorporate aspects of ergonomics are as important as simple control systems. Figure 18 illustrates these likely trends. Thus, the role of electronics, software, and algorithms will continue to increase, placing even greater emphasis on collaboration between automotive and electronics technologies.

RELATED ARTICLES

Volume 2, Chapter 56
Volume 3, Chapter 84
Volume 4, Chapter 137
Volume 4, Chapter 138
Volume 4, Chapter 139
Volume 4, Chapter 141
Volume 5, Chapter 143
Volume 5, Chapter 147
Volume 5, Chapter 153
Volume 5, Chapter 154
Volume 5, Chapter 155
Volume 5, Chapter 156
Volume 5, Chapter 157
Volume 5, Chapter 159
Volume 5, Chapter 161
Volume 5, Chapter 164
Volume 5, Chapter 165
Volume 5, Chapter 168
Volume 5, Chapter 169

REFERENCES

Casey, R. (2008) *The Model T, A Centennial History*, The Johns Hopkins University Press, Baltimore.

CISPR 12: *Vehicles, boats, and internal combustion engine driven devices – radio disturbance characteristics – Limits and methods of measurement.*

CISPR 25: *Radio disturbance characteristics for the protection of receivers used on board vehicles, boats, and on devices – Limits and methods of measurement.*

Ina, O. (1998) Recent Intelligent Sensor Technology in Japan. In Automotive Electronics: The 1980s - A Collection of Landmark Technical Papers. SAE International, 831–839.

ISO7637-1: *Road vehicles – Electrical disturbances from conduction and coupling – Part 1: Definitions and general considerations.*

ISO7637-2: *Road vehicles – Electrical disturbances from conduction and coupling – Part 2: Electrical transient conduction along supply lines only.*

ISO11451: *Road vehicles – Vehicle test methods for electrical disturbances from narrowband radiated electromagnetic energy.*

ISO11452: *Road vehicles – Component test methods for electrical disturbances from narrowband radiated electromagnetic energy.*

Kato, M. (2010) *Automotive Electronics Illustrated*, Nikkei Business Publications, Inc, Tokyo.

Manger, H. (1998) Electronics Engine Controls in Europe. In Automotive Electronics: The 1980s - A Collection of Landmark Technical Papers. SAE International, 465–469.

Numazawa, A. (1998) Automotive Electronics in Passenger Cars. In Automotive Electronics: The 1980s - A Collection of Landmark Technical Papers. SAE International, 789–802.

Shockley, W. (1950) *Electrons and Holes in Semiconductors, With Applications to Transistor Electronics*, D. Van Nostrand, New York.

FURTHER READING

Kitano, T. (1998) The Status of Automotive Electronics in Japan. In Automotive Electronics: The Early Years - A Collection of Landmark Technical Papers. SAE International, 377–395.

Scholl, H. (1998) Electronics Applications to the Automobile by Robert Bosch. In Automotive Electronics: The Early Years - A Collection of Landmark Technical Papers: The Early Years. SAE International, 521–537.

Shah, P. (1998) Programmable Memory Trends in the Automotive Industry. In Automotive Electronics: The 1990s and Beyond - A Collection of Landmark Technical Papers. SAE International, 93–104.

Chapter 137
ECU Technologies from Components to ECU Configuration

Yukihide Niimi
DENSO Corporation, Kariya, Japan

1 INTRODUCTION

Since the 1960s, when the first radios and stereos equipped with transistors began to be installed in cars, electronics were seen very much as optional equipment. The full-scale introduction of electronics for vehicle control began in 1970 with the proposal of the Clean Air Extension Act (popularly referred to as the *Muskie Act* after its main sponsor, the United States Senator Edmund Muskie) to regulate harmful emissions (NO_x, HC, and CO) from the engine. To comply with this Act, there was a trend for conventional carburetor fuel supply systems to be replaced with electronic fuel injection systems, which used injectors to control the amount of fuel injected into the engine intake system. This fuel system was controlled by an electronic control unit (ECU).

As electronic fuel injection systems became the mainstream means of supplying fuel, engine control systems began to evolve steadily. Initially, control simply judged the cleanliness of emissions to determine the optimum fuel injection amount. Once ECUs were introduced, automakers realized that various controls could be performed efficiently. Electronic circuits evolved from analog to digital (microcomputer control), and various controls were added to the engine ECU. These included ignition timing control, knock control, idling speed control, transmission control, and so on. At the same time, electronic control also spread quickly to functions such as the air conditioner, ABS (antilock braking system), and airbags.

Currently, controls are categorized in accordance with the field, such as powertrain, air conditioning, body, and dynamic and safety controls. There are many different control system variations within these categories. This chapter describes the internal structure and parts used in ECUs, using the engine ECU as a typical example.

2 ECU OUTLINE

Most car electronics contain an electronic control computer called an *ECU* (Kato, 2010a). This ECU functions to process inputs from sensors and the like to drive motors or other kinds of actuators. ECUs communicate with each other using the in-vehicle local area network (LAN). ECUs contain sensor and switch input processing circuits, analog–digital (AD) conversion circuits, microcomputers, a power supply, output processing circuits, and a communication circuit for interacting with other ECUs. The main components are semiconductors. Figure 1 shows a typical

Encyclopedia of Automotive Engineering, print
Edited by David Crolla, David E. Foster, Toshio Kobayashi, and Nick Vaughan.
 ISBN: 978-0-470-97402-5
Also published in the *Encyclopedia of Automotive Engineering* (online edition)
DOI: 10.1002/9781118354179.auto213

Figure 1. Typical ECU.

printed circuit board type ECU and Figure 2 shows the general functional blocks.

Each block in the ECU has the following roles. The power supply provides a stable voltage to each block (5, 3 V, and the like). It is connected to the battery (12 V) in the engine compartment. The power supply block must be highly accurate as it also uses the reference voltage of the AD conversion circuit (see below).

The input processing circuit converts digital input signals into signal levels that can be inputted into the microcomputers.

The AD conversion circuit (also known as the *AD converter*) converts analog input signals into digital values that can be inputted into the microcomputers. The microcomputers calculate the control amounts from each input signal and control the output signals in accordance with the results of these calculations. The output processing circuit converts the output signals of the microcomputers into signal formats capable of driving the actuators, amplifies voltages, and so on. The communication circuit contains a communication driver that converts data output from the microcomputers into communication signals that comply with communication standards. It also includes a communication receiver that converts signals transmitted from other ECUs into signal levels that can be input to the microcomputers.

3 ECU INSTALLATION LOCATIONS AND OPERATION CONDITIONS

3.1 ECU installation locations

The simple term *ECU* covers a wide range of ECU types that differ depending on the installation location. The structure of each ECU is designed in accordance with its installation environment (Kato, 2010b).

Electronic control systems comprise three types of devices: (i) an ECU that governs the system control, (ii)

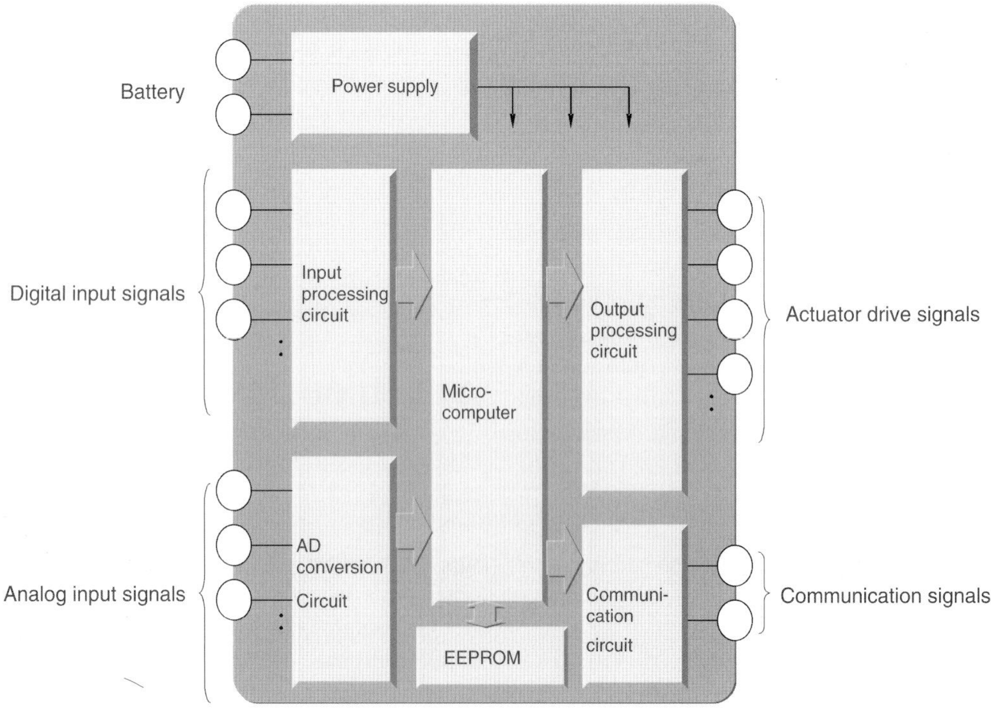

Figure 2. ECU functional block diagram.

sensor groups that input the states and information relevant to the control into the ECU, and (iii) actuator groups that ultimately achieve the targeted control. Of these device types, the sensors and actuators have to be installed in certain locations to achieve the function. In contrast, the installation locations of ECUs have changed over the course of history. When the electrification of vehicles started, electronic parts were particularly vulnerable to heat, humidity, and vibration. For this reason, ECUs were installed inside the occupant compartment. Subsequently, as the reliability of circuit board-based electronic parts and packaging technology for electronic parts improved, it became possible to install ECUs inside the engine compartment, despite its severe heat, humidity, and vibration, and even directly attach ECUs to the engine itself. By making the appropriate considerations at the design stage, it is now possible to install ECU hardware anywhere in the vehicle. This is known as the *free-installation concept*.

However, ECUs, sensors, and actuators have to be connected together electrically using wire harnesses. It is clearly preferable to ensure that these wire harness connections are as short as possible. Short wiring lengths have the following merits.

- lower material costs, enabling cheaper wire harnesses;
- less copper used, enabling vehicle mass reduction;
- lower electrical resistance, enabling thinner wires to be used;
- less susceptibility to the effects of induced noise caused by wiring impedance, which is generated depending on the frequency of the signal. Shorter wires also emit less noise.

For these reasons, the transmission ECU is installed as close to the transmission as possible. In some vehicles, it may actually be installed inside the transmission itself. Furthermore, a growing number of electronic parts related to engine control are installed directly on the engine. Ignition induction coils are installed on the plug hole to minimize ignition noise by making the high voltage wiring as short as possible. Impedance is a restriction for the electric driver unit (EDU) that is used to drive the injectors in a direct injection gasoline engine, because of the large current flow and considering operation speed. Accordingly, these units are installed close to the engine in the engine compartment with as short a wiring length as possible.

The installation location is often determined from a functional standpoint. For example, acceleration (G) sensors are located in multiple locations throughout the vehicle in the airbag system. As the airbag inflator has to deploy multiple airbags, it is located close to the center of the vehicle. In addition, the navigation system display has to be in a position that is easily visible, which restricts its installation location. Recently, as electronic circuits become more compact, motor control circuits are becoming extremely small. These compact controllers are now being installed and integrated into the actual motors, further driving reductions in size. Car electronics are installed in locations that facilitate the target function. Installation methods are categorized in accordance with the installation location (Figure 3). Each installation

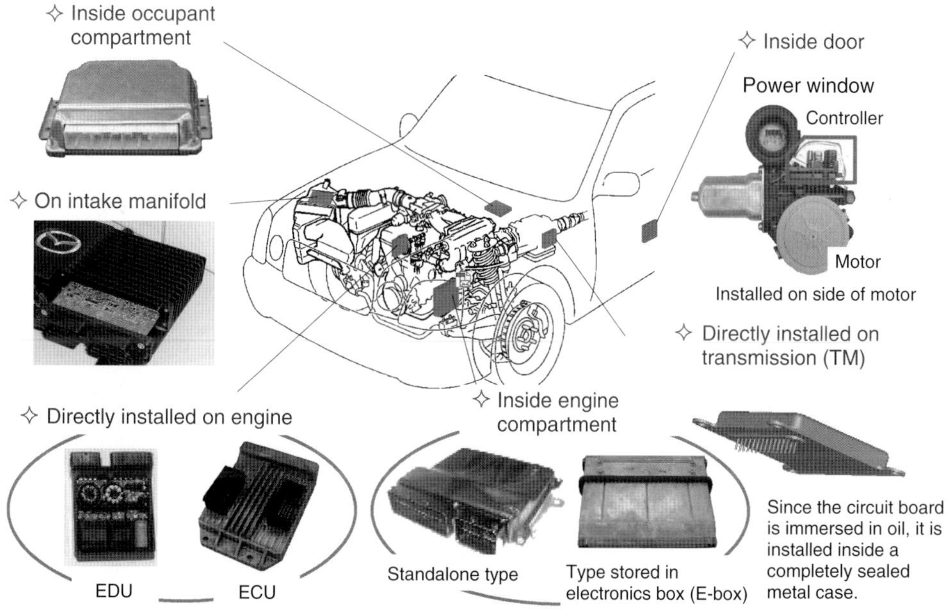

Figure 3. ECU installation location and method.

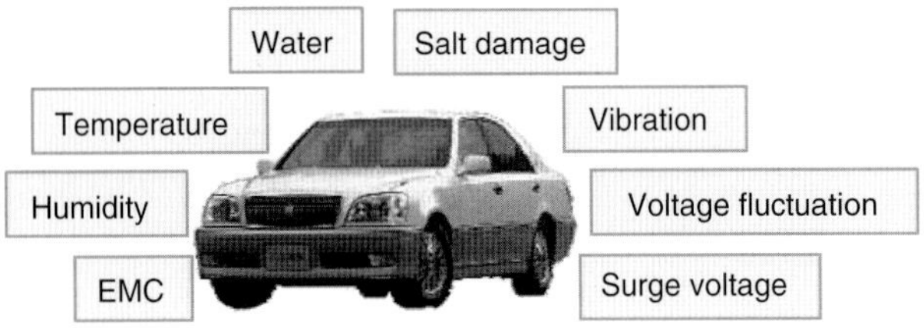

Typical maximum temperature exposure of parts in engine compartment

Location	Maximum temperature (°C)
Engine cooler	120
Engine oil	120
Transmission oil	150
Intake manifold	120
Exhaust manifold	650
Alternator intake air	130

Typical maximum humidity exposure of automotive parts

Location	Maximum humidity
Engine compartment (close to engine)	38°C 95% RH
Seats	66°C 80% RH
Close to side doors	38°C 95% RH
Close to dash board	38°C 95% RH
Floor sheet	66°C 80% RH
Rear deck	38°C 95% RH
Luggage compartment	38°C 95% RH

Typical maximum temperature exposure of parts in occupant compartment

Location	Maximum temperature (°C)
Top of dash board	120
Bottom of dash board	71
Floor	105
Rear deck	117
Headlining	83

Figure 4. Installation environments inside vehicle.

location has a different environment and ECUs are designed considering that environment. Figure 4 shows the typical environmental conditions of each installation location in a vehicle.

3.2 ECU operation conditions

Table 1 shows the different categories of operation conditions.

ECUs related to basic vehicle performance (the engine and dynamic ECUs) and ECUs related to occupant safety (the airbag ECU) operate under harsh conditions. In contrast, the operation conditions for the navigation system ECU are not set at the same level. This is because enabling an ECU to operate under harsher conditions than necessary generates excess cost and because the electronic circuits used in complex controls such as the navigation system are not yet able to operate in the harshest environments.

3.2.1 Operation conditions of parts used in ECU

Each part has different operating conditions (Kato, 2010b). Table 2 shows the temperature conditions for an electrolytic capacitor.

As the operation environment of automotive ECUs is harsher than those for home electronics, electrolytic capacitors are required to operate under more severe conditions. That difference is reflected in the higher purchase price.

Table 1. ECU operation conditions.

Application	Installation Location	ECU Type	Voltage (V)	Temperature (°C)	Vibration (G)
Engine	Engine compartment	Engine ECU	6–16	−30 to +95	4.4
Dynamics	Occupant compartment	ABS ECU	6–16	−30 to +80	4.4
Safety	Occupant compartment	Airbag ECU	6–16	−30 to +80	4.4
Body	Occupant compartment	Navigation system ECU	8–16	−30 to +80	4.4

Table 2. Operation conditions of electrolytic capacitor.

For Automotive Use		For Home Electronics
Engine ECU	Navigation System ECU	
−40 to 125°C	−40 to 105°C	−40 to 85°C

4 ECU EXAMPLES

4.1 Examples of printed circuit board-based ECUs

4.1.1 Engine ECU

The engine ECU is a standalone package and has only one internal control circuit board. As the number of control functions increased, some engine ECUs were developed with multiple boards. However, these functions have come to be collected together as integrated circuit (IC) devices, and modern engine ECUs have returned to a one-board configuration. Figure 5 shows an example of this type of ECU. It has a waterproof structure and its outside shape is designed to dissipate heat generated from the inside.

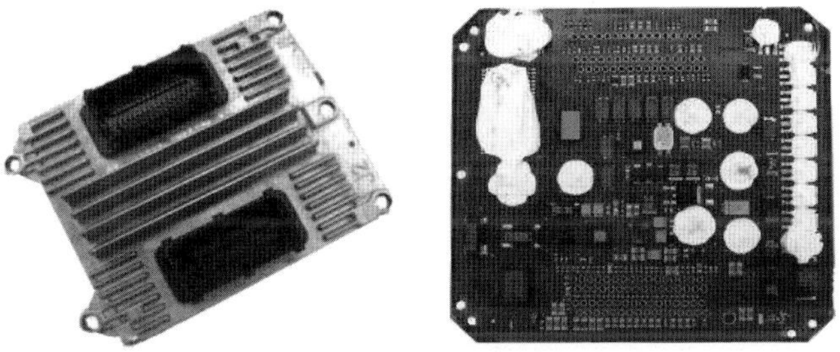

(a) (b)

The ECU case is provided with fins to improve heat dissipation. The white plastic portions in Figure (b) are made from a material called a thermally conductive gel, which is designed to improve heat transmission. These design points improve heat dissipation from the ECU and allow installation on the engine

Figure 5. Example of engine ECU. (a) Outside appearance of ECU. (b) Circuit board inside ECU.

4.1.2 Navigation system ECU

The navigation system ECU arranges components such as a display, DVD drive, and hard disc drive (HDD) in a compact package on the circuit board. Figure 6 shows a basic 2DIN type navigation system. Although it contains a built-in audio system and multiple circuit boards, the navigation system functions are collected on the same board. The navigation system uses a microprocessor with high functionality. To ensure efficient heat dissipation, a metal case is used that is punched with holes for ventilation.

4.2 Hybrid ECU

Hybrid ECUs are compatible with high functionality and large currents. In this type of ECU, bare chips are packaged on multilayered laminated ceramic circuit boards with high thermal conductivity. Figure 7 shows the example of a diesel engine hybrid ECU (Kato, 2010b). It contains a microcomputer, hybrid ICs, and power devices. Hybrid ECUs are usually adopted to meet requirements for size and weight reduction, and environmental resistance.

As shown in Figure 8, a hybrid ECU is structured with a control portion, power portion, large parts, and a connector. It uses multilayered laminated ceramic circuit control boards with high thermal conductivity, a bare chip type of chip scale package (CSP), conductive adhesive connections for the power ICs and chip parts, gold (Au) wire bonding, and silicon gel sealing.

Figure 9 shows an electric drive unit (EDU). A typical example of EDU application is as an injector drive controller. EDUs use a thick multilayered laminated ceramic circuit board. The ICs are bare packaged to help reduce the size of the circuit board. The largest parts on the

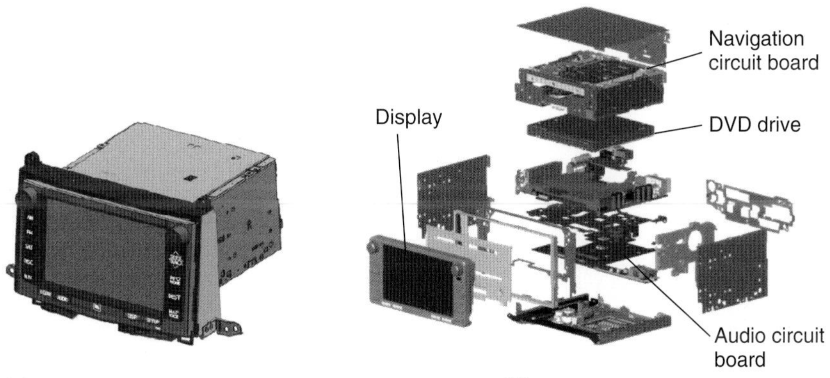

(a) (b)

Component parts such as the DVD drive and display, and control boards such as the navigation audio system boards are collected into a 2DIN space

Figure 6. Example of navigation system. (a) Outside appearance of ECU. (b) Internal configuration.

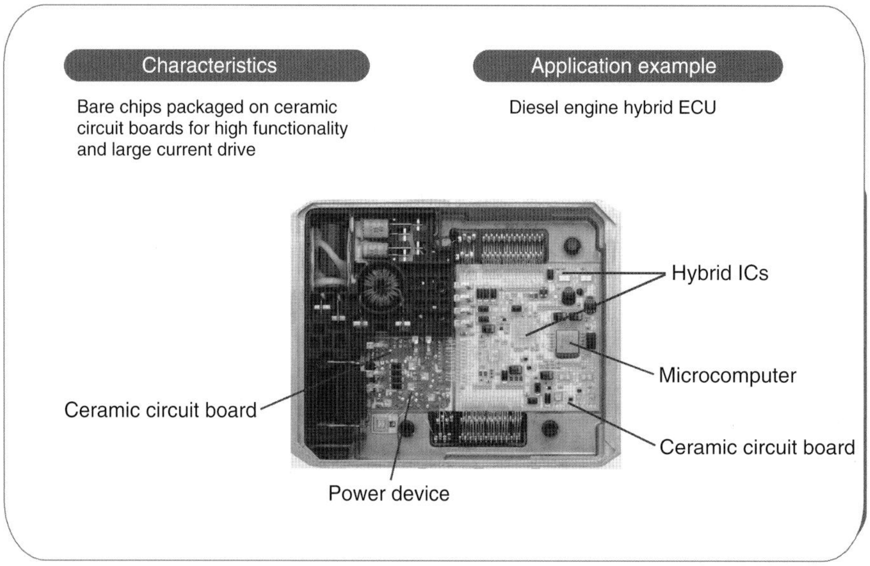

Figure 7. Characteristics and application example of hybrid ECU.

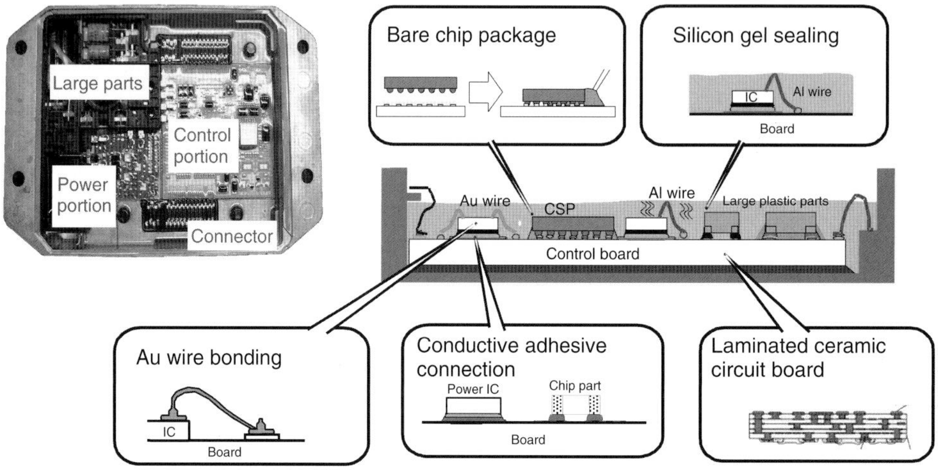

Figure 8. Hybrid ECU structure.

circuit board are the capacitors, which store accumulated energy for driving devices, and coils. EDUs are designed to be resistant against heat, vibration, and water. The bare chips are sealed in silicon gel to protect the bare packaged chips against the external environment.

5 ECU CONFIGURATION AND COMPONENT PARTS

5.1 ECU configuration

In broad terms, an ECU consists of a circuit board, a connector, and a case that protects these two portions from the external environment (Figure 10). Active, passive, and functional parts are arranged on the circuit board. Figure 11 shows the configuration of these parts. ECUs are increasingly being installed inside the engine compartment and on the engine itself, as well as in the occupant compartment. Figure 12 shows some typical examples of each.

5.2 Case

ECUs located inside the occupant compartment are open because water exposure is not considered likely. In contrast, ECUs installed inside the engine compartment or on the engine are designed with a waterproof case. The cases of ECUs attached to the engine are also designed to dissipate heat considering the harsh temperature environment. Most

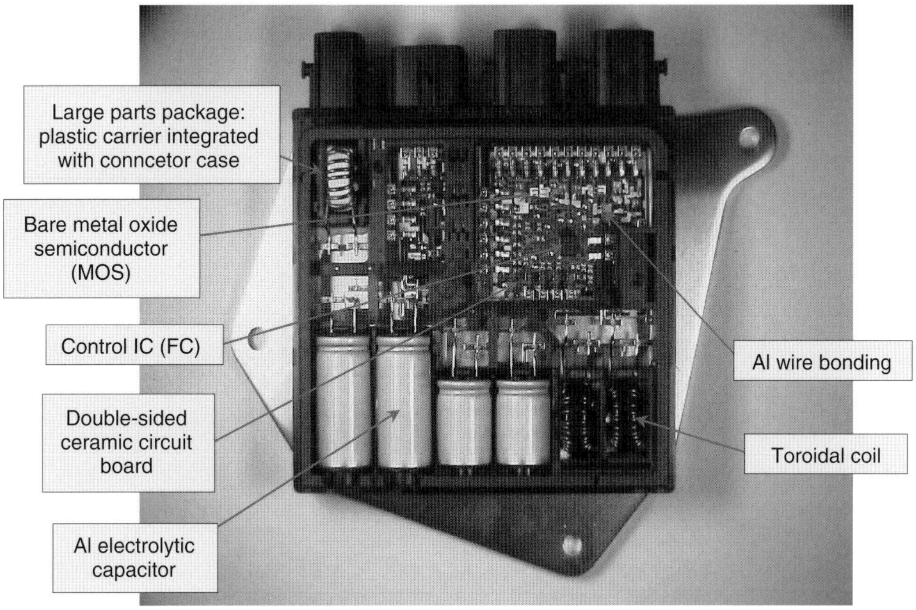

Figure 9. Electric driver unit.

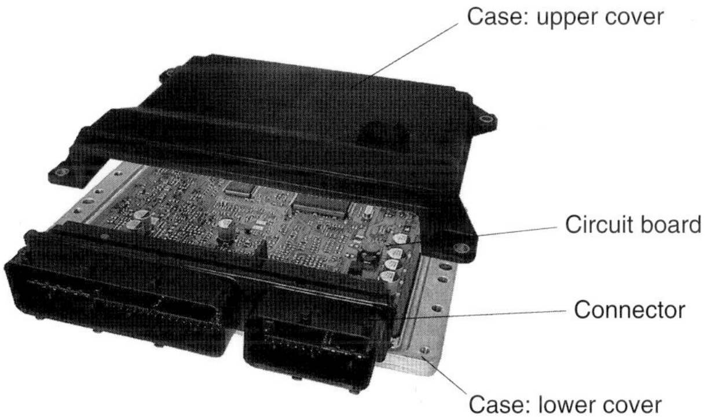

Figure 10. Configuration of typical ECU (waterproof type).

cases are configured in two sections with an upper and a lower part.

5.2.1 *Non-waterproof case*

The primary role of a non-waterproof case is to protect the circuits from the impact of external shocks. Non-waterproof cases that are not required to dissipate heat are often made of plastic. When heat dissipation (natural cooling) is required, metal cases are used, including Al die-cast, Al-stamped, and Fe-stamped cases.

5.2.2 *Waterproof cases*

Circuit assemblies in which the circuit board and connector are soldered together must be protected from water using upper and lower cases that have been designed accurately in accordance with the dimensional requirements. Accuracy is particularly required for the joining case surfaces, and shape around the connector, as these areas are closely related to waterproof performance. A waterproof sealing material is used to fill the gap between the cases. This material is generally a silicon gel with properties that do not change under temperatures as high as 150°C.

5.2.3 *Waterproof type high heat dissipating cases*

Cases installed in the engine compartment must be designed with resistance against heat, vibration, and water exposure. In addition to the waterproof design described earlier, the mechanical properties of the case material must be considered to withstand the vibration applied to the ECU from the engine. Most cases in engine compartment are also shaped to dissipate heat. This shape must be determined considering the surrounding airflow and recent developments use simulations to improve the efficiency of the shape design process.

5.3 Connector

Unlike conventional electrical products, which only have to perform a single function, ECUs are used for control based on signals measured by sensors with respect to a control target. ECUs identify the state of the control target, calculate the final control value with this built-in microcomputer, and transmit signals to drive actuators. Consequently, ECUs require a large number of signal

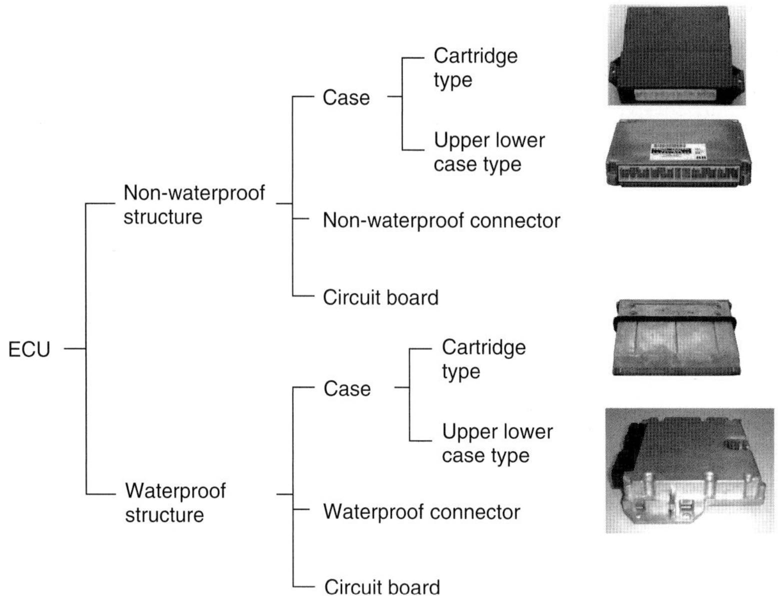

Figure 11. ECU component parts.

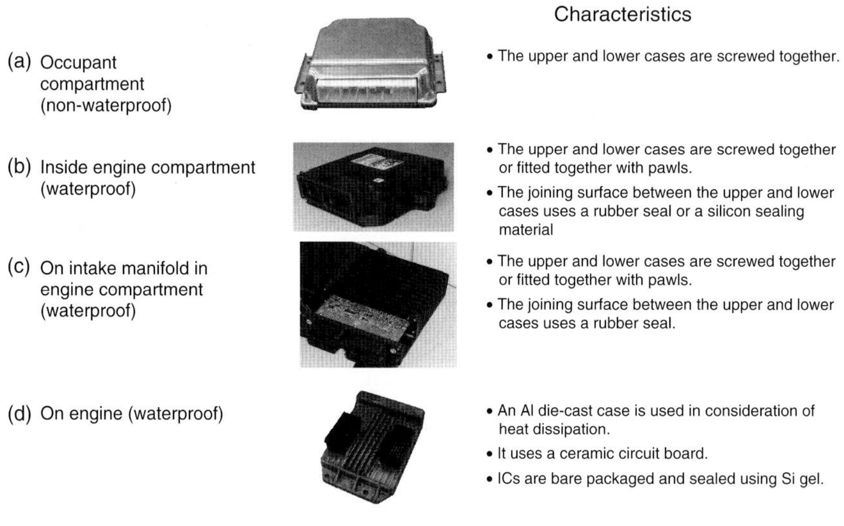

Figure 12. Appearance of ECU according to installation location.

terminals, such as sensor input terminals, actuator drive output terminals, power supply terminals, and terminals for communicating with other ECUs.

The part that collects these terminals into a single group is called the *connector* (Figure 13). The number of terminals in the connector is increasing as ECU functions become more sophisticated. Although advances in compact packaging technology have reduced the size of the circuit board, there has been no progress in making the connector smaller because of restrictions related to the diameter of the externally connected wires (i.e., the wire harnesses).

Waterproof and non-waterproof connectors are available. Waterproof connectors tend to be larger than non-waterproof connectors. Furthermore, non-waterproof connectors are divided into blocks of around 20 terminals. This is related to the maximum number of wires that can pass through a hole in the wall separating the engine and the occupant compartments (about 200). In contrast, waterproof connectors are designed with a split terminal layout, which is divided between wire harness groups used in the engine compartment and wire harnesses that transmit signals within the occupant compartment. In this

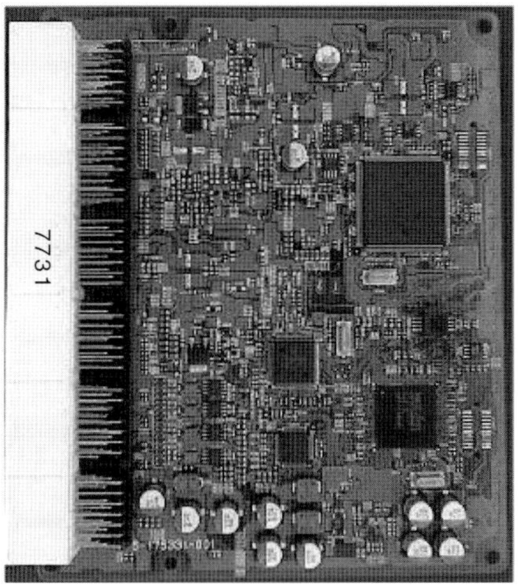

Non-waterproof type connector

Figure 13. Connector circuit board inside ECU.

case, a 200-terminal connector often has two blocks of about 140 terminals for transmitting signals to the engine compartment and about 60 terminals for the occupant compartment. Figure 14 shows the housing (circuit board side) and plug (vehicle wire harness side) of a waterproof and a non-waterproof connector.

5.4 Circuit board

Figure 15 shows the two types of circuit board used in an ECU, which are fabricated from epoxy resin and ceramic, respectively. Circuit parts are installed on the circuit board and the packaging format of these parts differs greatly between the circuit board types. More specifically, IC parts are bare packaged on ceramic circuit boards because this type of board is more expensive.

The type of the circuit board also affects the forms of the packaged parts. The use of through-holes to package parts has given way to the use of surface mounted devices (SMDs). This is because electronic parts are becoming smaller because of constant demands to install more functions within a smaller case. The merits of SMD can be outlined as follows.

- Compared to insertion, eliminating the through-hole enables more efficient use of the circuit board surface, increasing the available area for packaging. Therefore, a circuit board with the same number of parts can be made smaller.
- The soldered portion of the through-hole can be eliminated, reducing the size of the soldering land pattern and increasing the density of the wiring pattern.
- Eliminating the through-hole simplifies the circuit board fabrication process, which helps to reduce costs.

5.4.1 Epoxy resin printed circuit boards

Automotive applications widely use laminated circuit boards (Japan Institute of Electronics Packaging, 2006) based on glass epoxy resin. This type of circuit board is widely categorized into through-hole circuit boards (double-sided, 4-layer, and 6-layer), blind via hole (BVH) circuit boards (a type of through-hole circuit board), and build-up circuit boards that have a build layer on both sides and a conventional layered structure in the center (Figure 16). Build-up circuit boards that have a single build layer on the sides and a core 4-layer center structure are called *1-4-1 build-up circuit boards*. Build-up circuit boards are used in the navigation system and other systems that require high functionality with dimensional restrictions, as well as parts where small size is important. 1-4-1 build-up circuit boards have also begun to be adopted in engine

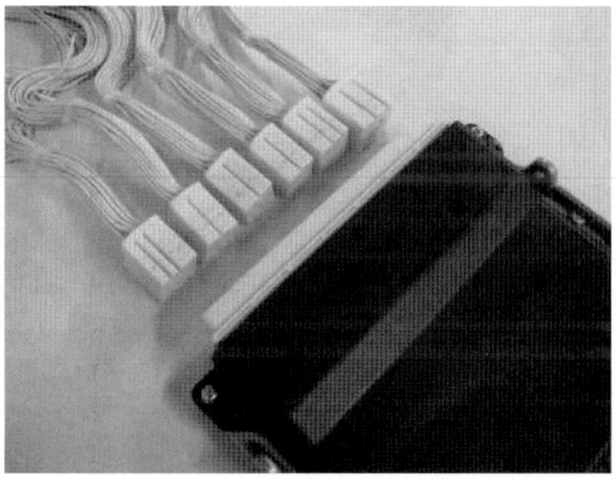
(a)

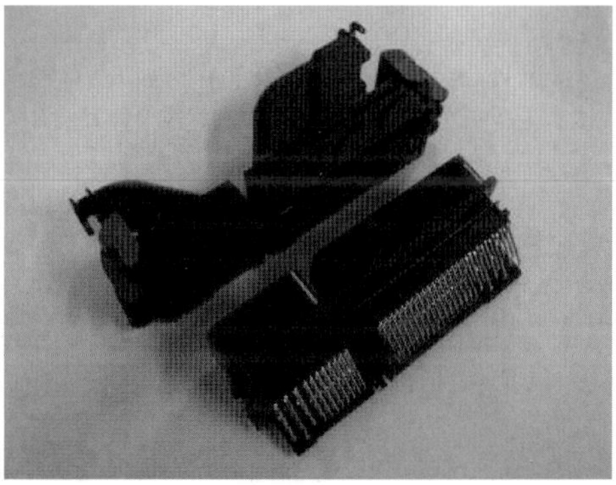
(b)

Figure 14. (a) Non-waterproof and (b) waterproof connectors.

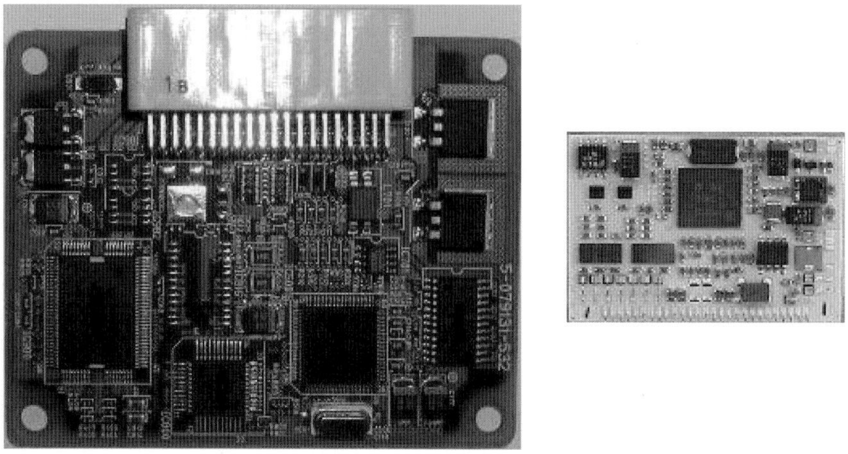

The photo shows the same circuit packaged onto an epoxy resin and a ceramic circuit board. Ceramic circuit boards are considerably smaller because the ICs are bare packaged

Figure 15. (a) Epoxy resin and (b) ceramic circuit boards.

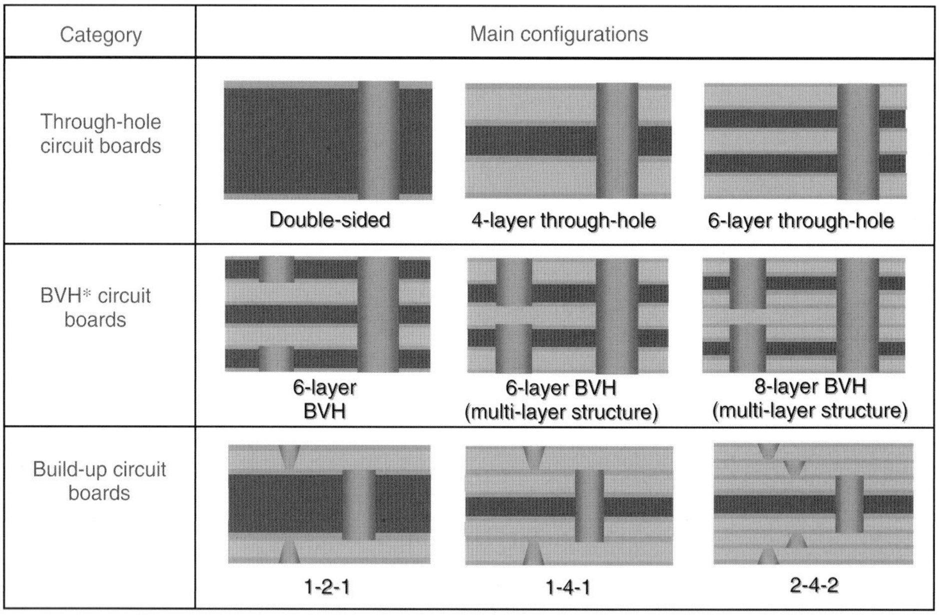

Category
Main configurations
Through-hole circuit boards
Double-sided
4-layer through-hole
6-layer through-hole
BVH* circuit boards
6-layer BVH
6-layer BVH (multi-layer structure)
8-layer BVH (multi-layer structure)
Build-up circuit boards
1-2-1
1-4-1
2-4-2

* BVH: blind via hole

Figure 16. Categories and layer configurations of epoxy resin circuit boards.

ECUs installed in the engine compartment. Through-hole circuit boards are particularly widely used in ECUs that have severe restrictions on size.

5.4.2 Ceramic circuit boards

Types of ceramic circuit boards include single-sided, double-sided, thick multilayered laminated, and sheet-laminated. The type used depends on the application. Single-sided to thick multilayered laminated ceramic circuit boards are widely used in control circuit boards for driving actuators. These control circuit boards also include built-in power devices for driving the actuators, with an emphasis on heat dissipation performance and size reduction. Sheet-laminated circuit boards are used in ECUs that operate under harsh temperatures and vibrations, such as those directly installed in the engine. These circuit boards include bare packaged ICs such as microcomputers and power devices. The configuration of sheet-laminated circuit boards is useful in achieving high wiring densities and size and weight reduction.

5.5 Categories of circuit parts

The parts (Japan Institute of Electronics Packaging, 2006) installed on a circuit board can be categorized in terms of function and shape. From the standpoint of function, these parts can be divided into general electronic and functional parts (Table 3). General electronic parts can be further categorized into passive parts, functional parts, connection parts, and conversion parts. Functional parts include individual semiconductors, ICs, and hybrid ICs. From the standpoint of shape, categories include inserted, surface mounted, and bare chip parts (Figure 17).

5.5.1 Inserted parts

These parts include axial lead parts, radial lead parts, irregular lead parts, and various types of packages, such as single inline packages (SIPs), dual inline packages (DIPs), and pin grid arrays (PGAs).

5.5.1.1 Axial lead parts. These parts have a transverse shape with two lead wires in the direction of the device. Lead wires are generally made of copper with a round cross section and use solder plating. These parts are inserted onto the circuit board and the lead legs are bent into an "L" shape. Typical axial lead parts include carbon film resistors, cylindrical ceramic capacitors, diodes, and so on.

5.5.1.2 Radial lead parts. These parts have a vertical shape with two or three lead wires protruding in parallel. Radial lead parts may be formed by processing the leads of axial lead part. As radial lead parts are created to match the pitch of the insertion holes in the printed circuit board, there is no need to bend the lead legs after insertion. In addition, the vertical shape of these parts means that a narrow interval can be set between leads, allowing denser packaging than with axial lead parts. More radial lead part types are available than axial lead parts, and these include Al electrolytic capacitors, transistors, ceramic filters, crystal oscillators, and the like.

5.5.1.3 Irregular lead parts. These are parts that cannot be categorized as either axial or radial lead parts. The shape, layout, and number of wires can be freely set, which means that there are various types of irregular lead parts. Typical examples include connectors, switches, variable resistors, and the like. Transistors and other devices that have an amplification action include ICs, power devices, and so on.

5.5.2 Surface mounted parts

These can be categorized into general electronic parts and active parts. General electronic parts can be further categorized by shape into flat, cylindrical, and irregular types. Flat parts are currently the mainstream type of surface mounted parts. Active parts can be categorized according to package type. These include mini-mold packages, small outline packages (SOPs), quad flat packages (QFPs), ball grid arrays (BGAs), and CSPs.

Table 3. Circuit part function categories.

Part Categories		Functions	Applications
General electronic parts	Passive parts	Controls voltage and current without changing the input signal characteristics	Resistors, capacitors, etc.
	Functional parts	Changes the input signal characteristics such as the frequency and time axis	Crystal oscillators, LC filters, etc.
	Connection parts	Mutually connects and switches parts and circuit devices	Switches, connectors, etc.
	Conversion parts	Converts input signals into different forms of energy	Speakers, sensors, etc.
Active parts	Individual semiconductors	Has active functions such as input signal amplification control, and memory. Standalone devices	Transistors, diodes, LEDs, etc.
	ICs	Collects and integrates multiple semiconductors.	Analog ICs, microcomputers, etc.
	Hybrid ICs	Collects active, passive, and film devices on a printed circuit board, and has active functions	Thick film hybrid ICs, thin film hybrid ICs, etc.

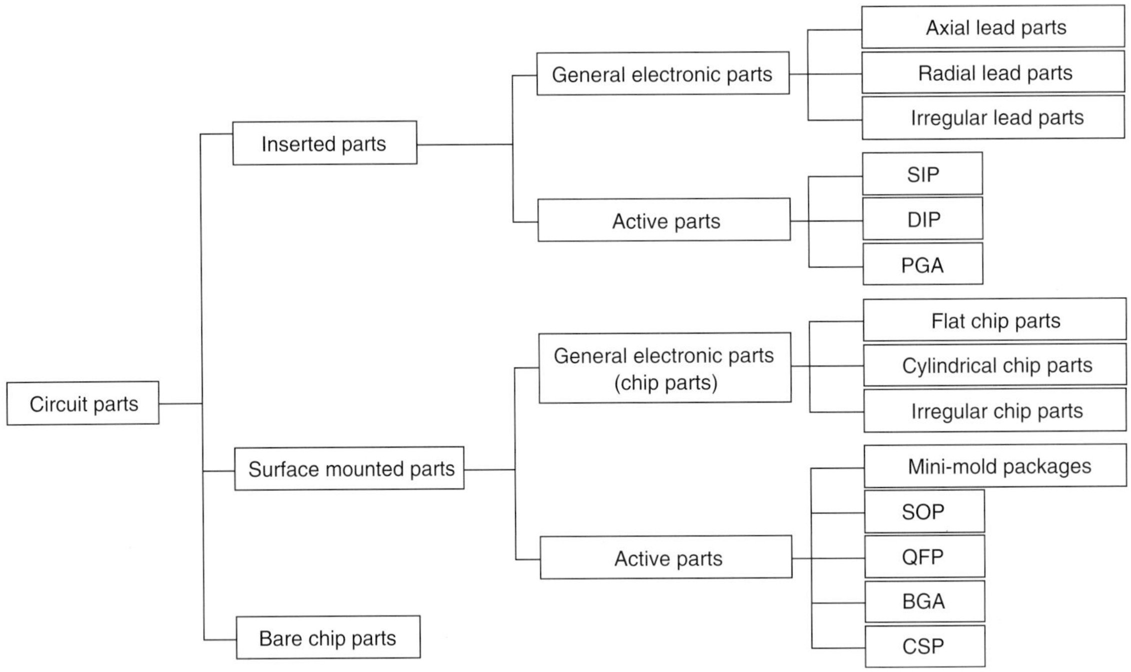

Figure 17. Circuit part shape categories.

5.5.2.1 Chip parts. Surface mounted general electronic parts are called *chip parts*. Flat chip parts are rectangular. These parts generally have no leads and have portions referred to as *terminals* that are located on the bottom or side surface of the solid shape. Fixed resistors, ceramic capacitors, laminated inductors, and the like use printed or laminated ceramics materials, which are formed to match the shape of the chip. These are formed into the flat shape by molding or enclosing the rectangular chip in a case. Typical flat chip parts include tantalum electrolytic capacitors, molded wound inductors, and the like.

Unlike flat chip parts, cylindrical chip parts have no directionality and can be used effectively in a package. However, the cylindrical shape is a disadvantage when packaging on a surface with various other parts. For this reason, there are not many parts with this shape. Irregular chip parts are those parts that are neither flat nor cylindrical. Typical examples include Al electrolytic capacitors, inductors, and the like.

5.5.3 Surface mounting semiconductor packages

5.5.3.1 Mini-molded packages. These packages enclose individual semiconductors such as transistors and diodes. The lead wires protrude to the outside of the package.

5.5.3.2 Small outline packages. SOPs are compact DIPs, with the leads processed for surface mounting. The lead pitch for automotive applications is between 1.27 and 0.5 mm, but consumer applications may have a lead pitch of 0.4 or 0.3 mm.

5.5.3.3 Quad flat packages. QFPs have lead wires that protrude in four directions from the package. QFPs are designed for high density packages with more leads than SOPs. The lead pitch for automotive applications is between 0.8 and 0.5 mm, but consumer applications may have a lead pitch of 0.4 or 0.3 mm.

5.5.3.4 Ball grid arrays and chip scale packages. Providing connection terminals for more than 500 pins on a QFP results in a large package. A countermeasure for this is to provide terminals on the back surface, opposite to the package front surface where the semiconductor devices are installed. The name BGA is derived from the round solder balls of these terminals. CSPs have the same structure as BGAs, but have a smaller package of virtually the same size as the semiconductor devices. CSPs attached with solder balls in the rewiring semiconductor fabrication process performed are sold under the names wafer process packages (WPPs) or wafer level packages (WLPs).

6 ECU AND MICROCOMPUTER TRENDS

6.1 ECU evolution

Figure 18 illustrates how ECUs have evolved. Figure 18a (Kato, 2010a) shows an electronic fuel injection device ECU with an analog circuit, which was used before the practical application of microcomputers. This is an analog ECU formed by discrete devices and used in 1973. This evolved into ECUs comprised of analog ICs, such as that shown in Figure 18b. This ECU is from the DIP era of ICs. Figure 18c is a digital engine ECU configured around a microcomputer. The custom microcomputers used in 1978 adopted software that was programmed using a 12-bit assembly language. The program used read-only memory (ROM), which was built in as a wafer during the microcomputer fabrication process. The photo shows the most complex engine ECU installed in 1983 models. Figure 18d is a modern powertrain ECU with two 32-bit microcomputers. Excluding the connectors, all the ECU parts used in 2006 were mounted on the surface. The software is written in C language. It uses flash memory and can read information from external terminals. As this shows, in addition to the scale of ECUs, the component parts, package shapes, and software languages have all evolved with the times.

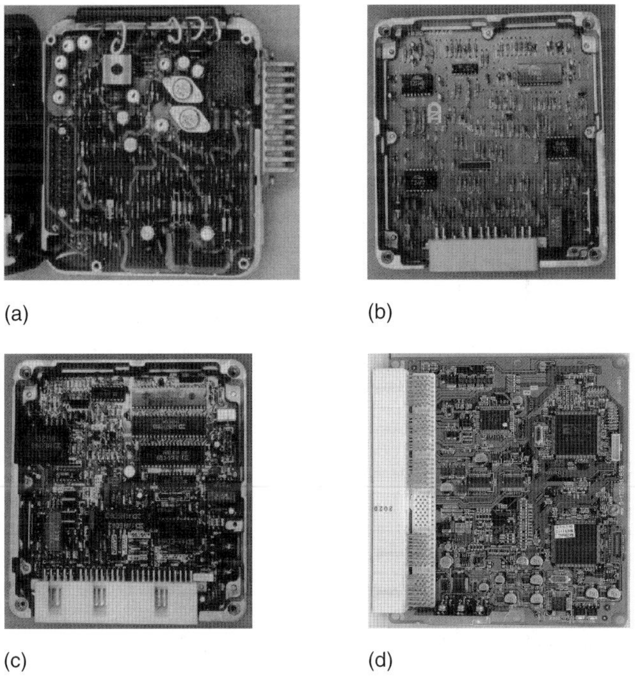

Figure 18. ECU evolution. (a) Analog ECU using discrete devices (1973 model). (b) Analog ECU using custom ICs (1979 model). (c) Digital 12-bit microcomputer (1983 model). (d) Digital 32-bit microcomputer (2006 model).

6.2 Microcomputer evolution

6.2.1 *Function*

This section describes the basic configuration of a microcomputer used in an ECU (Figure 19).

As the configuration of automotive microcomputers is no different from general-purpose types, a detailed description of general microcomputer characteristics is omitted in this section.

ROM has been adopted in recent years for regulatory compliance, and memory in modern microcomputers is configured to allow changes (i.e., flash memory) in block units within the ROM.

The timer controller performs time and time interval control. These devices have a compare function that changes an existing output when the set interval matches the internal timer value, and a capture function that memorizes the input time interval of external signal edges. The compare function is used to determine injection signal output timings and the like. The capture function is used to measure the engine speed signal input timing in combination with the interruption controller.

The communication interface is used to interact with input/output expansion ICs inside the ECU and with other ECUs. A growing number of microcomputers have an in-built CAN (controller area network) function for compatibility with the in-vehicle LAN.

6.2.2 *Microcomputer evolution*

The ROM capacity of microcomputers has rapidly increased to comply with new and more stringent regulations. This trend is likely to continue in the future, necessitating further miniaturization and increases in speed. Microcomputer types have also changed from 8- to 16-, to 32-bit configurations to meet requirements for greater processing speeds. As the pace of development has speeded up, the time to market introduction has grown shorter. In accordance with this trend, the type of ROM has switched from mask ROM to one-time programming (OTP) to flash ROM. Figure 20 shows an example of an engine control microcomputer. The details are described elsewhere.

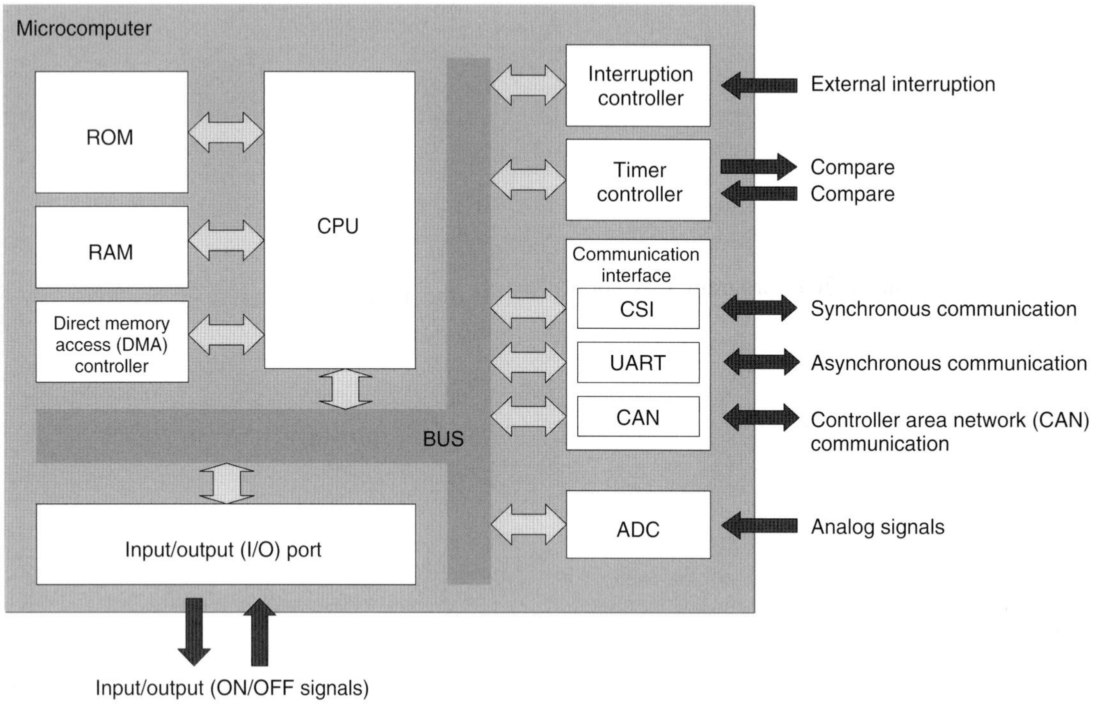

Figure 19. Internal microcomputer configuration.

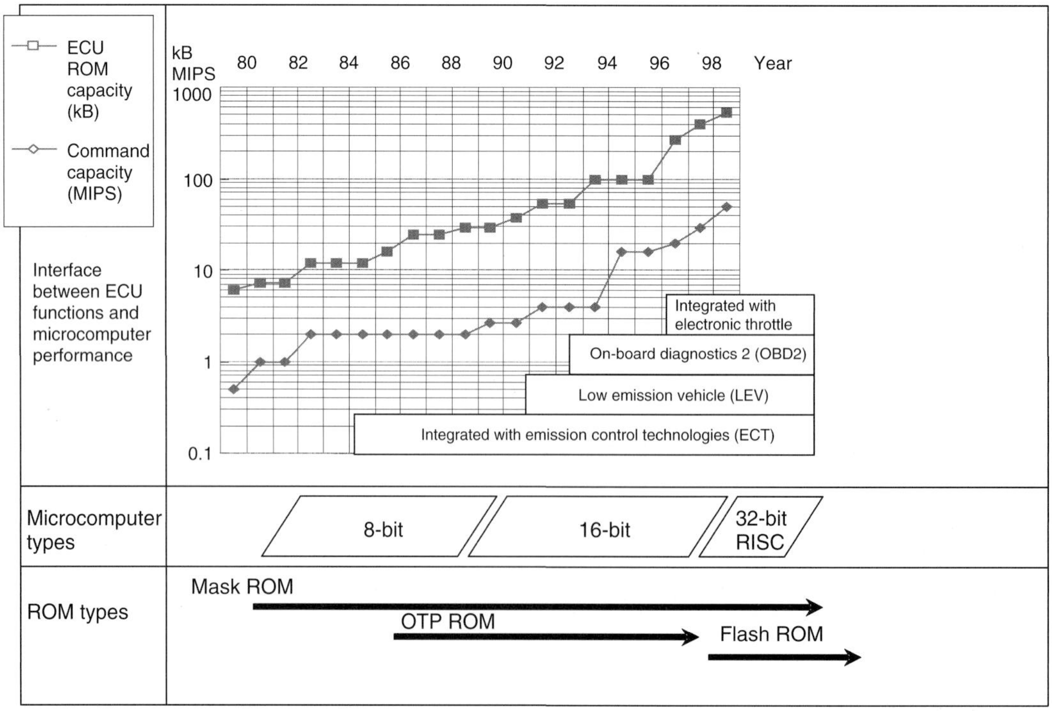

Figure 20. Engine control microcomputer evolution.

6.2.3 *Software trends*

Engine control by microcomputer began in the period from 1970 to 1980 in response to emissions regulations. Since then, microcomputers have spread to the transmission, airbags, navigation system, automatic air conditioner, and so on. As car electronics particularly the software is responsible for adding value to modern vehicles, this trend is likely to continue in the future. In the 1970s, the length of an engine control ECU program was about 4,000 lines of code. As a result of the recent explosive growth of electronic controls, which is rooted in the growing demands to improve the environment by reducing emissions and fuel consumption and to enhance vehicle safety through systems such as the airbags and seatbelts, a luxury vehicle may contain more than 80 ECUs that are larger in scale and far more sophisticated than ever before. Excluding the navigation system ECU, the total amount of program (software) code now exceeds 7 million lines, leading to a huge jump in developmental work hours.

6.3 Other ICs

Custom ICs designed to combine required functional blocks, such as operational amplifiers, comparators, and power supplies, and semi-custom ICs that make changes to existing blocks have also been developed. In the same volume, custom ICs are most expensive, followed by semi-custom ICs, and the lowest cost general-purpose ICs. The specifications of custom and semi-custom ICs are generally not disclosed to parties other than the customer. Another type of IC mixes digital and analog circuits on a single semiconductor chip.

In accordance with the configuration of the ECU, various single-chip custom ICs have been adopted that combine power devices with function blocks such as the power supply and in-vehicle LAN. As power devices are a cause of heat generation, these custom ICs use power packaging with a high heat dissipation performance instead of standard packaging to reduce the size of the ECU.

7 THE FUTURE OF ECU TECHNOLOGIES

This chapter described ECUs, focusing on their installation environment, operation conditions, common structures, component parts, and circuit parts, using the engine ECU as a typical example. ECUs contain sensor and switch input processing circuits, AD conversion circuits, microcomputers, a power supply, output processing circuits, and a communication circuit for interacting with other ECUs. This chapter detailed the parts provided on the circuit boards of ECUs, categorized in accordance with function and shape. It also included a brief outline of ECU trends.

In the past, ECUs operated independently, but ECUs today communicate with each other, and this has evolved into a networked system of ECUs. Recently, cooperation and integration with network applications have been seen. In the future, these systems will develop into cooperative autonomous systems that autonomously respond to the user's needs. So, as you can see, ECUs will continue to become more complex and larger in scale.

In addition, the ability to develop ECUs for which software and hardware in a short time frame is becoming crucial. It also goes without saying that functional safety must be considered at all times. We are in an age in which it is necessary to create frameworks for developing ECUs that can respond to a diverse range of needs.

RELATED ARTICLES

Volume 4, Chapter 136
Volume 4, Chapter 138
Volume 4, Chapter 139
Volume 4, Chapter 141
Volume 4, Chapter 142
Volume 5, Chapter 143
Volume 5, Chapter 159

REFERENCES

Japan Institute of Electronics Packaging (2006) *Handbook printed circuit technology*, 3rd edn, Nikkan kogyo Shinbun, Tokyo.

Kato, M. (2010a) *Automotive Electronics: Systems*, Nikkei Business Publications, Inc., Tokyo.

Kato, M. (2010b) *Automotive Electronics: Basic Technologies*, Nikkei Business Publications, Inc., Tokyo.

Chapter 138

Diversification of Electronics and Electrical Systems and the Technologies for Integrated Systems

Yukihide Niimi
DENSO Corporation, Kariya, Japan

1 INTRODUCTION

Until recently, cars were a collection of mechanical parts. Electronic control began in the 1970s with the engine, and now extends to the brakes, steering, suspension, instrument cluster, climate control, airbags, door locks, power windows, and the navigation system. In fact, virtually every device used in a modern vehicle is subject to electronic control. The human body works as a metaphor for an electronic control system. Sensors play the roles of the eyes and ears, actuators replace the hands and feet, and the brain becomes the electronic control unit (ECU). Signals measured by the sensors are transmitted to ECUs, where they are used to calculate control signals. The signals to and from the ECU are transmitted via wire harnesses and the like. An ECU consists of hardware such as a microcomputer, peripheral ICs, a power supply, printed circuit board, and connectors, as well as software that controls the calculation procedure.

A typical example of electronic control is the electronic control system in a gasoline engine. The engine ECU functions as the brain to calculate the fuel injection amount and ignition timing. As shown in Figure 3, the engine control ECU is connected with the in-vehicle network to exchange information with other ECUs. For example, the engine coolant temperature is transmitted to the instrument cluster for display on the temperature gage. Another example is in vehicles with adaptive cruise control (ACC), which functions to maintain a set distance with the vehicle ahead. If the vehicle-to-vehicle distance increases, the in-vehicle network is used to request the engine ECU to accelerate. In addition, when the cooling capacity of the climate control is insufficient in hot weather while the vehicle stops, the climate control ECU transmits a signal to the engine ECU to increase the engine speed. For details of automotive control systems, see Robert Bosch GmbH (2008, 2011), Denton (2004), Jurgen (1999), Mizutani (1992).

Figure 1 illustrates the change from independent standalone vehicle systems for engine control, brake control, climate control, and the like to integrated vehicle control systems that perform complex functions in cooperation with multiple control systems. There is also a growing trend for coordinated control with infrastructure outside the vehicle. Conventional vehicle electronics systems consisted of collections of individual ECUs installed separately to perform independent control. However, to achieve the type of complex and versatile control required by modern vehicles, these ECUs are being connected to the in-vehicle network (Figure 3). As the number of connected ECUs

Encyclopedia of Automotive Engineering, print © 2015 John Wiley & Sons, Ltd.
Edited by David Crolla, David E. Foster, Toshio Kobayashi, and Nick Vaughan.
This article is © 2015 John Wiley & Sons, Ltd. ISBN: 978-0-470-97402-5
Also published in the *Encyclopedia of Automotive Engineering* (online edition)
DOI: 10.1002/9781118354179.auto214

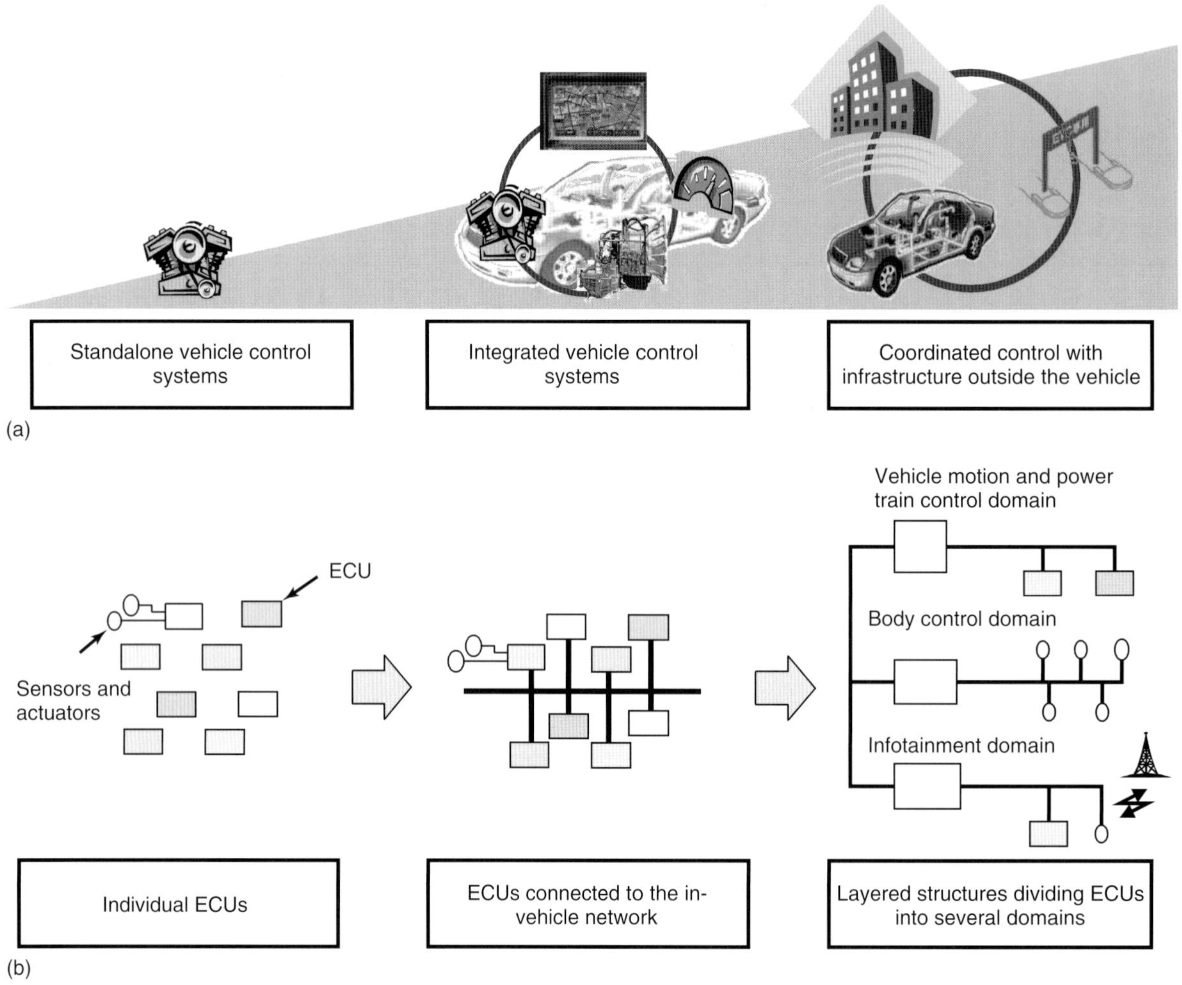

Figure 1. Evolution of automotive electronics systems. (a) Vehicle control systems. (b) Vehicle electronics systems.

increases, it is likely that layered structures will be created that divide these ECUs into several domains (Schäuffele, 2005a).

In the 1970s, the length of an engine control ECU program was about 4000 lines of code. As a result of the recent explosive growth of electronic controls, which is rooted in the growing demands to improve the environment by reducing emissions and fuel consumption and to enhance vehicle safety through systems such as the airbags and seat belts, a luxury vehicle may contain more than 80 ECUs that are larger in scale and far more sophisticated than ever before. Excluding the navigation system ECU, the total amount of program (software) code now exceeds 7 million lines. Figure 2 shows the trends for ECUs installed in luxury vehicles since 1995 (Schäuffele, 2005b).

Consequently, it is becoming more difficult to secure enough space to install electronic system components, particularly ECUs. As a result, ECUs are being made smaller and integrated with sensors and actuators. The commonization and standardization of parts, including software for configuring electronic control systems is also becoming more prevalent to help achieve major improvements in development efficiency.

The control systems of modern vehicles mutually connect the control functions of the power train, chassis, and body (i.e., the fundamental driving, turning, and stopping functions of a vehicle) with service functions that support the driver through information and communication. By doing so, modern vehicles are safer, more comfortable, easier to use, and more environmentally friendly than ever before. The scale of such systems is likely to increase in the future in response to the needs of society, which means that their functions will become even more sophisticated and versatile. New control, information, and communication technologies will be required to achieve such large-scale systems, and it will be necessary to design the optimum systems while ensuring high quality and reliability for both hardware and software.

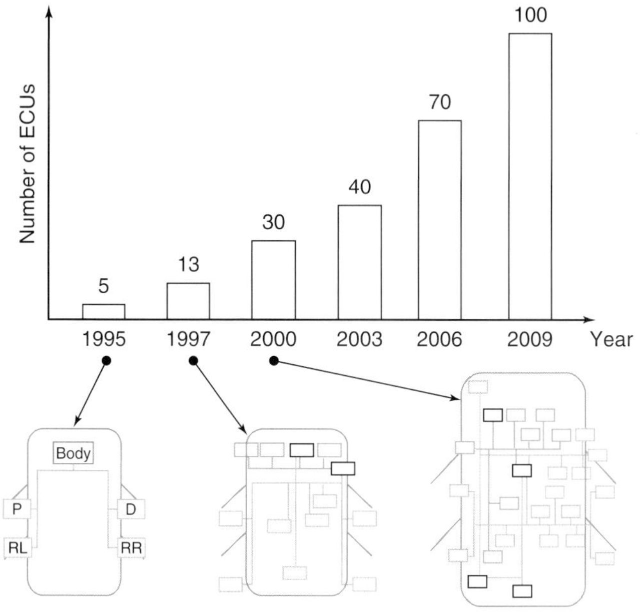

Figure 2. Number of ECUs installed in luxury vehicle.

2 LAYERING, STRUCTURING, AND FUNCTIONAL ALLOCATION

As shown in the example in Figure 3, the ECUs in a modern vehicle are connected to an in-vehicle network. The free exchange of data between the ECUs as well as between the sensors and actuators enables the vehicle system to determine which ECUs should be allocated with which functions from the standpoint of overall optimization. This stands in contrast to the conventional method in which the processes performed by the ECUs were automatically determined for each function.

However, as the number of ECUs in a vehicle increases, it is becoming more difficult to secure enough installation locations. Individual functional ECUs of standard functions that have no variations depending on the region of sale or vehicle grade are gradually giving way to integrated control. The integration of similar or mutually closely related functions is one way of reducing the number of ECU cases, ECU power supplies, component devices, connectors, and wire harnesses required. In other words, reducing these parts is a way of reducing costs. Under this strategy, nonstandard functions, new functions that have not yet found universal application, and functions for certain regions or vehicle grades are not subject to integration. These remain carried out by individual functional ECUs.

As the number of functions in each vehicle increases, another issue that has emerged is how to organize functional relationships and reconfigure the functional structure in the vehicle as a whole. In other words, this means determining how to most efficiently distribute functions throughout a vehicle.

Figure 4 analyzes each type of support function installed in a vehicle. This example shows three layers (the objective control layer, state control layer, and equipment control layer). The relationship between the three layers is as

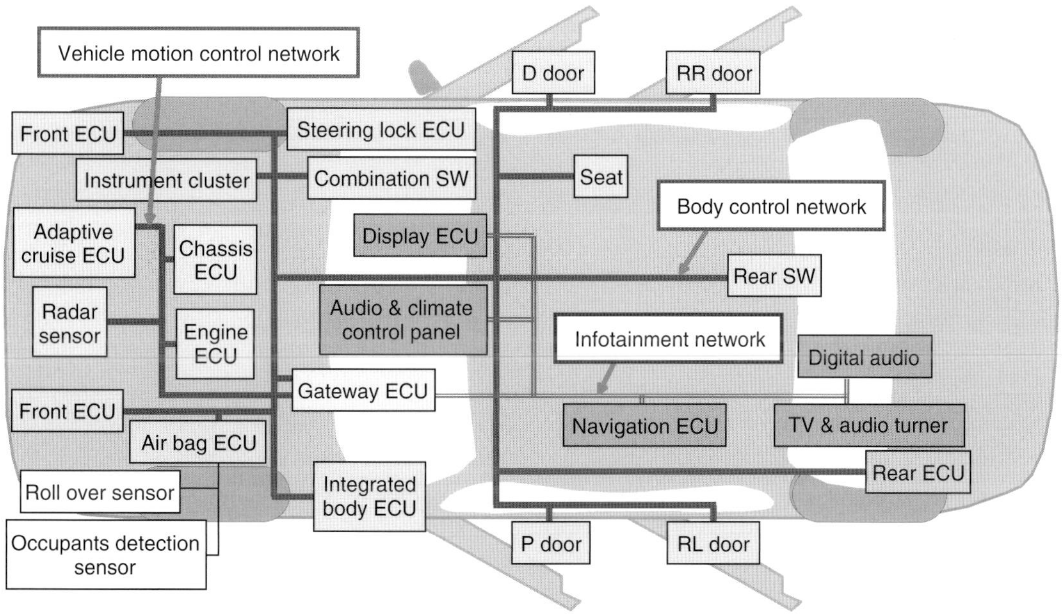

Figure 3. Example of in-vehicle network.

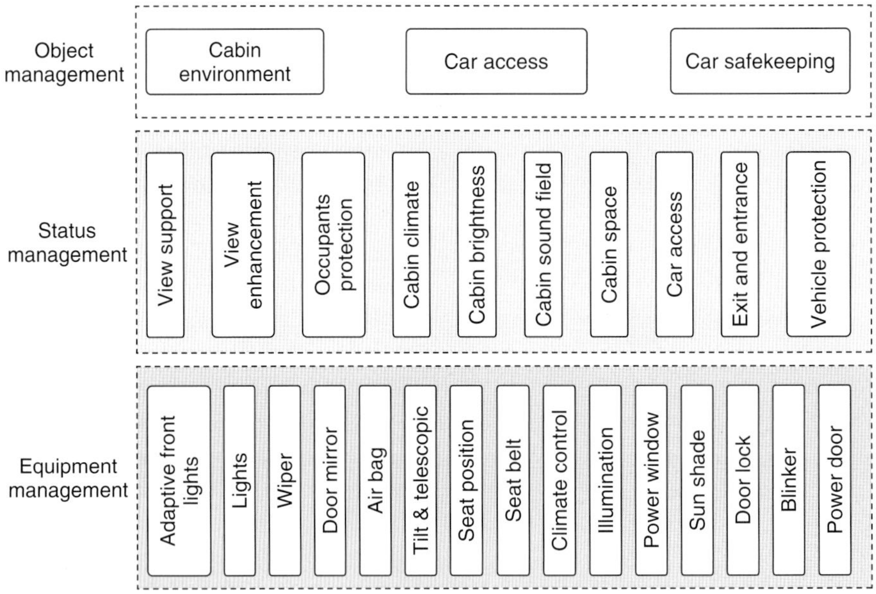

Figure 4. Example of layering of support functions.

	Role of layers	Interfaces of layers
Objective management layer	To obtain target value of required state to achive objective	Interface of objective management and state management: target value of required state
State management layer	To select the optimum equipment for achieving the state requested by the objective management, and to request drive to that equipment	Interface of state management and equipment management: target value of equipment control
Equipment management layer	To drive the actuators based on the required equipment control	

Figure 5. Role and interfaces of layers.

follows. The upper layer acts as the objective for the lower layer, and the lower layer acts as the means for the upper layer. As Figure 5 shows, the role of the objective control layer is to judge the required state and to set a target value that achieves the objective. The role of the state control layer is to select the optimum device for achieving the state requested by the objective control, and to request drive to that device. The role of the device control layer is to drive the actuators based on the required device operation.

For example, electronic keys are one type of support function that helps the driver enter the vehicle. If the driver approaches the vehicle carrying something bulky, the security control detects the key, temporarily deactivates the warning mode system, unlocks the door, and illuminates the interior with foot lamps. The driver can then open the door simply by gripping the door handle. Then, when the driver acts to close the door, the system drives the door closer and sets the driver's seat to the optimal position. Other examples are the climate control that controls the temperature and humidity of the vehicle interior to ensure the comfort of the occupants, and the view support functions such as the windshield wipers, mirrors, and glass defoggers.

Figure 1 illustrated that the scale of electronics systems will continue growing, particularly in luxury vehicles.

Layering and structuring is a key theme in these vehicles. In the future, a domain controller is likely to be provided for each system to enable sophisticated process function integration of these systems. Another likely trend is the integration of input/output processes with sensors and actuators, regardless of vehicle type. As these trends progress, it is also likely that the in-vehicle network will become layered with small-scale local networks between ECUs, sensors, and actuators, networks within domains for each system or domain, and a backbone network between domain controllers (Bertram *et al.*, 1998; Dieterich and Schröder, 1998; Lang *et al.*, 2008; Stolz, Kornhaas, and Krause, 2010).

3 INTRODUCTION OF COMMON STANDARDIZED PARTS AND REUSE

At present, one of the largest issues in vehicle development is how to establish an efficient process for designing the myriad of electronics systems concurrently and in a short space of time. This process has to cover a range of vehicle types from mass-market vehicles, which contain only an electronic instrument cluster and engine control ECU, to luxury vehicles that have large-scale electronics systems. This is achieved by maximizing the reuse of parts that have already been designed and developed through commonization and standardization, and newly designing only those parts that are required to achieve new functions. These two types of parts can then be combined to create versatile automotive electronics systems.

A structure of commonized and standardized parts and technologies may be called an *electronics platform*. Vehicle platforms generally refer to shared body and chassis designs. For computers, although the term generally relates to hardware or operating systems (OSs), the meaning and objective are the same from the standpoints of organizing systems, structures, commonized and standardized parts, and technologies to enable the efficient combination of a wide range of variations.

3.1 ECU software structure

This section describes the structure of the software in the ECU as a practical example. Figure 6 shows the relationship between the hardware and software in an ECU, and the software structure.

The electronic control system in an ordinary vehicle consists of sensors, actuators, and ECUs. Each ECU contains hardware such as a microcomputer, peripheral ICs, a power supply, and connectors, as well as software that performs the control procedure.

Conventionally, software was created in a structure that fitted with the characteristics of each individual ECU. More recently, software has been divided into two major categories: the application portion and the common portion, which does not rely on the vehicle type or function. Both are increasingly adopting structures that group together reusable parts. In principle, this structure can be used to form real-time OS, local area network (LAN) communication, device driver, and diagnostic driver software that allows common use across functions (such as engine control, brake control, steering control, and so on), vehicle models, and microcomputer types. It also facilitates the reuse of application component units.

In addition, creating the software for each ECU using the same structure may also enable the reuse of particular application components across ECUs. For example, application components installed in the engine ECU may be used in the climate control ECU. When adding a new function, only the application component that achieves that function has to be created. As a result, the function can be placed in the software platform with minimal changes to the platform, enabling fast and efficient development.

3.2 Other electronic platforms (commonized and standardized parts and technologies)

Obviously, the hardware parts of an ECU also require some form of platform. In the initial phase of electronic system adoption, it was acceptable to make optimal designs for each specific application. However, today, it is becoming increasingly important to achieve the standardization and common use of virtually all ECU hardware parts, such as microcomputers, power supply ICs, communication circuits, connectors, passive components, and so on.

Figure 3 illustrated that modern vehicles have an in-vehicle network system that connects ECUs by communication and exchanges data over the communication paths. This system enables cooperative control of vehicle behavior to assist the driver. For example, by sharing sensor data and using the navigation ECU to predict corners in advance, the engine ECU can command the transmission to generate engine braking by automatically changing gear. This in-vehicle network is utilized by something called a *communication platform*. This refers to a standard configuration of ECU groups connected to the network, a standard communication procedure, standard drivers, and standard interface signals for communicating between the ECUs.

This type of coordinated navigation system control means that the functional structure of the vehicle electronics system is becoming more complex. For example, in the case of longitudinal vehicle motion control, requests for engine

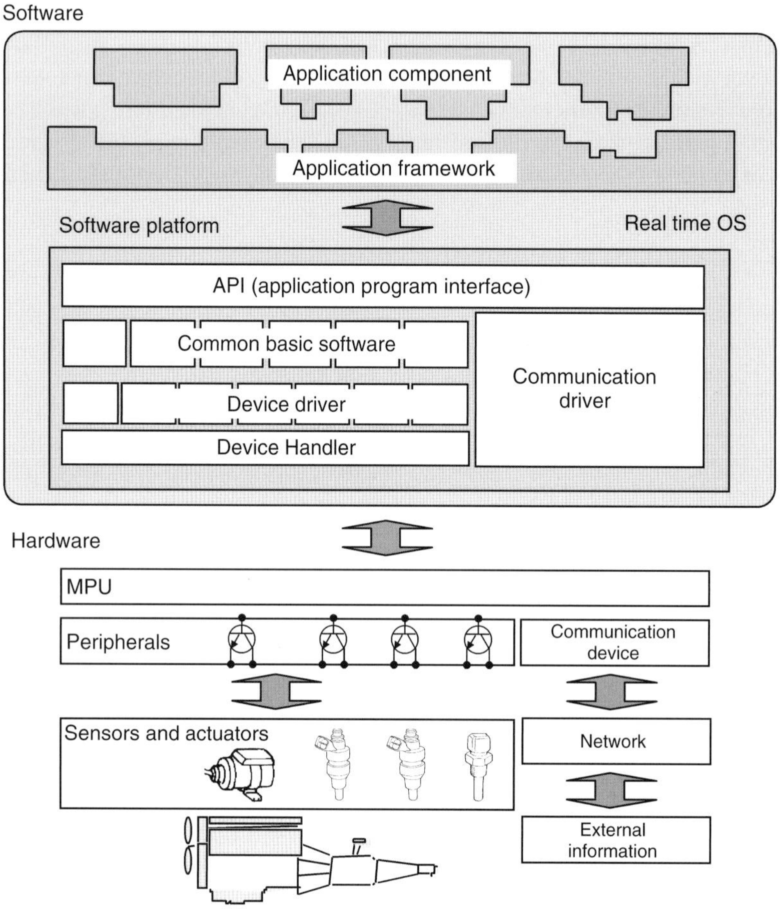

Figure 6. Software structure.

braking do not involve simply the conventional aspects of acceleration and braking. Other ECUs that are relevant to this control include the ACC ECU, which uses millimeter wave sensors and cameras to maintain a set distance to the vehicle ahead, the vehicle stability control (VSC) ECU that ensures stable vehicle handling on rough or dangerous road surfaces, the parking assist ECU that helps the driver parallel or reverse park, the precrash ECU that detects the potential of collision with objects, and the like. Torque commands from all these ECUs are transmitted through the in-vehicle network. Arbitration control prioritizes these signals is an indispensable part of the system and is an important element in the control platform.

Figure 7 shows a vehicle motion control platform. If the vehicle ahead is moving away, the ACC ECU will issue an acceleration request. Similarly, if the vehicle-to-vehicle distance is narrowing, the ECU will issue a deceleration request. The vehicle motions control platform factors in requests from other ECUs (e.g., a deceleration request from the precrash ECU will be prioritized) and calculates a target torque by prioritizing and judging information about the vehicle state to determine how the vehicle should react. This torque is then distributed to the engine and brakes. If gradual deceleration is required, it will reduce the engine request. If sudden deceleration is required, it will issue a braking torque request to the brakes.

The vehicle motion control platform also facilitates the reuse of functions such as these with other functions. By adopting the control structure shown in Figure 7, the high-speed ACC function (i.e., its control logic) developed for use on highways can also be used in principle without modification in vehicles with six-cylinder gasoline or diesel engines, or in hybrid or electric vehicles, as well as in the original four-cylinder gasoline engine vehicle. The development of new functions, such as full-speed ACC that includes start and stop control, can be separated from the development of other engines and carried out independently at the same time. Control platforms are also required for prioritizing vehicle behavior in the lateral and vertical directions as well as longitudinally (AUTOSAR, 2009).

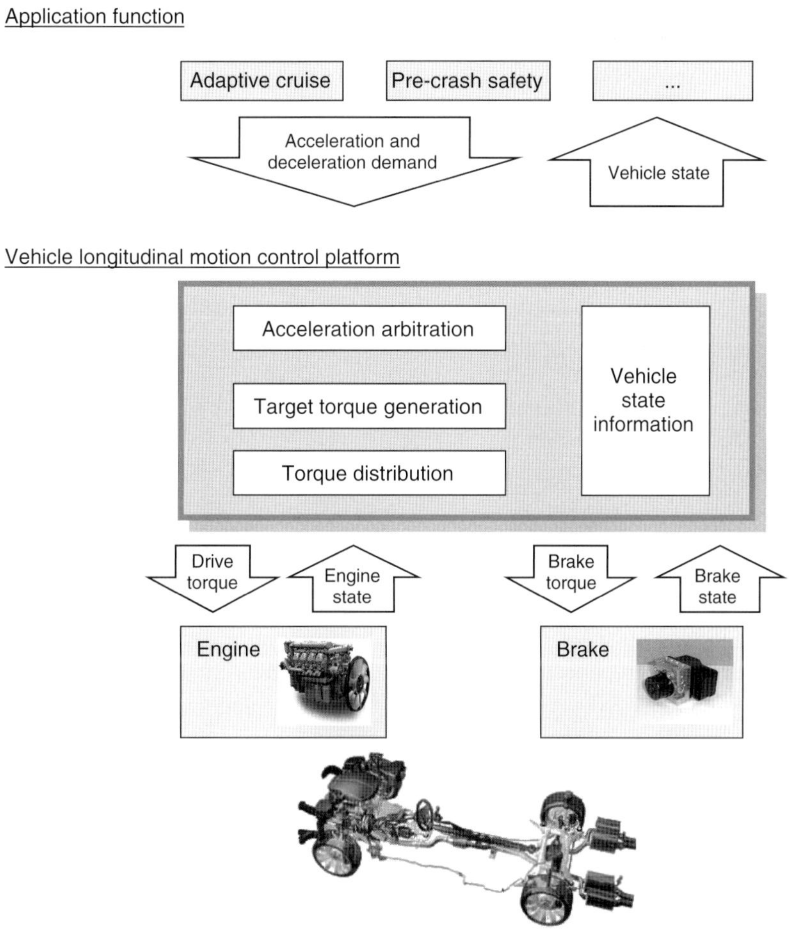

Figure 7. Structuring of vehicle motion control.

Another example of a control platform is the system for prioritizing displays of information to the driver. Since a vehicle only has one driver, it is not possible for that person to simultaneously process large amounts of information. A luxury vehicle has functions to display more than 10 different types of warnings as well as the operation states of various devices. This control platform prioritizes the displays in accordance with the constantly changing driving conditions and shows them in the proper sequence.

This section has described the details of electronic platforms (i.e., commonized and standardized parts and technologies) through various examples. In addition, electronic platforms are a wide-ranging concept in development that also encompasses vehicle power supply systems, types of vehicle control structures, and tools for designing ECUs. These platforms are also seen as the key concept for efficiently developing high quality automotive electronics systems that will continue evolving into the future. All electronic platforms have an interdependent relationship, and it will be necessary to find optimal solutions for the best way to combine such platforms into automotive electronics systems.

4 INTERNATIONAL STANDARDIZATION

The parts and technologies that are being popularized and standardized as described above contain many examples that require standardization on a global basis, irrespective of the industry standards of each country. A typical example is communication procedures, which are a basic technology for constructing the in-vehicle network. This has been the subject of international standardization efforts for some time. These efforts have recently spread to in-vehicle OS, ECU software structures, common basic software, the languages used for specifications, functional safety design methods, design and development tools, evaluation methods, and so on.

Conventionally, standardization in this field has been carried out internationally through the International Organization for Standardization (ISO) and International Electrotechnical Commission (IEC). Although these organizations dominated standardization efforts in the past, the speed of technological and market changes means that recent efforts have mainly been carried out through consortiums that make contracts with the main stakeholders to eliminate time-consuming deliberations on a country level. Although consortiums are in charge of establishing new standards that require speedy response, items that have legal and certification ramifications are still carried out by the ISO. These include the operation and maintenance of established standards, functional safety design methods, interfaces between ECUs and failure diagnostics tools, and procedures to dispose of airbags in scrapped vehicles.

As a typical example, the following sections describe the approach of functional safety design methods, which are formulated by AUTOSAR (automotive open system architecture) and established as ISO standards. Other main consortiums are LIN (local interconnect network) (http://www.lin-subbus.org/) and MOST (http://www.mostcooperation.com/home/index.html), which are related to communication procedures. OSEK (http://portal.osek-vdx.org/) deals with in-vehicle OS related to software, GENIVI (http://www.genivi.org/) works to reduce costs and to shorten development lead times in the fields of multimedia, telematics, and human machine interfaces, and the Association of Standardization of Automation and Measuring Systems (ASAM, http://www.asam.net/) is related to calibration and measurement.

4.1 AUTOSAR (automotive open system architecture)

In 2003, this consortium was founded, centering on German automakers and several major parts suppliers. AUTOSAR is working to standardize software structures, common basic software, software design tools, interfaces between application components, and certification test methods. As of June 2009, there were a total of 158 participating members: nine companies with voting rights (BMW, Daimler, VW, Toyota, PSA, GM, Ford, BOSCH, and Continental), 85 premium associate members with proposal rights (including Honda, Nissan, Mazda, DENSO, Hitachi, NEC, Fujitsu, and Renesas), another 85 associate members, and 8 development members. The work of AUTOSAR is being carried out through 26 working packages.

AUTOSAR has specified the common basic software and proposals for globally standardized specifications that transcend individual functional ECUs, vehicle types, manufacturers, and countries have been made with regard to in-vehicle network drivers such as CAN (controller area network) drivers, LIN drivers, and FlexRay drivers, systems services such as the ECU State Manager and Watchdog Manager, and software models including those related to input/output. For details of AUTOSAR activities, see AUTOSAR (2008), Ataide (2007), and Heinecke *et al.* (2004).

4.2 Functional safety standards

ISO 26262, an international standard for road vehicle functional safety was published in 2011. The standard covers automotive electronic systems relating to safety, and includes subjects such as definitions of concepts for determining safety levels such as the automotive safety integrity level (ASIL), risk assessment and other safety evaluation methods, and approaches to safe lifecycles.

ISO 26262 consists of 10 parts (Gräter, 2011). Part 1 begins with a definition of terminology, part 2 covers the requirements of organizations and people related to safety functions, and part 3 describes the procedure for developing functional safety concepts after setting safety targets. Parts 4–6 describe development and design procedures for systems, hardware, and software (in that sequence). Part 7 covers procedures related to production and market servicing. Part 8 describes guidelines for configuring and managing products developed to support these procedures, guidelines for managing modifications, guidelines for verification, guidelines for managing documentation, guidelines for certifying tools, guidelines for certifying components, and the like. Part 9 deals with guidelines related to ASIL. Finally, part 10 describes reference information to be used as guidelines for parts 1–9.

5 CONCLUSIONS

This chapter described the layering and structuring methods used in basic development of complex and large-scale vehicle electronics systems of today, the standardization and reuse of common parts in these systems, and activities to develop related international standards. Layering and structuring were outlined through a description of software structures and a motion control platform for longitudinal vehicle behavior. International standardization was covered in sections about AUTOSAR and functional safety standards. We expect these technologies will contribute to the higher level of improvement of the function and the performance in the field of fuel economy and vehicle safety in the near future.

RELATED ARTICLES

Volume 5, Chapter 143
Volume 5, Chapter 151
Volume 5, Chapter 159
Volume 5, Chapter 160
Volume 5, Chapter 161
Volume 5, Chapter 162
Volume 5, Chapter 163
Volume 5, Chapter 164
Volume 5, Chapter 165
Volume 5, Chapter 166
Volume 5, Chapter 167
Volume 5, Chapter 168
Volume 5, Chapter 169
Volume 5, Chapter 170

REFERENCES

AUTOSAR (2009) *Explanation of Application Inter-faces of the Chassis Domain*, http://www.autosar.org/download/R4.0/AUTOSAR_EXP_AIChassis.pdf (accessed 10 October 2013).

AUTOSAR (2008) *Technical Overview*, http://autosar.org/download/AUTOSAR_TechnicalOverview.pdf (accessed 10 October 2013).

Ataide, F.H. (2007) Automotive open system architecture—concept, benefits and challenges. SAE Paper 2007-01-2928.

Bertram, T, Bitzer, R., Mayer, R. and Volkart, A. (1998) CARTRONIC—an open architecture for networking the control systems of an automobile. SAE Paper 980200.

Denton, T. (2004) *Automobile Electrical and Electronic Systems*, 3rd edn, SAE International, USA.

Dieterich, K. and Schröder W. (1998) CARTRONIC—an ordering concept for future vehicle control systems. SAE Paper 98C011.

Gräter, A. (2011) Safety of electric vehicles during their life cycle.. *ATZ Autotechnology*, **11**, 12–17.

Heinecke, H., Schnelle, K., Fennel, H., *et al.* (2004) Automotive open system architecture—an industry-wide initiative to manage the complexity of emerging automotive E/E architectures. SAE Paper 2004-21-0042.

Jurgen, R.K. (1999) *Automotive Electronics Handbook*, 2nd edn, McGraw-Hill, USA.

Lang, H., Döricht, M., Preis, H., Spiegelberg, G. and Gombert, B. (2008) Vehicle architecture integration as an answer to the automotive challenges. SAE Paper 2008-01-0572.

Mizutani, S. (1992) *Car Electronics*, Sankaido Co., Ltd., Japan.

Robert Bosch GmbH (2008) *Automotive electrics and electronics*, 5th edn, Wiley.

Robert Bosch GmbH (2011) *Automotive handbook*, 8th edn, Wiley.

Schäuffele, J. (2005a) *Automotive Software Engineering Principles, Processes, Methods and Tools*, SAE International, USA, pp. 123–125.

Schäuffele, J. (2005b) *Automotive Software Engineering Principles, Processes, Methods and Tools*, SAE International, USA, pp. 16–17.

Stolz W., Kornhaas, R., and Krause, R. (2010) Domain control units—the solution for future E/E architecture? SAE Paper 2010-01-0686.

FURTHER READING

Kato, M. (2010a) *Automotive Electronics: Systems*, Nikkei Business Publications, Inc., Japan.

Kato, M. (2010b) *Automotive Electronics: Basic Technologies*, Nikkei Business Publications, Inc., Japan.

ECU General

Chapter 139

Microcomputers and Related Technologies: Enlargement of Software Size, Algorithms, Architectures, Hierarchy Design, Functional Decomposition, and Standardization

Hideaki Ishihara
DENSO Corporation, Kariya, Japan

1 INTRODUCTION

Automotive electronics have been advancing rapidly in the successful pursuit of environmental friendliness, safety, comfort, and convenience. These advancements have been achieved through the adoption of microcomputers, which now total to approximately 100 in each luxury car. Since the late 1970s, microcomputer technologies have achieved enhanced performance, affinity with real-time processing, high fault-tolerance, and the like. However, the consequent enlargement of software size has resulted in the need not only for new software architectures such as hierarchical design by functional decomposition but also for international standardization [such as OSEK (open systems and the corresponding interfaces for automotive electronics) and AUTOSAR (automotive open system architecture)]. In the twenty-first century, there is public awareness of the need for the creation of more green and dependable automotive systems, driving the need for the reduction of CO_2 emissions. There is also a general interest in achieving high road safety levels. From these points of view, this chapter describes the histories and future directions in regard to both microcomputers and software in automotive electronics, while considering semiconductor technologies and their applications.

Encyclopedia of Automotive Engineering, print © 2015 John Wiley & Sons, Ltd.
Edited by David Crolla, David E. Foster, Toshio Kobayashi, and Nick Vaughan.
This article is © 2015 John Wiley & Sons, Ltd. ISBN: 978-0-470-97402-5
Also published in the *Encyclopedia of Automotive Engineering* (online edition)
DOI: 10.1002/9781118354179.auto217

2 MICROCOMPUTERS

2.1 History of automotive microcomputer applications

The automobile and the microcomputer are two of the greatest developments of the twentieth century. The automobile enabled universal freedom of movement and the breakthrough of the microcomputer-enabled complex functions conceived by people and plays a main role in modern systems. Microcomputers have come to be used in almost every industry. This was triggered by the development of the integrated circuit (IC) after the invention of the transistor by John Bardeen, Walter H. Brattain, and William B. Shockley at Bell Laboratories, which had the effect of turning large computers into compact and portable devices. Present microcomputers generally contain not only microprocessors but also input/output circuits and memory devices all within a single semiconductor chip. A microprocessor consists of an arithmetic unit and a control unit, which functions as a central part of microcomputer circuits. The terms *central processing unit* (*CPU*), *processor*, and *micro processing unit* (*MPU*) are also used as general synonyms for a microprocessor. In 1971, Intel Corporation created the 4004, which was the world's first commercially available microprocessor. It had a 4-bit processing capability and a clock speed between 500 and 741 kHz. A total of 2300 transistors were integrated on a chip with an area of $3\,\text{mm} \times 4\,\text{mm}$ and it was manufactured using 10-μm semiconductor processes. The 4004 was developed as a joint venture between a Japanese electronics engineer called *Masatoshi Shima* and Intel Corporation based on a request from the Japanese company Busicom for use in an electronic calculator. Subsequently, microprocessor bit lengths evolved from 8 to 16 bits. From the beginning of the 1990s, 32-bit microprocessors also started to become available in certain embedded systems (Kato *et al.*, 2010).

When implementing an electrical function, a key difference between the microcomputer and the hard logic is the length of the development period between the software and the hardware. Software functions can be changed in a period from several hours to several days. However, hard logic takes at least several months (i.e., at least 100 times as long) to change as different semiconductors have to be prototyped and evaluated. Furthermore, the number of software designers can be increased substantially at short notice as software education and training is not subjected to serious physical limitations. These are the main reasons why microcomputers have spread so rapidly in vehicles (Kato *et al.*, 2010).

The operating voltage of microcomputers was generally 5 V, whereas the semiconductor manufacturing process was 0.5 μm or larger. Subsequently, since the beginning of the twenty-first century, the operating voltage has been decreasing, reaching 3.3 V at 0.35 μm, 2.5 V at 0.25 μm, and approximately 1 V at 90 nm. However, the operation voltage of most peripheral functions embedded in microcomputers remains 5 or 3.3 V for reasons related to sensor and actuator interfaces.

As automotive electronics have evolved, the number of microcomputers has increased to dozens in an ordinary passenger vehicle and more than a hundred in a luxury vehicle. Unlike personal computers, automotive microcomputers generally use a single chip in which the input and output circuits are integrated onto a semiconductor chip. This is to satisfy the various performance requirements of a vehicle, such as the built-in memory capacity, high real-time capabilities, strong electromagnetic compatibility (EMC), and wide operating temperature and voltage ranges. This has been enabled by the explosive evolution of semiconductor IC technology in the second half of the twentieth century. The aim of automotive microcomputers is not simply to achieve the basic moving, turning, and stopping performance aspect of the vehicle. Microcomputers are also installed in core portions of safety devices such as airbags, millimeter wave radar, and the like, anti-theft, power window, and other body control devices, and intelligent transport systems (ITS) devices such as electronic toll collection (ETC), and navigation systems. These functions can now be cooperatively controlled through the local area network (LAN) (Ishihara, 2009; Tsuda, 2007).

The first use of a microcomputer in a vehicle was for engine control, prompted by the introduction of stricter emissions regulations. Microcomputers were adopted by General Motors (GM) in 1976 and by Ford in 1977 for engine controls, such as ignition timings. The automotive industry in Japan was also researching microcomputers in engine controls at virtually the same time. In 1980, Nippondenso Co., Ltd. (the current Denso Corporation) adopted a single-chip microcomputer-based engine control with a 12-bit microprocessor and an 8-input interrupt function. These microcomputers had superior processing capabilities, real-time performance, and operating temperature range than the 8-bit microcomputers provided by Intel at that time. For a long time after then, the bit length of automotive microprocessors became a logical target for developers. However, 8-bit microprocessors became the predominant microcomputer since the second half of the 1980s because of their sufficient performance for all aspects of engine control by establishing a 16-bit address range capable of pointing to locations in memory. Although mainstream modern microprocessors are 32-bit, some 8-bit devices are still used in low-cost systems (Kato *et al.*, 2010).

Microcomputers entered widespread use over the second half of the 1980s. Figures 1 and 2 show two examples: the 1-bit microcomputer made by Denso for low-end electronic systems and the 8-bit microcomputer (ND8) made by Denso with a multithread structure to avoid runaway.

2.2 Requirements of automotive microprocessors

Non-multimedia automotive control systems require *hard real-time performance*. This refers to an absolute permissible maximum run time for individual software within the application. For microprocessors such as personal computers, the cache and other means are effective ways of

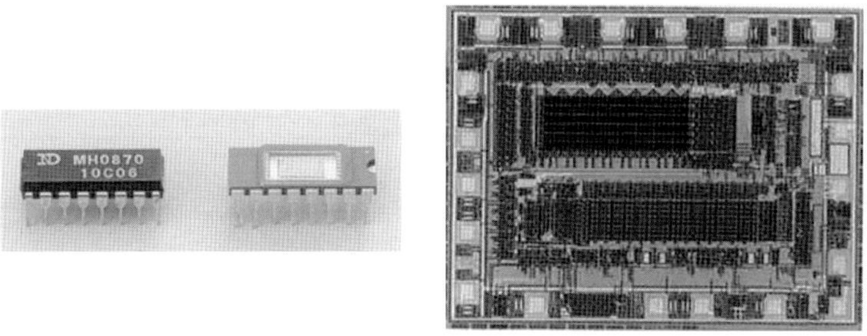

Figure 1. 1-Bit microcomputer made by Denso. (Reproduced by permission of H. Ishihara/Denso. © Denso Corporation.)

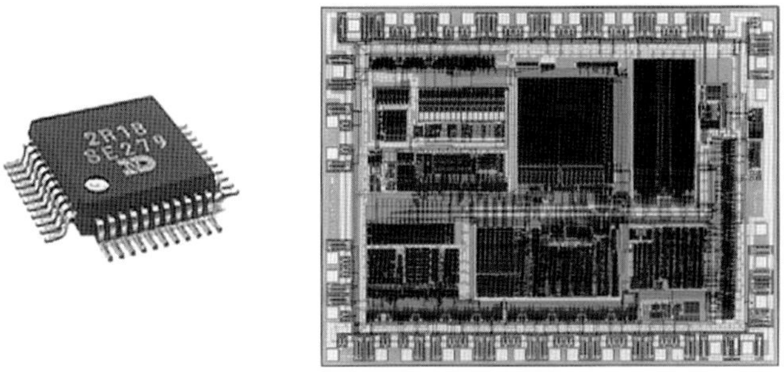

Figure 2. 8-Bit microcomputer (ND8) made by Denso. (Reproduced by permission of H. Ishihara/Denso. © Denso Corporation.)

improving performance. However, these do not satisfy the hard real-time requirements and therefore cannot be used in automotive microcomputers. Instead, memory devices are embedded in the chip to improve performance. As there is a limit to the capacity of memory that can be embedded in a chip, one key item for selecting a microcomputer is the efficiency of translating the software to machine code. This is also known as an *evaluation item* called *code size efficiency* (Figure 3) (Ishihara, 2007, 2009).

In 1995, Denso developed the Nippondenso reduced instruction set computer (NDR) machine, a 32-bit reduced instruction set computer (RISC) optimized for automotive controls. Using the RISC architecture of a general-purpose register machine (Figure 4), the development closely emphasized the relationship between the instruction set architecture and C compiler to enhance affinity with automotive control software. As a result, it was possible to greatly improve the code size efficiency in comparison with other microprocessors (Figure 5) (Ishihara, 2007; Kawamoto *et al.*, 2001).

Structure of microprocessors

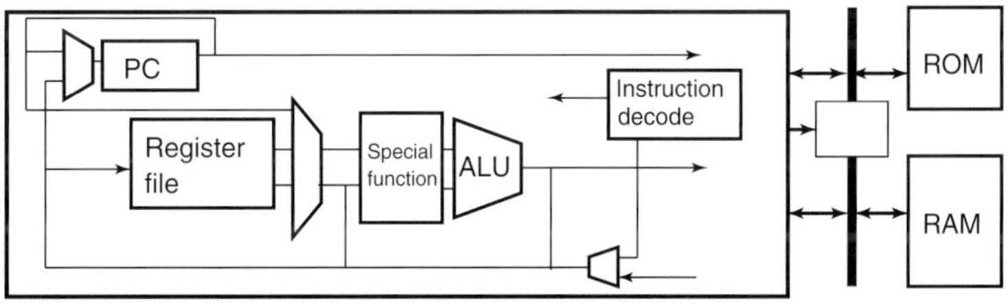

(a)

Evaluation items

	Evaluation items	Cost	Real time	Reliability	Power consumption
Technology	Small gate count (compact design)	✓			✓
	Fast response to interruption		✓	✓	
	Code size efficiency	✓	✓	✓	✓
	Special instruction and compiler optimization	Memory size		Memory devices should be embedded in the chip	

(b)

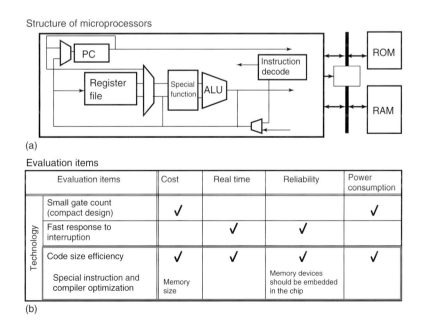

Figure 3. (a) Structure and (b) evaluation items of microprocessors.

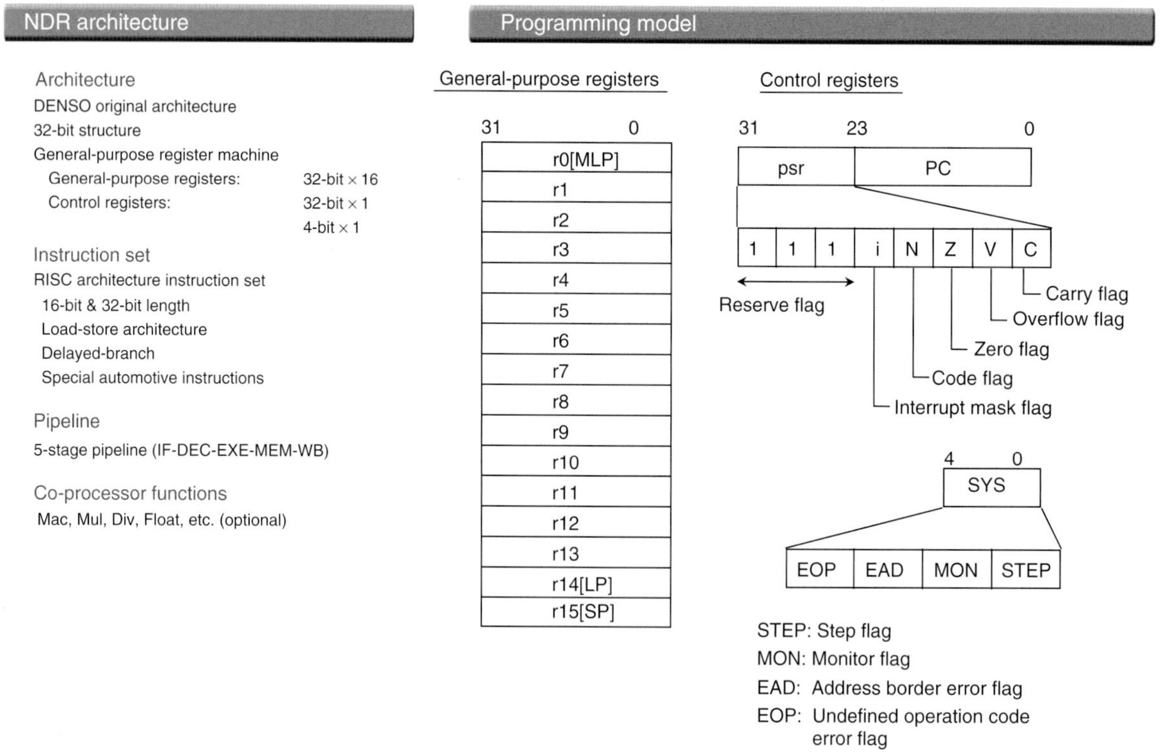

Figure 4. RISC architecture of general-purpose register machine.

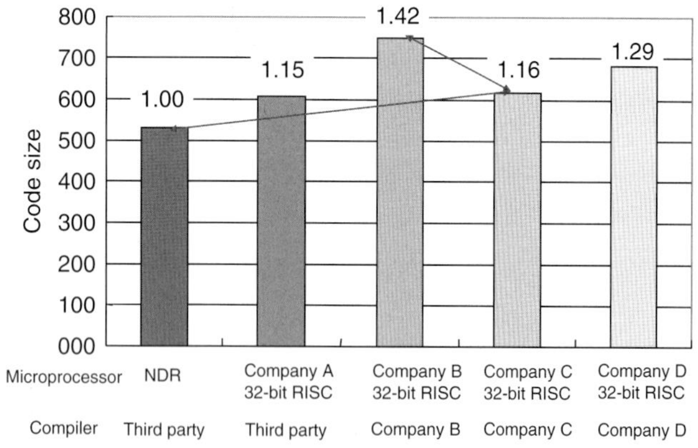

Figure 5. Affinity with automotive control software.

2.3 Internal structure of single-chip microcomputer

Figure 6 shows an example microcomputer structure.

In addition to the microprocessor, the integration of functions for special applications on single chips include embedded memory, interrupts (INT), timers, inputs [analog-to-digital converters (ADCs) and input capture], outputs [pulse width modulation (PWM)], communication functions [universal asynchronous receiver-transmitters (UARTs), serial peripheral interfaces (SPIs), and controller area networks (CANs)], and the like. Timers are used for the time management of tasks that are indispensable for real-time performance. Cooperative operations with INT are used to start periodic tasks and to measure the interval between events. ADCs are used to convert analog values from sensors to digital values that can be used by the microcomputer. PWM is capable of generating output waveforms with a modulation protocol that changes the duty ratio of the pulse wave and is used in motor control and for simple data transmission. SPI is characterized by clock-synchronized communication and is used for communication between the multiple peripheral ICs on the ECU (engine control unit) circuit board. CAN is used for communicating between ECUs. Input capture is used to measure the pulse width and interval. Other embedded functions may also include the voltage regulator required to operate the microcomputer, low voltage reset, and a watchdog timer. Recent years have also seen greater demands for extremely compact implementation as microcomputers are now being embedded within sensors and actuators. As an example of this trend, Figure 7 shows the microcomputer embedded in a power window motor (Ishihara, 2009; Ookura, 2005).

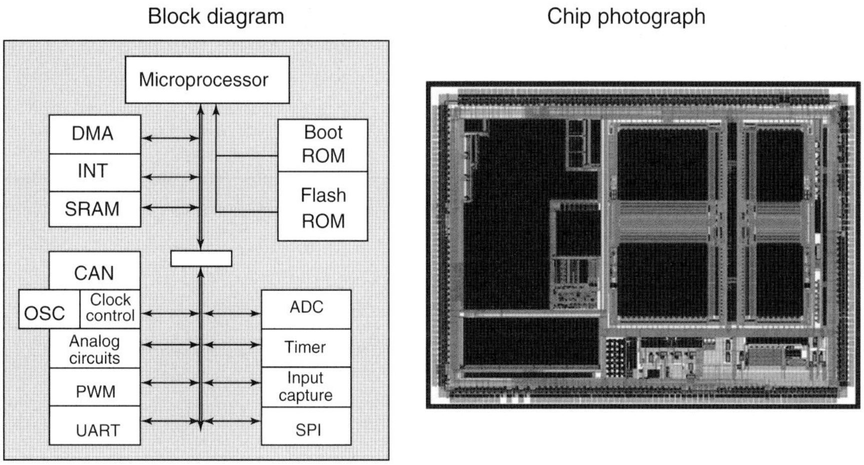

Figure 6. Example 32-bit microcomputer (body electronics).

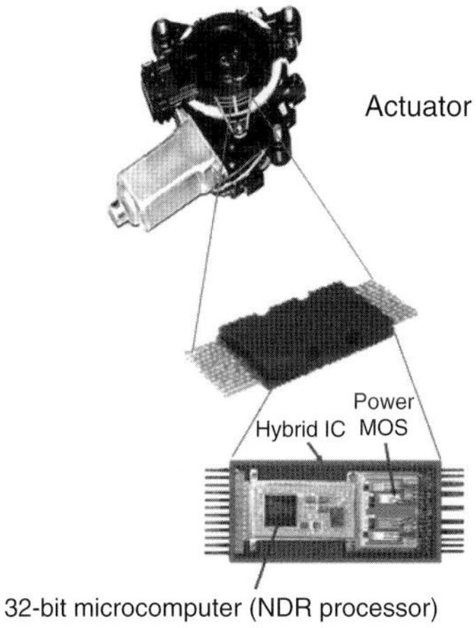

Figure 7. Example microcomputer (power window).

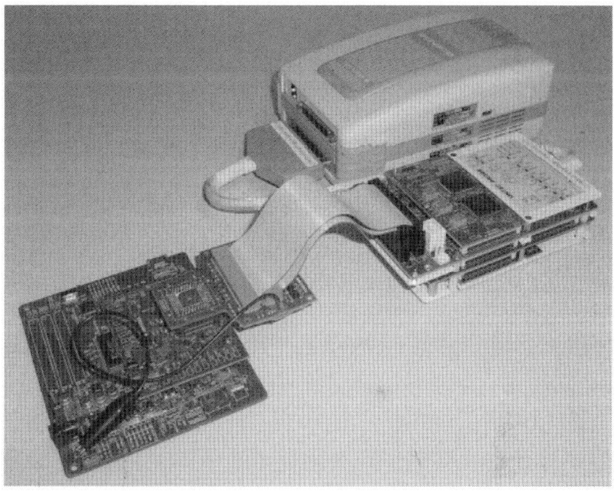

Figure 8. Example ICE.

2.4 Role of the software development environments

Use of the software development environment is an indispensable part of developing software to be embedded in microcomputers. When this software is developed in C language, the C programs are translated into assembly language by the C compiler and then further translated into machine language by an assembler and linker. The software is then written into the embedded memory of the microcomputer by a ROM (read-only memory) writer. The software is debugged by an in-circuit emulator (ICE) and/or RAM (random access memory) monitor, which confirms that the software operates according to expectations while monitoring the internal information of the microcomputer. Figure 8 shows an example ICE. As the integration density of semiconductors increases, the functions of the software development environments may be embedded into the microcomputer to confirm the software operation more efficiently.

2.5 Future prospects of automotive microcomputers

The amount of software is likely to increase dramatically in the future in response to requirements for greater environmental performance, safety, comfort, and convenience. Consequently, microprocessor performance will have to be enhanced to allow these processing requirements to be performed effectively. In the 30 years since 1980, microprocessor performance has improved by more than 1000 times. Moore's law is used by the world's semiconductor industries, computer industries, and research institutions as a guideline for making predictions regarding semiconductor and computer technologies. Gordon E. Moore, the cofounder and Chairman Emeritus of Intel Corporation, first proposed the law in 1965. Since then, Moore's law has been widely discussed among researchers, resulting in its use as a guideline for the prediction of future trends in both semiconductor and computer industries. This rule of thumb indicates that the numbers of transistors on an IC will double approximately every 18 or 24 months. This has been expected to be a driving force behind the progress of microprocessor performance. However, future trends remain unclear. As current developments are approaching their limits with respect to Moore's law in semiconductor technologies, the field of computer engineering is trending toward multi-core technologies that increase the number of microprocessors on silicon. However, multi-core technologies compatible with the hard real-time processing requirements of vehicles is still at a nascent stage around the world (Ishihara, 2007, 2009).

Hard real-time refers to systems in which the processing value switches to zero immediately after a preset time restriction (i.e., a deadline) is not satisfied. These are different from soft real-time systems in which processing values gradually fall over the course of time (such as personal computers and other information devices). In other words, automotive electronics resembles those used in the aviation field and are complex temporal protection systems. However, as the current focus of multi-core technology is on increasing average and optimum performance, these systems may even crash unless the minimum performance is guaranteed. In addition, as the number of microprocessors on a chip has increased, communication between microprocessors has also increased, which makes satisfying the fundamental temporal protection requirements unpredictable. Consequently, further technological innovation is strongly required. Figure 9 shows an example of a multi-core microcomputer with enhanced temporal protection performance (Ishihara, 2007, 2009).

Furthermore, many multi-core technologies are being researched that increase the number of processors on silicon from several tens to several hundreds. This is reputed to be generally suitable for applications that require high parallel performance, such as image processing. Multi-core technologies featuring more than 100 processors have already been applied for image recognition in automotive electronics.

In contrast, automotive microcomputers are trending strongly toward single-chip solutions. However, as technologies in ICs become increasingly advanced, it will be more difficult to integrate heterogeneous devices. Additionally, system in package (SiP) technologies that use through-hole vias are becoming more widely adopted. In the future, when the adoption of single-chip technologies is difficult or small amounts of a wide range of products are required, more fields are likely to request SiP solutions that combine standardized semiconductor chips.

3 SOFTWARE

Software is a key component of automotive systems. This section describes the fundamental details and characteristics of software for automotive electronics.

3.1 Role and positioning of software

The shape and form of software is invisible even under close examination of the ECU, which is the main component of automotive electronics. The software or programs are located within the internal ROM and RAM of the microcomputer. These are interpreted by the microprocessor and comprise rows of instructions to be operated and data

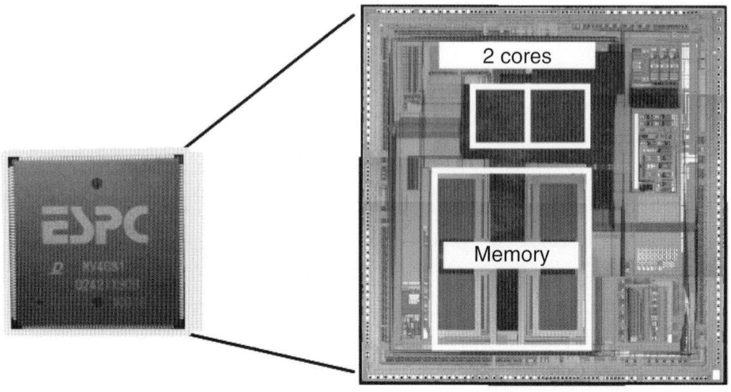

Figure 9. Example multi-core.

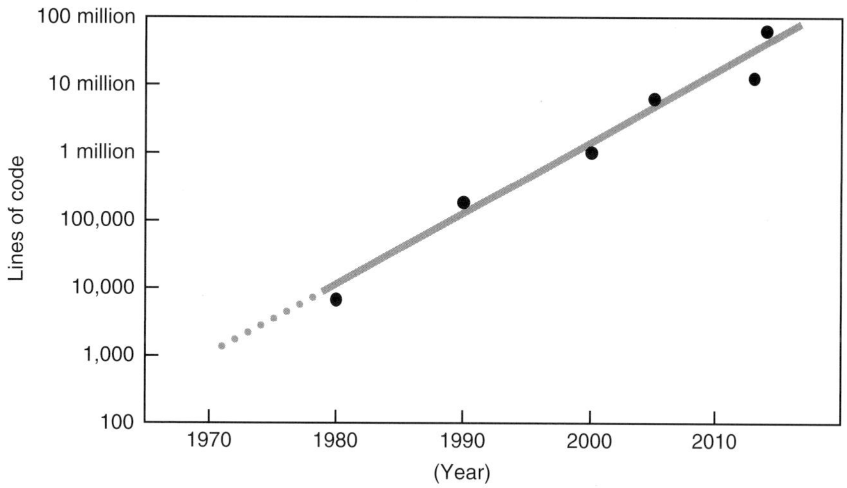

Figure 10. Trends for software installed per vehicle. (Estimated from trends of Denso and other companies.)

groups that represent the operation target. Software is the language used for comparison with the hardware and it defines how the hardware functions.

3.2 Increasing amount of software in automotive electronics

Initially, the adoption of microcomputers began in the 1970s with the development of engine controls. Subsequently, this trend began to spread to products used in body electronics, safety systems, and information systems.

In accordance with this growth, the number of microcomputers per vehicle increased to dozens in an ordinary passenger vehicle and more than a hundred in a luxury vehicle. From the standpoint of software, the lines of code increased from $<$10,000 in the 1980s to more than 10 million in 2010. This number is likely to continue growing (Figure 10) (Kato *et al.*, 2010).

3.3 Algorithms

The functions of the microprocessor embedded in a microcomputer include inputting and processing external data, and outputting the results. The procedure and timing for performing this data input/process/output sequence is determined by the algorithm. As a computer operates in accordance with the instructed program, the algorithm must be written correctly to ensure that the computer operates as intended. Historically, although Euclid's algorithm is regarded as the first mathematical algorithm, other types of algorithms based on state transition rules, physical models, and hardware restrictions are also in use. In automotive electronics, the means and procedures for key data processing are often regarded as the specifications. The software is implemented to realize the algorithm as concrete processing based on the specifications. Even if the same result is obtained, the algorithm used can result in substantial differences in the execution time required for microprocessor and the amount of data required for performing the work. The terms *software* and *program* are often used synonymously but actually represent slightly different nuances (Figure 11). A program is a set of instructions for computer processing, whereas *software* refers to all the documentation required to develop the programs in addition to the actual programs themselves (Kato *et al.*, 2010).

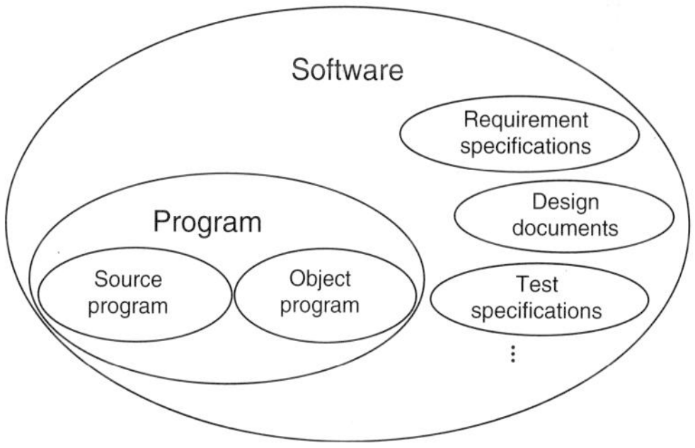

Figure 11. Software and program.

3.4 Software architecture, functional decomposition, and hierarchical design

Software architecture refers to the basic component, framework elements of the software, and the relationship between these elements (i.e., the input/output characteristics and the

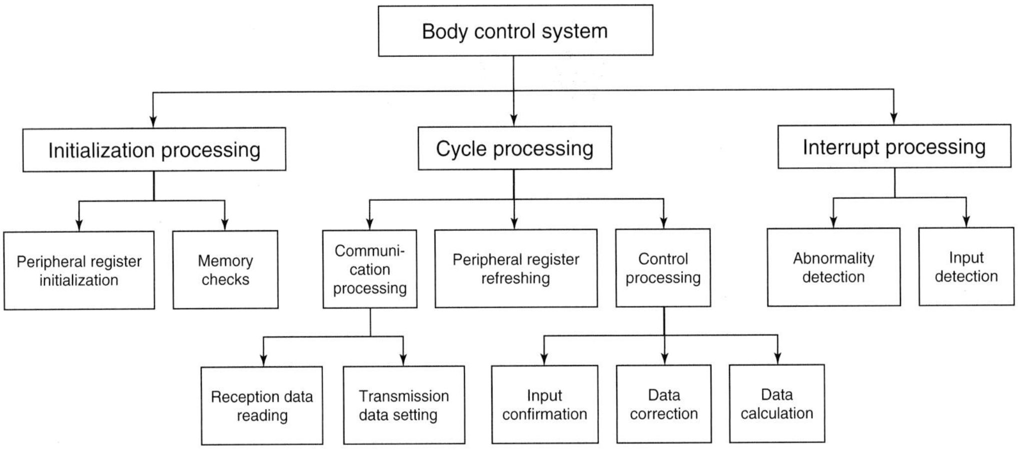

Figure 12. Example of hierarchical design in body electronics.

way one element uses another or is used). The software architecture can be used for mutual understanding and communication between software designers. In addition, the design policy is determined in the initial period of the development to ensure that the subsequent development steps proceed smoothly.

Hierarchical design by functional decomposition is used to clarify the component elements of software. In this design method, the software is divided into functional units that are assembled in a hierarchical manner (Figure 12). Hierarchical design enhances serviceability and quality by clarifying the design process.

3.5 Program categories

Programs can be categorized into system and application programs (applications). System programs can be further categorized into operating systems (OSs), device drivers, and middleware (Figure 13).

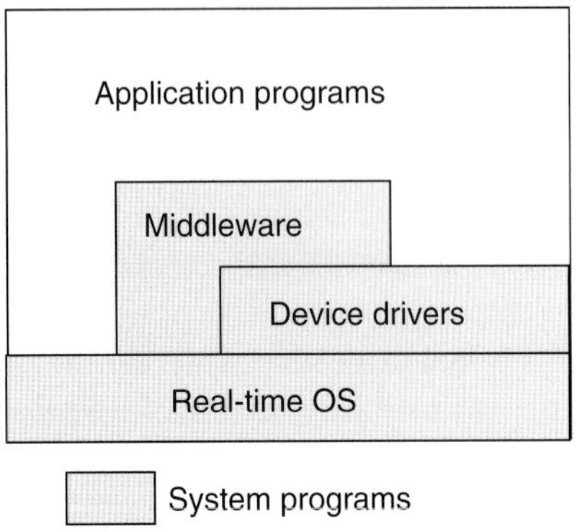

Figure 13. Categories of program.

Small-scale automotive systems are configured with interrupt processing programs, simple schedulers, and applications to reduce the processing time and memory overhead required by an OS. The OS is the fundamental program for ensuring that the computer resources are used efficiently. Automotive applications generally use a real-time operating system (RTOS). The basic functions of an RTOS include scheduling (task management), synchronization, and exclusive control. An RTOS with restricted functions may be referred to as a *kernel*. Under RTOS control, program units instructed by the microprocessor are called *tasks*. Task states are defined as run, ready, and wait. In addition, each task has a priority attribute. The RTOS transits to the run state of the task with the highest priority within the tasks that are in the ready state. This control is called the *scheduling function*. The wait state occurs when waiting for some kind of phenomenon to occur or when a resource cannot be used because it is being used by another task. The wait state continues until that phenomenon occurs or the object resource is released (Figure 14) (Kato *et al.*, 2010).

The program portion that quickly carries out the minimum functions required of the interrupt processing is called the *interrupt handler*. This is regarded as a separate unit from the tasks. In the event that multiple tasks are operating the same resources, the exclusive control acquires those resources to secure the operation and excludes the operation of other tasks (Figure 15) (Kato *et al.*, 2010).

The program that directly controls the hardware to allow the microcomputer to operate external data input and output is called the *device driver*. The device driver function is often achieved by a combination of the interrupt handlers and tasks. The aim of the device driver is to

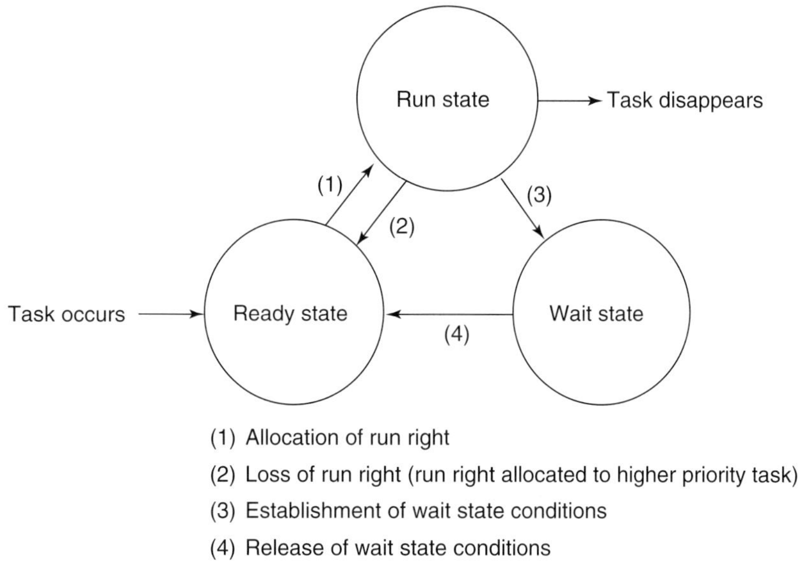

Figure 14. Task state transition.

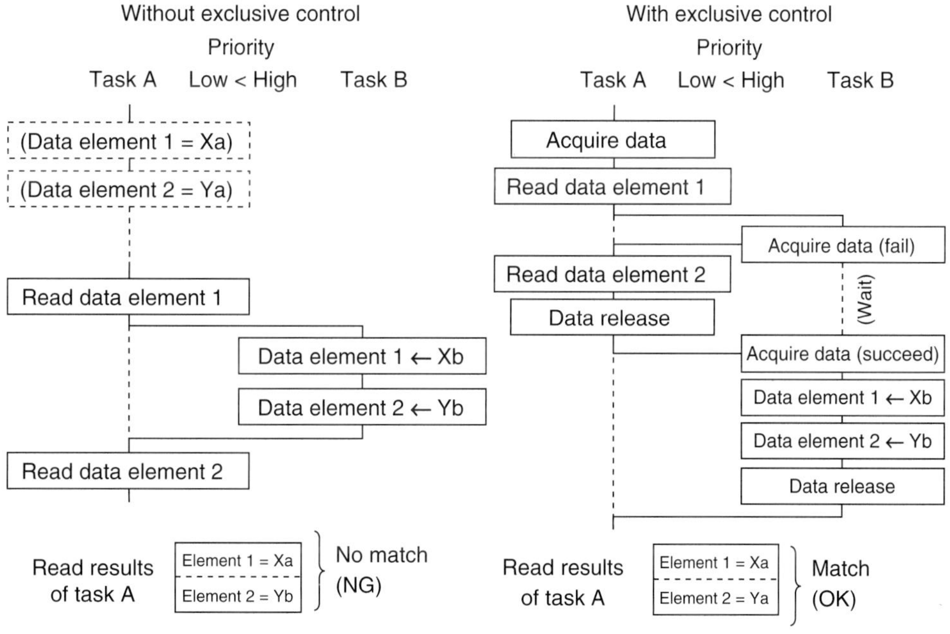

Figure 15. Securing of data operation by exclusive control.

isolate and conceal hardware differences and hardware control details from the applications. Isolating the hardware controls from the applications enables those applications to be reused with different hardware. Device drivers with the same hardware can also be used in common by different applications. In addition, the term *middleware* refers to *software* that is positioned between the RTOS and the applications.

3.6 Software development process

The series of activities involved in the development of software is referred to as the *software development process* or the *software process*. Each unit of these activities that occurs during the software process is called a *phase*. A *process model* is the general and abstract term for a software process. A

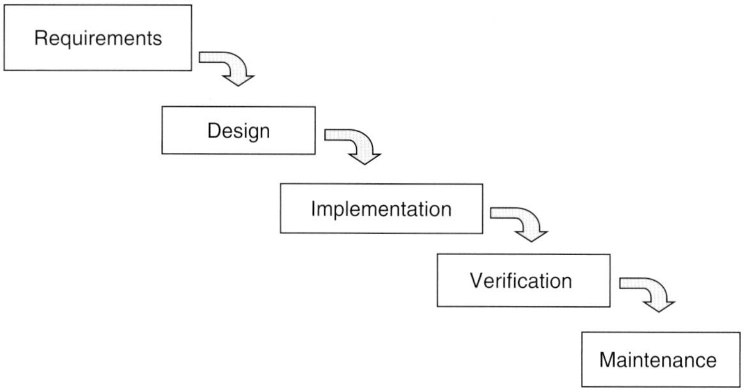

Figure 16. Waterfall model.

classic and well-known process is the waterfall model (Figure 16).

In the waterfall model, development is viewed as flowing downward through several phases throughout the software development process. In this model, the development moves on to the next phase only when the preceding phase is completed and perfected. The benefit of this process is the ease of quality control. If a problem is identified in the previous phase, it is fixable by reverting to the former phase. Therefore, it might be assumed that perfection within the given phase should ensure the perfection of the end product. However, this is often not the case. If the model itself has a problem that can only be revealed during development, perfecting each phase fails to produce the desired end result. Nevertheless, this development process is still widely used because of its ease of implementation.

The most popular process model employed in the automotive software development is the V-model, which is a modified version of the waterfall model. The V-model consists of the coding phase and the verification phase, in which the validation process is subdivided. As seen in Figure 17, the verification phase is divided into unit testing, integration testing, and system testing. Each of these serves the function of testing mounting, designing, and requirement analysis. The V-model shares weaknesses similar to the waterfall model but is superior because it emphasizes the verification phase. This ensures superior quality control of the software and is the most widely used model in the automotive industry.

3.7 Characteristics of automotive control software

The first characteristic of automotive control software is its hard real-time properties. Failure of an automotive control system to process a program within a certain deadline can have severe consequences for the vehicle.

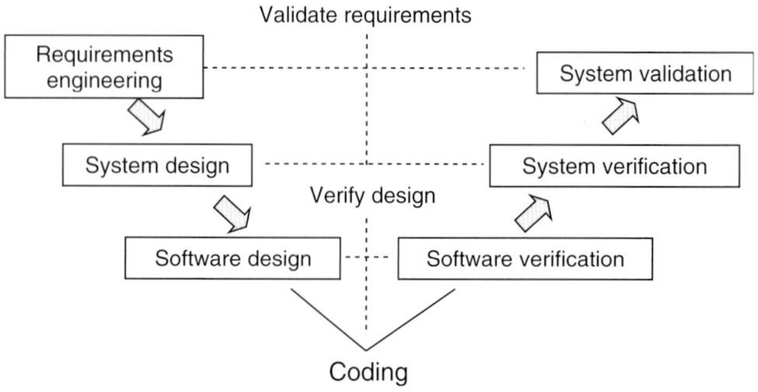

Figure 17. V-model.

The second characteristic is the importance of saving resources. As many automotive systems use single-chip microcomputers, it is crucial to implement the required functions as programs within the capacity of the chip. For this reason, there are strict requirements to save resources.

The third characteristic is the requirement for high quality. As software increases in scale and complexity, activities to ensure quality are becoming more difficult and are growing in importance.

4 ADOPTION OF ELECTRONIC SYSTEM PLATFORMS

4.1 Hierarchy of electronic systems

As automotive electronic systems have become more widespread, large numbers of ECUs, sensors, and actuators have been connected using automotive LAN. Consequently, compared with the early 1990s when the processes to be performed by the ECUs were determined for each function, there is now a greater need to determine the allocations of the role of each ECU from the standpoint of overall system optimization.

As the number of ECUs installed in a single vehicle increases, it is becoming clearly more difficult to secure adequate installation locations. Standard functions that do not rely on model variations (i.e., the intended region of sale, such as Asia, North America, Europe, and the like, and the vehicle grade) are being incorporated into integrated ECUs. In contrast, new nonstandard functions that are still in development as well as functions for certain regions and vehicle grades only are regarded as stand-alone functions.

As a result, function hierarchies have been created. This has generated an issue of function allocation, that is, which functions should be allotted to which hardware (i.e., ECUs, sensors, and actuators). A case of function allocation, in which multiple software and hardware items are defined in a hierarchy of related functions, is described later (Kato *et al.*, 2010). For example, the electronic key is a function that facilitates the driver entering the vehicle. Even if the driver is carrying a heavy package, the electronic key detects weak radio waves emitted by the key when the driver approaches the vehicle, temporarily disengages the alarm mode, unlocks the doors, and illuminates the courtesy lights at the driver's feet. This system then drives the door opener when the driver touches the door handle. After the driver enters the vehicle and starts to shut the door, the system drives the door closer and adjusts the seat to the optimal preset position (Kato *et al.*, 2010).

The air conditioning may also include functions that maintain the most comfortable environment in the vehicle for the driver by controlling the temperature and humidity in the occupant compartment, in addition to supporting the driver's field of vision through the windshield wipers, mirrors, and rear window defogger (Kato *et al.*, 2010).

Future automotive electronic systems, especially in luxury vehicles, will probably become more hierarchical and structured. It is also likely that ECUs with sophisticated processing functions will become combined, while creating sensors and actuators integrated with standard input and output processing, regardless of the vehicle model. As a result, automotive network hierarchies have been created with backbone and subnetwork systems.

4.2 Standardization and reuse of electronic parts

There is a need for approaches that efficiently develop and design various automotive electronic systems in parallel and in a short period of time. This can be accomplished by standardizing and reusing as many parts as possible, and newly developing only the parts required for new systems. Such approaches must be capable of constructing various systems quickly and easily.

This approach of adopting standard and reused parts is also called an *electronic platform*. Although vehicle platforms tend to refer to the body and chassis and *computer platforms* generally refer to the hardware and OSs, the purpose of enabling the simple construction of multiple variations is the same (Kato *et al.*, 2010).

Design is no longer focused on optimizing individual functions. Current trends are clearly highlighting the need to standardize and reuse many hardware parts, such as microcomputers, power supply ICs, communication circuits, connectors, passive devices, and the like.

4.3 Example of platform adoption

Modern vehicles use network systems to connect multiple ECUs to allow dynamic cooperative control to support the driver. For example, sensor data can be shared with corner prediction by the navigation system ECU to request the engine ECU to generate engine braking by automatically changing gear.

The longitudinal motion of the vehicle can be controlled by commands sent over the network. In addition to conventional accelerator and brake controls, engine and brake requests can be transmitted by adaptive cruise control (ACC) systems, which maintain a set vehicle-to-vehicle distance using millimeter wave sensors and/or cameras, vehicle stability control (VSC) that ensures stable vehicle behavior and maneuverability on poor road surfaces, parallel and reverse parking assist systems, precrash systems that detect the risk of a collision in advance, and so on. Such commands require arbitration systems that apply priorities to each signal (Kato *et al.*, 2010). These systems are not limited to vehicles with four-cylinder gasoline engines and should be applicable in theory with six-cylinder, diesel, hybrid, and electric vehicles.

5 INTERNATIONAL STANDARDIZATION (OSEK AND AUTOSAR)

International standardization activities include those led by the International Organization for Standardization (ISO) and the International Electro Technical Commission (IEC). However, a growing number of consortiums have been set up under contract with the main organizations to respond to the speed of changes in technology and the market. However, although the consortiums are responsible for establishing new standards that require a speedy approach,

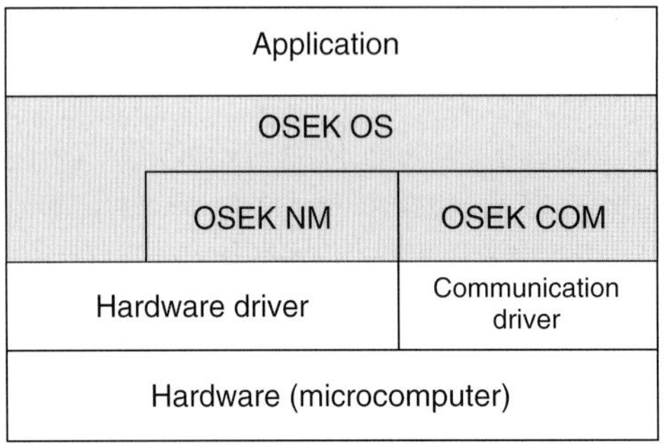

Figure 18. Software structure of OSEK/VDX.

items related to the maintenance, management, regulatory approval, and certification of these standards are still performed by the ISO, IEC, and the like.

The following sections describe some examples of international standardization for software. As development costs increase because of the multiplying types and scale of software, activities to standardize software and their use have been prompted by the European automotive industry, leading to the determination of an RTOS for automotive control called the *OSEK/VDX* (*vehicle distributed executive*) specifications. OSEK/VDX comprises of RTOS specifications (OSEK OS), communication specifications within and between ECUs (OSEK COM), and network management specifications (OSEK NM) (Figure 18). OSEK/VDX has been certified as ISO 17356 as an international standard for automotive OS.

The AUTOSAR consortium was established in 2003 for the purpose of international standardization. It is composed of more than 100 organizations including automotive manufacturers, suppliers, and tool vendors, divided into four types of membership consisting of core members, premium members, associate members, and development members. AUTOSAR specifications have been formulated based on the OSEK/VDX specifications and defined with extremely extended functions for realizing standardization, scalability, transferability, integration, maintainability, and software upgrades in the fields of automotive electrics and electronics (E/E) architectures. The AUTOSAR run time environment (RTE) provides a software application interface consisting of system services, memory services, communication services, I/O hardware abstraction, and complex drivers (Figures 19 and 20). These enable application programs to be used independently from the different hardware configurations (Kato *et al.*, 2010).

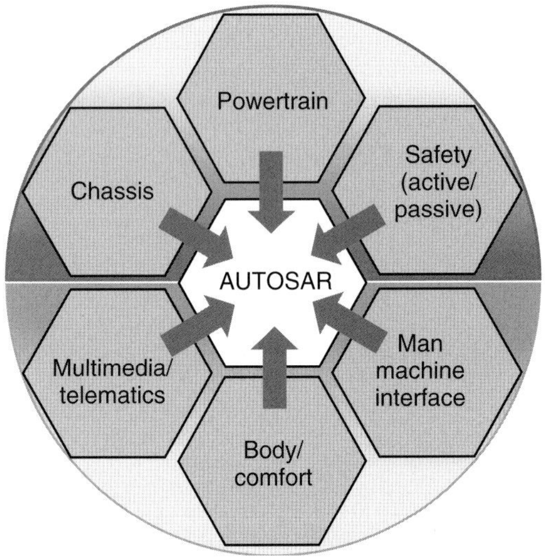

Figure 19. Field of applications in AUTOSAR.

This type of platform-based development may even form a key technological element with the potential to change the vertical integration business model into an open architecture business model.

6 SUMMARY AND CONCLUSION

This chapter has described the histories and future directions of automotive microcomputers, software, algorithms, architectures, hierarchy design, functional decomposition, and international standardization (OSEK and AUTOSAR), while considering semiconductor technologies and their applications in automotive electronics. These issues will continue to become increasingly important in the realization of further green and dependable automotive systems.

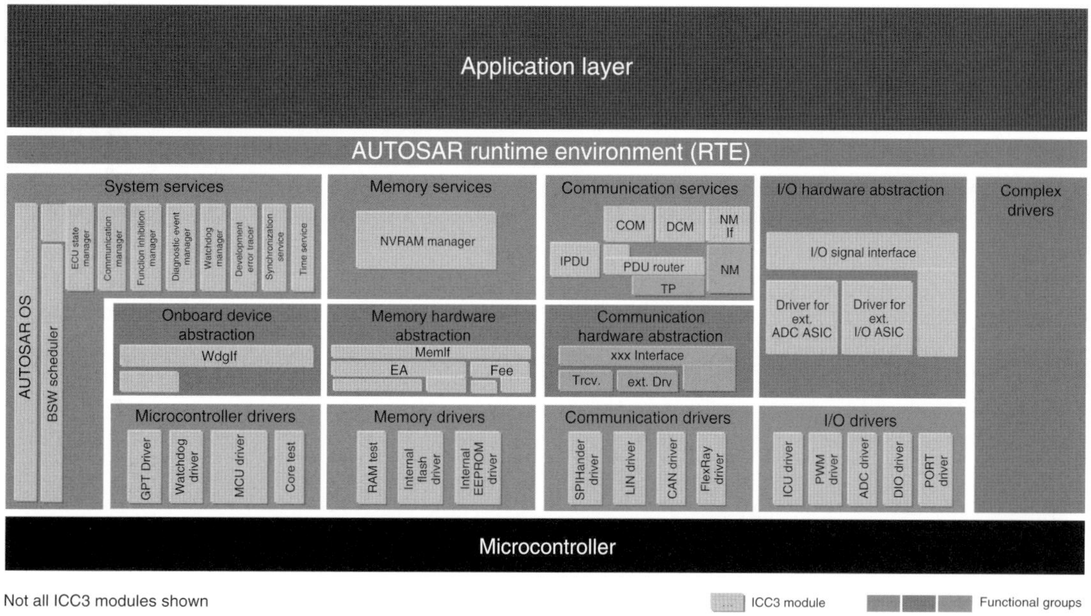

Figure 20. Software structure of AUTOSAR.

In the future, the progress of microcomputer performance and the evolution of software technologies are likely to result in advanced sensing technologies, super high speed data processing, higher integration, more efficient power electronics, and greater progress in high frequency applications such as millimeter wave and wireless functions. In addition to the need to advance these technologies, standardization across the industry will also be an important issue for pursuing development efficiency.

REFERENCES

Ishihara, H. (2007) An Overview of Automotive Electronics and Future Requirements for Microprocessors. *Microprocessor Forum 2007*, Keynote Speech.

Ishihara, H. (2009) Current issues and future prospects of automotive semiconductors. *Journal of Society of Automotive Engineers of Japan*, **63** (1), 65–70.

Kato, M., *et al.* (2010) *Illustrated Car Electronics in Two Volumes*, Nikkei Business Publications, Inc., Tokyo.

Kawamoto, K., *et al.* (2001) A single chip automotive control LSI using SOI bipolar complimentary MOS double-diffused MOS. *Japanese Journal of Applied Physics*, **40**, 2891–2896.

Ookura, K. (2005) Semiconductor Technology for Automotive Electronics. *Denso Technical Review*, **10** (2).

Tsuda, K. (2007) Denso's Automotive MCUs. *Microprocessor Report*, MPR Newsletter.

Chapter 140

Junction Blocks Simplify and Decrease Networks When Matched to ECU and Wire Harness

Takezo Sugimura, Kaoru Sugimoto, and Mamichi Tsuyuki

The Furukawa Electric Co., Ltd., Hiratsuka, Japan

1 INTRODUCTION

In this chapter, there is a description of the growth in the number of electronic control units (ECUs) and wire harnesses with the change in vehicle systems over time and the introduction of vehicle installation local area network (LAN). The number of ECUs on a vehicle has increased with new introductions and extension of vehicle systems year by year. Integration has been applied to control this ECU increase.

Along with this, the number of wire harnesses and their weight have increased, too. In addition, with the introduction of multiplex communication control, the computerization of the system and in-vehicle LAN has been introduced. Reduction of the amount of wire harnesses has been achieved by the introduction of this in-vehicle LAN.

Encyclopedia of Automotive Engineering, print © 2015 John Wiley & Sons, Ltd.
Edited by David Crolla, David E. Foster, Toshio Kobayashi, and Nick Vaughan.
This article is © 2015 John Wiley & Sons, Ltd. ISBN: 978-0-470-97402-5
Also published in the *Encyclopedia of Automotive Engineering* (online edition)
DOI: 10.1002/9781118354179.auto233

2 CAR ELECTRONICS

2.1 Transition of car electronics

In the 1950s, motorized body electric equipment has progressed, with the introduction of, for example, starters, wipers, power windows, and air-conditioning.

In the 1960s, automotive electronics still had single-control-function devices, such as voltage/current regulators and igniters. Later, in the 1970s, it became possible to control complex functions such as anti-lock brakes (ABS), air-conditioning, and fuel supply control by the advent of the microcomputer.

Since the 1980s, with the improved performance of the semiconductor, the reduction of cost and size in automotive electronics has progressed rapidly. After that, the installation of versatile electronic control for various individual systems made the wide expansion of applications possible (Figure 1).

In the late 1990s, engine, brakes, and power-integrated control systems were mounted on to HEVs (hybrid electric vehicles), which support an energy regeneration function.

Moreover, in recent luxury cars, vehicle lane-keeping functions, with integrated engine, brakes, steering control, and cooperative driver assistance systems that utilize infrastructure and information technology, have appeared with the aim of improved safety and comfort.

In addition, electric vehicles (EVs) and fuel cell vehicles that have no internal combustion engines have been developed with full-by-wire systems that have no mechanical backups at all. The aim is to introduce a system of full-by-wire (The Institute of Electrical Engineers of Japan, 2006, pp. 8–9).

			1970	1975	1980	1985	1990	1995	2000	2005	2010	2015
Powertrain		Power						Methanol		Hydrogen		
							Electric vehicle		Fuel cell vehicle			
			ICE (carburetor)		ICE (injection)			Hybrid				
		AT						CVT				
			3 speed mechanical AT			4 speed electronic controlled AT		5 speed electronic controlled AT	6 speed electronic controlled AT	7 speed electronic controlled AT		
		Throttle	Cable				Electronic controlled throttle					
		AWD	Mechanical			Viscous		Electric torque distribution mechanism				
Chassis		Steering				Speed sensitive power steering			Electric power steering (large car)			
								Electric power steering (medium car)				
					Hydraulic power steering		Electric power steering (small car)					
		Supension	Mechanical		Electronic controlled air suspension			Electronic controlled hydraulic	Electronic controlled magnetic fluid		(linear motor)	
		Brake					Traction control			Integrated chassis control system		
							Brake assist		Collision mitigation brake			
			Hydraulic		ABS			ESC				
		Cruise assist							Night vision monitoring / parking assist			
								Adaptive cruise control	Lane keeping assist			
					Cruise control (vacuum)			Electronic controlled cruise control				
		Occupant protection				Driver's seat airbags	Passenger seat airbags	Side airbags	Occupant detection sensor			
BODY	Air conditioning	Air conditioning	Cooler	Air conditioning	Auto air conditioning				Occupant detection air conditioning			
		Radiator cooling fan	Mechanical fan		Electric fan			Variable electric fan				
		Rear seat air conditioning					MPV (cooling)		MPV (heating and cooling)			
		Compressor	Mechanical compressor					Electric compressor	Hybrid compressor		CO_2 electric compressor	
	Entertainment & information	AUDIO	Cassette tape				CD		MD			
			AM Radio		AM/FM Radio					Digital radio		
		NAVI						Telematics	Rear seat entertainment			
							CD NAVI	DVD NAVI	HDD NAVI			
		Instrument panel	Mechanical Mater				Electronic controlled mater		Multi information display mater			
	Light & sight	Head light						Adaptive head light				
			Tungsten		Halogen			HID			LED	
		Defogger		Rear defogger					Front deicer		EHW	
		Side window	Mechanical	Power window		Key-off operation		Jam prevention				
	Other	Door lock	Mechanical		Electric door lock		Keyless entry		Smart entry			
		Mirror	Mechanical		Electric mirror	Retractable	Heated mirrors		Auto retractable			
		Seat	Mechanical			Power seat/heated seat		Memory seat	Memory seat with tilt /telescopic			
		Trunk	Mechanical					Power sliding door	Power tail gate & power trunk lid			

Figure 1. Transition in car electronics. (Reproduced from Institute of Electrical Engineers of Japan.)

Currently, by-wire systems such as throttle-by-wire, brake-by-wire, and shift-by-wire have been put to practical use. However, the complete control only using the by-wire technologies has not been achieved. For example, brake-by-wire keeps a safety mechanism in combination with a mechanical transmission. Steering-by-wire has still not been put to practical use, but progress of motorizing the steering is moving forward with electric power steering (EPS) and active-steering. The installation of the steering-by-wire system has started on many vehicles. For complete steering-by-wire, the most important developmental challenges are the establishment of reliability, development of legislation, and a dedicated communication protocol for practical use. These are progressing, and it is expected that market introduction will occur in the near future. By the introduction of by-wire systems, the mechanical transmission mechanism can be eliminated from the operation of the vehicle, and thus the flexibility of automotive design will be significantly increased. Many benefits such as space, power consumption, lighter weight, fuel consumption, and cost reduction can be expected. For this reason, the next generation of cars will be built with a system only using full-by-wire.

3 INCREASE OF ECUS

3.1 History

3.1.1 Outline

Undergoing this transition of car electronics, the number of ECUs and circuit wire harnesses has undergone rapid changes. An example of the growth of the number of ECUs mounted on a luxury car is shown in Figure 2.

By increasing the systems installed in a given model of car, at every full model cycle, the number of vehicle ECUs tends to increase. The number of ECUs is increasing in luxury vehicles more than smaller cars. In the example of the luxury car shown in Figure 2, the number of ECUs exceeded 70 in 2005, whereas in 2010, 100 or more ECUs are installed. The reason for the ECU increase is due to the addition of comfort and safety features at an acceptable price and environmental protection functions since 2000 (Figure 2).

In classifying ECUs by their functions, such as "HEV/EV system," "safety system," "information systems," "body electronics system," and "power train system," every function tends to increase in the number of ECUs.

3.1.2 Power train

Power train systems have one or two ECUs for the transmission and the engine. The number of ECUs depends on

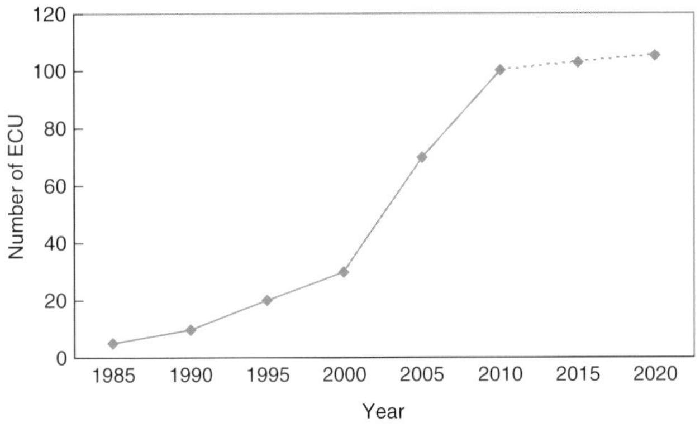

Figure 2. Transition of the number of ECUs (luxury car).

whether the function is integrated or independent. In recent years, for the realization of high engine efficiency, various systems have become electric-electronic. Processing by the engine ECU only is possible if the electronic control is simple, such as adjusting value timing, but if the aim is to control a complex motion, such as electric controlled variable valve timing and lift, to avoid strain on the engine ECU, a dedicated valve train ECU has begun to be installed.

An automatic stop and start control system, which is basically controlled by the engine ECU, may be performed by an increasing number of dedicated ECUs to accommodate factors such as supply voltage fluctuation.

3.1.3 Body system

Body system ECUs include air-conditioning, meters, smart key, latches, body control module (BCM), and so on. Basically, each ECU is installed individually. In some cases such as BCM, some control functions are integrated, for example, lighting system such as head lamps, door locks, wipers, and interior lamps. On the other hand, by functional integration to the BCM, there are increased problems in the number of circuit wire harnesses and need for increased installation space because of the larger ECU. For this reason, the case for mounting a dedicated ECU depends on how many functions for control are needed. For example, ECUs mounted on the doors, the engine, or around the steering. It is considered that there will be an increase in separated BCM functions controlled by ECUs in distributed areas.

3.1.4 Information system

ECUs of information systems, such as navigation, audio, telematics, and digital TV, have dedicated ECUs for each system, because there are many system configuration options. In the future, the addition of various information and services, such as communication between car and car and car and road transport infrastructure, will increase the number of ECUs further.

3.1.5 *Safety system*

The ECUs for safety systems, such as airbag, electronic stability control (ESC), vehicle camera systems, radar systems, and peripheral monitoring, have, because there are many system configuration options, dedicated ECUs mounted for each system. In the future, the addition of advanced safety functions will require an increase of ECUs.

3.1.6 *Hybrid electric vehicle/electric vehicle*

In the ECU systems of HEVs/EVs, the number of ECUs differs from that of an EV to a mild-HEV or a strong-HEV. Typically, dedicated ECUs are mounted for motor control, HEV/EV management, inverter control, and battery management.

3.2 Problems of increased ECUs

While increased ECUs can deliver benefits, a significant problem is that the space where the ECU can be mounted in the vehicle is limited. Usually, the place is in the vehicle cabin, such as the foot space under the driver's seat or passenger's seat, as a good temperature and/or vibration condition is required. On the other hand, expanding the cabin living space in the quest to comfort leaves less space for the ECU. Thus, it is difficult to keep the space for the ECU and this has become a serious problem.

In order to solve these problems, reducing the number of ECUs by integration is underway. For example, in the case of BCM, some ECUs belonging to the body electronics function are integrated into one ECU. In the case of safety systems such as the driving support ECU, the system is integrated with the face-direction-detecting ECU, the millimeter-wave radar ECU, and the stereo-camera ECU. In the case of the information system, an ECU is mounted in the center display module that is integrated with the audio and navigation ECU.

This ECU integration trend is expected to expand more and more in future generations of vehicle. In the future, the integration of ECUs will proceed beyond the function field, such as information systems and safety systems with the system detecting the field around the car.

3.3 Predictions for the future

Beyond the integration of ECUs, integrating power lines and signal lines that were necessary for the individual ECU is becoming possible; the effect is a reduction in overlapping circuits on wire harnesses.

Thus, as was indicated in Figure 2, the number of ECUs on a vehicle will tend to decrease as existing systems integration proceeds. However, this reduction effect is offset by the increase of ECUs with new systems. Therefore, the trend in the number of ECUs may remain flat until around 2020, or it may only increase slightly (Tanokura, Shindou, and Okubo, 2006, p. 115).

4 HISTORY OF WIRE HARNESSES

4.1 Historical trends in number and weight

An example of the trend in the number of circuits mounted on a luxury car is shown in Figure 3. The trend in wire harness mass for the same luxury car is shown in Figure 4.

For environmental control, safety, and comfort, electrical systems in cars have been increasing in functionality and circuits per vehicle wire harness every year.

Moreover, the number of wire harnesses will also increase as new electrical systems are added, and ECUs and multiple types of electrical equipment such as switches, sensors, valves, and motors are connected in a one-to-one connection.

In the past, when new electrical systems have been added, the number of circuits increased to a dozen at most, but the electrical systems advanced in recent years, which mean that 30–60 extra circuits have been added in one system.

4.2 Problems of increased wire harness

Although the space for mounting the vehicle wire harness is limited, owing to an increase in the number of these circuits, the problem of finding a suitable location for the

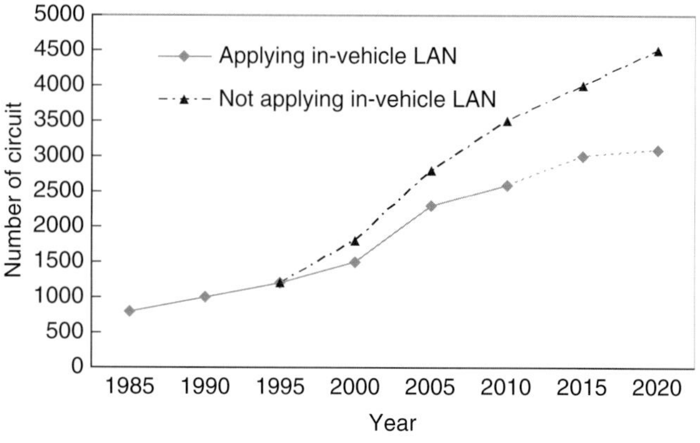

Figure 3. Transition of the circuits (luxury car).

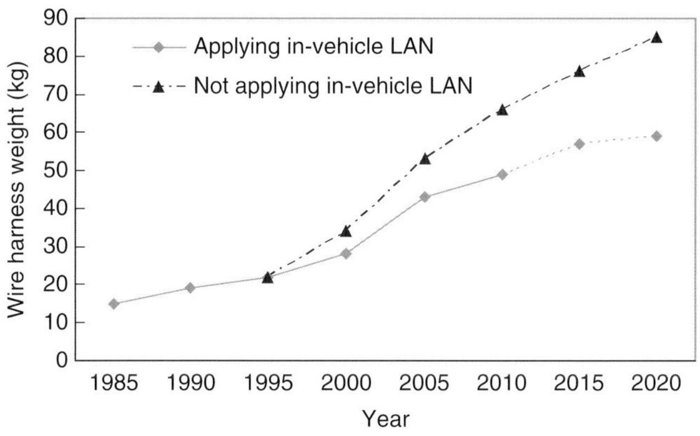

Figure 4. Transition of the wire harness weight (luxury car).

wire harness has arisen. In addition, the increase in the number of circuit wire harnesses has led to an increase in the vehicle weight and a deterioration in the ease of assembly work of the vehicle. As is required to improve the fuel efficiency of automobiles because of the rise of environmental concerns in recent years, the reduction in the weight of the entire car is being promoted and therefore lighter wire harnesses are also required.

4.3 Introduction of in-vehicle LAN

In order to solve these problems, the technology of in-vehicle LAN had begun to be introduced from the 1980s. Figure 3 shows the number of circuit wire harness introduced with the presence or absence of in-vehicle LAN. Comparing the 1990s to the present, the circuits increased from 1000 to 2600 circuits (applying in-vehicle LAN), but the increase of the number of circuits is estimated to be up to about 3500, without the introduction of in-vehicle LAN.

Figure 4 shows the mass growth of the wire harnesses in the presence or absence of in-vehicle LAN. Comparing the 1990s to the present, wire harnesses have increased from 15 to 49 kg (applying in-vehicle LAN), and the increase in weight of wire harnesses is estimated to up to about 66 kg if there is no application of in-vehicle LAN.

In this way, even though new electrical systems have been added each year, by introducing the vehicle LAN mode, it is possible to suppress the increase in the number of circuits and wire harness mass.

4.4 Rearranging of power supply distribution and the wiring

Generally, there are two roles for the junction block. The first is the storing/fixing/protecting the safety equipment such as a fuse, the relay, or the control of the power supply ON/OFF by the relay or rearranging the power supply distribution according to load and the joint circuit of the power supply. Secondly, there is a role for an organizing BOX wire harness to split the wire harness, to make for easier assembly in the vehicle. Duplication of circuit power lines and signal lines can be reduced by the aggregation of components for safe distribution and function, including power in the junction block, thus realizing a reduction of wire harnesses.

4.5 Decentralization of the power supply

In recent years, with the increase in the number of electrical systems for environmental, safety, and comfort purposes, there has been a rise in the number of circuit wire harnesses. Architecture has been considered to be a solution to this problem, by the distribution of the power source to the junction block for each area (i.e., a decentralization of the power supply). Specifically, a junction block placed on the engine, where the electrical equipment is concentrated, in the driver's seat and passenger's seat sections, the instrument panel section, or the center cluster section. Such decentralization of power reduces the wire harness power delivery (e.g., circuit lighting or a ground circuit). In addition, further lightweight wire harnesses can be used by minimizing the total wire length of power lines and electrical equipment cable leading to further weight reduction. Moreover, the power distribution architecture with in-vehicle LAN connections between the junction blocks for each area makes it possible to share power control information. The purpose of this power distribution method has the effect of reducing power by efficient distribution and management of ordered power, avoiding enlargement or concentration of the wire harness. In addition, ordered power distribution architecture contributes to the standardization of power supply arrangements: the standardization of power distribution (The Institute of Electrical Engineers of Japan, 2006, p. 18). It follows that junction blocks will have the role of power management and distribution control functions.

5 NETWORKS BETWEEN ECUS

5.1 Histories

Figure 5 shows an example of network deployment and the time of application LAN technology started to be introduced into cars from the 1980s. Many automobile manufacturers adopted in-vehicle LAN interfaces using their own original

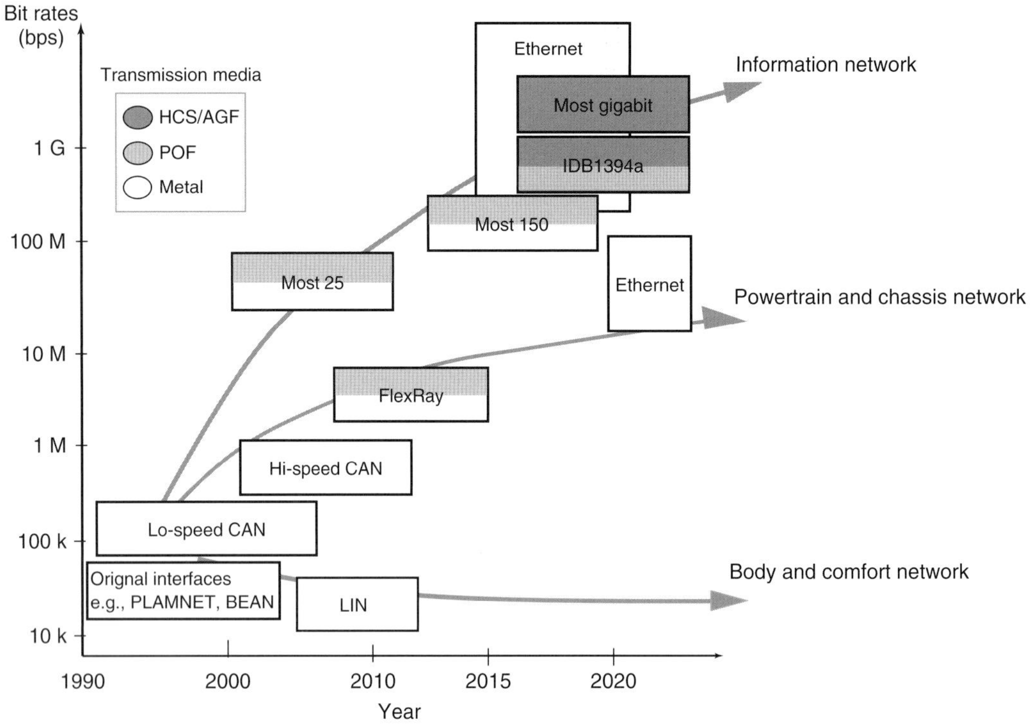

Figure 5. Network deployment and transition period.

standards developed independently of each other. Introduction of the automotive in-vehicle LAN began with the body control system. Examples are Chrysler "C2D", GM "J1850VPW," in the year 1990, Daimler "CAN," BMW "I-BUS" and "K-BUS," Toyota "BEAN," Mazda "PLAMNET," Honda "MPCS," and Nissan "IVMS." In many of these original interfaces, the data transmission speed was about 10 kbps.

5.2 Standardization of in-vehicle LAN

Not only does the wire harness continue to increase in number of systems installed in vehicles, but also there is a problem with the lack of data transmission speed in the standard interfaces previously adopted. From around 2000, world automobile manufacturers started to adopt the controller area network (CAN), which is a single interface for standard in-vehicle LAN.

In this way, CAN has become the mainstream in-vehicle LAN, although other LAN interface standards have also appeared since 2000. Those are not developed for each automobile manufacturer, as was the case in the 1980s, but standardized organizations such as Tier1 suppliers, automobile manufacturers, and semiconductor manufacturers participate and formulate optimal in-vehicle LAN standards for each vehicle system.

5.3 LAN structure

As shown in Figure 5, several kinds of standard LAN interface are used according to the difference of data transmission speed, purpose, or reliability.

Classification according to the in-vehicle LAN domain can be divided into three systems: control systems, body systems, and information systems. CAN and FlexRay are control systems, CAN and LIN are for body electronics systems, and MOST is for the information system; all are used widely as the typical in-vehicle LAN.

In this way, communication goes ahead through multi-channel LAN in the car, which is made up of plural subnetworks and used properly according to the domain. Currently, a typical luxury car may have multichannel networks composed of a LIN local network of multiple channels, MOST subnetworks, and nine channels of CAN subnetworks. Transmission of information between these subnetworks has been realized by a dedicated gateway ECU; its gateway function is set up to connect with the ECU's multiple plural subnetworks.

5.4 Problems of high functionality of electronic control

So far, with the high function of cars, LAN has been applied to cooperation or integrated controlling functions, by sharing actuator and sensor information between ECUs. In future, equipment related to the precrash safety advanced safety features developed in order to prevent a car accident, which will have wide applicability in the next generation of cars, and if cooperative control of vehicle infrastructure and road traffic by intelligent transport systems (ITSs) becomes widely used, the importance of in-vehicle LAN systems connecting each ECU is expected to progress further.

On the other hand, MOST and CAN have limits with this transmission speed; the maximum transmission speed of CAN is 1 Mbps and MOST25 is 25 Mbps. It follows that there can become a difficult situation transmitting by MOST25 or CAN, with an increased number of ECUs or growth in complexity of control, when the amount of data increases dramatically.

For example, control systems for the basic function of vehicle running required the improvement in motor control, to enhance vehicle safety, for which LAN interfaces need higher reliability and a faster real-time response. In addition, in information systems, LAN interface standards require emphasis on communication speed and a large amount of data transmission, with increased application of environmental monitoring cameras and simultaneous streaming of audio and video that this will mandate (Tanokura, 2011, p. 11).

5.5 Next-generation standard LAN interface

To compensate for these weaknesses and new requirements, various next-generation standard LAN car interfaces are being developed. In control systems, the standard interface equivalent to the next generation of "CAN" is "FlexRay," which has better speed and reliability, and guarantees a response at a given time, with a maximum transmission speed of up to 10 Mbps, and has begun to be adopted in some luxury cars. Other standardizations, such as the "automotive Ethernet" as a long-term measure, are being developed.

In information systems, the equivalent to next-generation "MOST" are "MOST150" with up to 150 Mbps transmission speed, "IDB1394" with a maximum transmission speed of 800 Mbps, and the "Ethernet" with a maximum transmission speed of 1 Gbps, as the standards for transmitting voice or video data.

The "Ethernet" has been developed as the standard to encompass both control and information systems in the long term.

In this way, the transmission speed of the next-generation standard LAN interface is moving in the direction of large capacity/high speed communications. In addition, using a standard LAN interface of the next generation, it will be possible to interact faster with much more information. This will be needed unless the number of ECUs does not decrease dramatically, using multichannels with multiple subnetworks (Noumi *et al.*, 2007, p. 20). However, it is thought that the future LAN constitution in the car will be multichannel and layered according to domain, with CAN, LIN, MOST, and Ethernet.

6 CONCLUSION

The introduction to the function of the junction box, the in-vehicle LAN, and integration of multiple ECUs discussed in this chapter has become indispensable in the solution of issues such as the increase in the number and weight of the wire harness circuits to transmit power and signals with the increase in number and ECUs as new electrical systems that are mounted on a vehicle. To build the overall architecture of the automobile electrical system by combining these solutions, to be developed further are the methods that deliver lightweight wire harnesses and reduced numbers of ECUs and the number of electrical systems for an increasingly complex future vehicles.

REFERENCES

Institute of Electrical Engineers of Japan (2008) Management technology of automotive electric power supply systems. Technical Report No. 1121. The Institute of Electrical Engineers of Japan, Japan.

Investigation Committee on Next Generation Automotive Electric Power Systems (2006) Roadmap of next generation automotive electric power systems. Technical Report No.1049. The Institute of Electrical Engineers of Japan, Japan.

Noumi, K., Nishihashi, S., Ishikawa, Y., and Takahashi, J. (2007) Approaches to the development of the in-vehicle LAN system. *FUJITSU TEN Technical Journal*, **49**. Fujitsu Ten, Japan.

Tanokura, Y. (2011) Toyota Eyes Adoption of Ethernet for On-Board LAN. *Nikkei Electronics* (June 27).

Tanokura, Y., Shindou, T., and Okubo, S. (2006) Improving the Lexus. *Nikkei Electronics* (December 18).

Chapter 141
ECU Design and Reliability

Joji Yoshimi
DENSO Corporation, Kariya, Japan

1 INTRODUCTION

There are increasing calls to make vehicles more environmentally friendly, safe, convenient, and comfortable. To meet these demands, onboard engine control units (ECUs) have grown more sophisticated in terms of function and performance. As a consequence, the scale of hardware and software has dramatically increased compared to the 1970s when the widespread use of automotive products with electronic applications first began. Drivers are also more sensitive to quality and now demand ECUs with higher reliability. This chapter describes the use of simulations for efficiently designing a large-scale ECU and methods for evaluating whether the ECU achieves a required level of quality.

2 ECU DESIGN

With the advancement of onboard electronics technology in recent years, luxury vehicles are now installed with around 100 ECUs, which cooperate with each other through an onboard LAN connection (Ookura, 2005). In the future, ECUs will have more functions that will also become increasingly sophisticated, and the interaction between the different functions is likely to become more complex.

Encyclopedia of Automotive Engineering, print © 2015 John Wiley & Sons, Ltd.
Edited by David Crolla, David E. Foster, Toshio Kobayashi, and Nick Vaughan.
This article is © 2015 John Wiley & Sons, Ltd. ISBN: 978-0-470-97402-5
Also published in the *Encyclopedia of Automotive Engineering* (online edition)
DOI: 10.1002/9781118354179.auto215

2.1 ECU design process

Figure 1 shows a general design flow for an ECU. The requirement definition step consists of defining all requirements throughout the ECU life cycle relating to reliability, regulations, and the ECU manufacturing process, in addition to defining requirements for the functions and performance demanded. This can reduce reworking at subsequent steps.

At the step for allocating hardware and software functions, a combination of hardware and software is selected that meets the requirements while minimizing the ECU price and development costs, and various specifications are then created. However, the specifications are often modified many times before mass production of the ECU starts. Any need for hardware changes or software structure changes lengthens the development period. Therefore, creating specifications capable of adapting to any likely requirement changes that can be anticipated at such time helps suppress the length of the development period.

At the hardware design and software design steps, the defined requirements are incorporated into design activities based on the technologies held by each ECU manufacturer.

At the hardware/software consolidated testing step, it is confirmed that the ECU satisfies all the defined requirements. Finally, the ECU is installed in a vehicle and system verification is performed, thus completing the ECU design process.

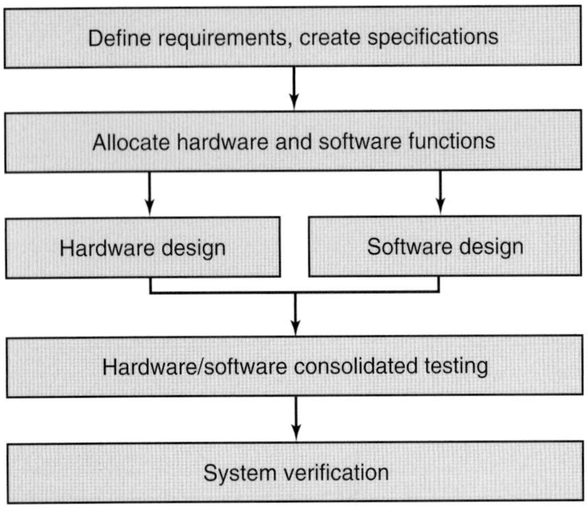

Figure 1. General design flow for ECU.

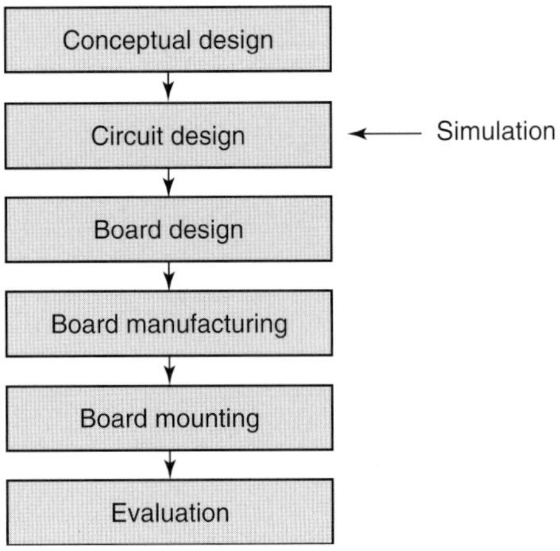

Figure 2. Design flow for printed circuit board.

2.2 ECU design support tools

As mentioned earlier, the ECU has grown in scale and become more complex. Designing and evaluating such an ECU by hand as in the past would require an enormous amount of man-hours and is not realistic. Therefore, the design support tools described below are gaining more widespread use.

2.2.1 ECU hardware development

Figure 2 shows the design flow for a printed circuit board. The process begins with conceptual design, and once the conceptual design is fixed, the process moves on to circuit design using circuit design tools. Next, operation verification is performed using an analysis tool (simulator). After circuit design, board design tools are used to set the layout and wiring for electronic components (board design), thus completing the printed circuit board design process (Kato, 2010).

2.2.2 Software development using virtual environments

Software used to be developed using actual chips. However, recent advances in simulation technology have enabled the co-validation of hardware (mainly microprocessors) and software by simulation (Keating and Bricaud, 2000). This virtual environment has the advantages of allowing software development to start before the trial production of actual chips with excellent observability. However, an amount of development time corresponding to roughly several hundred to several thousand times more than that needed when using actual chips is required (Keating and Bricaud, 2000). Accordingly, there is a trade-off relationship between the speed and precision of the simulation, which means that model development well suited to the simulation purpose and verification items is key.

2.2.3 Software verification using HILS

Instead of using actual automotive components, hardware-in-the-loop simulation (HILS) mimics (simulates) electric signals from such components (Figure 3). Software verification on the ECU is performed using signals originating from the HILS, without requiring the elaborate environment of the actual vehicle. Software-in-the-loop simulation (SILS) methods that build a simulation environment with only software on a computer and use no hardware such as an ECU or actuator have also been proposed (Kato, 2010).

2.3 Future of design support tools (CAE)

This section describes the prospects of high level design and model-based design as applications of design support tools for onboard electronics.

2.3.1 Onboard electronic system development using high level design technology

Owing to recent advances in onboard electronics, ECUs have increasingly sophisticated functions and the interaction between functions has grown more complex. In the future, this trend is expected to accelerate. As an example, the design and verification for an integrated control that requires cooperation among ECUs will become considerably more complicated. Therefore, a new ECU design environment that addresses broader concepts than individual optimization on a component level as in the past

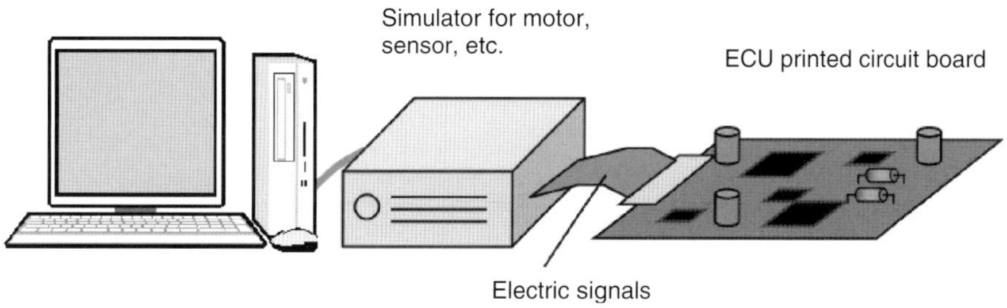

Figure 3. Example of HILS environment.

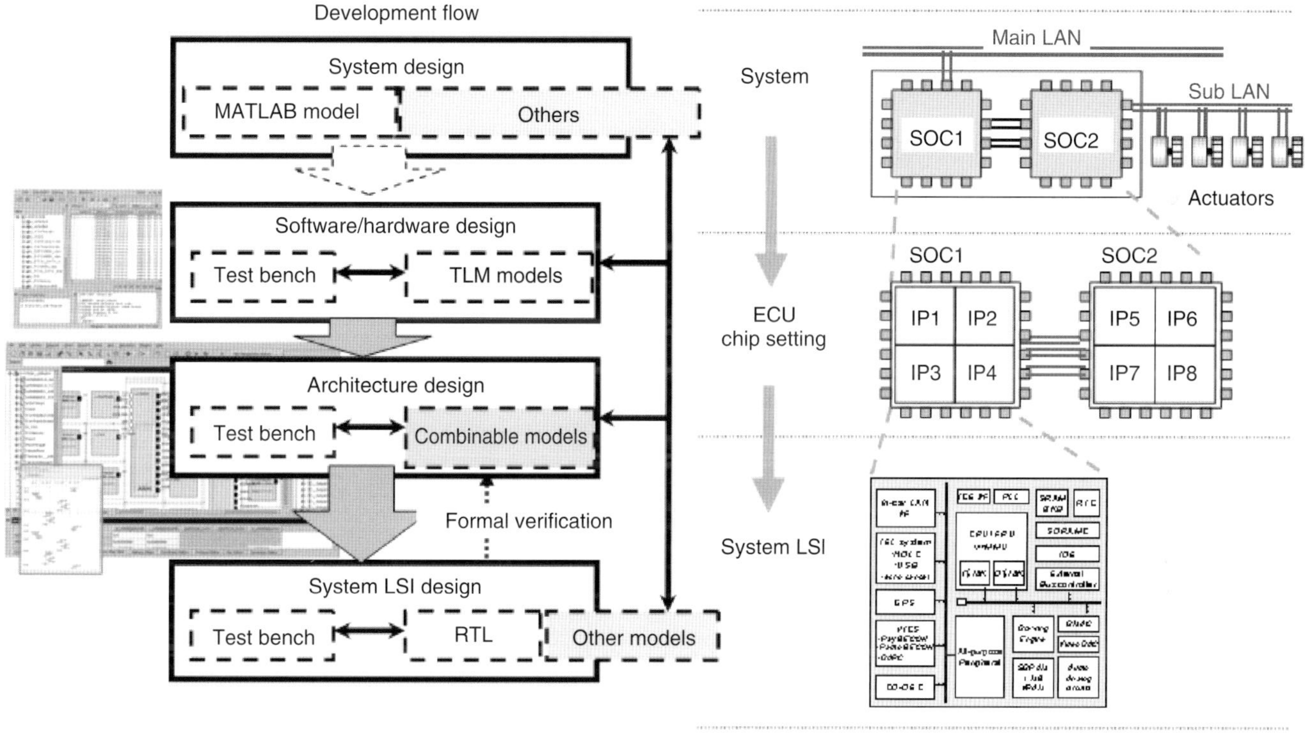

Figure 4. High level design and applicable layers.

is anticipated. One such environment gaining attention is electronic system level (ESL) design (high level design) for ECUs. In this field, the Open SystemC Initiative (OSCI) has standardized the design language SystemC and module interface TLM 2.0.

Figure 4 shows a design flow and applicable layers that apply high level design to an onboard electronic system. High level design has broad applicability ranging from onboard system operations to ICs. However, the pursuit of more precise verification lowers the abstraction level and reduces the simulation speed. Therefore, the verification items must be clarified and the abstraction level set for each applicable layer (Kato, 2010). Creating models of functional components and accelerating the speed of simulation including microprocessors are important elements in ESL design for ECUs.

The advantages of high level design in ECU design are listed below (Kato, 2010).

High level design can be considered as one type of SILS that does not require the actual device. However, high level design includes the concepts of hardware such as microprocessors and enables a system design that takes hardware into account.

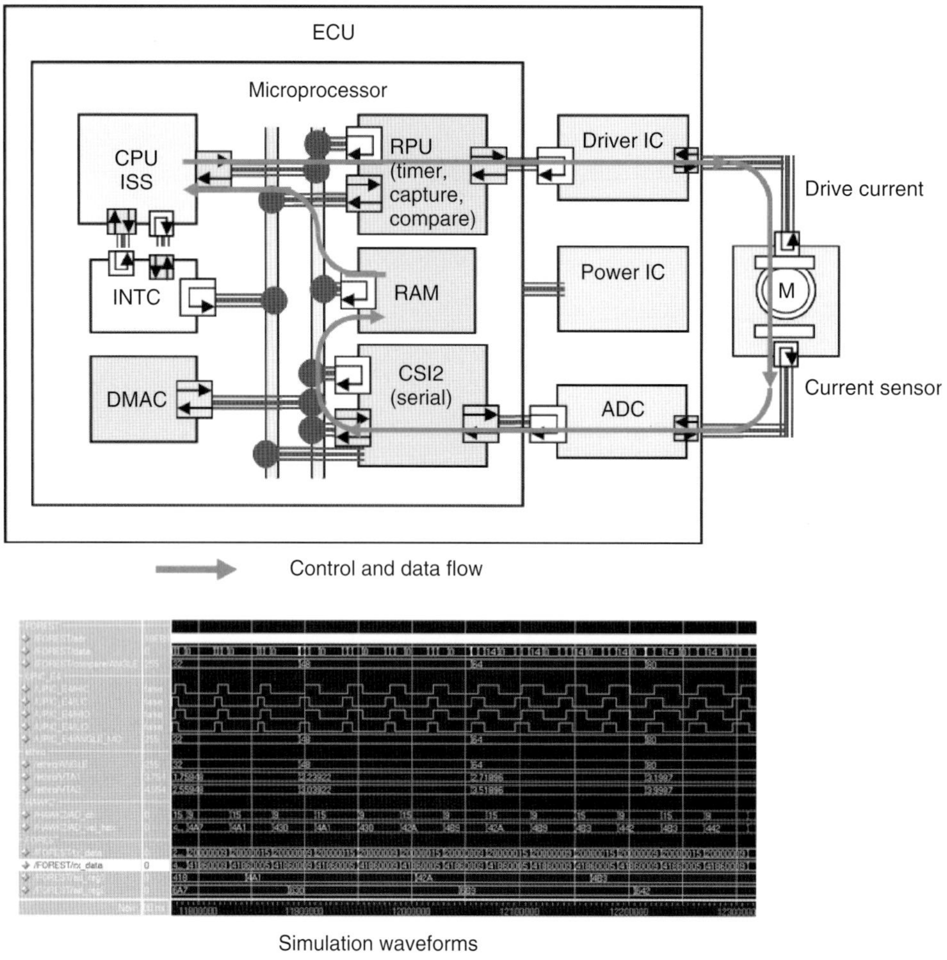

Figure 5. Example of high level design applied to ECU.

Instead of design that simply improves on the conventional ECU, various case studies of ECU architecture are possible in the virtual environment of high level design.

ECU design in the past involved mainly desk calculations, and it was necessary to incorporate a margin of error to achieve the required performance. Performance can be evaluated in a quantitative manner with high level design, which enables suitable design targets to be set and leads to cost reductions.

Problems can be discovered earlier by proceeding with verification in the virtual environment before making the actual device.

Advance software development is possible with the use of the microprocessor virtual environment.

Figure 5 shows an example of an ECU model. As the ECU, in addition to the microprocessor (CPU + peripheral circuits), models of the power IC and driver IC, as well as models of sensors and actuators connected to the ECU are required.

2.3.2 *Model-based design (linked to algorithm development)*

Physical states such as temperature and rotation angle are converted into sensor electric signals, and a feedback control is executed for actuators including the engine and motors based on these electric signals. This type of control development exists for many onboard systems.

The development process consists of developing algorithms first, and then designing the control hardware and building the system. MATLAB or the like is often used for algorithm development. An example of the flow of a model conversion that incorporates algorithms developed using MATLAB into a hardware model is described below.

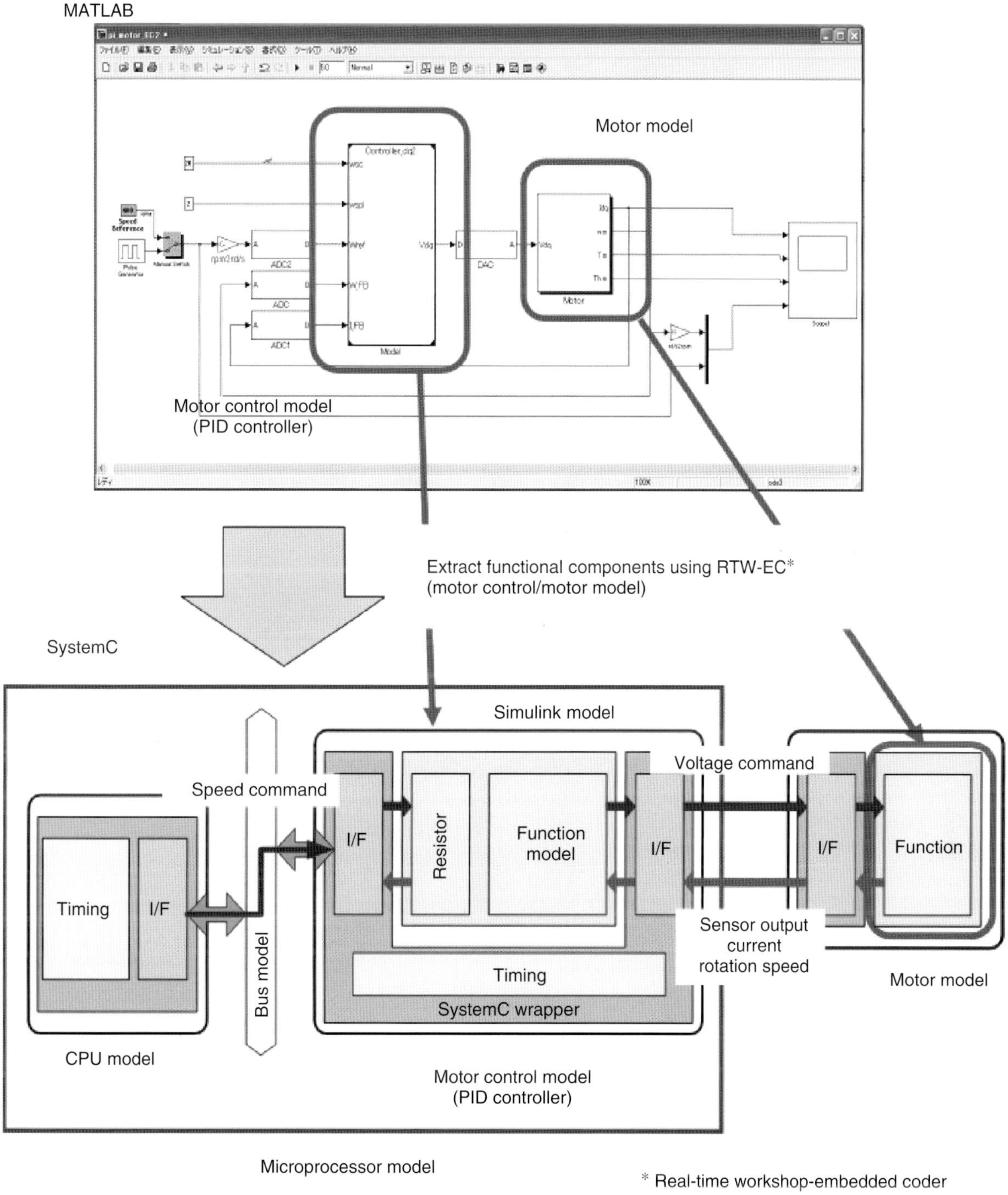

Figure 6. Example of conversion from MATLAB to SystemC model.

1. Algorithm development using MATLAB
2. Analog signal discretization
3. Function output in C code
4. Add interface and time control to C code and use as SystemC code
5. Use SystemC code circuit as high level design model

Although problems with simulation speed remain, tool vendors have recently developed simulation technology that combines MATLAB and high level design tools. Further advances in technology and application to onboard electronic systems are anticipated (Figure 6) (Kato, 2010).

3 ECU RELIABILITY

To ensure ECU reliability, first, the targeted quality and life are specified. Next, the environment in which the ECU will be used is studied. Finally, reliability testing that applies stresses corresponding to the use environment must be performed to confirm that the ECU satisfies the quality and life targets (Reliability Engineering Association of Japan, 1997).

A number of ECU failure patterns are represented by a bathtub curve in Figure 7. The failure rate tolerance and product service life shown in Figure 7 must satisfy the targeted quality and life.

The bathtub curve has an early failure period, a chance failure period, and a wear failure period based on shifts in the failure rate. The early failure period contains the highest probability of manufacturing and design defects, which means that there are many failures at the start of the period but the number gradually decreases. Screening by burn-in is effective for ensuring that defective products from this period do not enter the market (Renesas Electronics Corp, 2010). Burn-in is a method of applying stress to the ECU and ECU components over a set period to sort good and defective devices. This is explained in Figure 8 using an S-N curve.

Defective devices are normally distributed to a location away from good devices. Only defective devices fail after a fixed time (tB) following application of a stress (SB), so defective devices along with good devices can be stratified. Screening consists of focusing on the failure mode of the applicable device, and finding and eliminating design weaknesses and potential faults ahead of time through the

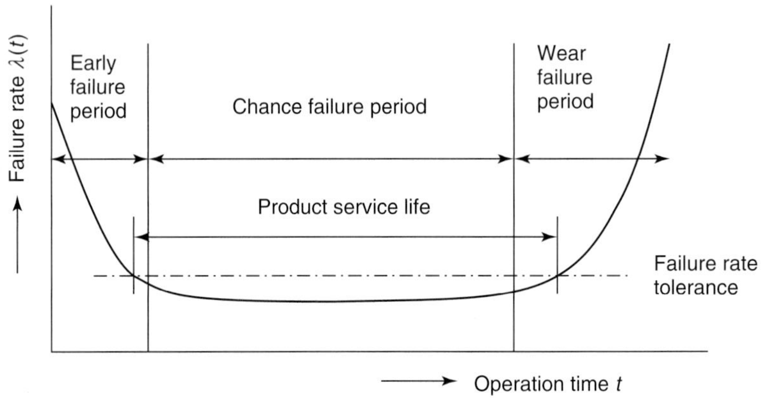

Figure 7. Bathtub curve.

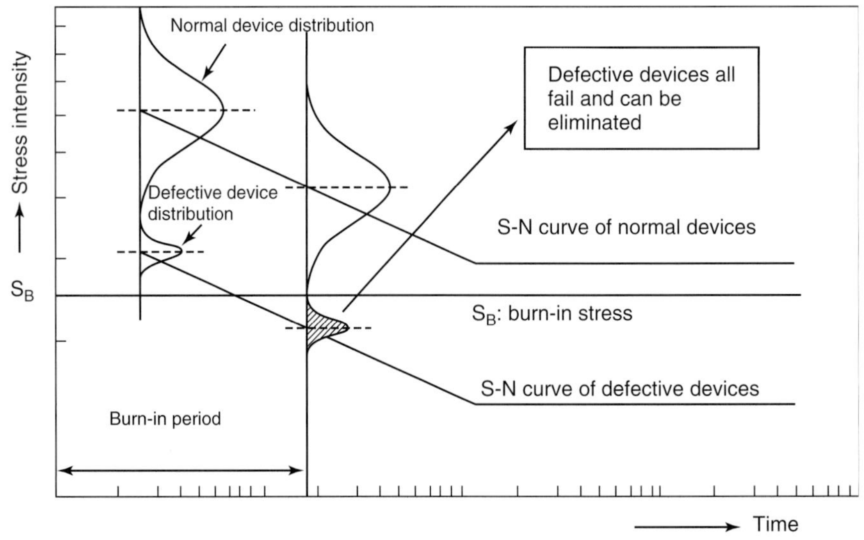

Figure 8. Burn-in.

appropriate application of stress. For effective screening, a stress type and application time capable of exposing a failure in the defective device must be selected. However, applying excessive stress to shorten the screening time or applying stress over a long period to reliably eliminate defective devices may adversely affect the characteristics of good devices or cause failures. Therefore, it is important to select the appropriate conditions.

In the chance failure period, failures occur for any random reason and there is a fairly steady failure rate. The wear failure period is a period in which failures increase due to deterioration over time from product use. The only ways to reduce the failure rate in the chance failure period and the wear failure period are by understanding various types of failure mechanisms and continuing steady efforts to improve tolerance against failure. In particular, countermeasures for wear-related failures should ensure that a large number of defects do not occur within the warranty period.

3.1 ECU environment

Vehicles must be safe and comfortable under various environments and continue to operate normally until they are scrapped. The quality of onboard ECUs must also remain high over a long period under harsh conditions such as heat, humidity, vibration, noise, and static electricity. Table 1 shows examples of environment conditions for an onboard ECU (Kato, 2010). The values listed in the table are only examples, and the ECU mounting environment in the vehicle must be accurately understood in the course of ECU development. These values ensure the validity of the reliability assessment described later.

3.1.1 Temperature

While the engine is operating, the temperature in the engine compartment can increase until 100°C or more. The temperature is also influenced by weather conditions, and may increase until 90°C on the top of the dashboard and until 65°C in the trunk when parked under the hot sun. In addition, the temperature may alternate between high (during the day) and low (at night), and sharp temperature increases may occur due to self-heating during load driving.

Table 1. Examples of environment conditions for onboard ECU.

Temperature	−30 to 110°C: engine compartment
	−30 to 80°C: vehicle cabin
Humidity, water	95% RH or more, water immersion
Vibration	20G, 20–200 Hz: engine compartment
	4.4G, 20–200 Hz: vehicle cabin
Electricity	Power voltage fluctuation: 6–24 V
Vibration	Surge voltage from induction load
	Electromagnetic damage, static electricity
Electricity	Dust, salt, grease

3.1.2 Humidity and water

This environment refers not only to outdoor humidity but also to the condensation that occurs when devices cooled by the air conditioner are subjected to outside air with high humidity, which may leak inside the circuit board and cause electrical or other corrosion. Water immersion even in the vehicle cabin must also be considered due to the possibility of spilled drinks and water entering the cabin during car washing.

3.1.3 Vibration

Driving is always accompanied by vibration, and some regions may reach until 20 G depending on the road surface and running conditions. Therefore, trouble caused by vibration must be taken into account.

3.1.4 Electricity

Various loads are connected to the vehicle battery, and the power voltage varies depending on their on/off state. At cold starts in particular, there are considerable changes in power voltage, which may range from a nominal 12 V to approximately 6 V. Operation of induction loads such as motors, solenoids, and relays, as well as current interruptions, may be accompanied by induction noise in other devices or parallel wire harnesses due to electromagnetic or electrostatic coupling.

3.2 ECU failure examples

Understanding the failure modes that occur in the ECU environment is essential to securing ECU reliability. This is because the accelerated testing described later focuses on failure modes to calculate an acceleration coefficient, and proper testing cannot be performed if the failure modes cannot be anticipated. Figure 9 shows an example of failures that occur when humidity, heat, hot/cold stresses, and vibration are applied as typical stresses. Most of these failures are exposed after a number of years in a normal use environment, so countermeasures at the design stage and sufficient evaluations are critical.

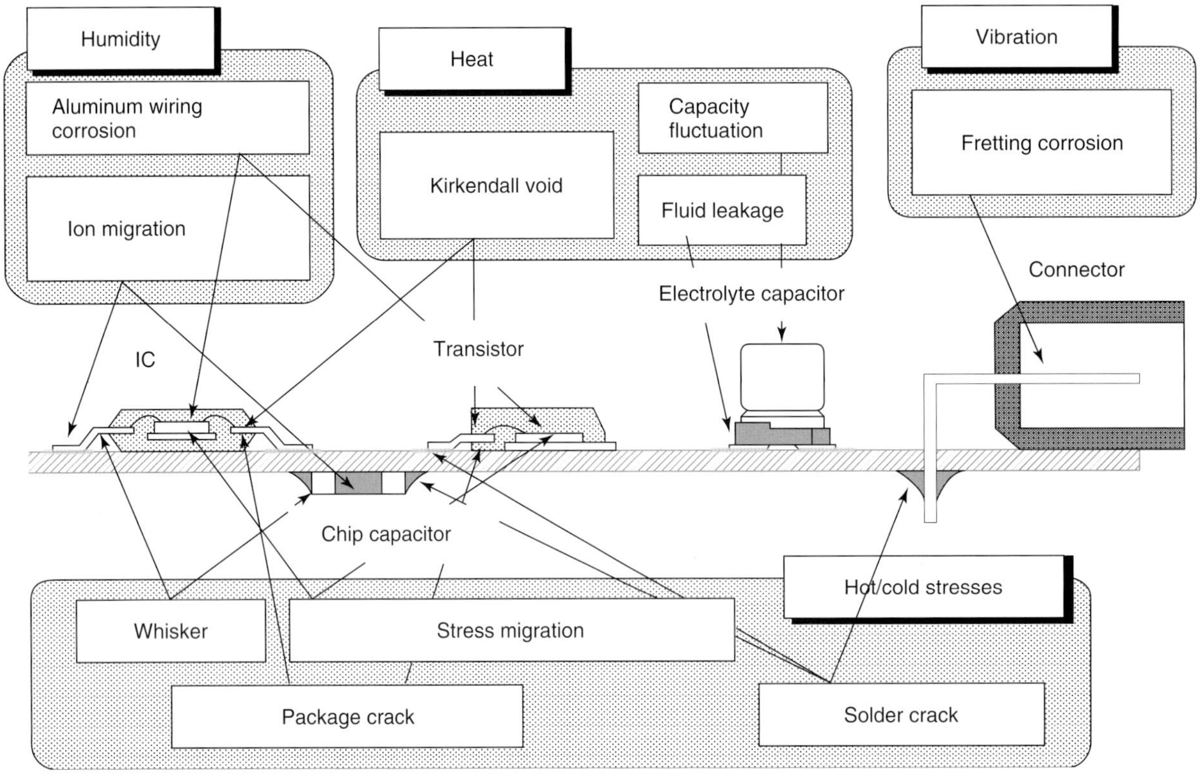

Figure 9. Examples of ECU failures caused by various stresses.

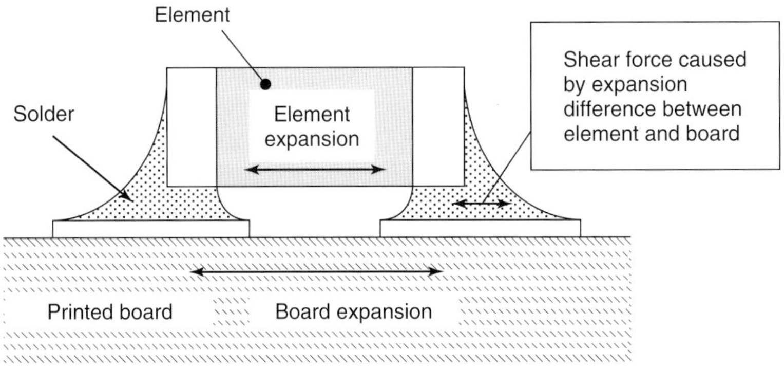

Figure 10. Solder crack mechanism.

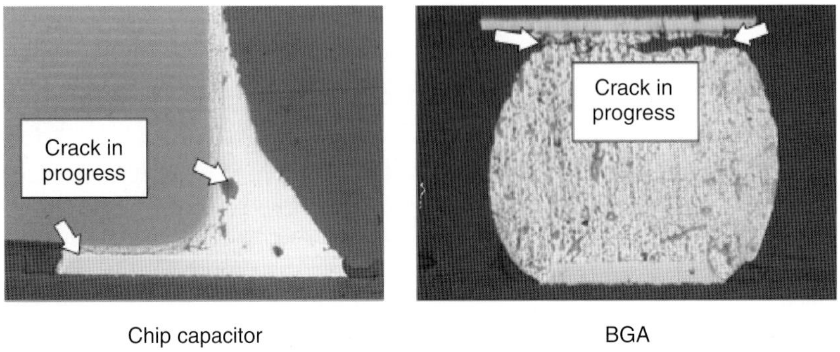

Figure 11. Solder crack examples.

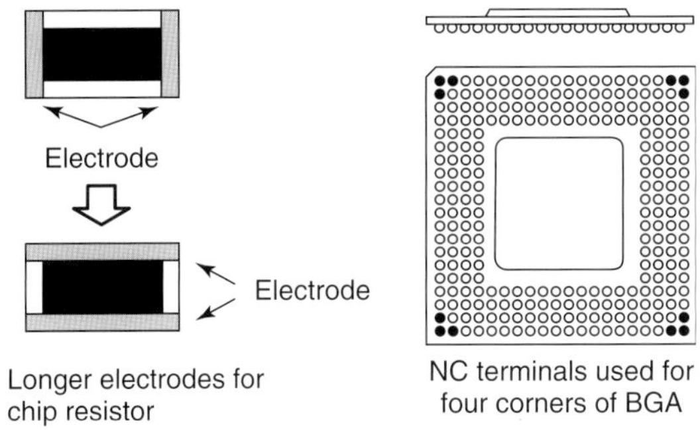

Figure 12. Solder crack countermeasures.

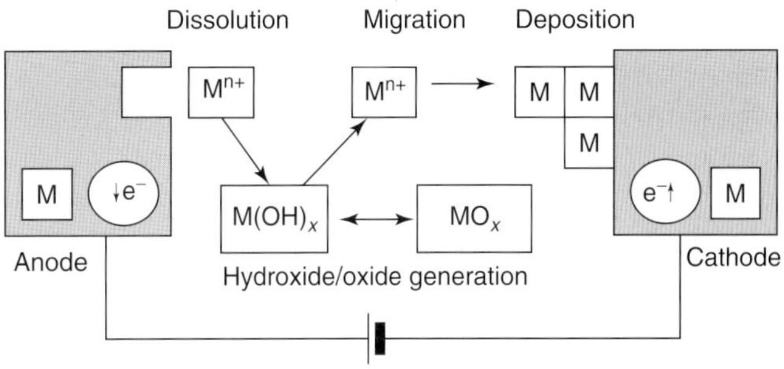

Figure 13. Ion migration mechanism.

3.2.1 Solder crack

The printed circuit board and its mounted electronic components have different coefficients of thermal expansion. Therefore, temperature fluctuations may produce stress on bonded sections (Figure 10). Such stress is absorbed by the solders, but with repeated heating and cooling, the solders may experience plastic deformation and exhibit minute cracks. The cracks may gradually grow as shown in Figure 11, and ultimately lead to an open defect.

For an ECU used in an environment with large temperature differences, to reduce hot/cold stresses, components with longer electrodes to shorten the interval between electrodes are adopted. For a BGA package IC, countermeasures such as using NC terminals without internal connections for the corner terminals subject to the maximum stress are adopted to ensure that functionality is not affected even if cracks occur (Figure 12) (Kato, 2010).

3.2.2 Ion migration

Ion migration is a phenomenon in which metal that includes electronic material ionized by the action of the electric field migrates between electrodes and is then reduced to metal again and deposited, as shown in Figure 13. The ion migration phenomenon is accompanied by visible tree-like branching called dendrites. An example is shown in Figure 14. The dendrites contain metal and are generated by the normal environment of the onboard ECU, such as temperature, humidity, electric field (voltage), and impurities (e.g., ions, contamination, and dust). These environment elements act in complex ways to form dendrites that cause insulation degradation in circuits and components.

For cases in which the ECU is used in an environment with high humidity and there are only narrow intervals of several millimeters between electrodes, countermeasures include avoiding the use of Ag, Pb, Sn, and Cu materials, blocking the ion migration path by covering the space between electrodes with a moisture-proof material or gel resistant to moisture absorption, and reducing the halogen concentration (Kato, 2010).

3.2.3 Whiskers

Whiskers are a phenomenon in which needle-like or nodule-like metal single crystals grow on a metal surface. Whiskers are known to occur from Sn plating and Zn plating. Whisker formation may lead to defects caused by short-circuiting between terminals of the circuit components. Figure 15 shows examples of whiskers found on an electrode of a chip capacitor and an IC lead in a temperature cycling test. Known causes of whiskers include formation by stress acting on the plating film, which occurs due to different coefficients of thermal expansion, intermetallics and their

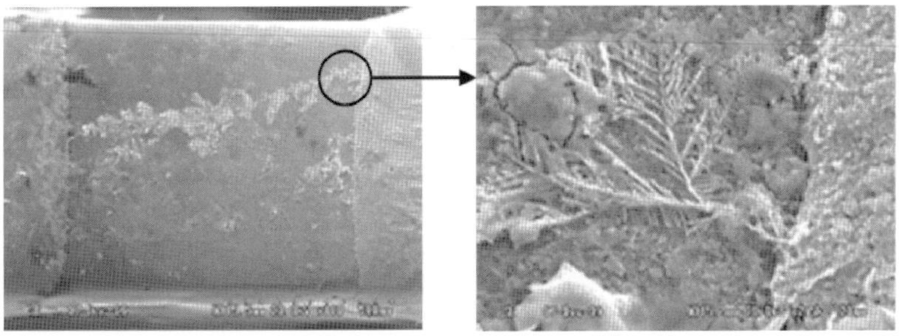

Figure 14. Dendrite formed between electrodes of ceramic capacitor.

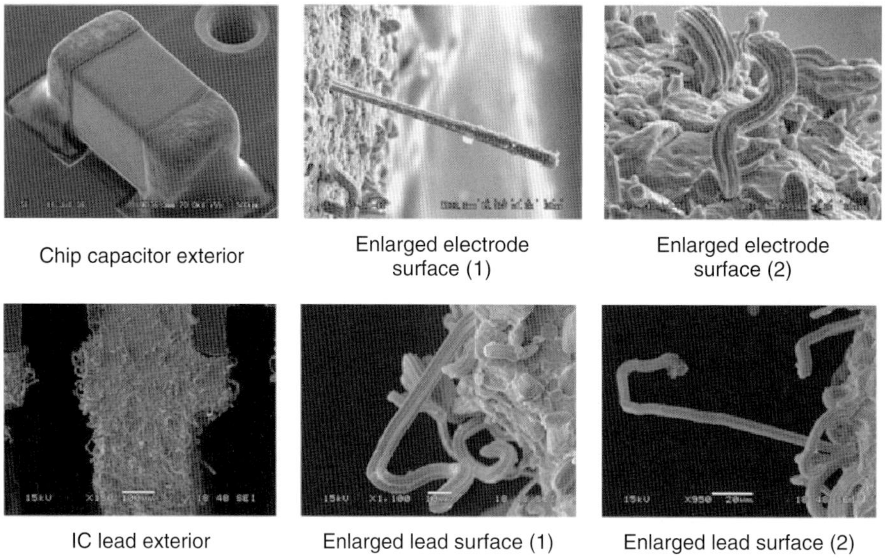

Figure 15. Whisker examples.

spread, galvanic corrosion, external stress, and the like. However, the mechanism is still not clear (Kato, 2010).

Countermeasures that have proved effective include applying a heat treatment (annealing treatment) after plating to reduce residual stress and adding a second metal to the Sn plating. Sn plating containing Pb was once often used in the industry. However, plating containing Pb can no longer be used because of the RoHS Directive, and reliability issues involving tin whiskers have arisen from the use of lead-free compliant components (Yoshino, Sanji, and Iguro, 2007). Therefore, whisker countermeasures for Sn plating are gaining attention again.

3.2.4 *Kirkendall voids*

Leaving the IC in a high temperature state or operating the IC in a continuous manner produces a solid phase reaction at the bonding portions of gold bonding wires and aluminum wiring. This may form a compound of gold and aluminum as a consequence and generate a disconnection defect as shown in Figure 16.

Interdiffusion at the Au–Al bond is accompanied by the formation of various types of metal compounds as shown in Figure 17. Voids known as Kirkendall voids may occur in intermetallics that are weaker than aluminum and gold, which could produce cracks after repeated cycles of heating and cooling and lead to fractures. Countermeasures include not using substances that accelerate the Kirkendall phenomenon such as bromine in the IC package resin and delaying the progress of the Kirkendall phenomenon by including palladium in the gold wires (Kato, 2010).

3.2.5 *Aluminum wiring corrosion*

Resin-sealed packages are widely used for semiconductor devices. However, external moisture may penetrate the package and corrode the aluminum wiring of the semiconductor chip, resulting in a failure (Figure 18).

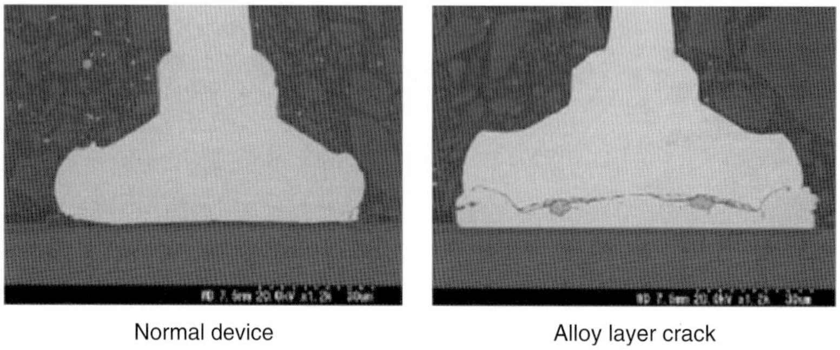

Figure 16. Kirkendall void.

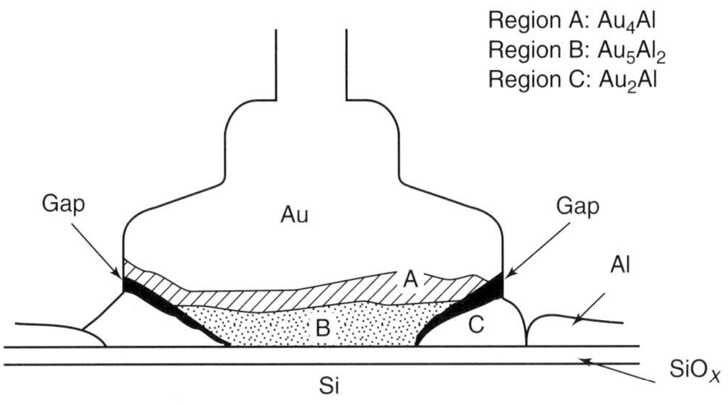

Figure 17. Alloy layers of bonding portion.

When left in dry air, aluminum develops a stable aluminum oxide (Al_2O_3) that acts as a protective film. However, an aluminum hydroxide ($Al(OH)_3$) soluble in acid and alkali is formed in the presence of moisture. If ions of impurities such as P^{3-}, Cl^-, F^-, and Na^+ are also present, corrosion occurs at exposed sections of aluminum wiring, bonding pads, pin holes in the chip protective film, cracks, and other locations. Countermeasures include adopting a highly moisture-resistant passivation film (chip protective film), lowering the moisture absorbency of the resin, and improving resin adhesion (Kato, 2010).

3.3 Reliability assessment

Before the ECU is mounted in a vehicle and shipped to market, it is evaluated by various tests and any design or manufacturing defects are corrected. To ensure effective reliability testing, it is important to accurately understand the vehicle environment and, using the accelerated testing approach described later, to select test conditions that correlate to stresses the ECU will likely receive once on the market, so that the extent of the margin with respect to the targeted life can be correctly grasped.

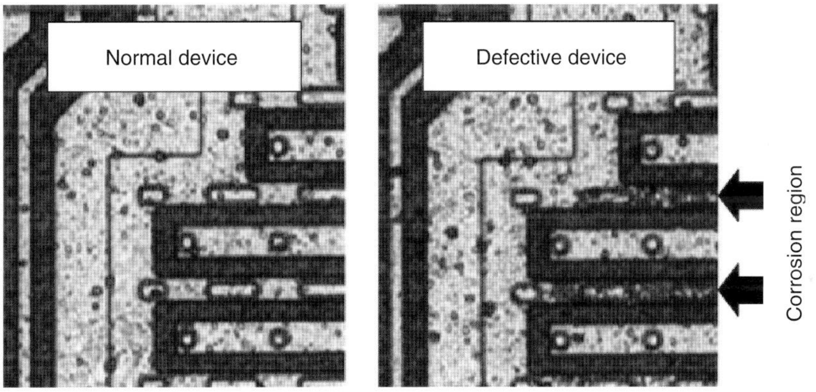

Figure 18. Aluminum wiring corrosion.

Table 2. Main reliability test items.

Test Item		Purpose
Life test	High temperature operation test	Evaluate ECU durability when subjected to electrical and thermal stress over a long period
	High temperature/high humidity operation test	Evaluate ECU durability when operating under a high temperature, high humidity atmosphere
	Humidity cycle test	Evaluate ECU durability in a temperature-varied environment that repeatedly cycles between high and low temperatures
	High humidity shelf test	Evaluate ECU durability when left in a high humidity environment
	High temperature/high humidity shelf test	Evaluate ECU durability when left in a high temperature, high humidity environment
Strength test	Load dump test	Evaluate ECU resistance to positive surge voltage applied with battery terminals removed
	Field decay test	Evaluate resistance to negative surge voltage from an alternator induction load
	RF immunity test	Evaluate ECU resistance when left under an intense electric field atmosphere
	Electrostatic test	Evaluate ECU resistance to static electricity during handling
	Vibration test	Evaluate ECU mechanical strength in a vibration environment
	Salt spray test	Evaluate ECU corrosion resistance when left in a saline environment

3.3.1 Reliability test items

Examples of general tests include ISO, IEC, SAE, and other standards. If the ECU will be used within the ranges assumed by these standards, the reliability of the ECU can be confirmed according to the test conditions set in the standards. However, car and parts manufacturers often set their own individual test items and evaluation conditions as internal standards to quickly respond to changes in onboard environments and achieve individual quality targets. Table 2 shows examples of test items (intended as only a partial list and not comprehensive). Many life tests overlap with test items performed for individual electronic components, and strength tests have many test items that concern the ECU used in an onboard environment such as the load dump test and the field decay test.

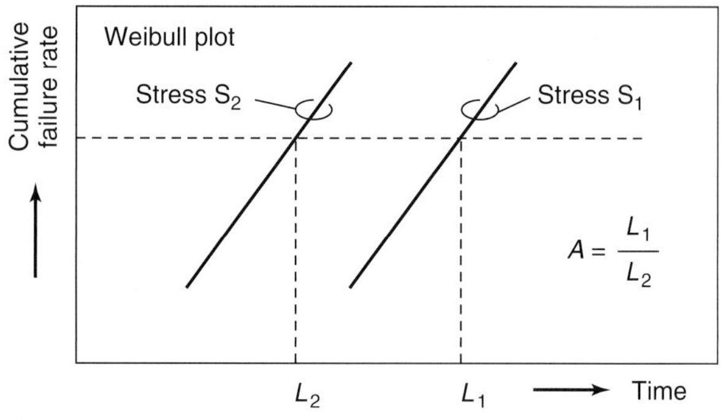

Figure 19. Calculation of acceleration coefficient.

3.3.2 Accelerated testing

Utilizing the fact that applying a large amount of stress to a product shortens the life of the product, accelerated testing is used to perform a reliability assessment in a limited period of time by testing under stricter conditions than use conditions and environment conditions in the market. The acceleration coefficient A is expressed as a ratio of the test times at which a stress S_1 used as a reference and a stress S_2 used as an acceleration condition correspond to the same failure rate as shown in Figure 19.

Table 3. Representative acceleration models.

Acceleration Factor	Acceleration Model/Life Equation	Notes
Temperature	Arrhenius model $L = A\bullet\exp\left(\frac{E_a}{k\bullet T}\right)$	- Electromigration - Change in oxide film over time (TDDB) - Al electric field capacitor life L: life A: constant E_a: activation energy k: Boltzmann constant T: absolute temperature
Electric field	E model $L = A\bullet\exp(-\beta\bullet E)$ $1/E$ model $L = A\bullet\exp\left(\frac{\gamma}{E}\right)$	- Gate oxide film TDDB β, γ: acceleration factor E: electric field
Temperature difference	Modified Coffin–Manson model $L = A\bullet f^m\bullet\Delta T^{-n}\bullet\exp\left(\frac{E_a}{k\bullet T_{max}}\right)$ Temperature difference accelerated model $L = A\bullet\Delta T^{-n}$	- Solder thermal fatigue m, n: constant ΔT: temperature difference f: frequency T_{max}: maximum temperature - Al slide - Interlayer cracks
Humidity	Absolute water vapor pressure model $L = A\bullet V_p{}^{-n}$ Relative humidity model $L = A\bullet(\mathrm{RH})^{-n}\bullet\exp\left(\frac{E_a}{k\bullet T}\right)$	- Al corrosion Vp: absolute water vapor pressure RH: relative humidity
Current	Black model $L = A\bullet J^{-n}\bullet\exp\left(\frac{E_a}{k\bullet T}\right)$	- Electromigration J: current density

Table 4. Activation energy examples.

Failure Mode	Failure Mechanism	Activation Energy (eV)
Threshold voltage shift at MOSFET gate	Ionic contamination	1.0–1.4
	Slow trapping	1.0–1.5
Current leakage	Generation of inversion layer	0.5–1.0
	Channel effect	0.5
Lowered hfe	Accelerated ion migration caused by moisture	0.8
Al wiring disconnection	Al wiring corrosion	0.5–1.0
	Al electromigration	0.4–0.7
Short-circuit	Breakdown of oxide film	0.3–1.1

3.3.3 Acceleration model

Table 3 shows representative acceleration model examples (Kato, 2010), and Table 4 shows examples of activation energy used in acceleration equations (Japan Electronics and Information Technology Industries Association, 2011). Using an acceleration equation enables infinite acceleration in desk calculations, but the failure mode may change depending on the stress conditions. Therefore, consideration must be given to setting the acceleration conditions within a range where the failure mode does not change.

RELATED ARTICLES

Volume 4, Chapter 137
Volume 4, Chapter 139
Volume 5, Chapter 159

REFERENCES

Japan Electronics and Information Technology Industries Association. (2011) *EDR-4708: Guideline for LSI Reliability Qualification Plan*.

Kato, M. (2010) *Automotive Electronics Illustrated*, Nikkei Business Publications, Inc., Tokyo.

Keating, M., Bricaud, P., and Nihon Synopsys G.K., Mentor Graphics Japan Co., Ltd. (2000) *Reuse Methodology Manual For System-On-A-Chip Designs*, Maruzen Publishing Co. Ltd, Tokyo.

Ookura, K. (2005) Semiconductor Technology for Automotive Electronics, *Denso Technical Review*, **10**(2).

Reliability Engineering Association of Japan (1997) *Reliability Handbook*, Union of Japanese Scientists and Engineers.

Renesas Electronics Corp (2010) *Semiconductor Reliability Handbook Rev.0.50*.

Yoshino, M., Sanji, M., and Iguro, S. (2007) Whisker Generation Mechanism of Lead-free Soldered Joints, *Denso Technical Review*, **12**(2).

Chapter 142

Manufacturing: An Introduction to Production Technology, Quality Assurance, and SCM

Naoki Ueda
DENSO Corporation, Kariya, Japan

1 INTRODUCTION

The production of automotive electronic control units (ECUs) has many requirements. First, ECUs with various forms and structures must be produced for a wide range of purposes in addition to the engine ECU. Electronics have been applied to many automotive controls, which means that vehicles are now equipped with a number of ECUs (Kato, 2010a). As a result, it is necessary to create structures that can be mounted near desired locations based on ECU function, as well as to secure installation space and reduce vehicle weight. In addition to the need to reduce size and weight, there are demands for mountability at locations with harsh environments in terms of temperature, humidity, moisture, and the like. The ECU mounting position varies depending on function and purpose. The engine ECU is often mounted inside the engine compartment because many signals originate from the engine compartment and also because the role of this ECU is to control the engine. Body system ECUs often receive signals from inside and around the vehicle cabin and are usually mounted inside the occupant compartment as a consequence. As it is critical that the combination graphic display and car navigation system can be easily recognized and operated by the driver, the ECUs for these are placed near the instrument panel. A location near the transmission is desirable for the transmission ECU. The motor ECU is preferably mounted near the motor, and integration of the ECU and its control object is also being studied.

Thus, the form and structure of the ECU varies depending on differences in usage and the installation environment. An ECU in the occupant compartment often uses a low-cost, resin-printed board with high flexibility. An ECU with a resin-printed board is often used in the engine compartment as well, because the reliability of parts and the mounting technology of parts have been improved and structure of heat dissipation also have been improved. However, for severe environments such as direct mounting on the engine, a hybrid ECU that uses a ceramic substrate capable of withstanding this type of environment is adopted (Kato, 2010b).

As explained earlier, there are ECUs with various functions, specifications, forms, and structures, and the production technology for each ECU is appropriately selected in consideration of quality and cost.

Second, a stable and high-quality ECU must be provided as necessary to various users while adapting to the increased sophistication and diversification of specifications. Therefore, it is important to ensure stable supply performance and quality over the entire supply chain, starting with material and part suppliers and ending with the manufacturer and the user.

Encyclopedia of Automotive Engineering, print © 2015 John Wiley & Sons, Ltd.
Edited by David Crolla, David E. Foster, Toshio Kobayashi, and Nick Vaughan.
This article is © 2015 John Wiley & Sons, Ltd. ISBN: 978-0-470-97402-5
Also published in the *Encyclopedia of Automotive Engineering* (online edition)
DOI: 10.1002/9781118354179.auto216

Third, environmentally friendly products and production methods are required. There have been demands for soldering using a solder that does not contain lead (lead-free soldering) as part of measures to reduce environmentally hazardous substances in recent years. This is a huge switch for electronic products that have been manufactured with lead soldering for many years (Japan Institute of Electronics Packaging, 2006). Going lead-free has also changed the solder melting point and wettability. Various countermeasures were required in addition to material improvement, such as improving component heat resistance, component plating, circuit board artwork (AW) design, production equipment, and process controls. In the home electronics industry, the switch to lead-free has rapidly progressed because of the Restriction of Hazardous Substances (RoHS) Directive effective from July 2006 (Official Journal of the European Union L 234/44, 2011). The automotive industry is similarly subject to the End-of-Life Vehicles (ELVs) Directive, in which the European Commission has set lead-free requirements for printed circuit boards in new vehicles to be sold from 2016 (Official Journal of the European Union L 85/3, 2011) and the automotive industry is starting to take action.

Figure 1. Example of a printed board engine ECU.

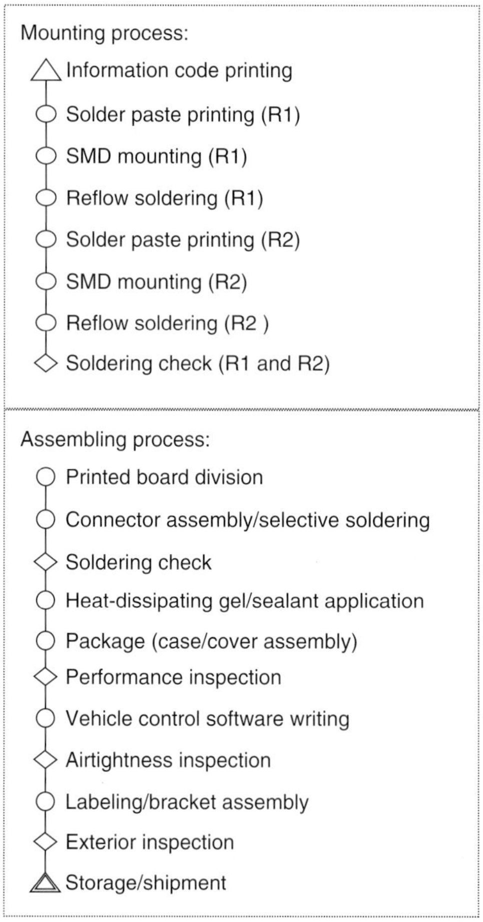

Figure 2. Example of the manufacturing process for a waterproof engine ECU.

2 ECU PRODUCTION TECHNOLOGY AND QUALITY ASSURANCE

Using the example of an engine ECU among the various types of ECUs, this section summarizes an example of the manufacturing process for an ECU with a resin-printed board, and an example of the manufacturing process for a hybrid ECU with a ceramic substrate to be utilized in harsh environments such as direct mounting on the engine. Approaches to and methods for production technology and quality assurance from a production standpoint and examples of countermeasures against part variations are illustrated (Kato, 2010b; Japan Institute of Electronics Packaging, 2006).

2.1 Manufacturing process for printed board ECU

The manufacturing process for a printed board ECU is explained in the following text. The engine ECU shown as an example in Figure 1 is configured from approximately 500 parts, including a microprocessor, transistor, resistor, capacitor, connector, resin-printed board, case, and cover. The manufacturing process can be roughly divided into a mounting process and a subsequent assembling process. Figure 2 shows an outline of the processes.

Features of the automotive engine ECU manufacturing process include a selective soldering process because the engine ECU has a large connector that is a through hole device (THD), and include a high reliability inspection process. An ECU without a waterproof structure or with certain mounting specifications also requires an additional process for coating the entire printed board or required sections with a drip-proof material to resist condensation and other humid environments. A detailed description of this process is omitted here as the main focus is on the manufacturing process for an ECU with a waterproof case structure.

2.1.1 Mounting process

Soldering is performed to mount electronic components on a printed board. To meet demands for reduced size and weight, surface mounting technology (SMT) is mainly used to mount the electronic components on both surfaces of the printed board (Japan Institute of Electronics Packaging, 2006).

2.1.1.1 Information code printing. An information code such as QR (quick response) code (ISO/IEC18004, 2006), which is a two-dimensional barcode, is used to print information such as the serial number and type number on the printed board with a laser. The information is utilized as automatic setup information for equipment and traceability, and within the assembly process for each product.

2.1.1.2 Solder paste printing. A pastelike solder material for bonding electronic components is printed on the printed board using a screen printer (Figure 3).

2.1.1.3 SMD mounting. Various types of surface mount devices (SMDs) are mounted on the printed solder paste material by a mounting machine. The code information is read to set the mounting program that corresponds to the product type and automatically perform mounting. Multiple suction nozzles are moved in succession to mount several tens of electronic components per second at high speed (Figure 3). Equipment that requires replenishing of electronic components is automatically displayed on a board, and a worker supplies the components. This operation takes advantage of the barcode information and prevents mistakes in the supply of electronic components.

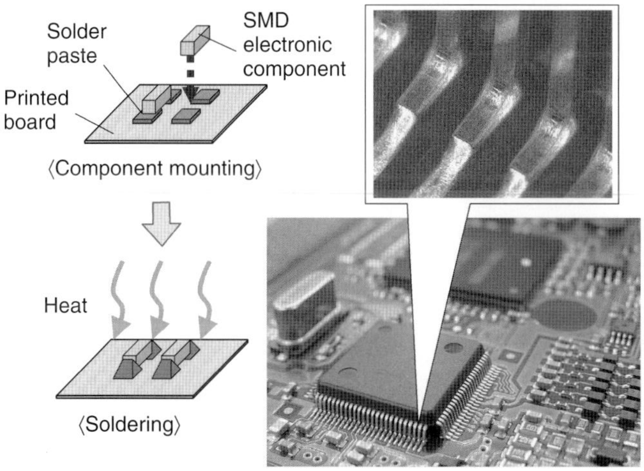

Figure 3. Solder printing, SMD mounting, reflow soldering.

2.1.1.4 Reflow soldering. Soldering is performed by melting the solder paste at approximately 230°C in a reflow (heating) oven (Figure 3). The temperature of the entire circuit board is controlled to maintain a uniform temperature and ensure a temperature that is below the upper temperature limit of the electronic components and that also melts the solder on all the components with different heat capacities. This temperature range has narrowed because of recent lead-free countermeasures, which has led to stricter temperature controls. This process completes the mounting of electronic components on one surface of the printed board. Once mounting on one surface is complete, electronic components are mounted on the other surface using the same process described earlier.

2.1.1.5 Soldering check. Next, the appearance of all soldered locations (roughly 1500 points) and the mounting condition of the components are automatically checked by an automatic viewing (optical inspection) device (Figure 4).

The principle behind the method of confirming the solder shape in a color image is explained here. The component and solder condition are judged based on the condition of this image. The component mounting process is subject to a meticulous process control to ensure that defects do not occur. However, as several hundred tiny electronic components are soldered all at once, there are rare cases of abnormal bonding as shown in Figure 4 (IPC-A-610E-2010, 2010). Therefore, all electronic components are checked so that defects are not passed on to downstream processes. The results, after the check, are fed back and used in the process control for the mounting process and to make further improvements. The checked printed board is then transferred to the subsequent assembling process.

There have been strong demands in recent years for higher functioning and smaller automotive ECUs, as well as smaller semiconductor packages with more pins. The commonly used quad flat package (QFP) form cannot satisfactorily meet these demands, and the ball grid alley (BGA) has been increasingly adopted. With this package, the solder-bonded sections are not visible after mounting on the circuit board, which means that inspection with the optical automatic viewing device used earlier is not possible. As a consequence, even tighter controls and even

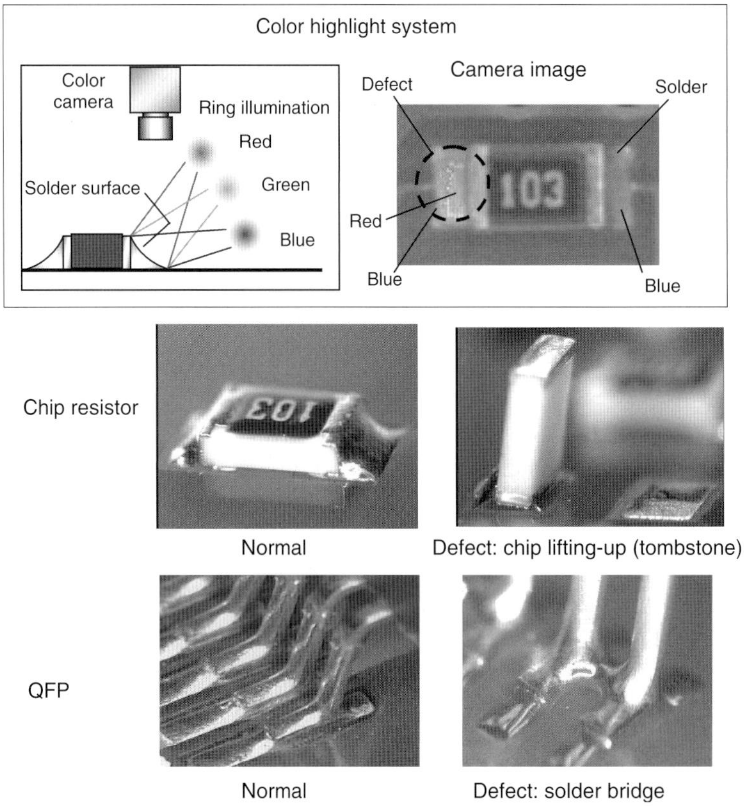

Figure 4. Soldering check by automatic viewing device.

better built-in quality from the component manufacturer up to the mounting process are required. After solder paste printing in the mounting process, an automatic viewing device is used to inspect that the solder paste has been normally printed. In addition, the solder-bonded sections may be checked by X-ray in the initial production as well as in the prototype stage (Figure 5).

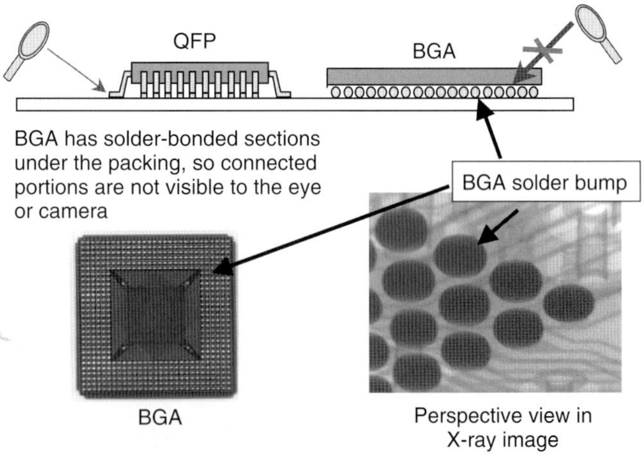

Figure 5. BGA and X-ray check.

2.1.2 Assembling process

2.1.2.1 Printed board division. In the mounting process, one sheet has multiple boards and the boards are cut and divided by a router or the like.

2.1.2.2 Connector assembly/selective soldering. The connector is inserted and assembled to the through hole of the printed board. For automotive ECUs such as the engine ECU, this connector alone is often a THD that is joined using a through hole. Also, as there are many pins, selective soldering is often used to enable soldering of all pins at once (Japan Institute of Electronics Packaging, 2006). First, the printed board is coated with flux to facilitate soldering. Next, it is preheated and then is on a soldering nozzle, after which melted solder is injected from below to perform selective soldering. The amount and range of the flux coating, the solder immersed state, and so forth greatly influence bonding, and process conditions are precisely set and tightly controlled for each type (Figure 6).

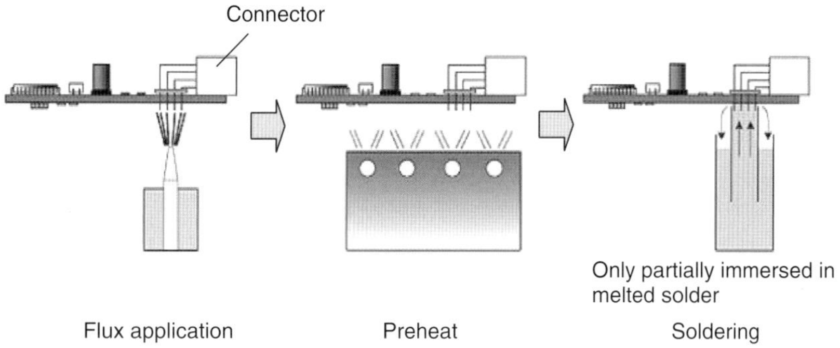
Connector
Only partially immersed in melted solder
Flux application
Preheat
Soldering

Figure 6. Selective soldering.

2.1.2.3 Soldering check. Similarly to the mounting process, the solder condition is checked by an automatic viewing device.

2.1.2.4 Heat-dissipating gel/sealant application. Heat from the electronic components must be dissipated to prevent excessive heating of the electronic components. Therefore, a heat-dissipating gel (silicone) is quantitatively applied to correspond with the layout of the electronic components on the case. To prevent water from entering the ECU interior, a sealant (silicone) is applied for hermetically sealing the outer periphery of the connector and case (Figure 7).

2.1.2.5 Package (case/cover assembly). A case and cover are attached to the printed board to package the printed board (Figure 8).

2.1.2.6 Performance inspection. To check the several hundred electronic components, as well as their assembling quality and operation as an ECU, a performance inspection is performed in which quasi-signals such as vehicle speed signals and engine speed signals are input to determine whether a set output signal can be acquired (Figure 9).

Inspections at high and low temperatures that estimate the temperature environment when mounted in a vehicle may be performed to detect any defects in the temperature characteristics of the components.

2.1.2.7 Vehicle control software writing. Programs for controlling the vehicle are automatically written based on ECU type. The programs are written in time to create an ECU capable of engine control.

2.1.2.8 Airtightness inspection. Hermetic performance is inspected to check whether the sealant applied for a

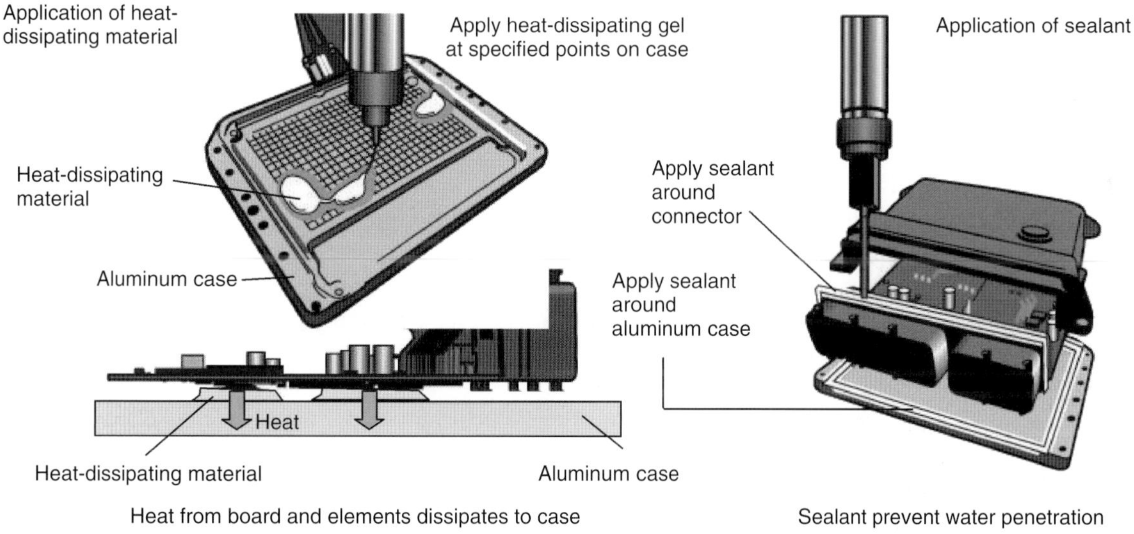
Application of heat-dissipating material
Apply heat-dissipating gel at specified points on case
Heat-dissipating material
Aluminum case
Heat
Heat-dissipating material
Aluminum case
Heat from board and elements dissipates to case
Application of sealant
Apply sealant around connector
Apply sealant around aluminum case
Sealant prevent water penetration

Figure 7. Heat-dissipating material, sealant application.

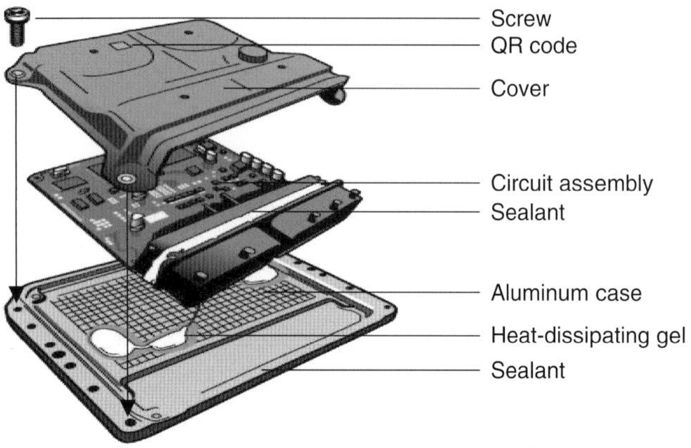

Figure 8. Package (case/cover assembly).

hermetic seal can withstand the installation environment in the vehicle.

2.1.2.9 Labeling/bracket assembly. Labels for each type are printed and attached. Brackets for attachment to the vehicle are assembled.

2.1.2.10 Exterior inspection. The exterior of the fully assembled product is inspected, including the name, labels, and the like. This completes production of the ECU.

2.1.2.11 Storage/shipment. Following completion of the inspection, the product is matched to an order and shipped in accordance with the order.

The preceding process and production technology examples are only brief summaries, and the processes may be differently ordered, omitted, or added to, depending on the product.

As mentioned earlier in one of the processes, each process uses the information code for the production progress management of each product, process controls, automation of in-process switching of product types, and the like. A traceability system is also employed so that history information can be traced to ensure thorough quality control. Although various inspections as mentioned earlier are implemented to prevent the outflow of any products with defects, quality is built into each process as a general rule.

For example, process capability is normally evaluated using the process capability index Cp (Cp = standard width/6σ, where σ is the standard deviation). However, the product data average may deviate from the standard median, and Cpk (Cpk = (data average − limit)/3σ) must be considered in this case. Cp and Cpk are used in combination to evaluate processes and implement necessary improvements. To ensure that excessive stress is not applied to electronic components or bonded portions during processing and assembling, the strain applied to the product in each process is evaluated and the positional relationships between components are reflected in the board design. Necessary improvements are also made to the processes themselves. Specific examples include the screw position and the locations of board divisions.

This section introduced some activities to ensure quality. As described earlier, stable manufacturing and supply is only possible after first building high-quality processes.

2.2 Manufacturing process for hybrid ECUs

Figure 10 shows an example of a hybrid ECU that uses a ceramic substrate. The ECU is directly mounted on the engine and used as an engine ECU. The ECU is configured from approximately 150 parts, including a microprocessor, transistor, resistor, capacitor, connector, ceramic substrate, case, and cover (Kato, 2010b).

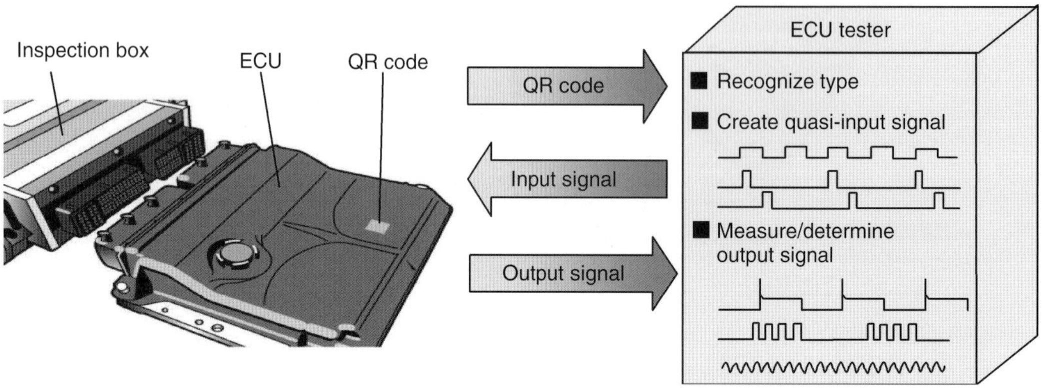

Figure 9. Performance inspection.

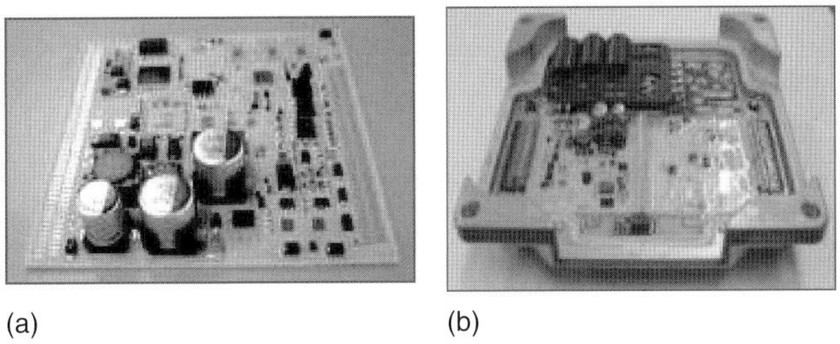

Figure 10. Example of hybrid ECU directly mounted on engine. (a) Photo of exterior of components mounted on ceramic substrate. (b) Exterior of ECU product.

Directly mounting the ECU on the engine requires high heat resistance and a small size. To increase heat resistance, a ceramic substrate is used for the circuit board, and semiconductors are mounted on the circuit board by bare chip mounting.

The production technology for a hybrid ECU is outlined in the following text. Figures 11 and 12 illustrate examples of the process flow.

First, a solder material is applied to required locations on the ceramic substrate using a screen printing method. Parts such as semiconductor bare chips and chip capacitors are mounted on the solder material, and reflow soldering is performed at approximately 230°C. Flux contained in the solder material is subsequently cleaned. Fluorocarbon cleaning solvents have been used in the past as the cleaning solution, although water-based cleaning solutions are now employed in consideration of the global environment. Semiconductor bare chip components are mounted with an electrically conductive adhesive, after which gold (Au) wire bonding and aluminum (Al) wire bonding are performed to create electrical connections.

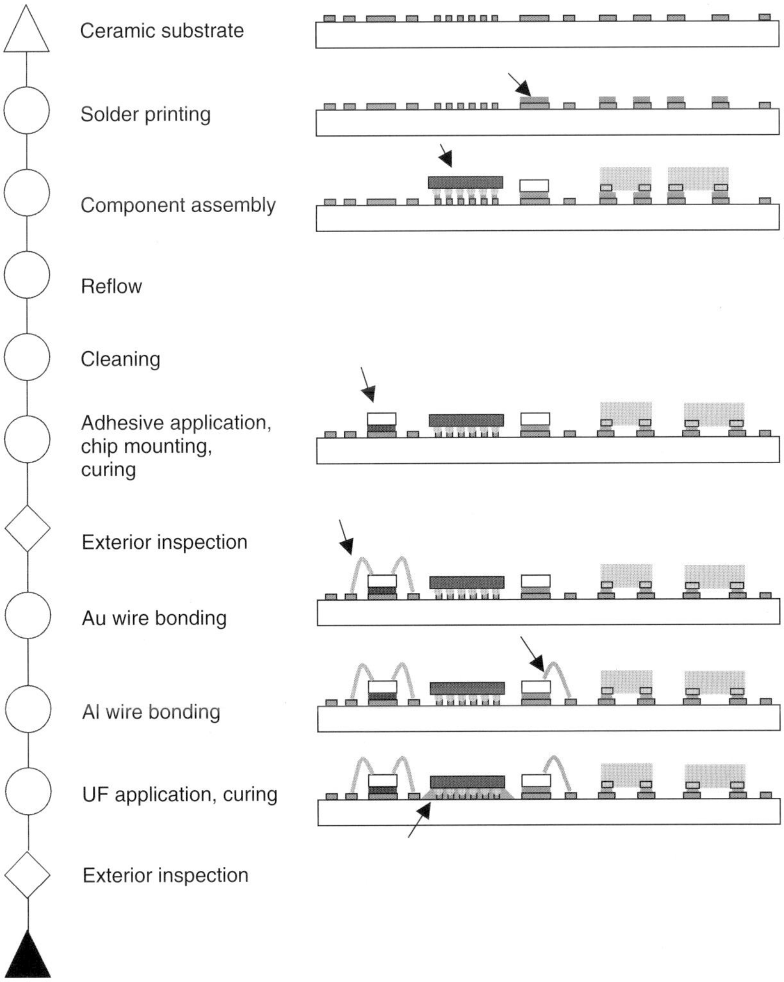

Figure 11. Example of manufacturing process flow for hybrid ECU (circuit board process).

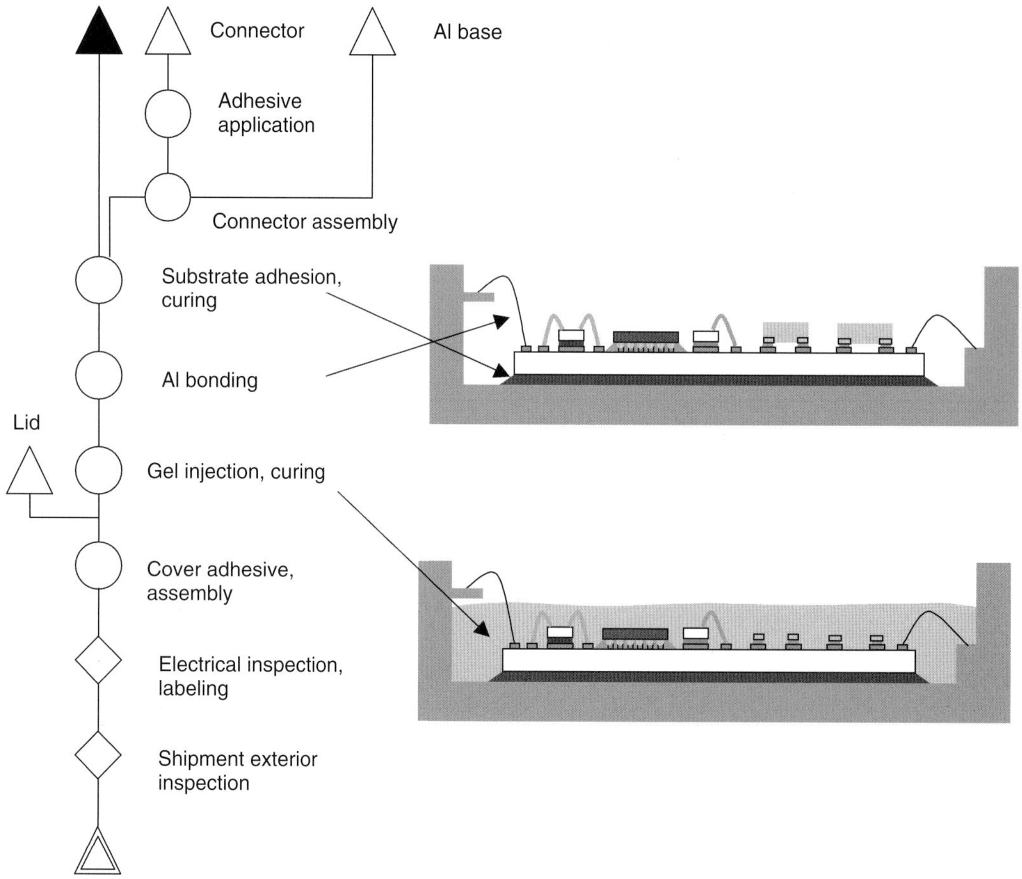

Figure 12. Example of manufacturing process flow for hybrid ECU (assembling process).

To improve the reliability of the solder-bonded portions of some components, an epoxy-based resin material [underfill (UF) material] is applied to reinforce the soldering. This completes the manufacture of one circuit board. An exterior inspection is performed at this stage, and the circuit board is then transported to the downstream process. Note that, although not described in detail here, the ECU shown in Figure 10 is configured from three boards.

The circuit board with assembled components is closely attached using adhesive to components to which a base made of metal (generally Al) and connector components are already adhered. Next, these are covered with a potting material (silicone-based gel) to protect the semiconductors and bonding wires, and a lid is then adhered. Following an electrical inspection, the product is shipped.

2.3 Comparison of hybrid ECU and printed board ECU

This section briefly compares the hybrid ECU that uses a ceramic substrate and the commonly used printed board ECU, while focusing on the merits of the hybrid ECU in particular. The hybrid ECU is advantageous in terms of having a small size, as well as good heat resistance and environment resistance (Table 1).

From the standpoint of downsizing, the hybrid ECU has an edge because of easy bare chip mounting. The printed board ECU uses molded ICs including QFPs, and the practical area required for mounting these components is, therefore, approximately five times larger than that needed for bare chip mounting. In addition, the hybrid ECU can adopt structures well suited for downsizing, for example, direct printing of resistors on the substrate and full-thickness via holes.

With regard to heat resistance, the hybrid ECU uses a ceramic substrate, that is, a sintered compact, and is extremely stable with respect to heat. Ceramic components mounted on the substrate have a matching coefficient of thermal expansion for reduced stress on connected regions and high reliability. The environment resistance of the hybrid ECU enables installation in high-temperature regions subject to large vibrations.

Table 1. Comparison of hybrid ECU and printed Board ECU.

			Hybrid ECU	Printed board ECU
Downsizing	IC		Bare chip	Molded IC
	Resistance		Direct printing resistor + chip resistor component	Chip resistor component
	Board design		Via hole	Through hole + via hole
Heat resistance	Board glass-transition point		Approximately 1600°C (firing temperature)	Approximately 140°C (FR4: glass epoxy board)
	Coefficient of thermal expansion	Board	7 ppm/°C	16 ppm/°C
		Components	6–9 ppm/°C	6–9 ppm/°C
Environment resistance	Use environment		Direct mounting on engine	Mounting in occupant compartment or engine compartment
	Vibration resistance		30G	5G

3 SUPPLY CHAIN MANAGEMENT

Automotive ECUs are configured from a number of materials and parts, including electronic components such as microprocessors, transistors, resistors, and capacitors, as well as printed boards, connectors, cases, covers, solders, sealants, screws, and brackets. In addition, those individual materials and parts are composed of raw materials and other parts and materials. Their quality and supply support ECU production, and the management, that is, supply chain management (SCM), of the material and part supply network is extremely critical to achieve high-quality and stable production.

In the automotive industry, consolidation of the supply of raw materials and parts has advanced thus far based on the pursuit of lower costs and higher quality. Partial modularization has similarly progressed. Furthermore, there are parts that demand lower costs and higher quality of both the ECU and components, which can only be achieved by surveying the supplier processes and the entire ECU production process, and considering all processes and inspections as a whole. Instead of a pyramid supply configuration, it may be discovered that

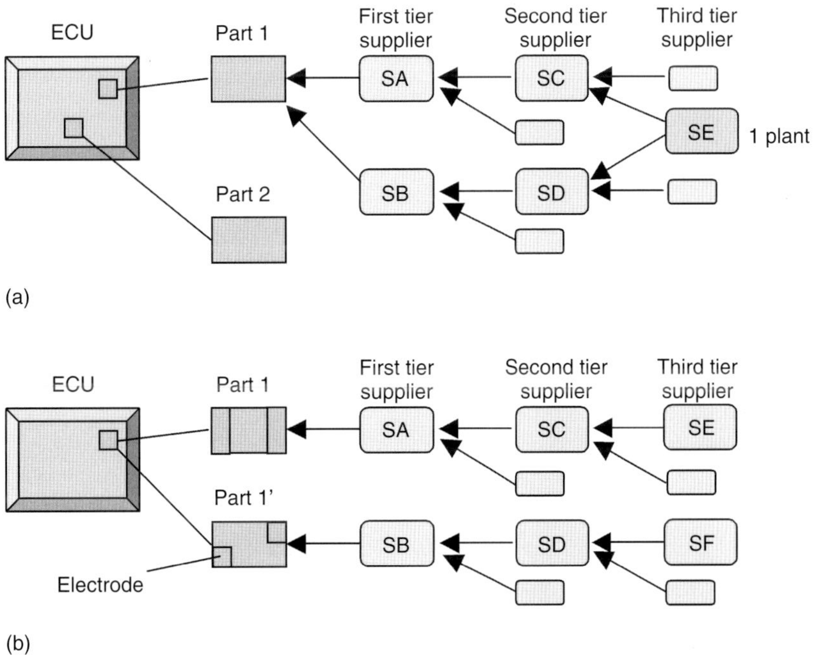

Figure 13. Supply chain. (a) Example of common raw material supplier. (b) Example of different suppliers and electrode geometry.

some materials or parts are restricted to a limited number of suppliers.

Supply risk management has become extremely important because of large-scale disasters, company risks, and other factors and is gaining more attention. As the ECU is configured from a large number of materials and parts as described earlier, it cannot be produced if the supply of even one part is interrupted. Although there are parts that can be easily substituted or are not directly linked to product competitiveness, materials and parts that sustain the competitiveness of the product in terms of quality, performance, and cost cannot be easily replaced. For this reason, restructuring and visualization of the supply chain, particularly in consideration of risk management, is necessary.

An in-depth and comprehensive examination of the supply chain must be performed. For example, as shown in Figure 13a, in the case of two first-tier suppliers, there should be no problem with the two first-tier suppliers purchasing materials and parts to manufacture their own parts from different second-tier suppliers. However, the fact that an underlying third-tier supplier supplies raw material to both second-tier suppliers and also has only one plant may be overlooked. If the third-tier supplier is affected by a disaster, the supply of raw material could be stopped. Moreover, this could happen again at any time in the future.

Parts with the same exterior shape and function cannot necessarily be easily interchanged. For example, as shown in Figure 13b, even parts with the same exterior shape and functions may have different electrodes. In this case, the geometry of the electrodes (lands) on a printed board may correspond to one part but not the other. If the other part is mounted to the noncorresponding printed board and soldered, a mounting defect such as part rotational displacement may occur. If possible, the geometry of the electrodes of printed board (lands) should be designed in advance so that either part can be mounted to printed board without any problems. This involves AW design technology, and in a wider sense, production technology.

Simply prioritizing risk avoidance by increasing inventory results in unnecessary costs and reduces product competitiveness. Instead of thinking strictly in terms of the quality, cost, and supply of materials and parts, the supply chain should be considered and strategically studied at the stages of design, part development/selection, and material development/selection.

4 CONCLUSION

There are ECUs with various functions, specifications, forms, and structures and appropriate production technology for each ECU is selected considering required quality and cost. This chapter takes the engine ECU as a representative example and gives an overview of production technology for printed board ECUs and hybrid ECUs made of ceramic substrate. It also summarizes approaches and methods for quality assurance taken during production of these.

It is important to ensure stable supply performance and quality over the entire supply chain, starting with material and part suppliers to the manufacturer itself and the users. Supplier chain management is outlined also in view of significant supply issues that have surfaced in recent years.

In the near future, integration of the ECU and its control object, that is, electro-mechanical integration, will increasingly progress. Further development of connection technology and size reduction technology will be required, together with improved production technology to achieve these goals.

RELATED ARTICLES

Volume 4, Chapter 137
Volume 4, Chapter 141
Volume 5, Chapter 159

REFERENCES

IPC-A-610E-2010 (2010) Acceptability of electronic assemblies, association connecting electronics industries.

ISO/IEC18004 (2006) Information technology—automatic identification and data capture techniques—QR code 2005 bar code symbology specification.

Japan Institute of Electronics Packaging (2006) *Handbook printed circuit technology*, 3rd edn, Nikkan kogyo Shinbun, Tokyo.

Kato, M. (2010a) *Automotive Electronics: Systems*, Nikkei Business Publications, Inc., Tokyo.

Kato, M. (2010b) *Automotive Electronics: Basic Technologies*, Nikkei Business Publications, Inc., Tokyo.

Official Journal of the European Union L 234/44 (2011) Commission Decision of 8 September 2011, amending, for the purposes of adapting to technical progress, the Annex to Directive 2002/95/EC of the European Parliament and of the Council as regards exemptions for applications containing lead or cadmium, September 10, 2011.

Official Journal of the European Union L 85/3 (2011) Directives, Commission Directive 2011/37/EU of 30 March 2011, amending Annex II to Directive 2000/53/EC of the European Parliament and of the Council on end-of-life vehicles, March 31, 2011.